Intel's 80486 32-bit microprocessor, courtesy Intel Corporation

DIGITAL MICROELECTRONICS

DIGITAL MICROELECTRONICS

HALDUN HAZNEDAR

Texas Instruments
Semiconductor Group
Design Automation Division, Houston, Texas

THE BENJAMIN/CUMMINGS PUBLISHING COMPANY, INC.

Redwood City, California   Menlo Park, California   Reading, Massachusetts
New York   Don Mills, Ontario   Wokingham, U.K.   Amsterdam
Bonn   Sydney   Singapore   Tokyo   Madrid   San Juan

To my grandparents
and
my parents, Nükhet and Mazhar Haznedar

Executive Editor: Alan Apt
Sponsoring Editor: Mark McCormick
Associate Editor: Mary Ann Telatnik
Senior Production Supervisor: Bonnie B. Grover
Production and Art Coordinator: Stacey C. Sawyer, Sawyer & Williams
Text Designer: Eleanor Mennick
Cover Designer: Rodolphe M. Zehntner, The Belmont Studio
Text Art: Rolin Graphics, Inc.
Copyeditor: Nicholas Murray
Proofreader: Anita Wagner
Compositor: Arizona Composition Service, Inc.

Library of Congress Cataloging-in-Publication Data

Haznedar, Haldun.
Digital microelectronics / Haldun Haznedar.
p. cm.
Includes bibliographical references and index.
ISBN 0-8053-2821-1
1. Digital electronics. 2. Microelectronics. I. Title.
TK7868.D5H36 1991
621.381—dc20 90-46494
CIP

2345678910 -MA- 95 94 93 92 91

The Benjamin/Cummings Publishing Company, Inc.
390 Bridge Parkway
Redwood City, California 94065

PREFACE

Audience

The primary objective of this book is to serve as the text in digital microelectronics courses for advanced undergraduate students in electrical and computer engineering by providing insight into the analysis and design of integrated circuits. The student, after reading this book, is sufficiently prepared to function proficiently in a more advanced graduate-level course on digital IC design and in industry. The broad variety of topics and the depth of discussion also make this text a valuable supplement to the continuing education of practicing engineers in the aforementioned fields. The technical level of the material requires prior knowledge of fundamentals of logic design and of analog electronics, which is usually acquired through junior-level electronics courses.

Coverage

The presentation is balanced between practical circuit design problems and theoretical rigor. Realistic applications of the commercially produced digital devices and real-life problems encountered in designing digital systems are included in the book. Besides a comparative study of available logic families, technologies such as advanced Schottky, advanced low-power Schottky, high-speed CMOS, and advanced CMOS logic are treated in detail. Among the other topics covered are VLSI circuits such as semiconductor memories and programmable logic devices, digital regenerative circuits, and data converters. A subject that is neglected by electronics textbooks is the design of the logic systems in noisy environments. A whole chapter is devoted to types and sources of electrical noise in board-level digital circuits, types of failures, and design considerations that may help minimize noise susceptibility. The material covered in this book is more than enough for a one-semester three-hour course. Thus, course pacing and relevant material selection can satisfy various curricular needs.

A comprehensive coverage of modern digital integrated circuits necessitates some understanding of the semiconductor materials used to fabricate these chips and the physical mechanisms of the microelectronic devices employed in these circuits for switching purposes. Treated in depth are the integrated circuit versions of those semiconductor devices that form the basis of modern digital microelectronic circuits: *pn*-junction and Schottky-barrier diodes (Chapter 2), metal-oxide-semiconductor field-effect transistor (Chapter 3), and bipolar junction transistor (Chapter 5). The principles underlying the operation, analysis, and design of these devices are explained and their large-signal transient behavior in switching circuits are discussed in detail. Energy band diagrams are used in visualizing their internal status under various biasing conditions.

Computer Simulation

Since circuit simulation is critical to the design and optimization of a circuit, extensive use of the de facto industry-standard SPICE simulator is made to model the devices as well as to predict the performance of integrated circuit designs. It is employed in many examples and problems to generate the voltage transfer characteristics or switching responses of digital circuits for the reader to acquire an appreciation of the values used for model parameters and to gain familiarity with the tool itself to utilize it in the analyses of more complicated circuits.

Examples/Problems

The worked examples, incorporated in the body of the text to illustrate the concepts, use practical parameter values. The end-of-chapter problems constitute an important component of the learning process. Most of them reinforce topics of prime importance and extend the basic concepts introduced in the text. They range from drill in analysis to computer simulation employing SPICE to design issues. The rest are numerical exercises that provide the student with an intuitive feel for magnitudes and dimensions of key parameters encountered in real devices and circuits. Furthermore, programs written in Pascal are included to simulate hand-analysis.

Chapter Outline

Chapter 0 introduces the basic concepts. The basic properties of digital circuits are summarized and their performance characteristics such as power dissipation, propagation delay, speed-power product, noise margins, and fan-out are defined.

Chapter 1 reviews the properties of semiconductor materials necessary for a basic understanding of the operation of microelectronic devices. It also formulates the mathematical equations needed to characterize the mechanism of current flow.

Chapter 2 deals with semiconductor diodes. Being the most fundamental of all semiconductor devices, the *pn*-junction diode is especially essential to our study. Its static and dynamic properties are described. Schottky-barrier diode is also introduced. The SPICE diode model is presented.

Chapter 3 starts with the description of the MOS capacitor. Based on the concepts already presented for the two-terminal structure, it then becomes easy to understand the operation and modeling of the four-terminal MOS transistor. Short-channel effects such as channel length modulation and velocity saturation are studied. The SPICE model for the MOSFET is described.

Chapter 4 introduces the CMOS technology. The metal-gate and silicon-gate CMOS technologies and different series of CMOS subfamilies are introduced. The basic CMOS inverter, its voltage transfer characteristic and dynamic behavior, and the design and analysis of CMOS NAND and NOR gates are among the subjects treated in this chapter.

Chapter 5 studies the bipolar transistor and the basic BJT inverter. Even though the bipolar transistor is employed under large-signal conditions in the switching mode in digital circuits, its operation still passes through the forward active region when the transistor is switched into saturation from the off mode or vice versa. Therefore, charge control model for all modes of operation are developed in this chapter. The BJT inverter, its static and dynamic characteristics, and the bipolar transistor as a charge-controlled device are all discussed in detail. Various time intervals during switching between states are considered. The SPICE model of the bipolar transistor is discussed.

Chapter 6 deals with a detailed study of the bipolar logic circuit families. Comprehensive analysis of the basic NAND gate is made for each and every TTL subfamily. Performance characteristics of each subfamily are considered. Another bipolar family, ECL, is also treated in this chapter. Coverage of the emitter-coupled inverter is followed by the discussions of 10K, 100K, and 10KH series. Finally, the I^2L is introduced.

Chapter 7 presents detailed performance comparison of logic families. Interfacing ECL and TTL as well as CMOS and TTL are treated in this chapter. The open collector and three-state devices, as well as buffers and drivers in different logic families, are treated in detail.

Chapter 8 introduces the digital regenerative circuits. Digital circuits with feedback connections, existence of which leads to functional blocks for sequential circuits, are discussed. Various types of astable, bistable, and monostable multivibrators as well as Schmitt triggers are presented.

Chapter 9 deals with D/A and A/D converters as well as S/H circuits. Several techniques to convert a digital signal to analog form are explained. For analog to digital conversion, various approaches like successive approximation, dual-slope, counting, and subranging are covered.

Chapter 10 discusses the digital memory and other VLSI circuits. Random access type semiconductor memories such as mask programmable ROMs, PROMs, EPROMs, E^2PROMs, and dynamic and static read-write memories are treated in detail. Available memory chips using different technologies, their internal organization, read and write operations, sense amplifiers, alpha-induced soft errors, and new trends in high-density memory chips are considered. In addition, semicustom VLSI circuits such as EPLDs, E^2PLDs, LCAs, PLAs, PALs, and GALs are covered.

Chapter 11 considers the design of logic systems in noisy environments. Types and sources of electrical noise such as transmission line reflections, crosstalk, and electromagnetic interference are classified. Design considerations such as printed circuit board layout, power supply distribution, decoupling, using bypass capacitors, grounding and shielding techniques, electrostatic discharge and its effects are discussed.

The book concludes with three appendices. *Appendix A* reviews the information available in typical TTL and CMOS logic data sheets of commercial and military ICs as provided by the manufacturers. *Appendix B* goes over the integrated circuit fabrication technology covering both the CMOS and bipolar IC fabrication processes. *Appendix C* lists basic physical constants and properties of silicon.

The traditional practice of covering bipolar devices and logic families before MOSFETs and CMOS has not been followed. Following a chapter on the *pn*-junction diode, the book moves directly to the MOSFET since it is the backbone of modern VLSI circuits. Nevertheless, the chapters are so written as to permit some variation in the order in which they can be assigned. Thus, Chapters 5 and 6 can be read right after Chapter 2 if the course emphasis is on bipolar devices.

Acknowledgments

It is a pleasure to acknowledge here the help I received from various sources during the development and production of this book. I wish to thank the people at the Benjamin/Cummings Publishing Company who brought their expertise to this project. Its staff have been unfailingly helpful throughout the long process of preparing the manuscript. Above all, this project is the brainchild of my acquisition editor Craig S. Bartholomew who instilled motivation and enthusiasm. To him, I give my grateful thanks. For their wise counsel and expert professional advice, warm thanks are also due to my former and present editors, Mark McCormick, Mary Ann Telatnik, Alan R. Apt, and Stacey C. Sawyer who have persisted with me in this endeavor through many drafts and revisions.

Since I began writing the book five years ago, I have drawn on the comments of numerous colleagues who reviewed the manuscript at various stages of its development and contributed to its improvement with their feedback and constructive criticism. Special acknowledgment should go to T. V. Blalock, University of Tennessee; J. E. Dalley, University of Utah; R. G. Deshmukh, Florida Institute of Technology; L. W. Eggers, California State University at Los Angeles; W. J. Helms, University of Washington; H. Z. Massoud, Duke University; J. P. Palmer, California Polytechnical State University at Pomona; M. Soma, University of Washington; S. K. Vadhva, California State University at Sacramento; K. Watson, Texas A & M University; and C. R. Zimmer, Arizona State University.

This book could have never been written without the encouragement of my family. To my wife Binnur and my daughter Bengü goes my heartfelt appreciation. Their patient understanding enabled me to devote the long hours necessary to complete this project. I also wish to express my gratitude to my parents and my sister Şeyma for their support from afar and for their belief in me.

HALDUN HAZNEDAR
October, 1990

CONTENTS

0. PROLOGUE

Digital vs. Analog

There are two ways to represent the numerical values of quantities that we constantly deal with in everyday life: *analog* and *digital.* Analog quantities can vary over a *continuous* range of values, while digital representation of quantities is *discrete* in nature. A *digital system* manipulates data that are composed of a finite number of discrete elements and produces results that are also composed of discrete elements. In contrast, an *analog system* manipulates data that are represented in a continuous form and produces continuous results.

Digital systems are easier to design than analog systems because the *switching* circuits do not require exact values of voltage or current; only the range (*high* or *low*) in which they fall is important. Furthermore, in digital systems information storage is easy because certain switching circuits can hold information as long as necessary. Their accuracy is also greater; they can handle as many digits of precision as needed simply by adding more switching circuits. The precision of analog systems is usually restricted because voltage and current values are directly dependent on the circuit component values.

A digital system can easily be programmed so that its operation is controlled by a set of stored instructions. Even though it is also possible to program analog systems, the variety and complexity of operations are severely limited. Digital circuits are less affected by noise because exact values are not important. Moreover, in passing through each circuit, noise is attenuated as the desired logic signals are restored to full amplitude. Thus, in digital systems noise does not accumulate from one logic circuit to another, as it does in analog systems.

While their use of high-value capacitors, precision resistors, transformers, and inductors has prevented analog systems from employing fully the advantages of high-density *integrated-circuit* (*IC*) technology, digital circuitry has already achieved a very high degree of integration. However, the fact that the real world is mainly an analog system is a major drawback for digital techniques. Thus, to take advantage of these techniques when dealing with analog inputs and outputs, the inputs are first quantized and converted to digital form, then the digital information is processed, and finally the digital outputs are converted back to analog forms. Both types of conversion require *data converters*. Naturally, this conversion process between different forms of information requires extra time and adds complexity and expense. However, the advantages of digital techniques far outweigh these shortcomings.

Digital Circuits

Digital systems use switches to represent the information that is being processed in *binary* form. We can arbitrarily define an *open* switch to represent *logical* **0** and a *closed* switch to represent *logical* **1**. In electronic digital systems, binary information is usually represented by voltages that are present at the inputs or outputs of various circuits. Typically, the *logical states* are described by two nominal voltage levels, the higher of which, corresponding to the power *supply* voltage, V_{CC}, represents **1** in *positive logic* while the *ground* (i.e., 0 V) represents **0**. However, as stated before, the circuits are actually designed to respond to input voltages and to produce output voltages that fall within prescribed voltage ranges.

Since a digital circuit obeys a certain set of logic rules, it is also called a *logic circuit*. Almost all the digital circuits used in digital systems are ICs. Several fabrication technologies are used to produce digital ICs, the most common being the *transistor-transistor logic* (*TTL*) and *complementary metal-oxide semiconductor* (*CMOS*). The former uses the bipolar transistor as its switching element, whereas CMOS employs the MOS transistor.

A logic circuit with one or more inputs and outputs is an interconnection of several logic devices designed to perform a desired function. The outputs of a *combinational* logic circuit at any particular time are each a function of the combination of inputs at that time, regardless of the previous combination of inputs. Therefore, the circuit does not possess any memory. The basic building blocks of combinational circuits are *gates*. A logic circuit with a memory is called a *sequential* circuit, the output of which at any particular time is a function of both the inputs and the state of the circuit at that time. The state in turn depends on what has happened to the circuit previously. The basic building blocks of sequential circuits are *flip-flops*.

Entire systems containing thousands of components can be fabricated on a single chip. The term *microelectronics* refers to the design and fabrication of these high-density ICs. The levels of integration include *small-scale integration* (*SSI*), with about 1 to 10 gates per chip; *medium-scale integration* (*MSI*), with 10 to 100 gates per chip; *large-scale integration* (*LSI*), with 100 to 1000 gates per chip; *very-large-scale integration* (*VLSI*), with 1000 to 100,000 gates per chip; and *ultra-large-scale integration* (*ULSI*), with more than 100,000 gates per chip.

The design of any given circuit, whether it be on a printed circuit board or at the IC level, involves an ongoing iteration among the setting of system performance specifications and circuit parameters as well as the selection of basic materials and the fabrication aspects of design. Thus, a systems approach to digital electronics design requires a designer to gain a broad appreciation of all the important factors that enter into digital electronics by developing an understanding of materials science and physical electronics, the characteristics of the building-block elements to be used in the design, and the process of combining these elements into complete circuits and systems. This approach also requires the designer to consider all the performance objectives that affect the design process. Technical requirements, cost requirements, and manufacturing capabilities must all be taken into account to produce an optimum solution to a design problem.

CAD/CAE Tools

Computer-aided design and engineering (*CAD/CAE*) tools are essential to the development of all digital circuits and to the configuration of IC building-block subsystems and systems. Computer simulations are especially important in IC design, because the circuit cannot be breadboarded, and its first prototype is prohibitively expensive. Moreover, computer simulation lets one perform *what if?* experiments that are not feasible with a physical prototype.

This text makes extensive use of the best-known and most widely used circuit simulator, SPICE, the *S*imulation *P*rogram with *IC E*mphasis. Its dc and transient analysis capabilities are particularly important for the study of digital circuits.

The version chosen for illustration in this text is PSPICE, by MicroSim Corporation. This commercial package was the first version of SPICE that ran on a personal computer. It incorporates all of the features of the generic public-domain software and embodies several additions as well. The original SPICE, designed to produce output on a line printer, does not offer any graphics capability. A **.PLOT** command generates graphs that use alphanumeric characters. On the other hand, one of the most useful options PSPICE offers is its graphics postprocessor program, called PROBE, which is a *software oscilloscope*, allowing the user to view any waveform in the system on the screen and/or to output to a printer or plotter. A classroom version, restricted to ten transistors and one hundred nodes (suitable for most educational purposes), is available at no cost from MicroSim, the software developer.

Mathematical Foundation

The mathematical foundation of all logic design work is the *Boolean algebra*, developed by George Boole, a Scottish mathematician, in the last century.[1] Among many existing Boolean algebras, the two-valued Boolean algebra, also called the *switching algebra*, with the set of input symbols $P = \{\mathbf{0}, \mathbf{1}\}$, guides the logic-level analysis and design of digital circuits.

Briefly, a Boolean algebra is a *closed*[2] mathematical system consisting of a set, P, of at least two distinct elements and two binary operators ($+$ and $\cdot$) that satisfy the following axioms:

---

1. George Boole. *An Investigation of the Laws of Thought*. London: Macmillan, 1854; reprinted, Dover Publications: New York, 1958.
2. That is, the result of operating on any group of members of P is itself a member of P.

A1. *Operators are commutative.* For every $A, B \in P$:
$A + B = B + A$,
$A \cdot B = B \cdot A$

A2. *Operators are associative.* For every $A, B, C \in P$:
$(A + B) + C = A + (C + B)$,
$(A \cdot B) \cdot C = A \cdot (B \cdot C)$

A3. *Operators are distributive over each other.* For every $A, B, C \in P$:
$A + (B \cdot C) = (A + B) \cdot (A + C)$,
$A \cdot (B + C) = A \cdot B + A \cdot C$

A4. *Unique elements* **0** *and* **1** *exist.* For every $A \in P$:
$A + \mathbf{0} = A$,
$A \cdot \mathbf{1} = A$

A5. *The complement of each element exists.* For every $A \in P$:
$A + \overline{A} = \mathbf{1}$,
$A \cdot \overline{A} = \mathbf{0}$

Figure 0.1 illustrates the three basic operations in switching algebra using logic gates. The *logical product*, with the symbol · usually omitted, is performed by an *AND* gate. The symbol + corresponds to *logical sum* and is implemented by an *OR* gate. The unary complementation operation, *not*, is realized by an *inverter*. The latter is a misnomer since there is no division or cancellation operation and hence no inversion in Boolean algebra.

Properties of Digital Circuits

The four most important characteristics, in addition to cost, in selecting a logic family are the *propagation delay*, *fan-out*, *power dissipation*, and *noise immunity*.

The *propagation delay* time of a switching circuit such as an inverter defines how fast the circuit output responds to changes at its input. It is measured between the

Figure 0.1 The three basic operations: (a) logical product and the AND gate; (b) logical sum and the OR gate; and (c) complementation and the NOT gate.

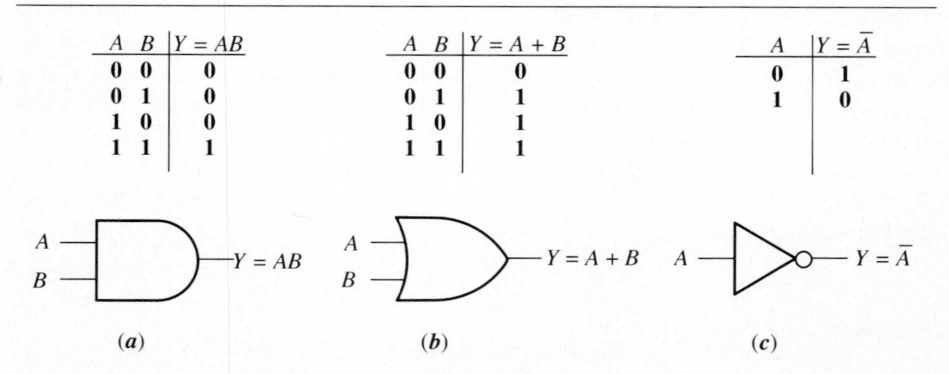

A	B	$Y = AB$
0	0	0
0	1	0
1	0	0
1	1	1

A	B	$Y = A + B$
0	0	0
0	1	1
1	0	1
1	1	1

A	$Y = \overline{A}$
0	1
1	0

50% points of the input and the output waveforms at both the leading and trailing edges, as illustrated in Figure 0.2 for an inverting circuit.

The turn-on propagation delay time, t_{PLH}, where the subscripts L and H indicate the low-to-high transition of the output, is defined as the time between the 50% points on the input and output voltage waveforms with the output changing from the logical **0** level to the logical **1** level. Similarly, the turn-off propagation delay time, t_{PHL}, is the time interval between the 50% points on both waveforms with a high-to-low transition at the output. The average propagation delay time is defined as

$$t_{PD} \equiv \frac{(t_{PLH} + t_{PHL})}{2} \tag{0.1}$$

Next, consider the input and output current specifications for a given logic circuit. The high-level input current, I_{IH}, is defined as the current into an input when a high-level voltage is applied to that input. The current into an input when a logical **0** voltage is applied to that input is designated by I_{IL} and called the low-level input current. Similarly, the high-level output current, I_{OH}, is the current into an output with the input conditions that will establish a logical **1** at the output. Finally, I_{OL} is the low-level output current when the applied input conditions establish a low level at the output. Note that the current out of a terminal will be given a negative value.

The term *fan-out* describes the maximum number of load gates, N, that can be connected to the output of the driving logic circuit without preventing the output from reaching guaranteed high and low voltage levels. Therefore, it is equal to either I_{OL}/I_{IL} or I_{OH}/I_{IH}, whichever calculation produces the lower natural number:

$$N \equiv \min\left[\left|\frac{I_{OL}}{I_{IL}}\right|, \left|\frac{I_{OH}}{I_{IH}}\right|\right] \tag{0.2}$$

Figure 0.3 illustrates the fan-out of a gate for both logical levels.

A logic circuit designer must also know the amount of power consumed by the circuit being designed to be able to determine the amount of current that must be

Figure 0.2 Input and output voltage waveforms of a logic inverter, illustrating the propagating delay times.

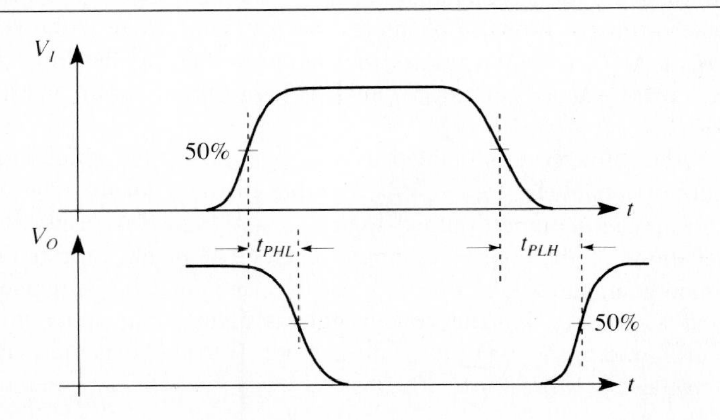

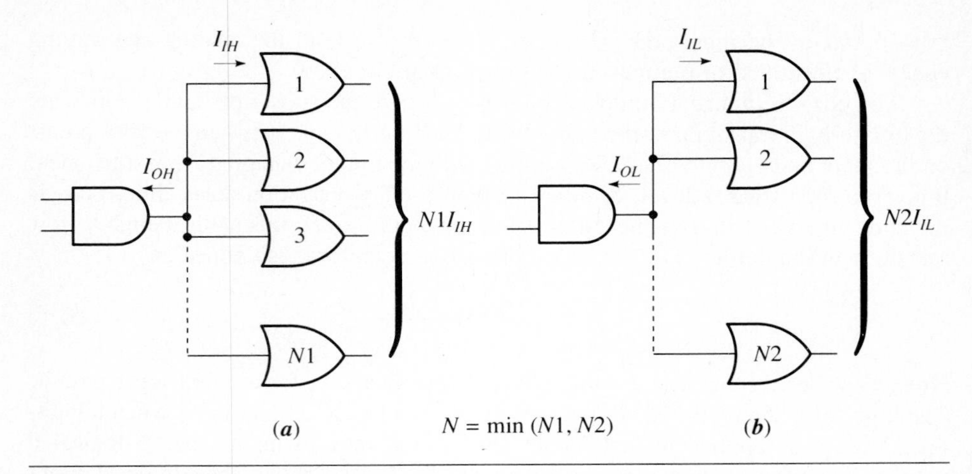

Figure 0.3 Illustration of the fan-out of a gate for (a) logical **1**, and (b) logical **0** levels.

provided by the power supply. There are two components of the power dissipated in a logic circuit. The *static power dissipation* is the power consumed while the circuit is not changing states. The *dynamic power dissipation* is the power consumed while the circuit is switching states.

The categorization of the logic families strictly on the basis of power consumption or propagation delay is inconclusive as far as overall system performance is concerned. Therefore, to provide a means of measuring circuit efficiency, a *speed-power product* efficiency index (*SP*), obtained by multiplying the average propagation delay by the circuit static power dissipation, has been developed and is often used as a figure of merit for the integrated circuits. It is usually given in picojoules, that is, in 10^{-12} W · s. The lower the SP figure for a logic family, the more effective the logic family is. One would be interested in the ideal case of high-speed performance combined with low power dissipation. Usually, though, these two requirements are in conflict because the circuit delay increases as the supply current is decreased to reduce power dissipation.

Another principal property of interest for a digital circuit is the *voltage transfer characteristic* (*VTC*), which relates the output voltage to the input voltage under static (i.e., steady-state), conditions. Such a characteristic for an inverter is shown in Figure 0.4.

The *transition region* is defined by $V_{IH} - V_{IL}$, and the difference between the two output voltage levels, $V_{OH} - V_{OL}$, signifies the *logic swing*. The subscripts I, O, H, and L designate the input, output, logical **1**, and logical **0**, respectively, such that V_{IH}, for example, is the high-level input voltage. The manufacturers usually provide the minimum guaranteed values for V_{OH} and V_{IH}, and the maximum guaranteed values for V_{IL} and V_{OL}. Note that the voltage gain is greater than unity in the transition region. Furthermore, the magnitude of the slope of the VTC is unity at points where the input voltage is either V_{IL} or V_{IH}.

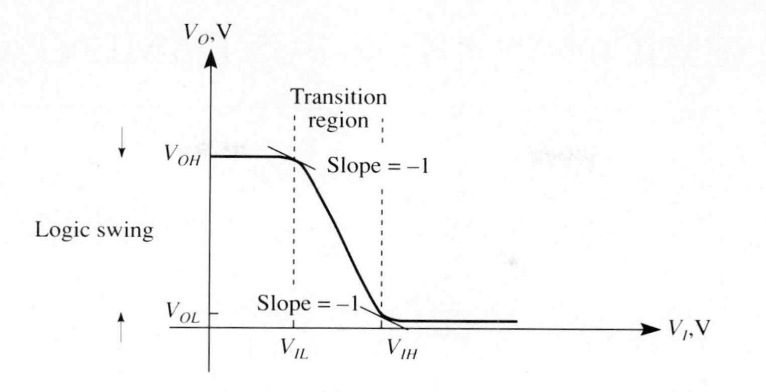

Figure 0.4 Voltage transfer characteristic of an inverting circuit.

The high- and low-level *noise margins* are defined as

$$NM_H \equiv V_{OH} - V_{IH} \tag{0.3a}$$

$$NM_L \equiv V_{IL} - V_{OL} \tag{0.3b}$$

These margins are best illustrated by the use of the logic-level diagram shown in Figure 0.5. Note that a large logic swing or a small transition region will improve the noise margins.

Figure 0.5 Logic-level diagram.

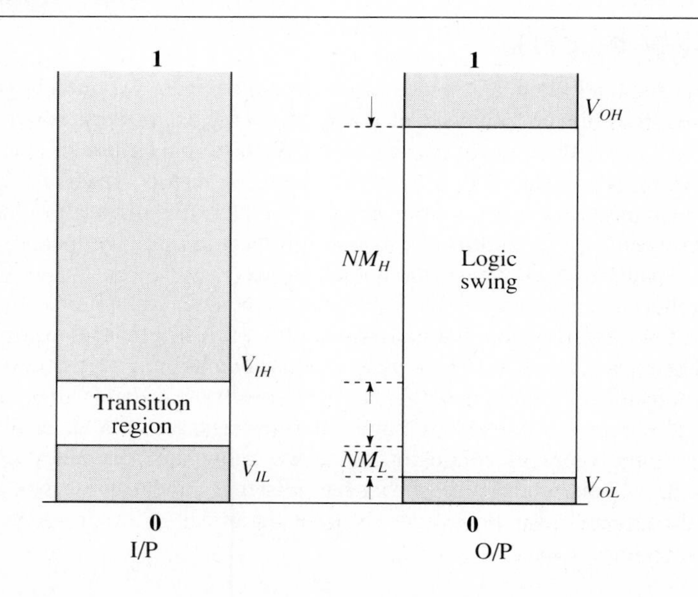

1. SEMICONDUCTOR FUNDAMENTALS

A review of the physics of semiconductor devices is especially fundamental for those future chapters in which the static and dynamic characteristics of semiconductor diodes and of bipolar and MOS transistors are discussed.

In this chapter, the varying ability of materials to carry an electric current is explained by studying the energy-band structure of solids. Our attention is mainly concentrated on semiconductors. Factors such as energy gap, doping densities, and temperature that determine carrier concentrations are studied, and current-producing mechanisms in semiconductors are also discussed. Finally, we develop a set of differential equations whose solutions give the minority carrier distributions.

1.1 Theory of Semiconductors

The Energy-Band Model

To understand the performance of electronic devices, we must begin with a discussion of the allowed behavior of electrons in solids. A very convenient way of portraying the conduction of current in conductors makes use of the *energy-band model*, which also leads us to differentiate between *metals*, *semiconductors*, and *insulators*. Quantum mechanics are utilized to develop this model for electrons. From this description we can identify the material properties that give metals, semiconductors, and insulators their electrical characteristics.

Quantum mechanics state that a particle is characterized not only by its mass, m, but also by a wave function that is derived from a three-dimensional, time-dependent differential equation called *Schrödinger's equation*. Solving this equation leads to a significant result: the energy levels available to electrons in solids are a series of bands of allowed energies separated by bands of forbidden gaps. We can distinguish between the three types of conductors once we know the occupancy of the various energy bands. This model also shows the existence of the two types of carriers in semiconductors and insulators, and describes the effect of impurities on the conduction of semiconductors.

In one type of energy-band structure, the two important bands are the *valence band* and the *conduction band*. The valence band is the range of energy levels of a solid crystal in which lie the energies of the valence electrons that bind the crystal together. It is the highest energy band containing electrons at absolute zero temperature. Since electrons always seek the lowest energy levels available, all energy levels below the valence band are completely filled at 0 K. The conduction band is the range of energy levels in which the electrons can move as free particles, giving rise to electrical conduction. It is located at higher energies than the valence band and is empty at 0 K. The two bands are separated by an *energy gap* E_g, also known as the *forbidden gap*, the width of which represents the amount of energy in electron volts (eV) that a valence electron must acquire to be able to break its covalent bond and become a free electron. Hence, when the thermal energy of a valence electron exceeds its average value by a sufficient amount for the electron to traverse the energy gap, there will be free electrons in the conduction band to provide conduction. This type of structure is found in semiconductors and insulators. These bands are illustrated in Figure 1.1a.

In another type of structure, the valence band is partially filled at 0 K, although it contains a large amount of electrons and quantum states. The valence electrons can acquire energy within the band. Thus, the valence and the conduction bands are the

Figure 1.1 (a) Illustration of the allowed energy bands and the forbidden energy gap in a semiconductor or insulator; (b) Fermi distribution function for an intrinsic semiconductor; (c) energy bands of a metal.

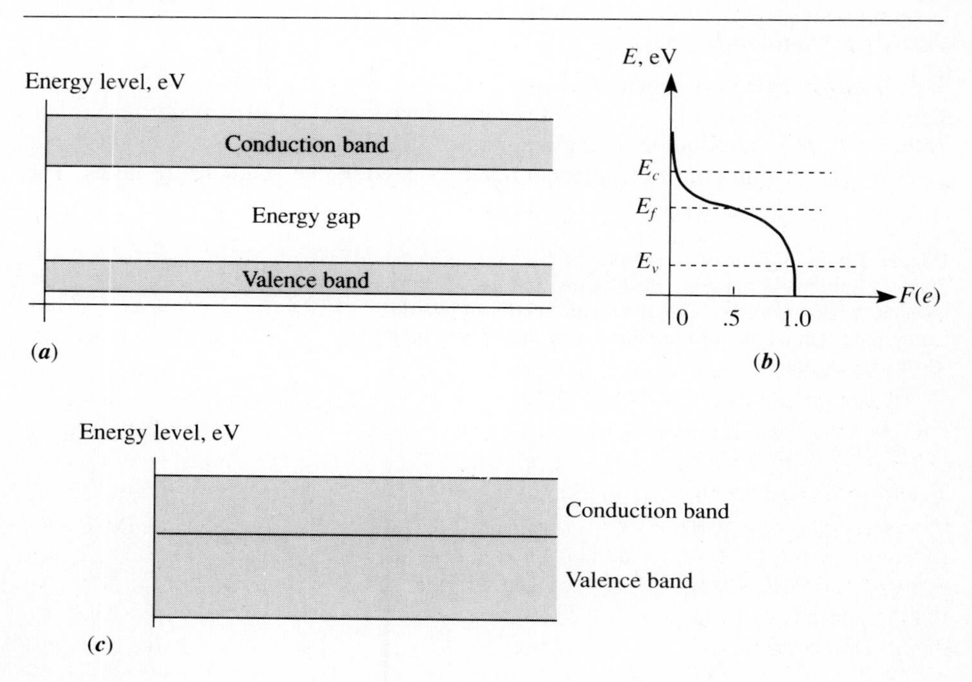

same. Materials having this energy-band structure are called *metals*. This structure is shown in Figure 1.1c.

Therefore, in both semiconductors and insulators the valence band at absolute zero temperature is completely full and separated from the conduction band by a forbidden gap. In insulators this gap is sufficiently large to allow very few carriers into the conduction band, even at reasonable temperatures. The semiconductors differ from the insulators in having a smaller value of E_g, so that at room temperature a much larger number of electrons appears in the conduction band. They have energy gaps on the order of 1 eV, while band gaps of insulators average about 10 eV.

Hole Conduction

One important feature of conduction in semiconductors and insulators is the conduction due to so-called *positively charged* particles, or *holes*. As the temperature is increased above 0 K, the electrons pass through the energy gap to fill the conduction band and leave some levels in the valence band empty, allowing the remaining electrons to acquire an incremental energy and a net velocity, which in turn result in additional current. The current due to electrons in the conduction band is proportional to the number of electrons in the conduction band, whereas the current carried by the electrons in the valence band is proportional to the vacancies in the valence band that behave like positively charged particles. Hence, the conductive process seems to be described by two oppositely charged particles. However, one should note that the hole is actually an abstract notion to explain a physical occurrence in a simple way. In reality, electrons are the only charge carriers; they are either free electrons in the conduction band or bound electrons in the valence band.

Intrinsic Semiconductors

A pure, single-crystal semiconductor, in which all the electrons in the conduction band have been thermally excited from the valence band, is called an *intrinsic semiconductor*. The simplified bonding diagram of Figure 1.2a depicts the four valence electrons in the outer orbit of each atom being shared by the neighboring atoms. The

Figure 1.2 (a) Valence-band model of silicon; (b) valence-band model of silicon containing phosphorus; the fifth electron is not needed to fill a bond. (c) Valence-band model of silicon containing boron; one of the bonds with the surrounding host atoms is not filled.

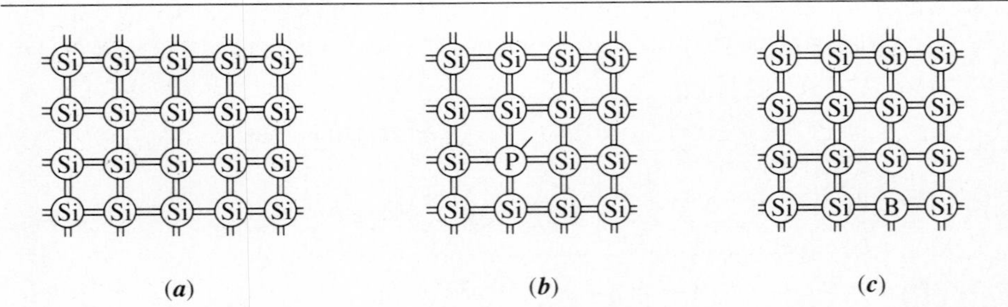

eight lines connected to each atom show not only the contribution of four shared electrons by an atom but also the accepted electrons from adjacent atoms.

The intrinsic concentration of holes, p_i, in the valence band is equal to the intrinsic concentration of electrons, n_i, in the conduction band. In *thermal equilibrium*, the temperature is uniform in space and does not vary with time. In addition, no electric field is being impressed upon the semiconductor, so that the motion of the holes and electrons is completely random. The intrinsic carrier concentrations in thermal equilibrium are designated as p_{io} and n_{io}, and again

$$p_{io} = n_{io} \tag{1.1}$$

since the holes in the valence band result from the thermal promotion of valence electrons to the conduction band.

Effect of Impurities and Change in Energy-Band Structure

Semiconductors whose electrical properties are dominated by impurities are called *extrinsic semiconductors*. These are the materials most often used in device and integrated-circuit fabrication. The presence of an impurity atom leads to a different solution to the aforementioned Schrödinger equation, which results in a different energy spectrum for an electron in the crystal.

Impurity atoms with an excess of valence electrons are called *donors*. If the density of the impurity atoms added is small compared to the density of the host semiconductor, then their presence introduces a discrete energy level, E_d, in the forbidden energy gap just below the bottom edge of the conduction band. This energy level, otherwise empty at 0 K, is now occupied by an electron.

Consider an element—for example, phosphorus—from Group V of the periodic table, with five valence electrons. Suppose such an atom is substituted for one of the host silicon atoms, as illustrated in Figure 1.2b. Since the fifth electron is weakly bound to the phosphorus ion and requires only .044 eV to break free, compared to the energy of 1.12 eV needed to free an electron from an Si-Si bond, these atoms are much more easily *thermalized* than the host semiconductor atoms to donate conduction electrons, while maintaining four valence electrons. Therefore, donors produce conduction electrons without corresponding valence-band holes and preserve charge neutrality because the *ionized* donor is now positively charged. In this way, free electrons can be produced at much lower temperatures than those required to produce them from broken silicon bonds. Even very small concentrations of impurities have profound effects on the semiconductor's conductivity. The slight change due to donors in the energy states of the four electrons is shown in Figure 1.3a, where the bound-state energy level, E_d, is called the *donor energy level*.

Example 1.1

Intrinsic silicon at room temperature contains about 10^{10} carriers/cm^3, and the atomic density of silicon is approximately 10^{22} atoms/cm^3, which corresponds to one carrier per 10^{12} silicon atoms. The addition of 10^{16} phosphorus atoms/cm^3 translates to an impurity

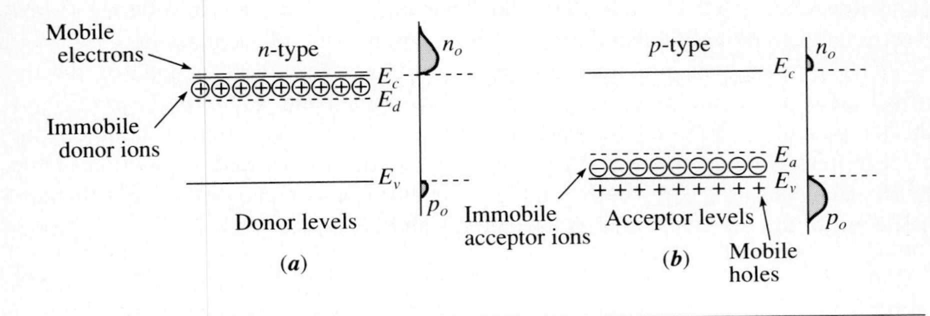

Figure 1.3 Energy levels in extrinsic semiconductors and carrier concentrations.

concentration of 10^{-6}. Therefore, as each phosphorus atom gives up its fifth electron, the carrier concentration and conductivity increases a millionfold.

Intrinsic generation of free electrons involves the generation of an equal number of holes. In impurity generation, on the other hand, as in the ionization of the phosphorus atom, there is no hole involved. Therefore, such impurities cause conduction by only one carrier type. In a semiconductor containing Group V impurities, the charge carriers are predominantly negatively charged, and hence such a semiconductor is said to be *n-type*. The electron concentration is larger than the hole concentration; that is, electrons are the *majority* carriers, and holes are the *minority* carriers.

Impurity atoms with a deficiency of valence electrons are called *acceptors*. Thus, they tend to capture electrons, producing excess valence-band holes in the semiconductor but producing no conduction-band electrons. The single energy level necessary to capture an electron is called the *acceptor energy level*, E_a, and is shown in Figure 1.3b. E_a is above the top of the valence band ready to accept an electron from this band to create a hole.

Consider substituting a Group III element such as boron into a lattice position of a silicon crystal. Having only three valence electrons, boron cannot fill all the bonds with the surrounding silicon atoms, as illustrated in Figure 1.2c. An energy of .045 eV is required to move a neighboring bound electron into the empty B-Si bond position, thus causing a hole in the Si-Si bonds. A boron atom is ionized, and its presence produces a *p-type* semiconductor because the majority carriers now are positively charged holes.

Over a wide range of temperature, donors and acceptors are almost totally ionized. The former contribute conduction electrons, and the ion sites have a net positive charge; the latter produce holes and carry a net negative charge. In extrinsic semiconductors, the numbers of conduction electrons and valence holes are not equal. The number of holes in a *p*-type semiconductor is normally much greater than the number of free electrons. For *n*-type semiconductor, the reverse is true.

There are two basic reasons for using Group III or Group V elements as impurities:

1. Their concentrations can be readily controlled.
2. Their small ionization energy allows all impurities to be ionized at temperatures far below room temperature.

On the other hand, when a donor level lies near the valence band, the conductive properties of the semiconductor are variable due to the continuing ionization of the impurities at and beyond room temperature. Elements from other groups, when used as donor or acceptor impurities, introduce energy levels in the energy gap far away from either band edge, thereby resulting in higher ionization energy requirements compared to those for the elements in Group III and Group V. The ionization energies of various impurities are listed in Table 1.1. Note that gold gives rise to more than one impurity level in both silicon and germanium.

As the temperature is increased far above room temperature, the intrinsic generation starts to dominate, and the semiconductor again behaves as an intrinsic material. For Si, the onset of this phenomenon is at about 250° C.

Charge-Carrier Concentrations

The number of available carriers determines the electrical properties of semiconductors and is found from the number of allowed states as well as from the probability that an energy level, E, is occupied by conduction electrons. This probability is given by the *Fermi distribution function*:

$$f(E) = \frac{1}{e^{(E-E_f)/kT} + 1} \tag{1.2}$$

where E_f is called the *Fermi level*, T is the absolute temperature in degrees Kelvin, and the quantity k is a universal constant, called *Boltzmann's constant*. Its value is 1.38×10^{-23} J/K so that at 300 K, $kT = 4.1 \times 10^{-21}$ J or .026 eV. The Fermi level is a function of the temperature, type of material, and impurities present, although for any system in thermal equilibrium, it is independent of time and spatial position. It is defined as the energy at which the probability of occupation by an electron is exactly .5.

The distribution function is symmetrical with respect to E_f, as illustrated in Figure 1.4. For $E_f > E$ at 0 K, $f(E)$ is unity, which indicates that all energy levels below E_f are occupied, and all energy levels greater than E_f are empty. Furthermore, at $E_f = E$ for $T > 0$ K, the probability of occupancy is always .5. Finally, the function can be approximated by

$$f(E) \approx e^{-(E-E_f)/kT}$$

for all energy levels higher than $E_f + 3kT$.

For intrinsic semiconductors, the probability of an electron occupying one of the allowed states in the conduction band is small. By contrast, the probability of an electron occupying one of the large number of allowed states in the valence band is almost unity. This is illustrated in Figure 1.1b. As can be seen, E_f is located near the middle of the energy gap.

Table 1.1 Ionization Energies of Various Impurities in Si and Ge

Impurities	Ionization energies (eV)	
	Si	Ge
Group III		
Boron	.045	.010
Aluminum	.057	.010
Gallium	.065	.011
Indium	.160	.011
Group V		
Phosphorus	.044	.012
Arsenic	.049	.013
Antimony	.039	.010
Bismuth	.069	.000
Gold (donor)	.350	.050
Gold (acceptor)	.540	.040
		.150
		.200

Figure 1.4 The Fermi probability density function (a) as $T \to 0$ K, and (b) generalized $T > 0$ K plot.

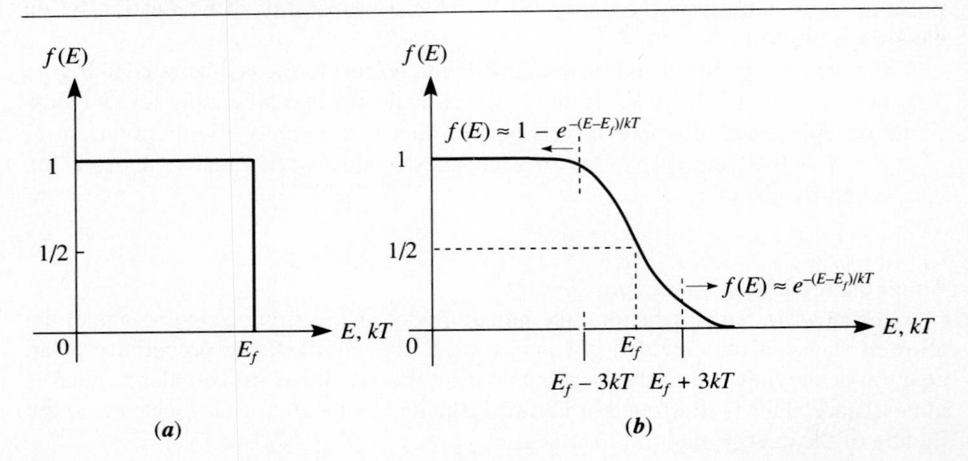

In addition to charges, electrons and holes in a semiconductor carry masses that are different from the electron mass, m_o, in free space. The *effective mass*, m_e, of a carrier depends on the energy band it occupies as well as the momentum of the carrier and the direction of the applied electric field. If the effective masses of electrons and holes are equal, the *intrinsic Fermi level*, denoted by E_i, is located at the middle of the gap, specifying the *midgap energy* of the semiconductor.

For an *n*-type material doped with a donor concentration N_d at absolute zero temperature, the valence band is full, the conduction band is empty, and N_d donor electrons/cm^3 reside in donor states. As the temperature is increased to room temperature, suppose that N_d donor electrons are all excited into the conduction band. Some holes will also exist, arising from the thermal excitation of valence electrons out of the valence band and into the conduction band. Therefore, we will have

$$n_o = N_d + p_o \tag{1.3}$$

where the subscript *o* indicates that the concentration has its thermal equilibrium value.

In thermal equilibrium, new electron-hole pairs are generated as fast as old ones are annihilated by the *recombination* process, as will be discussed later. To bring about a thermal equilibrium balance between thermal generation and recombination, it can be shown that

$$p_o n_o = n_{io}^2 \tag{1.4}$$

for a given temperature. Equation (1.4) states that when impurities are introduced to increase one carrier type, the other carrier type is decreased. Therefore, we have

$$p_o = \frac{n_{io}^2}{n_o} \tag{1.5}$$

Normally, the doping is such that $N_d >> n_{io}$. Since the minority concentration, p_o, is always smaller than n_{io}, we obtain in *n*-type

$$n_o \approx N_d \tag{1.6}$$

Then by Equation (1.5), we find

$$p_o \approx \frac{n_{io}^2}{N_d} \tag{1.7}$$

An exact solution would proceed by first substituting Equation (1.5) into (1.3),

$$n_o = N_d + \frac{n_{io}^2}{n_o}$$

and then by solving the quadratic formula to obtain

$$n_o = \frac{N_d}{2} + \sqrt{\left(\frac{N_d}{2}\right)^2 + n_{io}^2} \tag{1.8}$$

Similarly, in a p-type material doped with an acceptor concentration N_a, the acceptors are neutral at absolute zero temperature. Suppose that at 300 K all acceptors accept electrons from the valence band to produce N_a holes/cm^3. Thus,

$$p_o = N_a + n_o \approx N_a \tag{1.9}$$

and from Equation (1.5),

$$n_o = \frac{n_{io}^2}{p_o} \approx \frac{n_{io}^2}{N_a} \tag{1.10}$$

An exact solution is obtained as follows:

$$p_o = \frac{N_a}{2} + \sqrt{\left(\frac{N_a}{2}\right)^2 + n_{io}^2} \tag{1.11}$$

A usually accepted expression for the intrinsic carrier concentration is given in terms of the band-gap energy, E_g, as

$$\begin{aligned} n_i &= \sqrt{N_c N_v}\, e^{-E_g/2kT} \\ &= N_c e^{-(E_c - E_i)/kT} \\ &= N_v e^{-(E_i - E_v)/kT} \end{aligned} \tag{1.12}$$

where E_v is the potential energy of the top of the valence band, and E_c is the potential energy associated with the bottom of the conduction band, so that $E_c - E_v \equiv E_g$, and N_v and N_c are called the *effective densities* of allowed energy states at E_v and E_c, respectively. Both N_v and N_c depend on $\sqrt{T^3}$, although the exponential term in Equation (1.12) dominates. Values for N_v and N_c as well as for n_i at room temperature are given in Table 1.2.

Therefore, considering Equation (1.4), the carrier concentrations are exponentially related to the band-gap energy of the material, which in turn is a slowly varying function of the temperature. Hence, the calculations based on Equation (1.12) are slightly different from the measured values.

As long as the Fermi energy, E_f, does not approach within several kT of either energy-band edge, the concentrations are given by

$$\begin{aligned} n &= N_c e^{-(E_c - E_f)/kT} \\ &= n_i e^{(E_f - E_i)/kT} \end{aligned} \tag{1.13}$$

Table 1.2 Values for N_c, N_v, and n_i per cm^3 for Si and Ge at 300 K

	Si	Ge
N_c	2.9×10^{19}	8.8×10^{18}
N_v	1.1×10^{19}	6.2×10^{18}
n_i	1.5×10^{10}	2.4×10^{13}

and

$$p = N_v e^{-(E_f - E_v)/kT}$$
$$= n_i e^{(E_i - E_f)/kT} \quad (1.14)$$

Example 1.2

To find the Fermi level at room temperature for Si doped with 10^{15} phosphorus atoms per cm^3, we first note that $N_d >> n_i$, so that $n \approx N_d = 10^{15}\ cm^{-3}$, and $p = 2.25 \times 10^5\ cm^{-3}$. Then, substituting Equation (1.6) into (1.13) and solving for E_f, we obtain

$$E_f = E_c - kT \ln\left(\frac{N_c}{N_d}\right) = E_c - .265\ \text{eV}$$

Current Flow in a Solid

The two basic mechanisms that lead to the flow of charges in semiconductors are a potential gradient, which results in a *conduction current* described by Ohm's Law, and a concentration gradient due to a nonuniform distribution of mobile carriers, which leads to a *diffusion current*. The conduction current is also called the *drift current*. It results from electrical forces producing directed motion of electric charges. A useful form of Ohm's Law that includes explicit parameters of the conducting medium is given in terms of the current density, **J**, and electric field intensity, $\mathcal{E}$, as

$$\mathbf{J} = \sigma\mathcal{E} = \left(\frac{1}{\rho}\right)\mathcal{E} \quad (1.15)$$

where σ is the electrical conductivity, and ρ is the resistivity of the material. From Equation (1.15), it is still not possible to determine the reason for the conductivity of a metal being much larger than that of an insulator. Hence, to gain more insight into the nature of σ, we first consider Newton's Second Law,

$$\mathbf{F} = m\mathbf{a}$$

and the expression for the force **F** on an electron in a solid due to an applied field $\mathcal{E}$,

$$\mathbf{F} = -q\mathcal{E}$$

where q is the magnitude of electron charge. Equating these two equations, we obtain the differential equation

$$\frac{-q\mathcal{E}}{m_e} = \frac{d^2x}{dt^2} \quad (1.16)$$

where x is the presumed direction of the current flow. Solving Equation (1.16) results in

$$\frac{-q\mathcal{E}t}{m_e} = v - v_{th} \tag{1.17}$$

where v_{th} is the average thermal velocity, having a value on the order of 10^7 cm/sec at room temperature. Hence, a velocity component is superimposed upon the random thermal velocity to cause a net displacement of the electrons. The resulting average net velocity is called the electron *drift velocity* $\mathbf{v}_{de}$, and its magnitude is given by

$$v_{de} = \frac{v - v_{th}}{2} \tag{1.18}$$

Using Equation (1.17), we get

$$v_{de} = -\frac{1}{2}\left(\frac{qT_e}{m_e}\right)\mathcal{E} \tag{1.19}$$

where T_e is the average time between collisions, called the *mean free time*. The factor 1/2 in Equation (1.19) is eliminated in a more accurate model when proper averaging of T_e over the distribution of random electron velocities is included. The quantity within the parentheses in (1.19) is called the *electron mobility*, μ_e

$$\mu_e \equiv \frac{qT_e}{m_e} \tag{1.20}$$

Mobility is a measure of the ease of carrier motion within a semiconductor crystal. It is the rate of movement of the carrier (in cm/s) per unit electric field (1 V/cm). Thus, the standard unit for μ is $cm^2/V \cdot s$. A low value means the carriers are suffering a large number of collisions with lattice atoms, while a large mobility implies relative ease in carrier movement. In major semiconductors μ_e is always greater than the *hole mobility* μ_h for a given doping level and temperature simply because it is easier to move a free electron in the conduction band than it is to move a bound electron into a hole in the valence band. It is not unreasonable to talk about hole mobility, because the holes can be made to move by the application of a voltage. Their movement is brought about by the movement of valence-band electrons from hole to hole. Table 1.3 lists the observed doping dependence of the electron and hole mobilities in silicon at room temperature. Below 10^{14} cm^{-3}, the carrier mobilities are independent of the doping concentration. For dopings above this value, the mobilities decrease as the concentrations increase.

For all practical purposes, the following empirical relationships can be employed to describe the doping density dependence of the mobilities at 300 K:

$$\mu_e = 88 + \frac{1252}{1 + (6.984 \times 10^{-18})N_a} \tag{1.21a}$$

$$\mu_h = 54.3 + \frac{407}{1 + (3.745 \times 10^{-18})N_d} \tag{1.21b}$$

Table 1.3 Room-Temperature Carrier Mobilities ($cm^2/V \cdot s$) in Si as a Function of the Doping Density

N_a or N_d (cm^{-3})	μ_e	μ_h
10^{14}	1358	492
2×10^{14}	1357	489
5×10^{14}	1352	484
10^{15}	1345	477
2×10^{15}	1332	465
5×10^{15}	1298	438
10^{16}	1248	406
2×10^{16}	1165	363
5×10^{16}	986	291
10^{17}	801	233

Another factor that affects mobility is the temperature. As energy is added to the crystal structure with increased temperature, some of this energy is used to excite the electrons to the conduction band, while the rest is given to the atomic cores, causing them to oscillate in and out of position. That is to say, the thermal energy of a system leads to deviations of the atoms from their positions. The resulting collision with electrons and holes is called *lattice scattering*, which increases with increasing temperature because of the increased lattice vibration. These vibrations are considered in quantum mechanical terms as discrete particles called *phonons*, which, like electrons, have actually the dual nature of wave and particle. A phonon's oscillation frequency is quantized corresponding to a discrete energy. At any rate, increased scattering gives rise in turn to a shorter mean free time and hence impedes carrier motion, resulting in a lower mobility.

Consider now the current flowing in a conductor of cross-sectional area A. The volume of an element of the conductor of length Δx is given as $A\Delta x$. Then, the charge, ΔQ, within this element is $-nqA\Delta x$, where n is the number of electrons per unit volume of mobile electrons available. The time, Δt, needed for them to traverse Δx is $\Delta x / v_{de}$. Therefore, the current density for any conductor can be written in terms of the electron drift velocity as

$$\mathbf{J}_e = \frac{I_e}{A} = \frac{\Delta Q}{A\Delta t} = -qn\mathbf{v}_{de}$$

Therefore, using Equation (1.20), we obtain

$$\mathbf{J}_e = qn\mu_e \boldsymbol{\mathcal{E}}$$

so that the electron conductivity is given by

$$\sigma_e = qn\mu_e \tag{1.22}$$

which is the expression we have been seeking for the conductivity in terms of the material parameters.

When both holes and electrons are present, the currents they carry become additive. Consider an electric field applied in a semiconductor containing n electrons/cm^3 and p holes/cm^3, with carrier mobilities μ_e and μ_h, respectively. The holes have the same direction as the electric field. This results in a positive hole conduction current density given by

$$\mathbf{J}_h = qp\mathbf{v}_{dh} = qp\mu_h \boldsymbol{\mathcal{E}} = \sigma_h \boldsymbol{\mathcal{E}}$$

where $\mathbf{v}_{dh}$ is the hole drift velocity, and σ_h is the hole conductivity. Therefore, both current densities have the same direction, so that the total current density expression for a semiconductor becomes

$$\mathbf{J} = q(p\mu_h + n\mu_e)\boldsymbol{\mathcal{E}} = (\sigma_h + \sigma_e)\boldsymbol{\mathcal{E}} = \sigma\boldsymbol{\mathcal{E}}$$

where

$$\sigma = q(p\mu_h + n\mu_e) \tag{1.23}$$

is the total conductivity. In many extrinsic semiconductors the drift current due to the minority carriers can be neglected because of their very small number.

A local concentration of particles in a portion of a volume causes the particles to spread out until they are eventually distributed uniformly throughout the available space. Similarly, if the holes and the electrons in a semiconductor are not uniformly distributed, their random thermal motion tends to reduce the nonuniformity. This process is referred to as *diffusion*.

A mathematical expression for the particle flux density, F, developed to describe the diffusive flow (i.e., the rate at which particles flow through a unit area), is given by

$$F = -D\nabla\zeta \tag{1.24}$$

where ζ is the particle concentration, ∇ is the vector gradient operator, and D is the *diffusion coefficient*, which depends on the type of particles diffusing as well as the medium in which they flow. For the diffusion of electrons in one dimension, Equation (1.24) takes the form

$$F_e = -D_e \frac{dn}{dx}$$

where n is the nonuniform electron concentration and D_e is the diffusion coefficient of the electrons in the particular medium considered. To describe the diffusion of holes, we will have

$$F_h = -D_h \frac{dp}{dx}$$

where p is the nonuniform hole concentration, and D_h is the diffusion coefficient of the holes in the medium being considered.

In the presence of a concentration gradient, the electric current density is given by

$$J_e = -qF_e = +qD_e \frac{dn}{dx}$$

for electrons, and

$$J_h = qF_h = -qD_h \frac{dp}{dx}$$

for holes. Therefore, the total current in a semiconductor is

$$\mathbf{J} = \mathbf{J}_{e(\text{drift})} + \mathbf{J}_{e(\text{diffusion})} + \mathbf{J}_{h(\text{drift})} + \mathbf{J}_{h(\text{diffusion})}$$

or

$$\mathbf{J} = qn\mu_e \mathcal{E} + qD_e \frac{dn}{dx} + qp\mu_h \mathcal{E} - qD_h \frac{dp}{dx}$$

Arranging the terms on the right side, we obtain

$$\mathbf{J} = q(n\mu_e + p\mu_h)\mathcal{E} + q\left(D_e \frac{dn}{dx} - D_h \frac{dp}{dx}\right) \tag{1.25}$$

which is different from the expression for the current density in a metal, where $\mathbf{J} = qn\mu_e \mathcal{E}$. Semiconductors can sustain an electric field and carrier concentration gradients, while metals cannot.

The preceding discussion also shows that the conduction current is proportional to the carrier concentration, while the diffusion current is proportional to the gradient of concentrations. It is obvious from Equation (1.25) that the electric field in a semiconductor includes not only the applied electric field but also an internal electric field produced by the carrier distribution due to doping concentration gradients.

Thermal Equilibrium in a Nonuniformly Doped Semiconductor

Consider an n-type semiconductor sample with a nonuniform concentration, n_o. For a material in thermal equilibrium, the total net current is zero. Moreover, since there is no externally applied electric field, the current of the conduction electrons must also be zero. On the other hand, the unbalanced charge distribution also creates a *built-in* electric field $\mathcal{E}_o$ causing a conduction flow of electrons whose corresponding current density is expressed as $qn_o\mu_e\mathcal{E}_o$. Hence, in equilibrium the electron concentration must vary in such a way that the diffusion set up by the nonuniformity is just balanced by the conduction flow due to $\mathcal{E}_o$. Thus,

$$qn_o\mu_e\mathcal{E}_o + qD_e \frac{dn_o}{dx} = 0 \tag{1.26}$$

In Equation (1.13), the temperature, T, the Fermi level, E_f, and N_c are constant so that E_c is the only variable on the right side. Also note that E_c is the potential energy for the conduction electrons. Hence, its gradient is the negative of the force $-q\mathcal{E}_o$ that is being applied to the conduction electrons. Therefore, we obtain

$$\frac{dE_c}{dx} = -(-q\mathcal{E}_o) = q\mathcal{E}_o \tag{1.27}$$

Taking the derivative of both sides in Equation (1.13) with respect to x, we get

$$\frac{dn_o}{dx} = \left(-\frac{1}{kT}\right) n_o \frac{dE_c}{dx} \tag{1.28}$$

Substituting Equation (1.27) into (1.28), and (1.28) into (1.26), simplifying and rearranging (1.26) lead to

$$\left(\frac{kT}{q}\right) \mu_e = \phi_T \mu_e = D_e \tag{1.29}$$

where $\phi_T \equiv kT/q$ is called the *thermal voltage* and is equal to 25.9 mV at 300 K. Carrying out the development for a p-type semiconductor would result in

$$\left(\frac{kT}{q}\right) \mu_h = \phi_T \mu_h = D_h \tag{1.30}$$

Equations (1.29) and (1.30) are called the *Einstein relations*, which become invalid when the impurity concentration becomes too large. Experimentally, the value for the concentration is found to be around $10^{19}/\text{cm}^3$ in silicon before these relations fail.

To find the value of the internal electric field intensity $\mathcal{E}_o$ that should balance out the effect of the diffusion in equilibrium, we solve Equation (1.26) for $\mathcal{E}_o$ in one dimension and use Equation (1.29) to get

$$\mathcal{E}_o = \left(-\frac{\phi_T}{n_o}\right) \frac{dn_o}{dx} \tag{1.31}$$

Equation (1.31) expresses the built-in field in terms of the majority carrier concentration. It states that this internal field exists only if there is a nonuniform carrier concentration.

Recombination-Generation

In addition to drift and diffusion, there is one more type of carrier action occurring inside semiconductors. It is called the *recombination-generation* process. *Recombination* is a process in which electrons and holes are annihilated whereas in *generation*, electrons and holes are created. There are a number of different ways in which carriers can be created and destroyed within a semiconductor. In *photogeneration*, for instance, process light with an energy greater than E_g impinges upon the semiconductor and excites electrons from the valence band into the conduction band, thereby creating conduction electrons and valence holes. *Direct thermal generation*, as already mentioned, is identical to photogeneration, except the excitation is due to the thermal energy present within the material. In *direct thermal recombination*, an electron and hole moving in the lattice wander into the same spatial vicinity, annihilating each other. Finally, *indirect thermal recombination-generation*, which is the dominant way to thermally create or annihilate carriers in the semiconductors, takes place at special locations known as *R-G centers* or *traps*. These centers are lattice defects and unintentional impurities, and they are present even in device-quality materials. Figure 1.5 illustrates an important property of the R-G centers, using the energy-band model. These traps introduce an allowed energy level E_T near the center of the band gap. As shown on the left-hand side of Figure 1.5, an electron travels to the vicinity of an R-G center, loses energy, and is trapped. Subsequently, a hole comes along, is attracted to the electron, and annihilates the electron and itself within the center. Obviously, it is also possible for the hole to be captured first.

The right-hand side of Figure 1.5 shows the opposite process, namely, *indirect thermal generation*. An electron is first thermally excited to the E_T level, leaving a hole behind in the valence band. Then it absorbs additional thermal energy and jumps into the conduction band, becoming a carrier and completing the creation process.

Thermal recombination-generation is a mechanism that stabilizes the carrier concentrations within a material, whereas photogeneration acts so as to create an excess of carriers and occurs when the semiconductor is illuminated. Recombination-generation affects current flow in semiconductors by controlling the carrier concentrations involved in drift and diffusion.

The creation of excess carriers in semiconductors is called *injection*. Let us define excess carrier concentrations Δp and Δn that measure the extent of departure from thermal equilibrium, as follows:

$$\Delta p \equiv p - p_o \tag{1.32a}$$

$$\Delta n \equiv n - n_o \tag{1.32b}$$

where n and p are carrier concentrations under arbitrary conditions. The injection of excess carriers is considered to be small if

$$\Delta p \ll n_o,\ n \approx n_o \text{ in an } n\text{-type material,}$$

$$\Delta n \ll p_o,\ p \approx p_o \text{ in a } p\text{-type material.}$$

If this condition, referred to as *low-level injection*, is met, then the net effect of an indirect recombination-generation process is characterized mathematically by specifying the time rate of change in the perturbed minority carrier concentration as

$$\left.\frac{\partial p}{\partial t}\right|_{\text{R-G}} = -\frac{\Delta p}{\tau_h} \text{ for holes in an } n\text{-type material} \tag{1.33a}$$

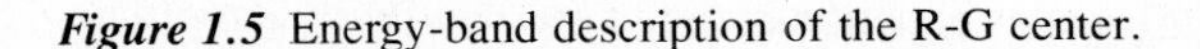

Figure 1.5 Energy-band description of the R-G center.

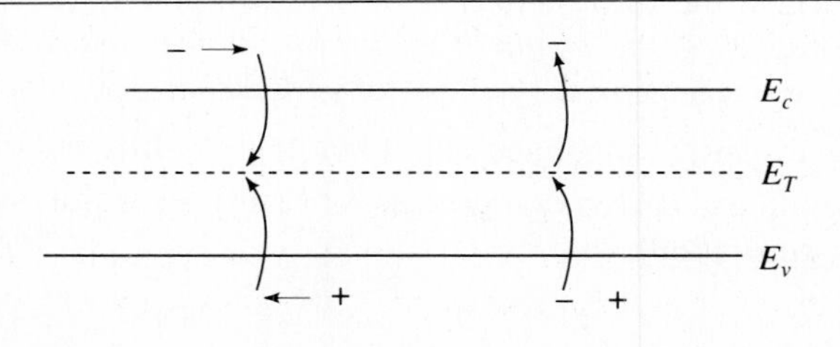

$$\left.\frac{\partial n}{\partial t}\right|_{\text{R-G}} = -\frac{\Delta n}{\tau_e} \text{ for electrons in a } p\text{-type material} \tag{1.33b}$$

where the time constants τ_h and τ_e are called the *excess minority carrier lifetimes* in n-type and p-type materials, respectively. They can be interpreted as the average time an excess minority carrier will *live* in the midst of majority carriers. The fact that these important parameters depend on poorly controlled R-G center concentration rather than the carefully controlled doping parameters N_a and N_d is the main reason not to have relevant facts and information about them. Fabrication procedures, the type of the semiconductor, the presence of various impurities such as intentional introduction of gold, and the degree of crystalline perfection as measured by the number of R-G centers greatly affect the values of τ_h and τ_e. Equation Set (1.33) shows that the thermal recombination-generation process is essentially controlled by the excess minority carrier concentration.

The following empirical relationship associates the minority carrier lifetimes for both types of material to the impurity concentration:

$$\tau = \frac{5 \times 10^{-7}}{1 + (2 \times 10^{-17})N} \tag{1.34}$$

The appendix at the end of this chapter lists a program that, given the impurity concentration, calculates various semiconductor parameters at room temperature for both types of silicon material.

Example 1.3

Assume a doped Si material at 300 K with $N_d = 10^{14}$ cm^{-3} and subject to a perturbation with $\Delta p = \Delta n = 10^9$ cm^{-3}. To determine if we have low-level injection, we note that it is an n-type material, so that the majority carrier concentration $n_o \approx N_d = 10^{14}$ cm^{-3} while the minority carrier concentration $p_o \approx n_{io}^2/N_d = 2.25 \times 10^6$ cm^{-3}. Thus, $n = n_o + \Delta n \approx n_o$ remains essentially unperturbed. Also, $\Delta p = 10^9$ cm^{-3} $\ll n_o = 10^{14}$ cm^{-3}.

Therefore, low-level injection conditions prevail. Note that $\Delta p \gg p_o$; that is, the minority carrier concentration increases over its thermal equilibrium value by many orders of magnitude.

Continuity Equations

Next, we develop a set of differential equations, called the *continuity equations*, whose solutions give the distributions of carrier concentrations both in space and time. They are based on the *continuity principle*, which states that in a specified region of space, any change in the number of particles depends upon

1. The flow of particles into and out of that region (drift and diffusion)
2. The creation and destruction processes of particles within the region (recombination-generation)

We use this principle to develop the continuity equations and determine the state of a semiconductor system by taking into account the combined effect of the individual types of carrier action.

The overall change in the minority carrier concentration per unit of time, then, is given by

$$\frac{\partial n}{\partial t} = \left.\frac{\partial n}{\partial t}\right|_{\text{drift}} + \left.\frac{\partial n}{\partial t}\right|_{\text{diffusion}} + \left.\frac{\partial n}{\partial t}\right|_{\substack{\text{indirect}\\ \text{thermal R-G}}} + \left.\frac{\partial n}{\partial t}\right|_{\substack{\text{other}\\ \text{processes}}}$$

for p-type material, and by

$$\frac{\partial p}{\partial t} = \left.\frac{\partial p}{\partial t}\right|_{\text{drift}} + \left.\frac{\partial p}{\partial t}\right|_{\text{diffusion}} + \left.\frac{\partial p}{\partial t}\right|_{\substack{\text{indirect}\\ \text{thermal R-G}}} + \left.\frac{\partial p}{\partial t}\right|_{\substack{\text{other}\\ \text{processes}}}$$

for n-type material.

To derive the one-dimensional drift and diffusion components of the continuity equation, we consider either current component $J_\eta(x)$ entering the interval between x and $x + dx$ of a conductor of cross-sectional area A. The net rate of change in the number of particles within the volume Adx is then given by

$$\left.\frac{\partial \zeta}{\partial t}\right|_{\substack{\text{drift or}\\ \text{diffusion}}} (Adx) = \left[\frac{J_\eta(x)A}{-q}\right] - \left[\frac{J_\eta(x + dx)A}{-q}\right]$$

Expanding the second term in a Taylor's series results in

$$J_\eta(x + dx) = J_\eta(x) + \left(\frac{J_\eta(x)}{x}\right) dx + \cdots$$

Therefore,

$$\left.\frac{\partial \zeta}{\partial t}\right|_{\substack{\text{drift or}\\ \text{diffusion}}} = \frac{1}{q}\frac{\partial J_\eta(x)}{\partial x}$$

In three dimensions, then, we have

$$\frac{\partial n}{\partial t} = \left(\frac{1}{q}\right) \nabla \cdot \mathbf{J}_e - \frac{\Delta n}{\tau_e} + \left.\frac{\partial n}{\partial t}\right|_{\text{other}} \qquad p\text{-type material} \qquad (1.35a)$$

$$\frac{\partial p}{\partial t} = \left(-\frac{1}{q}\right) \nabla \cdot \mathbf{J}_h - \frac{\Delta p}{\tau_h} + \left.\frac{\partial p}{\partial t}\right|_{\text{other}} \qquad n\text{-type material} \qquad (1.35b)$$

The first term on the right-hand side of Equation Set (1.35) states that if more carriers drift and/or diffuse into a given section of the semiconductor than out of that section, then there will be a change in the carrier concentrations within that region. When these equations are solved, we observe that the excess carrier concentrations decay exponentially to their equilibrium values when the external stimulus for carrier production is removed. Although the proof is beyond the scope of our discussion, we should note that the electric field induced by any carrier concentration gradient has a very small effect on the minority carrier flow, so that the conduction term can be dropped. This means that the minority current density is due entirely to diffusion. This results in further simplification of the continuity equations to yield, in one dimension:

$$\frac{\partial n}{\partial t} = D_e \frac{\partial^2 \Delta n}{\partial x^2} - \frac{\Delta n}{\tau_e} + \left.\frac{\partial \Delta n}{\partial t}\right|_{\text{other}} \qquad p\text{-type material} \qquad (1.36a)$$

$$\frac{\partial p}{\partial t} = D_h \frac{\partial^2 \Delta p}{\partial x^2} - \frac{\Delta p}{\tau_h} + \left.\frac{\partial \Delta p}{\partial t}\right|_{\text{other}} \qquad n\text{-type material} \qquad (1.36b)$$

Here, we used Equations (1.32) and (1.33) and assumed that the equilibrium minority carrier concentrations were not a function of position, so that $\partial\zeta/\partial x = \partial\Delta\zeta/\partial x$. It turns out that, in general, to examine only the behavior of minority carriers is sufficient to derive anything else in semiconductor devices. Since the equilibrium carrier concentrations are not a function of time either, we finally arrive at the *excess* minority carrier continuity equations:

$$\frac{\partial \Delta n}{\partial t} = D_e \frac{\partial^2 \Delta n}{\partial x^2} - \frac{\Delta n}{\tau_e} + \left.\frac{\partial \Delta n}{\partial t}\right|_{\text{other}} \qquad p\text{-type material} \qquad (1.37a)$$

$$\frac{\partial \Delta p}{\partial t} = D_h \frac{\partial^2 \Delta p}{\partial x^2} - \frac{\Delta p}{\tau_h} + \left.\frac{\partial \Delta p}{\partial t}\right|_{\text{other}} \qquad n\text{-type material} \qquad (1.37b)$$

Summary

Charge carriers that are responsible for the electricity conduction in the semiconductor are the electrons in the conduction band and holes in the valence band. In thermal equilibrium, there is no net charge flow and hence no current as the particles move in a random motion with a temperature-dependent thermal velocity. In their motion, electrons and holes meet, recombine, and disappear as free charges. At absolute zero temperature, no electron-hole pairs are produced and the semiconductor acts as an insulator.

The distribution of electrons among permissible states in a system at thermal equilibrium is described by the Fermi distribution function. It defines a Fermi level that is 50% occupied at all temperatures. It also predicts that all states below the Fermi level are 100% occupied at absolute zero temperature.

When a semiconductor crystal is doped with Group III or V impurities, it becomes an extrinsic semiconductor. Donor and acceptor impurities are incorporated into the semiconductor lattice to produce n- and p-type materials, respectively. In the former, each impurity atom creates a permissible state near the top of the forbidden gap. This state is originally occupied. At room temperature, all impurity atoms ionize. Each atom contributes one electron to the conduction band and generates a positive ion bound to the crystal lattice. Electrons outnumber holes. In p-type semiconductors, each impurity atom creates an empty permissible state near the bottom of the gap. At room temperature, all ionized impurity atoms receive one electron each from the crystal's valence band, resulting in holes and negative ions bound to the crystal lattice. Holes outnumber electrons.

Disturbance of thermal equilibrium by nonthermal sources of energy introduces excess carriers into the semiconductor, raising the total concentration of one or both of the carriers above the equilibrium value. The two most common mechanisms of charge flow in semiconductors are drift and diffusion. The application of an electric field causes carriers to acquire a drift velocity in a direction parallel to the electric field, establishing a conduction current. At low and moderate fields the drift velocity is proportional to the electric field, with the constant of proportionality being called the *mobility*, which depends on the impurity concentration and temperature. The conductivity of a material depends on the carrier concentrations and mobilities. It links current density and electric field as expressed by Ohm's law.

Diffusion is the net movement of carriers due to a concentration gradient of carriers. The diffusivity (i.e., the diffusion coefficient), is the constant of proportionality between the flux of carriers and the concentration gradient. As minority carriers diffuse, they travel a distance characterized by the diffusion length before they recombine with majority carriers. The time taken for an excess minority carrier to recombine with a majority carrier is called the minority carrier lifetime. The diffusion length is related to the lifetime via the diffusion coefficient.

The continuity equation associates the rate of change of carrier concentration with spatial differences in current and temporal changes in carrier concentration. It is mainly used in finding carrier concentration profiles from which currents can be computed.

Appendix 1A: Pascal Program for Silicon Parameters

Given only the impurity concentration, the following program, developed in Turbo Pascal, Version 5, calculates the following parameters at 300 K for both types of silicon material:

- μ minority carrier mobility
- D diffusion coefficient
- τ minority carrier lifetime
- L diffusion length
- σ_T total conductivity
- ρ resistivity of the semiconductor

```
program semiconductor_parameters;

uses crt;
var stat: char;

procedure semi_par;

var N, Dh, De, ue, uh, Le, Lh, t, se, sh : real;

begin {semi_par}
  textbackground(red);
  clrscr;
  writeln;
  writeln;
  writeln ('    *****************************************************************');
  writeln ('    ***********     SILICON SEMICONDUCTOR PARAMETERS      ***********');
  writeln ('    *****************************************************************');
  writeln;
  writeln;
  writeln ('        Given the impurity concentration, this program calculates');
  writeln ('             the following semiconductor parameters at 300 K:');
  writeln;
  writeln ('               Dh, De, Lh, Le, μh, μe, τh, τe, σh, σe, rho');
  writeln;
  writeln;
  writeln ('    Enter the concentration in units of per cm3 in the form x.(x)Exx');
  writeln;
  write   ('    Enter N: '); readln(N);

  ue:=88+(1252/(1+6.984E-18*N));
  uh:=54.3+(407/(1+3.745E-18*N));
  Dh:=0.02586*uh;
  De:=0.02586*ue;
  t:=5E-7/(1+2E-17*N);
  Lh:=sqrt(Dh*t);
  Le:=sqrt(De*t);
  se:=1.6021892E-19*(N*ue+2.25E20*uh/N);
  sh:=1.6021892E-19*(N*uh+2.25E20*ue/N);
```

```
  clrscr;
  writeln ('    *****************************************************');
  writeln ('    * An n-type silicon with an impurity concentration of *');
  writeln ('    * Nd = ', N:9 ,' will have the following parameters:  *');
  writeln ('    *****************************************************');
  writeln ('       Minority hole mobility,      µh =',uh:7:2,' cm²/V.s ');
  writeln ('       Diffusion coefficient,       Dh =',Dh:7:2,' cm²/s   ');
  writeln ('       Minority hole lifetime,      τh =',1E9*t: 7:2,' ns  ');
  writeln ('       Diffusion length,            Lh =',1E4*Lh:7:2,' µm  ');
  writeln ('       Total conductivity,          σT =',se:7:2,' (Ω.cm)-1');
  writeln ('       Resistivity,                rho =',1/se:7:2,' Ω.cm  ');
  writeln ('    *****************************************************');
  writeln;
  writeln ('    *****************************************************');
  writeln ('    * A p-type silicon with an impurity concentration of  *');
  writeln ('    * Na = ', N:9 ,' will have the following parameters:  *');
  writeln ('    *****************************************************');
  writeln ('       Minority electron mobility,  µe =',ue:7:2,' cm²/V.s ');
  writeln ('       Diffusion coefficient,       De =',De:7:2,' cm²/s   ');
  writeln ('       Minority electron lifetime,  τe =',1E9*t:7:2, ' ns  ');
  writeln ('       Diffusion length,            Le =',1E4*Le:7:2,' µm  ');
  writeln ('       Total conductivity,          σT =',sh:7:2,' (Ω.cm)-1');
  writeln ('       Resistivity,                rho =',1/sh:7:2,' Ω.cm  ');
  writeln ('    *****************************************************');
end; {semi_par}

begin {main program}
  stat := 'Y';
  while (stat =  'Y') or (stat = 'y') do
    begin
    semi_par;
    writeln;
    write  ('    Do you want to change the impurity concentration? ');
    stat:= readkey;
    end
end. {main program}
```

References

1. D. L. Pulfrey and G. Tarr. *Introduction to Microelectronic Devices.* Prentice-Hall, Englewood Cliffs, NJ: 1989.
2. M. Zambuto. *Semiconductor Devices.* McGraw-Hill, New York: 1989.
3. E. S. Yang. *Microelectronic Devices.* McGraw-Hill, New York: 1988.
4. D. H. Navon. *Semiconductor Microdevices and Materials.* Holt, Rinehart & Winston, New York: 1986.
5. R. A. Colclaser and S. Diehl-Nagle. *Materials and Devices for Electrical Engineers and Physicists.* McGraw-Hill, New York: 1985.
6. S. M. Sze. *Semiconductor Devices, Physics and Technology.* John Wiley & Sons, New York: 1985.
7. R. F. Pierret. *Semiconductor Fundamentals.* Addison-Wesley, Reading, MA: 1983.
8. H. E. Talley and D. G. Daugherty. *Physical Principles of Semiconductor Devices.* The Iowa State University Press, Ames, IO: 1976.

PROBLEMS

For the following problems, assume a room temperature of 300 K and complete ionization of impurity atoms at that temperature.

1.1 Determine the carrier concentrations and Fermi level for a silicon wafer with $N_d = 10^{15}\ \text{cm}^{-3}$.

1.2 Find the resistivity of an *n*-type Si with $N_d = 10^{16}\ \text{cm}^{-3}$.

1.3 The kinetic energy of electrons in an *n*-type semiconductor sample with uniform donor concentration in thermal equilibrium is given by

$$\frac{m_e v_{th}^2}{2} = \frac{3kT}{2}$$

a. Calculate the mean free time of an electron having a mobility of $1000\ \text{cm}^2/\text{V}\cdot\text{s}$.
b. Find the mean free path, that is, the distance traveled by an electron between collisions. Assume $m_e = .2369 \times 10^{-30}$ kg.

1.4 For a *p*-type Si with $N_a = 10^{17}\ \text{cm}^{-3}$, the minority carrier distribution is given as

$$n_p(x) = 10^6(1 - x) + 7 \times 10^9\ \text{cm}^{-3}, \qquad 0 \leq x \leq 1\ \text{cm}$$

Determine the following;
a. The electron flux density
b. The diffusion current density at $x = .5$ cm.

1.5 In a particular semiconductor, if the Fermi level is 275 meV below the edge of the conduction band, what is the probability of finding an electron in an energy kT above that edge?

1.6 For a semiconductor sample, the probability of finding electrons in a state kT above the conduction edge is e^{-11}. Find the location of the Fermi level with respect to E_c.

1.7 An electric field of 60 V/cm, applied across a homogenous n-type Si sample, moves holes injected into the semiconductor a distance of 1 cm in 50 μs. Find v_{dh} and D_h.

1.8 Which semiconductor whose energy-band diagram is shown in the accompanying diagram will have the greatest intrinsic carrier concentration?

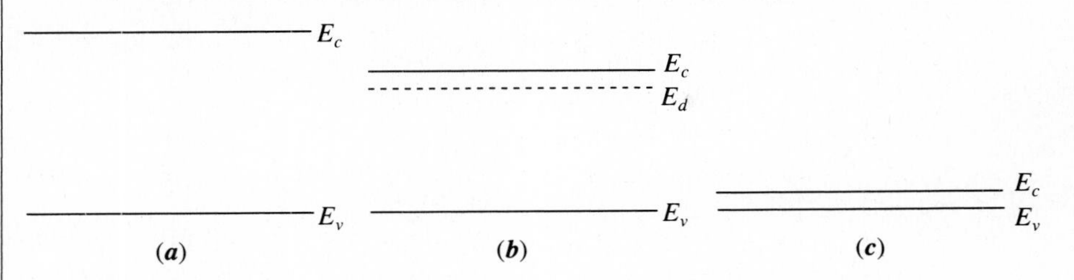

1.9 What special conditions must exist in the physical systems defined by the following modified forms of the electron excess minority carrier continuity equations?

a. $D_e \dfrac{d^2 \Delta n_p}{dx^2} - \dfrac{\Delta n_p}{\tau_e} = 0$

b. $\dfrac{d^2 \Delta n_p}{dx^2} = 0$

c. $\dfrac{d \Delta n_p}{dt} = -\dfrac{\Delta n_p}{\tau_e}$

1.10 A piece of Si with $N_a = 10^{-15}$ cm^{-3} and $\tau_e = 1$ μs is first illuminated for a time $t \gg \tau_e$ with light that generates $G_{L0} = 10^{16}$ electron-hole pairs per cm$^3 \cdot$ s uniformly throughout the volume of the material. At time $t = 0$, the light intensity is reduced, making $G_L = G_{L0}/2$ for $t \geq 0$.
a. Investigate if we have low-level injection.
b. Determine $\Delta n_p(t)$ for $t \geq 0$.

1.11 A uniformly donor-doped silicon wafer is suddenly illuminated at $t = 0$. Assuming $N_d = 10^{16}$ cm^{-3}, $\tau_h = 1$ μs, and a light-induced creation of 10^{18} electrons and holes per cm$^3 \cdot$ s throughout the semiconductor, find $\Delta p_n(t)$ for $t > 0$.

2. SEMICONDUCTOR DIODES

This chapter is concerned with the semiconductor diodes. An understanding of the physical properties of the *pn* junction is particularly essential to the study of digital electronics because it is most fundamental of all the semiconductor devices. Schottky diodes, which are used in some TTL logic subfamilies to improve switching speed, will also be introduced.

2.1 The *pn*-Junction Diode

So far, the concentrations and motions of charge carriers in semiconductors have been studied separately. Now it is time to start considering what happens when *p*- and *n*-type materials are joined together to form a single semiconductor crystal called the *pn junction*, whose circuit symbol is shown in Figure 2.1. The following discussion assumes a *pn* junction with an abrupt change of impurity at the junction.

The *pn* junction is a continuous crystalline lattice in which the impurity doping is changed from donor dominance in one portion to acceptor dominance in the other. The *p* region has holes as the majority mobile carriers; in the *n* region, electrons are the majority carriers. The boundary is called the *metallurgical junction*. The current-voltage characteristic of the *pn* junction is due almost entirely to the behavior of the holes and electrons in the regions near this junction.

The pn *Junction in Thermal Equilibrium*

Before they are joined, the *n*-type piece has many more electrons than holes, and the *p*-type piece has many holes and very few electrons. Still, both are electrically neutral. In the *n*-type, the ionized donors and the few holes exactly balance the electron charge. In the *p*-type, the ionized acceptors and the few electrons exactly balance the hole charge.

The instant the two types are brought together, there will be large concentration gradients at the boundary. The electron concentration changes abruptly from a large

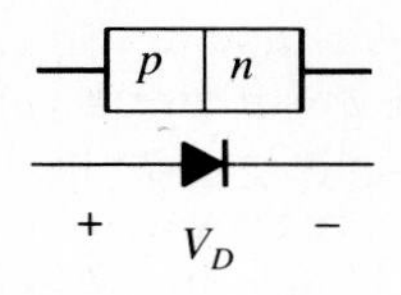

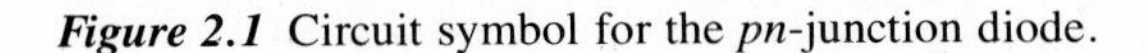

Figure 2.1 Circuit symbol for the *pn*-junction diode.

value in the *n*-type material to a very small value in the *p*-type while the hole concentration change is exactly the reverse. The result is the diffusion of holes from *p* to *n* and electrons from *n* to *p*. As the holes leave the *p* material, they leave behind negatively charged and immobile acceptor ions. Being surrounded by a large number of holes as they enter the *p* region, the electrons recombine, leaving the acceptor ions without adjacent positive charges to maintain charge neutrality. Consequently, near the *pn* boundary the *p* region will have a net negative charge. Similarly, the *n* region will take a net positive charge near the boundary, as shown in Figure 2.2a. These charges create a *built-in* electric field $\mathcal{E}(x)$ across the boundary, directed from the *n* region to the *p* region and counteracting the diffusion of holes and electrons. The force exerted by this field pushes some holes back toward the *p* region, thereby caus-

Figure 2.2 (a) Carrier concentrations and charge distributions at equilibrium for a *pn* junction; (b) current components at equilibrium in the depletion region.

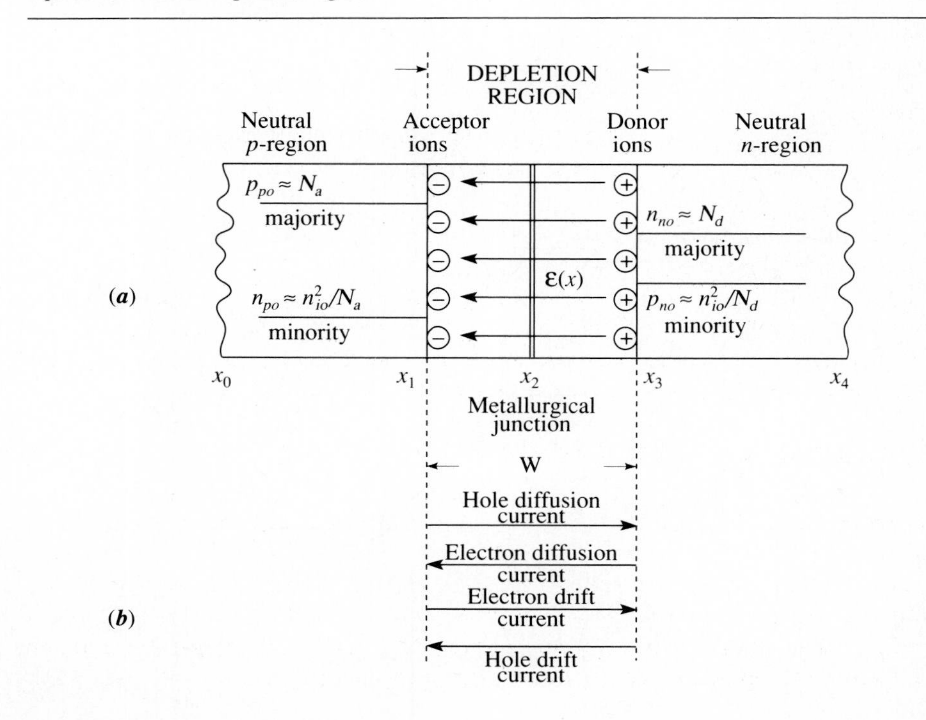

ing a hole drift current from the n region to the p region, and some electrons back toward the n region, creating an electron drift current from the p region to the n region. These drift currents, naturally, reduce the original diffusion flow rates. The current components in this region are shown in Figure 2.2b. Eventually, the charge imbalance becomes sufficient for the electric field to be able to reduce the net hole and electron flows across the boundary to zero, and thermal equilibrium exists. Away from the junction, the free electrons and holes compensate the immobile ionic charges to maintain charge neutrality in the *bulk* regions.

Figure 2.3 illustrates the energy-band diagram under equilibrium conditions. The slope of the energy band edges, E_c and E_v, is proportional to the magnitude of the electric field intensity $\mathcal{E}(x)$ in the depletion region.

Note that the existence of an electric field implies a potential difference across the boundary, called the *built-in* or *barrier potential* ϕ_o. The *height* of this barrier equals the change in electron potential energy across the barrier and is called the *barrier energy* $E_{Bo} = q\phi_o$.

To obtain an expression for the built-in potential, we note that the electric field can be expressed as $\mathcal{E}(x) = -d\phi_o/dx$. Using Equation (1.31) to eliminate $\mathcal{E}(x)$ in this expression, we get

$$d\phi_o(x) = \phi_T \frac{dn_o}{n_o} \tag{2.1}$$

We integrate both sides of Equation (2.1) from x_1 to x_3. Observing the boundary values for the electron concentrations from Figure 2.2a, we get

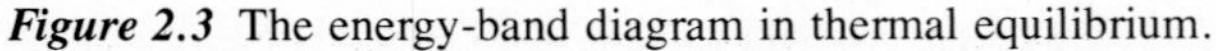

Figure 2.3 The energy-band diagram in thermal equilibrium.

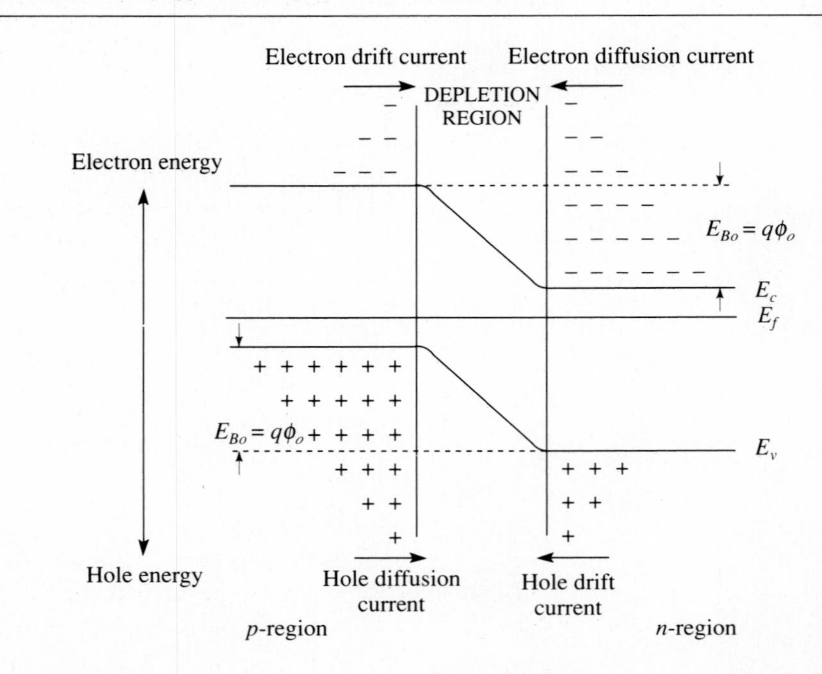

$$\int_{x1}^{x3} d\phi_o(x) = \phi_T \int_{n_{po}}^{n_{no}} \frac{dn_o}{n_o} \tag{2.2}$$

where the subscripts n and p denote the n- and p-type regions, respectively, so that n_{no} is the equilibrium majority carrier concentration in the n region, and n_{po} designates the equilibrium minority carrier concentration in the p region. Then

$$\phi_o = \phi_T \ln\left(\frac{n_{no}}{n_{po}}\right) \tag{2.3}$$

Using Equations (1.6) and (1.7) results in

$$\phi_o = \phi_T \ln\left(\frac{N_a N_d}{n_{io}^2}\right) \tag{2.4}$$

which shows that the built-in potential is a function of the impurity concentrations on either side of the junction as well as the temperature. To produce a current, the carriers must overcome this potential barrier. Rearranging Equation (2.3) leads to

$$n_{po} = n_{no} e^{-\phi_o/\phi_T} \tag{2.5a}$$

Similarly, for the hole concentrations we obtain

$$p_{no} = p_{po} e^{-\phi_o/\phi_T} \tag{2.5b}$$

Equation Set (2.5) relates the electron and hole concentrations on either side of the junction for the equilibrium condition.

The region surrounding the metallurgical junction is called the *space charge layer* (*SCL*) because of the net charge that exists within it. It is also called the *depletion region*, because the electric field $\mathcal{E}(x)$ depletes the region of mobile carriers. Note that the SCL itself has overall neutrality. The negative charge on the p side of the boundary exactly matches the positive charge on the n side because the original pieces are neutral.

Example 2.1

We use Equation (2.4) to find the barrier potential across an Si junction at room temperature when the p region has (a) $N_a = 10^{15}/\text{cm}^3$; (b) $N_a = 10^{17}/\text{cm}^3$, and N_d changes as follows:

	$N_a = 10^{15}/\text{cm}^3$	$N_a = 10^{17}/\text{cm}^3$
N_d, cm^{-3}	ϕ_o, V	ϕ_o, V
10^{16}	.635	.754
10^{17}	.695	.814
10^{18}	.754	.874
10^{19}	.814	.933
10^{20}	.874	.993

Therefore, the greater the doping on either side, the greater the barrier potential.

The pn *Junction under Forward Bias*

With a forward voltage V_D applied across the junction, the potential barrier is reduced. Thus, one should expect an increase in the flow of mobile carriers across the junction, resulting in *excess minority carriers*. A good assumption at lower current levels for modern devices is that the electric field in the bulk n and p regions is approximately zero, so that no voltage drops exist across them.

Figure 2.4 shows the effects on the energy-band diagram of applying a positive potential to the p region with respect to the n region. Here, the p region is arbitrarily chosen to remain fixed, while the n region is shifted upward in electron energy by qV_D, as a result of which the slope of the band edges is reduced as compared to thermal equilibrium. This implies that the magnitude of the electric field is reduced. The reduced barriers for the diffusion of holes and electrons result in increases in the respective diffusion-current components over the thermal equilibrium value. A large number of holes and electrons have energies greater than the barrier height, $q(\phi_o - V_D)$. On the other hand, the drift-current components of both holes and electrons remain the same as the equilibrium values because a change in barrier height does not affect the number of minority carriers. The carrier supply is limited by thermal generation. Therefore, the net effect of forward bias is a huge increase in the diffusion-current components, while the drift components remain fixed.

Next, we find an expression for the total diode current in terms of the applied

Figure 2.4 The energy-band diagram under forward bias.

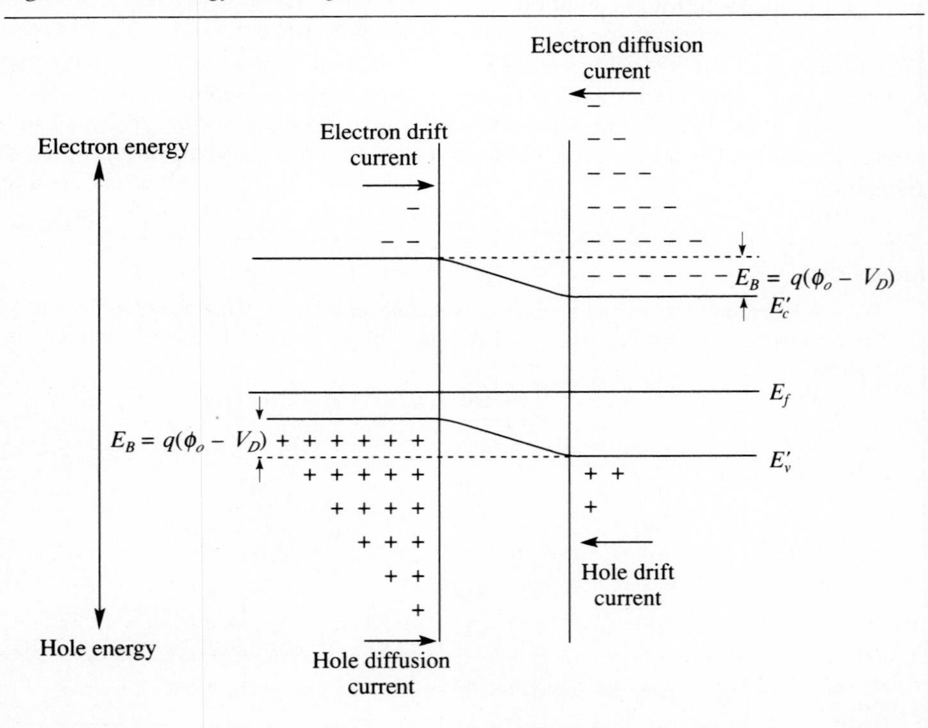

voltage, V_D. Consider first the n material. For the case of constant forward-bias voltage, the steady state means zero time dependence; that is, $\partial \Delta p / \partial t = 0$, so that Equation (1.37b) reduces to a linear, second-order differential equation:

$$0 = D_h \frac{d^2 \Delta p_n(x)}{dx^2} - \frac{\Delta p_n(x)}{\tau_h} \tag{2.6}$$

The first boundary condition, $p_n(x_3)$, is produced by the arrival of holes from the p material, which have enough energy to surmount the barrier. The expression is similar to the one in Equation (2.5b), with ϕ_o replaced now by $\phi_o - V_D$, assuming the ohmic voltage drops along the neutral portions of both regions are negligible, so that all the voltage applied appears across the SCL. Hence, using Equation (2.5b) and assuming that $p_p(x_1) \approx p_{po}$ under low-level injection, we have

$$\begin{aligned} p_n(x_3) &= p_p(x_1) e^{-(\phi_o - V_D)/\phi_T} \\ &= p_{po} e^{-(\phi_o - V_D)/\phi_T} \\ &= p_{no} e^{V_D/\phi_T} \end{aligned} \tag{2.7}$$

This characterization applies not only throughout the entire p material but also within the SCL, because there is little recombination due to the low electron concentration.

For the second boundary condition, we will assume that for $x >> x_3$ in the n region, the hole concentration approaches the thermal equilibrium value. This is called the *long diode approximation*.

Solving Equation (2.6) by standard methods then yields

$$p_n(x) = p_{no} + K_1 e^{-x/\sqrt{D_h \tau_h}} + K_2 e^{+x/\sqrt{D_h \tau_h}} \tag{2.8}$$

The length constant given by $L_h \equiv \sqrt{D_h \tau_h}$ is called the minority carrier *diffusion length*. It represents the average distance that holes diffuse during their lifetime, τ_h, that is, before their number decreases by a factor of $1/e$ as a result of recombination.

In order to satisfy the second boundary condition, we set $K_2 = 0$, and apply Equation (2.7) to get

$$p_n(x) = p_{no} + p_{no}(e^{V_D/\phi_T} - 1) \exp\left(-\frac{x - x_3}{L_h}\right) \tag{2.9}$$

which is plotted in Figure 2.5. Obviously, the concentration is now greater than the thermal equilibrium level, due to the excess minority carriers. Since the minority current density is due entirely to diffusion, the hole current in the n material can be expressed as

$$I_h(x) = AJ_h = -qAD_h \frac{dp_x(x)}{dx} \tag{2.10}$$

where A is the cross-sectional area of the n region in cm^2. Taking the derivative of Equation (2.9) with respect to x and substituting it in Equation (2.10), we obtain the hole current as a function of distance:

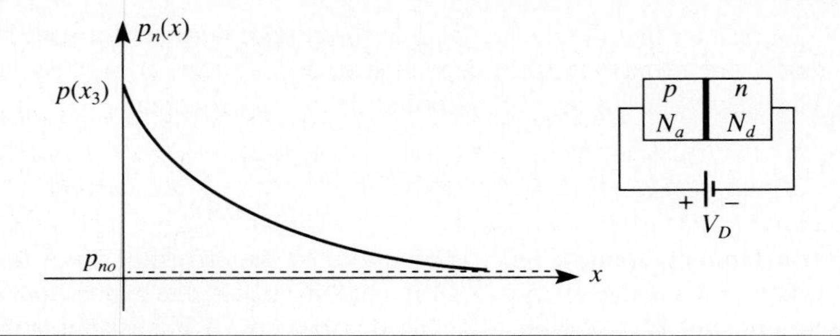

Figure 2.5 Minority carrier concentration in the n-region under forward bias.

$$I_h(x) = qAD_h \frac{p_{no}}{L_h}(e^{V_D/\phi_T} - 1) \exp\left(-\frac{x - x_3}{L_h}\right) \tag{2.11}$$

As expected, the hole current decays to zero as it traverses the n region. At the edge of the neutral n region, where $x = x_3$, the hole current is

$$I_h(x_3) = qAD_h \left(\frac{p_{no}}{L_h}\right)(e^{V_D/\phi_T} - 1) \tag{2.12}$$

Since there is little recombination-generation process within the SCL, the preceding equation is valid there, too.

When we analyze the behavior of electrons in the p region in the manner of the preceding discussion, we end up with an expression for the electron current at $x = x_1$:

$$I_e(x_1) = qAD_e \left(\frac{n_{po}}{L_e}\right)(e^{V_D/\phi_T} - 1) \tag{2.13}$$

where L_e is the diffusion length for the minority electrons in the p region. Equation (2.13) also applies within the SCL and at $x = x_3$.

The sum of Equations (2.12) and (2.13) yields the *ideal diode equation*, giving the total diode current as

$$\begin{aligned} I_D &= qA\left(\frac{D_h p_{no}}{L_h} + \frac{D_e n_{po}}{L_e}\right)(e^{V_D/\phi_T} - 1) \\ &= I_S\,(e^{V_D/\phi_T} - 1) \end{aligned} \tag{2.14}$$

where

$$\begin{aligned} I_S &\equiv qA\left(\frac{D_h p_{no}}{L_h} + \frac{D_e n_{po}}{L_e}\right) \\ &= qA\left(\frac{D_h}{L_h N_d} + \frac{D_e}{L_e N_a}\right) n_{io}^2 \end{aligned} \tag{2.15}$$

is called the *reverse saturation current* and varies linearly with the two minority carrier concentrations, and hence inversely with the doping concentrations N_d and N_a. Therefore, the more lightly doped side of the *pn* junction produces a larger number of minority carriers, that is, the largest current component. At room temperature, I_S is less than 1 pA in small-area silicon diodes. In practice, I_S is somewhat less than the value given by Equation (2.15) due to the actual voltage drops across the neutral regions. In addition, it is strongly dependent on temperature. Figure 2.6 illustrates the total current as a function of the applied voltage V_D, and the change in carrier concentrations in both bulk regions from those in thermal equilibrium as a result of forward bias is shown in Figure 2.7.

Temperature Behavior of the pn *Junction*

It is obvious from Equation (2.14) that the electrical characteristics of the *pn* junction are temperature-sensitive. Furthermore, the saturation current is also dependent on temperature. To show this, we consider Equation (1.12) and the subsequent discussion to express the square of the intrinsic carrier density in thermal equilibrium as

$$n_{io}^2 = CT^3 e^{-E_g/kT}$$

where C is a constant. Therefore, this temperature dependence affects the saturation current of Equation (2.15), too. Thus, the approximate fractional change in I_S with temperature is given by

$$\frac{1}{I_S}\frac{dI_S}{dT} = \frac{3}{T} + \frac{E_g}{kT^2} \tag{2.16}$$

For silicon at room temperature, we have

$$\frac{dI_S}{I_S} = 47.4\,\frac{dT}{T}$$

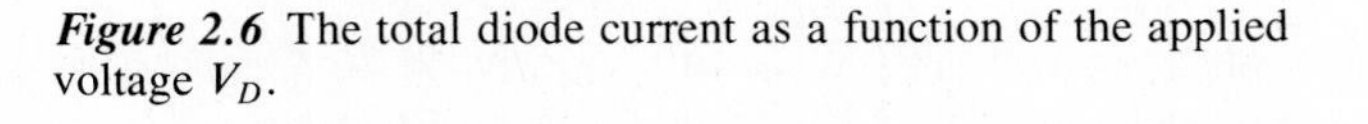

Figure 2.6 The total diode current as a function of the applied voltage V_D.

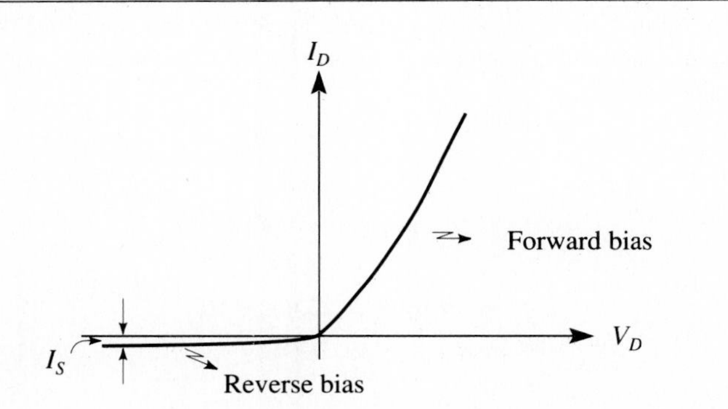

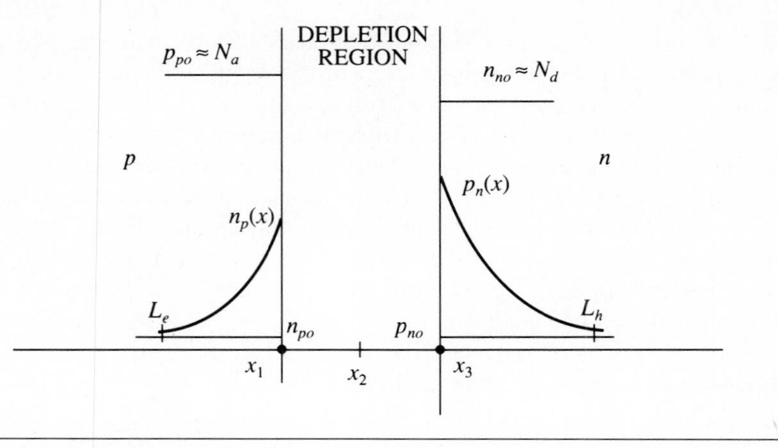

Figure 2.7 Change in carrier concentrations from those in thermal equilibrium as a result of forward bias.

The pn *Junction under Reverse Bias*

In reverse bias, V_D is a negative value, and the polarity of the voltage is positive on the n-region. The effect of the reverse bias on the energy-band diagram, as compared to the thermal equilibrium, is illustrated in Figure 2.8. The n region is arbitrarily shown to be shifted downward by $-qV_D$. The slope of the energy-band edges in the depletion region is increased, implying a larger electric field for the reverse bias. Also note that the applied reverse-bias voltage adds to the built-in potential because $V_D < 0$. The increase in barrier height leads to reduced number of holes, p_p, and electrons, n_n, that have sufficient energy to traverse the depletion region. However, the drift-current components of both types of carriers remain at their equilibrium values. Therefore, the net result is a constant reverse current, small in magnitude and negatively directed from the n region to the p region.

Example 2.2

Given an Si pn junction with $A = 10^{-4}$ cm^2, $N_a = 10^{17}$/cm^3, $N_d = 10^{15}$/cm^3, $\tau_e = .1$ μs and $\tau_h = 1$ μs, we can easily find $D_e = 20.75$ cm^2/s, $D_h = 12.35$ cm^2/s, $L_e = 1.44 \times 10^{-3}$ cm, $L_h = 3.51 \times 10^{-3}$ cm, $\phi_o = .695$ V, and $I_S = 1.32 \times 10^{-14}$ A. Now, describing the diode characteristics by Equation (2.14), the diode current is found for various values of V_D as

V_D, V	I_D, A
-5	$-I_S$
$-.1$	-1.29×10^{-14}
$.2$	2.98×10^{-11}
$\phi_o = .695$	5.95×10^{-3}
1	774(!)

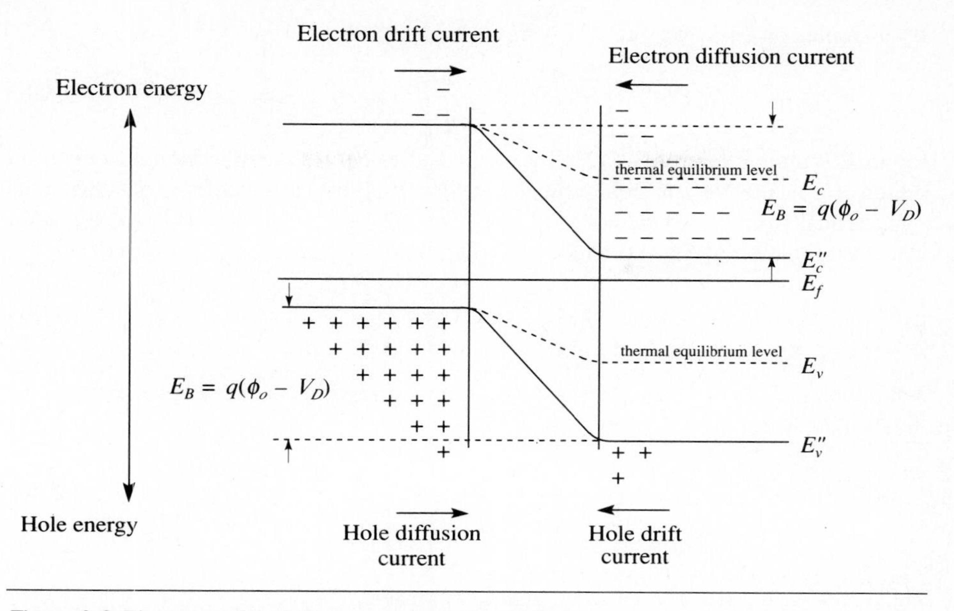

Figure 2.8 The energy-band diagram under reverse bias.

Depletion Region Width

The width of the depletion region in thermal equilibrium is determined by the number of immobile charges that must be uncovered to produce the barrier voltage. Hence, considering Equation (2.4), the width also depends on the doping concentrations on each side of the junction. In order to derive an expression for the width, W, of the space-charge layer, we first examine the variations of the electric field and barrier potential with distance within the region. The depletion region is as shown in Figure 2.2a. The total acceptor and donor charges are numerically equal because the diode is neutral.

All flux lines originating on donor ions terminate on acceptor ions. We may assume, ignoring the flux fringing, that the flux lines are parallel to the x-axis and point in the negative x-direction. If we also assume that the mobile carriers are few compared to the donor and acceptor ion concentrations N_d and N_a, respectively, then the net charge to the left of an imaginary plane at some position x between x_1 and x_2 is given by $-qAN_a(x - x_1)$. Gauss's Law states that the net electric flux through any closed Gaussian surface is equal to the net charge inside the surface divided by the electric permittivity of the medium, or

$$\oint \mathcal{E} \cdot d\mathbf{A} = \frac{q_{in}}{\epsilon_s} \tag{2.17}$$

Equation (2.17) yields an expression for the electric field intensity $\mathcal{E}(x)$:

$$\mathcal{E}(x) = \frac{-qN_a(x - x_1)}{\epsilon_s} \qquad x_1 \leq x \leq x_2 \tag{2.18a}$$

By a similar process, we get

$$\mathcal{E}(x) = -\frac{qN_d(x_3 - x)}{\epsilon_s} \qquad x_2 \le x \le x_3 \tag{2.18b}$$

Figure 2.9 plots Equation Set (2.18) as well as the charge density within the depletion region. Here, the electric field intensity at the metallurgical junction is designated as $-\mathcal{E}_m$. Thus, from (2.18) we have $|x_2 - x_1| = \epsilon_s \mathcal{E}_m / qN_a$ and $|x_3 - x_2| = \epsilon_s \mathcal{E}_m / qN_d$, whose summation yields the width of the depletion region:

$$W = |x_3 - x_1| = \frac{\epsilon_s \mathcal{E}_m}{q}\left(\frac{1}{N_a} + \frac{1}{N_d}\right) \tag{2.19}$$

Again using $\mathcal{E}(x) = -d\phi_o/dx$, we obtain an expression for the barrier potential ϕ_o in terms of W as

$$\phi_o = -\int_{x_1}^{x_3} \mathcal{E}(x)dx = \frac{\mathcal{E}_m W}{2} \tag{2.20}$$

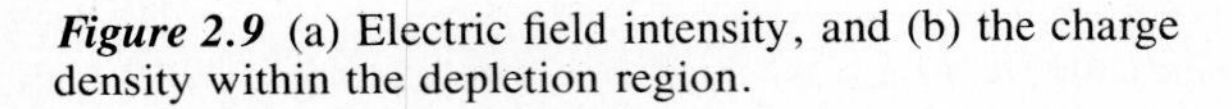

Figure 2.9 (a) Electric field intensity, and (b) the charge density within the depletion region.

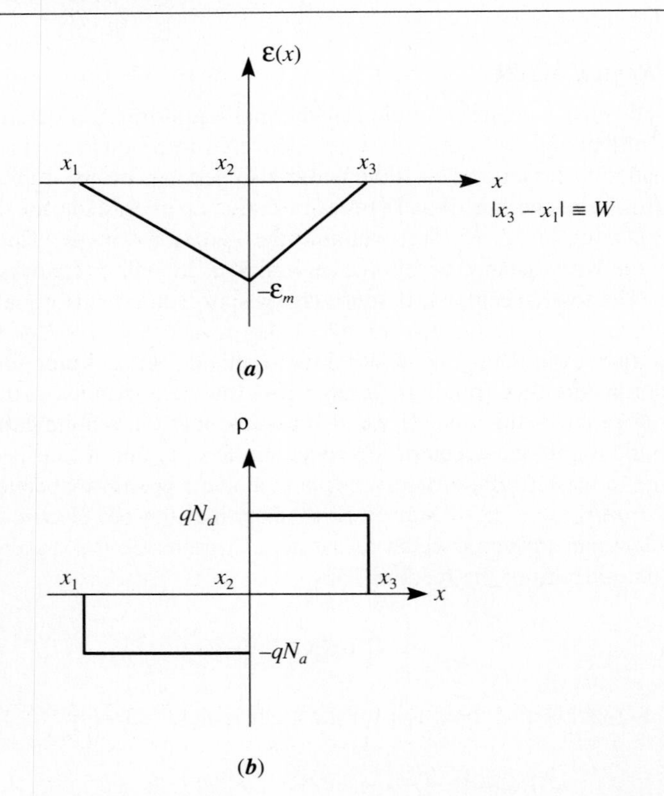

Now, we use Equation (2.19) to eliminate $\mathcal{E}_m$ from (2.20) to get

$$\phi_o = \frac{q}{2\epsilon_s} \frac{N_a N_d}{N_a + N_d} W^2 \tag{2.21}$$

Note that in a so-called *one-sided junction*, in which one side of the junction is much more heavily doped than the other, for example $N_a >> N_d$, W is approximated, using Equation (2.21), as

$$W \approx \sqrt{\frac{2\epsilon_s \phi_o}{q N_d}} \tag{2.22}$$

In one-sided junctions most of the depletion region width appears on the more lightly doped side of the metallurgical junction.

When a voltage, V_D, is applied across the *pn* junction, the barrier potential can be replaced with $\phi_o - V_D$, resulting in

$$W = \sqrt{\frac{2\epsilon_s}{q} \left(\frac{N_a + N_d}{N_a N_d} \right) (\phi_o - V_D)} \tag{2.23}$$

This equation is not applicable for forward-bias voltages equal to or above ϕ_o. However, V_D may not reach ϕ_o because the diode current I_D as given by Equation (2.14) becomes large enough to destroy the diode. Application of a reverse-bias voltage gives rise to an increase in the width of the depletion region.

The Transition Capacitance

The depletion region contains very few mobile carriers, so that it acts like an insulator with a dielectric constant ϵ_s of the semiconducting material. Since the carriers diffuse through, the region is not actually a perfect insulator. In any case, the *pn* junction can be visualized as having conducting sections, the bulk *p* and *n* regions, separated by a dielectric depletion region. Figure 2.10a represents the *pn* junction with the conducting bulk regions acting as the plates of a capacitor, although the structure is somewhat different from that of a typical parallel-plate capacitor. The dipoles in the depletion region have their positive charge in the *n* part of the region and their negative charge on the *p* side. On the other hand, in a typical parallel-plate capacitor, the separation between the charges in the dipoles is much less, and they are distributed more homogeneously throughout the dielectric, as shown in Figure 2.10b. Also, the electric field direction in the *pn* junction is independent of the polarity of the applied potential, which is not true for the parallel-plate capacitor. Hence, the *transition capacitance*, C_t, is given by

$$C_t = \frac{\epsilon_s A}{W} \tag{2.24}$$

We already know that the width W varies as the junction voltage is changed. Therefore, the capacitance varies inversely with $\sqrt{\phi_o - V_D}$. We see that by using Equation (2.23) in (2.24) to get

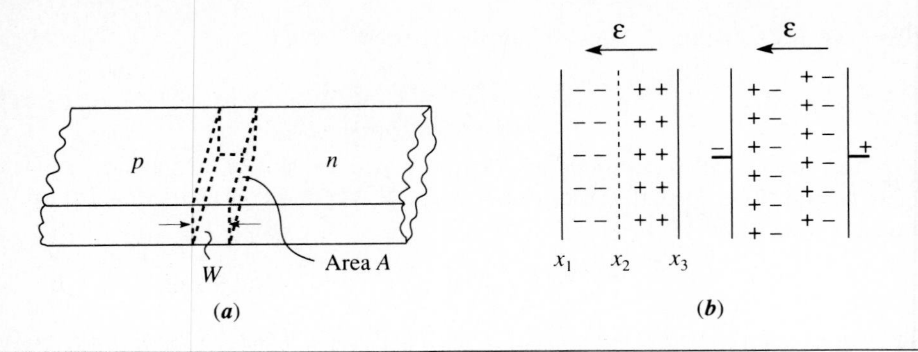

Figure 2.10 (a) The *pn* junction with bulk regions acting as the plates of a parallel-plate capacitor while the depletion region acts as a dielectric; (b) depletion region of a *pn* junction compared to a parallel-plate capacitor.

$$C_t = A\sqrt{\frac{q\epsilon_s}{2}\frac{N_aN_d}{N_a + N_d}}(\phi_o - V_D)^{-1/2}$$

$$= \frac{C_{to}}{\sqrt{1 - (V_D/\phi_o)}} \tag{2.25}$$

where

$$C_{to} \equiv A\sqrt{\frac{q\epsilon_s}{2\phi_o}\left(\frac{N_aN_d}{N_a + N_d}\right)} \tag{2.26}$$

is the capacitance at equilibrium, that is, when $V_D = 0$. As V_D approaches ϕ_o, C_t becomes large; if V_D is negative, then C_t decreases. Typical values of C_t for digital integrated circuits are around .1 pF, measured at 0 V. Notice that Equation (2.25) is not valid for values of V_D equal to or above ϕ_o. The capacitance does not become a short circuit when the applied voltage equals the barrier potential. One reason is that our assumption that there are very few or no mobile carriers within the depletion region is not appropriate anymore. Moreover, the applied voltage appears across the depletion region only if there is no voltage drop across the bulk regions of the device. This is true only when there is no (or an extremely small) current through the device, which is not the case for the forward-biased diode.

The transition capacitance C_t, as given by Equation (2.25), is actually a small-signal parameter. Over the voltage ranges used in digital circuits, it is more reasonable to define an average linear capacitance, $C_{(av)}$, to eliminate the voltage-dependent effect of the nonlinear transition capacitance C_t. $C_{(av)}$ will require the same change in charge as C_t for a transition between two voltage levels, V_r and V_f. Therefore,

$$C_{(av)} = \frac{\Delta Q}{V_f - V_r}$$

where

$$\Delta Q = \int_{V_r}^{V_f} C_t(V_D)dV_D = \int_{V_r}^{V_f} C_{to}\left(\frac{1}{\sqrt{1 - V_D/\phi_o}}\right) dV_D$$

so that

$$C_{(av)} = -\frac{2C_{to}\phi_o}{V_f - V_r}\left(\sqrt{1 - \frac{V_f}{\phi_o}} - \sqrt{1 - \frac{V_r}{\phi_o}}\right) \tag{2.27}$$

Figure 2.11 illustrates the increase in the transition capacitance under forward-bias conditions, which peaks around barrier potential, and the decrease beyond that point as V_D increases.

For most circuits, one can consider the maximum forward- and reverse-bias limits to be ϕ_o and $-k\phi_o$, respectively, where the constant k is chosen such that $k\phi_o$ is equal to the maximum reverse bias. In this case, Equation (2.27) reduces to

$$C_{(av)} = \frac{2C_{to}}{\sqrt{1 + k}}$$

For $k = 1$, $C_{(av)} = \sqrt{2}C_{to} \approx 1.41\ C_{to}$, while for $k = 2$, $C_{(av)} \approx 1.16\ C_{to}$, and for $k = 3$, $C_{(av)} = C_{to}$. This shows that the average value does not vary drastically from the zero-bias value. The appendix at the end of this chapter lists a Pascal program that, given the doping densities in both regions and the applied voltage, computes various diode parameters at room temperature.

Example 2.3

A *one-sided pn* junction, with the p side much more heavily doped than the n side, is denoted by the symbol p^+n. For a p^+n junction in thermal equilibrium at 300 K, with $N_d = 10^{16}/\text{cm}^3$, $N_a = 10^3 N_d$, and $\epsilon_s = 11.9\epsilon_o = 1.053 \times 10^{-12}$ F · cm^{-1}, we calculate ϕ_o, W, and $\mathcal{E}_m$ by using Equations (2.4), (2.22), and (2.20):

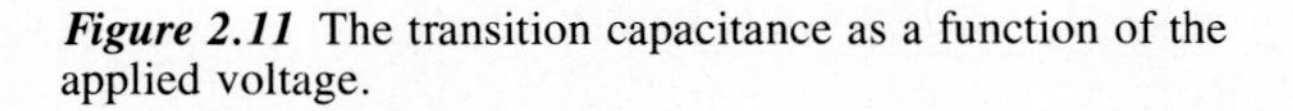

Figure 2.11 The transition capacitance as a function of the applied voltage.

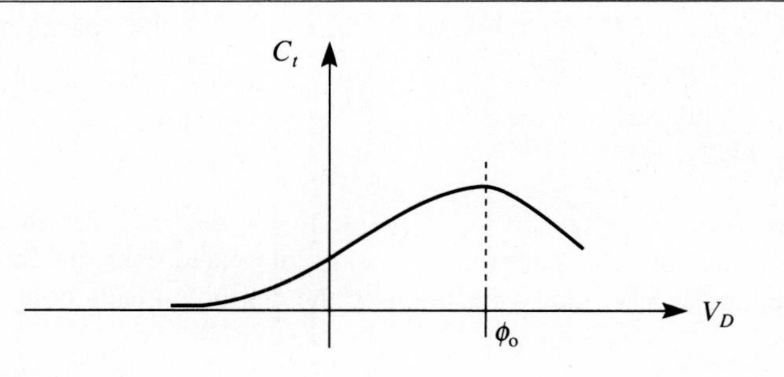

$$\phi_o = .879 \text{ V}$$

$$W = .336 \text{ } \mu\text{m}$$

$$\mathcal{E}_m = 5.23 \times 10^6 \text{ V/m}$$

Note that by using Equation (2.22) to eliminate ϕ_o, the maximum value of the electric field intensity can also be written as

$$\mathcal{E}_m = \frac{qN_d W}{\epsilon_s}$$

Next, denoting the transition capacitance per unit area in thermal equilibrium as C_{toa}, and using Equation (2.26) for a p^+n junction, we find

$$C_{toa} = \sqrt{\frac{q\epsilon_s N_d}{2\phi_o}} = 307 \text{ } \mu\text{F/m}^2$$

Charge-Control Model

In order to provide insight into the diode behavior under time-varying conditions, we will now evaluate minority currents at the edges of the SCL in terms of charge distributions in the neutral regions and then combine them to obtain the total current. This approach is called the *charge-control model*.

In Figure 2.2a, consider the volume in the neutral portion of the n region, that is, between x_3 and x_4. As far as the hole current is concerned, the continuity principle says that

(*the rate of change of hole charge in the n region*)
= (*the hole current entering at* $x = x_3$) − (*the hole current exiting at* $x = x_4$)
+ (*the effects of the recombination-generation process within the region*).

To translate this into a symbolic equation, assuming low-level injection and no photogeneration, we use Equation (1.35) and proceed as follows:

$$\frac{\partial \Delta p(x, t)}{\partial t} = \frac{-1}{q}\frac{\partial J_h}{\partial x} - \frac{\Delta p(x, t)}{\tau_h} \tag{2.28}$$

We multiply Equation (2.28) by $+qAdx$, where $+q$ is the charge on a hole, and integrate over the region to get

$$\frac{d}{dt}\left[qA\int_{x_3}^{x_4} \Delta p(x, t)dx\right] = -A\int_{J_h(x_3)}^{J_h(x_4)} dJ_h - \frac{qA}{\tau_h}\int_{x_3}^{x_4} \Delta p(x, t)dx \tag{2.29}$$

Now, we note that the total charge, Q_h, due to the excess holes in the n region is simply the excess hole concentration $\Delta p(x, t)$ integrated over the entire volume between $x = x_3$ and $x = x_4$, and multiplied by the charge on each hole. Therefore, we have

$$Q_h \equiv +q \int_{x_3}^{x_4} p(x, t) A dx \tag{2.30}$$

Thus, Equation (2.29) can be rewritten as

$$\frac{dQ_h(t)}{dt} = -A[J_h(x_4) - J_h(x_3)] - \frac{Q_h(t)}{\tau_h}$$

Also, $J_h(x_4)$, the hole current density at $x = x_4$, is typically zero, for the n region is normally long enough to ensure that almost all the holes entering at $x = x_3$ recombine before they reach $x = x_4$. Therefore, setting $J_h(x_4) = 0$ and approximating $AJ_h(x_3)$ as the total hole current $i_h(t)$, we get

$$\frac{dQ_h(t)}{dt} = i_h(t) - \frac{Q_h(t)}{\tau_h} \tag{2.31a}$$

which reveals that the rate of change of charge storage is equal to the current minus that lost to recombination. In other words, the current $i_h(t)$ supplies holes to the neutral n region at the rate the holes are being lost by recombination plus the rate the stored charge increases.

Undertaking a similar analysis for the behavior of electrons in the p region leads to

$$\frac{dQ_e(t)}{dt} = i_e(t) - \frac{Q_e(t)}{\tau_e} \tag{2.31b}$$

The two current components in Equation Set (2.31) are used to evaluate the total diode current, excluding a capacitive term. Thus, we get

$$\begin{aligned} i_D(t) &= i_h(t) + i_e(t) \\ &= \frac{d(Q_h + Q_e)}{dt} + \frac{Q_h}{\tau_h} + \frac{Q_e}{\tau_e} \end{aligned} \tag{2.32}$$

Equation (2.32) is a fundamental equation relating the diode current to the excess minority carrier charges. It is very useful, especially in the analysis of switching transients in diodes and bipolar junction transistors. Under time-independent (i.e., static) conditions, the first term on the right-hand side is reduced to zero, which indicates that the total diode current is the sum of the recombination currents in the two neutral regions.

The Diffusion Capacitance

Under a forward bias, the *pn* junction exhibits a capacitive effect much larger than the transition capacitance. This capacitive behavior is due to the injected charge stored near the junction outside the depletion region and is modeled by a small-signal capacitance C_d, called the *diffusion capacitance*.

Assuming that the p side is much more heavily doped than the n side, so that the total diode current is entirely due to holes, the total charge storage in steady state due to minority carrier injection across the junction is given by

$$Q_T(t) \approx Q_h(t) = \tau_h i_h(t) \approx \tau_h i_D(t) = \tau_h I_S(e^{V_D/\phi_T} - 1)$$

Then C_d is given by

$$C_d \equiv \frac{dQ_T(t)}{dV_D} = \tau_h I_S \frac{e^{V_D/\phi_T}}{\phi_T} = \tau_h g_d = \frac{\tau_h}{r_d} \tag{2.33}$$

where

$$g_d \equiv \frac{di_D(t)}{dV_D} = I_S \frac{e^{V_D/\phi_T}}{\phi_T} \tag{2.34}$$

is called the *diode dynamic conductance*, and $r_d \equiv 1/g_d$ is the *diode dynamic resistance*. For a reverse bias, g_d is very small and r_d is very large, so that C_d is negligible compared to C_t.

The small-signal model for the *pn* junction is depicted in Figure 2.12. It consists of the dynamic resistance r_d in parallel with the transition and diffusion capacitances.

Example 2.4

For the diode of Example 2.1, we use Equations (2.25) and (2.33) to determine C_t and C_d for various values of applied voltage:

V_D, V	C_t, pF	C_d, pF
−5	.38	0
0	1.09	0
.5	2.05	123

As V_D increases, both capacitances increase, but with forward bias, C_d dominates; the reverse is true under reverse bias.

Turn-Off Transient

Suppose a *pn* junction is connected in the circuit of Figure 2.13 with a step driving voltage. A differential equation rather than an algebraic equation is required to predict the dynamic response of this circuit. When a forward-biased *pn* junction diode is switched to the reverse-biased condition, the excess charge stored in the *n* and *p* regions must be removed before the diode can support the reverse voltage.

Consider a one-sided junction with the *p* side much more heavily doped than the

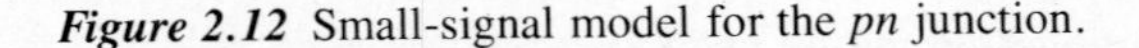

Figure 2.12 Small-signal model for the *pn* junction.

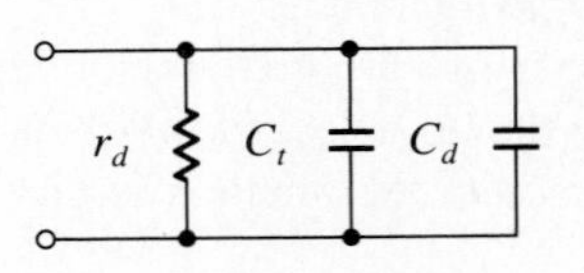

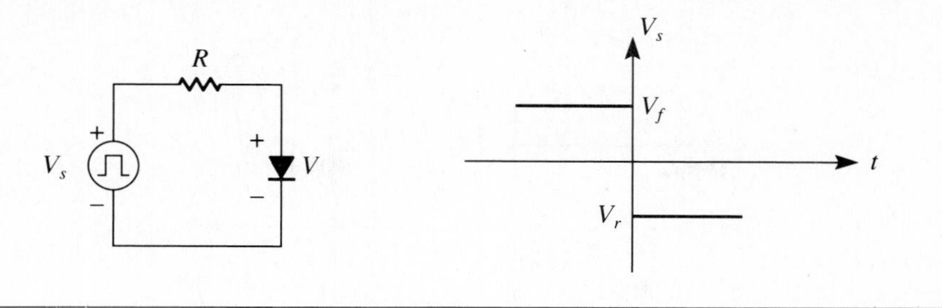

Figure 2.13 A simple diode circuit with a step driving voltage.

n side. We need only to take into account the charge stored in the n region in the form of the distribution of the excess-hole concentration. In this case, we use (2.31a), where $Q_h(t)$ represents the charge due to the excess holes in the n side of the diode. In Figure 2.13, we assume that at $t = 0^-$, a steady-state forward current I_f flows through the circuit, and at $t = 0^+$, the current becomes I_r instantaneously. Suppose that the ohmic voltage drops, as usually accounted for by a series resistance r_s, can be ignored, so that no distinction is necessary between the diode applied voltage V_D and the diode junction voltage V_j. We proceed with the analysis by writing Equation (2.31a) as

$$I_r = \frac{dQ_h(t)}{dt} + \frac{Q_h(t)}{\tau_h} \tag{2.35}$$

At $t = 0^-$, the stored charge is $Q_h(0^-) = I_f\tau_h$. As long as an infinite current is not available, the stored charge will have the same value at $t = 0^+$.

The complementary solution for Equation (2.35) can be shown to be $K_1 e^{-t/\tau_h}$. Since the left-hand side of Equation (2.35) is a constant, the particular function must be a constant K_2, too. Substituting this into (2.35) yields $K_2 = I_r\tau_h$. Thus, the complete solution is given by

$$Q_h(t) = K_1 e^{-t/\tau_h} + I_r\tau_h$$

which must be equal to $I_f\tau_h$ at $t = 0^+$. Hence, we obtain the following solution:

$$Q_h(t) = (I_f - I_r)\tau_h e^{-t/\tau_h} + I_r\tau_h \tag{2.36}$$

The diode junction voltage cannot immediately become V_r to permit the current to be zero because it cannot change without the stored charge $Q_h(t)$ changing. This dependence can be explained by considering Equations (2.9) and (2.30) along with Figure 2.2, and letting $x_4 \to \infty$. Then, we see that $Q_h(t)$ is proportional to $e^{V_D/\phi_T} - 1$, which in turn implies that the voltage is logarithmically related to $Q_h(t)$. Therefore, the junction voltage remains positive as long as $Q_h(t)$ is positive. Figure 2.14 shows the charge, voltage, and current transients of a pn-junction diode. As is obvious from Figure 2.14a, Equation (2.36) is valid until almost all the excess hole charges stored in the n region have been removed. The time needed for this process is called the *storage time* t_s and is found from Equation (2.36) by setting $Q_h(t_s) = 0$ and solving for t_s:

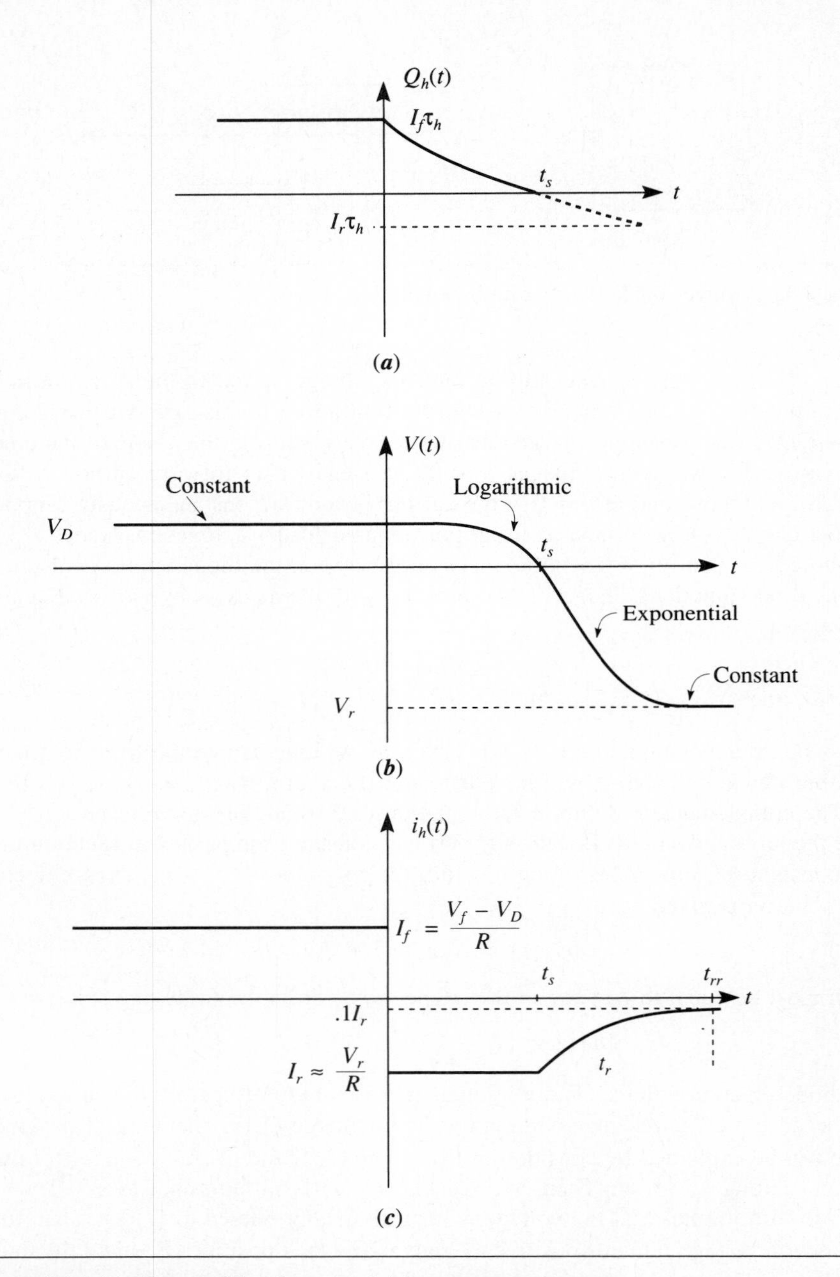

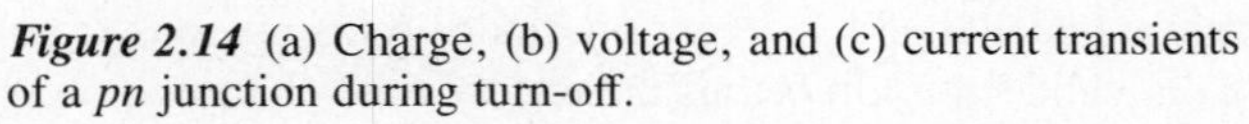

Figure 2.14 (a) Charge, (b) voltage, and (c) current transients of a *pn* junction during turn-off.

$$t_s = \tau_h \ln\left(\frac{1 - I_f}{I_r}\right) \quad (2.37)$$

During this period, the junction is flooded with injected minority carriers, so that its impedance is negligibly small compared to R, which then limits the current. The storage time indicates how long it takes for the diode to start acting like a reverse-biased device and is a factor that limits the speed of digital integrated circuits. Hence, it is used as a figure of merit in switching applications. Typical storage time of a switching diode ranges from .5 ns to several nanoseconds, whereas it ranges from 1 μs to several milliseconds for power rectifiers.

An examination of Equation (2.37) reveals that t_s decreases if τ_h and I_f are decreased, which has the same effect as decreasing $Q_h(0^+)$. The storage time is also reduced by increasing I_r, which corresponds to pulling more charge per second from the n region.

The control of lifetime is accomplished to a certain extent by introducing extra impurities, like gold, to produce additional energy levels in the forbidden gap, as discussed in Chapter 1. These levels increase the rate of combination needed to get rid of the stored charge faster when the diode is being switched from forward to reverse bias.

After a time t_s, a sufficient number of holes either flow out of the n region or are recombined so that the junction gradually returns to its high impedance state. Once the storage interval is completed, and $Q_h(t)$ is zero for all practical purposes, the charge-control equation is not valid anymore. Thus, the excess minority carrier charges lose control over the junction voltage, whereupon V_j tends toward V_r gradually during the *recovery time* t_r, as shown in Figure 2.14c. A new current component emerges from the transition capacitance C_t, so that the diode current is now governed by an RC decay, where R is the circuit resistance. This current is given by

$$i_D(t) = C_t \frac{dV_j(t)}{dt}$$

The recovery is completed when the current that flows is the saturation current I_S. The *reverse recovery time* t_{rr} is defined as the time necessary for the diode current to recover to $.1 I_r$ after it has been switched off. Thus,

$$t_{rr} = t_s + t_r \quad (2.38)$$

The general nature of the switching response can be explained by considering the excess minority concentration profiles in both bulk regions during transient turn-off, as illustrated in Figure 2.15. The initial concentrations before the device is switched are shown in Figure 2.15a. The total excess minority carrier charge stored in the junction is the area under each curve above p_{no} and n_{po}. Figure 2.15b depicts the time progression of the excess charge removal with a constant current I_r, which corresponds to a constant slope of the carrier concentration at the depletion region edges.

Figure 2.15c illustrates that actually $Q_h(t_s) = Q_e(t_s) \neq 0$. However, to assume that it is zero leads to a larger value and hence a conservative estimate of t_s. At $t = t_s$, the junction voltage becomes zero. During t_r, a reverse voltage develops across

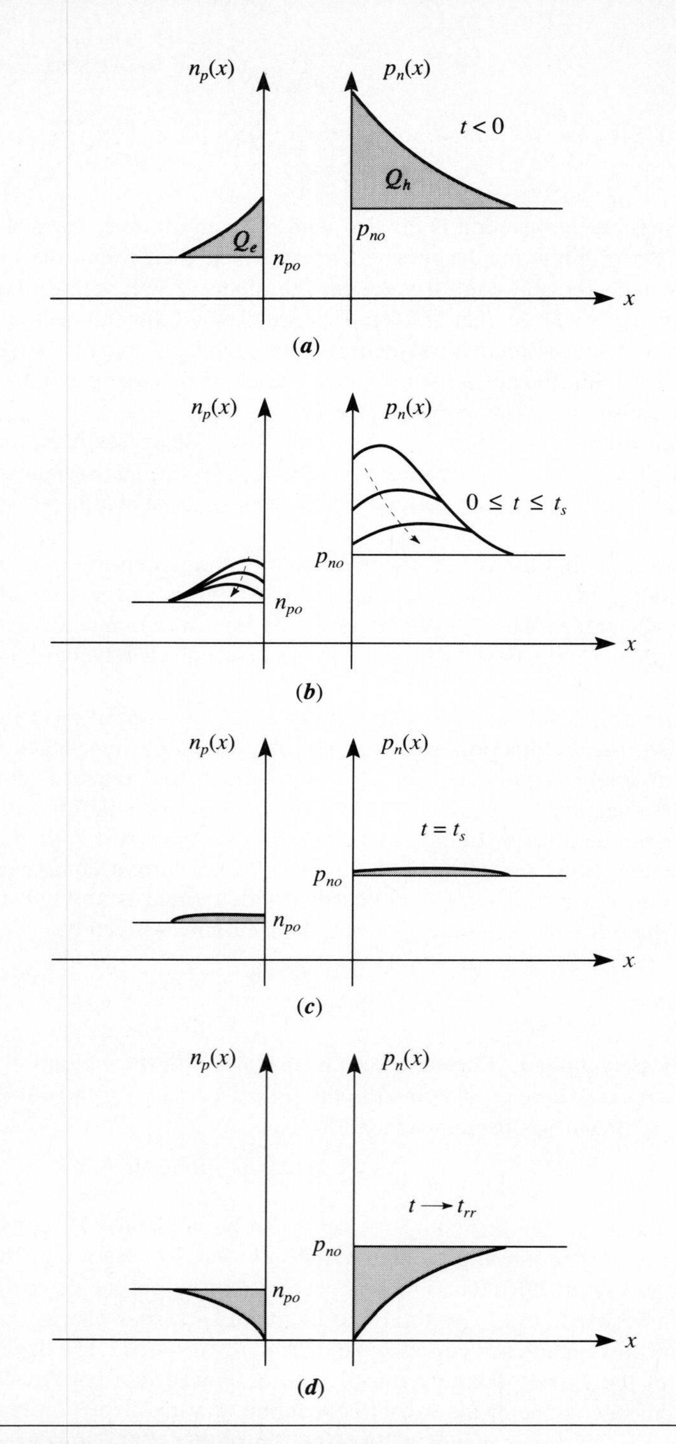

Figure 2.15 Excess minority concentration profiles during turn-off: (a) initial, (b) storage interval, (c) $t = t_s$, (d) recovery phase.

the junction. The absolute value of the current decreases rapidly towards the steady-state saturation value I_S. In the meantime, the total excess charge is eliminated, and the reverse bias causes the carrier deficit, as indicated in Figure 2.15d.

Turn-On Transient

Next, we will consider the case where the diode is pulsed with a current step from $i_D(t) = 0$ to I_f at $t = 0$, as shown in Figure 2.16a. Again, if we assume a one-sided junction with the p side much more heavily doped than the n side, then the total current is essentially due to the holes injected into the n region at $x = x_3$. Since this current is constant, the slopes of both hole and electron concentrations are constant, as depicted in Figure 2.16b.

Applying Equation (2.31a), we see that the rate at which the charge builds up is due to the current I_f minus that used up by recombination:

$$\frac{dQ_h(t)}{dt} = I_f - \frac{Q_h(t)}{\tau_h}$$

Figure 2.16 Turn-on transient: (a) current, (b) minority carrier concentrations.

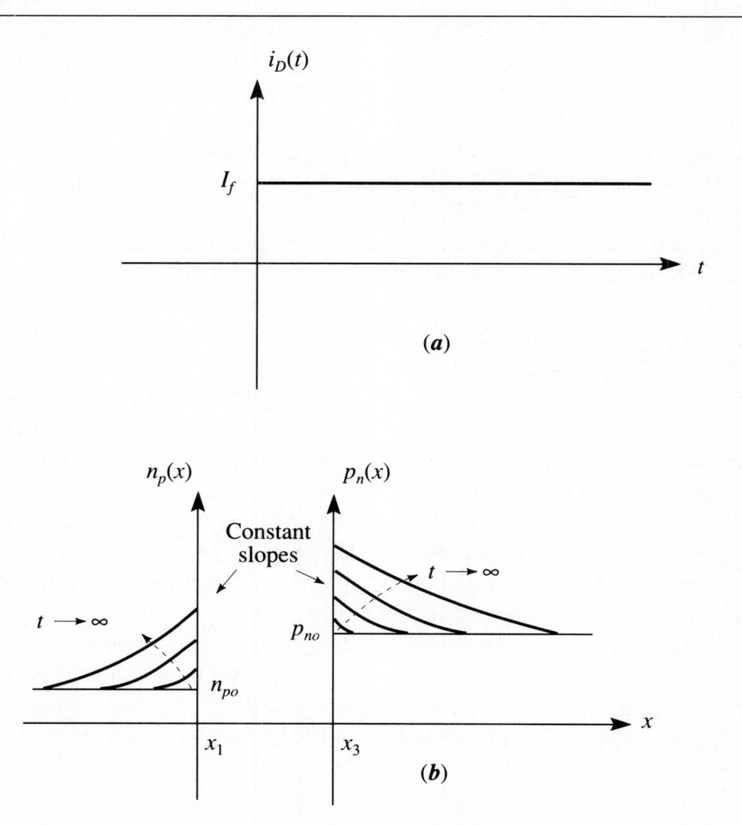

The solution obtained is

$$Q_h(t) = I_f \tau_h (1 - e^{-t/\tau_h}) \tag{2.39}$$

which means that $Q_h(\infty) = I_f \tau_h$.

2.2 The Schottky-Barrier Diode

From the discussion of switching transients so far, we see that a diode with small charge storage and capacitance is needed to improve the switching speed in a digital circuit. A *Schottky-barrier diode* (*SBD*) is often used in place of a *pn* junction for this reason. It is a metal-semiconductor junction formed by evaporating a small metallic contact onto a lightly doped *n*-type material to obtain a current-voltage relationship identical to Equation (2.14) but with a larger value of I_S that results in smaller *cut-in*, that is, turn-on, ($\approx$ .3 V) and forward ($\approx$ .4 V) voltage drops as compared to those of a *pn* junction for the same current level. This contact has a characteristic potential barrier ϕ_B, called the *Schottky barrier*, which is not a function of the semiconductor doping but depends only on the two materials brought together.

In a junction diode, the current is controlled by the diffusion of minority carriers. The applied voltage establishes a distribution of minority carriers in both sides of the device. On the other hand, in a Schottky diode, the applied voltage controls the flow of the majority carriers across the border, so that it does not store minority charge during forward-bias operation, resulting in faster switching times for the turn-off transient. Therefore, even though in both cases the equilibrium between currents is disturbed by the lowering of the barrier at the junction with a forward bias, the currents in an SBD are due to the *emission* of majority carriers rather than to *diffusion* of minority carriers.

We already know that the valence and conduction bands in a metal overlap, making a large number of allowed energy states available to be filled by free electrons injected into the conduction band. The energy necessary to remove an electron from the Fermi level to the vacuum level, where the material can no longer exert any force on the electron, is defined as the *work function*. Now, if the semiconductor has a higher Fermi level before contact, as shown in Figure 2.17a (and as is usually the case), corresponding to a different work function than the metal, then the electrons in its conduction band are in energy states above those of the electrons in the conduction band of the metal. Thus, upon contact, the electrons from the semiconductor will *spill over* into the metal and fill the energy states of the metal. This will give rise to fixed, positive ionized donors in the lightly doped n^- region, resulting in an upward *band bending* near the surface and hence a depletion layer in the semiconductor, as diagrammed in Figure 2.17b, and forming a junction potential, ϕ_o.

The corresponding built-in electric field opposes further electron flow from the semiconductor and eventually becomes sufficient to sustain a balance, at which time the thermal equilibrium is established. Moreover, the Fermi level of the system is aligned, because the semiconductor's Fermi level is now lowered by

$$q\phi_o = q(\phi_m - \phi_s)$$

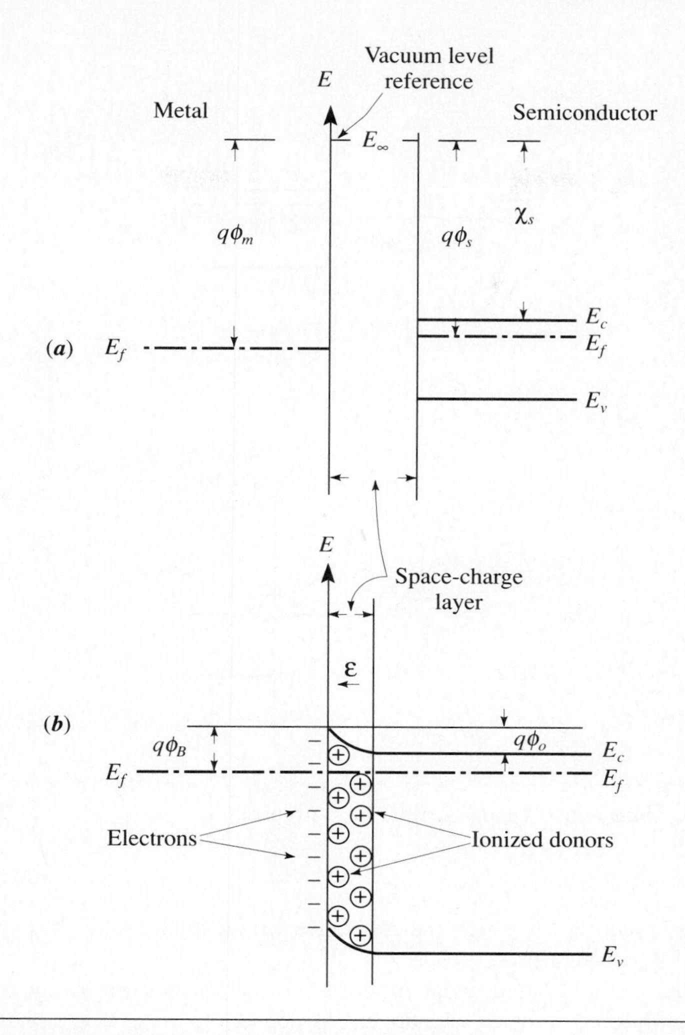

Figure 2.17 The energy-band diagrams of a Schottky barrier with $\phi_m > \phi_s$: (a) before contact; (b) after contact at thermal equilibrium.

where ϕ_m and ϕ_s, with $\phi_m > \phi_s$, are the original potential differences in the metal and semiconductor, respectively, between the Fermi and vacuum levels. The Schottky barrier, on the other hand, is given by

$$q\phi_B = q(\phi_m - \chi_s)$$

where χ_s is the *electron affinity* that specifies the energy required to release an electron from the bottom of the conduction band to the vacuum level. Therefore, due to the presence of the Schottky barrier, the contact is a *rectifying junction* as opposed to a

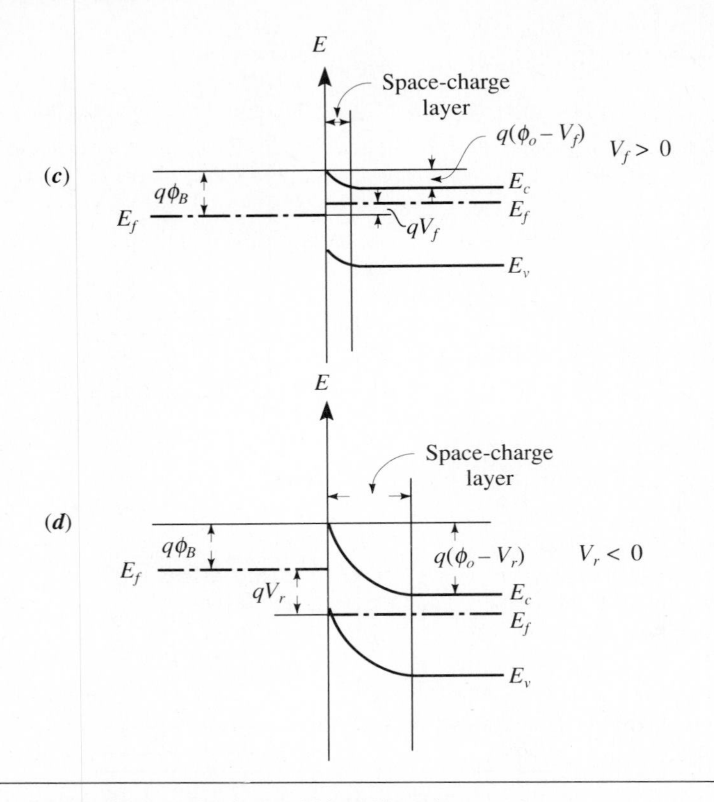

Figure 2.17 *(continued)* (c) under forward bias, (d) under reverse bias.

linear *ohmic contact*, for which the resistance would be the same, regardless of the direction of the current flow.

With the metal positive under forward bias V_f, the built-in potential is reduced to $\phi_o - V_f$ while ϕ_B remains unchanged, as depicted in Fig. 2.17c. This reduction on the semiconductor side increases the number of electrons having sufficient energy to traverse the junction. In fact, these *hot electrons* have energies much greater than the Fermi energy in the metal. Note that the current flow in an SBD is entirely due to the majority carrier electrons compared to the minority injected carriers that carry the current in a forward-biased *pn* junction. Thus, while the switching speed of a *pn*-junction diode is limited by the minority carrier storage effect, only the RC time constant associated with the charging time of the transition capacitance has an effect on the speed of an SBD.

Under an external reverse bias V_r, the metal electrons and semiconductor holes are forced toward the barrier by the applied field. However, only a few of them have enough energy to overcome the barrier potential, constituting a small leakage current. Furthermore, electrons in the semiconductor are repelled from the junction, thereby widening the depletion layer. This is depicted in Figure 2.17d.

2.3 SPICE Diode Model

The model used for the *pn*-junction diode in SPICE is depicted in Figure 2.18. In fact, the SPICE diode model is also applicable to Schottky-barrier diodes. The ohmic resistance, dc characteristic of the device, and the charge storage are modeled by the linear resistor r_s, nonlinear current source I_D, and the charge-storage element Q_D, respectively. The value of I_D follows the ideal diode equation,

$$I_D = I_S(e^{V_D/n\phi_T} - 1)$$

where the parameter n, called the *emission coefficient*, is equal to 1 in almost all cases, but can be somewhat larger than 1 for Schottky diodes. The extrapolated intercept of the graph of log (I_D) vs. V_D, as shown in Figure 2.19, estimates the saturation current I_S, while the slope of the characteristic in the ideal region of operation, corresponding to the straight line in Figure 2.19, determines n.

In practice, the diode current deviates from the ideal exponential characteristic at both higher and lower levels of bias. The curvature at low currents (<1 nA) is basically caused by the thermal recombination-generation of carriers in the depletion region. At low current levels, this process is more evident, due to the higher percentage of carriers lost by trapping. At high currents (>1 mA), the deviation is due to the bulk ohmic effect of the neutral regions as well as high-level injection effects. The former yields a larger terminal voltage for V_D at any given value of I_D. This fact is used to determine the value of r_s from the deviation of the actual diode voltage from the ideal exponential characteristic at a specific current. A typical value of r_s is 10 Ω.

When the minority carrier concentrations at the junction boundaries approach those of majority carriers, the observed diode current stops obeying the ideal diode equation, but follows a modified form:

$$I_D = I_S e^{V_D/2n\phi_T}$$

Since SPICE's emphasis is on integrated circuits, which usually do not develop such large currents in normal operation, high-level injection is not included in the SPICE

Figure 2.18 The SPICE diode model.

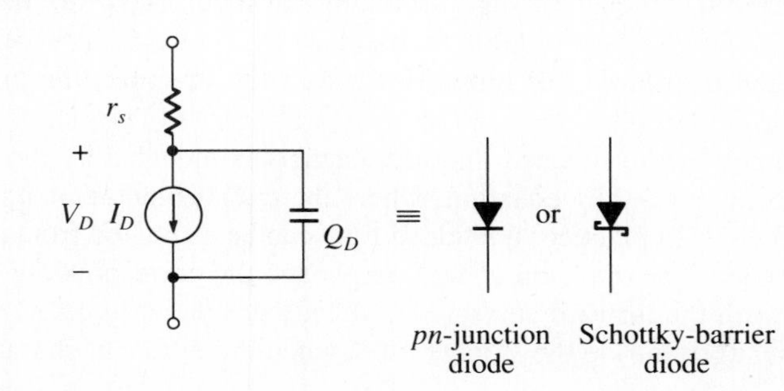

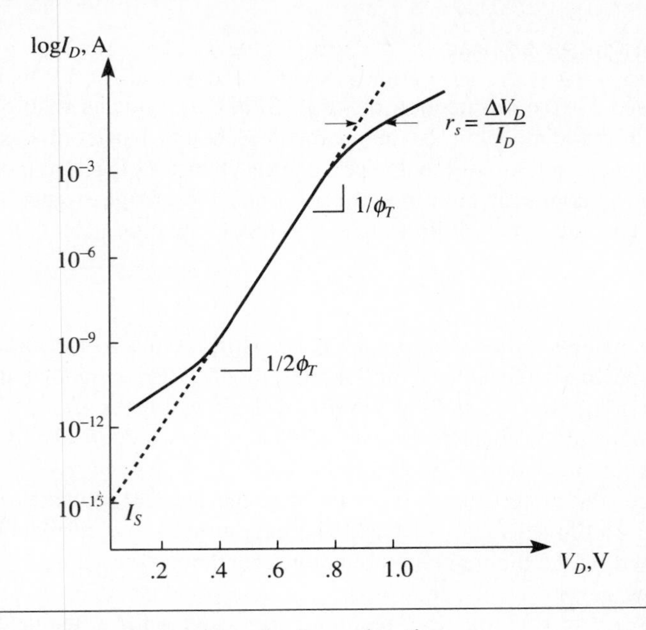

Figure 2.19 Log current versus voltage for the *pn*-junction device.

diode model. Instead, the effects of both ohmic resistance and high-level injection are modeled by r_s, which can be set to limit the exponential effect of the diode equation.

The charge Q_D is determined by

$$Q_D = \tau_T I_D + C_{to} \int_0^{V_D} \left[1 - \frac{V}{\phi_o} \right]^{-m} dV$$

As seen from the above equation, the charge-storage element models two distinct effects in the diode. Charge storage in the depletion region is described by the parameters C_{to}, ϕ_o, and m, where the latter is known as the *gradient coefficient*. It results from the spatial dependence of $n(x)$ in solving Poisson's equation across the space-charge layer. So far, we have assumed it to be .5, since we considered only the abrupt junction in which the change of impurity at the junction is a step. This is true in current fabrication techniques of digital circuits. In older ones, the transition from uniform *n*-type doping to *p*-type doping took place over a finite distance. In this type of linearly graded distribution of impurities across the junction, the grading coefficient, m, is taken to be .33.

Charge storage due to injected minority carriers is modeled by the first term on the right side of the preceding equation, where the transit-time parameter τ_T is equal to τ_e, τ_h or their sum. In practice, the transit time can be estimated from pulsed delay-time measurements. The diffusion charge and hence the corresponding diffusion capacitance vary with the forward current. As already discussed in detail, the diffusion charge manifests itself during the storage time, when the diffusion charge in the junction is discharged for the diode to switch off. During the recovery time, as the junction becomes reverse-biased, the transition capacitance dominates, and the total capacitance is taken to be the sum of these capacitances. Thus, the charge-storage element can be equivalently defined by a voltage-dependent capacitor with two components as

$$C_D = \frac{dQ_D}{dV_D} = \tau_T I_S \frac{e^{V_D/n\phi_T}}{n\phi_T} + \frac{C_t}{\left(1 - \frac{V}{\phi_o}\right)^m}$$

We already know that the second term on the right side of this equation is not valid for values of V equal to or above ϕ_o. For forward biases beyond some fraction of ϕ_o, as set by the parameter **FC**, SPICE calculates the transition capacitance as

$$C_t = C_{to}(1 - \mathbf{FC})^{-(1+m)}\left[1 - \mathbf{FC}(1 + m) + \frac{mV}{\phi_o}\right], \qquad V > \mathbf{FC}\phi_o$$

A complete listing of the parameters required to model the diode is given in Table 2.1. The temperature dependence of the saturation current is defined by the parameters **EG** and **XTI**; the reverse breakdown is described by an exponential increase in the reverse diode current and is determined by the positive parameters **BV** and **IBV**. The optional [(area) value] in the input statement scales **IS**, **RS**, **CJO**, and **IBV** and defaults to 1.

Table 2.1 SPICE Diode Model Parameters

Name	Parameter name	Unit	Default
IS	Reverse saturation current	A	1.0E-14
RS	Ohmic resistance	Ω	0
N	Emission coefficient		1
TT	Transit time	s	0
CJO	Zero-bias transition capacitance	F	0
VJ	Built-in potential	V	1
M	Grading coefficient		.5
EG	Energy gap	eV	1.12
XTI	Saturation current temperature exponent		3
KF	Flicker noise coefficient		0
AF	Flicker noise exponent		1
FC	Forward bias nonideal transition capacitance coefficient		.5
BV	Reverse breakdown voltage	V	∞
IBV	Reverse breakdown current	A	1.0E-10

Example 2.5

Given $R = 1\ \text{K}\Omega$, $C_{to} = 3$ pF, $\phi_o = .7$ V, and $\tau_T = 5$ ns for the circuit of Figure 2.13 with a 50-ns input pulse of 5 V, the following is the corresponding SPICE input file:

```
SIMPLE DIODE CIRCUIT
VIN 1 0 PULSE 0 5 0 1P 1P 50N
R 1 2 1K
D 2 0 SWITCH
.MODEL SWITCH CJO 3PF VJ .7 TT 5N
.TRAN .5N 100N
.PROBE
.END
```

From the resulting current and voltage waveforms plotted in Figure 2.20, we observe that $t_s = 11$ ns and $t_r = 8$ ns.

As its name implies, the SPICE program is mainly developed to simulate integrated circuits. The focus was on the accurate simulation of circuits containing many small and fast devices. On the other hand, it works equally well for discrete components.

Figure 2.20 SPICE simulation for the simple diode circuit of Figure 2.13: (a) current through, and (b) voltage across the *pn* junction.

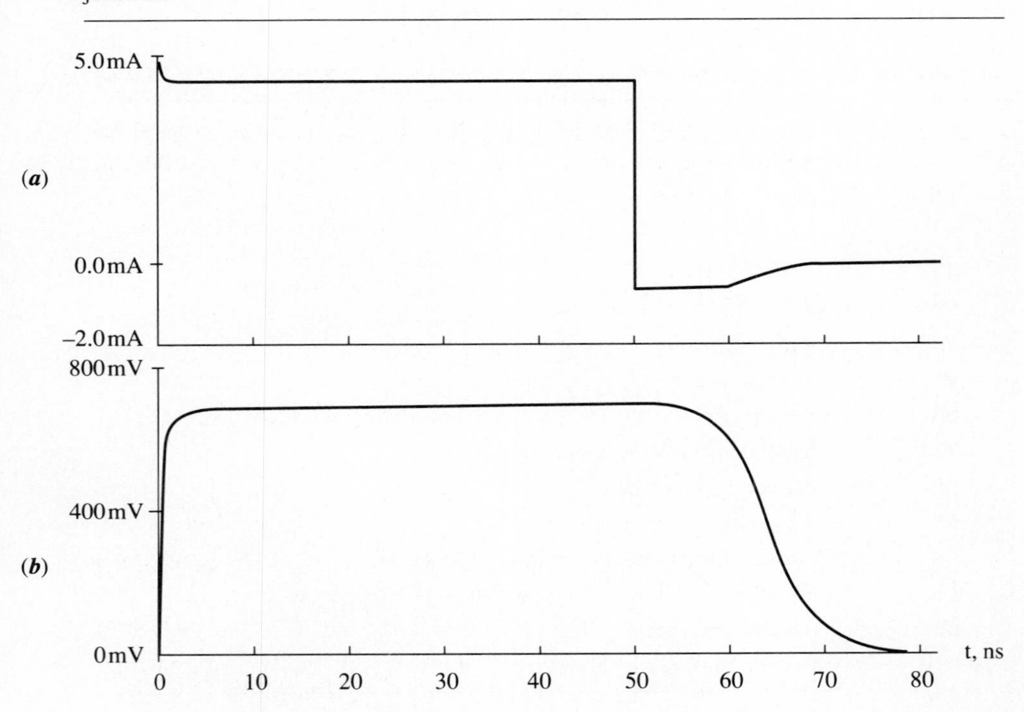

However, due to the IC emphasis, the default values of some parameters are not optimum for discrete devices.

One should note that the default value for all parasitic resistances and capacitances in **.MODEL** statements of switching devices is zero. This implies that if **RS** and **CJO** are not specified, the diode will not have ohmic resistance or transition capacitance. With **RS** = 0, the circuit may have nothing to limit the forward current through a diode, which can be large enough to cause numerical problems.

Without a specified transition capacitance and with **TT** = 0, the diode will have a zero switching time. A transient analysis run will then try to make a transition in zero time, which in turn will cause the program to make the internal time step increasingly smaller until it will collapse and report a transient convergence problem.

Summary

When a *pn* junction is formed, the junction interface is no longer a surface delimiting the crystal. The continuity effectively merges the two original crystals into one larger structure, as a result of which the potential barrier that initially prevented the carriers from leaving each bulk region is now lowered so that the majority carriers are free to diffuse through the junction interface into the next region, where they become minority carriers and recombine, thereby forming a depletion region on both sides of the metallurgical junction. The diffusion fades out when the Fermi levels align in equilibrium, at which time the diffusion current is balanced by the conduction current generated by the built-in electric field. Thus, an electrostatic potential difference exists across the junction at thermal equilibrium. This built-in potential as well as the depletion region width and transition capacitance depend on doping densities on both sides of the junction.

Application of a forward bias disturbs the equilibrium, decreases the width of the depletion region, lowers the potential barrier, and increases majority carrier diffusion. Thus, the current balance is destroyed, and the diode current increases exponentially with applied voltage. In the *p* region, the current is brought about essentially by holes drifting toward the junction. In the vicinity of the junction, due to recombination, electrons increasingly carry a larger portion of the current. Finally, within the *n* region, electrons are the primary carriers of the current. Application of a reverse bias widens the depletion region and raises the junction potential barrier. Only a small reverse saturation current flows through the device.

The small-signal equivalent circuit for a forward-biased *pn*-junction diode is a parallel combination of a resistor, a transition capacitance, and a diffusion capacitance. The latter is the dominant component, but under reverse-bias conditions, the transition capacitance becomes the major parameter. These capacitors play an important role in determining the frequency response and switching speed of the diode. Over the voltage ranges encountered in digital circuits, an average linear capacitance is employed to eliminate the nonlinearity of the small-signal capacitances.

When a contact is made between a metal and a lightly doped *n*-type semiconductor, with the latter's Fermi level above that of the metal, an emission of electrons from the semiconductor to the metal generates a depletion region similar to that of a one-sided *pn* junction without involving minority carriers. The Schottky-barrier diode characteristics are similar to the *pn*-junction diode *I-V* curves, with larger reverse saturation currents and smaller cut-in voltages. Since they do not exhibit minority carrier storage effects, SBDs have faster switching transients than *pn*-junction diodes.

Appendix 2A: Pascal Program for *pn*-Junction Diode Parameters

Given the doping densities at both bulk regions and the applied voltage across the *pn* junction, the following program developed in Turbo Pascal, Version 5, calculates the following parameters:

ϕ_o = built-in potential,

W = depletion region width,

$\mathcal{E}_m$ = maximum electric field intensity in the depletion region,

$x_N \equiv |x_3 - x_2|$, depletion region width on the n side,

$x_P \equiv |x_2 - x_1|$, depletion region width on the p side,

ρ_n = net charge density in the n region, and

ρ_p = net charge density in the p region.

```pascal
program diode_parameters;

uses crt;
var stat : char;

procedure dio_par;

var Na, Nd, Vd, W, phio, Em, xN, xP, c_n, c_p : real;

begin; {dio_par}
  textbackground(red);
  clrscr;
  writeln;
  writeln;
  writeln ('    ****************************************************************');
  writeln ('    **********    SILICON PN JUNCTION DIODE PARAMETERS    ***********');
  writeln ('    ****************************************************************');
  writeln;
  writeln ('    *     Given doping densities Na, Nd, the applied voltage Vd,     *');
  writeln ('    *     this program calculates the following diode parameters:   *');
  writeln;
  writeln ('    φo  = built-in potential,');
  writeln ('    W   = depletion region width,');
  writeln ('    Em  = maximum electric field intensity in the depletion region,');
  writeln ('    xN  ≡ x3-x2, depletion region width on the n-side,');
  writeln ('    xP  ≡ x2-x1, depletion region width on the p-side,');
  writeln ('    c_n = net charge density in the n region,');
  writeln ('    c_p = net charge density in the p region,');
  writeln;
  writeln ('    Enter the concentrations in atoms per cm-3 in the form x(.x)Exx');
  writeln ('    Enter the applied voltage in volts in the form .xx');
  writeln;
  write   ('    Enter Na ') ; readln(Na);
  write   ('    Enter Nd ') ; readln(Nd);
  write   ('    Enter Vd ') ; readln(Vd);
```

```
  phio:=0.02586*ln(Na*Nd/sqr(1.5E10));
  W:=sqrt((2*1.036E-12/1.602E-19)*(Na+Nd)/(Na*Nd)*(phio-Vd));
  Em:=-2*phio/W;
  xN:=Na*W/(Nd+Na);
  xP:=Nd*W/(Nd+Na);
  c_n:=1.602E-19*Nd;
  c_p:=-1.602E-19*Na;

  clrscr;
  writeln;
  writeln;
  writeln ('    *******************************************************************');
  writeln ('    * An Si pn junction with doping densities of Na = ', Na:8, ' cm-3, *');
  writeln ('    *   Nd = ', Nd:8, ' cm-3, and an applied voltage of Vd = ', Vd:4:2, ' V    *');
  writeln ('    *         will have the following parameters:                     * ');
  writeln ('    *******************************************************************');
  writeln;
  writeln ('                  φo  = ',phio:4:2,' V');
  writeln ('                  W   = ',1E7*W:5:2,' nm');
  writeln ('                  Em  = ',Em:6:0,' V/cm');
  writeln ('                  Xn  = ',1E7*Xn:5:2,' nm');
  writeln ('                  Xp  = ',1E7*Xp:5:2,' nm');
  writeln ('                  c_n = ',c_n:9,' C/cm3');
  writeln ('                  c_p = ',c_p:9,' C/cm3');
end; {dio_par}

begin; {main program}
  stat := 'Y';
  while (stat = 'Y') or (stat = 'y') do
    begin;
    dio_par;
    writeln;
    write('    Do you wish to change tne input data? ');
    stat:=readkey;
    end
end. {main program}
```

References

1. D. L. Pulfrey and G. Tarr. *Introduction to Microelectronic Devices.* Prentice-Hall, Englewood Cliffs, NJ: 1989.
2. M. Zambuto. *Semiconductor Devices.* McGraw-Hill, New York: 1989.
3. P. Antognetti and G. Massobrio, eds. *Semiconductor Device Modeling with SPICE.* McGraw-Hill, New York: 1988.
4. E. S. Yang. *Microelectronic Devices.* McGraw-Hill, New York: 1988.
5. D. H. Navon. *Semiconductor Microdevices and Materials.* Holt, Rinehart & Winston, New York: 1986.
6. R. A. Colclaser and S. Diehl-Nagle. *Materials and Devices for Electrical Engineers and Physicists.* McGraw-Hill, New York: 1985.
7. S. M. Sze. *Semiconductor Devices, Physics and Technology.* John Wiley & Sons, New York: 1985.
8. G. W. Neudeck. *The PN Junction Diode.* Addison-Wesley, Reading, MA: 1983.
9. H. E. Talley and D. G. Daugherty. *Physical Principles of Semiconductor Devices.* The Iowa State University Press, Ames, IO: 1976.
10. L. W. Nagel. "SPICE2, A Computer Program to Simulate Semiconductor Circuits." *ERL Memorandum ERL-M520,* University of California, Berkeley: May 1975.

PROBLEMS

For the following problems, assume a room temperature of 300 K.

2.1 Assuming $N_d = 10^{16}$ cm^{-3}, find $\Delta p(x_3)$ for V_D over the range .1 to .7 V in steps of .1 V. Plot $\Delta p(x_3)$ vs. V_D using a software tool such as Lotus 1-2-3 or MathCAD.

2.2 Given $N_a = 10^{17}$ cm^{-3} and $N_d = 10^{15}$ cm^{-3}, sketch plots of carrier concentration versus distance for an Si *pn* junction.

2.3 For the *pn* junction of Problem 2.2, find the resistivities of bulk regions.

2.4 What is the value of the diode voltage required for the reverse current in a *pn* junction to reach 90% of its saturation value?

2.5 What will be the effect on the forward voltage of a *pn* junction if it is shunted by a second identical diode while being supplied with a constant current I?

2.6 Determine the change in barrier height of a *pn*-junction diode when the doping on the *n* side is changed by a factor of (a) 100, and (b) 1000, while the doping on the *p* side remains constant.

2.7 To keep a semiconductor material from being degenerate, its Fermi level must be at least $3kT$ away from either band edge. What should be the maximum value of ϕ_o if an abrupt *pn* junction is not to have degenerate bulk regions?

2.8 Given $\tau_h = 1\ \mu$s, $R = 1$ KΩ, $V_f = 2$ V, $V_r = -20$ V and $V_D = .65$ V at $t = 0$, find the storage time of the diode in Figure 2.13.

2.9 a. Determine t_{rr} of the diode in Figure 2.13 if $I_f = 10$ mA, $I_r = -20$ mA and $R = 500\ \Omega$. The diode is characterized by $I_S = 2 \times 10^{-15}$ A, $C_t = 5$ pF when it is reverse-biased, and $N_a = 100N_d = 10^{19}$ cm^{-3}.
b. Verify the result in (*a*) with SPICE.

2.10 Given an Si *pn* junction with a cross-sectional area $A = 10^{-4}$ cm^2, $N_a = 10^{17}$ cm^{-3}, $N_d = 5 \times 10^{15}$ cm^{-3}, $\tau_h = 100$ ns, and $\tau_e = 10$ ns, calculate the following.
a. Leakage current due to holes
b. Leakage current due to electrons
c. Reverse saturation current, I_S
d. Injected minority carrier currents and excess carrier concentrations at 0 and 1 μm into the bulk regions if $V_D = \phi_o/2$
e. Minority carrier concentrations at the edge of the depletion region if $V_D = -\phi_o/2$
f. Value of the applied voltage at which the low-level injection assumption fails, by taking 1/10 as the limit of the carrier ratio
g. Transition capacitance at $V_D = 0$, $-\phi_o/2$, and -10 V
h. Diffusion capacitance at $V_D = \phi_o/2$

2.11 Consider two diodes with $I_{S1} = 10^{-14}$ A and $I_{S2} = 100 I_{S1}$ connected in series. Calculate the diode current I and the voltage across each diode if the applied voltage is 1 V.

2.12 a. For the input waveform and circuit shown in the diagram, sketch $V_D(t)$ and $I_R(t)$ assuming $W >> L_h$, $I_S = 10^{-15}$ A, $\tau_h = 10$ ns, and neglecting any diode capacitance.
b. Verify the result in (*a*) with SPICE.

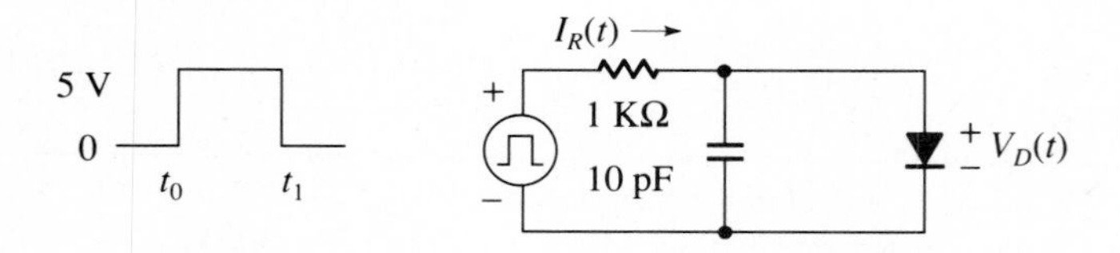

2.13 The following are provided for a *pn* junction: $N_a = 10^{15}$ cm^{-3}, $N_d = 10^{17}$ cm^{-3}, and $A = (40 \times 40)$ μm^2. Calculate the following.
a. ϕ_o, C_{to}, and $C_{(av)}$ as V changes from -10 V to .5 V,
b. W and C_t for $V = -10$ V and .5 V, and
c. I_S.

2.14 For a *pn* junction with $N_a = 10^{17}$ cm^{-3}, $N_d = 10^{16}$ cm^{-3}, and $A = (50 \times 50)$ μm^2, find the following.
a. W, C_t, and $\mathcal{E}_m$ for $V = -5$ V and .5 V, and
b. $C_{(av)}$ as V changes from -5 to .5 V.

2.15 Given an Si *pn* junction with $N_a = 10^{17}$ cm^{-3} and $N_d = 10^{15}$ cm^{-3} in thermal equilibrium, do the following:
a. Calculate ϕ_o.
b. Determine W.
c. Plot $\mathcal{E}(x)$.

2.16 An Si *pn* junction is doped such that $E_f = E_c - E_g/4$ on the *n* side and $E_f = E_v - 2kT$ on the *p* side with $A = 10^{-3}$ cm^2. Find ϕ_o under thermal equilibrium.

2.17 Define the transition capacitance per unit area as $C_{ta} \equiv C_t/A$. Assuming $N_a >> N_d$, show that the doping profile N_d for a one-sided junction can be expressed as

$$N_d = \left(\frac{C_{ta}^3}{q\epsilon_s}\right)\left(\frac{dC_{ta}}{dV_D}\right)^{-1}$$

2.18 a. For a one-sided junction, show that

$$\frac{1}{C_{ta}^2} = \left(\frac{2}{q\epsilon_s N}\right)(\phi_o - V_D)$$

where N represents the minority carrier concentration in the lightly doped side.

b. Plot $(10^{-18} \cdot C_{ta}^2)$ vs. V_D for values of N over the range 10^{14} to 10^{17} cm^{-3}, with the impurity concentration in the heavily doped side fixed at 10^{19} cm^{-3}.

2.19 Using the result of Problem 2.18, plot $(10^{-18} \cdot C_{ta}^2)^{-1}$ vs. V_D is the slope is 1.2×10^{18} and $\phi_o = .7$ V. Find the impurity concentration in the heavily doped side. Is the one-sided assumption justified?

2.20 For an SBD with $\phi_B = .66$ V and $N_d = 10^{16}$ cm^{-3}, find the following.

a. ϕ_o

b. W

c. $\mathcal{E}_m$

2.21 Consider an Si *pn* junction with $N_a = 10^{19}$ cm^{-3}, $N_d = 10^{16}$ cm^{-3} and $A = 1$ mm^2. Also given is an SBD with same N_d and cross-sectional area having $I_S = 8.5 \times 10^{-7}$ A and $\phi_B = .66$ V.

a. Compute the currents in both devices when $V_D = .3$ V.

b. What bias is needed for the *pn* junction to have the same current value as that of the SBD?

c. Find the small-signal equivalent parameters for both devices corresponding to the current value found in (*a*) for the SBD.

d. Comment on their switching performances by computing their time constants at the particular operating point considered in (*c*).

2.22 Using SPICE, compare the switching performances of the diodes in Problem 2.21. Model each diode by the parameters `IS`, `CJO`, `VJ`, and `TT`.

3. THE MOS TRANSISTOR

This chapter is an introduction to one of the most popular logic series: the *C*omplementary-symmetry *M*etal-*O*xide *S*emiconductor, or, simply, the *CMOS*, which is made up of *p*- and *n*-channel enhancement MOS devices, and is discussed in detail in Chapter 4. The *PMOS* (*p-channel MOSFET*) was initially more popular due to the technological difficulties of fabricating *NMOS* (*n-channel MOSFET*) devices in IC form. Today, most of the problems have been overcome, and the NMOS has replaced the PMOS in digital electronics. The latter is now used only in conjunction with the former to obtain the CMOS, which has recently become much more widely used than the NMOS for LSI and VLSI circuits.

Unlike bipolar-junction transistors, discussed in Chapter 5, the operation of the *field-effect transistor* (*FET*) depends only on the flow of the majority carriers. There are two types of FETs. The switching device used in CMOS technology, the MOSFET, is a four-terminal device in which the lateral current flow is controlled by an externally applied vertical electric field. The FETs fabricated with bipolar-junction technology are called *JFETs*. Since the former is the basic building block of the CMOS, we will concentrate on this device.

3.1 The MOS Capacitor

Before investigating the MOS transistor, we will first consider the electrical properties of the less complicated, two-terminal MOS structure shown in Figure 3.1a. Using the vacuum energy level as the reference, and assuming that the energy bands are flat and there is no work-function difference between the metal and the semiconductor, Figure 3.1b shows the ideal electron energy-band diagrams of the metal, oxide, and semiconductor as three separate parts. When they are brought into contact under thermal equilibrium, as depicted in Figure 3.1c, the Fermi levels line up, and the metal electrode, called the *gate*, and the *p*-type bulk semiconductor, called the *substrate*, form the plates of a capacitor with dielectric SiO_2 sandwiched in between. Note that

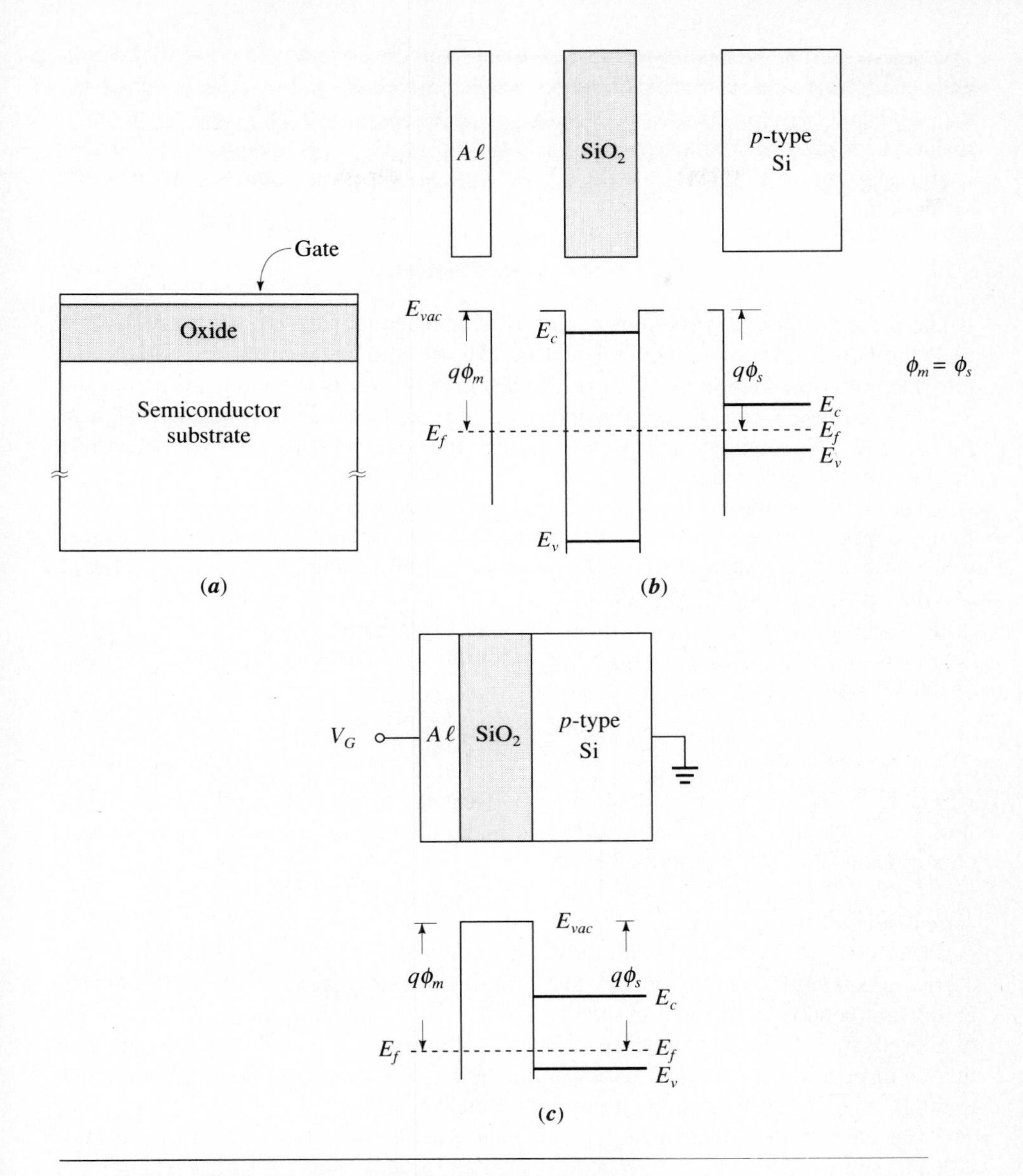

Figure 3.1 The MOS capacitor: (a) structure and idealized energy-band diagram (b) before contact, (c) after contact.

there is no charge transfer upon contact due to the above assumption on work functions.

As an arbitrary voltage is applied across the MOS capacitor, charges appear in the semiconductor, all of which are practically contained within a region adjacent to its top surface. Outside this region we can safely assume that the substrate is neutral.

The potential drop across the region is defined from the surface to a point in the bulk but outside that region, and is called the *surface potential*, ϕ_s. An electric field $\mathcal{E}_{ox}$ is established between the plates, causing a displacement of mobile carriers near the surface of each plate, which in turn leads to two space-charge layers. The induced charge per unit area (C/cm^2) at the silicon surface Q_s, using Gauss's Law, is given by

$$-Q_s = \epsilon_{ox}\mathcal{E}_{ox} = \epsilon_s\mathcal{E}_s \tag{3.1}$$

where ϵ_{ox} and ϵ_s are permittivities of the oxide and semiconductor, respectively, and $\mathcal{E}_s$ is the field at the semiconductor surface. The second equality above is due to the fact that the normal component of the displacement vector **D** is continuous across the Si-SiO_2 interface when there is no interface charge. Note that the negative sign in Equation (3.1) is required, because Q_s is not on the positively charged metal electrode but lies on the semiconductor surface.

The field penetrates into the semiconductor, modifies the energy-band diagram, forms a space-charge region, and establishes a potential barrier beneath the surface. Being an excellent insulator, the oxide does not permit conduction between the metal and the semiconductor, so that the bulk semiconductor has a constant Fermi level as in thermal equilibrium, even under an external bias. Neglecting the voltage drop in the metal plate, the applied voltage V_{GB}, with the subscript B representing the *body*, or the substrate, is expressed as

$$V_{GB} = V_{ox} + \phi_s \tag{3.2}$$

where $V_{ox} = |\mathcal{E}_{ox}x_{ox}|$ is the potential drop across the charge-free oxide whose thickness is x_{ox}. Depending on the polarity and magnitude of V_{GB}, three different surface conditions can exist, as described below.

1. The application of a negative voltage at the gate attracts holes to the oxide-semiconductor interface, causing the majority carrier concentration below the silicon surface to be greater than the equilibrium hole density in the bulk, and leading to the carrier *accumulation* at the surface. The resulting negative surface potential produces an upward bending of the energy-band diagram, as illustrated in Figure 3.2a, leading to a larger $E_i - E_f$ near the surface. Using Equations (1.13) and (1.14), we note that there will be a lower electron density and a higher hole density at the surface than at the bulk, and the surface conductivity will increase.

The gate oxide capacitance per unit area is essentially independent of dc bias voltage, and is given by

$$C_{ox} = \frac{\epsilon_{ox}}{x_{ox}} \ \text{F/cm}^2 \tag{3.3}$$

2. Now assume that a small but positive V_{GB} is applied to the electrode. Then, the surface potential is positive, and the energy band bends in the opposite direction, as diagrammed in Figure 3.2b. The value of $E_i - E_f$ is decreased, and holes are depleted from the vicinity of the Si-SiO_2 interface, leaving behind negatively charged immobile acceptor ions to create a *depletion* region.

The space-charge of the depletion region acts as a capacitor:

$$C_d = \frac{\epsilon_s}{x_d} \tag{3.4}$$

where x_d is the depletion region width and is a function of the bias voltage V_{GB}. C_d is in series with C_{ox}, so that the overall capacitance is given by

$$C_{mos} = \frac{1}{\dfrac{1}{C_{ox}} + \dfrac{1}{C_d}}$$

$$= \frac{C_{ox} C_d}{C_{ox} + C_d} \tag{3.5}$$

An expression for ϕ_s in terms of the space-charge width x_d is obtained by first considering the Poisson equation:

$$\frac{d\mathcal{E}}{dx} = \frac{\rho}{\epsilon_s} \tag{3.6}$$

where ρ is the net space-charge density in a semiconductor, which is given by the algebraic sum of the charge-carrier densities and ionized impurity concentrations and can be expressed as

$$\rho = q\left[p(x) + N_d - \{n(x) + N_a\}\right]$$

because an ionized donor atom has a fixed positive charge, and an ionized acceptor atom has a fixed negative charge. Here, N_d and N_a represent the ionized donor and acceptor concentrations in the semiconductor, respectively, while $p(x)$ and $n(x)$ correspond to the mobile hole and electron carrier densities at a distance x from the oxide-semiconductor interface. However, in the depletion region, the free carriers are negligible due to the presence of a strong electric field, so that Equation (3.6) reduces to

$$\frac{d\mathcal{E}_s}{dx} = -\frac{qN_a}{\epsilon_s} \tag{3.7}$$

for a p-type body.

Solving Equation (3.7) for $\mathcal{E}_s$ and substituting into

$$\phi_s \equiv -\int_0^{x_d} \mathcal{E}_s \, dx$$

yields

$$\phi_s = \frac{qN_a x_d^2}{2\epsilon_s} \tag{3.8}$$

From Equations (3.1), (3.2), and (3.3), we have

$$V_{GB} = -\frac{Q_s}{C_{ox}} + \phi_s \tag{3.9}$$

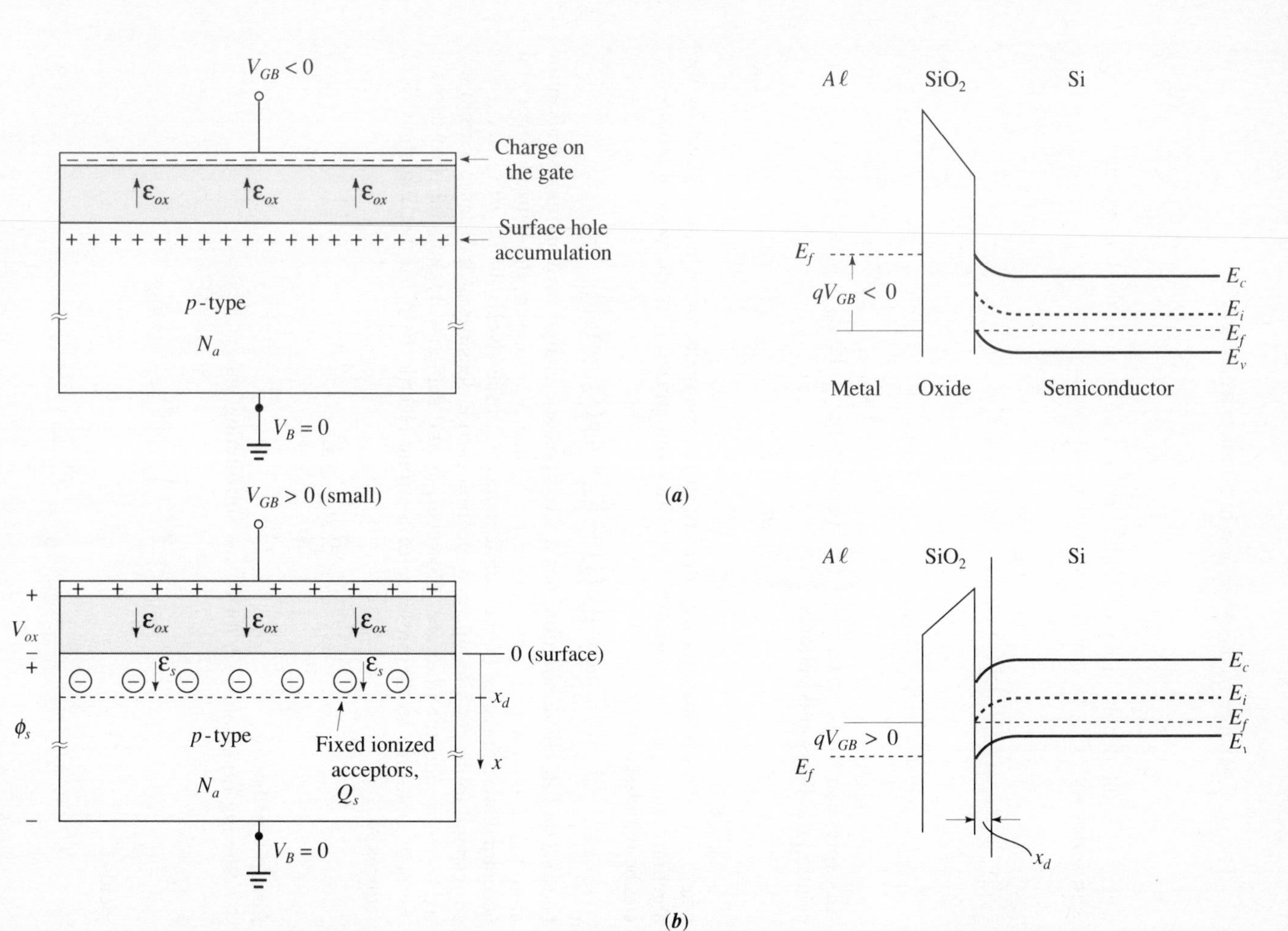

$V_{GB} < 0$
Charge on the gate
ε_{ox}
Surface hole accumulation
p-type
N_a
$V_B = 0$
Aℓ
SiO_2
Si
E_f
$qV_{GB} < 0$
E_c
E_i
E_f
E_v
Metal
Oxide
Semiconductor
$V_{GB} > 0$ (small)
V_{ox}
ϕ_s
ε_s
0 (surface)
x_d
x
p-type
N_a
Fixed ionized acceptors, Q_s
$V_B = 0$
Aℓ
SiO_2
Si
$qV_{GB} > 0$
E_f
E_c
E_i
E_f
x_d

(*a*)

(*b*)

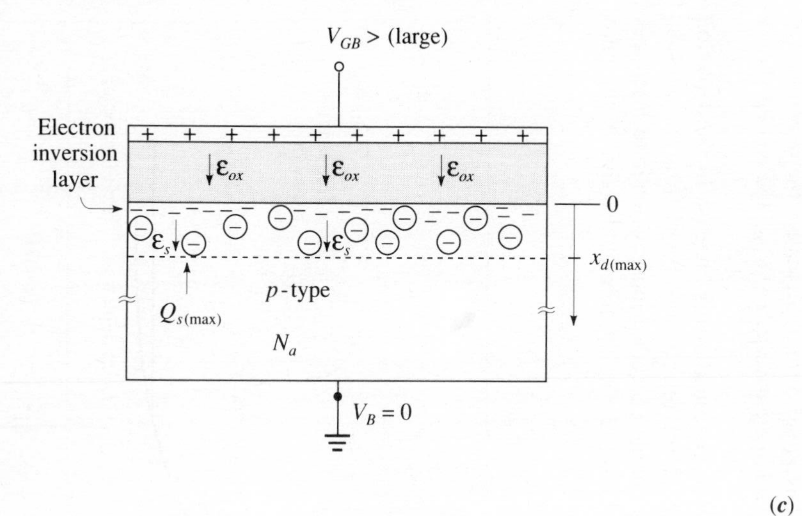

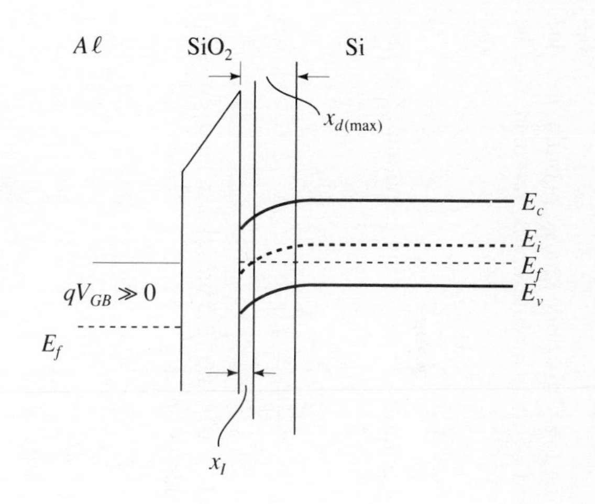

(*c*)

Figure 3.2 Energy-band diagrams of an ideal MOS capacitor: (a) accumulation, (b) depletion, (c) inversion.

where Q_s, as defined in Equation (3.1), is the total immobile charge per unit area due to the acceptor ions that have been stripped of their mobile holes, and can also be expressed as

$$Q_s = -qN_a x_d$$

with the negative sign signifying the charge polarity. Substituting Equation (3.8) into (3.9) and solving for x_d yields

$$x_d = \frac{\epsilon_s}{C_{ox}} \left(\sqrt{1 + \frac{2V_{GB}C_{ox}^2}{q\epsilon_s N_a}} - 1 \right) \tag{3.10}$$

Finally, substitution of Equations (3.4) and (3.10) into (3.5) results in an expression of the MOS capacitance under the carrier depletion as

$$C_{mos} = \frac{C_{ox}}{\sqrt{1 + \left(\frac{2C_{ox}^2}{q\epsilon_s N_a} \right) V_{GB}}} \tag{3.11}$$

The ratio C_{mos}/C_{ox} is plotted in Figure 3.3 as a solid curve. Note that the depletion approximation of Equation (3.7) is not accurate near zero bias, because the transition between the depletion and neutral regions is not abrupt. Actually, the depletion capacitance at zero bias is known as the *flatband capacitance*, C_{FB}, and can be shown to be equal to $\sqrt{qN_a\epsilon_s/\phi_T}$ at zero bias.

C_{mos}/C_{ox} decreases as V_{GB} is increased because of the increase in the depletion layer width under the oxide as a result of the rejection of holes from the Si-SiO$_2$ interface. This leaves behind negatively charged acceptor ions, which in turn reduces C_d and hence the ratio according to Equation (3.5). The minimum value of the ratio

Figure 3.3 A plot of C_{mos}/C_{ox} as a function of the gate voltage.

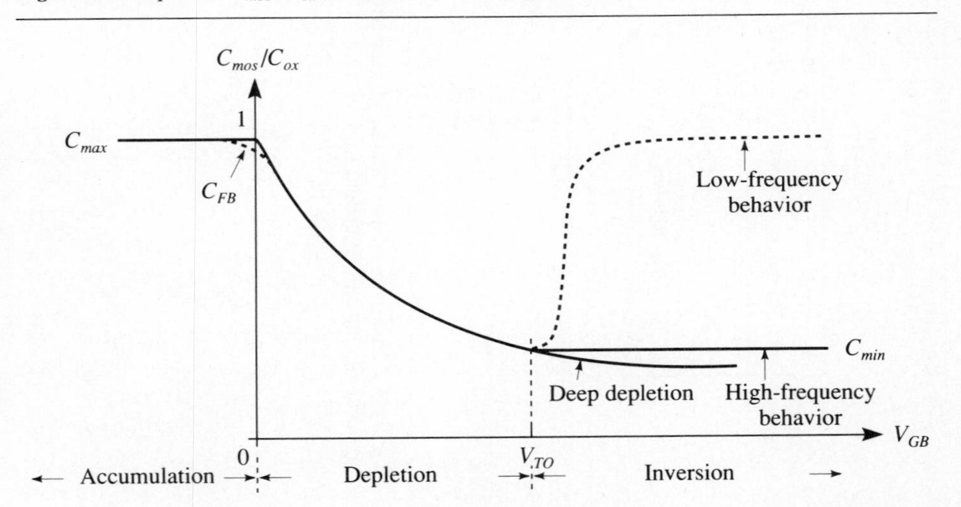

occurs when $x_d = x_{d(max)}$. Additional applied positive gate charge leads to the collection of mobile minority electrons in the p-type silicon at its interface with the oxide.

3. If a larger positive bias is applied, the downward bending of the band is accentuated. When the midgap energy E_i crosses over the constant Fermi level near the Si surface, as depicted in Figure 3.2c, an extremely thin *inversion layer* of width x_I, typically in the order of 100 Å or 10 nm, is formed in which the electron density is greater than the majority hole density due to the additional collection of minority carriers from the p-type semiconductor substrate. The inversion layer now becomes n-type, thereby inducing a pn junction under the metal electrode.

Due to the exponential nature of Equation (1.13), a small increase of ϕ_s produces a large increase of electrons at the surface, because the increase of ϕ_s, in parallel to an increase in V_{GB}, is added to $E_f - E_i$. Then the inversion layer acts like a narrow n^+ layer. Notice, however, that the ionized impurities in this inversion layer are negatively charged acceptors, not positively charged donors as found in an n-type material. All induced charge will be in the inversion layer for a large positive applied voltage, while the depletion-layer charge will remain constant. As we shall see below, inversion will be responsible for current conduction when the two-terminal MOS structure under consideration becomes part of a MOS transistor and the surface potential is raised above $2|\phi_f|$, where

$$|\phi_f| \equiv \frac{|E_i - E_f|}{q} = \phi_T \ln\left(\frac{N}{n_i}\right) \tag{3.12}$$

is the equilibrium electrostatic potential in a semiconductor, also known as the *Fermi potential*. Here, N is either N_a or N_d, depending on the type of the substrate.

Equation (3.11) is valid until the inversion-layer charge becomes significant, after which the capacitance starts to diverge from (3.11) in one of two ways, depending on the frequency of the applied voltage. As we know from the preceding discussion, the incremental positive voltage increases the silicon surface potential, so that the holes are depleted and the depletion layer is widened; hence more and more negative fixed charge is accumulated at the edge of the p-type semiconductor, giving rise to two capacitors in series. In *inversion* and at very low frequencies, the electron-hole pairs can be generated fast enough for the depleted holes at the edge of the depletion region to be replenished by the generated holes. In the meantime, the generated electrons will be attracted by the field and accumulate at the SiO_2 interface. This implies that all electric field lines generating from additional positive charges placed on the gate will be able to terminate on these electron charges, and the total capacitance will again be C_{ox}. Therefore, the characteristic in Figure 3.3 will follow the dotted curve.

If the frequency is high, however, the large inversion-layer charge cannot keep up with the rapidly changing voltage, because it is isolated from the outside world by the oxide on top and the depletion region below. The electron concentration can then only be changed by the slow process of thermal recombination-generation. The inversion-layer width eventually reaches a maximum value of $x_{d(max)}$ and remains at this value even if the gate voltage is increased further. Therefore, the depletion capacitance, as given by Equation (3.4), becomes a constant, which in turn results in a constant MOS capacitance, as illustrated in Figure 3.3 with solid lines. We will subsequently see that communication with the outside world is provided by the *source* and *drain* regions in a MOS transistor, so that the inversion-layer charge can be supplied or removed externally.

If V_{GB} is increased very rapidly from accumulation toward inversion, there may not be time for an inversion layer to form at the Si surface. The capacitance, instead of flattening out at the inversion value of C_{min}, follows the depletion curve and continues to drop below C_{min} because the minority carriers cannot be generated fast enough to balance the changing gate voltage; thus, the depletion layer grows indefinitely past its $x_{d(max)}$ value to compensate for the increased gate charge. This is called the *deep depletion* condition.

A Pascal program written to extract various information from a CV-plot appears at the end of this chapter.

Threshold Voltage

The turning point of the low-frequency capacitance curve in Figure 3.3 is called the *threshold voltage* V_{T0}, with the subscript 0 signifying a grounded bulk electrode. At this critical voltage, the inversion layer is formed significantly, causing a rapid increase in charge for higher gate voltages. It is defined, using Equation (3.9), as

$$V_{T0} \equiv -\frac{Q_{s(max)}}{C_{ox}} + \phi_{si} \tag{3.13}$$

where the bulk charge $Q_{s(max)}$ is due to the ionized acceptor atoms when $x_d = x_{d(max)}$, and is given as

$$Q_{s(max)} = -qN_a x_{d(max)}$$

In Equation (3.13), ϕ_{si} is the surface potential corresponding to the onset of *strong inversion* and is given by

$$\phi_{si} \approx 2\phi_f \tag{3.14}$$

the exact value being dependent on oxide thickness and substrate doping. It can also be expressed as

$$\phi_{si} = 2\phi_f + m\phi_T$$

where m is a parameter chosen to fit the experimental data. For example, if $x_{ox} =$ 500 Å and $N_a = 10^{15}$ cm^{-3}, then $m = 6$ and the surface potential is about 150 mV above $2\phi_f$.

At this point, we can find an expression for the maximum layer width by setting $\phi_s = \phi_{si}$ in Equation (3.8):

$$x_{d(max)} = \sqrt{\frac{2\epsilon_s(2\,|\phi_f|)}{qN_a}} \tag{3.15}$$

Therefore, the value of gate voltage V_{GB} needed to produce strong inversion (i.e., V_{T0}) seems to consist of two components. The first term on the right side of Equation (3.13) is needed to offset the depletion layer charge $Q_{s(max)}$, while the second component corresponds to the voltage needed to produce band bending for strong inversion, in which case the total charge per unit area below the oxide (i.e., at the surface) is given by

$$Q_{si} = Q_I + Q_{s(max)} \tag{3.16}$$

$$= Q_I - qN_a x_{d(max)} = Q_I - \sqrt{2q\epsilon_s N_a (2\,|\phi_f|)}$$

where Q_I is the charge density per unit area due to the electrons in the inversion layer. Now, substituting Equations (3.9) and (3.13) into (3.16) produces

$$Q_I = -C_{ox}(V_{GB} - V_{TO}) \tag{3.17}$$

which expresses the inversion-layer charge in terms of the gate voltage above the threshold voltage. The actual graph of the inversion-layer charge as a function of gate bias is plotted in Figure 3.4. Equation (3.17) is a good approximation for $V_{GB} > V_{si}$, where V_{si} is the break-point corresponding to the onset of strong inversion in Figure 3.4. Below V_{si}, experimental results cannot be described by a simple equation.

Equation (3.13) is not accurate because we have assumed so far that the energy-band diagram is flat when the gate voltage is zero. This is not true in practice due to the *difference* in work functions between the gate material and bulk silicon, and also due to the thin layer of positive parasitic oxide charge Q_{ox} formed at the oxide-silicon interface as SiO_2 is grown thermally on an Si substrate. This undesirable fixed charge can be up to .16 fC/μm^2 in modern devices. It cannot be altered by the gate bias.

In a MOS structure, we consider *modified* work functions of materials because we deal with energies required to bring an electron from the respective Fermi level in the metal and semiconductor to the oxide conduction-band edge instead of the

Figure 3.4 Inversion-layer charge as a function of gate bias.

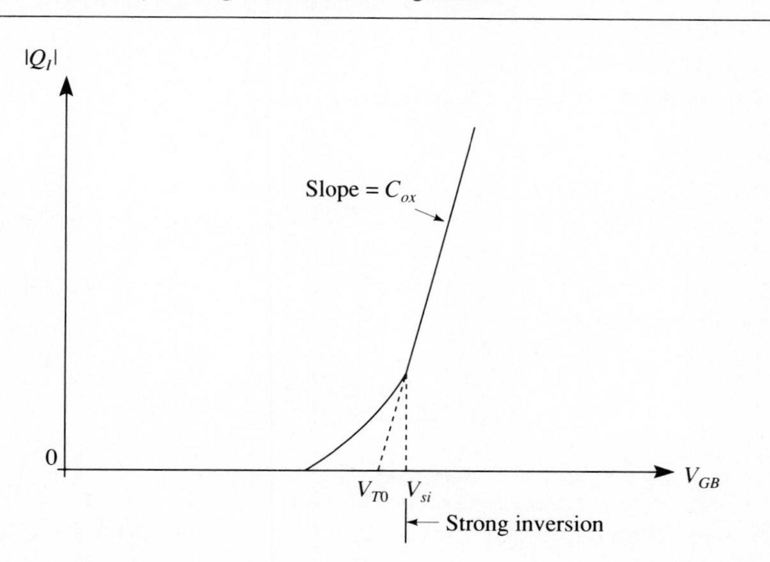

vacuum level. For the metal-SiO_2 system, the difference is the SiO_2 *electron affinity* of $\chi = .9$ V. The corresponding energy-band diagrams for the structure before and after the contact with zero gate bias are shown in Figures 3.5a and 3.5b. Note the band bending in the semiconductor to satisfy the requirement of constant Fermi level under thermal equilibrium. The lower electron energy near the surface and the posi-

Figure 3.5 Energy-band diagrams of an MOS structure and band bending: (a) before contact, (b) after contact at $V_{GB} = 0$, and (c) with $V_{GB} = V_{GB1}$.

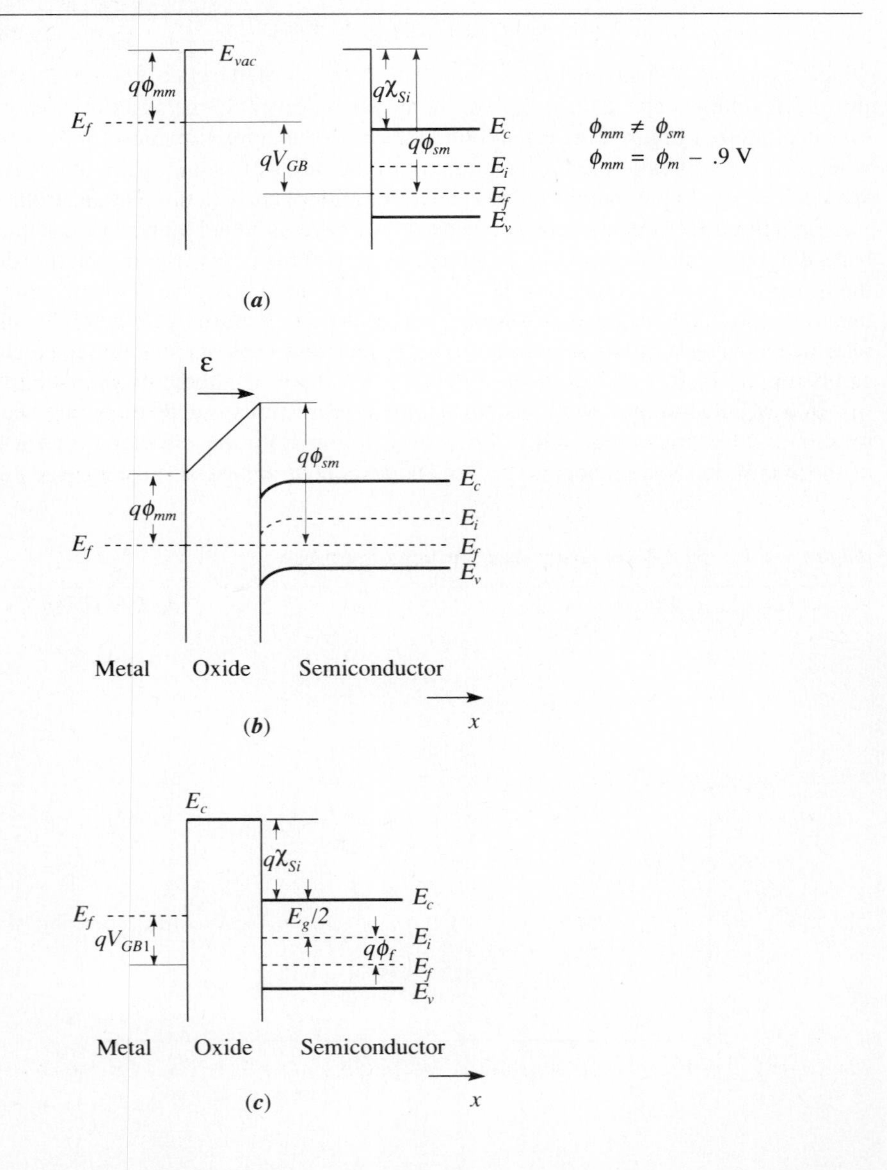

tive slope of the band edges imply the existence of an electric field directed in the positive x-direction that forces holes from the surface, so that the surface is somewhat depleted of holes and tends toward n-type. Furthermore, the positive and constant energy slope in the oxide indicates that there is an electric field there, too. These fields are built in under thermal equilibrium, similar to the built-in field of a *pn* junction.

Now, the gate voltage necessary to eliminate this bending, and hence the electric field in the silicon, is the difference of the modified work functions. The energy-band diagram for an *Aℓ-p*-type silicon structure with $V_{GB1} = \phi_{mm} - \phi_{sm}$, where the second subscripts on the right side signify the *modification*, is illustrated in Figure 3.5c. The modified work function for the $A\ell$-SiO_2 system is 3.2 V. Thus, the gate voltage V_{GB1} required to eliminate the band bending is given by

$$V_{GB1} = 3.2 - \left(\frac{\chi_{Si}}{q} + \frac{E_g}{2q} + \phi_f\right) = -.61 - \phi_f$$

where $\chi_{Si} = 3.25$ eV is the modified electron affinity for silicon. For an n-type substrate the gate voltage would be

$$V_{GB1} = -.61 + \phi_f$$

In the latest MOSFETs, however, the metal has been replaced by polycrystalline silicon so heavily doped that it is degenerate, and the Fermi level of the gate electrode is essentially pinned at either one of the band edges. Therefore, even though the electron affinity remains constant, the work-function difference changes with the substrate impurity concentration.

For a MOS system with a p^+ poly gate whose Fermi level is pegged at E_v, the work-function difference is given by

$$qV_{GB1} = (E_{c(oxide)} - E_v) - (E_{c(oxide)} - E_{f(Si)})$$

where $E_{c(oxide)}$ is the SiO_2 conduction band edge, and $E_{f(Si)}$ is the Fermi level of the substrate. Using Equation (1.13) for an n-type substrate, we have

$$E_{f(Si)} = E_c - kT \ln\left(\frac{N_c}{N_d}\right)$$

while for a p-type substrate, we have

$$E_{f(Si)} = E_v + kT \ln\left(\frac{N_v}{N_a}\right)$$

Therefore, for a system with a p^+ poly gate and an n-type substrate, we obtain

$$V_{GB1} = \frac{E_g}{q} - \phi_T \ln\left(\frac{N_c}{N_d}\right)$$

For a system with a p^+ poly gate and a p-type substrate, the work-function difference is expressed as

$$V_{GB1} = \phi_T \ln\left(\frac{N_v}{N_a}\right)$$

Similarly, it can easily be shown that the work-function difference for an n^+ poly/n-type substrate system is

$$V_{GB1} = -\phi_T \ln\left(\frac{N_c}{N_d}\right)$$

while for an n^+ poly/p-type substrate structure, it takes the form

$$V_{GB1} = -\frac{E_g}{q} + \phi_T \ln\left(\frac{N_v}{N_a}\right)$$

In deriving the last two expressions, it has been assumed that the Fermi level coincides with the conduction-band edge of the heavily doped polysilicon gate.

The charge at the Si-SiO_2 interface also modifies the gate voltage required to achieve the flatband condition. It makes a negative contribution in the amount of $V_{GB2} = -Q_{ox}/C_{ox}$. Combining the effects of the work-function difference and the fixed oxide charge then yields the total gate voltage required to achieve the flatband condition, called the *flatband voltage*, which is given by

$$V_{FB} \equiv V_{GB1} + V_{GB2} = \phi_{mm} - \phi_{sm} - \frac{Q_{ox}}{C_{ox}} \tag{3.18}$$

The flatband voltage causes a shift of the CV-plot away from $V_{GB} = 0$. Thus, C_{FB} is now the value at V_{FB}, not at zero-bias.

Therefore, Equation (3.13) is modified as

$$V_{T0} = V_{FB} - \frac{Q_{s(max)}}{C_{ox}} + \phi_{si} \tag{3.19}$$

For an $A\ell$-p-type silicon MOS capacitor, V_{GB1} is negative, Q_{ox} is positive, $Q_{s(max)}$ is negative, and ϕ_f is positive. For an $A\ell$-n-type silicon MOS capacitor, $Q_{s(max)}$ becomes positive and ϕ_f negative. Equation (3.19) does not give an exact value for V_{T0} in practical cases due to such factors as the possible variations of the body doping near the oxide interface and of the oxide thickness and dielectric constant due to changes in process, and the inexact control of the oxide charge. Its usefulness lies especially in predicting the threshold voltage changes as a function of dimensions and dopings.

As we will see later, the enhancement-mode n-channel MOSFET requires a positive threshold voltage for proper switching operation. Since Q_{ox} is positive, and the magnitude and sign of V_{GB1} depend on the gate and substrate materials, V_{FB} may be a negative number, which in turn may lead to a negative V_{T0}. To ensure a positive threshold voltage, an additional step in the wafer-fabrication process, known as the *threshold adjustment implant*, provides an acceptor ion implantation into the substrate that is characterized by an ion dose D_I (ions/cm^2), giving rise to an additional term qD_I/C_{ox} on the right-hand side of Equation (3.19):

$$V_{TO} = V_{FB} + 2\phi_f + \left(\frac{1}{C_{ox}}\right)\sqrt{2q\epsilon_s N_a(2\,|\phi_f|)} + \frac{qD_I}{C_{ox}} \tag{3.20}$$

If donor ions were implanted, V_{TO} would shift by the same amount but in the opposite direction. In that case, D_I is taken to be a negative quantity because implanting donor ions lowers the threshold voltage. Employment of ion implantation to control V_{TO} makes it possible to set different threshold voltages for different transistors on the same wafer.

For *body bias*, that is, for nonzero values of the bulk voltage V_B, the bulk depletion charge is modified as

$$Q_{s(max)} = -\sqrt{2q\epsilon_s N_a(2\,|\phi_f| + V_B)}$$

to yield the following expression for the change in the threshold voltage:

$$\Delta V_T = V_T - V_{TO} = \frac{\sqrt{2q\epsilon_s N_a}}{C_{ox}}\left(\sqrt{2\,|\phi_f| + V_B} - \sqrt{2\,|\phi_f|}\right)$$

where V_T is the threshold voltage with the body bias. Therefore, we have

$$V_T = V_{TO} + \gamma\left(\sqrt{2\,|\phi_f| + V_B} - \sqrt{2\,|\phi_f|}\right)$$

where

$$\gamma \equiv \frac{\sqrt{2q\epsilon_s N_a}}{C_{ox}} \tag{3.21}$$

is called the *body-bias coefficient*. Its unit is $\sqrt{V}$.

Example 3.1

To calculate the threshold voltage for an *Aℓ-n*-type silicon MOS capacitor characterized by $N_d = 10^{15}\ \text{cm}^{-3}$, $x_{ox} = 125$ nm, $Q_{ox} = 5 \times 10^{-8}\ \text{C/cm}^2$, and $V_{GI} = -.3$ V, we first find the oxide capacitance and the Fermi potential as

$$C_{ox} = \frac{\epsilon_{ox}}{x_{ox}} = 2.83 \times 10^{-8}\ \text{F/cm}^2$$

$$\phi_f = -\phi_T \ln\left(\frac{N_d}{n_i}\right) = -.29\ \text{V}$$

to get

$$\phi_{si} = -(2\phi_f + 6\phi_T) = -.74\ \text{V}$$

$$Q_{s(max)} = 1.6 \times 10^{-8}\ \text{C/cm}^2$$

Now, using Equations (3.18) and (3.19), we find the threshold voltage as

$$V_{TO} = -3.3\ \text{V}.$$

3.2 The MOS Transistor

The logic circuits can be implemented using only MOSFETs requiring no resistors because these transistors can also be connected as resistor loads. MOSFETs are either of the p- or the n-channel type. The circuit symbols of both types are shown in Figure 3.6. They can be either in *depletion* or *enhancement* mode. The CMOS family uses the latter mode because it can be turned on and off with the same polarity voltage.

Static Characteristics of the MOSFET

Figure 3.7 shows the physical structure of the n-channel enhancement type MOSFET, the NMOS. It is a four-terminal device consisting of a p-type semiconductor substrate in which two n^+ regions, the drain and source, are formed. The substrate is a single-crystal silicon wafer and is utilized to physically support the final circuit. The insulator, also called the *field* region, should not permit conduction between separate transistor regions. Since a typical NMOS is symmetrical, the source and drain are interchangeable. The metal contact on the oxide is called the gate. One important device parameter is the *channel length*, L. As shown in Figure 3.7, it is defined as the distance between the two metallurgical n^+p regions. Other basic parameters in determining the electrical characteristics of a MOSFET are the *channel width*, W, the oxide thickness, x_{ox}, and the substrate doping concentration N_a or N_d.

On-chip MOS devices have input resistances in the range of $10^{17}\ \Omega$ to $10^{18}\ \Omega$, which is typical of gate oxides. Even when the resistance is reduced by the shunt resistance of package leads at inputs and outputs or by the shunt junction leakage of input protective devices, these values still allow MOS logic stages to be directly coupled, without the need for separate isolation.

The analysis of digital CMOS circuits does not necessarily require the inclusion of body-bias effects. The basic CMOS logic can be biased with $V_T = V_{T0}$; to simplify the notation, therefore, V_T will be used from now on to denote the threshold voltage, even if $V_T = V_{T0}$.

At zero gate bias, that is, when no voltage is applied at the gate, the only current that can flow is the reverse leakage current from source to drain. The fact that the enhancement MOS device is inherently off when $V_{GS} = 0$ is especially helpful for improving the logic implementation of MOS circuits. Consider an ideal logic element as a simple switch which has zero resistance when closed and infinite resistance when

Figure 3.6 Circuit symbols for MOSFETs: (a) n-channel, (b) p-channel.

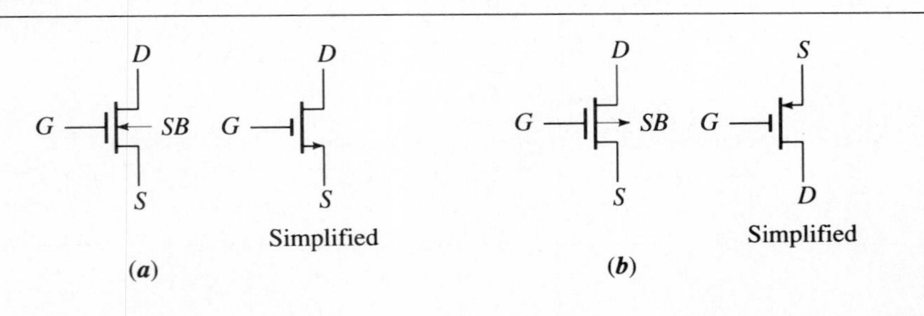

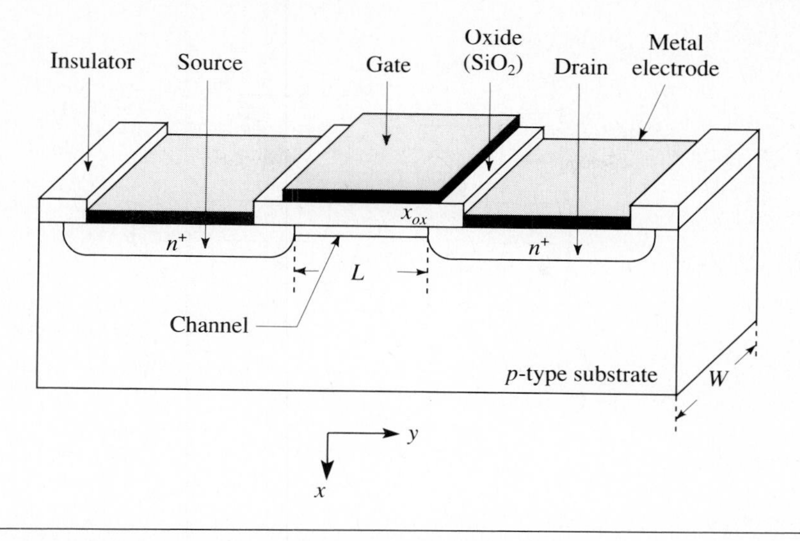

Figure 3.7 The physical structure of an n-channel enhancement type MOSFET.

opened and requires no energy to open or close. The MOSFET can closely simulate this ideal switch because its open circuit resistance is nearly infinite, its closed resistance is in the order of a few hundred ohms at most, and it typically requires less than a picoJoule/s to turn on or off.

Using the source contact as the voltage reference, when a positive gate voltage V_{GS} is applied, with $V_{DS} = V_B = 0$, electrons are attracted from the p-type substrate and accumulate at the surface beneath the oxide layer. When $V_{GS} < V_T$, the p-type substrate is depleted, as shown in Figure 3.8a, for two reasons. The regions around the n^+ source and drain deplete because of the pn^+ junctions formed with the body, while the depletion underneath the gate is due to the electric field penetration from the oxide into the semiconductor. This *field effect* is the greatest influence on the drain-source conduction properties of the MOSFET. Under zero bias, due to the symmetry of the structure, one cannot distinguish between the source and drain. The source is defined to be the n^+ region at the lower potential after the terminal voltages are applied.

After a certain threshold value, V_T, has been reached, an *n channel* (i.e., a conduction path) is established between the source and drain. Thus, the n^+ regions are electrically connected, as illustrated in Figure 3.8b. However, the absence of a driving voltage prevents a drain current, I_D, from flowing.

The Output Transfer Characteristics

In the preceding brief discussion of channel formation, we assumed that the drain-source voltage was zero. Usually, V_{DS} is a nonzero value. Consider the case shown in Figure 3.9a, where $V_{GS} < V_T$. In this *cutoff* mode of operation, no channel exists, and $I_D \approx 0$.

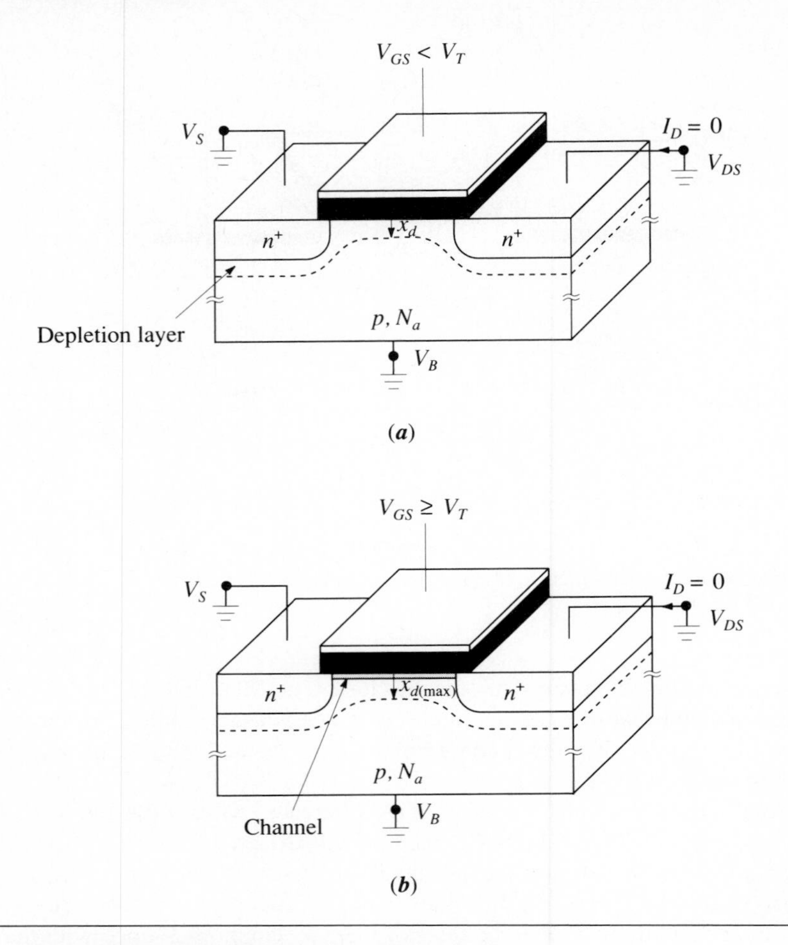

Figure 3.8 The basic MOSFET channel formation with $V_{DS} = 0$: (a) depletion region formation, (b) channel formation.

When the gate-source voltage is increased so that $V_{GS} - V_T > 0$ and $V_{DS} > 0$, the drain current starts flowing through the device by creating a channel and establishing an electric field parallel to the surface in the channel due to the potential difference in n^+ regions. Now, the MOSFET is in the *active* mode of operation. In this channel, or inversion layer, the minority carrier electron concentration is higher than the equilibrium majority carrier hole concentration. Due to their high conductivity, many electrons flow through the inversion layer from source to drain, creating a current from drain to source. The modulation of the channel conductance is possible by varying the gate voltage, resulting in changes of the drain current, which is considered to be positive flowing into the drain electrode.

Depending on the relative value of V_{DS} with respect to $V_{GS} - V_T$, the transistor exhibits two distinct characteristics as far as current flow is concerned. When V_{DS} is small, further increase of V_{GS} *enhances* the already created channel and hence

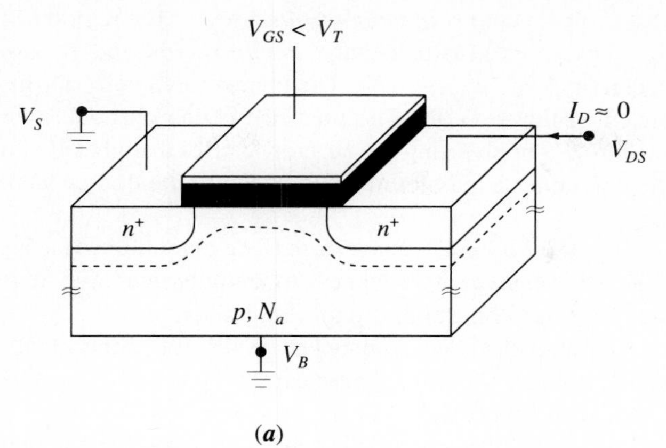

(*a*)

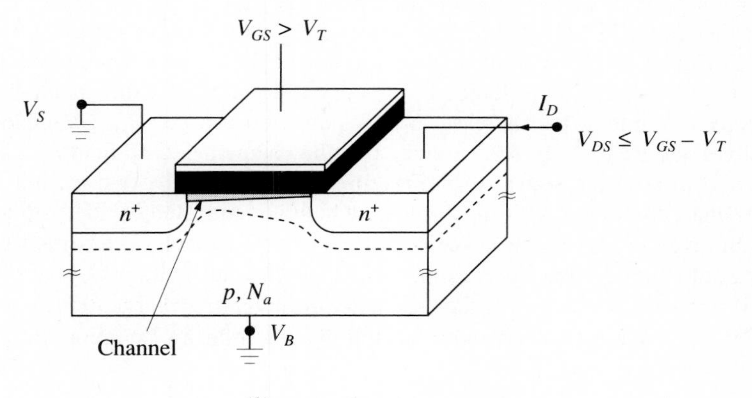

(*b*)

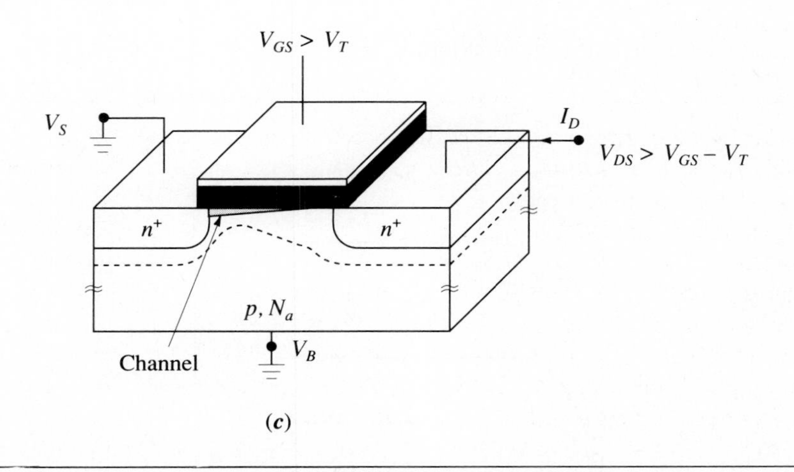

(*c*)

Figure 3.9 Basic modes of operation of a MOSFET: (a) cutoff, (b) nonsaturated, (c) saturated.

decreases its resistance. This is the *voltage-controlled-resistance* region of operation, where $V_{DS} < V_{GS} - V_T$. It is also called the *ohmic region* and is shown in the I_D-V_{DS} output characteristics of Figure 3.10. The channel geometry in this *nonsaturated* mode is depicted in Figure 3.9b. The presence of the channel voltage $V(y)$, due to V_{DS} and opposing V_{GS} in inverting the surface, causes a slight tilt at the bottom edge of the channel. It assumes a maximum value of V_{DS} at the drain end and reduces to zero at the source.

If $V_{DS} > V_{GS} - V_T$, the MOSFET is *saturated*. As diagrammed in Figure 3.9c, the channel experiences a *pinchoff*; its thickness essentially decreases to zero at the pinchoff point before reaching the drain. Figure 3.11 illustrates the I_D-V_{GS} transfer characteristics for an n-channel enhancement-type MOSFET. Simplified schematic diagrams of an NMOS structure showing charge carriers and fixed ions under various biasing conditions appear in Figure 3.12.

To study the output transfer characteristics, we will reduce the actual three-dimensional system down to a one-dimensional current flow problem that can be solved analytically and will still yield results that agree well with experimental measurements. To derive an approximate expression for the drain current in the ohmic region, assume that the threshold voltage, V_T, is constant throughout the channel, which is not exactly true due to the increase in bulk depletion charge, $Q_{s(max)}$, created by the channel voltage $V(y) > 0$. Also assume that the channel electric field $\mathcal{E}_y$, established by V_{DS} and driving the source-to-drain current in the negative y-direction in Figure 3.13, is small compared with the field $\mathcal{E}_x$, which is controlled by the gate voltage to induce an inversion layer and is perpendicular to the channel. At a certain distance y from the source and along the channel, the gate-channel voltage is given as $V_{GS} - V(y)$. If this value exceeds V_T, then the electron inversion charge density induced at this point under the gate is given by the following similar to Equation (3.17):

$$Q_I(y) = -C_{ox}[V_{GS} - V(y) - V_T] \tag{3.22}$$

Figure 3.10 The I_D-V_{DS} transfer characteristic for an n-channel enhancement-type MOSFET.

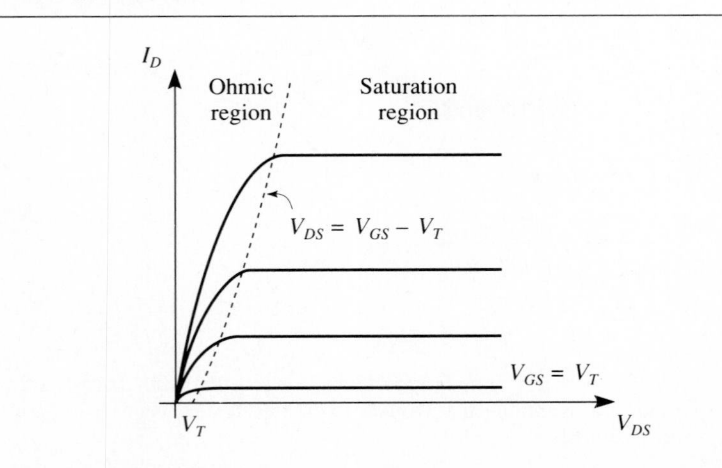

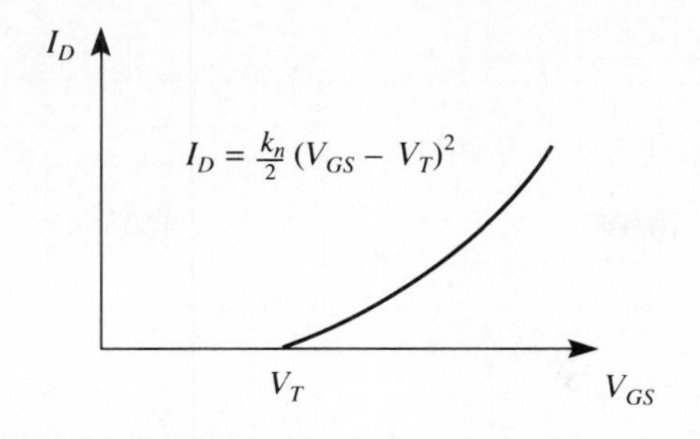

Figure 3.11 The I_D-V_{GS} transfer characteristic for an n-channel enhancement-type MOSFET.

where C_{ox} is the gate capacitance per unit area, as defined in Equation (3.3). The conductivity of the channel at position y is given by Equation (1.22) as

$$\sigma_e(x) = q\mu_e n(x)$$

where μ_e $(\mathrm{cm^2/V \cdot s})$ is the mobility of electrons in the inversion layer, known as the *surface mobility*. As seen in Table 1.3, the mobility values for electrons are 2.5 to 3 times as much as those for holes; thus, for the same silicon area, the current drive capability of an n-channel transistor is at least 2.5 times that of a PMOS. With larger gate fields (i.e., at higher gate voltages), however, the electrons encounter more collisions with the oxide-semiconductor interface as they drift along the channel, leading to a lower surface mobility. In fact, for a given temperature, the surface mobility is found to be practically independent of doping density as long as $N < 10^{16}\ \mathrm{cm^{-3}}$. It is rather a function of an average normal electric field in the inversion layer.

The channel conductance, g_d, can be expressed as

$$g_d \equiv \left.\frac{\partial I_D}{\partial V_{DS}}\right|_{V_{GS}=const} = \left(\frac{W}{L}\right)\int_0^{x_I} \sigma_e(x)\, dx \tag{3.23}$$

where x_I is the thickness of the inversion layer. Substitution of the above expression for conductivity into Equation (3.23) yields

$$g_d = \mu_e \left(\frac{W}{L}\right)\int_0^{x_I} qn(x)\, dx \tag{3.24}$$

The integral on the right-hand side of Equation (3.24) corresponds to the total charge per unit area in the channel, that is, $Q_I(y)$. Consequently, the channel resistance dR of a differential channel increment of length dy, using Equation (3.23), is obtained as

$$dR = -\frac{dy}{g_d L} = -\frac{dy}{W\mu_e Q_I(y)}$$

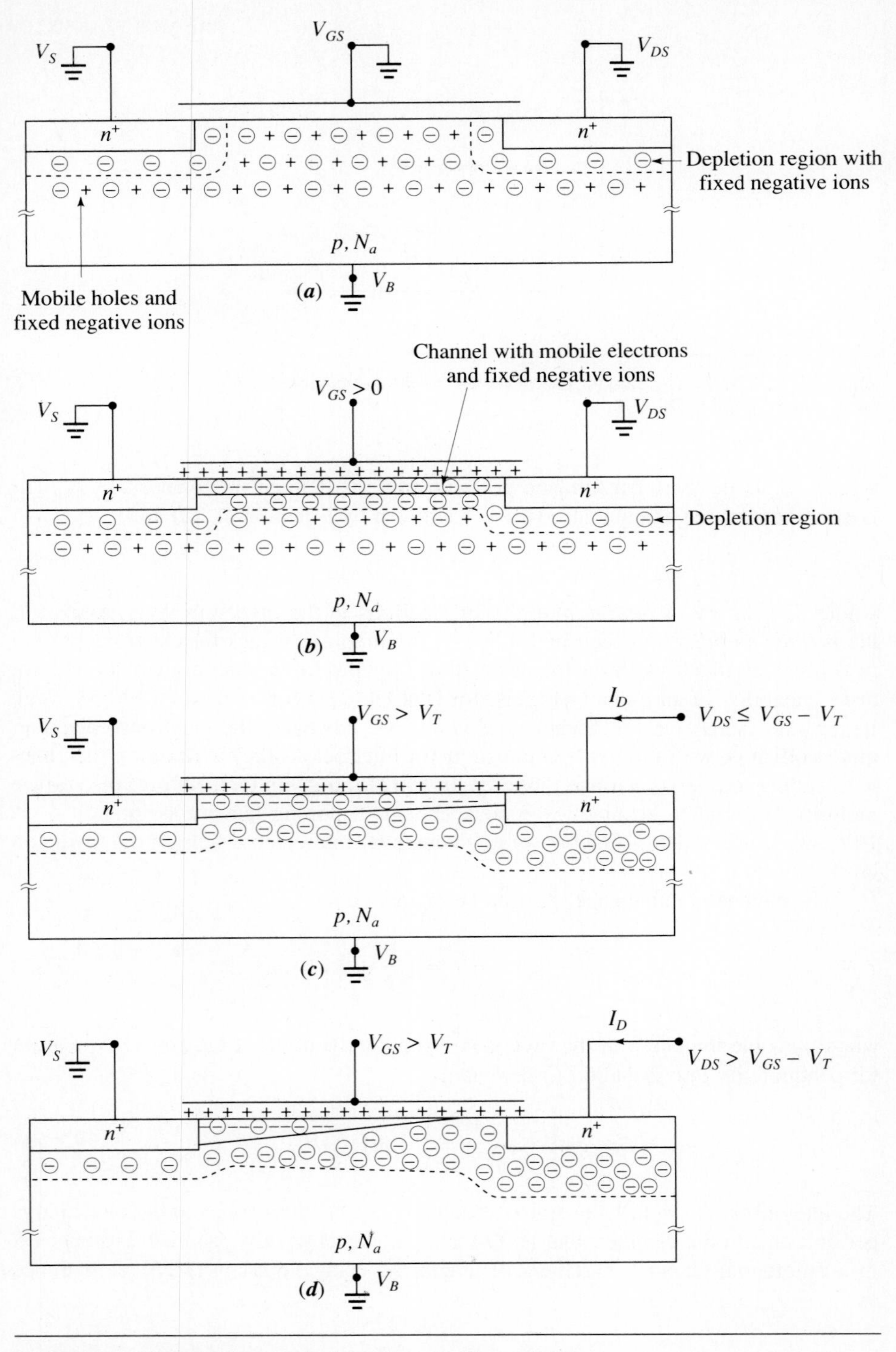

Figure 3.12 (a) Zero bias on all electrodes; (b) $V_{GS} > 0$, $V_{DS} = 0$; (c) nonsaturated mode; (d) saturated mode.

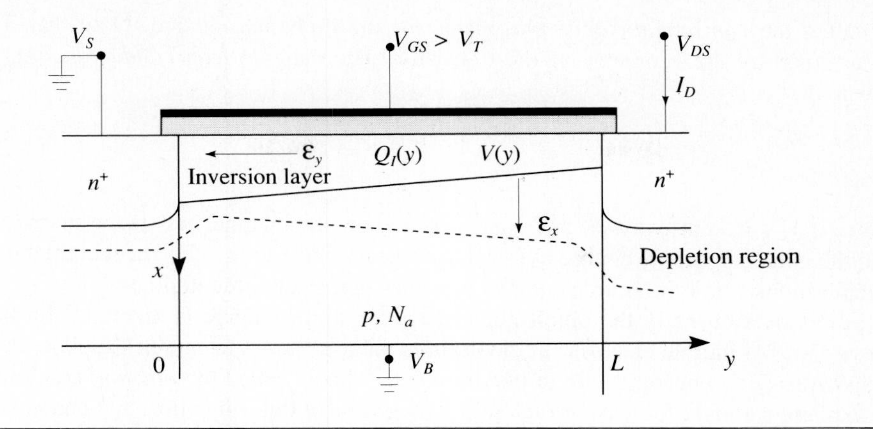

Figure 3.13 MOSFET geometry used in drain current derivation in the ohmic region.

where the minus sign is used to compensate for the negative value of $Q_I(y)$. The differential voltage drop along the length of the channel dy is then given by

$$dV(y) = I_D dR = -\frac{I_D dy}{W\mu_e Q_I(y)} \tag{3.25}$$

Note that the drain current I_D is independent of y. Substituting Equation (3.22) into (3.25) and rearranging, we obtain

$$I_D dy = W\mu_e C_{ox}[V_{GS} - V(y) - V_T]\, dV(y) \tag{3.26}$$

Integrating both sides of Equation (3.26) over the entire length of the channel, we get

$$I_D \int_0^L dy = W\mu_e C_{ox} \int_0^{V_{DS}} [V_{GS} - V(y) - V_T]\, dV(y)$$

Therefore, the I_D-V_{DS} output characteristics of a nonsaturated MOSFET are described by the parabolic relationship

$$I_D = (\mu_e C_{ox})\left(\frac{W}{L}\right)\left[(V_{GS} - V_T)V_{DS} - \frac{1}{2}V_{DS}^2\right]$$

$$= k_n\left[(V_{GS} - V_T)V_{DS} - \frac{1}{2}V_{DS}^2\right] \tag{3.27}$$

where the gain constant

$$k_n \equiv \mu_e C_{ox}\left(\frac{W}{L}\right) \tag{3.28}$$

is called the *transconductance parameter* for the n-channel device. Note that it is determined by the geometry of the transistor. The channel *transconductance* g_m is given as

$$g_m \equiv \left. \frac{\partial I_D}{\partial V_{GS}} \right|_{V_{DS}=const} = k_n V_{DS} \tag{3.29}$$

Now, if V_{GS} is kept constant and V_{DS} is increased, the channel resistance increases, and the induced channel charge $Q_I(y)$ decreases near the drain. The channel thickness remains constant at the source end and becomes narrower at the drain end, giving rise to a nonlinear curve in the ohmic region. As the drain voltage is increased further, the number of mobile electrons at the drain is reduced, and the channel thickness and the charge $Q_I(y)$ become zero at the drain end. This is called the *pinchoff* condition, as explained above. Increasing the drain voltage above this value does not change the shape of the channel. Therefore, above pinchoff, the drain current remains essentially constant for a given value of the gate voltage, and the device approximates a *voltage-controlled current source*. This region of operation is called the *saturation region*, where Equation (3.27) no longer applies.

The loci of points between the two regions give the I_D-V_{GS} characteristic, as in Figures 3.10 and 3.11, which are found by substituting $V_{DS(sat)} = V_{GS} - V_T$ in Equation (3.27). Therefore, in the saturation region we have

$$I_{D(sat)} = \frac{1}{2} k_n (V_{GS} - V_T)^2 \tag{3.30}$$

which manifests the enhancement nature of the device. For an ideal MOSFET in the saturation region, the inversion-layer charge and channel conductance are zero, and the transconductance is obtained, using Equation (3.30), as

$$g_{m(sat)} = k_n (V_{GS} - V_T) \tag{3.31}$$

The operation and characteristics of the p-channel enhancement-mode MOSFET are similar to those of the n-channel device except for a reversal of polarity of all currents and voltages. The I_D-V_{GS} characteristic for a p-channel MOSFET is shown in Figure 3.14.

Figure 3.14 The I_D-V_{GS} transfer characteristic for a p-channel enhancement-type MOSFET.

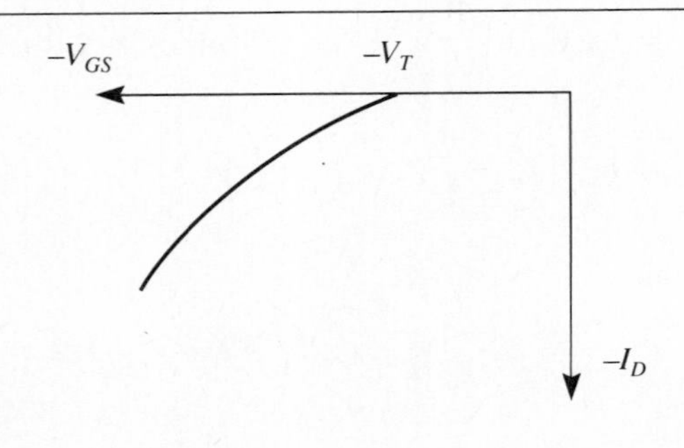

One should note here that, in deriving the current-voltage equations, we considered only the drift of carriers under an electric field. This can be seen by noting that $\mathcal{E}_y(y) = -dV(y)/dy$ and rewriting Equation (3.26) as

$$I_D = \mu_e W Q_I(y)\mathcal{E}_y(y)$$

If a diffusion-current component were included, the current expression would be

$$I_D = W\left[\mu_e Q_I(y)\mathcal{E}_y(y) + D_e \frac{dQ_I(y)}{dy}\right]$$

where D_e is the diffusion coefficient for electrons, as described in Chapter 1. Therefore, by excluding this contribution from our approximation, we assumed that $Q_I(y)$ is a slowly varying function of y, corresponding to a gradual change in the channel properties, which may not be valid in *short-channel* devices. Even in a large-channel device, the current is dominated by carrier diffusion under *weak inversion*, in which $\phi_s < 2|\phi_f|$. Since this current flows for voltages $V_{GS} < V_T$, it is termed the *subthreshold* current. Therefore, under this condition, I_D is made up of both subthreshold and reverse leakage currents.

Example 3.2

Consider the n-channel enhancement MOSFET shown in Figure 3.15a. For a constant positive $V_{GS} = 8$ V, a 12 V potential difference between the source and drain is distributed

Figure 3.15 The n-channel enhancement MOSFET and the formation of the inversion layer.

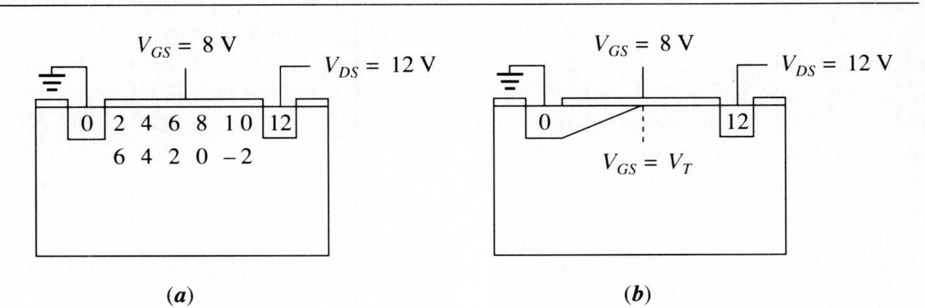

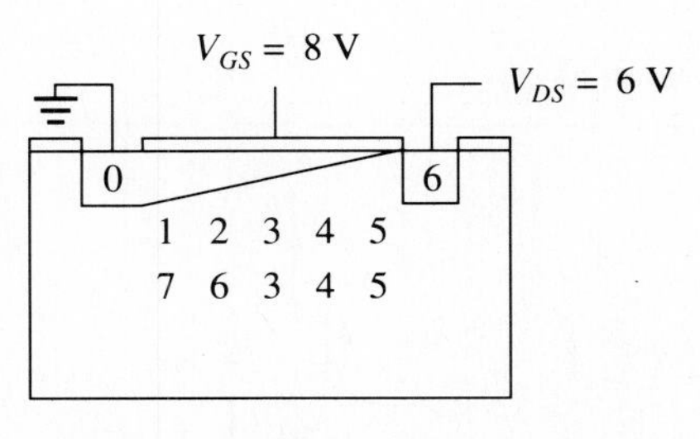

uniformly across the substrate. The potential difference between the gate and any point on the substrate beneath is also indicated in the second row of numbers.

If the threshold voltage $V_T = 2$ V, the channel will be partially inverted. Figure 3.15b shows that the channel thins down as the gate-channel potential decreases. The channel is not inverted when $V_{GS} < V_T$.

The channel extends as V_{DS} is reduced until $V_{GS} - V_{DS} > V_T$, at which point the channel is completed, and the transistor goes into saturation. This is illustrated in Figure 3.15c.

Example 3.3

The transistor shown in Figure 3.16 has $k_n = .2$ mA/V^2 and $V_T = 1$ V. To find an expression for I_y as a function of V_{DS}, we first note that, since $V_{DS} = V_{GS} > V_{GS} - V_T$, the transistor is operating above pinchoff. Hence, it is in the saturation region. Thus, the current I_y can be found by using Equation (3.30) as

$$I_y = I_{D(sat)} = \frac{1}{2} k_n (V_{GS} - V_T)^2 = .1(V_{DS} - 1)^2 \text{ mA}$$

Small-Signal Equivalent Circuit

A sufficiently accurate small-signal equivalent circuit of a MOSFET at high frequencies is shown in Figure 3.17. The gate-to-source capacitance C_{gs} is usually much larger than the gate-to-drain capacitance C_{gd}, which provides an undesirable feedback between the input and the output. Although C_{gd} is typically less than 1 pF for a high-frequency MOSFET, it is still included in the model because it is in parallel with C_{gs} due to the Miller effect, resulting in a high input capacitance. The value of the drain-to-source capacitance C_{ds} is comparable to that of C_{gs}. A more detailed discussion of these capacitances is presented in Section 3.3. Using Equations (3.27) and (3.30), we obtain the channel conductance as

$$g_d = k_n(V_{GS} - V_T - V_{DS}) \quad \text{in the ohmic region,}$$

$$= 0 \quad \text{in the saturation region} \qquad (3.32)$$

Figure 3.16 NMOS in saturation.

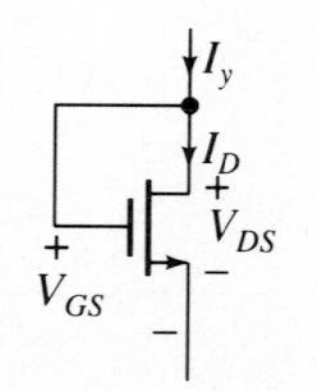

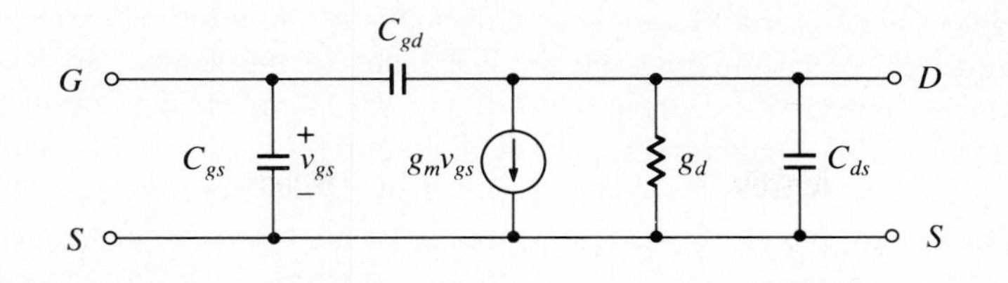

Figure 3.17 High-frequency, small-signal equivalent circuit of the MOSFET transistor.

The *unity gain bandwidth* of a MOSFET is defined as

$$\omega_T \equiv \frac{g_m}{C_G} = \frac{g_m}{(C_{gs} + C_{gd})} \tag{3.33}$$

where C_G is the total gate capacitance. Using Equations (3.28) and (3.31), and assuming that $C_G \approx C_{ox}WL$, Equation (3.33) reduces to

$$\omega_T = k_n \frac{(V_{GS} - V_T)}{(C_{gs} + C_{gd})}$$

$$= \mu_e \frac{(V_{GS} - V_T)}{L^2} \tag{3.34}$$

for the saturated case. Thus, to obtain high-frequency operation, it is necessary to have a short channel length and high carrier mobility. High gate-voltage values also improve the speed, but they increase the power consumption.

Short-Channel Effects

A MOS device is considered to be short when the channel length is the same order of magnitude as the depletion-layer widths of the source and drain junctions. As the channel length L is reduced to increase both the operation speed and the number of components per chip, the so-called *short-channel effects* arise. Today, devices with minimum feature length in the submicron ($<1\ \mu$m) region are already available.

Similar to Equation (2.22), the expressions for the drain and source junction widths are

$$W_D = \sqrt{\left(\frac{2\epsilon_s}{qN_a}\right)(V_{DS} + \phi_{si} + V_{SB})}$$

and

$$W_S = \sqrt{\left(\frac{2\epsilon_s}{qN_a}\right)(\phi_{si} + V_{DB})}$$

respectively, where V_{SB} and V_{DB} are source-to-body and drain-to-body voltages. When the depletion region surrounding the drain extends to the source, so that the two depletion layers merge (i.e., when $W_S + W_D = L$), *punchthrough* occurs: the gate voltage loses control of the current, and the drain current rises sharply. A current can then flow even at gate-source voltages lower than the flatband voltage. If the drain voltage is increased, the electrostatic potential at points between the source and drain increases and the barrier to electrons decreases, leading to *drain-induced barrier-lowering effect*. Punchthrough can be minimized with thinner oxides, larger substrate doping, shallower junctions, and obviously with longer channels. It does not cause permanent damage as long as there is no local melting.

As the channel length becomes smaller due to the lateral extension of the depletion layer into the channel region, the longitudinal electric field component $\mathcal{E}_y$ increases, and the surface mobility becomes field-dependent. Since the carrier transport in a MOSFET is confined within the narrow inversion layer, and the *surface scattering*, that is, the collisions suffered by the electrons that are accelerated toward the interface by $\mathcal{E}_x$, causes reduction of the mobility, the electrons move with great difficulty parallel to the interface, so that the average surface mobility, even for small values of $\mathcal{E}_y$, is about half as much as that of the bulk mobility.

The performance of short-channel devices is also affected by *velocity saturation*, which reduces the transconductance in the saturation mode. At low $\mathcal{E}_y$, the electron drift velocity v_{de} in the channel varies linearly with the electric field intensity, as given by Equation (1.19). However, as $\mathcal{E}_y$ increases above 10^4 V/cm, the drift velocity tends to increase more slowly, and approaches a saturation value of $v_{de(sat)} = 10^7$ cm/s around $\mathcal{E}_y = 10^5$ V/cm at 300 K.

From the discussion in Chapter 1, the corresponding drain current is

$$I_{D(sat)} = Wv_{de(sat)} \int_0^{x_I} qn(x)\, dx = Wv_{de(sat)} \left| Q_I \right|$$

From Equations (3.17) and (3.29), then, we have

$$I_{D(sat)} = WC_{ox} v_{de(sat)} (V_{GS} - V_T)$$

corresponding to a constant transconductance

$$g_m = WC_{ox} v_{de(sat)}$$

which in turn defines the maximum gain possible for a MOSFET. The transconductance is independent of the gate and drain bias as well as the channel length. Note also that the drain current is limited by velocity saturation instead of pinchoff. This occurs in short-channel devices when the dimensions are scaled without lowering the bias voltages.

Under velocity saturation, the unity gain bandwidth is found, using Equation (3.33), as

$$\omega_T = \frac{v_{de(sat)}}{L}$$

Example 3.4

For a silicon-gate NMOS transistor with $W = 20\ \mu m$, $L = 1\ \mu m$, $\mu_e = 800\ cm^2/V \cdot s$, $C_{ox} = 100\ nF/cm^2$, and $V_T = 1$ V, the saturated drain current and transconductance values for the long-channel case at $V_{GS} = 5$ V are obtained as

$$I_{D(sat)} = \frac{1}{2} k_n (V_{GS} - V_T)^2 = (8 \times 10^{-4})(16) = 12.8 \text{ mA}$$

and

$$g_{m(sat)} = k_n (V_{GS} - V_T) = (16 \times 10^{-4})(4) = 6.4 \times 10^{-3} \text{ S}$$

On the other hand, in the case of velocity saturation, we get

$$I_{D(sat)} = WC_{ox} v_{de(sat)} (V_{GS} - V_T) = (20 \times 10^{-6})(10^{-7})(10^7)(4) = 80\ \mu A$$
$$g_{m(sat)} = WC_{ox} v_{de(sat)} = 20 \times 10^{-6} \text{ S}$$

Note that both parameters are reduced drastically.

Another undesirable short-channel effect, especially in NMOS, occurs due to the high velocity of electrons in the presence of high longitudinal fields that can generate electron-hole (e-h) pairs by *impact ionization*, that is, by impacting on silicon atoms and ionizing them.

It happens as follows: normally, most of the electrons are attracted by the drain, while the holes enter the substrate to form part of the parasitic substrate current. Moreover, the region between the source and drain can act like the base of an *npn* transistor, with the source playing the role of the emitter and the drain that of the collector. If the aforementioned holes are collected by the source, and the corresponding hole current creates a voltage drop in the substrate material of the order of .6 V, the normally reverse-biased substrate-source *pn* junction will conduct appreciably. Then electrons can be injected from the source to the substrate, similar to the injection of electrons from the emitter to the base. They can gain enough energy as they travel toward the drain to create new e-h pairs. The situation can worsen if some electrons generated due to high fields escape the drain field to travel into the substrate, thereby affecting other devices on a chip.

Another problem, again related to high electric fields, is caused by so-called *hot electrons*. These high-energy electrons can enter the oxide, where they can be trapped, giving rise to *oxide charging* that can accumulate with time and degrade the device performance by increasing V_T and affect adversely the gate's control on the drain current.

Even though the threshold voltage, as described in Equation (3.20) for the MOS capacitor, naturally ignores the presence of the n^+ source and drain regions of a MOSFET structure, it is still fairly accurate in describing large MOS transistors because the depletion widths of the source and drain *pn* junctions can be considered to occupy a negligible portion of the channel length L. However, the expression collapses when applied to small-geometry MOSFETs.

Equation (3.20) assumes that the bulk depletion charge is only due to the electric field created by the gate voltage, while the depletion charge near the n^+ source and drain regions is actually induced by the *pn* junction band bending. Therefore, the amount of bulk charge the gate voltage supports is overestimated, leading to a larger V_T than the actual value. This is especially true in a short-channel MOSFET. The electric flux lines generated by the charge on the MOS capacitor gate electrode terminate on the induced mobile carriers in the depletion region just under the gate. For short-channel MOSFETs, on the other hand, some of the field lines originating from the source and drain electrodes terminate on charges in the channel region, as diagrammed in Figure 3.18. Thus, less gate voltage is required to cause inversion. This implies that the fraction of the bulk depletion charge originating from the *pn* junction depletion and hence requiring no gate voltage, must be subtracted from the V_T expression.

A simple geometric model to determine the amount of threshold voltage shift when $V_{DS} = 0$ is depicted in Figure 3.19a, in which the bulk charge that is relevant to V_T exists in a trapezoidal volume instead of a rectangular one. Let us introduce the effective channel length

$$L_{eff} \equiv L - \Delta L$$

as the *average* length of the depletion along the channel, where ΔL is the lateral extent of the depletion charge due to the *pn* junctions on both ends of the channel. Then the portion of the bulk depletion charge that contributes to V_T can be expressed as

$$\left(\frac{L_{eff}}{L}\right) Q_{s(max)} = \left(\frac{1 - \Delta L}{L}\right) Q_{s(max)} \tag{3.35}$$

Next, to find an expression for ΔL, we assume that (1) the *pn*-junction depletion regions extend a distance $x_{d(max)}$ into the substrate, and (2) the source and drain edges are quarter-circular arcs, each having a radius equal to the junction depth x_j. From Figure 3.19b,

Figure 3.18 Illustration of charge sharing in short-channel devices.

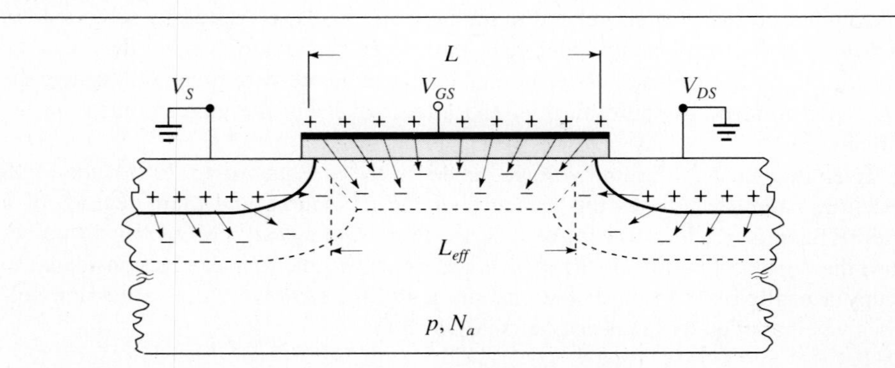

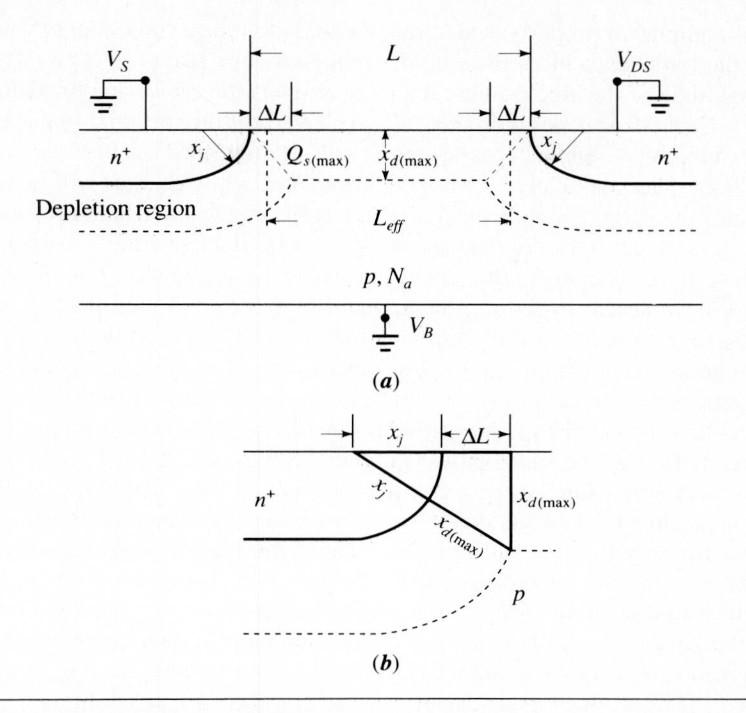

Figure 3.19 A simple model to calculate V_T in a short-channel device: (a) geometry, (b) ΔL calculation.

$$x_{d(max)}^2 + [x_j + \Delta L]^2 = [x_j + x_{d(max)}]^2$$

Solving for ΔL yields

$$\Delta L = \sqrt{x_j^2 + 2x_j x_{d(max)}} - x_j$$

Substituting this into Equation (3.35) and using that expression in the third term in Equation (3.20), we obtain a complete expression for the threshold voltage, including the short-channel effects:

$$V_T = V_{FB} + 2\phi_f + \gamma\sqrt{2\,|\phi_f|}\left[1 - \frac{x_j}{L}\left(\sqrt{1 + \frac{2x_{d(max)}}{x_j}} - 1\right)\right] + \frac{qD_I}{C_{ox}} \tag{3.36}$$

Hence, because short-channel effects in a small-geometry MOSFET degrade device performance, they should be minimized, but without trading off the long-channel behavior of the device. One way of minimizing them is by simply reducing all dimensions and voltages by a scaling factor $\alpha > 1$, so that the electric field in the channel is held constant.

In *constant-field scaling*, the basic assumption is that a device is scaled in all three dimensions by $1/\alpha$, so that areas are scaled by $1/\alpha^2$. Let us assume that the substrate doping density, N, of the MOSFET is scaled by α.

To accomplish a properly scaled reverse-biased voltage throughout the channel, all operating voltages and the threshold voltage must be scaled by $1/\alpha$. The maximum magnitude of the electric field intensity remains the same, and breakdown will not occur. Capacitances per unit area, C_a, are inversely proportional to distances, so they are scaled by α, while capacitances $C = C_a A$ are scaled by $1/\alpha$.

The body-bias coefficient, γ, is scaled by $1/\sqrt{\alpha}$, while charges will be scaled by $1/\alpha^2$. Since the surface mobility is independent of impurity concentration as long as $N < 10^{16}\ \text{cm}^{-3}$ and does not change under constant-field scaling, the drain current is scaled by $1/\alpha$. Therefore, the power dissipation is scaled by $1/\alpha^2$. Since device areas have been scaled by $1/\alpha^2$, the device density per unit area will be scaled by α^2, so that the power per unit of chip area will actually not be scaled.

The rate of change of charging capacitances, $dV/dt = I/C$, is not scaled. These capacitances, however, now only need to be charged to voltages scaled down by $1/\alpha$, and hence the time needed to charge them is scaled by $1/\alpha$, which in turn implies a faster circuit. Hence, the speed-power product will be reduced by $1/\alpha^3$.

Now, scale the width of the metal and polysilicon lines that form the interconnections and gates by $1/\alpha$. It may also be necessary to reduce the height of these lines, because very thin but tall lines may cause fabrication problems. Thus, scale the height by $1/\alpha$, too. Then, the current density in these lines will be scaled by α because the cross-sectional area of the lines is scaled by $1/\alpha^2$. This is unwelcome because the increased current density can lead to what is known as *electromigration*, in which metal atoms move along the direction of electron flow, leading to a possible failure. For $A\ell$ lines, the current density should not exceed $1\ \text{mA}/\mu\text{m}^2$.

The resistance of interconnection lines, being proportional to the length and inversely proportional to the cross-sectional area, is scaled by α. The parasitic capacitances of these lines to the substrate is scaled by $1/\alpha$, so that the corresponding time constant does not change. Note that long lines may nullify the speed advantage gained by scaling. Furthermore, the voltage drop across them is not scaled, which implies that a larger fraction of the available voltage that has been scaled by $1/\alpha$ is now wasted across these lines. For these reasons, the height of the interconnection lines is reduced not by $1/\alpha$ but less drastically.

Adherence to established chip interface requirements means that fixed voltage levels cannot be scaled. Decreasing device dimensions while keeping the voltages unchanged is referred to as *constant-voltage scaling*. In this scheme, W, L, and N are scaled as before. However, if the oxide thickness is scaled by the same factor, the resulting field can be prohibitively high because voltages are kept constant. To lessen this problem somewhat, the oxide thickness is usually scaled less drastically. To avoid the extreme cases of constant-field or constant-voltage scaling, compromise rules are generally used.

3.3 SPICE MOSFET Model

SPICE provides three levels of a MOSFET model that differ from each other in the formulation of the transistor's I-V characteristic. The **LEVEL** = 1 model is illustrated in Figure 3.20 for an n-channel MOSFET. For a PMOS, the current source as well as the polarities of the terminal voltages and of the two substrate junctions, neither

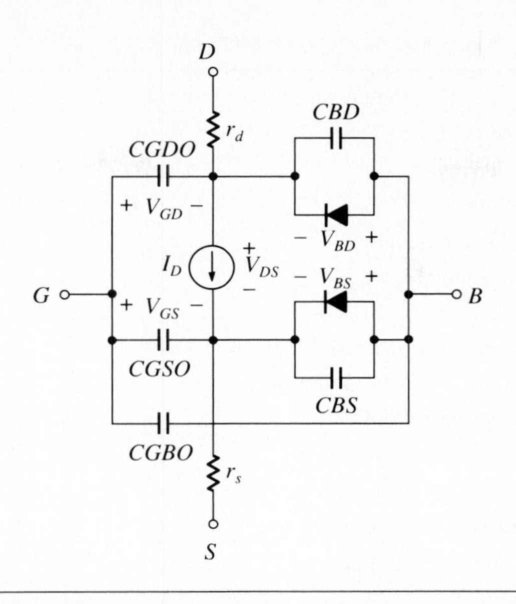

Figure 3.20 The SPICE MOSFET model, **LEVEL1**.

of which is normally forward-biased, are reversed. All SPICE parameters are listed in Table 3.1.

The dc equations used in the basic model are those developed in this chapter. The analytic **LEVEL** = 2 is a geometry-based model in which detailed device physics have been employed to define its equations; the empirical **LEVEL** = 3 model relies more on measured characteristics of the device. However, depending on the parameters specified, mixed models can be utilized by the designer for calculating threshold voltage and drain current.

As in the discussion in Section 3.1, with the source-to-body voltage, V_{SB}, taking the place of V_B, we obtain an expression for the threshold voltage for the MOS transistor as

$$V_T = V_{TO} + \gamma(\sqrt{2\,|\phi_f| + V_{SB}} - \sqrt{2\,|\phi_f|}) \tag{3.37}$$

Given the process parameters oxide thickness x_{ox} (**TOX**), surface mobility μ_e or μ_h (**UO**), and substrate doping concentration N_a or N_d (**NSUB**), SPICE calculates the five electrical parameters **PHI**, **VTO**, **GAMMA**, **KP**, and **LAMBDA** to determine the dc characteristics of the MOSFET. User-specified parameters, however, always override.

Parameter Extraction

In SPICE, the threshold voltage V_{TO} is positive for enhancement-mode devices and negative for depletion-mode devices. For an NMOS operating in the forward region with $V_{SB} = 0$, it is the value of V_{GS} where the channel begins to conduct. It can be determined from a graph of $(\sqrt{I_D})$ vs. $(V_{GS} = V_{DS})$ by extrapolating to zero drain current. The circuit arrangement and the resulting characteristics for different values of V_{SB} are shown in Figure 3.21. Therefore, the *extrapolated threshold voltage*

Table 3.1 SPICE MOSFET Model Parameters

Name	Parameter name	Unit	Default
LEVEL	Model type (1, 2 or 3)		1
L	Channel length	m	
W	Channel width	m	
LD	Lateral diffusion (length)	m	0
WD	Lateral diffusion (width)	m	0
TOX	Oxide thickness	m	∞
XJ	Metallurgical junction depth (2 and 3)	m	0
VMAX	Maximum drift velocity (2 and 3)	m/s	0
TF	Ideal forward transit time	s	0
IS	Bulk junction saturation current	A	1E-14
JS	Bulk junction saturation current/area	A/m^2	0
KP	Process transconductance parameter	A/V^2	2E-5
VTO	Zero-bias threshold voltage	V	0
PHI	Fermi potential	V	.6
PB	Bulk junction potential	V	.8
GAMMA	Body-bias coefficient	$\sqrt{V}$	0
LAMBDA	Channel-length modulation (1 and 2)	V^{-1}	0
THETA	Mobility modulation (3)	V^{-1}	0
UCRIT	Mobility degradation critical field (2)	V/cm	1E4
RD	Drain ohmic resistance	Ω	0
RS	Source ohmic resistance	Ω	0
RG	Gate ohmic resistance	Ω	0
RB	Bulk ohmic resistance	Ω	0
RDS	Drain shunt resistance	Ω	∞
RSH	Drain, source diffusion sheet resistance	Ω/□	0
CBD	Bulk-drain zero-bias transition capacitance	F	0
CBS	Bulk-source zero-bias transition capacitance	F	0
CJSW	Bulk junction zero-bias sidewall capacitance/length	F/m	0
CGSO	Gate-source overlap capacitance/channel width	F/m	0
CGDO	Gate-drain overlap capacitance/channel width	F/m	0
CGBO	Gate-bulk overlap capacitance/channel length	F/m	0
CJ	Bulk junction zero-bias bottom capacitance/area	F/m^2	0
UO	Surface mobility	$cm^2/V \cdot s$	600
NSS	Surface state density (2 and 3)	cm^{-2}	0
NFS	Fast surface state density (2 and 3)	cm^{-2}	0
NSUB	Substrate doping density N_a or N_d	cm^{-3}	0

continued

Table 3.1 *continued*

Name	Parameter name	Unit	Default
MJ	Bulk junction bottom grading coefficient		.5
MJSW	Bulk junction sidewall grading coefficient		.33
FC	Bulk junction forward-bias capacitance coefficient		.5
NEFF	Channel charge coefficient (2)		1
KF	Flicker noise coefficient		0
AF	Flicker noise exponent		1
XQC	Fraction of channel charge attributed to drain (2 and 3)		1
ETA	Static feedback on threshold voltage (3)		0
KAPPA	Saturation field factor (3)		.2
UEXP	Exponential coefficient for mobility (2)		0
UTRA	Transverse field coefficient (2)		0
DELTA	Width effect on threshold voltage (2 and 3)		0
TPG	Gate material type (2 and 3): + 1 = opposite of substrate, − 1 = same as substrate, 0 = aluminum		

Figure 3.21 Body-bias effects: (a) circuit arrangement, (b) $\sqrt{I_D}$ vs. V_{GS} in saturation for different values of V_{SB}.

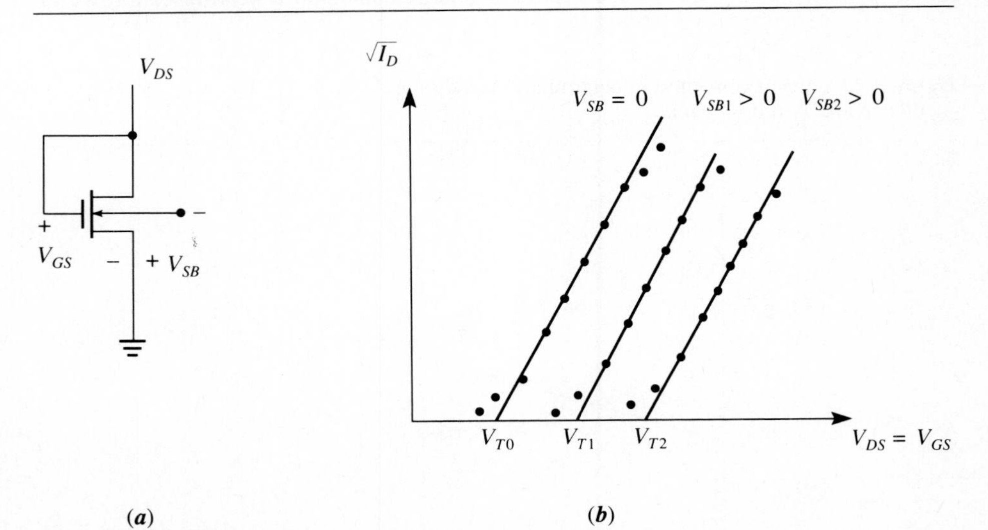

is actually not the voltage parameter obtained from the production measurement, which usually corresponds to a gate voltage necessary to achieve a certain drain current for a saturated transistor of specified W and L.

If V_T values obtained from Figure 3.21 are now plotted vs. $\sqrt{2\,|\phi_f| + V_{SB}}$, assuming a value for $2\phi_f$ around .6 V, γ can be determined, as depicted in Figure 3.22. The value of $2\phi_f$ needs to be chosen such that the points will fall on a straight line. Then, using Equation (3.37), the slope of the straight line yields the body-bias coefficient, γ.

The *process transconductance parameter* **KP** ($= \mu C_{ox}$) is also determined by the slope of the aforementioned characteristic, while the *channel-length modulation parameter*, λ (**LAMBDA**), can be estimated from curve tracer measurements of I_D vs. V_{DS}. The latter has the same role that the parameter V_A has for the BJT. Since the depletion layer at the drain widens as V_{DS} increases, in effect reducing the channel length, $I_{D(sat)}$ is not completely independent of V_{DS} and increases as drain voltage increases. An empirical correction to the actual drain current in saturation region yields

$$I_{D(sat)} = \frac{1}{2} k_n (V_{GS} - V_T)^2 (1 + \lambda V_{DS})$$

The modulation parameter actually has little effect on the operation of digital MOS circuits. The tilt due to channel-length modulation is depicted in Figure 3.23. One circuit arrangement to determine λ is shown in Figure 3.24.

MOSFET Capacitances and the SPICE Input Data

1. Overlap Capacitances. The cross section of the large-signal MOSFET capacitance model illustrating the lumped-element capacitances and the top view geometry needed for calculations is shown in Figure 3.25. During fabrication, even in the so-called *self-aligned* process, those processing steps that require heating of the wafer

Figure 3.22 Graphical method to determine the body-bias coefficient and Fermi potential.

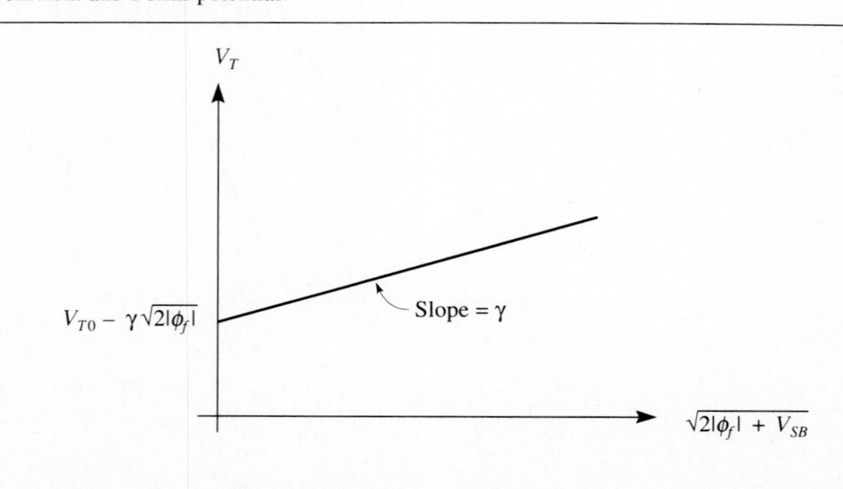

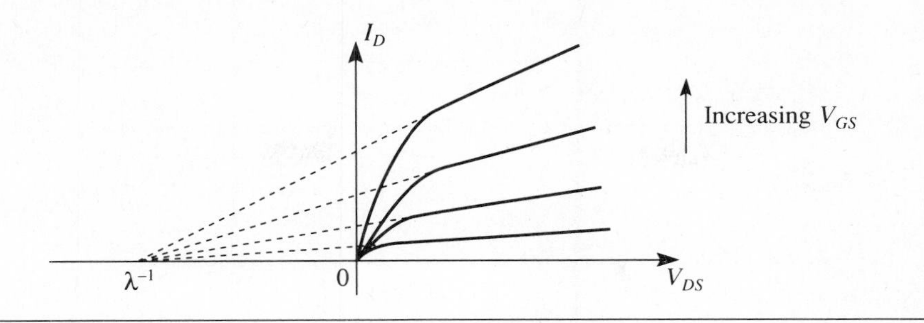

Figure 3.23 Channel-length modulation effects in MOSFET characteristics.

cause the n^+ source and drain regions to diffuse in a direction parallel to the surface of the wafer and toward each other under the gate electrode. This unavoidable *lateral diffusion* leads to an overlap of the nominally self-aligned gate above these regions so that the *effective channel length* L_{eff} becomes shorter than the *physical gate length* **L**, even though these lengths are usually mentioned interchangeably. It also gives rise to the parasitic and almost linear parallel-plate *overlap* capacitors C_{ols} and C_{old}, with the oxide acting as a dielectric. Due to the symmetry of the device, the overlap distance on each side is entered as **LD** on the SPICE **.MODEL** card, so that **L** is decreased by twice **LD** to get L_{eff}. Typically, **LD** is less than 1 μm. Obviously, the circuit designer has no control over this parameter. As a matter of fact, for the inversion layer to make contact with both n^+ regions, we need to have a nonzero value for **LD**. Also, **W** is decreased by twice **WD** to get the effective channel width. With both source and drain having a width of **W**, the overlap capacitances can be expressed in terms of the gate oxide capacitance per unit area as

$$C_{ols} = C_{ox}\mathbf{W} \cdot \mathbf{LD} = C_{old}$$

Therefore, **W** is the only layout parameter that can be controlled by the designer. SPICE requires these capacitances to be entered as overlap capacitances *per unit gate width*:

$$\mathbf{CGSO} = C_{ox}\mathbf{LD} = \mathbf{CGDO}$$

Figure 3.24 Circuit arrangement to determine channel-length modulation parameter.

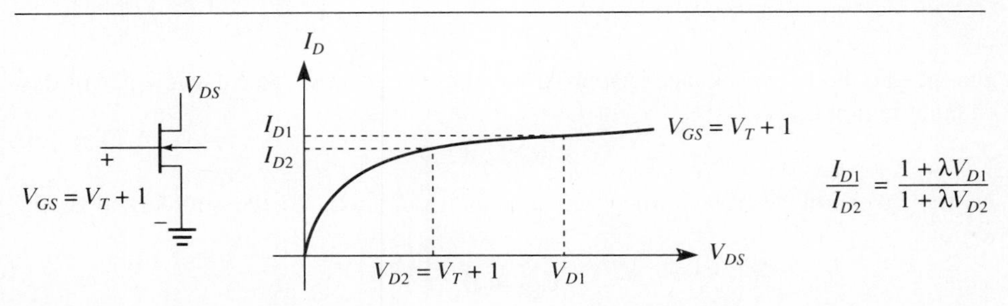

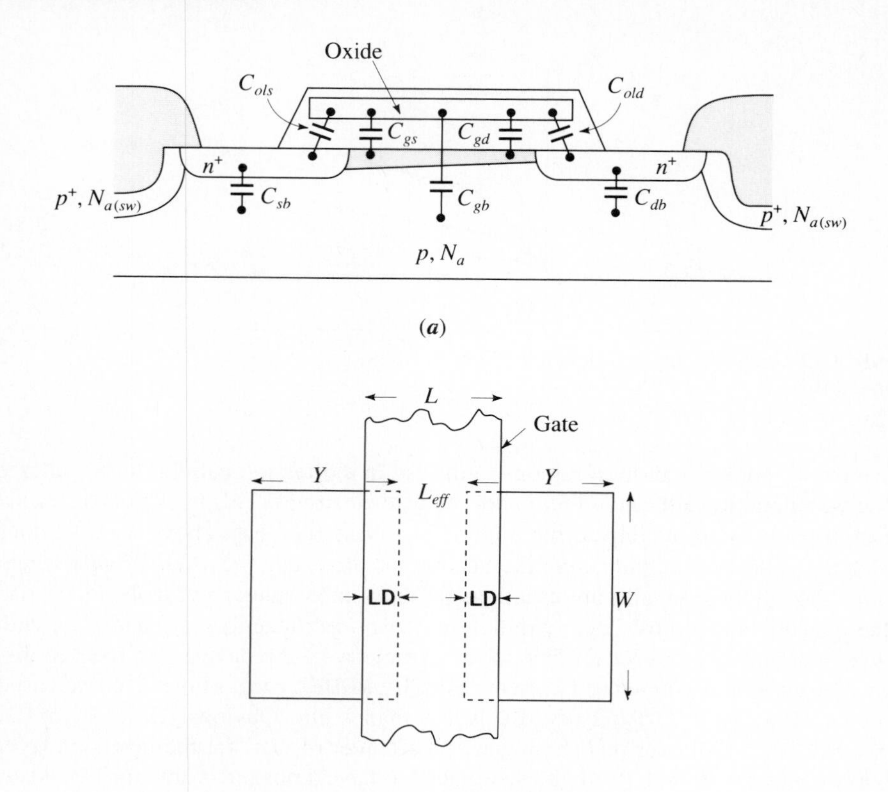

Figure 3.25 MOSFET capacitances: (a) model, (b) top-view geometry.

2. *Intrinsic Capacitances.* In Figure 3.25, the capacitors denoted by C_{gs}, C_{gd}, and C_{gb} represent *intrinsic* gate-source, gate-drain, and gate-body capacitances, respectively. The first two parameters are really the gate-to-channel capacitances as seen between the gate and respective n^+ region. Since the formation of the channel is voltage-dependent, these parameters are nonlinear.

Figure 3.26 illustrates gate capacitances in the three regions of operation. In cutoff, with $V_{GS} < V_T$, no inversion layer exists. Thus,

$$C_{gs} = C_{gd} = 0$$

and the gate-body capacitance is approximated by the intrinsic gate capacitance of the channel region as

$$C_{gb} = C_{ox}WL_{eff}$$

As the inversion layer is formed and acts as a conductor to the source and drain regions,

$$C_{gb} = 0$$

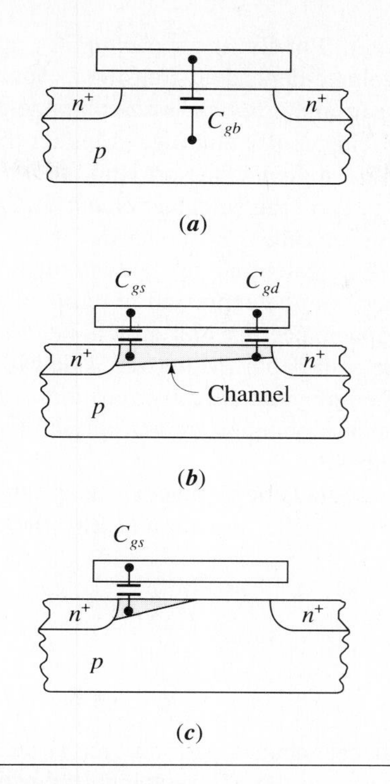

Figure 3.26 MOSFET gate capacitances in (a) cutoff, (b) ohmic region, and (c) saturation.

because the body electrode is shielded from the gate by the channel formed. When the channel extends across the entire transistor and the device is in the ohmic region,

$$C_{gs} \approx \frac{1}{2} C_{ox}WL_{eff} \approx C_{gd}$$

When the channel is pinched off and the device is saturated, the capacitances are approximated as

$$C_{gs} \approx \frac{2}{3} C_{ox}WL_{eff}$$

and

$$C_{gd} \approx 0 \approx C_{gb}$$

Combining the intrinsic capacitances with overlap contributions, the total gate capacitance is given by

$$C_G = C_{ox}\mathbf{W} \cdot \mathbf{L} = C_{gs} + C_{gd} + C_{gb} + C_{ols} + C_{old}$$

To enable SPICE to calculate the gate capacitances, **TOX** must be entered on the **.MODEL** card.

3. Transition Capacitances. Finally, the capacitors C_{sb} and C_{db} in the model of Figure 3.25 represent the voltage-dependent transition capacitances, each of which is composed of a *sidewall* part and a *bottom-wall* component. This distinction is made because the sidewall capacitance per unit area is higher than that of the bottom wall due to the heavily doped p^+ regions that surround the n^+ sidewalls near the surface outside the channel area to form the so-called *channel stop* areas, which are utilized in isolating devices from each other, as shown in Figure 3.25. The p^+ field doping of $N_{a(sw)}$ is around ten times greater than the body doping, N_a. The bottom-wall component of both capacitances can be expressed in terms of a capacitance per unit area in F/m^2; the sidewall capacitance per unit area is integrated over the sidewall area because it is not constant with depth. Moreover, sidewalls are not plane, but cylindrical. Therefore, SPICE employs an effective zero-bias *sidewall capacitance per unit length*, **CJSW**, in F/m that is obtained by multiplying the sidewall capacitance per unit area by junction depth **XJ**.

The calculation of these transition capacitances is similar to that of the transition capacitance of a *pn* junction. Using Equation (2.25), the total bottom capacitance is then given by

$$C_{bottom} = \frac{\mathbf{CJ}\,WY}{\sqrt{1 + \dfrac{V_r}{\mathbf{PB}}}}$$

where **CJ** is the zero-bias capacitance per unit area at the bottom of the n^+ drain or source region, WY is the area of the n^+/p-type bulk junction, V_r is the magnitude of the reverse-bias junction voltage, and $\mathbf{PB} = \phi_o$ is the built-in junction potential as given in Equation (2.4). SPICE assumes the same value of the latter for bottom and sidewall junctions. V_r is either V_{SB} or V_{DB}. In SPICE, WY is replaced by AD, AS on the input card for drain and source diffusion areas.

With $N_{a(sw)} \approx 10N_a$, the sidewall capacitance per unit area is higher than **CJ** by the square root of 10. Hence, the sidewall capacitance per unit length can be taken as

$$\mathbf{CJSW} = \sqrt{10}\mathbf{CJ} \cdot \mathbf{XJ}$$

if not specified. The total sidewall capacitance is then obtained as

$$C_{sw} = \frac{\mathbf{CJSW}(W + 2Y)}{\sqrt{1 + \dfrac{V_r}{\phi_{o(sw)}}}}$$

where $W + 2Y$ is the sidewall perimeter length because the high value of sidewall capacitance is not present along the edge of diffusion bordering on active channels, and $\phi_{o(sw)}$ is the sidewall built-in junction potential. However, for simplicity the entire perimeter, as specified by PD, PS for the drain or source diffusion perimeters on the input card, is multiplied by **CJSW**. The total transition capacitance is computed by summing C_{bottom} and C_{sw}. Finally, note that SPICE offers an alternative to specify zero-bias transition capacitances. They can be set by **CBD** and **CBS**, which are absolute values in farads. Then, as in the discussion in Section 2.3 for the diode model, the parameters **MJ**, **MJSW**, and **FC** must also be specified.

Example 3.5

Suppose that the following data are available for the MOSFET of Figure 3.25b: $W = Y = 10\ \mu\text{m}$, $L = 6\ \mu\text{m}$, $V_T = 1$ V, **KP** $= 30\ \mu\text{A/V}^2$, $\gamma = .4\ \sqrt{\text{V}}$, $N_a = 10^{15}\ \text{cm}^{-3}$, $N_d = 5 \times 10^{19}\ \text{cm}^{-3}$, $x_{ox} = 75$ nm, $x_j =$ **LD** $= 1\ \mu$m. To prepare the SPICE device and **.MODEL** cards, we first need to calculate the capacitances per unit area in the form needed for input to SPICE. First, from Equation (2.6), we have

$$\phi_o = \textbf{PB} = .86\ \text{V}$$

Then, as in Example 2.3, but this time with $N_d \gg N_a$, we get from Equation (2.26)

$$C_{toa} = \textbf{CJ} = 98.97\ \mu\text{F/m}^2$$

Therefore,

$$\textbf{CJSW} = \sqrt{10}\textbf{CJ} \cdot \textbf{XJ} = 313\ \text{pF/m}$$

Also, using Equation (3.3), we have $C_{ox} = 460\ \mu\text{F/m}^2$ and hence

$$\textbf{CGSO} = \textbf{CGDO} = 460\ \text{pF/m}$$

The source and drain perimeters and areas are found as

$$\text{PS} = \text{PD} = 4 \times 10 = 40\ \mu\text{m}$$

$$\text{AS} = \text{AD} = 10 \times 10 = 100\ \mu\text{m}^2$$

The SPICE input data are then as follows:

```
M⟨name⟩ ⟨(drain) node⟩ ⟨(gate) node⟩ ⟨(source) node⟩ ⟨(substrate) node⟩
+ ⟨(model) name⟩ AS = 100P AD = 100P PS = 40U PD = 40U
.MODEL ⟨(model)name⟩ NMOS VTO 1 KP 30E-6 GAMMA .4 NSUB 1E15
+ TOX 75N XJ 1U LD 1U CJ 32U CJSW 102P
+ CGSO 460P CGDO 460P PB .86
```

Example 3.6

The circuit of Figure 3.27a is used to compare the output characteristics of an n-channel MOSFET as generated by SPICE **LEVEL 1** and **LEVEL 2** models. Four basic physical parameters describe identical transistors. The more accurate **LEVEL2** model predicts considerably smaller currents as V_{DS} is increased, as depicted in Figure 3.27b.

```
COMPARISON OF NMOS ID-VDS CHARACTERISTICS
VDS 1 0
VGS 2 0
M1  1 2 0 0 N1
M2  1 2 0 0 N2
.MODEL N1 NMOS TOX 50N NSUB 1E16 L 10U W 10U
.MODEL N2 NMOS LEVEL 2 TOX 50N NSUB 1E16 L 10U W 10U
.DC VDS 0 5 .1 VGS 0 5 1
.PROBE
.END
```

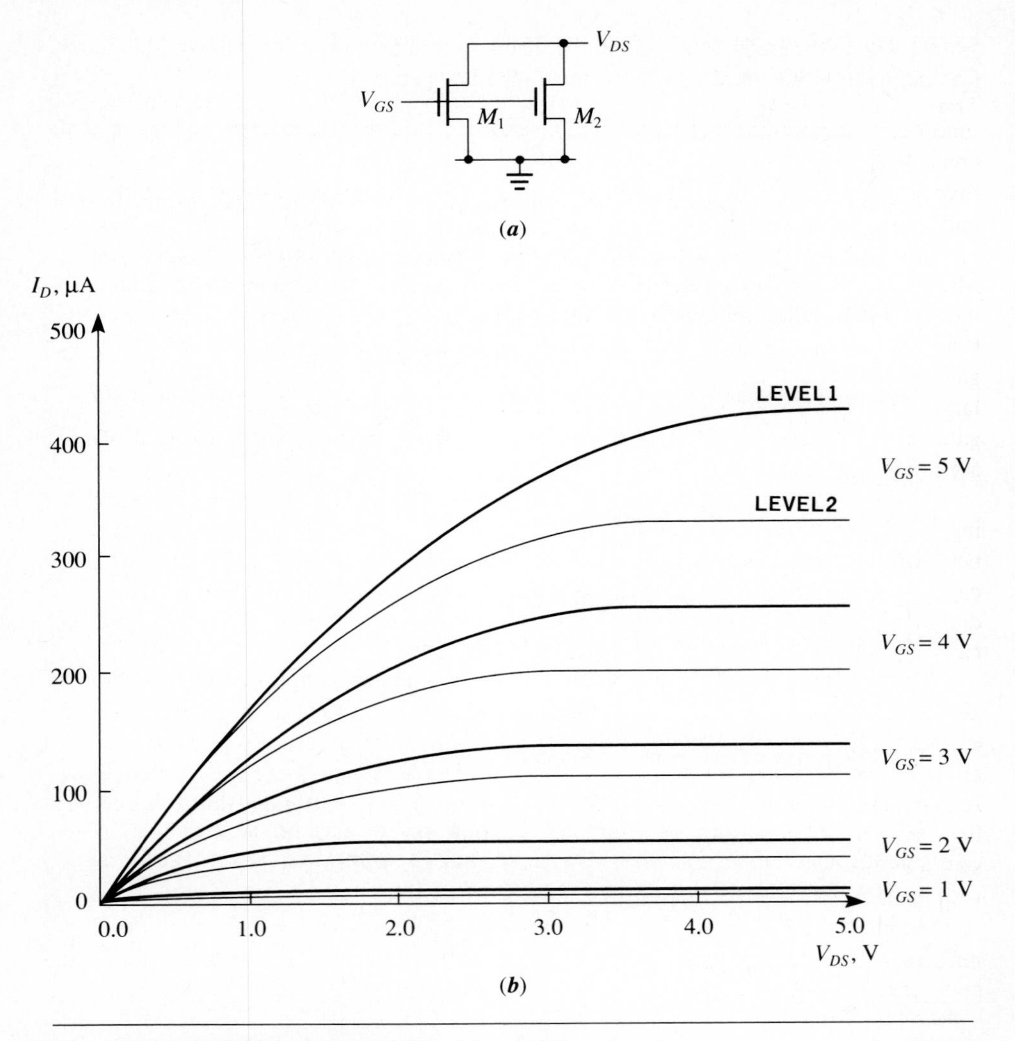

Figure 3.27 SPICE simulation: Output characteristics as predicted by **LEVEL1** and **LEVEL2** models, (a) the circuit, and (b) the waveforms.

Summary

Being a less complicated structure than the MOSFET, the MOS capacitor is an important tool to help us understand the electrical properties of the interface between the oxide and semiconductor substrate as a bias voltage between the metal or polysilicon gate on the oxide and the semiconductor is applied to control the energy-band conditions in this critical region.

Depending on the polarity and magnitude of the gate bias, three regions of operation are possible. A negative gate bias with respect to the substrate results in the accumulation of majority carriers, leading to no conduction. A low positive bias with $V_{GB} < V_T$ causes depletion very much like that in the space-charge layer of a *pn*-junction diode. Still, there is no conduction. A high positive bias, $V_{GB} \geq V_T$, increases the width of the space-charge layer and results in the formation of a thin inversion layer of majority carrier type in the surface region of the semiconductor, and hence conduction.

Under depletion and inversion, the small-signal capacitance per unit area for the MOS capacitor is represented by two capacitors in series, one due to the oxide and the other due to the space-charge region. The MOS capacitor has interesting characteristics as far as the frequency of the capacitance measurement as a function of dc gate bias is concerned. For instance, the capacitance measured at low frequencies is larger than that measured at high frequencies for a *p*-type substrate with a positive gate bias. A single CV-plot that can be obtained from a simple test structure conveys a wealth of information.

The enhancement-type MOSFET is essentially a MOS capacitor with neighboring heavily doped regions that serve as the source and drain. One important difference between the dynamic response of the two devices is that in the MOS capacitor, the carriers in the channel are thermally generated, while in the MOSFET the source and drain regions deliver an immediate supply of carriers. Consequently, the latter device can respond instantly to any change in the gate voltage.

The threshold voltage V_T is the minimum gate bias required for inversion under zero drain voltage and zero flatband voltage. However, a nonzero flatband voltage is caused by a variety of factors, such as the metal-semiconductor work-function difference and the fixed oxide charge. The MOSFET has two distinct regions of operation. At low drain voltages, the device is in the ohmic region, and the drain current is linearly proportional to the drain voltage. At high drain voltages, the device is in the saturation region, and the drain current essentially remains constant, even if the drain voltage increases.

In a MOSFET, short-channel effects are encountered when the channel length is reduced to increase the speed of the circuit and chip density. These effects include punchthrough, drain-induced barrier lowering, velocity saturation, impact ionization, and hot electrons.

Appendix 3A: Pascal Program for CV Analysis

The following simple program written in Turbo Pascal, Version 5, calculates the oxide thickness, maximum depletion-layer width, doping density, and flatband capacitance, given only the maximum and minimum values of the capacitance from a CV curve. The maximum value occurs in accumulation and is equal to C_{ox}; the minimum value is taken from the flat portion of the high-frequency plot. The voltage corresponding to the calculated flatband capacitance is the flatband voltage and can be read off the plot.

```
program CV_plot;

uses crt;
var Cox, Cmin, Xox, Xdmax, N, Cfb : real; stat : char;

begin {main program}
  stat := 'Y';
  while (stat =  'Y') or (stat = 'y') do
    begin
    clrscr;
    textbackground(red);
    writeln;
    writeln;
    writeln ('   ****************************************************************');
    writeln ('   ***********       MOS CAPACITOR CV PLOT ANALYSIS      ***********');
    writeln ('   ****************************************************************');
    writeln;
    writeln;
    writeln ('       Given the maximum and minimum values of the capacitance   ');
    writeln ('           this program calculates the following parameters:   ');
    writeln;
    writeln ('                        Xox, Xdmax, N, Cfb');
    writeln;
    writeln;
    write   ('       Enter Cox,  nF/cm²: '); readln (Cox);
    write   ('       Enter Cmin, nF/cm²: '); readln (Cmin);

    Xox := 3.4531E-4/Cox;
    Xdmax := 1.0537E-3*(1/Cmin - 1/Cox);
    N := 1E11;
    while not (abs(N - 681305.11*ln(6.67E-11*N)/sqr(Xdmax)) < 0.01) do
      begin
      N := 681305.11*ln(6.67E-11*N)/sqr(Xdmax);
      end;
    Cfb := 1/(1/(2.554E-6*sqrt(N)) + 1/Cox);
```

```
    clrscr;
    writeln;
    writeln;
    writeln ('    **************************************************************');
    writeln ('    * A MOS capacitor with Cox = ', Cox:3:0, ' nF/cm² and Cmin = ', Cmin:3:0, ' nF/cm² *');
    writeln ('    *              will have the following parameters:          *');
    writeln ('    **************************************************************');
    writeln;
    writeln ('    Oxide thickness,                  Xox = ', Xox*1E7:6:2, ' nm');
    writeln ('    Maximum depletion region width, Xdmax = ', Xdmax*1E7:6:2, ' nm');
    writeln ('    Impurity concentration,             N = ', N:9, ' cm-3');
    writeln ('    Flatband capacitance,             Cfb = ', Cfb:6:2, ' nF/cm²');
    writeln;
    write   ('    Do you want to change the input data? ');
    stat:= readkey;
  end;
end. {main program}
```

Note that the iterative method of *successive substitution* has been utilized to find the impurity concentration.

References

1. D. L. Pulfrey and G. Tarr. *Introduction to Microelectronic Devices*. Prentice-Hall, Englewood Cliffs, NJ: 1989.
2. M. Zambuto. *Semiconductor Devices*. McGraw-Hill, New York: 1989.
3. P. Antognetti and G. Massobrio, eds. *Semiconductor Device Modeling with SPICE*. McGraw-Hill, New York: 1988.
4. D. A. Hodges and H. G. Jackson. *Analysis and Design of Digital Integrated Circuits*. 2d ed. McGraw-Hill, New York: 1988.
5. J. P. Uyemura. *Fundamentals of MOS Digital Integrated Circuits*. Addison-Wesley, Reading, MA: 1988.
6. E. S. Yang. *Microelectronic Devices*. McGraw-Hill, New York: 1988.
7. Y. P. Tsividis. *Operation and Modeling of the MOS Transistor*. McGraw-Hill, New York: 1987.
8. D. H. Navon. *Semiconductor Microdevices and Materials*. Holt, Rinehart & Winston, New York: 1986.
9. R. A. Colclaser and S. Diehl-Nagle. *Materials and Devices for Electrical Engineers and Physicists*. McGraw-Hill, New York: 1985.
10. S. M. Sze. *Semiconductor Devices, Physics and Technology*. John Wiley & Sons, New York: 1985.
11. D. G. Ong. *Modern MOS Technology: Processes, Devices, and Design*. McGraw-Hill, New York: 1984.
12. J. Mavor, M. A. Jack, and P. B. Denyer. *Introduction to MOS LSI Design*. Addison-Wesley, London: 1983.
13. R. F. Pierret. *Field Effect Devices*. Addison-Wesley, Reading, MA: 1983.
14. L. W. Nagel. "SPICE2, A Computer Program to Simulate Semiconductor Circuits," *ERL Memorandum ERL-M520*, University of California, Berkeley: May 1975.

PROBLEMS

For the following problems, assume a room temperature of 300 K. Use $\epsilon_s = 11.9\epsilon_o = 11.9 \times 8.85 \times 10^{-14}$ F/cm, and $\epsilon_{ox} = 3.9\epsilon_o$.

3.1 A MOSFET is characterized by $V_{T0} = .8$ V, $\gamma = .4\sqrt{\text{V}}$, $2\phi_f = .6$ V, **KP** $= 20\ \mu\text{A}/\text{V}^2$, $\lambda = 0$.

a. Determine the aspect ratio, W/L, needed for a current flow of $I_D = .1$ mA if the device is biased with $V_{GS} = 2.2$ V, $V_{DS} = 4$ V, and $V_{SB} = 2$ V.

b. Calculate the total gate capacitance C_G if $W = L = 2.5\ \mu$m.

3.2 For an enhancement-mode NMOS transistor, the following parameters are provided: $V_{T0} = .7$ V, $\gamma = .35\sqrt{\text{V}}$, $C_{ox} = 7 \times 10^{-8}$ F/cm^2, $\lambda = 0$, $k_n = 40\ \mu\text{A}/\text{V}^2$.

a. Write a program to plot V_T as a function of V_{SB} for $0 \leq V_{SB} \leq 5$ V.
b. Plot $\sqrt{I_D}$ as a function of V_{GS} for $0 \leq V_{GS} \leq 5$ V, and $0 \leq V_{SB} \leq 3$ V, if $V_{DS} = 5$ V.

3.3 If $N_a = 10^{15}$ cm^{-3}, and the channel surface mobility μ_e is 1/2 the bulk value, determine **KP** for the following:
a. $x_{ox} = 75$ nm
b. $x_{ox} = 50$ nm
c. $x_{ox} = 25$ nm

3.4 Given $N_a = 10^{15}$ cm^{-3}, $x_{ox} = 750$ Å, $V_{GB1} = -.8$ V, $Q_{ox} = 2.5 \times 10^{-8}$ C/cm^2, $D_I = 5.3 \times 10^{11}$ cm^{-2} for a MOSFET, compute the following:
a. V_{T0}
b. γ
c. V_T when $V_{BS} = 1$ V, 2 V, and 3 V

3.5 The NMOS shown in the diagram is characterized by the following parameters: $N_{a(sw)} = 10N_a = 10^{16}$ cm^{-3}, $N_d = 10^{19}$ cm^{-3}, $x_{ox} = 75$ nm, **XJ** = **LD** = .5 μm. Compute the following:
a. **CGSO**, **CGDO**, **CJSW**
b. Total gate capacitance C_G
c. Total transition capacitance of the source region as V_{SB} changes from $-.3$ V to -5 V

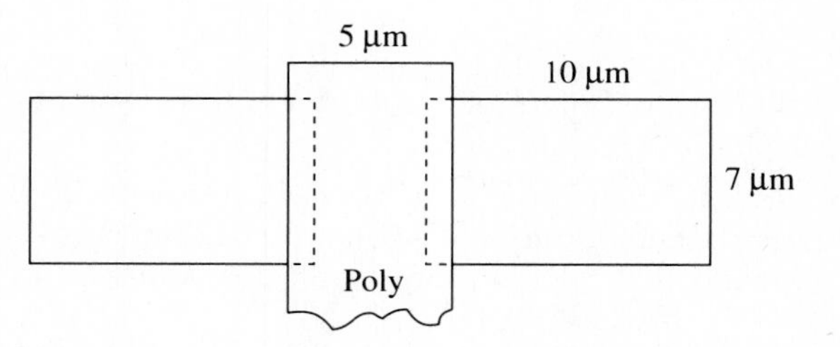

3.6 Given a p-type silicon substrate with $N_a = 10^{15}$ cm^{-3}, find the following:
a. The surface band bending at strong inversion ϕ_{si}
b. Depletion-layer width $x_{d(max)}$
c. Total bulk charge $Q_{s(max)}$
d. V_{T0}, assuming $V_{FB} = 0$, if an oxide of 75 nm is grown and a metal gate is deposited

3.7 For an MOS capacitor with $x_{ox} = 100$ nm on a p-type Si substrate with $N_a = 10^{15}$ cm^{-3}, compute the capacitance at
a. $f = 1$ Hz and $V_{GB} = 10$ V, and
b. $f = 1$ MHz and $V_{GB} = 10$ V.

3.8 An MOS structure consists of an n-type substrate with $N_d = 10^{15}$ cm^{-3}, $x_{ox} = 75$ nm, and an $A\ell$ contact. If $V_T = -2.5$ V, and $V_{GB1} = -.3$ V, find Q_{ox}, assuming $D_I = 0$.

3.9 Given $N = 5 \times 10^{15}$ cm^{-3}, $C_{ox} = 30$ nF/cm^2, $Q_{ox} = 8 \times 10^{-8}$ C/cm^2, $V_{GB1} = -.44$ V for gold and p-type silicon and $V_{GB1} = -.28$ V for aluminum and n-type silicon. Compute V_{T0} for an
a. $A\ell$-n-type MOS device, and an
b. Au-p-type MOS device.

3.10 Would the maximum layer width $x_{d(max)}$ differ for Ge from that of its value for Si if N_a is the same?

3.11 For a polysilicon-p-type MOSFET with $N_a = 10^{15}$ cm^{-3}, $Q_{ox} = q10^{10}$ C/cm^2, and $x_{ox} = 25$ nm, calculate the following:
a. Ideal V_{T0} with $V_{FB} = 0$
b. The minimum capacitance as the strong inversion takes place at V_T
c. Actual V_{T0} with $V_{GB1} = -1$ V
d. Boron dose required to increase V_{T0} found in (c) to 1 V.

3.12 If $V_{GB2} = -1$ V for a MOS structure whose oxide thickness is 100 nm, find the number of parasitic charges per μm^2 at the Si-SiO$_2$ interface.

3.13 Show that for an NMOS transistor at the edge of saturation, the electric field along the channel is given by

$$\mathcal{E}_y = \frac{V_{DS(sat)}}{2L\sqrt{1 - \frac{y}{L}}}$$

where L is the channel length.

3.14 For an NMOS transistor with $V_{GS} = 5$ V, $V_{DS} = 4$ V, $x_{ox} = 125$ nm, and $L = 2$ μm, determine the following:
a. $\mathcal{E}_{ox}$ at the source, where $\mathcal{E}_{ox}(y)$ is the field in the gate oxide in the x-direction at any point y
b. $\mathcal{E}_{ox}$ at the drain
c. $\mathcal{E}_y$ halfway between the source and drain when the transistor is at the edge of saturation with $V_{DS(sat)} = 5$ V

3.15 To investigate the body effect on the threshold voltage of the n-channel MOSFET N1 of Example 3.6, use SPICE to generate the transistor's I_D-V_{GS} characteristics in the range $0 \le V_{GS} \le 5$ V for a fixed $V_{DS} = .1$ V. Change V_{BS} from 0 to -5 V in steps of -1 V. Determine V_{T0}.

3.16 Use SPICE to demonstrate the body effect on the drain current of the NMOS in Problem 3.15. Generate the I_D-V_{DS} characteristics in the range $0 \le V_{DS} \le 5$ V for $V_{BS} = 0$ and $V_{BS} = -5$ V with $V_{GS} = 5$ V. Comment on the result.

3.17 Use SPICE to examine the effect of the substrate doping on the body-bias effect by generating the I_D-V_{GS} characteristics in the range $0 \le V_{GS} \le 5$ V and observing the change in V_T. Consider a circuit with three n-channel MOSFETs connected in parallel. The transistors are identical with the exception of the doping density in the substrate. Change **NSUB** from 10^{15} to 10^{17} cm^{-3} in powers of 10. Otherwise use parameters of Problem 3.15 for all transistors. Assume $V_{DS} = .1$ V.

3.18 Repeat Problem 3.17, but this time examine the effect of oxide thickness on body effect. Set the substrate impurity concentration at 10^{16} cm^{-3} and change **TOX** from 50 nm to 100 nm in steps of 50 nm.

3.19 Use SPICE to study the effect of the flatband voltage on the I_D-V_{GS} characteristics in the range $0 \le V_{GS} \le 3$ V. Consider a circuit with three n-channel MOSFETs connected in parallel. The transistors are identical with the exception of the gate material type that is specified by **TPG**. Use the parameters of Problem 3.15 for all transistors. Also, include **NSS** $= 2 \times 10^{10}$ charges/cm^2 to represent the fixed oxide charge. Set $V_{DS} = .1$ V and $V_{BS} = -1$ V.

3.20 To illustrate the effect of channel-length modulation on the drain current of a short-channel MOSFET, use SPICE to simulate a circuit with two *n*-channel MOSFETs connected in parallel. The parameters for both transistors will be the same as those used in Problem 3.15, but the channel length and width for one of them will be reduced to 1 μm. Use **LEVEL2** and generate the I_D-V_{DS} characteristics for $0 \leq V_{DS} \leq 5$ V and $V_{GS} = 3$ V.

3.21 To investigate the limiting effect of velocity saturation on the drain current of a short-channel NMOS, use **LEVEL2** SPICE modeling to generate I_D-V_{DS} characteristics of two transistors connected in parallel. The transistors have the same parameters as those in Problem 3.15, but their channel length has been reduced to 1 μm. Sweep V_{DS} from 0 to 5 V and set $V_{GS} = 5$ V. Specify **VMAX** $= 2 \times 10^6$ m/s for only one of the transistors.

3.22 To analyze the short-channel effects on the threshold voltage, use SPICE to simulate a circuit with two *n*-channel MOSFETs connected in parallel. The parameters for both transistors will be the same as those used in Problem 3.15, but the channel length and width for one of them will be reduced to 1 μm. Use **LEVEL2** and generate the I_D-V_{GS} characteristics for $0 \leq V_{GS} \leq 1$ V and $V_{DS} = .1$ V.

3.23 For all six combinations of gate material and substrate type, find expressions for the work-function difference V_{GB1} in terms of the doping density in the substrate, but independent of the effective density of states N_v or N_c. Use Lotus 1-2-3 or MathCAD to plot V_{GB1} as a function of $10^{14}\ \text{cm}^{-3} \leq N \leq 10^{18}\ \text{cm}^{-3}$ for all combinations.

3.24 The program listed at the end of this chapter can be modified to provide more information about the MOS structure under consideration.

a. Add appropriate lines to include the calculation of the depletion region charge, $Q_{s(max)}$.
b. Use the result of Problem 3.23 to incorporate the calculation of V_{GB1} into the program.
c. Prompt the user to read off the CV-plot the flatband voltage once the program outputs the flatband capacitance and make the program use this information to calculate (1) the fixed oxide charge Q_{ox} and (2) the threshold voltage V_T.

4. THE CMOS FAMILY

The CMOS logic family, as compared to bipolar families, has been relatively slow to emerge as a mainstream technology. Developed by RCA in the early sixties, CMOS has, until a decade ago, remained in the background. It consists of the older *metal-gate CMOS* and the newer *silicon-gate CMOS* subfamilies.

4.1 Metal-Gate vs. Silicon-Gate Processing

In this section we compare the two basic types of CMOS technologies without going into the details of fabrication processes, which are discussed in Appendix B.

Metal-Gate CMOS Subfamilies

The conventional metal-gate CMOS ICs used a fabrication technology in which metal comprised the gate electrode on the MOSFETs. To provide both channels for CMOS, usually, a *p*-type *well* is formed into *n*-substrate as shown in Figure 4.1. The gate electrode is metal, specifically, aluminum. In a metal-gate process, since the source and drain areas are diffused before the gate is defined, the gate needs to overlap the source and drain regions in order to compensate for a possible misalignment between masking operations. In this way, the presence of a continuous channel from source to drain is assured. However, this process not only enlarges the layout area for the circuit, but the thin gate oxide in the overlap region also produces a large overlap capacitance between the gate and drain, which impedes the circuit's performance because the oxide capacitance per unit area C_{ox} is inversely proportional to x_{ox}. Furthermore, the metal mask is extended beyond the side of the gate to prevent exposure of the gate oxide in case the metal mask is misaligned.

Another shortcoming of the metal-gate process is the increase in the area of the device due to the introduction of the *guard rings*. These are heavily doped regions of the same type as that of the well or the substrate and are usually embedded near the boundary of the regions. They are used to avoid leakage paths in the field regions

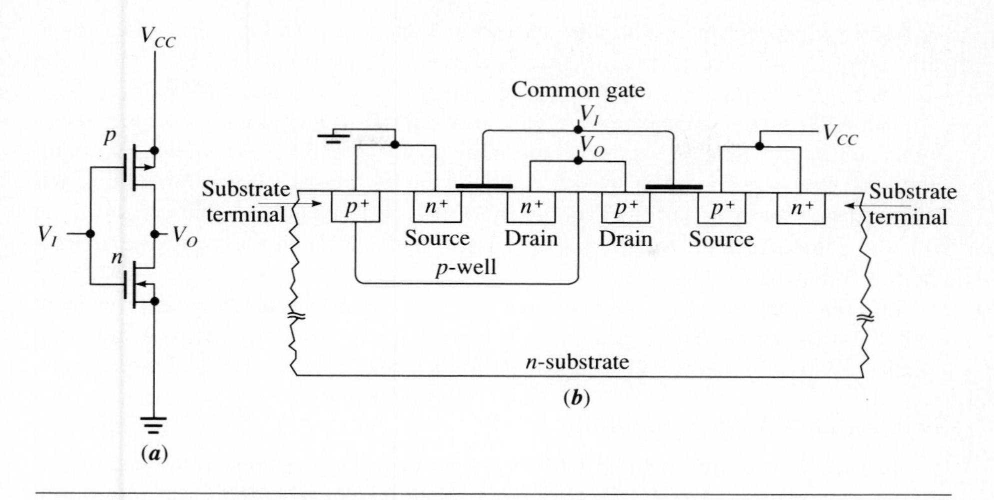

Figure 4.1 (a) The CMOS inverter, and (b) its simplified cross section.

and to lower the probability of a *latch-up* due to the undesirable parasitic transistors under metal lines. Figure 4.2 shows the cross section of a metal-gate inverter with inherent capacitances.

Yet, the metal-gate CMOS subfamily satisfies the requirements of those applications where low power consumption, wide power supply range, and high noise immunity are more important than high speeds. Therefore, initially, for applications that required high speeds, the system designers had to trade the advantages of CMOS

Figure 4.2 The cross section of a metal-gate CMOS inverter with the parasitic capacitances.

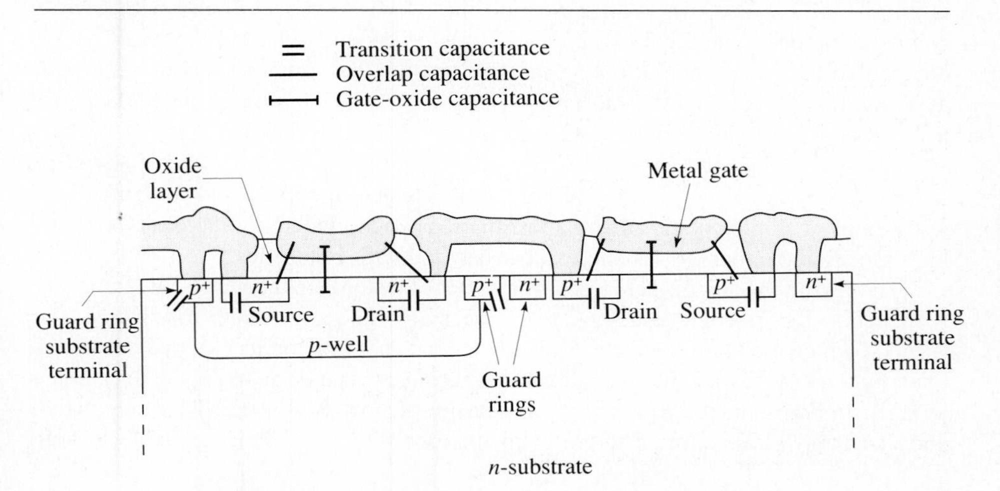

for faster switching times. The introduction of high-speed CMOS subfamilies in the early eighties provided switching speeds comparable to Schottky TTL subfamilies without having to trade off most of the advantages of the original CMOS.

The 4000 series, introduced in 1968, was the oldest and already obsolete series in this subfamily. Since it was slow and more costly than TTL for equivalent circuit functions, it was restricted to specific applications. Series 4000B was a modified and improved version of the original subfamily. It had buffered outputs that enhanced the sink and source current capabilities to ± 1 mA and improved the switching characteristics somewhat.

Introduced in the early seventies, the 54C/74C series was a further improvement over the 4000 series. The distinguishing feature of this series is that it is pin- and function-compatible with TTL integrated circuits, making interchangeability easier.

Silicon-Gate CMOS Subfamilies

The recent breakthroughs in CMOS technology have resulted in the emergence of new subfamilies. The 54HC/74HC *High-speed CMOS* (*HCMOS*) was introduced in 1982. In addition to enjoying high noise immunity, low power dissipation, and a substantial range of operating voltages, which are characteristics of all CMOS devices, this series can operate at speeds comparable to Schottky TTL subfamilies thanks to the use of silicon-gate processing technology. It is also buffered like the 4000B series.

The 54HCT/74HCT series is primarily used to interface TTL output signals to 54HC/74HC inputs. The input transistor geometries were changed to make these devices TTL compatible. Therefore, though 74HCT still provides considerable savings over TTL, its power consumption is higher compared to the equivalent 74HC device.

In 1985, Fairchild was the first to make the next step forward in the evolution of the CMOS family by introducing the *advanced CMOS logic* (*ACL*). It is based on the fastest CMOS technology yet devised, called by the manufacturer *Fact*, for *F*airchild *a*dvanced *C*MOS *t*echnology. It utilizes a 2-μm isoplanar silicon-gate CMOS process to rival the speeds of the advanced Schottky subfamilies while retaining the advantages of CMOS. The two ACL series are the 54AC/74AC subfamily with CMOS compatible inputs, and the 54ACT/74ACT line of logic with both TTL- and CMOS-compatible inputs and outputs. A year later, in late 1986, other manufacturers, such as Texas Instruments, RCA, and Philips, began introducing advanced CMOS parts, too. Table 4.1 lists the supply voltage ranges of all CMOS subfamilies.

The gate-electrode the HCMOS technology utilizes is a layer of polycrystalline silicon, referred to as *polysilicon*, or *poly*, in this text. It is a deposited layer of silicon with grains of single-crystal silicon a few microns wide, which, unlike metal, endures high processing temperatures and can be oxidized with impurities.

The reason the silicon-gate transistor switches faster than the metal-gate transistor is related to the inherent parasitic capacitances and the gain of both transistors. The speed at which a CMOS device can switch depends not only on its external capacitive load, but on how fast its internal parasitic capacitances can be charged and discharged as well. In addition, the gain is also a measure of how well a transistor can charge and discharge a capacitor. Therefore, to increase the speed, it is desirable to both

Table 4.1 Operating Voltage Ranges of CMOS Subfamilies

Subfamily	Voltage range (V)
4000	3–15
4000B	3–15
74C	3–15
74HC	2–6
74HCT	4½–5½
74AC	2–6
74ACT	2–6

decrease the parasitic capacitances and increase the transistor gain. This is what has been accomplished in silicon-gate HCMOS technology, as explained below.

Figure 4.3 shows the cross section of a silicon-gate inverter with parasitic capacitances. There are three major ways in which silicon-gate processing reduces the parasitic capacitances:

1. A thick layer of field oxide recesses below the surface of the silicon substrate and prevents the current flow by completely isolating each transistor from the others, thereby eliminating the need for the guard rings.
2. Shallower junction depths decrease the size of the parasitic diodes between the p^+ regions and the n-substrate, as well as the n^+ regions and the p-well. Since the capacitance is proportional to diode area, the reduction in diode areas results in reduced transition capacitance.
3. The *self-aligned* gate process used in this technology, in which the polysilicon is deposited over the gate oxide before the source and drain implants are made, so that the gate serves as a mask for these diffusions, greatly reduces the gate overlap and hence the gate-to-source and gate-to-drain capacitances associated with it.

The use of polysilicon also increases the gain of the transistors used in the HCMOS device because polysilicons may be etched to finer line widths than metal, resulting in transistors with shorter gate lengths. Using Equation (3.28), we see that a decrease in gate length will cause an increase in current gain capability, which allows the MOSFET to charge a capacitor more rapidly. The silicon-gate structure in the HCMOS features a 3-μm gate length, while the 4000 series had a gate length of 7 microns. Moreover, although a thinner gate oxide increases C_{ox}, it also increases the gain, k.

Finally, V_T is lower in a silicon-gate CMOS, so that there is more current available in the saturation region, as seen from Equation (3.30), and the problem of latch-up has virtually been eliminated by a combination of process enhancements and layout techniques. Therefore, the net result in using silicon-gate processing is a considerable increase in the switching speed of the device.

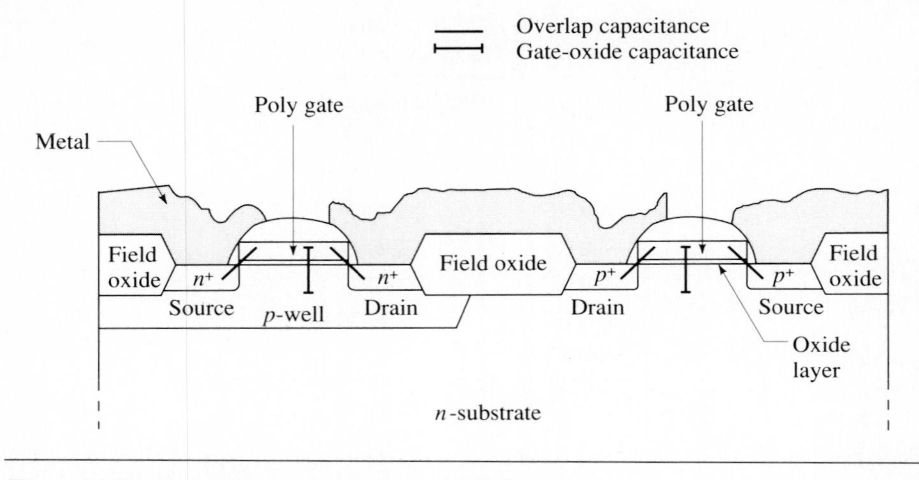

Figure 4.3 The cross section of a silicon-gate CMOS inverter with the parasitic capacitances.

The Latch-Up Problem

As mentioned, especially the metal-gate CMOS process produces certain parasitic *npn* transistors. When these transistors are improperly biased, a *latch-up* phenomenon presents a near short-circuit condition across the power supply. This causes large currents to flow that destroy the MOS device.

To understand this problem, we first need to discuss briefly the basic operation of a *silicon-controlled rectifier* (*SCR*) which is a *pnpn* device with three terminals, called the *anode*, the *cathode*, and the *gate*, as illustrated in Figure 4.4. The cross-coupled two-transistor equivalent circuit of the SCR is shown in Figure 4.5. An increase in the gate current I_G causes an increase in the collector current I_{C2} which in turn increases I_{C1}, resulting in a further increase of the base current I_{B2}. The regenerative positive feedback action turns the device on permanently even after the gate current is removed, leading to self-destruction.

The formation of the parasitic transistors in the silicon is depicted in Figure 4.6a; the equivalent electrical circuit is shown in Figure 4.6b. Here, R_s is the substrate resistance and R_w represents the well resistance. If a positive impulse is applied at node G (e.g., when the power is turned on), the voltage across R_w will induce a

Figure 4.4 The silicon-controlled rectifier (SCR).

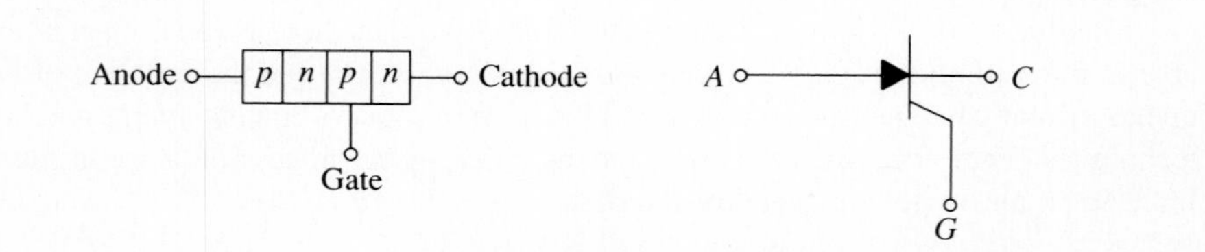

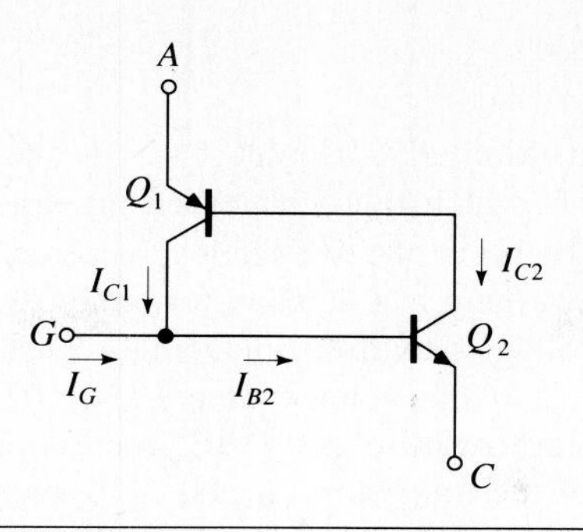

Figure 4.5 The two-transistor equivalent of the SCR.

Figure 4.6 (a) Parasitic bipolar transistors in a CMOS, and (b) equivalent electrical circuit.

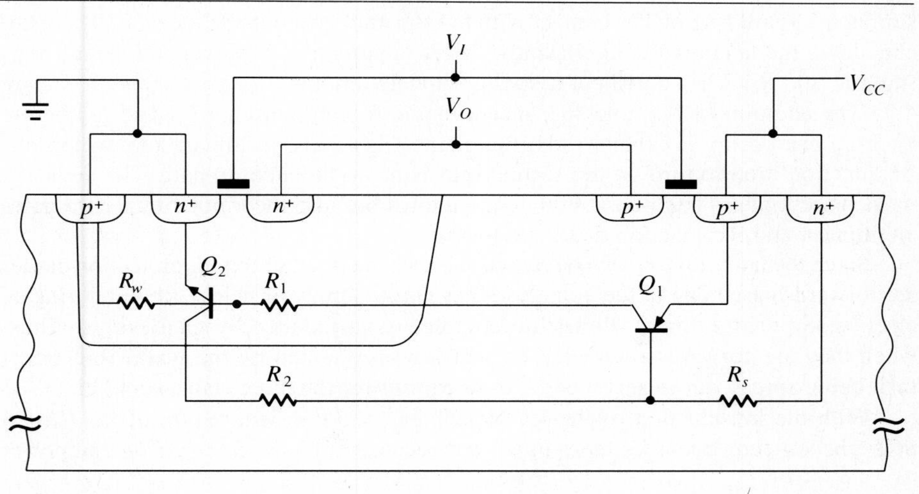

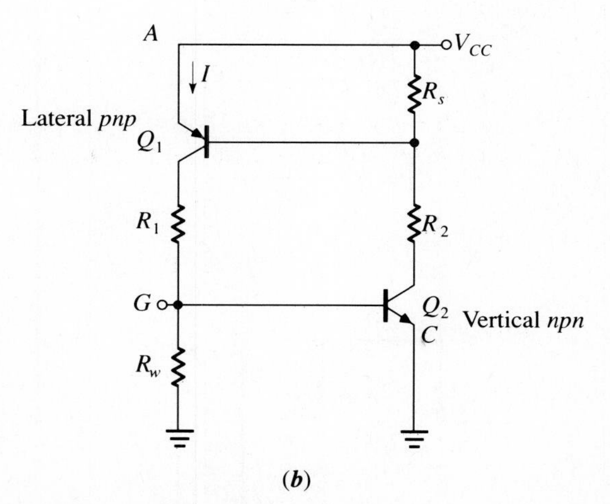

positive feedback current. Similarly, if point A surges above V_{CC}, due to a phenomenon called *inductive kick*, which can occur when the inputs are switched, a current will flow in R_s, again inducing a positive feedback current. Finally, during dynamic switching in CMOS, when there is a demand for large current for a very short duration, the power supply voltage may drop, inducing a latch-up.

In all these situations, if the voltage drop IR_s or IR_w, where I is the induced current, is large enough to forward-bias the BE junction of one of the parasitic transistors, and if the gain of the transistors β_1 and β_2 is such that $\beta_1\beta_2 \geq 1$, then the latch-up takes place. The vertical *npn* is of fairly high current gain, from 100 to even 500, while the lateral *pnp* is of low gain, from .01 to 1. In a metal-gate CMOS, the values of R_s and R_w are reduced as much as possible by using guard rings, which are low-resistivity connections, to supply voltages built around the CMOS *p*-channel and *n*-channel transistors. The contact to the well, in the form of a p^+ diffusion, is placed between any *n*-channel drain and the edge of the *p*-well, as shown in Figure 4.2. A similar n^+ guard ring makes contact with the substrate just outside the well. The latter cuts down the lateral gain significantly. Most importantly, however, the guard rings shunt R_s and R_w. The equivalent transistor model with guard rings is shown in Figure 4.7. The addition of R_{pgr} and R_{ngr} increases the forcing currents I_{C1} and I_{C2} before latch-up can occur. In other words, the guard rings make it difficult to develop one BE junction drop to turn on the second transistor in the latch-up pair. The device is most susceptible to latch-up at high temperatures because the bipolar transistor gains are highest and BE junction drops are lowest.

Since the latch-up can also occur when either the input or output protection diodes are forward-biased due to the voltage values above supply or below ground during an electrostatic charge, these diodes are completely surrounded by guard rings. Thus, when they are forward-biased, the current flow is shunted by the guard ring rather than being spread out to serve as the base current for the other transistors.

With the introduction of the HCMOS logic, the transient nature of the CMOS SCR phenomenon becomes more important because of signal-line *ringing* and power

Figure 4.7 Equivalent transistor model with guard rings.

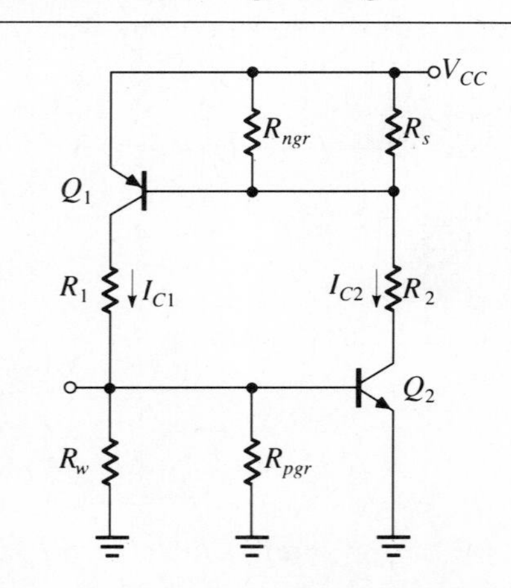

supply transients. If the trigger is a short pulse, then the peak value of the pulse current that triggers the SCR can be much larger than the static dc trigger current because of the poor frequency characteristics of the SCR.

For short noise pulses of smaller than 5 μs, the peak current required to latch up a device depends on the duty cycle of the pulses. At these speeds, the latch-up is caused by the average current. If the pulse widths become longer than 5 μs, then the latch-up current approaches the dc value even for low duty cycles. Hence, pulses lasting several microseconds may be long enough to appear as dc currents to the SCR. For example, for a 1-MHz 50% duty cycle pulse train applied to a device that otherwise latches with a 20 mA dc current, the typical peak current required would be about 40 mA, while for a 25% duty cycle it would be as high as 80 mA. On the other hand, for a slow repetition rate of, say, 2.5-KHz 10% duty cycle pulse train, a peak input current of around 35 mA would be sufficient to latch the device.

In typical high-speed systems, the noise spikes are only a few nanoseconds in duration and the average duty cycle is small. Therefore, even a device that is not designed to prevent latch-up will probably not latch up. However, in noisy applications such as automotive and relay drivers, where inductive loads are used, transients of several microseconds can be easily generated. In spite of that, as mentioned before, SCR latch-up in HCMOS is effectively controlled in IC level with advanced process and layout considerations.

4.2 The CMOS Inverter

An ideal logic family should dissipate no power, have zero propagation delay, and have noise immunity equal to 50% of the logic swing. The properties of complementary MOS begin to approach these ideal characteristics, as we will see below in detail.

The Operation

The basic CMOS circuit is the inverter shown in Figure 4.8. It utilizes two enhancement-type MOSFETs, namely, an n-channel *pull-down* and a p-channel *pull-up*. The power supplies for CMOS are either called V_{DD} or V_{CC}. The conventional MOS circuits use the former for drain supplies. This should not directly apply to CMOS because the supply is actually connected to the source. V_{CC} is borrowed from the TTL logic and is being used widely, especially since the introduction of the 74C series of CMOS.

In a CMOS system, V_{CC} corresponds to logical **1** and ground to logical **0** level. When $V_I = 0$ V, Q_1 is operating at $V_{GSn} = 0$. Hence, it is cut off and $I_{D1} = 0$. For the p-channel Q_2, however, the gate-to-source voltage $|V_{GSp}| = -V_{CC}$. If V_{CC} is larger than the threshold voltage of this transistor, then the transistor will have an inversion channel but no current will be drawn because $I_{D1} = I_{D2} = I_D = 0$. As shown in Figure 4.9, we will have $V_{OH} = V_{CC}$ at the operating point.

For $V_I = V_{CC}$, $|V_{GSp}| = 0$, the pull-up transistor is cut off, and $I_{D2} = 0$. If V_{CC} is larger than the threshold voltage of Q_2, then the n-channel transistor will have an inversion channel but still no current will be flowing through the two devices, and $V_{OL} = 0$. This is also shown in Figure 4.9. Therefore, in either extreme state, one

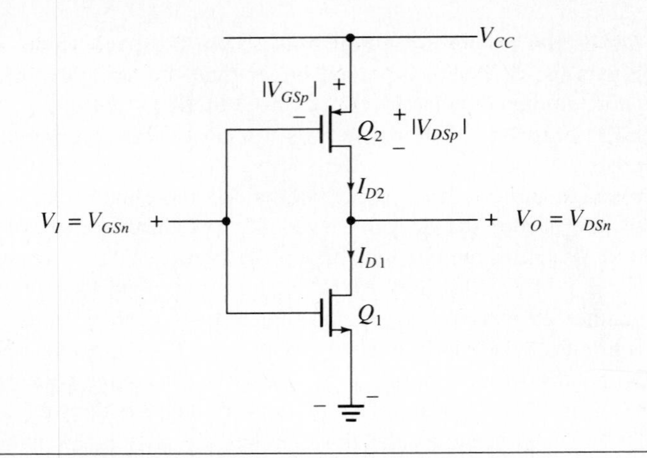

Figure 4.8 The basic CMOS inverter to be analyzed.

transistor in the series path from the supply to the ground is nonconducting. The only current which flows in either steady state is the extremely small leakage current, so that the static power dissipation is essentially zero.

An advantage of CMOS circuits over NMOS is that the former are *ratioless*. This means that the output voltage levels do not depend on the relative dimensions of the *n*- and *p*-channel devices, so that the transistors can be of minimum size. Due to the independence of logic voltage levels and device sizes, the relative sizes of the pull-down and pull-up transistors in a CMOS circuit are determined by other considerations. The equivalent resistance of the transistors in the conducting state, for instance, is primarily affected by the transistor sizes.

This implies that approximately equal capability to source or sink load current can be achieved in CMOS using device sizing. Thus, the symmetrical drive capability

Figure 4.9 The load curves for the driver Q_1 and the load Q_2 of the CMOS inverter.

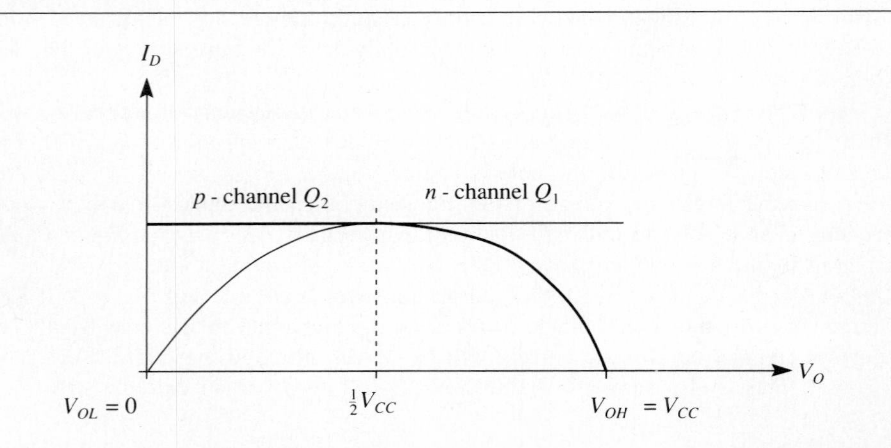

of CMOS allows comparable switching times at the output for both directions of the transition between logic states. Circuits employing *ratioed* logic, such as the NMOS, on the other hand, cannot provide symmetrical output drive because different pull-down and pull-up resistances are needed to obtain useful logic levels.

The Static Performance

The VTC for the CMOS inverter appears in Figure 4.10. As it shows, the characteristic has five distinct regions. The modes of operation of both transistors are indicated for each region. The following simplifying assumptions are made to obtain the characteristic:

1. The threshold voltage is the same for both transistórs; that is, $V_{Tn} = |V_{Tp}| = V_T$.
2. The gain constant of the transfer function $k = k_n = k_p$.
3. The saturation *I-V* characteristics of both transistors are ideal, that is, horizontal straight lines.
4. The gate oxides are grown at the same time, so that C_{ox} is the same for both devices.
5. $V_{CC} > 2V_T$.

Note from the above assumptions that the device *aspect ratios* are related by

$$\frac{(W/L)_p}{(W/L)_n} = \frac{\mu_e}{\mu_h} \approx 2.5$$

such that a minimum-area *symmetrical* CMOS inverter with equal rise and fall times will have $(W/L)_n \approx 1$ and $(W/L)_p \approx 2.5$.

For small values of input voltage V_I, with $V_I < V_T$ in region 1, the *n*-channel Q_1 is off, the *p*-channel Q_2 is in the ohmic region, and $V_O = V_{OH} = V_{CC}$. There will be no drain current until $V_I = V_T$. As soon as $V_I > V_T$, Q_1 will be in saturation. Q_2 does not saturate until

$$|V_{DSp}| = |V_{GSp}| - V_T \tag{4.1}$$

Since

$$|V_{DSp}| = V_{CC} - V_O \tag{4.2}$$

and

$$|V_{GSp}| = V_{CC} - V_I \tag{4.3}$$

substituting these values into Equation (4.1), we obtain the corresponding value of the output voltage as

$$V_O = V_I + V_T \tag{4.4}$$

Similarly, Q_1 saturates when

$$V_{DSn} = V_{GSn} - V_T$$

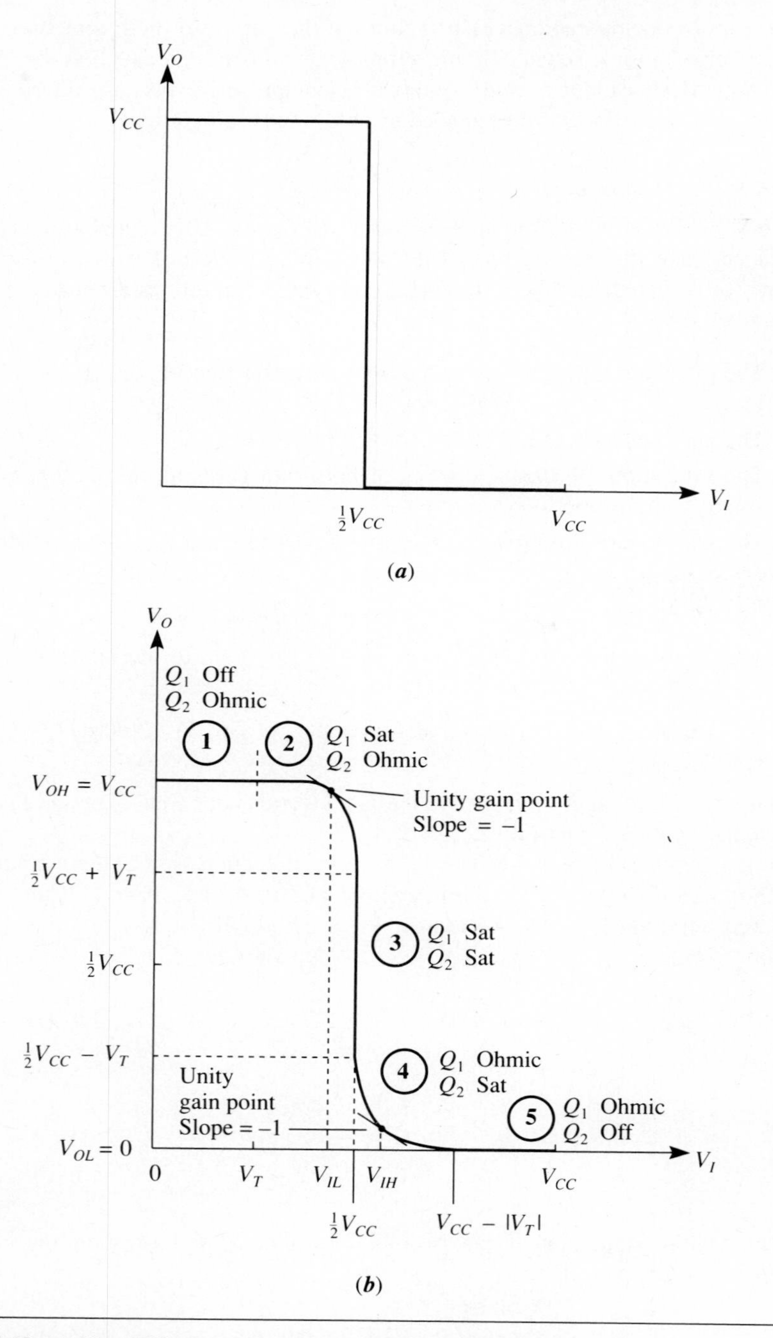

Figure 4.10 (a) The VTC of an ideal CMOS inverter; (b) VTC of a symmetric CMOS inverter with matched *p*- and *n*-channels.

But,

$$V_{DSn} = V_O$$

and

$$V_{GSn} = V_I$$

so that, when Q_1 saturates, the output has the value

$$V_O = V_I - V_T \tag{4.5}$$

Therefore, when both transistors are saturated in region 3, Equation (3.30) yields

$$I_{D1} = \frac{1}{2}k(V_I - V_T)^2 \tag{4.6}$$

and

$$I_{D2} = \frac{1}{2}k(V_{CC} - V_I - V_T)^2 \tag{4.7}$$

With no load being driven by the output terminal, the drain currents are equal to each other. Therefore, from the last two equations, we obtain

$$V_I = \frac{1}{2}V_{CC} \tag{4.8}$$

as the input value needed to saturate both transistors. Substituting Equation (4.8) into Equations (4.4) and (4.5) gives the corresponding output values as

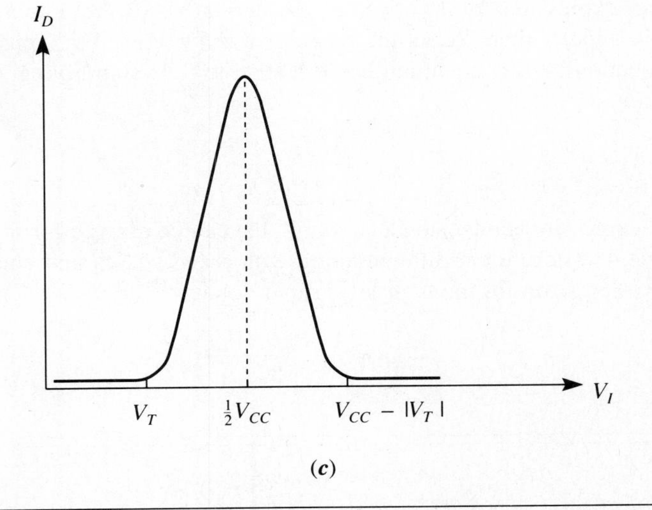

Figure 4.10 (*continued*) (c) The corresponding current flow.

$$V_O = \frac{1}{2} V_{CC} + V_T$$

when Q_2 saturates, and

$$V_O = \frac{1}{2} V_{CC} - V_T$$

when Q_1 saturates.

Between these two values of the output voltage, a constant current source is loading another, so that each transistor sees a very high load resistance. Notice that the inverter switches state at half the power supply voltage V_{CC}. Also, as seen in Figure 4.10 and due to the third assumption, the transition region is a vertical line, which implies an infinitely high gain. This desirable characteristic maximizes the noise immunity. In reality, as V_{DS} increases, I_{DS} also increases slightly, resulting in a finite slope in region 3.

Now, consider region 2 in Figure 4.10, in which Q_1 is saturated and Q_2 is in the ohmic region. Here, the expression for the drain current is given by Equation (3.27). Substituting Equations (4.2) and (4.3) into Equation (3.27) yields the expression for the load drain current in the ohmic region as

$$I_{D2} = \frac{1}{2} k[2(V_{CC} - V_I - V_T)(V_{CC} - V_O) - (V_{CC} - V_O)^2] \tag{4.9}$$

The saturation current for the driver is given by Equation (4.6). Equating (4.6) to (4.9), we derive an expression for V_O as a function of V_I in region 2:

$$V_O = V_I + V_T + \sqrt{(V_I + V_T)^2 + V_{CC}(V_{CC} - 2V_I - 2V_T) - (V_I - V_T)^2} \tag{4.10}$$

To derive an expression for V_{IL}, we set the derivative of the output voltage with respect to the input voltage equal to -1 and solve for $V_I = V_{IL}$. Thus, after differentiating Equation (4.10), equating the derivative to -1, simplifying, and rearranging, we obtain

$$V_{IL} = \frac{3V_{CC} + 2V_T}{8} \tag{4.11}$$

An analytic expression similar to Equation (4.11) can be derived for the output voltage in region 4 which, upon differentiating with respect to V_I and equating to -1, produces an expression for the high level input voltage:

$$V_{IH} = \frac{5V_{CC} - 2V_T}{8} \tag{4.12}$$

Example 4.1

Substituting typical values of $V_{CC} = 5$ V and $V_T = 1$ V into Equations (4.11) and (4.12) yields $V_{IL} = 2.125$ V and $V_{IH} = 2.875$ V so that the noise margins are found as

$$NM_H = V_{OH} - V_{IH} = \frac{3V_{CC} + 2V_T}{8} = 2.125 \text{ V}$$

and

$$NM_L = V_{IL} - V_{OL} = 2.125 \text{ V}$$

Note that increasing V_{CC} improves the noise immunity but also increases the power dissipation of the inverter.

When V_I is increased to the point where $V_I > V_{CC} - |V_T|$, Q_2 goes into cutoff, resulting in $I_{D1} = I_{D2} = 0$. Even though Q_1 is in the ohmic region of operation and has an established inversion layer, we have $V_O = V_{OL} = 0$ V in region 5. As illustrated in Figure 4.10c, no current flows between the supply and ground as long as $V_I < V_T$ or $V_{CC} - V_I < V_T$. Significant current flow exists only during switching between the two logical states. Table 4.2 summarizes the characteristics associated with the five regions.

To find a general expression for the inverter threshold, $V_{th} = V_I = V_O$, with $k_n \neq k_p$ and $V_{Tn} \neq |V_{Tp}|$, note that the equality occurs when both transistors are saturated so that

$$k_n(V_{th} - V_{Tn})^2 = k_p(V_{CC} - V_{th} - |V_{Tp}|^2)$$

Therefore, we have

$$V_{th} = \frac{V_{Tn} + \sqrt{k_p/k_n}\,(V_{CC} - |V_{Tp}|)}{1 + \sqrt{k_p/k_n}} \tag{4.13}$$

The Dynamic Performance

Consider the capacitively loaded inverter shown in Figure 4.11a. Capacitive loading can be due to the input capacitance of another gate and the stray capacitance of the interconnecting wires. Assume that the transistors have identical but complementary characteristics and threshold voltages. A pulse is being applied at the input.

At $t = 0^-$, just before the input pulse appears, the capacitor C_L has been charged

Table 4.2 Summary of the Basic CMOS Inverter Operation

Region	Input	NMOS	PMOS	Output*
1	$0 \leq V_I \leq V_T$	Cutoff	Ohmic	5V
2	$V_T \leq V_I \leq ½\, V_{CC}$	Saturated	Ohmic	$V_I + \sqrt{15 - 6V_I} + 1$
3	$V_I = ½\, V_{CC}$	Saturated	Saturated	$3½ \text{ V} \leq V_O \leq 1½ \text{ V}$
4	$½V_{CC} \leq V_I \leq V_{CC} - \vert V_T\vert$	Ohmic	Saturated	$V_I - \sqrt{6V_I - 15} - 1$
5	$V_{CC} - \vert V_T\vert \leq V_I$	Ohmic	Cutoff	0

*V_{CC} = 5V, V_T = 1V

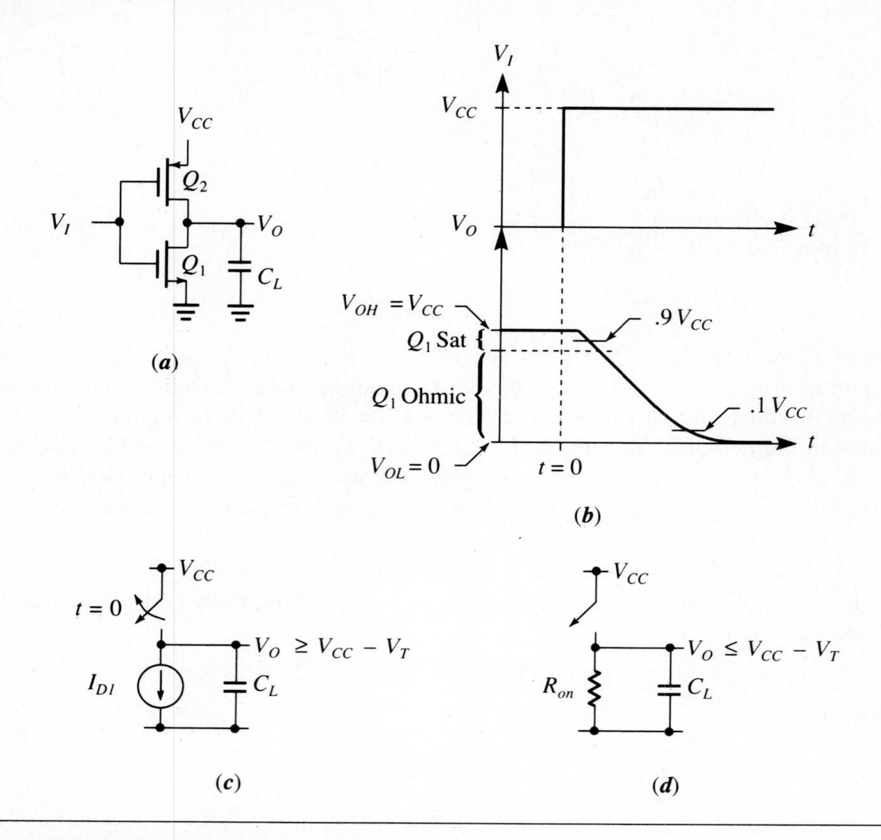

Figure 4.11 The transient behavior of the inverter: (a) capacitive loading, (b) input and output waveforms, (c) discharging circuit during t_{f1}, and (d) discharging circuit during t_{f2}.

to $V_{OH} = V_{CC}$. At $t = 0$, V_I rises to V_{CC} which turns off the pull-up transistor, Q_2, and saturates the pull-down transistor, Q_1. The capacitor discharges through Q_1 during a finite fall time. Therefore, although the static power dissipation is extremely low in both logical states, the dynamic power dissipation during transitions between the states will be appreciably higher due to the charging and discharging of the capacitive loads.

Due to symmetry in the circuit, the rise and fall times of the output waveform will be equal. Hence, we consider here only the turn-off time. Referring to Figure 4.11b, we see that during the fall time, the initial voltage change across the capacitive load is a ramp, that is, C_L first discharges linearly until the point where $V_O = V_{CC} - V_T$ due to the current source characteristics of the pull-down transistor in saturation. This is followed by a rounding off, as output voltage approaches zero, due to the resistive characteristics of the transistor in the ohmic region.

Figure 4.11 also shows the equivalent circuits during turn-off time. The fall time t_f can be expressed as

$$t_f = t_{f1} + t_{f2}$$

where t_{f1} is the interval during which the output voltage drops from the 90% of its initial value of V_{CC} to $V_{CC} - V_T$, and t_{f2} is the time required for V_O to drop from $V_{CC} - V_T$ to 10% of its initial value.

During t_{f1}, using Equation (4.6), we have

$$I_D = \frac{1}{2} k(V_{CC} - V_T)^2 \tag{4.14}$$

Therefore, the drain current is constant and the output voltage falls linearly. The discharging operation of the capacitor is described by

$$I_D \Delta t = C_L \Delta V_O \tag{4.15}$$

where $\Delta V_O = .9V_{CC} - (V_{CC} - V_T)$, and $\Delta t = t_{f1}$. Substituting into Equation (4.15) for I_D from (4.14) yields

$$t_{f1} = \frac{2C_L(V_T - .1V_{CC})}{k(V_{CC} - V_T)^2} \tag{4.16}$$

Note that this expression is valid only if $V_T > .1V_{CC}$. Otherwise, the device is already in the ohmic region, and $t_{f1} = 0$.

After Q_1 enters the ohmic region, we can use Equation (3.27) to express the discharging current as

$$I_D = C_L\left(\frac{dV_O}{dt}\right) = \frac{1}{2} k[2(V_{CC} - V_T)V_O - V_O^2] \tag{4.17}$$

Rearranging terms in Equation (4.17) and integrating both sides, we get

$$t_{f2} = \frac{C}{k(V_{CC} - V_T)} \int_{V_{CC}-V_T}^{.1V_{CC}} \frac{dV_O}{[1/2(V_{CC} - V_T)]V_O^2 - V_O}$$

Using the identity,

$$\int \frac{dx}{ax^2 - x} = \ln\left(1 - \frac{1}{ax}\right)$$

we obtain

$$t_{f2} = \frac{C_L}{k(V_{CC} - V_T)} \ln\left[1 - \frac{2(V_{CC} - V_T)}{V_O}\right]\Bigg|_{V_{CC}-V_T}^{.1V_{CC}}$$

$$= \frac{C_L}{k(V_{CC} - V_T)} \ln\left[20\,\frac{V_{CC} - V_T}{V_{CC}} - 1\right] \tag{4.18}$$

Taking $V_T = .2V_{CC}$, Equation (4.16) reduces to

$$t_{f1} \approx \frac{.3125C_L}{kV_{CC}}$$

while Equation (4.18) becomes

$$t_{f2} \approx \frac{3.385C_L}{kV_{CC}}$$

so that the total turn-off time is found as

$$t_f \approx \frac{3.7C_L}{kV_{CC}} \tag{4.19}$$

As V_{CC} is increased, the inverter is forced to drive the load through a larger voltage swing. At the same time, however, the drive capability is increased, and, as is obvious from Equation (4.19), the switching times are improved. Thus, for a given capacitive load, the speed of the device increases as the power supply voltage increases.

The contributions to the output capacitance C_L in a CMOS inverter are shown in Figure 4.12. Naturally, only capacitors that experience a voltage change during switching need to be included. Therefore, the average values of the transition capacitances must be used as the output voltage changes from V_{OL} to V_{OH} and vice versa. Since the capacitors are in parallel, we have from Figure 4.12, for a fan-out of 1,

$$C_L \approx C_{GDn} + C_{GDp} + C_{dbp(av)} + C_{dbn(av)} + C_{line} + C_G$$

where C_{line} is the lumped-element approximation of the interconnecting wiring capacitance, and C_G is the total gate capacitance of the load, with both MOSFETs contributing capacitance that must be charged or discharged during switching between states. For short interconnect lines, C_G dominates in the expression above so that, if the output of the inverter is connected to the input of an identical inverter, the load can be approximated as

$$C_L \approx C_G = C_{Gn} + C_{Gp} = C_{ox}(W_pL_p + W_nL_n)$$

where the quantity in parantheses is the total area occupied by the n- and p-channel transistors. If the transistors are identical, the above approximation becomes

Figure 4.12 The contributions to the output capacitance in a CMOS inverter.

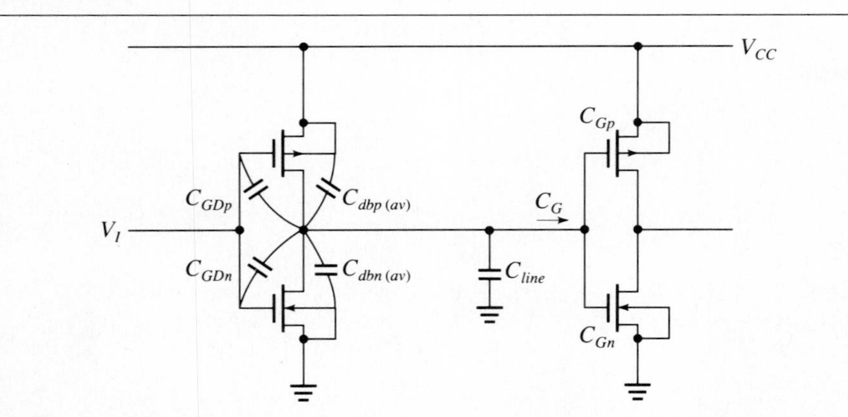

$$C_L \approx 2C_{ox}WL \tag{4.20}$$

Then, substituting Equations (3.28) and (4.20) into Equation (4.19), we find

$$t_f = t_r \approx \frac{7.4L^2}{\mu V_{CC}} \tag{4.21}$$

where t_r is the turn-on time, and μ is either μ_e or μ_h. In reality, since the electron mobility is about 2.5 times the hole mobility, there will be some asymmetry in switching speed for the turn-on and turn-off times. However, it can be removed by making the width of the p-channel transistor 2.5 times that of the NMOS.

In Equation (4.21), we define

$$\tau_e \equiv \frac{L^2}{\mu_e V_{CC}}$$

as the transit time for the electrons to pass through the n-channel transistor from the source to the drain, and

$$\tau_h \equiv \frac{L^2}{\mu_h V_{CC}}$$

as the transit time for the holes to pass through the p-channel transistor. Therefore, we have

$$t_f \approx 7.4\tau_e$$

$$t_r \approx 7.4\tau_h$$

The equivalent resistance of the circuit during t_{f2}, frequently referred to as the *on resistance*, is defined as

$$R_{on} \equiv \frac{V_{DS}}{I_D}$$

where I_D is given by Equation (3.27). Factoring out V_{DS} in that equation and assuming that

$$2(V_{GS} - V_T) = 2(V_{CC} - V_T) >> V_{DS}$$

we obtain

$$R_{on} = \frac{1}{V_{CC} - V_T} \tag{4.22}$$

Example 4.2

Consider a CMOS inverter with $k = .2\ \text{mA/V}^2$ and $V_T = 1$ V. To determine the turn-off time for a capacitive load $C_L = 5$ pF at $V_{CC} = 5$ V, and R_{on} during t_{f2}, we use Equation (4.19) to find $t_f = 18.5$ ns and Equation (4.22) to get $R_{on} = 1250\ \Omega$.

Example 4.3

A CMOS inverter is characterized by $k_n = 2.5k_p = 25\ \mu A/V^2$, and $V_{Tn} = |V_{Tp}| = 1$ V. The following SPICE file with $V_{CC} = 5$ V and $C_L = 50$ fF simulates the circuit to obtain the switching times:

```
BASIC CMOS INVERTER
M1   2 1 0 0 MN L = 4U W = 4U
M2   2 1 3 3 MP L = 4U W = 4U
CL   2 0 50F
VIN 1 0 PULSE 0 5 0 0 0 5N
VCC 3 0 5
.MODEL MN NMOS VTO  1 GAMMA .4 KP 2.5E-5
.MODEL MP PMOS VTO -1 GAMMA .4 KP 1E-5
.TRAN 1N 15N
.PROBE
.END
```

Figure 4.13 shows the resulting *nonsymmetric* switching, with $t_r = 1.48$ ns and $t_f = 3.7$ ns. Hand analysis, using Equation (4.19), would yield similar results.

Example 4.4

Consider the circuit described in Example 4.3, but now assume that $k_n = k_p = 25\ \mu A/V^2$. The SPICE file is modified as follows:

```
SYMMETRIC CMOS INVERTER
M1   2 1 0 0 MN L = 4U W = 4U
M2   2 1 3 3 MP L = 4U W = 10U
CL   2 0 50F
VIN 1 0 PULSE 0 5 0 0 0 5N
VCC 3 0 5
.MODEL MN NMOS VTO  1 GAMMA .4 KP 2.5E-5
.MODEL MP PMOS VTO -1 GAMMA .4 KP 1E-5
.TRAN 1N 10N
.PROBE
.END
```

Figure 4.14 illustrates the symmetric waveform at the output, with $t_r = t_f = 1.48$ ns.

The Power Consumption

There are basically two types of power consumption in a CMOS:

1. static (quiescent) power dissipation
2. dynamic power dissipation

The static power dissipation is ideally zero because there is no current flowing through the circuit at extreme states. In reality, though, there is a leakage current flowing across the reverse-biased diode junctions due to the inherent nature of semiconductors. Figure 4.15 shows the paths of the leakage current for an inverter when Q_1 is turned on. One path is through the reverse-biased drain junction of Q_2 and the turned-

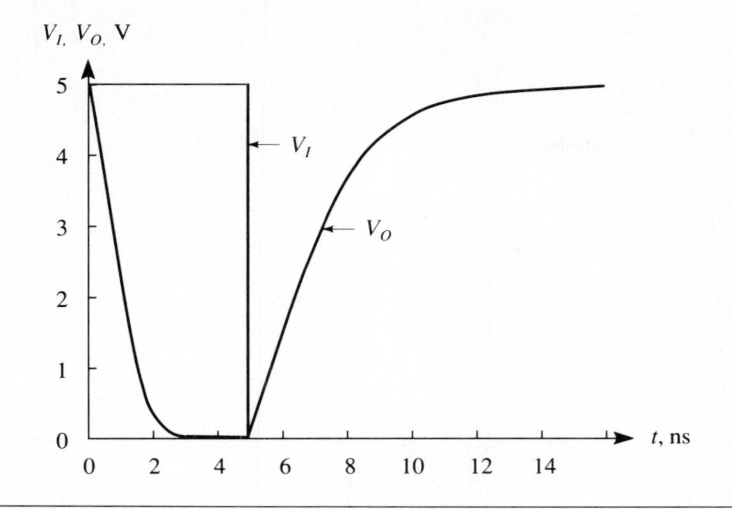

Figure 4.13 SPICE simulation: input and output voltage waveforms for the basic CMOS inverter.

on n-channel. The other is through the reverse-biased p-well-to-substrate junction, which is larger than the first component due to the larger well area and depletion width. Since these leakages are caused by thermally generated charge carriers, the number of these carriers and hence the power dissipation increases as the ambient temperature of the junction increases. The power dissipation due to this leakage current in the quiescent state is given by

Figure 4.14 SPICE simulation: input and output voltage waveforms for the symmetrical CMOS inverter.

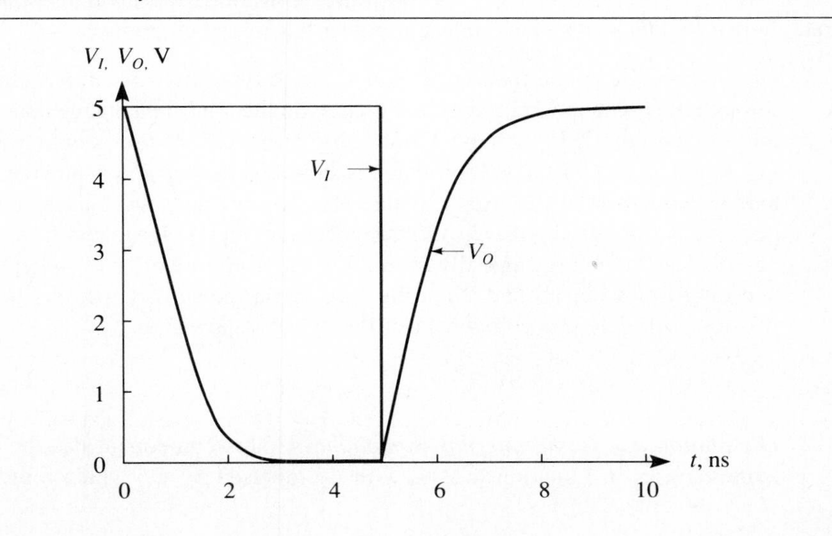

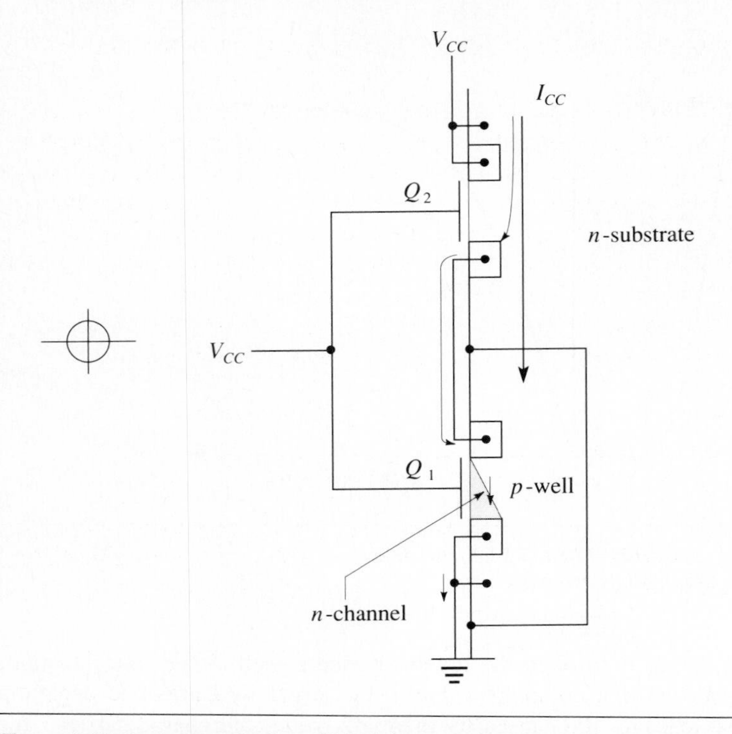

Figure 4.15 Leakage paths in a CMOS inverter that determine the static power dissipation.

$$P_S = V_{CC} I_{CC} \tag{4.23}$$

where I_{CC} is specified as the maximum quiescent supply current in data sheets.

The dynamic power dissipation is the power consumed during transitions between the two logical states. This transient power has three components:

1. *Dissipation due to the load capacitance.* Each time the capacitor is charged through the *p*-channel transistor, the charge on the load capacitance rises from almost zero to $C_L V_{CC}$; thus, the energy provided by the source will be $(C_L V_{CC}) V_{CC} = C_L V_{CC}^2$, half of which is stored on the capacitor, and the other half is dissipated in the load resistor, R_{off}. In the same way, each time C_L discharges through the driver on-resistance, the energy dissipated in R_{on} will be equal to the energy initially stored on C_L. If the inverter is switched on and off f times per second, then the sum of the power dissipations in both resistors will give this component of the dynamic power as

$$P_{D1} = C_L V_{CC}^2 f \tag{4.24}$$

2. *Dissipation due to the internal capacitances.* This component has the same expression as in Equation (4.24), with C_L replaced by C_I, which represents

the total internal parasitic capacitance of the device that is being charged or discharged every time a switching occurs.

3. *Dissipation due to current spiking* is the dc power consumed during switching due to a momentary low-impedance path for the supply current from the supply to the ground when both transistors are conducting simultaneously. If the rise time of the input signal were zero, there would be no current flow through the circuit. However, the input voltage ramps linearly from zero to V_{CC}, and hence spends a finite amount of time between the values V_T and $V_{CC} - V_T$, as seen in Figure 4.16. Note that the time spent in this region by the input increases as the power supply increases. Figure 4.16 also shows the corresponding switching current, whose average value over a period can be approximated as

$$I_{SW} = \frac{1}{2} I_{peak} \frac{t_r + t_f}{t_p}$$

where t_p is the period of the input signal. If the transistors match, then $t_f = t_r$. Using this and the fact that $t_p \equiv 1/f$, we obtain an expression for the average dc-transient power as

$$P_{SW} = V_{CC} I_{SW} = I_{peak} V_{CC} t_r f \tag{4.25}$$

Therefore, as seen from Equations (4.24) and (4.25), the power dissipation is directly proportional to the operating frequency. Note also that, although increasing V_{CC} increases the speed, it also increases the power consumption.

Usually, the last two components of the transient power are combined to define a capacitance C_{PD}, called the *power-dissipation capacitance* on data sheets for CMOS devices, which determines the no-load dynamic power consumption when used in the following expression:

$$P_{D(nl)} = C_{PD} V_{CC}^2 f \tag{4.26}$$

Figure 4.16 The input signal and the current spiking during switching between the states.

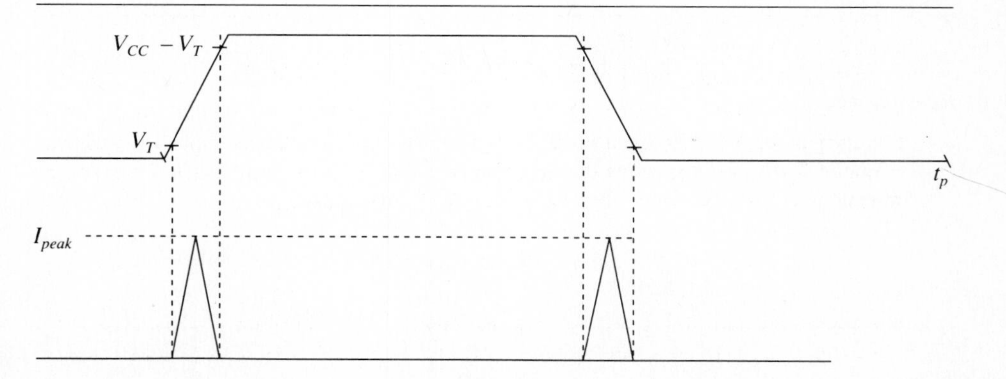

Therefore, the total power consumption of any CMOS device at any operating frequency is computed as

$$P_T = (C_L + C_{PD})V_{CC}^2 f + V_{CC}I_{CC} \tag{4.27}$$

Equation (4.27) includes both dc and ac contributions to the power usage. For example, the power-dissipation capacitance for HCMOS devices is calculated under worst-case conditions by choosing the worst path possible, as follows:

1. The power supply voltage is set to $V_{CC} = 5.5$ V.
2. The input signals are set up so as to switch as many outputs as possible.
3. The power supply current is measured and recorded at two different frequencies, f_1 and f_2. These are usually 200 KHz and 1 MHz, respectively.
4. Using Equation (4.27), the following simultaneous equations are solved:

$$P_{01} = C_{PD}V_{CC}^2 f_1 + I_{CC}V_{CC}$$

$$P_{02} = C_{PD}V_{CC}^2 f_2 + I_{CC}V_{CC}$$

to get

$$C_{PD} = \frac{(I_{01} - I_{02})}{V_{CC}(f_1 - f_2)}$$

where $I_{01} = P_{01}/V_{CC}$ and $I_{02} = P_{02}/V_{CC}$ are the values for the supply current at f_1 and f_2.

Example 4.5

The data sheet of the 74HCT00 2-input NAND gate gives the typical value for the power-dissipation capacitance, C_{PD}, as 20 pF with $C_L = 15$ pF at $V_{CC} = 5$ V, and the maximum quiescent supply current, I_{CC}, as 2 μA. To determine the total power consumption of this gate at $f = 1$ MHz we use Equation (4.27) to get

$$P_T = (15 + 20)(10^{-12})5^2 \times 10^6 + 5(2 \times 10^{-6}) = .885 \text{ mW}$$

Example 4.6

The current spike vs. the input voltage for the CMOS inverter of Example 4.2 is shown in Figure 4.10c. To determine the maximum current, I_{peak}, we note that it flows during transition between the states when $V_I = V_{CC}/2$. Therefore, using Equation (3.30), we get

$$I_{\text{peak}} = \frac{1}{2}k\left(\frac{1}{2}V_{CC} - V_T\right)^2 = .1\left(\frac{5}{2} - 1\right)^2 = .225 \text{ mA}$$

4.3 CMOS Logic Gates

In standard CMOS structuring, each input is connected to both NMOS and PMOS. The circuit of a two-input NAND gate is shown in Figure 4.17a. It consists of two n-channel transistors connected in series and two p-channel pull-up transistors connected in parallel. That NAND logic operation is satisfied can be demonstrated as follows. When both inputs are logical **0**, both PMOS transistors will be conducting, and the n-channel transistors will be off. Hence, the output voltage, Y, will be approximately V_{CC}, corresponding to logical **1**. Even if only one of the inputs is grounded, the corresponding n-channel MOSFET will be off, and the output will remain at **1**. When both inputs are tied to V_{CC}, the p-channel transistors will be off, and the n-channel devices will be conducting, so that the output voltage will be approximately 0 V, corresponding to a **0**.

A two-input NOR gate is shown in Figure 4.17b. Being the *dual* of a NAND gate, the NOR drivers are in parallel, while the PMOS transistors are in series. When at least one input is **1**, the corresponding p-channel device(s) will be off, and the n-channel transistor(s) will turn on, leading to a **0**. When both inputs are tied to ground, both p-channel devices will be conducting while both n-channels will be off. Therefore, the output voltage Y will be approximately V_{CC}.

The design of CMOS logic circuits focuses on establishing the desired VTC properties. As compared to the CMOS inverter, multiple-input gates are more complicated. The logic threshold voltage V_{th} assumes different values for different input switching combinations because the equivalent device resistances between the output node and supply or ground depend on the conducting modes of various transistors.

Figure 4.17 (a) The two-input CMOS NAND gate, and (b) the two-input CMOS NOR gate.

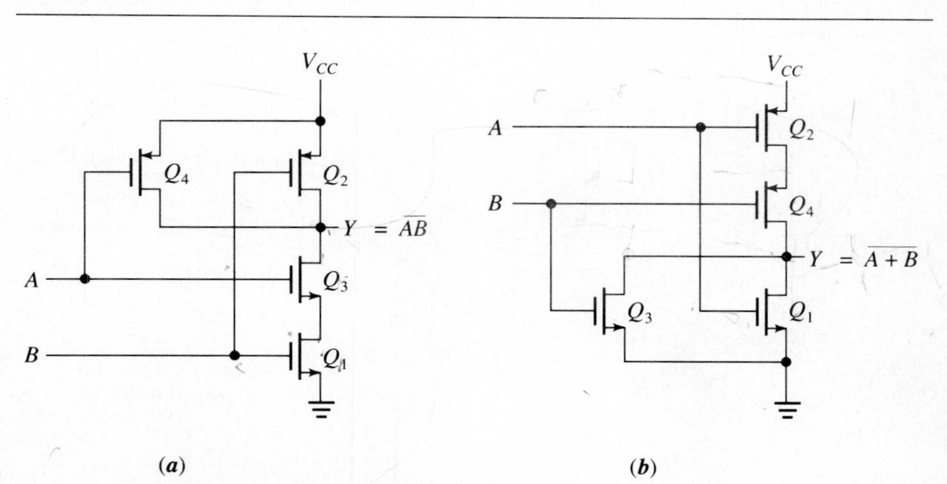

Design and Analysis of a Two-Input NAND *Gate*

Consider the two-input CMOS NAND gate of Figure 4.17a and assume that V_O is initially at $V_{OH} = V_{CC}$. There are three input combinations that can cause this:

1. $A = B = 0$

2. $A = 0, B = V_{CC}$

3. $A = V_{CC}, B = 0$

Therefore, three possible ways of increasing input voltages from zero up to the logical threshold voltage of V_{th} will induce output transitions of $V_O \rightarrow V_{th}$, giving rise to three distinct values of V_{th}. Assume that both n-channel devices have a transconductance of k_n, and both p-channel devices have a transconductance of k_p.

1. Suppose now that both inputs are switched simultaneously toward **1**. The corresponding circuit arrangement is depicted in Figure 4.18a. The NMOS gate voltages are $V_{GS1} = V_{th}$ and $V_{GS3} = V_{th} - V_{DS1}$. Substituting the output KVL equation,

$$V_{th} = V_{DS1} + V_{DS3}$$

in the latter equation reveals that $V_{GS3} = V_{DS3}$, which in turn implies that Q_3 is saturated. Furthermore, with $V_{GS1} > V_{GS3}$, Q_1 must be conducting in the ohmic region because a common current flows through both devices. This can be verified by the graphical representation of its operating point, as plotted in Figure 4.18b. Notice here that we ignore the presence of body-bias effects in Q_3. Equating the expressions for drain currents, we obtain

$$I_D = \frac{1}{2} k_n (V_{th} - V_{Tn} - V_{DS1})^2 = \frac{1}{2} k_n [2(V_{th} - V_{Tn}) V_{DS1} - V_{DS1}^2]$$

Figure 4.18 (a) Circuit voltages with the CMOS NAND inputs switching simultaneously, and (b) graphical representation of the Q_3 operating region.

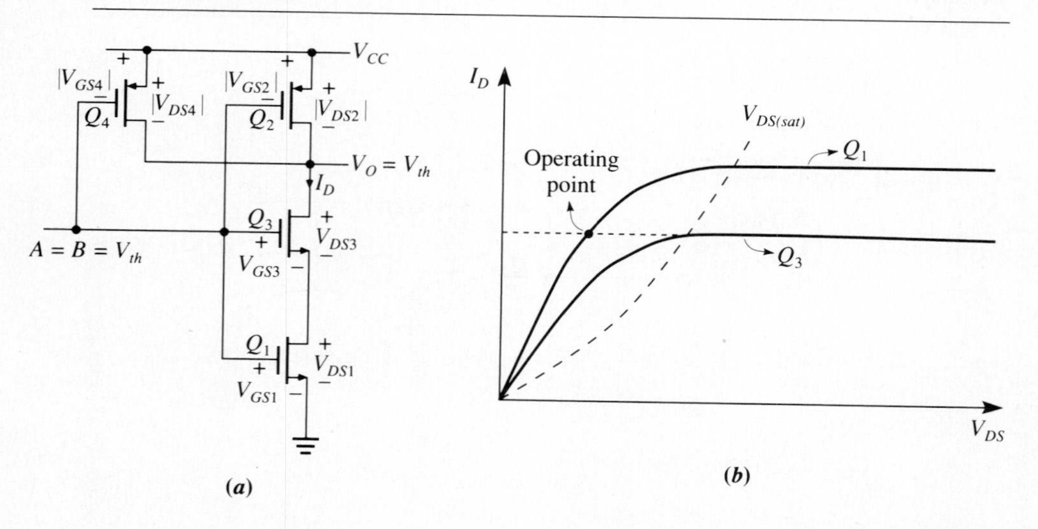

Eliminating V_{DS1} above and solving for V_{th}, we obtain

$$V_{th} = V_{Tn} + 2\sqrt{\frac{I_D}{k_n}} \tag{4.28}$$

On the other hand, the p-channel devices have

$$|V_{GS2}| = |V_{GS4}| = V_{CC} - V_{th} = |V_{DS2}| = |V_{DS4}|$$

so that both Q_2 and Q_4 are saturated with equal drain currents. Therefore, the total current drawn from the supply is

$$I_D = k_p\left(V_{CC} - V_{th} - |V_{Tp}|\right)^2$$

Substituting this into Equation (4.28) and rearranging give the threshold voltage as

$$V_{th} = \frac{V_{Tn} + 2\sqrt{k_p/k_n}\,\left(V_{CC} - |V_{Tp}|\right)}{1 + 2\sqrt{k_p/k_n}} \tag{4.29}$$

For a symmetrical NAND gate with $V_T = V_{Tn} = |V_{Tp}|$ and $k_n = k_p$, Equation (4.29) reduces to

$$V_{th} = \frac{2\,V_{CC} - V_T}{3}$$

which implies a nonsymmetrical VTC.

2. Now, suppose that initially $A = 0$ and $B = V_{CC}$, and the A input is then increased to V_{th}. This case of single-input switching is illustrated in Figure 4.19. The

Figure 4.19 The CMOS NAND gate with single input switching.

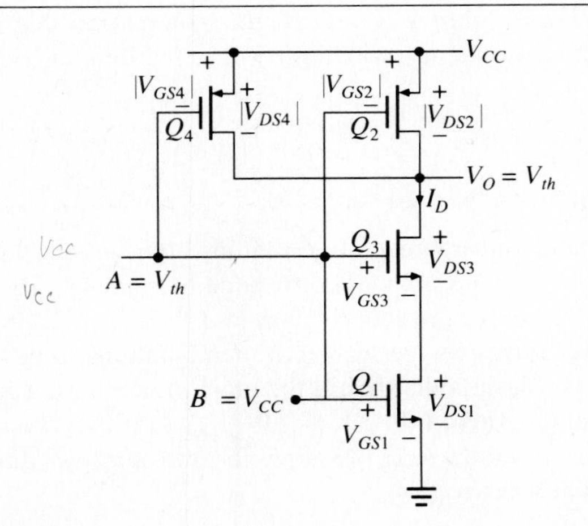

gate voltages of the NMOS devices are $V_{GS1} = V_{CC}$ and $V_{GS3} = V_{th} - V_{DS1}$. Using the output KVL equation yields

$$V_{GS3} = V_{DS3}$$

Thus, Q_3 is saturated while Q_1 is in ohmic region, so that

$$I_D = \frac{1}{2} k_n (V_{th} - V_{Tn} - V_{DS1})^2 = \frac{1}{2} k_n [2(V_{CC} - V_{Tn}) V_{DS1} - V_{DS1}^2] \quad (4.30)$$

On the other hand, since $|V_{GS2}| = 0$, Q_2 is in cutoff while Q_4 is in saturation. Hence, we also have

$$I_D = I_{D4} = \frac{1}{2} k_p (V_{CC} - V_{th} - |V_{Tp}|)^2 \quad (4.31)$$

Eliminating V_{DS1} in Equation (4.30), substituting Equation (4.31) into (4.30), and assuming that $V_{Tn} = |V_{Tp}| = V_T$ result in

$$[1 + 2(x + \sqrt{x})] V_{th}^2 - [(2 + 4x + 2\sqrt{x}) y + 2V_T + 2V_{CC}\sqrt{x}] V_{th} + z = 0 \quad (4.32)$$

where

$$x \equiv \frac{k_p}{k_n}$$

$$y \equiv V_{CC} - V_T$$

and

$$z \equiv 2xy^2 + 2(V_T + \sqrt{x}) y + V_T^2$$

If we further assume that $k_p = k_n$, Equation (4.32) can be solved to give

$$V_{th} = V_{CC} - .6V_T - .2\sqrt{5V_{CC}^2 - 10V_{CC}V_T + 4V_T^2}$$

The set of VTC curves in Figure 4.20 reflects the general trait that V_{th} is the largest for the case of simultaneous input switching. Thus, rearranging Equation (4.29) to determine the ratio,

$$\sqrt{\frac{k_p}{k_n}} = \frac{V_{th} - V_{Tn}}{2(V_{CC} - V_{th} - |V_{Tp}|)}$$

for a particular V_{th} can be one approach to designing a two-input NAND gate. If $V_{Tn} = |V_{Tp}|$, for instance, $V_{th} = V_{CC}/2$ will correspond to $k_n = 4k_p$. However, since the other switching cases will have threshold values less than $V_{CC}/2$, an alternate design procedure can be to take the average V_{th} for all combinations to be $V_{CC}/2$. In high-density CMOS designs, on the other hand, the most important concern is to employ minimum-size transistors, where $(W/L)_n = (W/L)_p$ describes the smallest allowed layout. Taking $\mu_e \approx 2\mu_h$, for the sake of simplicity, the corresponding ratio becomes $k_n/k_p \approx 1/2$, which in turn yields

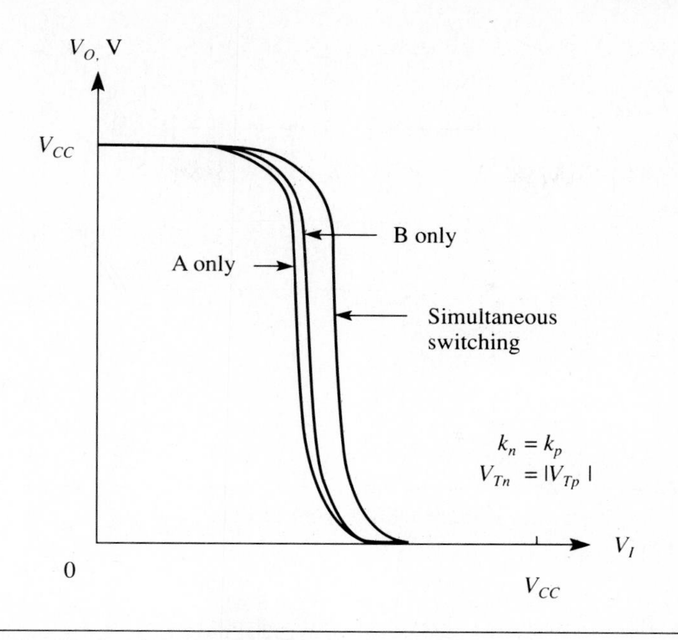

Figure 4.20 Switching properties for a two-input CMOS NAND gate.

$$V_{th} \approx \frac{1.414\,V_{CC} - .414\,V_T}{2.414}$$

for the case of simultaneous switching with $V_T = V_{Tn} = |V_{Tp}|$. We can conclude that the chip-area minimization leads to nonsymmetrical voltage transfer characteristics that depend on the supply voltage. Note also that designing a CMOS circuit with $k_n \neq k_p$ gives rise to nonsymmetrical switching times which should be taken into account in system timing.

Design and Analysis of a Two-Input NOR Gate

Consider the two-input CMOS NOR gate of Figure 4.17b and assume that V_O is initially at $V_{OH} = V_{CC}$, corresponding to **0** at both inputs. Then, as in the case of the NAND gate, there are three possible input combinations that can bring about the output transition $V_O \rightarrow V_{th}$. Each of these cases gives rise to a different value of V_{th}.

1. The circuit arrangement resulting from the simultaneous increasing of inputs from zero up to V_{th} is shown in Figure 4.21. Both NMOS devices are saturated with equal current levels because $V_{GS1} = V_{GS3} = V_{th} = V_{DS1} = V_{DS3}$. The total drain current is obtained as

$$I_D = k_n(V_{th} - V_{Tn})^2 \qquad (4.33)$$

where, as before, k_n describes both n-channel devices. Rearranging Equation (4.33) yields

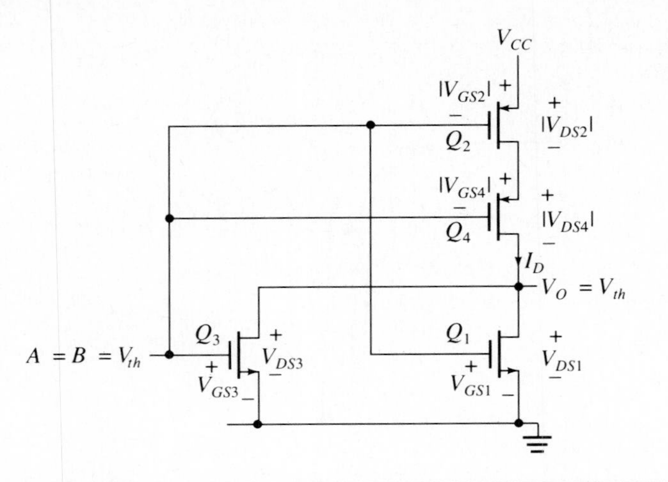

Figure 4.21 Circuit voltages with the CMOS NOR inputs simultaneously.

$$V_{th} = V_{Tn} + \sqrt{\frac{I_D}{k_n}} \tag{4.34}$$

On the other hand, the p-channel MOSFETs have $|V_{GS2}| = V_{CC} - V_{th}$ and $|V_{GS4}| = V_{CC} - V_{th} - |V_{DS2}|$. Substituting the output KVL equation

$$V_{CC} - V_{th} = |V_{DS2}| + |V_{DS4}|$$

into the latter equation reveals $|V_{GS4}| = |V_{DS4}|$, which in turn implies that Q_4 is saturated. Furthermore, with $|V_{GS2}| > |V_{GS4}|$, Q_2 must be conducting in the ohmic region. Equating the expressions for drain currents, we obtain

$$I_D = \frac{1}{2} k_p[2(V_{CC} - V_{th} - |V_{Tp}|)|V_{DS2}| - |V_{DS2}^2|]$$

$$= \frac{1}{2} k_p(V_{CC} - V_{th} - |V_{Tp}| - |V_{DS2}|)^2$$

with k_p being the device transconductance for the PMOS devices. Eliminating $|V_{DS2}|$ produces

$$I_D = \frac{1}{2} k_p(V_{CC} - V_{th} - |V_{Tp}|)^2$$

Substituting this into Equation (4.34) and rearranging, we get

$$V_{th} = \frac{V_{Tn} + \frac{1}{2}\sqrt{k_p/k_n}\,(V_{CC} - |V_{Tp}|)}{1 + \frac{1}{2}\sqrt{k_p/k_n}} \tag{4.35}$$

as the threshold voltage with simultaneous input switching. For $V_T = V_{Tn} = |V_{Tp}|$ and $k_p = k_n$, Equation (4.35) reduces to

$$V_{th} = \frac{V_{CC} + V_T}{3}$$

which also implies a nonsymmetrical VTC, as in the case of the two-input NAND gate.

2. Figure 4.22 depicts one case of single switching at the input with $A = 0$ while B is increased to V_{th}. Since $V_{GS1} = 0$, Q_1 is in cutoff. Meanwhile, Q_3 is saturated with $V_{GS3} = V_{th} = V_{DS3}$. Thus, the current is given by

$$I_D = \frac{1}{2} k_n (V_{th} - V_{Tn})^2 \tag{4.36}$$

The gate voltages of the PMOS transistors are $|V_{GS2}| = V_{CC}$ and $|V_{GS4}| = V_{CC} - V_{th} - |V_{DS2}|$. From the output KVL equation,

$$V_{CC} - V_{th} = |V_{DS2}| + |V_{DS4}|$$

it can easily be seen that $|V_{GS4}| = |V_{DS4}|$, so that Q_4 is saturated, with Q_2 conducting in the ohmic region. Consequently, equating the drain currents produces

$$I_D = \frac{1}{2} k_p \left[2(V_{CC} - |V_{Tp}|)\,|V_{DS2}| - |V_{DS2}^2| \right]$$

$$= \frac{1}{2} k_p (V_{CC} - V_{th} - |V_{Tp}| - |V_{DS2}|)^2 \tag{4.37}$$

Eliminating $|V_{DS2}|$ above and substituting Equation (4.37) into (4.36) yield a quadratic equation for the logic threshold voltage:

Figure 4.22 The CMOS NOR gate with single input switching.

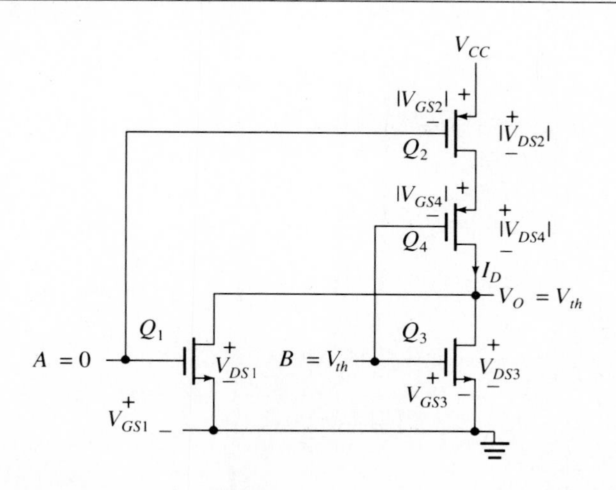

$$\left[1 + 2(x + \sqrt{x})\right]V_{th}^2 - 2V_{Tn}(2x + \sqrt{x})V_{th} - \left[\left(V_{CC} - |V_{Tp}|\right)^2 - 2xV_{Tn}^2\right] = 0 \tag{4.38}$$

where x is as defined before. For a symmetrical NOR gate, Equation (4.38) reduces to

$$V_{th} = .6V_T + .2\sqrt{5V_{CC}^2 - 10V_{CC}V_T + 4V_T^2}$$

The VTC curves in Figure 4.23 illustrate that the lowest V_{th} corresponds to simultaneous input switching, which is the opposite of the situation with the NAND gate.

To design a two-input NOR gate, Equation (4.35) can be rearranged to give

$$\sqrt{\frac{k_p}{k_n}} = \frac{2(V_{th} - V_{Tn})}{V_{CC} - V_{th} - |V_{Tp}|}$$

which in turn may be used to determine the device ratio for a particular V_{th}. For a design goal of $V_{th} = V_{CC}/2$ with $V_{Tn} = |V_{Tp}|$, the equation gives exactly the inverse of the result found for the NAND gate (i.e., $k_p = 4k_n$). However, since the other switching cases will have threshold values above $V_{CC}/2$, an alternate design procedure is to choose the average V_{th} for all combinations as $V_{CC}/2$. For minimum-size MOSFETs with all threshold voltages equal to V_T, the threshold voltage expression reduces to

$$V_{th} \approx \frac{.646V_T + .354V_{CC}}{1.354}$$

Figure 4.23 Switching properties for a two-input CMOS NOR gate.

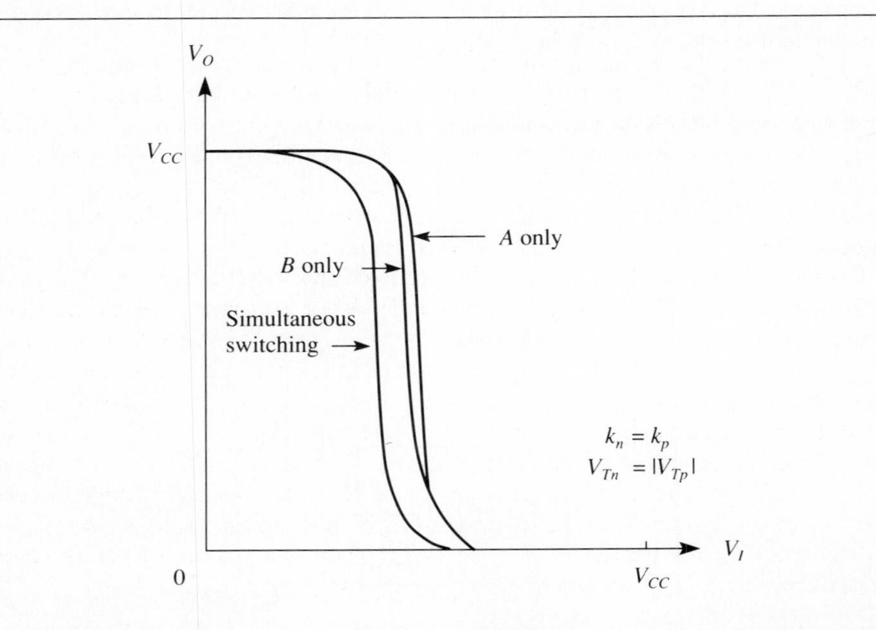

To compare the two-input CMOS NAND and NOR gates, let us assume that our design criterion is to set $V_{th} = V_{CC}/2$ with simultaneous input switching. This choice calls for a NAND gate with $k_n = 4k_p$, and a NOR gate with $k_n = k_p/4$. Again using $\mu_e \approx 2\mu_h$, a NAND design will have

$$\left(\frac{W}{L}\right)_p \approx \frac{1}{2}\left(\frac{W}{L}\right)_n$$

Similarly, a NOR design will require

$$\left(\frac{W}{L}\right)_p \approx 8\left(\frac{W}{L}\right)_n$$

To obtain the smallest layout geometry, the logical choice is to set $(W/L)_p = (W/L)_{min}$ for the NAND gate, where the latter aspect ratio represents the smallest MOSFET dimensions that a particular fabrication process permits. Thus,

$$\left(\frac{W}{L}\right)_n \approx 2\left(\frac{W}{L}\right)_{min}$$

for the n-channel devices of the NAND gate. On the other hand, the NOR gate should have $(W/L)_n = (W/L)_{min}$, so that its p-channel devices will be specified as

$$\left(\frac{W}{L}\right)_p \approx 8\left(\frac{W}{L}\right)_{min}$$

The number of MOSFETs in each gate being equal, the ongoing discussion demonstrates that NAND gates will consume much less area than NOR gates.

Propagation Delay in CMOS

As in the case of power consumption, the speed of a device is a function of the load capacitance. The internal structure also plays a role, so that propagation delay is not zero for zero load capacitance, and an offset term that is unique to the particular device type must be added. Another factor is the value of the power supply. Although they can operate over a 4-volt range to enable versatility to a certain extent, the performance of HCMOS series is, nevertheless, optimized for 5-volt operation. Consequently, lowering the power supply voltage results in increased propagation delays, as illustrated in Figure 4.24 for the 74HC00 two-input NAND gate. As V_{CC} is decreased from 5 V to 2 V, the delay increases by about two times. This is due to the dynamic power dissipation with increased voltage, as discussed above.

The manufacturers supply the propagation delays in data sheets only for a certain value of load capacitance C_L at a certain power supply voltage. In some applications, however, it is desirable to compute the propagation delay under specific load conditions. To calculate the expected delay, the rate of change of the propagation delay with the load capacitance, t_C, must be known. This value, also called the *capacitive delay variation constant*, is used to extrapolate the delay from the value provided in the data sheet to the actual value, C_{act}. Figure 4.25 plots t_C versus power supply voltage for HCMOS devices with standard output structures.

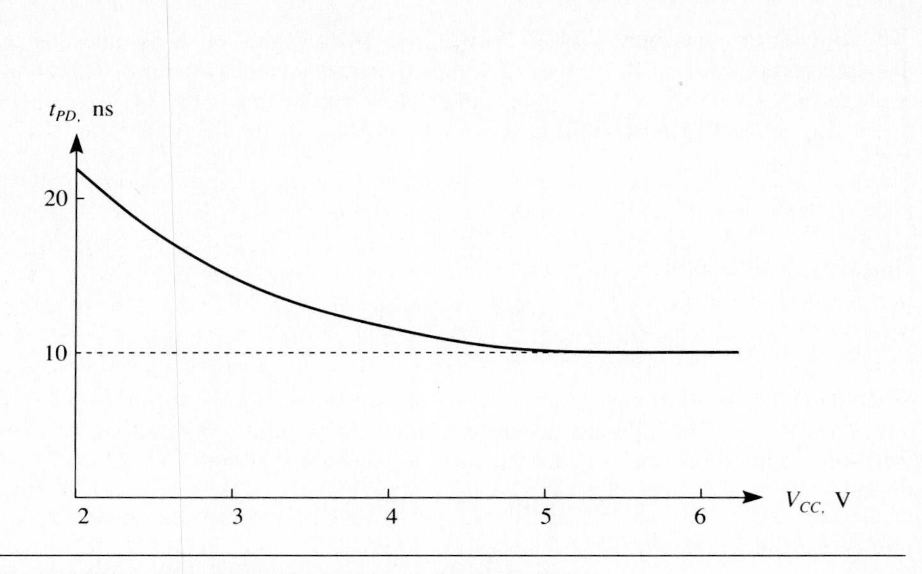

Figure 4.24 Typical propagation delay variation for 74HC00 with power supply voltage.

Figure 4.25 The capacitive delay variation constant t_C as a function of the power supply voltage for the 74HC series.

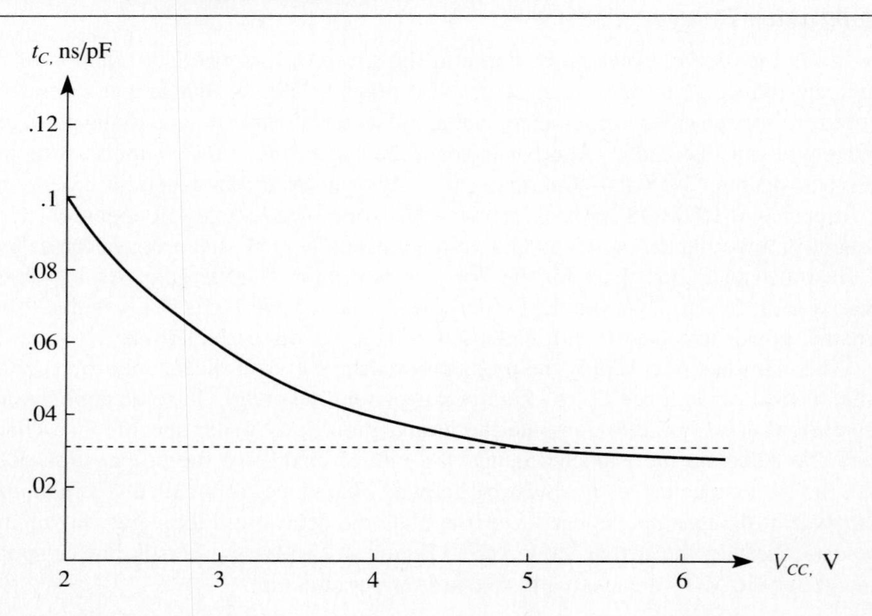

The following expression is used to compute the propagation delay at any load, C_{act}, and power supply, V:

$$t_{PD}(C_{act}, V) = (C_{act} - C_L)t_C + t_{PD}t_V \tag{4.39}$$

where C_L is the value of the load capacitance and t_{PD} is the corresponding propagation delay in nanoseconds as supplied by the data sheets for a certain value of power supply voltage. The parameter t_V is the variation of delay with the operating supply voltage, normalized to $V_{CC} = 5$ V, as shown in Figure 4.26 for the 74HC series. The first term on the right-hand side of Equation (4.39) is the difference in the delay between the actual load, C_{act}, and C_L. The second term is an offset to incorporate the dependence on the internal structure of the propagation delay, and yields the resultant delay at the desired supply voltage with respect to t_{PD}.

Changes in temperature also cause some change in speed. Both metal- and silicon-gate CMOS parts operate slightly slower at elevated temperatures. This is due to the dependence of the carrier mobility, which decreases with increase in temperature, which in turn decreases the transistor gain and increases the delay. For HCMOS, the propagation delay derates fairly linearly from 25°C at around $-.003/°C$, when operated at $V_{CC} = 5$ V with $C_L = 50$ pF. For example, at 125°C the delay increases about 30% from its value at 25°C.

Therefore, to calculate the expected device speed at any temperature, we have

$$t_{PD}(T) = [1 + .003(T - 25)]t_{PD}(25) \tag{4.40}$$

Figure 4.26 Normalized propagation delay variation constant t_V as a function of power supply voltage for 74HC series.

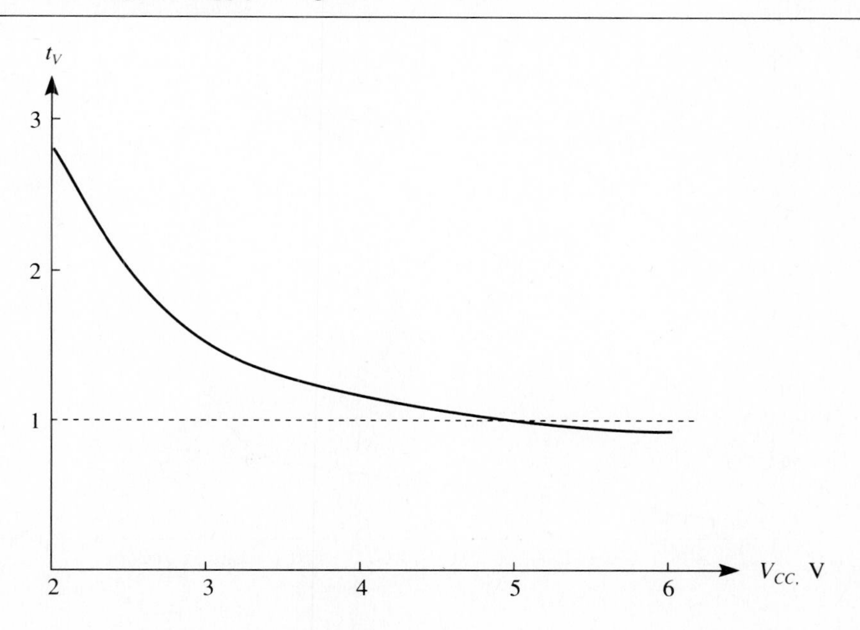

where $t_{PD}(T)$ is the delay at temperature T in °C, and $t_{PD}(25)$ is the delay at room temperature.

Figure 4.27 compares the HCMOS propagation delays to those of the 4000B and 74C series. At 5 V and with a load capacitance of 15 pF, the HCMOS parts are about ten times faster than the buffered 4000 series, and about five times faster than the unbuffered 74C. Note that the delay differential between metal-gate and silicon-gate CMOS increases as C_L increases.

Example 4.7

To calculate the propagation delay for the 74HC00 at $V_{CC} = 6$ V with a 15 pF load, we see from the data sheets that the typical propagation delay t_{PD} is 8 ns for $C_L = 15$ pF at $V_{CC} = 5$ V. Then, from Figure 4.26, we find $t_V = .8$ ns at $V_{CC} = 6$ V. Using Equation (4.39), we obtain

$$t_{PD}(15 \text{ pF}, 6 \text{ V}) = 0 + 8(.8) = 6.4 \text{ ns}$$

For a 100 pF load, from Figure 4.25, we have $t_C = 38$ ps/pF at $V_{CC} = 6$ V. Substituting into Equation (4.39), we get

$$t_{PD}(100 \text{ pF}, 6 \text{ V}) = .038(100 - 15) + 8(.8) = 9.63 \text{ ns}$$

Figure 4.27 Propagation delay vs. load capacitance for two-input NAND gates.

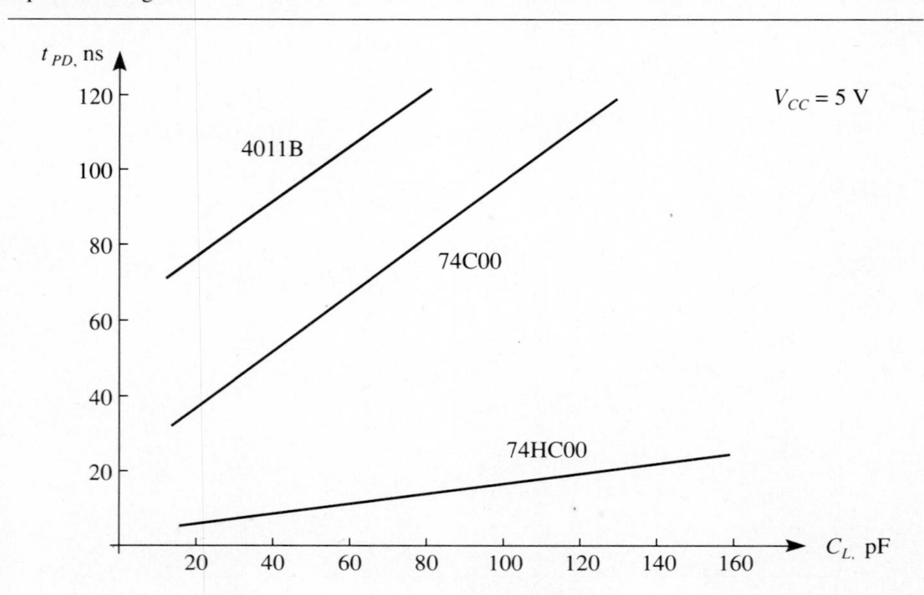

Noise Performance

As we already know, noise immunity is the maximum amount of voltage that can be applied to the input without causing the output to change state. Ideally, it should equal 50% of the logic swing, that is, $V_{CC}/2$. In reality, however, the logic swing at the output is given by $V_{OH} - V_{OL}$.

For standard CMOS circuits, the typical noise immunity does not change with voltage, and is 45% of V_{CC}. This means that a spurious input that is $.45\,V_{CC}$ away from the supply voltage value or ground will not propagate through the system. The existence of mismatch between the *p*- and *n*-channels does not allow the switching to occur at exactly $V_{CC}/2$, as discussed above in detail. Therefore, manufacturers specify the two limits of the transfer characteristics in data sheets. Since, depending on the operating point, MOSFETs are either voltage-controlled resistors or current sources, the transfer characteristic and hence the noise immunity of a CMOS device are determined by the parallel-series combination of the transistor impedances in conjunction with the input voltages, number of inputs, and the gate circuit configuration. For different input conditions, different combinations of the impedances of *n*-channel and *p*-channel transistors come into play. These different situations give rise to varying transfer characteristics, as a result of which NM_L decreases and hence NM_H increases (or vice versa) as the number of inputs at the logical **1** level increases.

Manufacturers also guarantee that metal-gate CMOS circuits have a 1-V dc noise margin over the full power supply and temperature range, as shown in Figure 4.28.

Figure 4.28 Guaranteed noise margins for metal-gate CMOS circuits as a function of V_{CC}.

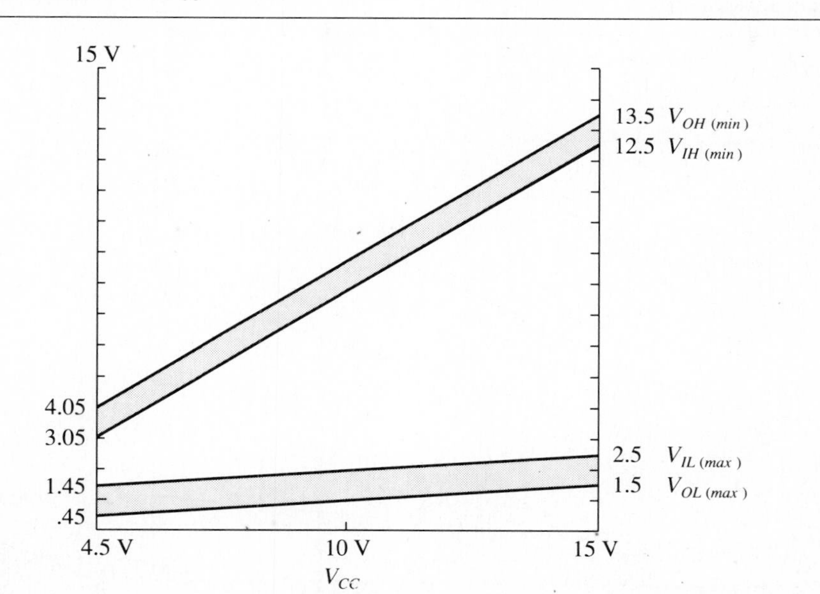

Silicon-gate CMOS series also enjoy the same wide noise margins. For these devices, V_{OH} is typically .1 V below V_{CC} and V_{OL} is .1 V above ground. The worst-case input and output voltages over the operating supply range are shown in Figure 4.29. Over the entire range, except when $V_{CC} = 2$ V, $V_{IH} = .7\,V_{CC}$ and $V_{IL} = .2\,V_{CC}$. The reason for the exception is that at low voltages the transistors' threshold voltages start playing significant roles.

The excellent noise immunity of the 74HC series can be seen from the transfer characteristic of the 74HC00 two-input NAND gate in Figure 4.30. It has a very sharp transition around 2.25 V that is very stable with temperature.

The 74HCT ICs have input buffers specifically designed to yield guaranteed TTL input levels. Hence, their noise characteristics are substantially different from those of 74HC devices. For the 74HCT series, the nominal transition voltage has been designed to be around 1.4 V, as compared to the 2.25-volt value for 74HC series. While the transition for the latter is set to offer optimal noise margins for both logic levels, the trip point set for the 74HCT degrades the **0**-level noise margin by almost a volt. This can be explained if one considers the fact that the 74HCT is actually designed to be interfaced with TTL. Figure 4.31 compares the noise performances of the 74HC and 74HCT series.

The metal-gate CMOS subfamilies (the 4000, 4000B, and 74C series) are highly immune to certain types of system noise. Although this is partly due to the nature of the CMOS, their relatively slow speeds reduce the self-induced supply noise and crosstalk, thus preventing the circuit from responding to short, externally-generated transient noise.

Figure 4.29 Guaranteed noise margins for the 74HC series as a function of V_{CC}.

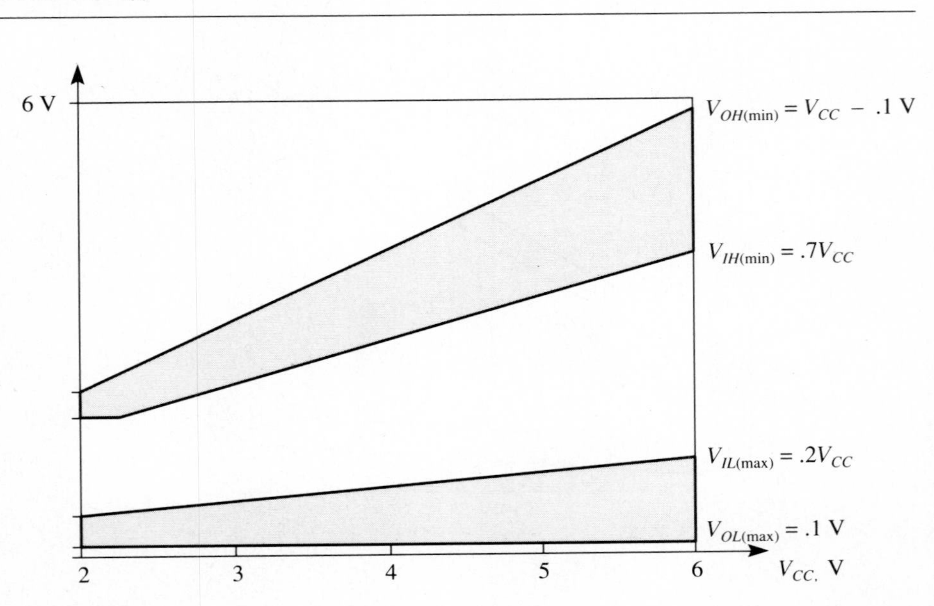

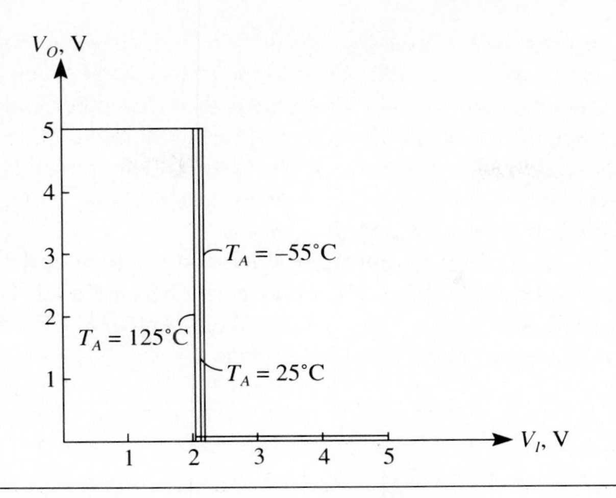

Figure 4.30 The VTC of 74HC00 quad two-input NAND gate.

On the other hand, in a high-speed silicon-gate CMOS, the *crosstalk*, induced supply noise, and noise transients become important factors that degrade the performance of the device. Being much faster than the older series of CMOS devices, the 74HC not only responds more quickly to these types of noise sources but tends to emphasize the parasitic interconnection capacitances, which in turn increase the self-

Figure 4.31 Comparison of guaranteed noise margins for the 74HC and 74HCT CMOS series at V_{CC} = 4.5 V.

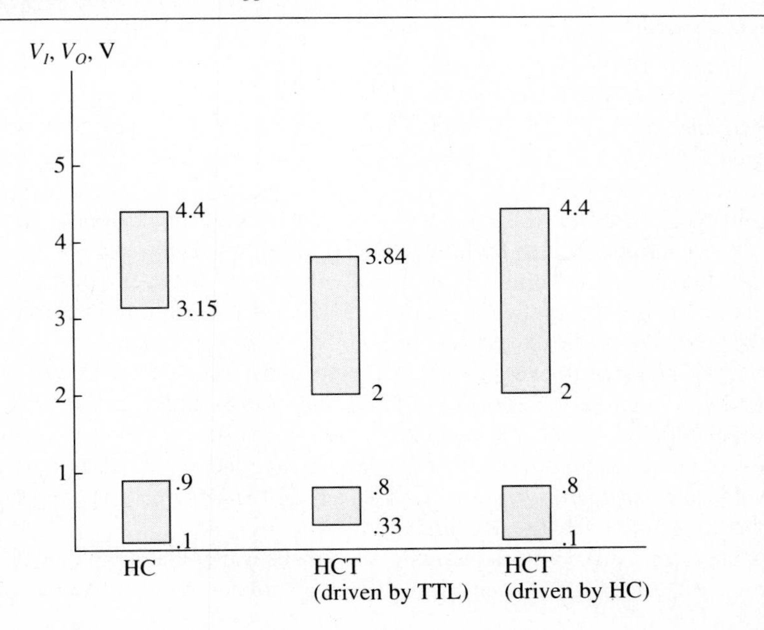

induced noise and crosstalk caused by capacitive or inductive coupling of extraneous voltages from one signal line to another or to the power supply line. However, since the amount of noise voltage coupled by capacitively induced currents is directly proportional to the output impedance of the driver, a high-speed CMOS device is much less susceptible to this type of noise simply because its output impedance is one-tenth that of a device in the 4000 series. Consequently, lower stray voltages are induced for a given amount of current coupling.

There is also *ringing* that results from signals propagating down improperly terminated transmission or signal lines. It occurs partly because a CMOS gate switches back and forth from a very high impedance to a very low one. Chapter 11 presents a more detailed treatment of noise in digital circuits.

Example 4.8

For the CMOS inverter of Example 4.2, to determine NM_L and NM_H when $V_{CC} = 5$ V, we consider the following two cases:

Case 1: Q_1 is off and Q_2 is saturated so that $V_O = V_{OH} = V_{CC} = 5$ V. The maximum value of input voltage for Q_1 to stay off is V_T. Therefore, $V_{IL} = V_T = 1$ V.

Case 2: Q_2 is off and Q_1 is saturated so that $V_O = V_{OL} = 0$ V. The minimum value of the input voltage for Q_2 to stay off is $V_{IH} = V_{CC} - V_T = 4$ V. Hence, we get

$$NM_L = 1 - 0 = 1 \text{ V}$$
$$NM_H = 5 - 4 = 1 \text{ V}$$

Input Protection

In a MOSFET, the extremely thin ($\approx$ 1000 Å with metal gates and less than 500 Å with silicon gates) SiO_2 layer beneath the gate, acting as an insulator, has a typical dielectric strength of around 10^7 V/cm. Defects in the gate oxide can easily reduce it below 3×10^6 V/cm, causing it to break down when the gate-to-source voltage, V_{GS}, exceeds 100 V. A typical input capacitance smaller than 5 pF in parallel with a typical 10^{12}-Ω resistance will form a very high input impedance, which will be very much inclined to build electrostatic charges. Thus, even a small amount of static charge accumulating on the input capacitance between source and gate may result in a large enough field to rupture the dielectric, resulting in permanent damage to the oxide due to the excessive current flow. A person walking across the laboratory can easily generate in excess of 12,000 volts under suitable conditions. If this person touches the input terminal of a MOSFET, the energy stored in the body capacitance will be sufficient to transfer enough voltage to the device to cause breakdown. The ambient relative humidity has an especially great effect on the amount of static charge developed because the moisture provides a useful leakage path to the ground, thereby reducing the static charge accumulation.

Therefore, protective networks, which make use of clamping diodes at the inputs and outputs of CMOS devices, are utilized to protect the gate oxide against *electro-*

static discharge (*ESD*) to a certain extent. The purpose of these diodes is to prevent the incoming signal from rising above the supply voltage V_{CC} by more than one forward diode voltage drop and from falling below ground voltage by the same amount. Figure 4.32 shows one such circuit, which is mainly used in metal-gate CMOS devices. The input protection circuit consists of a series isolation resistor R_s, whose typical value is around 250 Ω, and diodes D_1 and D_2, which clamp the input voltages as discussed above. R_s is used to limit the current flow when the input is subjected to a high-voltage static discharge. Diode D_1 connected to V_{CC} protects against positive transients, while D_2 shunts the negative surges. Diode D_3 is a distributed structure resulting from the diffusion fabrication of R_s and does not actually contribute to the protection of the gate. The ESD protection for the output consists of a diffused diode D_4 from the output to the power supply voltage. The other diodes are intrinsic.

The input and output protection for a silicon-gate CMOS inverter is shown in Figure 4.33. A polysilicon resistor is used to isolate the input diodes more effectively than the diode-resistor network used in the metal-gate CMOS. It slows down the input transient and partially dissipates its energy. The input diodes connected to the resistor clamp the input spike and prevent large voltages from appearing across the transistors. They are larger than those used in a metal-gate CMOS in order to be able to shunt greater current. The parasitic output diodes isolate the drains from the substrate and clamp large voltages that may appear across the output. Figure 4.34 shows the cross section of the inverter of Figure 4.33.

Fan-Out

Since the input current to a CMOS gate is extremely low for both logic states due to high input impedances, the fan-out for CMOS devices, in general, is unlimited for all practical purposes. However, if dynamic performance is taken into account, input capacitances deteriorate switching speeds at high frequencies, and the input rise time becomes a limiting constraint, as discussed in this section for the 74HC series.

For the HCMOS, a total of 15 pF—10 pF for the load and 5 pF of stray capac-

Figure 4.32 Typical input and output protections for metal-gate CMOS parts.

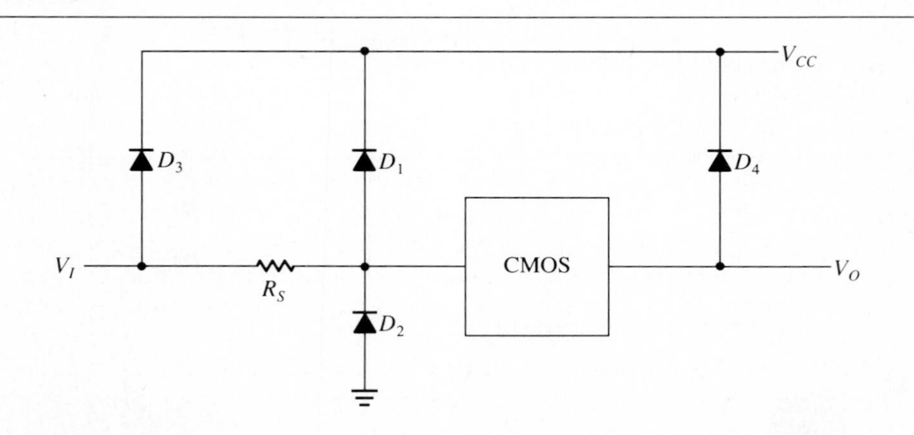

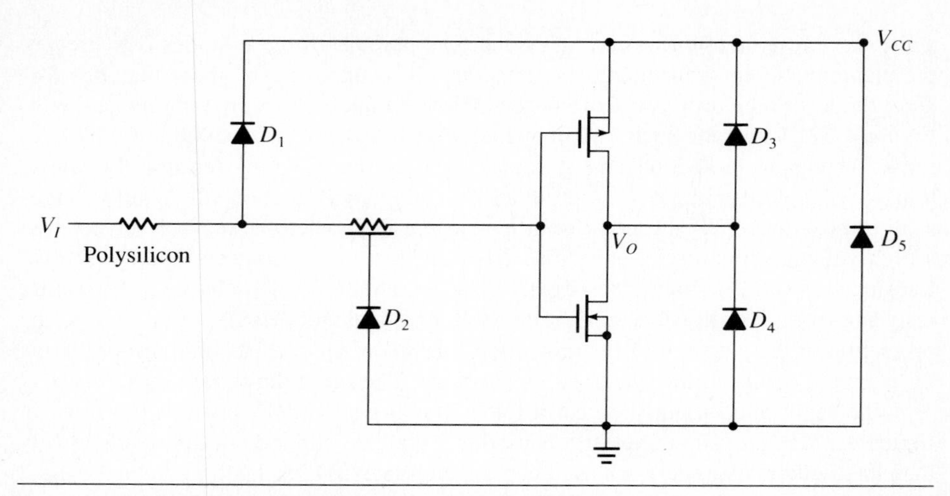

Figure 4.33 Typical input and output protections for a silicon-gate CMOS inverter.

itance—is assumed to be seen by the driver under worst-case conditions. The typical input capacitance is actually 3 pF. The input resistance, r_i, can then be approximated as

$$r_i = \frac{V_I}{I_I}$$

where V_I is the input voltage and I_I is the input current. V_I is equated to $V_{CC} = 6$ V for test conditions, and the typical value for I_I is taken as .1 nA. Therefore, the

Figure 4.34 Simplified cross section of the silicon-gate CMOS inverter with input and output protections.

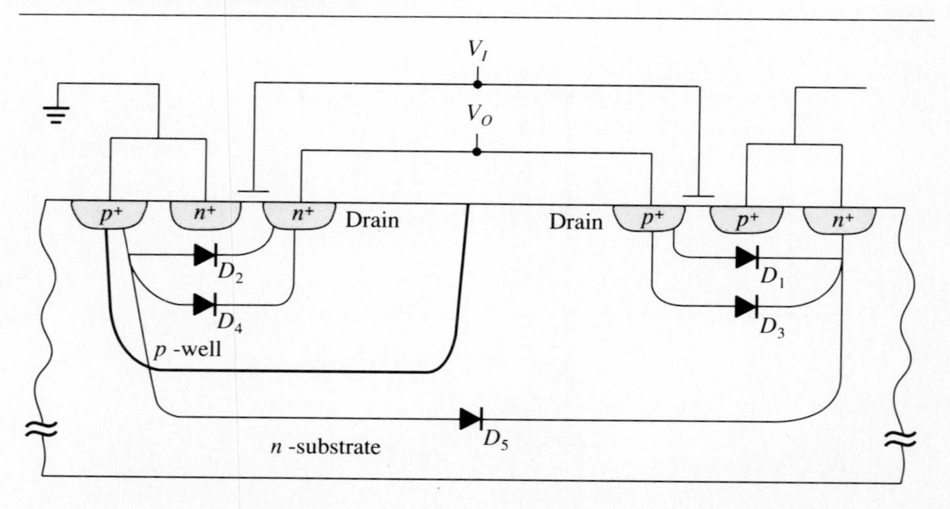

calculated input resistance is

$$r_i = \frac{6\text{ V}}{.1\text{ nA}} = 60\text{ M}\Omega$$

The maximum output resistance can be calculated in a similar manner using the following equation:

$$r_o = \frac{V_{CC} - V_{OH}}{I_{OH}}$$

where V_{OH} is typically given as 4.3 V at $V_{CC} = 4.5$ V, and $I_{OH} = 4$ mA. Hence, the maximum output resistance is approximately 50 Ω. Figure 4.35 illustrates the worst-case input and output models of the HCMOS during rise time.

As depicted in Figure 4.36, for a fan-out of n HCMOS devices, the total load capacitance will be $C_T = nC_L$ since the capacitances will all be in parallel. When the driver output switches from low to high level, the input capacitances of all driven devices must be charged up to V_{IH} within the maximum recommended rise time of 500 ns as specified in data sheets. Since $r_i >> r_o$, the voltage across the capacitors can be expressed as

$$v(t) = V_{OH}\,(1 - e^{-t/r_oC_T})$$

The value $V_{IH} = 3.15$ V will be reached at $t = t_t$, the time at which the transition occurs. Therefore,

$$V_{IH} = V_{OH}\,(1 - e^{-t_t/r_onC_L}) \tag{4.41}$$

Rearranging the terms and taking the natural logarithm of both sides, we get an expression for the fan-out n as

$$n = \frac{(-t_t/r_oC_L)}{\ln\,[1 - V_{IH}/V_{OH}]} \tag{4.42}$$

Substituting the numerical values, we obtain the maximum fan-out as $n = 505$.

Figure 4.35 The worst-case models of high-speed CMOS: (a) input, and (b) output.

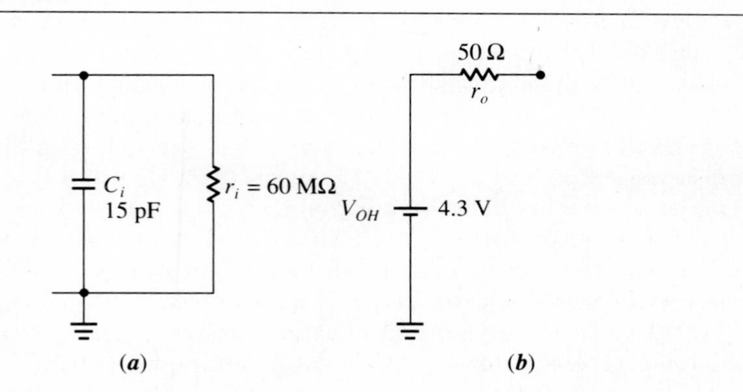

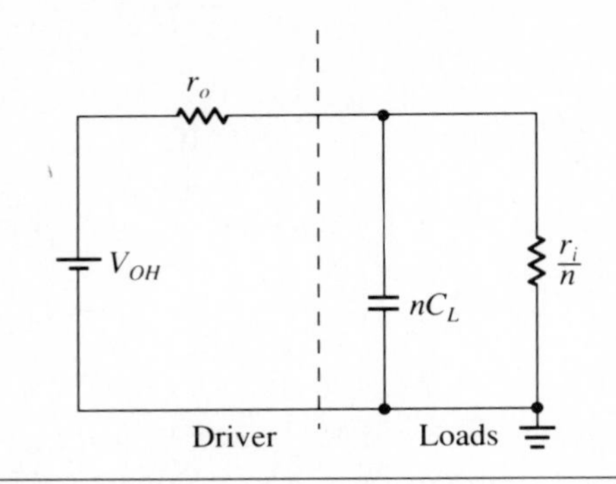

Figure 4.36 Model for the output stage of an HCMOS part during rise time, driving n loads.

Example 4.9

To determine the propagation delay within the driver caused by each HCMOS load added to the fan-out, consider Equation (4.41) and solve for $t = t_{PD}$ to yield the desired result:

$$t_{PD} = (-nr_oC_L)\ \ln\left[1 - \frac{V_{IH}}{V_{OH}}\right] \tag{4.43}$$

Substituting the values, we obtain

$$t_{PD} = 1\ \text{ns}$$

Therefore, each added load will increase the propagation delay by about 1 ns at $V_{CC} =$ 4.5 V which corresponds to approximately 67 ps/pF of added delay.

4.4 The Advanced CMOS Logic

The next step in the evolution of the CMOS family was the *advanced CMOS logic* (*ACL*). Featuring gate propagation delays of around 5 ns, as compared to a typical 8 ns for the high-speed CMOS, the ACL is the fastest CMOS logic yet available and matches the speed performance of the advanced Schottky bipolar subfamilies at CMOS power levels. Just as the HCMOS became an industry standard competing with LS TTL, the ACL is expected to challenge the AS/F/ALS subfamilies.

The output drive capability of ±24 mA, compared with ±4 mA for the HC/HCT, enables the subfamily to drive transmission lines while still generating voltage levels necessary to operate the loads safely. The TTL-compatible 54/74ACT series is designed to interface with TTL outputs operating with a $V_{CC} = 5\ \text{V} \pm 10\%$, but the devices are functional over the entire operating voltage range of 2–6 V. They have buffered outputs that will drive CMOS or TTL devices with no additional interface circuitry. The output characteristics for the 74AC00 two-input NAND gate with various power supply voltages are given in Figure 4.37.

All ACL outputs are buffered to ensure consistent output voltage and current specifications across the family. Two clamp diodes are internally connected to the

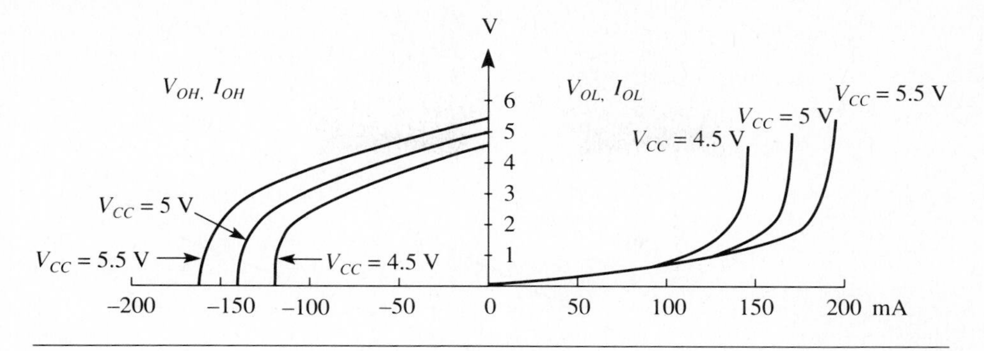

Figure 4.37 The output characteristics of the 74AC00 NAND gate with various supply voltages.

output pin to improve the impedance matching with other device inputs from the same series and to suppress the voltage overshoot and undershoot in noisy applications. The balanced output design allows for controlled edge rates and equal rise and fall times. The worst-case current values for the 74 line of commercial grade parts are given as $I_{OL} = 86$ mA and $I_{OH} = -75$ mA when $V_{CC} = 5.5$ V and $T_A = 85°$C. At room temperature, the parts can typically withstand dynamic current forced into or out of the outputs of over 450 mA, thereby virtually eliminating the latch-up problem.

For propagation delays, capacitive loading effects should be taken into account in addition to temperature and power supply effects. Figure 4.38 shows the effects of the capacitive loading on propagation delay times for the 74AC00. Normalized propagation delay as a function of the supply voltage for the ACL subfamily is illustrated in Figure 4.39. Finally, Table 4.3 summarizes the electrical characteristics of the HCMOS and ACL subfamilies.

The Pinout Controversy

Having reasoned that the latest CMOS subfamilies were too fast to perform properly using the traditional and familiar *dual-in-line package* (*DIP*) pinouts, Texas Instruments and several other manufacturers implement a new packaging scheme, leading to revised pinout patterns. On the other hand, other leading manufacturers, such as Fairchild and RCA, hold on to the conventional pinout setup.

In recent years, the glitches, reflections, and ringing induced on printed circuit boards and in logic packages by faster and faster TTL switching has become a major problem for logic system designers. Yet, the system noise generated by the bipolar logic was small enough to be readily overcome. However, CMOS devices with gate lengths of around a micron, which make them the equal of bipolar parts in both the speed and current drive capability, made the switching glitches worse in some cases, in spite of their much higher noise margins, resulting in the CMOS logic operating marginally in the very applications in which it was designed to excel, that is, in those requiring high speed and drive.

Contented that the device speeds outran the packaging technology, Texas Instruments, Signetics Corporation, and Signetics's parent, N. V. Philips, agreed to shorten the power and ground paths within the chips and packages by not only relocating the

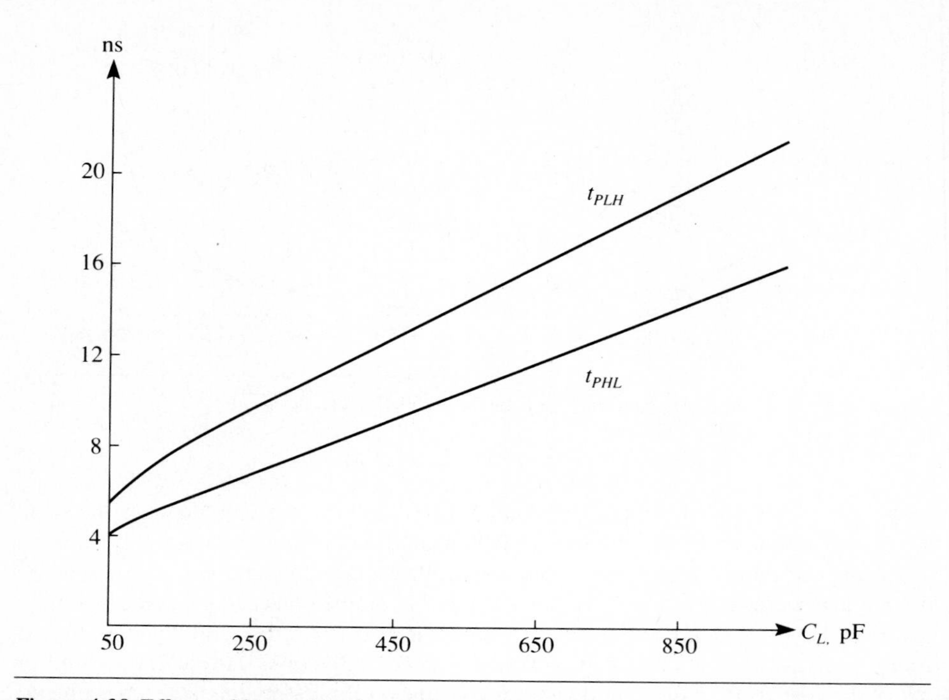

Figure 4.38 Effects of lumped load capacitance on propagation delay times of 74AC00 at V_{CC} = 5 V.

Figure 4.39 Normalized propagation delay as a function of the supply voltage for the ACL subfamily.

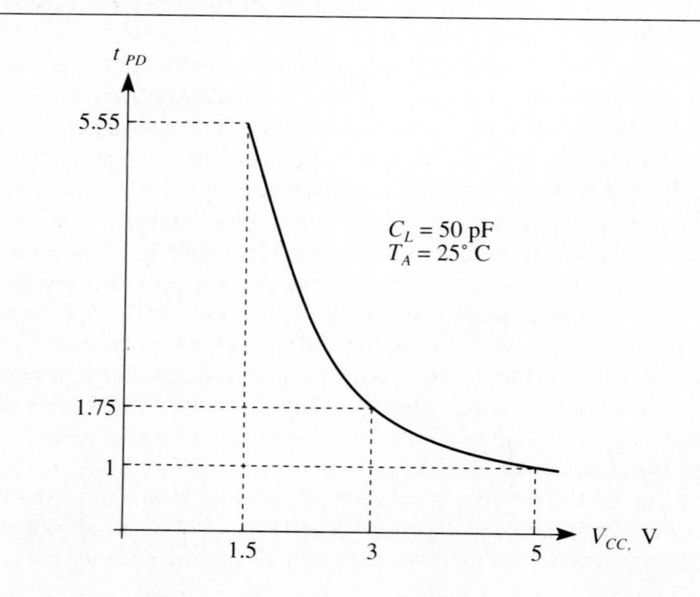

Table 4.3 Key Performance Parameters of Commercial Grade HCMOS and ACL Subfamilies at $V_{CC} = 4.5$ V

	HCMOS	ACL
V_{OH}, V	4.4	4.4
V_{IH}, V	3.15	3.15
NM_H, V	1.25	1.25
V_{OL}, V	.1	.1
V_{IL}, V	.9	1.35
NM_L, V	1.25	1.25
I_{OL}, mA	4	24
I_{IL}, μA	−1	−1
I_{OH}, mA	−4	−24
I_{IH}, μA	1	1
t_{PD}*, ns	8	3
Power/gate:		
static, μW	2.5	2.5
at 1 MHz*, mW	1.75	1.75
SP, pJ at 1 MHz	14	9

*$C_L = 50$ pF

corresponding pins but including multiple power and ground pins in larger packages as well. On the other hand, Fairchild and RCA argued that there was no need for another TTL or CMOS logic series because the logic was being absorbed into larger LSI and VLSI chips, leaving little *glue logic* for the future. By the mid-nineties, it is estimated that discrete logic will not be able to provide the cost and performance required. The push for more speed in discrete logic functions will probably be in *emitter-coupled logic* (*ECL*), which eliminates the glitch problems. Consequently, the latter group believe that the ACL will be the last logic line, so that it does not make sense to add a new packaging scheme. Moreover, they hold that glitches are a problem only in rarely encountered situations, and such a limited problem does not justify changing pinouts on the entire series of parts. Finally, they argue that the problems arise only when DIP packages are used. Moving to *surface mountable* units eliminates the glitches. Therefore, they opted to keep the edge speeds to 3 ns to get the subfamily around a number of board-level design problems like transmission effects, which arise when faster rates are used.

Besides new ground- and power-pin placements in the AC/ACT series, the new pinout scheme changes the input, output, and control-pin strategy. Texas Instruments calls the new arrangement a *flow-through architecture*. Similar schemes had already been proposed by Intel Corporation for large memory chips and could ironically be seen in certain high performance products such as Fairchild's 100K ECL and 54F/74F series. In this architecture, the power and input pins are located on the right side of the package, the ground and output pins are on the left, and the control pins are situated as the end pins. Therefore, in a 14-pin package, the ground pin is 4 and the power is on 11. The power pin moves to 12 on the 16-pin package if the chip has one or two outputs. If the 16-pin package has three or four outputs, then the grounds are on pins 4 and 5, and the supply voltage is on pins 12 and 13. The part-number designation system has also been altered to indicate the new pinout scheme. An *11* has been added to the numerical function indicator. For example, the two-input NAND gate is identified as 74AC1100.

Summary

Today's VLSI ICs almost exclusively employ CMOS gates. A CMOS gate uses an enhancement-mode NMOS for pull-down and an enhancement mode PMOS for pull-up in a complementary manner. These two types of devices necessitate two electrically isolated regions. The standard technique is to form an *n*-well in a *p*-type substrate or a *p*-well in an *n*-type substrate.

The CMOS is a fully restored logic. The output settles at either V_{CC} or ground. When a CMOS gate is in steady state, it carries only the leakage current of the off device, so that static power dissipation is negligible. However, the dynamic power loss during switching is directly proportional to the switching frequency. CMOS inputs have an extremely large impedance that draws essentially no current from the signal source. However, the fan-out is limited by the input capacitance. The driver must charge and discharge the parallel combination of each capacitive load, so that the switching time deteriorates in proportion to the number of loads being driven.

The dominant problem in CMOS circuits used to be device latch-up, in which an internal feedback due to the presence of parasitic bipolar junction transistors led to permanent loss of circuit operation. Improved design techniques and advances in device fabrication technology have eliminated this problem.

There are several different series in the digital CMOS family. Devices in the latest subfamily, ACL, attain speeds similar to that of the fastest TTL series while retaining the advantages of the CMOS logic: very low static power and high noise immunity.

References

1. J. P. Uyemura. *Fundamentals of MOS Digital Integrated Circuits.* Addison-Wesley, Reading, MA: 1988.
2. A. Mukherjee. *Introduction to nMOS & CMOS VLSI Systems Design.* Prentice-Hall, Englewood Cliffs, NJ: 1986.
3. N. Weste and K. Eshraghian. *Principles of CMOS VLSI Design, A System Perspective.* Addison-Wesley, Reading, MA: 1985.
4. Fairchild Digital Unit. *Fairchild Advanced CMOS Technology Logic Data Book.* So. Portland, ME: 1985.
5. V. Kulkarni. "DC Noise Immunity of CMOS Logic Gates." *Application Note 377,* National Semiconductor, Santa Clara, CA: July 1984.
6. L. Wakeman. "AC Characteristics of High-Speed CMOS." *Application Note 317,* National Semiconductor, Santa Clara, CA: June 1983.

7. L. Wakeman. "DC Electrical Characteristics of High-Speed CMOS Logic." *Application Note 313*, National Semiconductor, Santa Clara, CA: June 1983.
8. K. Karakotsios. "High-Speed CMOS Processing." *Application Note 310*, National Semiconductor, Santa Clara, CA: June 1983.
9. S. Calebotta. "CMOS, the Ideal Logic Family." *Application Note 77*, National Semiconductor, Santa Clara, CA: January 1983.
10. T. P. Redfern. "54C/74C Family Characteristics." *Application Note 90*, National Semiconductor, Santa Clara, CA: August 1973.

PROBLEMS

For the following problems, assume a room temperature of 300 K, $V_{CC} = 5$ V, and $V_T = V_{Tn} = |V_{Tp}| = 1$ V unless otherwise stated.

4.1 A CMOS network for any arbitrary combinational function can easily be derived by first splitting the circuit into two subcircuits as shown below. The top subcircuit Z, made up of p-channel devices, provides transmission paths for all input combinations for which the output f is logical **1.** Similarly, the bottom subcircuit, $\overline{Z}$, of n-channel devices provides paths for all input combinations for which f is **0.** Also note from DeMorgan's theorem that the complement of any Boolean expression is obtained by complementing the variables and by replacing the AND operations with the OR operations and vice versa. Hence, if f is specified as a sum-of-products form, we first derive the $\overline{Z}$ network from $\overline{f}$ in product-of-sums form and then use the aforementioned theorem to implement Z, except that it is not necessary to complement the input variables because the complementary p-channel transistors are employed for the Z subcircuit. Therefore, the dual of $\overline{Z}$ is actually used to realize Z. Verify this procedure by deriving the two-input NAND and NOR gates of Figure 4.17.

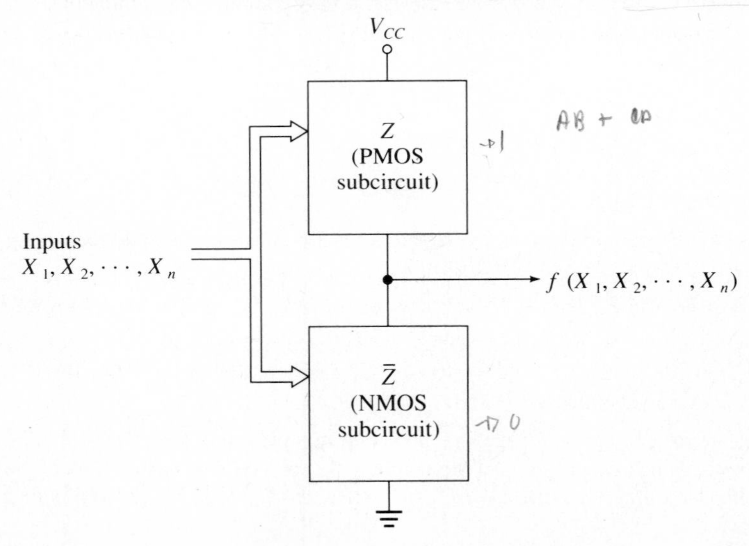

4.2 Using the procedure described in Problem 4.1, draw the schematic of a CMOS
a. AND-OR-INVERT gate, $f = \overline{ab + cd}$
b. OR-AND-INVERT gate, $f = \overline{(a + b)(c + d)}$

4.3 a. The majority function

$$M(a, b, c) = ab + bc + ca$$

represents the carry output bit from an adder stage, where c is the carry-in, and a and b are the input bits to the stage. Draw the schematic for the CMOS implementation of $\overline{M}$.

b. An interesting property of M is that it is self-dual, that is, interchanging the logical sum with logical product yields the same function:

$$M = (a + b)(b + c)(c + a)$$

Realize the CMOS implementation of $\overline{M}$, this time starting with the above expression.

4.4 Construct a truth table for the CMOS circuit shown in the diagram realizing the three-input logical function $Y(A, B, C)$.

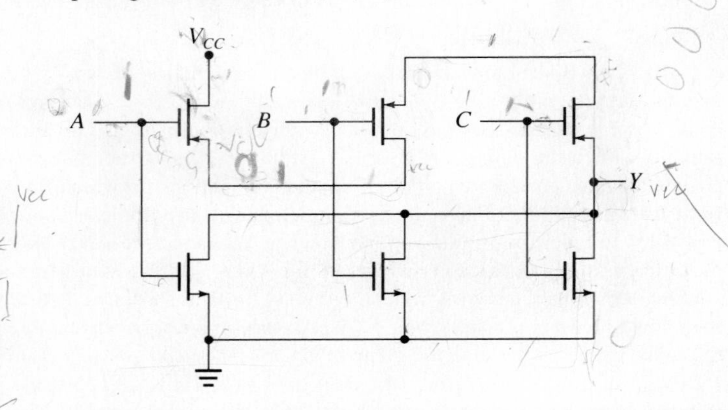

4.5 For a CMOS inverter whose transistors have identical but complementary characteristics and threshold voltages, use SPICE to find its VTC when the following are true:
a. $V_{CC} < V_T$
b. $V_{CC} = V_T$
c. $V_T < V_{CC} < 2V_T$
d. $V_{CC} > V_T$
e. $V_{CC} > 2V_T$

Choose appropriate model parameters for the transistors, and assume $V_T = 2$ V.

4.6 For the basic CMOS inverter, find the equivalent resistance, R_{off}, of the p-channel transistor in the logical **1** state, assuming that $2(V_{CC} - V_T) \gg |V_{DS}|$.

4.7 Consider a CMOS inverter with $k_p = 5k_n = .0625$ mA/V^2. A signal having linear ramp edges, with $t_r = t_f = 10$ ns, has been applied at the input. Find I_{SW} if the inverter is operated at 10 MHz.

4.8 Give the expression for the total power consumption of a 4-bit CMOS binary counter with a clock frequency f_o. Then compute the total power consumption of the following 4-bit binary counters with $f_o = 1$ MHz and $C_L = 50$ pF on each output:

a. 74C161, $I_{CC} = .05\ \mu A$, $C_{PD} = 95$ pF,
b. 74HC161, $I_{CC} = 8\ \mu A$, $C_{PD} = 90$ pF

4.9 For the basic CMOS inverter, assume $V_I = 5$ V, $N_a = N_d = 10^{15}$ cm^{-3} for substrate dopings, and $N_a = N_d = 10^{19}$ cm^{-3} for source and drain dopings. Also, $W/L = 100$, $x_{ox} = 125$ nm, and $I_S = 10^{-12}$ A for both devices.
a. Compute the output voltage.
b. If the inverter is loaded with an identical driver and its input is switched from 5 to 0 V, what is the total charge supplied to the load assuming that the Si-SiO$_2$ interface has a surface area of 10^{-3} cm^2? Which device supplies the charge?

4.10 For a CMOS inverter intended for SSI circuit applications with $(W/L)_p = 2(W/L)_n = 40$, and $\mu_e C_{ox} = 2\mu_h C_{ox} = 20\ \mu A/V^2$, do the following:
a. Find the maximum current that it can sink while V_O remains $\leq .2$ V.
b. Determine the peak current drawn from the supply during switching.
c. What are the maximum source and sink currents for which V_{OH} and V_{OL} deviate from the ideal rail-to-rail values by at most $.1V_{CC}$?
d. Suppose that the inverter is loaded by $C_L = 15$ pF. Compute the dynamic power at 2 MHz. What is the average current drawn from the supply?
e. Find t_{PD} for the load capacitance of part (d).

4.11 Suppose that, for the inverter of Problem 4.10, a 1-KHz sawtooth waveform at the input rises from 0 V to 5 V in 50 μs.
a. Use SPICE to plot the waveforms of the inverter current and output voltage.
b. Compute the total charge that flows in the switch in one cycle.
c. Determine the average switching current and power dissipation.

4.12 For a VLSI CMOS inverter with $(W/L)_p = 2(W/L)_n = 4$, $\mu_e C_{ox} = 2\mu_h C_{ox} = 20\ \mu A/V^2$, and $C_L = 100$ fF, determine the following:
a. t_{PLH}
b. t_{PHL}
c. t_{PD}
d. SP at 40 MHz

4.13 Suppose that $L = 2\ \mu$m and $(W/L)_n = 2$ for a certain CMOS technology. Find the width, W, of each of the NMOS and PMOS transistors and the total area occupied by all devices for the following:
a. A two-input NAND gate
b. A two-input NOR gate, so that equal output current-driving capability will be achieved in both directions. Take $\mu_e = 2\mu_h$.

4.14 Repeat Problem 4.13 for three-input NAND and NOR gates.

4.15 Determine the circuit-design parameters for an n-well CMOS process with the following specifications:

$$x_{ox} = 500\ \text{Å},$$
$$V_{T0n} = |V_{T0p}| = .75\ \text{V},$$
$$\mu_e = 2.5\mu_h = 580\ \text{cm}^2/\text{V}\cdot\text{s},$$
$$N_{d(sw)} = 5N_{d(well)} = 5 \times 10^{16}\ \text{cm}^{-3},$$
$$N_{a(sw)} = 5N_a = 5 \times 10^{15}\ \text{cm}^{-3},$$
$$\text{for } n^+ \text{ and } p^+ \text{ regions: } N_{d(n^+)} = N_{a(p^+)} = 10^{20}\ \text{cm}^{-3}.$$

The design parameters include body-bias coefficients and bulk Fermi potentials for both NMOS and PMOS, and all transition capacitances.

4.16 For a nonsymmetric CMOS inverter with $k_n = 2.5\ k_p$, and $L = W = 5\ \mu$m for both devices, determine the following:

a. V_{IL} and the corresponding output voltage
b. V_{IH} and the corresponding output voltage
c. V_{th}
d. Generate the VTC using SPICE with **GAMMA** $= .4\sqrt{\text{V}}$ for both devices to verify the results obtained in parts (*a*), (*b*), and (*c*).

4.17 Repeat Problem 4.16 for a symmetric inverter, with the only change being $W_p = 12.5\ \mu$m. Also calculate the current flow through the circuit when $V_I = V_{th}$.

4.18 a. Design a CMOS inverter that will produce $V_{th} = 2.5$ V for $\textbf{KP}_p = 16\ \mu\text{A/V}^2$, $\textbf{KP}_n = 40\ \mu\text{A/V}^2$, $V_{Tn} = .8$ V, and $V_{Tp} = -.7$ V.
b. Simulate the circuit in (*a*) using SPICE.

4.19 The layout for two cascade CMOS inverters is shown in the accompanying diagram. Each inverter employs a *p*-type substrate for the *n*-channel devices and an *n*-well region for the *p*-channel devices. The design parameters are given as follows:

For both devices:

$$C_{ox} = 70\ \text{nF/cm}^2$$
$$\textbf{LD} = .3 \times 10^{-4}\ \text{cm}$$
$$\textbf{GAMMA} = .4\sqrt{\text{V}}$$

For NMOS:

$\textbf{KP} = 40\ \mu\text{A/V}^2$
n^+ to *p*-substrate transition capacitances:
$C_{toa} = .1\ \text{fF}/\mu\text{m}^2$
$C_{t(sw)} = .11\ \text{fF}/\mu\text{m}$
$\phi_o = .88$ V
$\phi_{o(sw)} = .92$ V

For PMOS:

$\textbf{KP} = 16\ \mu\text{A/V}^2$
p^+ to *n*-well transition capacitances:
$C_{toa} = .03\ \text{fF}/\mu\text{m}^2$
$C_{t(sw)} = .36\ \text{fF}/\mu\text{m}$
$\phi_o = .94$ V
$\phi_{o(sw)} = .99$ V

For the driver inverter, do the following:

a. Find $C_{GDp(max)}$ and $C_{GDn(max)}$.
b. Separating the sidewall and bottom contributions, find the zero-bias values of C_{dbp} and C_{dbn}.
c. Excluding the interconnect and load gate capacitance values, determine C_L.
d. Determine the transient response by calculating t_r and t_f.
e. Examine the accuracy of the hand-calculated results of (*d*) by simulating this portion of the circuit using SPICE with the lumped-element C_L of (*c*) as an input parameter.
f. Generate another SPICE simulation, this time employing SPICE modeling of the gate and transition capacitances.

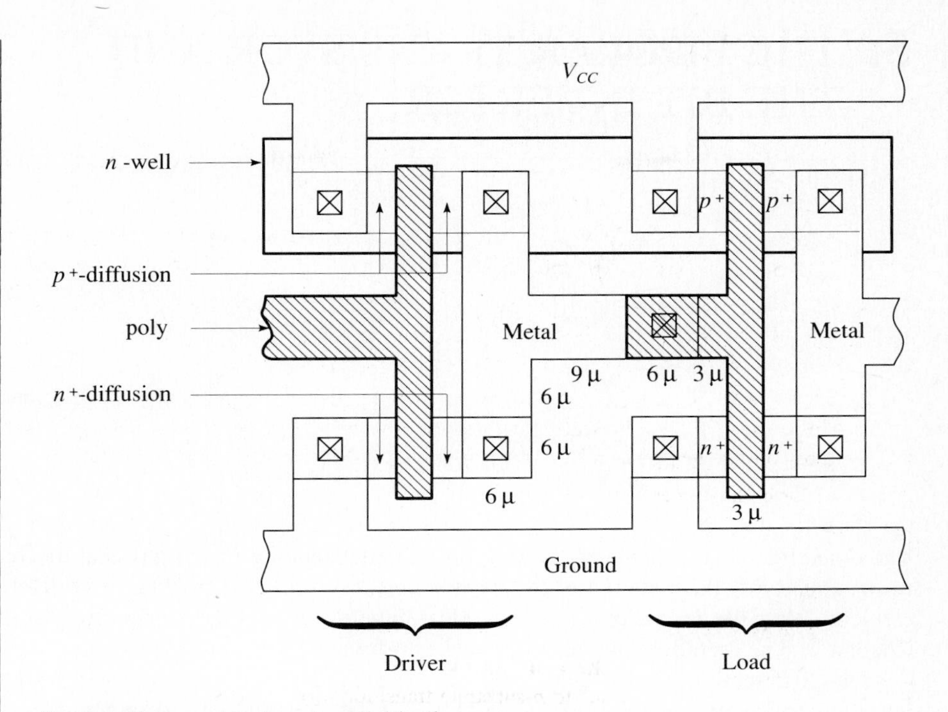

Overlap distance, **LD**, is not explicitly shown.

4.20 To incorporate the parasitic line interconnect capacitances in the circuit of Problem 4.19, assume $C_{m\text{-}f} = .04 \text{ fF}/\mu\text{m}^2$ and $C_{p\text{-}f} = .06 \text{ fF}/\mu\text{m}^2$ where $C_{m\text{-}f}$ is the parasitic interconnect capacitance of metal over the field oxide, and $C_{p\text{-}f}$ is that of poly over the field oxide. Then do the following:

a. Ignoring the regions where the metal overlaps with n^+, p^+, and poly, determine $C_{m\text{-}f}$ from the output of the driver to the metal-poly contact cut at the input of the load.

b. Calculate the input capacitance of the load as seen looking into the poly line.

c. Determine the sum of C_{line} and C_G to compare it to C_L as found in Problem 4.19.

d. Determine the transient response of the driver, using SPICE and including interconnect and load contributions.

5. THE BIPOLAR TRANSISTOR AND THE BJT INVERTER

The objective of this chapter is to develop an understanding of the physical mechanisms underlying the behavior of the *bipolar junction transistor* (*BJT*) as a switching element. We will make use of the basic ideas developed for the *pn*-junction diode in Chapter 2. The material covered is divided into four sections.

The first section introduces the BJT. Its physical description provides an insight into the performance of the transistor as a switching element. The various modes of operation of the transistor are presented and the charge-control equations are derived for each mode. The hybrid-pi equivalent circuit model is also developed. The second section describes the static and dynamic characteristics of the basic BJT inverter. Various BJT models used in circuit simulation programs such as SPICE are discussed in the third section. The fourth section presents important BJT design considerations for linear and digital applications.

5.1 Fundamentals of the Bipolar Junction Transistor

The bipolar junction transistor is a semiconductor device containing three doped regions of *p*- and *n*-type materials resulting in two *pn* junctions. Depending on the type of the narrow center material, there are two kinds of devices: *pnp* and *npn*. Since digital integrated circuits predominantly use *npn* transistors, this text concentrates on that type.

The central region is known as the *base*. The outer layers are labeled the *emitter* and *collector*. These regions are not interchangeable, because the emitter region is much more heavily doped than the collector. Their cross-sectional areas are also different. Conventional circuit symbols for *npn* and *pnp* transistors are shown in Figure 5.1, and the four possible modes of operation for the BJT are listed in Table 5.1. In general, the transistor is operated only in the forward active mode in linear or analog circuits. On the other hand, in digital circuits all four modes of operation may be

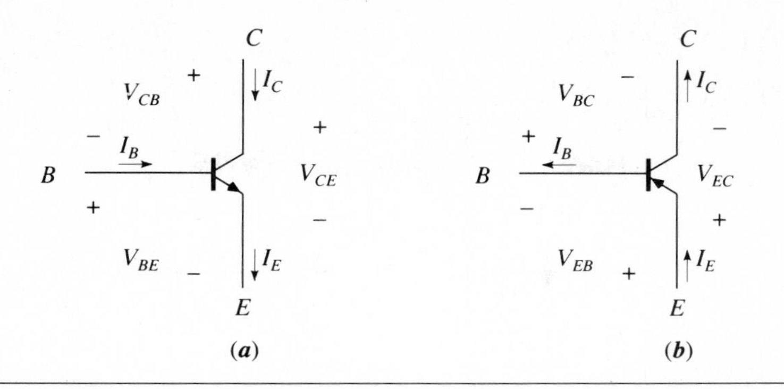

Figure 5.1 Circuit symbols with forward active region voltage and current polarities: (a) *npn*, (b) *pnp*.

involved. However, the forward active region of operation is one of primary concern in understanding the operation of the device.

Forward Active Mode

In the forward active mode, the *base-emitter* (*BE*) junction is forward-biased and the *base-collector* (*BC*) junction is reverse-biased. The energy-band diagrams for an *npn* transistor in thermal equilibrium and under forward active conditions are plotted in Figure 5.2. The *active area* of the base is the area between the depletion regions. In this quasi-neutral region the electric field is zero under low-level injection because the externally applied voltages appear essentially across these junctions. This simplifies the analysis, because the minority carrier current flow through the base layer has only a diffusion component. The field-dependent drift-current component is zero.

In forward active mode, the potential barrier for electrons entering the *p*-type base region from the emitter is lowered. Hence, the majority carrier electrons are allowed to be injected from the emitter into the base to become minority carriers. Also note that the barrier for holes in the base is lowered, and the majority carrier holes are injected from the base into the emitter. The combined effect is that the forward-biased BE junction creates a positive emitter current I_E.

The BC junction is reverse-biased, and the right-hand side of Figure 5.2b is lowered with respect to the base region, which effectively raises the barrier for holes

Table 5.1 Modes of Operation for the BJT

Emitter junction	Collector junction	Mode of operation
Forward	Reverse	Forward active
Reverse	Forward	Reverse active
Reverse	Reverse	Cutoff
Forward	Forward	Saturation

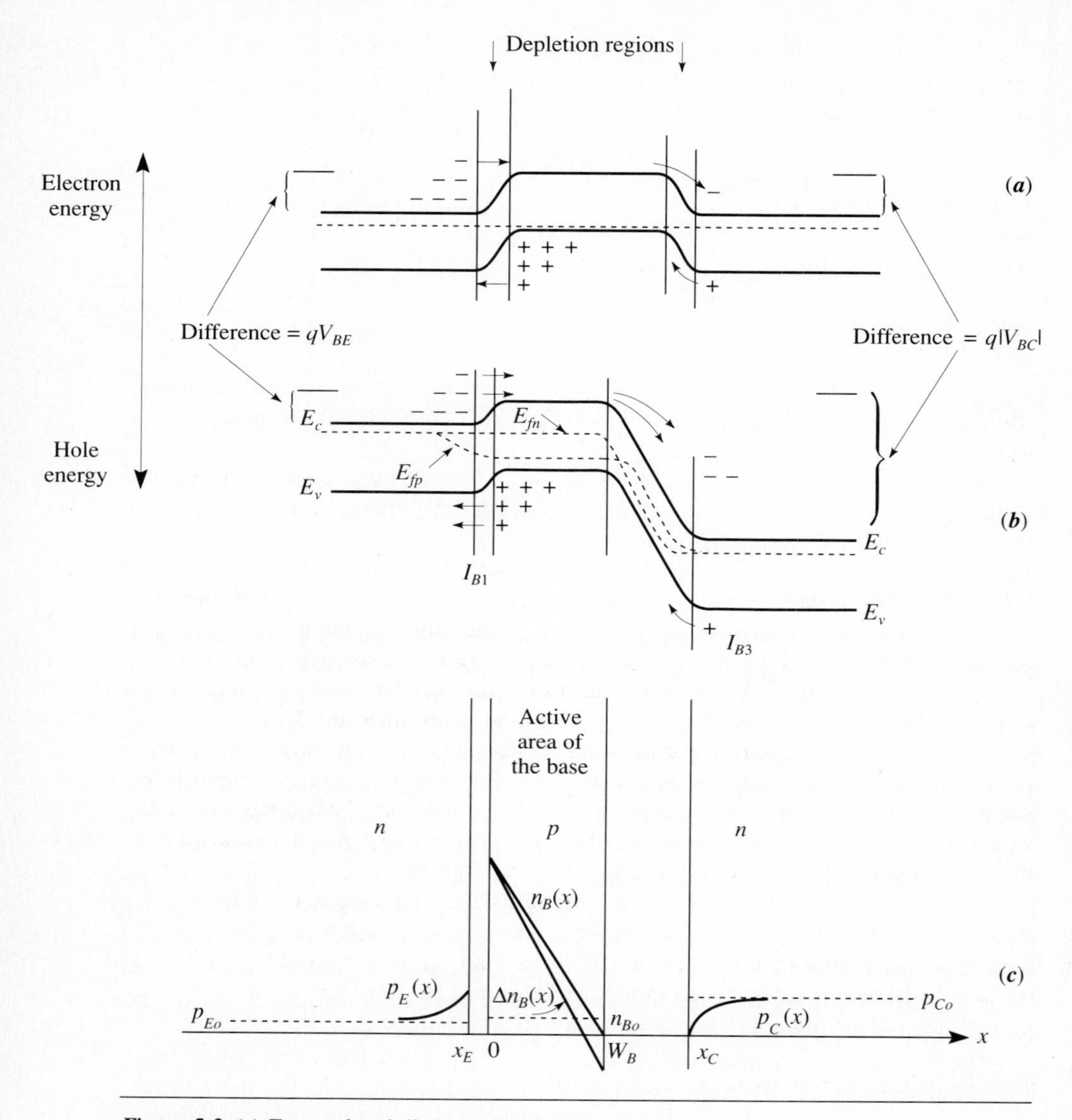

Figure 5.2 (a) Energy-band diagrams for an *npn* transistor in thermal equilibrium, and (b) under forward-bias conditions; (c) minority action profiles in the forward active mode.

in the base that may want to travel to the collector. Similarly, the electron barrier from collector to base is increased, thereby reducing any carrier flow. However, the electrons injected from the emitter into the base easily diffuse through the base and find a potential hill to slide down into the collector to be *collected*. The width W_B of the base active area is kept very narrow (<1 μm) compared to the minority carrier diffusion length in the base, L_B, so that most of the electrons injected into the base will be swept out of the base at the BC depletion region. If the recombination rate in the base region is very small, then the collector electron current I_{Ce} is slightly less than the emitter electron current I_{Ee} and may be written as

$$I_{Ce} = \alpha_T I_{Ee} \tag{5.1}$$

where the proportionality constant α_T, called the *base transport factor*, signifies the portion of the surviving electrons and depends on the doping and dimensions of the transistor.

The hole component I_{Ch} of the total collector current I_C arises from the thermally generated minority holes that are within one diffusion length of the BC junction edge and that fall down the potential hill.

As seen in Figure 5.2b, the BE junction depletion region is smaller than in equilibrium because of the foward bias, whereas the BC junction depletion region is wider, as expected.

Current components for an *npn* transistor in the forward active mode are depicted in Figure 5.3. There are basically three base current components to make up the total current going into the base. The first component $I_{B1} = I_{Eh}$ arises from excess holes injected across the forward-biased BE junction from the base into the active area of the emitter. I_{B2} corresponds to the holes that must enter the base to replace holes used up in recombining with electrons injected from the emitter, while $I_{B3} = I_{Ch}$ is as described above. The latter is nothing but a small, reverse-bias leakage current.

Another base current component can result from carrier generation and recombination in the emitter space-charge region. Being determined by the quality of the starting wafer, it is beyond the control of the device designer. However, in well-prepared modern transistors, in which the crystal defects and metallic impurity are reduced by a technique called *gettering*, which involves the heat treatment of the wafer at high temperatures for several hours to remove the oxygen below the surface, this component can be neglected for all practical purposes.

Thus, the total collector current is almost equal to the electron current passing through the base. Since the emitter doping is much greater than that of the base, the electron current across the BE junction is larger than the hole current I_{B1}. Hence, I_E is approximately equal to the electron current I_{Ee} injected into the base. The *emitter*

Figure 5.3 The current components for an *npn* transistor in the forward active mode.

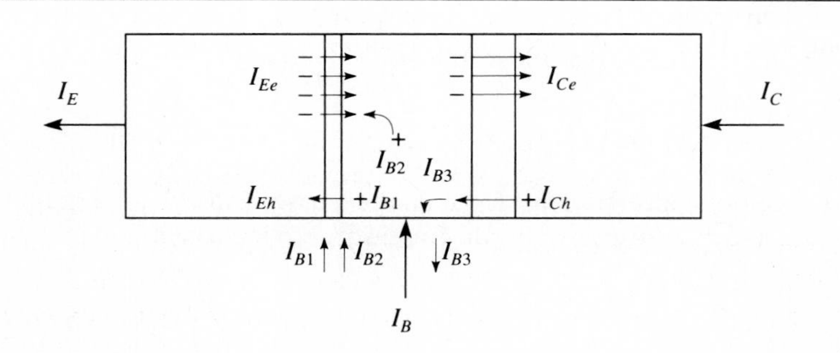

injection efficiency measures the injected electron current compared to the total emitter current:

$$\gamma \equiv \frac{I_{Ee}}{I_E} \tag{5.2}$$

Very few electrons are lost as they traverse the base because W_B is much smaller than L_B. For a well-designed transistor, both α_T and γ approach unity.

Therefore, the terminal currents are given by

$$I_B = I_{B1} + I_{B2} - I_{B3} \tag{5.3}$$

$$I_C = I_{Ce} + I_{Ch} = \alpha_T I_{Ee} + I_{Ch} = \alpha_T \gamma I_E + I_{Ch} \tag{5.4}$$

$$I_E = I_B + I_C = I_{Ee} + I_{Eh} \tag{5.5}$$

It is easier now to explain the large *current gain*

$$\beta_F \equiv \frac{I_C}{I_B} \tag{5.6}$$

of which the BJT is capable. It is large because the BE junction needs only a small hole current to provide a large electron current; that is, even a small base current can force this junction to become forward-biased and inject large numbers of electrons.

Using Equations (5.4)–(5.6), and neglecting I_{Ch}, the dc-current gain is given by

$$\beta_F = \frac{\alpha_F}{1 - \alpha_F} \tag{5.7}$$

where $\alpha_F \equiv \alpha_T \gamma = I_{Ce}/I_E$ is the *forward common-base current gain*.

Recalling from Chapter 2 that the minority carrier concentration at the edge of a depletion region is the thermal equilibrium value of the concentration multiplied by an exponential term, we get the excess minority carrier concentrations as

$$\Delta n_B(0) \equiv n_B(0) - n_{Bo} = n_{Bo}(e^{V_{BE}/\phi_T} - 1) \tag{5.8a}$$

$$\Delta n_B(W_B) \equiv n_B(W_B) - n_{Bo} = n_{Bo}(e^{V_{BC}/\phi_T} - 1) \tag{5.8b}$$

$$\Delta p_E(x_E) \equiv p_E(x_E) - p_{Eo} = p_{Eo}(e^{V_{BE}/\phi_T} - 1) \tag{5.8c}$$

$$\Delta p_C(x_C) \equiv p_C(x_C) - p_{Co} = p_{Co}(e^{V_{BC}/\phi_T} - 1) \tag{5.8d}$$

With the impurity concentrations at the emitter, base, and collector regions denoted by N_{dE}, N_{aB}, and N_{dC}, respectively, the minority carrier concentrations at equilibrium are given by

$$p_{Eo} = \frac{n_{io}^2}{N_{dE}}$$

$$n_{Bo} = \frac{n_{io}^2}{N_{aB}}$$

$$p_{Co} = \frac{n_{io}^2}{N_{dC}}$$

Note that since the exponential term in (5.8b) is approximately zero for a reverse-bias voltage of $V_{CB} >> \phi_T$, $\Delta n_B(W_B) \approx -n_{Bo}$. This indicates a *deficit* of excess electron concentration, which in turn implies that the total electron concentration there is approximately zero. Since n_{Bo} in Si BJTs is quite small, $\Delta n_B(W_B)$ itself can be taken as zero.

Using the boundary conditions given in the Equation Set (5.8), the minority carrier distributions can be obtained from the solution to the continuity equation. The widths of the neutral emitter and collector regions are much greater than the respective diffusion lengths, L_E and L_C. Therefore, the concentrations decay exponentially from their boundary conditions to the thermal equilibrium values. These are sketched in Figure 5.2c. Note that we assume negligible voltage drops across the neutral regions and low-level injection in the base region; that is, $p_{Bo} >> n_B(0)$. In reality, there are bulk resistors in the neutral regions in series with the BE and BC junctions, so that the terminal currents develop a voltage across these regions. The bulk emitter resistance r_E is usually much less than the bulk collector resistance r_C due to different impurity concentrations.

The minority carrier distribution in the neutral base region can be described for the case of constant forward-bias voltage by the field-free, steady-state continuity equation as

$$0 = D_B \frac{d^2 \Delta n_B(x)}{dx^2} - \frac{\Delta n_B(x)}{\tau_{BF}} \tag{5.9}$$

where D_B and τ_{BF} are the diffusion coefficient and the *effective minority carrier lifetime* in the neutral base region, respectively. The latter is a time constant that represents the combined effects of the base current I_{B1} injected into the emitter and base recombination current I_{B2}. The solution to Equation (5.9) is

$$n_B(x) = n_{Bo} + K_1 e^{x/L_B} + K_2 e^{-x/L_B} \tag{5.10}$$

where $L_B \equiv \sqrt{D_B \tau_{BF}}$.

Since the base is thin and lightly doped, x in (5.10) is small, and L_B is large, which means $n_B(x)$ and $\Delta n_B(x)$ are almost linear, as diagrammed in Figure 5.2c. This implies that both emitter and collector currents are proportional to $\Delta n_B(0)/W_B$, the slope of the electron distribution in the base.

Defining

$$\frac{d\Delta n_B(x)}{dx} \equiv \frac{\Delta n_B(0) - \Delta n_B(W_B)}{W_B}$$

and noting that $\Delta n_B(W_B) \approx 0$, we get

$$I_E \approx I_{Ee} = J_E A_E$$

$$= +qA_E D_B \left.\frac{d\Delta n_B(x)}{dx}\right|_{x=0}$$

$$\approx +qA_E D_B \frac{\Delta n_B(0)}{W_B} \tag{5.11}$$

where A_E is the emitter-junction cross-sectional area. On the other hand, an expression for the total excess minority carrier charge stored in the base, denoted by Q_F in the forward active mode, can be found as

$$Q_F = \left| -qA_E \int_0^{W_B} \Delta n_B(x)\, dx \right| \tag{5.12}$$

Using the linear approximation produces

$$Q_F = \frac{qA_E W_B \Delta n_B(0)}{2} \tag{5.13}$$

Combining Equations (5.11) and (5.12) to express Q_F in terms of I_C, we obtain

$$Q_F = \left(\frac{W_B^2}{2D_B}\right) I_E \approx \left(\frac{W_B^2}{2D_B}\right) I_C \tag{5.14}$$

where the quantity in parentheses is called the *mean forward transit time* τ_F of the electron carriers in the base:

$$\tau_F \equiv \frac{W_B^2}{2D_B} \tag{5.15}$$

For switching purposes, τ_F should be as short as possible, which implies that W_B should be small. A typical value for a high-speed transistor with $W_B = .5\ \mu\text{m}$ and $D_B = .001\ \text{m}^2/\text{s}$ would be .125 ns. Equation (5.14) allows us to describe the transistor as a *charge-controlled device* as opposed to describing it as a *current-controlled device* in the analog world, symbolized by Equation (5.7). Finally, using Equations (5.8a), (5.13), and (5.14), we can relate Q_F, I_E, and I_C to the base-emitter voltage, V_{BE}, as

$$Q_F = \left(\frac{1}{2} qA_E W_B n_{Bo}\right)(e^{V_{BE}/\phi_T} - 1) \tag{5.16}$$

$$I_E = \left(\frac{qA_E D_B n_{Bo}}{W_B}\right)(e^{V_{BE}/\phi_T} - 1)$$

$$= I_{ES}\left(e^{V_{BE}/\phi_T} - 1\right) \tag{5.17}$$

and

$$I_C = \alpha_F I_{ES}\left(e^{V_{BE}/\phi_T} - 1\right) \tag{5.18}$$

where

$$I_{ES} \equiv \frac{qA_E D_B n_{Bo}}{W_B}$$

is the reverse saturation current for the BE junction.

Next, we develop the charge-control model for the BJT in the forward active mode. At one side of the base region there is an entering electron current $I_{Ee}(0)$. At the other side there is an exiting electron current $I_{Ce}(W_B)$. Any difference between these two is essentially caused by the recombination process and is accounted for by the external circuit, which replaced the missing holes. Ignoring I_{B1} and I_{B3}, we have

$$i_B(t) = I_{B2} = I_{Ee}(0) - I_{Ce}(W_B) \tag{5.19}$$

Analogous to the discussion in Chapter 2, the continuity principle states that

(*the time rate of change of excess electron charge*)
= (*the electron current entering at x = 0*)
− (*the electron current exiting at x =* W_B)
+ (*the effects of the recombination-generation process within the region*).

In symbolic form, this yields

$$\frac{\partial \Delta n_B(x, t)}{\partial t} = \left(+\frac{1}{q}\right)\frac{\partial J_B}{\partial x} - \frac{\Delta n_B(x, t)}{\tau_{BF}} \tag{5.20}$$

where J_B is the electron current density in the neutral base region. Next, we multiply Equation (5.20) by $-qAdx$, where $-q$ is the charge on an electron and $A = A_E$, and integrate over the region to get

$$\left(\frac{d}{dt}\right)\left[-qA\int_0^{W_B} \Delta n_B(x, t)\, dx\right]$$

$$= -A\int_{J_B(0)}^{J_B(W_B)} dJ_B - \left(-\frac{qA}{\tau_{BF}}\right)\int_0^{W_B} \Delta n_B(x, t)\, dx \tag{5.21}$$

Using Equation (5.12) yields

$$\frac{dQ_F(t)}{dt} = -A\left[J_B(W_B) - J_B(0)\right] - \frac{Q_F(t)}{\tau_{BF}} \tag{5.22}$$

Finally, rearranging terms in Equation (5.22), and making use of (5.19), we obtain a first-order differential equation as

$$i_B(t) = \frac{dQ_F(t)}{dt} + \frac{Q_F(t)}{\tau_{BF}} \tag{5.23}$$

Equation (5.23) is a fundamental equation relating the base current to the excess minority carrier charge and incorporating not only the static but also the dynamic properties of the BJT. It shows that the base current supplies holes to the neutral base region at the same rate at which the holes are being lost by recombination plus the rate at which the stored charge increases. Note that, in steady-state, the transient term in Equation (5.23), $dQ_F(t)/dt$, is zero. Thus, we have from Equations (5.7), (5.14), (5.15), and (5.23)

$$\beta_F = \frac{\tau_{BF}}{\tau_F} \tag{5.24}$$

Therefore, a large ratio of minority carrier lifetime to base transit time is desirable to obtain a high value of β_F.

Solving Equation (5.23) for the constant base current $i_B(t) = I_B$ of Figure 5.4a yields

$$Q_F(t) = I_B \tau_{BF} + K e^{-t/\tau_{BF}}$$

To solve for the constant K, we assume that, at $t = 0$, the collector current and hence the excess minority carrier base charge $Q_F(0)$ are zero. Therefore, $K = -I_B \tau_{BF}$. Substituting this into the preceding equation yields

$$Q_F(t) = I_B \tau_{BF}(1 - e^{-t/\tau_{BF}}) \tag{5.25}$$

A plot of $Q_F(t)$ is shown in Figure 5.4b. Now, using Equation (5.14), the collector current is given by

Figure 5.4 Solution to Equation (5.23): (a) the constant excitation $i_B(t) = I_B$; (b) excess stored charge in the base, $Q_F(t)$; and (c) the corresponding collector current.

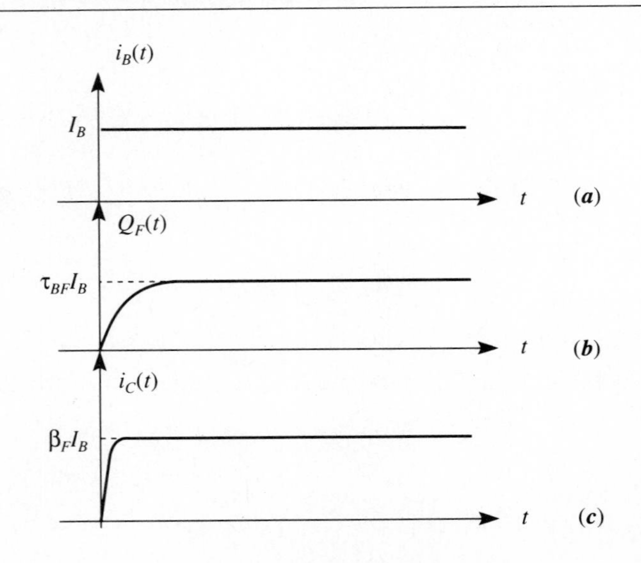

$$i_C(t) = I_B\left(\frac{\tau_{BF}}{\tau_F}\right)(1 - e^{-t/\tau_{BF}})$$
$$= \beta_F I_B(1 - e^{-t/\tau_{BF}})$$
$$= I_C(1 - e^{-t/\tau_{BF}}) \tag{5.26}$$

This equation indicates that the collector current rises from zero and levels off at a steady-state value of $I_C = \beta_F I_B$. The variation of $i_C(t)$ with time is diagrammed in Figure 5.4c. It is now apparent that it takes some time for the collector current to increase from its initial value of zero to its final value of $\beta_F I_B$. This transition as well as the parameters that characterize various aspects of the BJT's transient performance are treated in Section 5.2.

Finally, the emitter current is given by

$$i_E(t) = i_C(t) + i_B(t)$$
$$= \frac{Q_F(t)}{\tau_F} + \frac{Q_F(t)}{\tau_{BF}} + \frac{dQ_F(t)}{dt} \tag{5.27}$$

This equation is not complete because the effects of the transition capacitances are not taken into account. These capacitances are due to the charges stored in the depletion regions of the transistor, at the emitter and collector junctions. Therefore, whenever there is a voltage change across either of these junctions, we should also consider the charging currents to these regions. For the emitter junction, the charging current is simply given by

$$i_E(t) = C_{tE}\frac{dv_{BE}(t)}{dt} = \frac{dQ_{BE}(t)}{dt} \tag{5.28}$$

where C_{tE} is the transition capacitance, and Q_{BE} is the depletion charge associated with the BE junction. Similarly, for the BC junction

$$i_C(t) = -C_{tC}\frac{dv_{BC}(t)}{dt} = -\frac{dQ_{BC}(t)}{dt} \tag{5.29}$$

where C_{tC} is the transition capacitance, and Q_{BC} is the depletion charge associated with that junction. The negative sign is due to the fact that the conventional direction of the collector current is opposite to the forward current in the collector diode of the BJT. Therefore, we obtain the complete charge-control model in the forward active mode as

$$i_E(t) = \frac{Q_F(t)}{\tau_F} + \frac{Q_F(t)}{\tau_{BF}} + \frac{dQ_F(t)}{dt} + \frac{dQ_{BE}(t)}{dt} \tag{5.30a}$$

$$i_C(t) = \frac{Q_F(t)}{\tau_F} - \frac{dQ_{BC}(t)}{dt} \tag{5.30b}$$

$$i_B(t) = \frac{dQ_F(t)}{dt} + \frac{Q_F(t)}{\tau_{BF}} + \frac{dQ_{BE}(t)}{dt} + \frac{dQ_{BC}(t)}{dt} \tag{5.30c}$$

Saturation Mode

In the saturation mode of operation, both junctions are forward-biased. The emitter is more strongly biased. The electrons flow into the base region across both the emitter and collector junctions. Hence, for a given $i_C(t)$, the charge stored in the base is increased over its value in the forward active case, as illustrated in Figure 5.5. This distribution is obtained from a solution to the continuity equation, and assuming that $L_B >> W_B$, it can be approximated as being linear. Thus, the stored charge in the base is proportional to the area of the triangle plus that of the rectangle in Figure 5.5. The *critical base charge* Q_A that brings the transistor to the edge of saturation (*eos*) is represented by the triangle, and the collector current $I_{C(eos)}$ by the slope of the profile. Consequently, we have

$$Q_A = \tau_F I_{C(eos)} = \tau_{BF} I_{B(eos)} \tag{5.31}$$

where $I_{B(eos)}$ is the corresponding base current component.

As depicted in Figure 5.5, and in contrast to the forward active mode, the excess minority concentration is not zero at the edge of the BC junction because that junction is now forward-biased. The *overdrive base charge* Q_S stored in the base is represented by the shaded rectangle. Since this charge does not contribute to the slope of the concentration profile but rather arises from the pushing of more current into the base than is required to saturate the transistor, it does not lead to an additional collector current component. The *excess base drive* over and above $I_{B(eos)}$ that actually takes the transistor into saturation and gives rise to Q_S is denoted by I_{BS}:

$$I_{BS} \equiv I_B - \frac{I_C}{\beta_F}$$

Therefore,

$$Q_S = \tau_S I_{BS} \tag{5.32}$$

where the *storage time constant* τ_S determines the rate at which the excess saturation charge decays to zero. To find an expression for the instantaneous base current in

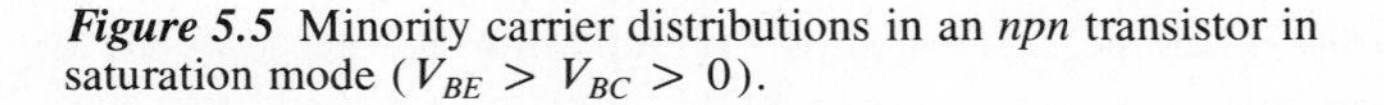
Figure 5.5 Minority carrier distributions in an *npn* transistor in saturation mode ($V_{BE} > V_{BC} > 0$).

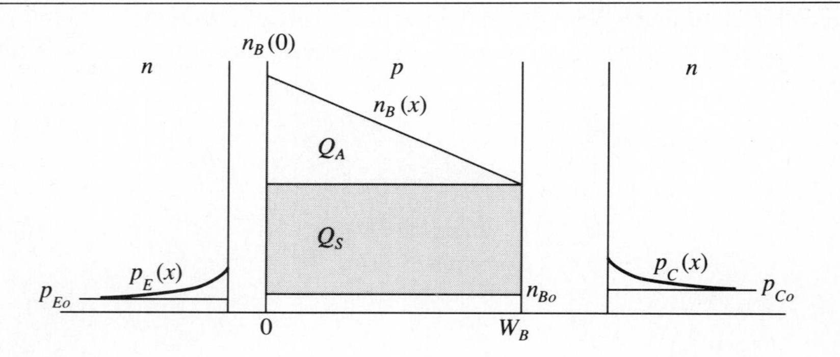

terms of the stored charge, we must account for the time rate of change of Q_S, too. Thus, we have

$$i_B(t) = \frac{dQ_S(t)}{dt} + \frac{Q_A}{\tau_{BF}} + \frac{Q_S(t)}{\tau_S} \tag{5.33}$$

to describe the charge-control model of the transistor in saturation. Note that dQ_A/dt is absent from (5.33) because the external circuit constrains the collector current and hence Q_A to a constant value. Also, since both junction voltages are constant, there is very little change in charges stored in the depletion regions, so that the transition capacitance effects can be neglected. We use (5.31) to eliminate τ_{BF} in (5.33) for the saturation mode charge-control equation to contain only one time constant. Hence, the instantaneous base current will be given by

$$i_B(t) = I_{B(eos)} + \frac{dQ_S(t)}{dt} + \frac{Q_S(t)}{\tau_S} \tag{5.34}$$

Reverse Active Mode

In this mode, the BC junction is forward-biased, and the BE junction is reverse-biased. This means that there is an injection of carriers from the collector into the base, and a collection of these carriers at the emitter junction. The profile of minority carrier distribution is shown in Figure 5.6. The excess minority carrier reverse base charge Q_R is represented by the shaded triangle. In general, $Q_F >> Q_R$. The reverse active mode charge-control equations can be obtained by an analogy with the forward active mode equations:

$$i_C(t) = -\frac{Q_R(t)}{\tau_R} - \frac{Q_R(t)}{\tau_{BR}} - \frac{dQ_R(t)}{dt}$$

$$i_B(t) = \frac{Q_R(t)}{\tau_{BR}} + \frac{dQ_R(t)}{dt}$$

Figure 5.6 Minority concentration profiles in the reverse active mode.

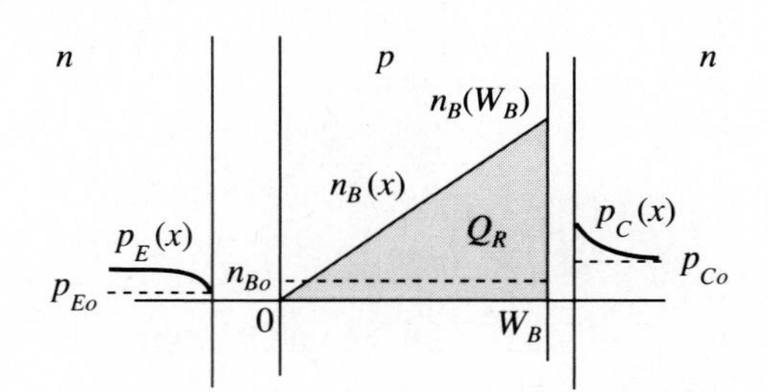

so that

$$i_E(t) = -\frac{Q_R(t)}{\tau_R}$$

Similar to (5.16), we have

$$Q_R = \frac{qA_C W_B n_{Bo}}{2}(e^{V_{BC}/\phi_T} - 1)$$

where A_C is the collector junction area. Also,

$$\beta_R = \frac{\alpha_R}{1 - \alpha_R} = \frac{\tau_{BR}}{\tau_R} \tag{5.35}$$

where β_R is the *reverse current gain* and α_R is the *reverse common-base current gain.* The latter is due to the ratio of the number of electrons collected at the emitter to the number emitted into the base by the collector at the forward-biased BC junction.

Therefore,

$$\alpha_R \equiv \frac{I_{Ee}}{I_{Ce}} \tag{5.36}$$

Cutoff Mode

In the cutoff mode, both the collector and emitter junctions are reverse-biased so that V_{BC} and V_{BE} are negative and very small. Hence, the magnitude of Q_R is negligible. The currents are given by

$$i_C(t) = -\frac{dQ_{BC}(t)}{dt}$$

and

$$i_B(t) = \frac{d(Q_{BC} + Q_{BE})}{dt}$$

Figure 5.7 depicts the junction polarities and minority carrier profiles of an *npn* transistor under all four modes of operation.

Hybrid-Pi Circuit Model

In this section, we develop a high-frequency, small-signal model of the BJT that relates to the physical processes in the transistor.

The total base current consists of a relatively large dc component and a small time-varying component:

$$i_B(t) = I_B + i_b(t)$$

where I_B is the dc term and i_b is the ac term. Similar expressions can be written for the collector current and junction voltages:

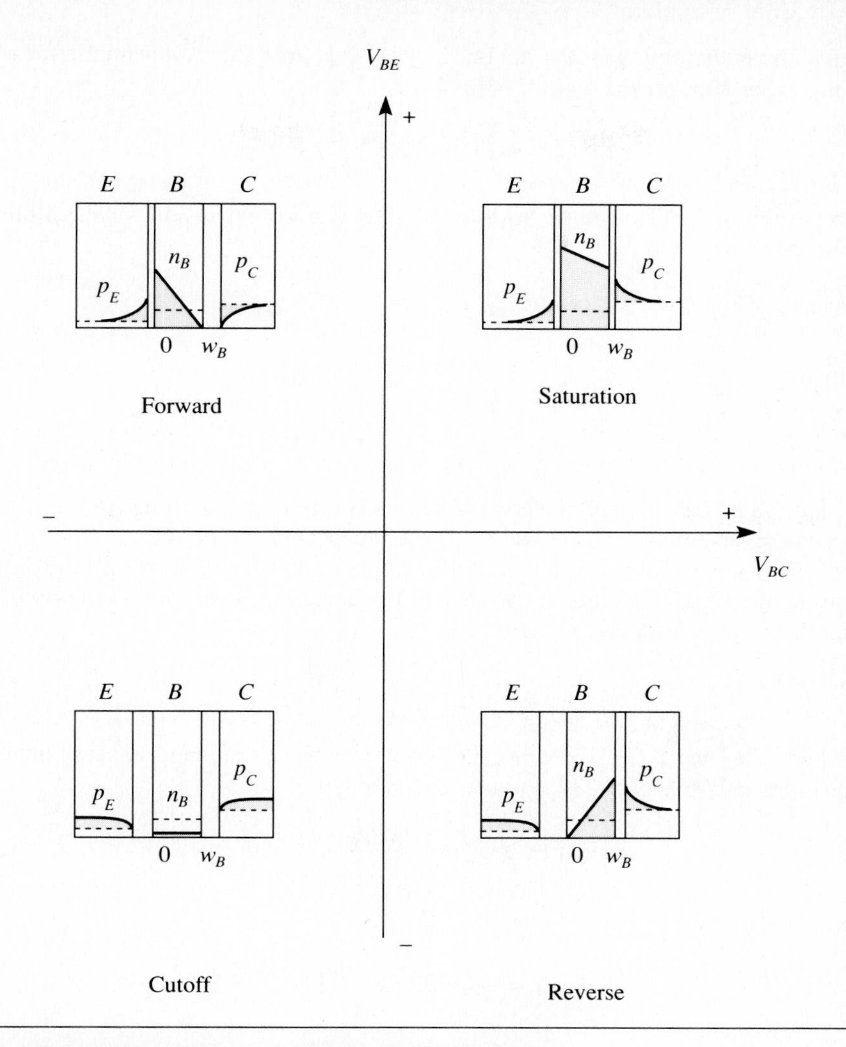

Figure 5.7 Junction polarities and minority carrier distributions of an *npn* transistor under all four modes of operation.

$$i_C(t) = I_C + i_c(t)$$

$$v_{BE}(t) = V_{BE} + v_{be}(t)$$

$$v_{BC}(t) = V_{BC} + v_{bc}(t)$$

Small-signal conditions apply when the ac components are small enough to be related by linear equations in spite of the nonlinear relationships among the total current and voltages. The permitted magnitude of v_{be} depends on the condition $\phi_T >> v_{be}$ because this is the only way the nonlinearity arising from the exponential term e^{v_{be}/ϕ_T} in the expression for the emitter current

$$i_E(t) = I_{ES}\left[\exp\left(\frac{V_{BE} + v_{be}}{\phi_T}\right) - 1\right]$$

can be removed. This can be seen by writing a series expansion for the aforementioned term as

$$e^{v_{be}/\phi_T} \approx 1 + \frac{v_{be}}{\phi_T}, \qquad \frac{v_{be}}{\phi_T} << 1$$

Hence,

$$i_E(t) \approx (I_{ES}e^{V_{BE}/\phi_T})\left(1 + \frac{v_{be}}{\phi_T}\right)$$

so that $i_E(t)$ and v_{be} are linearly related. Also note that as long as the collector remains reverse-biased, $i_B(t)$ and $i_C(t)$ are independent of $v_{BE}(t)$.

In a manner analogous to that of the discussion in Chapter 2, we get the diffusion capacitance C_d as the ratio of the change in charge stored in the base region to the change in voltage which caused it:

$$C_d \equiv \frac{\Delta Q_F}{\Delta v_{BE}} = \frac{\Delta Q_F}{v_{be}} \tag{5.37}$$

To find $i_b(t)$ and $i_c(t)$, we temporarily omit the transition capacitances, increment Equations (5.14) and (5.23), and use (5.37) to get

$$i_c(t) = \Delta i_C(t) = \frac{\Delta Q_F}{\tau_F} = \left(\frac{C_d}{\tau_F}\right) v_{be} \tag{5.38}$$

and

$$i_b(t) = \Delta i_B(t) = \frac{d\Delta Q_F}{dt} + \frac{\Delta Q_F}{\tau_{BF}}$$

$$= \frac{C_d dv_{be}}{dt} + \left(\frac{C_d}{\tau_{BF}}\right) v_{be} \tag{5.39}$$

The *transconductance* g_m of the transistor as a measure of the output response to the input signal is defined as

$$g_m \equiv \frac{di_c}{dv_{be}} \tag{5.40}$$

Now, using Equation (5.18), we note that $i_c(t) \approx I_C e^{v_{be}/\phi_T}$, which in turn can be expressed as $i_c(t) \approx I_C(1 + v_{be}/\phi_T)$ under steady-state condition of $v_{be} << \phi_T$, so that Equation (5.40) yields

$$g_m = \frac{|I_C|}{\phi_T} \tag{5.41}$$

and

$$C_d = \tau_F g_m \tag{5.42}$$

Referring next to Equation (5.39), the coefficient of the second term on the right-hand side is called the *input conductance* of the transistor, the reciprocal of which is designated as r_π:

$$r_\pi \equiv \frac{\tau_{BF}}{C_d} = \tau_{BF}\left(\frac{1}{\tau_F g_m}\right) = \frac{\beta_F}{g_m} \tag{5.43}$$

so that

$$\beta_F = h_{fe} = \frac{i_c}{i_b} = r_\pi g_m \tag{5.44}$$

Here, we assume that the dc current gain is equal to the ac current gain, h_{fe}. Now, Equations (5.38) and (5.39) can be rewritten as

$$i_c = g_m v_{be} \tag{5.45}$$

$$i_b = \left(\frac{1}{r_\pi}\right) v_{be} + C_d \frac{dv_{be}}{dt} \tag{5.46}$$

The resulting equivalent circuit modeling the transistor is shown in Figure 5.8a. Next, the transition capacitances are accounted for by simply connecting C_{tE} from B to E, and C_{tC} from C to B to obtain Figure 5.8b. Still, we have to incorporate the ohmic resistances in the base, emitter, and collector regions. Since the emitter is characteristically heavily doped to obtain a high emitter efficiency and thus produce a high current gain, and since the cross-sectional area is relatively large, the associated resistance is small. The collector region, on the other hand, is relatively lightly doped, but this region is so thin that its resistance is also small.

In two principal components of the base current, namely, I_{B1} and I_{B2}, the flow of holes is from the base contact into the narrow base region. Since the base is relatively lightly doped to increase γ and hence to obtain a high β_F, and since the cross-sectional dimension W_B is small to increase α_T, the base resistance is significantly large. This large resistance causes the BE junction voltage to differ from the terminal base-emitter voltage. Relabeling the B terminal of Figure 5.8b as B' and calling the external terminal B, we have a *base spread resistance*, r_x, connecting the internal and external base points, as illustrated in Figure 5.8c. Figure 5.8d shows the same complete model with C_d and C_{tE} combined into C_π, and with C_{tC} relabeled as C_μ to conform to the accepted convention. Typically, C_π is in the range of a few pF to a few tens of pF, and C_μ is in the range of a fraction of a pF to a few pF.

C_π may not be provided by the manufacturer. Rather, the behavior of h_{fe} as a function of frequency is given, and the value of C_π can be obtained from the *3 dB cutoff frequency*,

$$\omega_\beta = 2\pi f_\beta = \frac{1}{r_\pi(C_\pi + C_\mu)} \approx \frac{1}{r_\pi C_\pi} \tag{5.47}$$

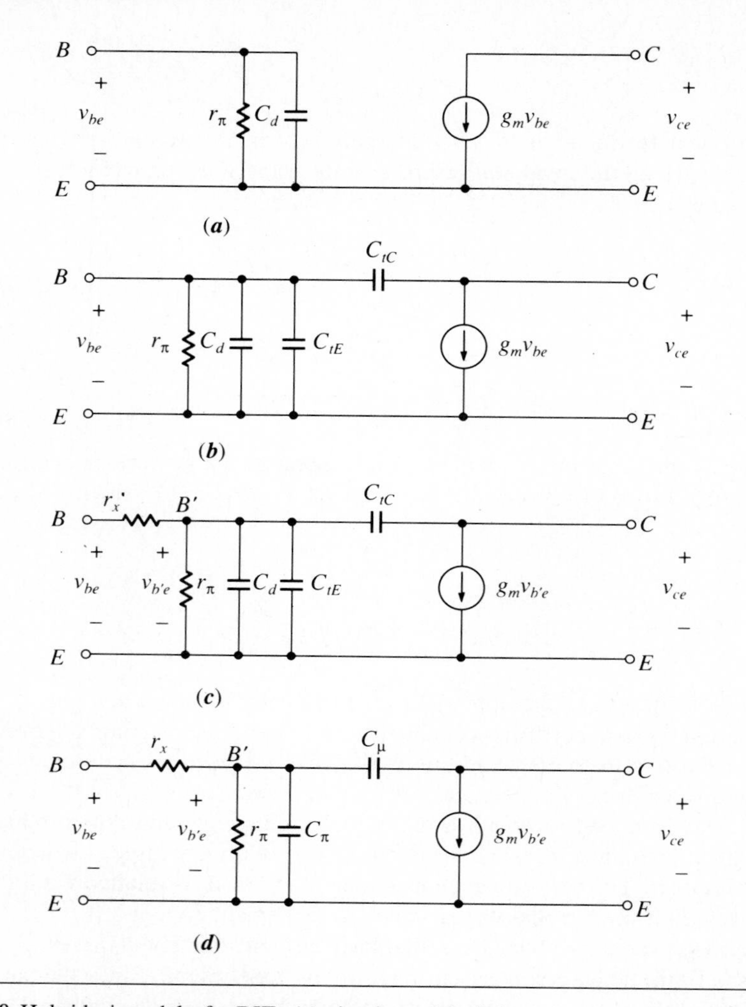

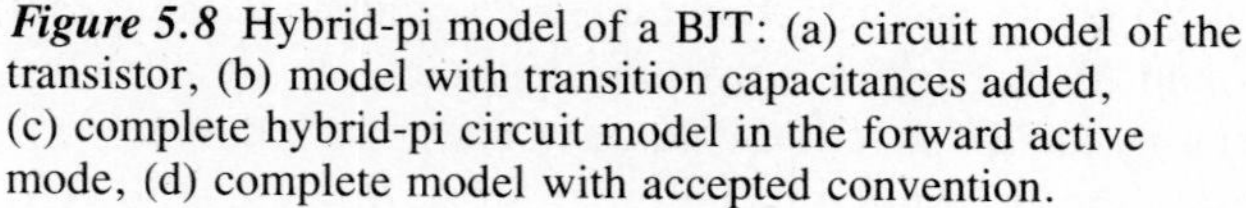

Figure 5.8 Hybrid-pi model of a BJT: (a) circuit model of the transistor, (b) model with transition capacitances added, (c) complete hybrid-pi circuit model in the forward active mode, (d) complete model with accepted convention.

which is defined as the frequency at which h_{fe} is $1/\sqrt{2} = .707$, or 3 dB off its dc value of β_F. Figure 5.9 shows a Bode plot for the current gain. As the frequency of the input signal is increased, C_π approaches a short circuit, and h_{fe} decreases. From the -6 dB/octave slope, it follows that the frequency at which h_{fe} drops to unity, called the *unity-gain cutoff frequency* or *unity-gain bandwidth*, is given by

$$\omega_T = 2\pi f_T = \beta_T \omega_\beta \tag{5.48}$$

and is usually specified on the data sheets of the transistor. Substituting Equations (5.44) and (5.47) into (5.48) leads to

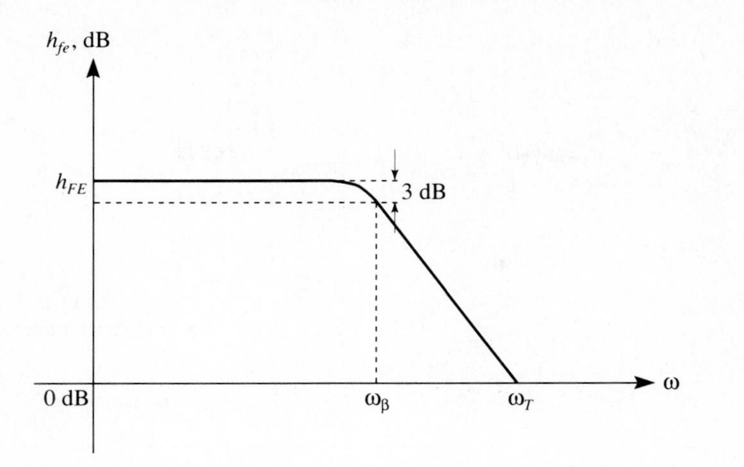

Figure 5.9 Bode plot for h_{fe}.

$$\omega_T = \frac{g_m}{C_\pi + C_\mu} \approx \frac{g_m}{C_\pi} \tag{5.49}$$

Finally, assuming $C_\pi \approx C_d$, we have

$$\omega_T = \frac{1}{\tau_F} \tag{5.50}$$

and

$$\omega_\beta = \frac{1}{\tau_{BF}} \tag{5.51}$$

The hybrid-pi model we have been discussing characterizes the transistor operation accurately only up to a frequency of about $.2\omega_T$. Other parasitic elements come into play at higher frequencies in such a way that the model needs to be refined to account for them.

5.2 The BJT Inverter

The Voltage Transfer Characteristic (VTC)

Next, we study the static transfer characteristic of the BJT inverter driving N similar circuits, shown in Figure 5.10a. Since V_{IL} is the maximum allowable logical **0** value, it is equal to the cut-in voltage, V_γ, of the transistor. Therefore, for input values less than V_{IL}, Q is off. The output voltage, on the other hand, is not equal to V_{CC} because a current flows through R_C and into the bases of N driven transistors that should be sufficient to drive each transistor into saturation. As N increases, however, each base current will decrease. To calculate the maximum allowable fan-out, we consider the worst-case situation, which happens when the driven gates draw maxi-

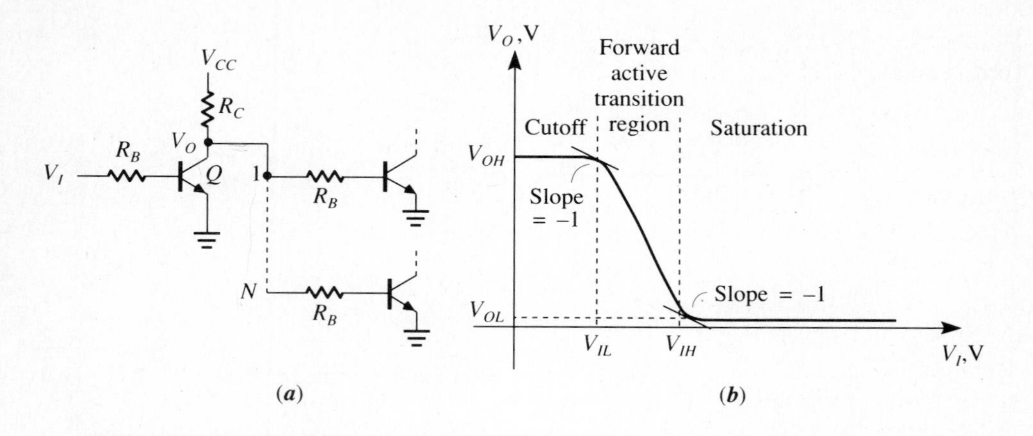

Figure 5.10 (a) The basic BJT inverter driving N similar gates, and (b) its VTC.

mum current through R_C. This means that the fan-out limitation occurs when the driving transistor is off and $V_{OH} = V_{IH}$, that is, $NM_H = 0$.

From Figure 5.10a, we see that if Q is off, then in each succeeding stage

$$I_B = \left(\frac{1}{N}\right)\left[\frac{V_{CC} - V_{BE(sat)}}{R_C + R_B/N}\right]$$

$$= \frac{V_{CC} - V_{BE(sat)}}{R_B + NR_C}$$

and

$$I_C = \frac{V_{CC} - V_{CE(sat)}}{R_C}$$

Here, we replaced each load transistor by a resistance, R_B/N, in series with $V_{BE(sat)}$, neglecting the transistor's input impedance of $\approx r_\pi$. For the transistors in these stages to be in saturation, $\beta_F I_B \geq I_C$ or

$$\frac{\beta_F R_C}{R_B + NR_C} \geq \frac{V_{CC} - V_{CE(sat)}}{V_{CC} - V_{BE(sat)}}$$

Solving for N, we get

$$N \leq \frac{\beta_F(V_{CC} - V_{BE(sat)})}{V_{CC} - V_{CE(sat)}} - \frac{R_B}{R_C}$$

which, neglecting $(V_{CE(sat)}/V_{CC})^2$, can be rewritten as

$$N \leq \beta_F\left(1 - \frac{V_{BE(sat)}}{V_{CC}}\right)\left(1 + \frac{V_{CE(sat)}}{V_{CC}}\right) - \frac{R_B}{R_C}$$

Multiplying out and neglecting the product term $(V_{BE(sat)}/V_{CC})(V_{CE(sat)}/V_{CC})$, we obtain an expression for the maximum fan-out:

$$N_{max} = \beta_{F(min)}\left(1 - \frac{.6}{V_{CC}}\right) - \frac{R_B}{R_C} \tag{5.52}$$

where we used $V_{CE(sat)} = .2$ V and $V_{BE(sat)} = .8$ V. The corresponding output high voltage is found to be

$$V_{OH} = V_{CC} - R_C\left(\frac{V_{CC} - V_{BE(sat)}}{R_C + R_B/N_{max}}\right)$$

As input voltage V_I exceeds V_{IL}, Q turns on and operates in the forward active mode, giving rise to the transition region of the gate transfer characteristic. Eventually, Q enters saturation when the value V_{IH} is reached at the input. Therefore, V_{IH} is the minimum input voltage just sufficient to saturate the transistor, and it is interpreted by the circuit as a logical **1**.

The input voltage, V_{IH}, is obtained as

$$V_{IH} = V_{BE(sat)} + I_{B(eos)}R_B$$

where

$$I_{B(eos)} = \frac{I_{C(eos)}}{\beta_F} = \frac{V_{CC} - V_{CE(sat)}}{\beta_F R_C}$$

so that

$$V_{IH} = V_{BE(sat)} + \left(\frac{R_B}{R_C}\right)\frac{V_{CC} - V_{CE(sat)}}{\beta_F}$$

Input voltages greater than V_{IH} drive the transistor deeper into saturation, and the output voltage decreases slightly below $V_{OL} = V_{CE(sat)}$. The resulting VTC is depicted in Figure 5.10b.

Example 5.1

For the simple BJT inverter circuit of Figure 5.11a, with $V_{CC} = 5$ V, $R_B = 10$ KΩ, $R_C = 1$ KΩ, $V_{CE(sat)} = .2$ V, and $\beta_F = 100$, the voltage transfer characteristic can readily be determined as follows:

If the input voltage is less than $V_\gamma \approx .5$ V, the transistor is off; hence $V_{IL} = .5$ V. We obtain the high-level input voltage as

$$V_{IH} = .8 + \left(\frac{10\ \text{K}\Omega}{1\ \text{K}\Omega}\right)\frac{(5 - .2)}{100} = 1.28\ \text{V}$$

Note that at the output we have $V_{OH} = 5$ V, and $V_{OL} = V_{CE(sat)} = .2$ V. Then the logic swing is given as $5 - .2 = 4.8$ V, and the transition region is obtained as $1.28 - .5 = .78$ V. Finally, the noise margins are found to be

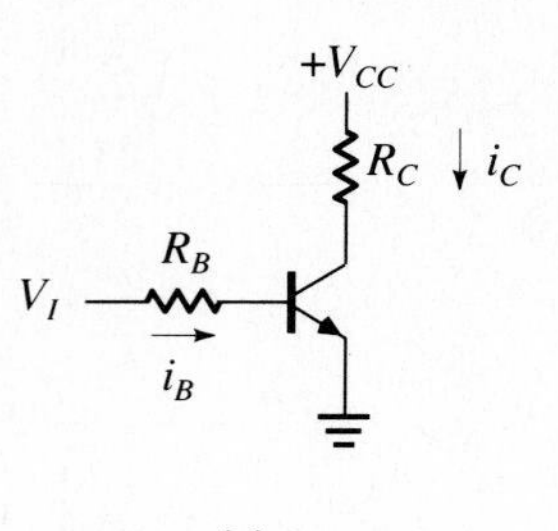

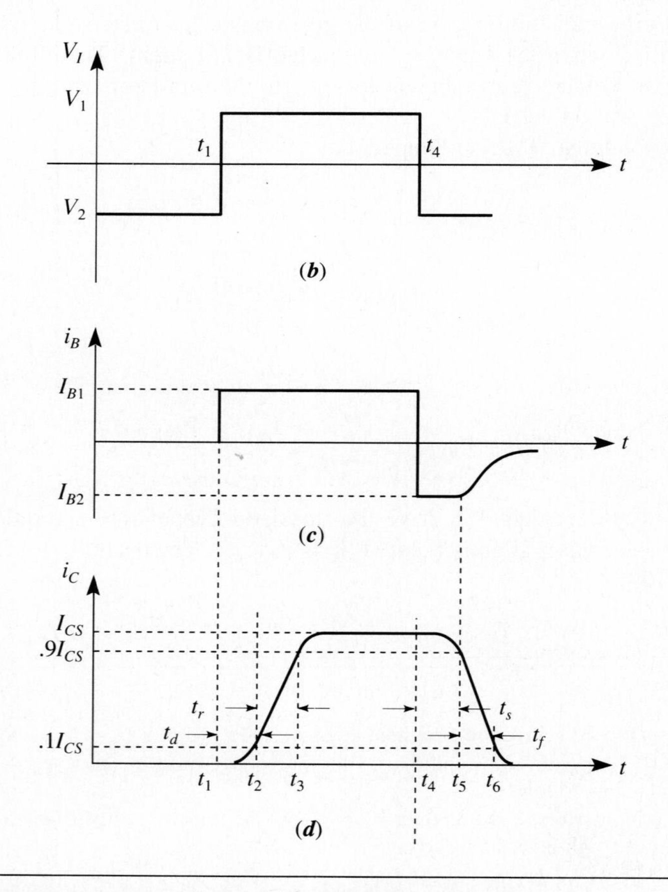

Figure 5.11 Switching-time waveforms: (a) the test circuit; (b) input waveform; (c) the base current; and (d) the output collector current.

$$NM_H = 5 - 1.28 = 3.72 \text{ V}$$

and

$$NM_L = .5 - .2 = .3 \text{ V}$$

The BJT Inverter Switching Times

Now we can demonstrate the significance of the charge-control approach in the calculation of the switching times for the BJT. For this purpose, we will make use of the basic BJT inverter as sketched in Figure 5.11a. The input voltage is a pulse that varies from a negative V_2 to a positive V_1 and back to V_2, as illustrated in Figure 5.11b. When $V_I = V_1$, a base current,

$$I_{B1} = \frac{V_1 - V_{BE(on)}}{R_B}$$

flows and saturates the transistor. When the input voltage is at V_2, initially a reverse base current,

$$I_{B2} = \frac{V_2 - V_{BE(sat)}}{R_B}$$

flows, which drops to zero eventually as the transistor is cut off. Figure 5.11c depicts the base current. Note that here I_{B1} and I_{B2} are unrelated to those in Equation (5.3). Although the switching times shown in Figure 5.11d and defined in Table 5.2 are for the output collector current, equivalent definitions in terms of the output voltage could also be used.

Table 5.2 Switching-Time Definitions

	Parameter	Definition
t_d	Delay time	Elapsed time between the application of an input base signal and the time when the output collector current rises to 10% of its final value.
t_r	Rise time	Time required for the collector current to rise from 10% to 90% of its final value.
t_{on}	Turn-on time	$t_{on} = t_d + t_r$
t_s	Storage time	Time between removal of the input and the time when the output falls to 90% of its initial saturated value $I_{C(eos)}$.
t_f	Fall time	Time for the collector current to fall from 90% to 10% of its initial saturated value.
t_{off}	Turn-off time	$t_{off} = t_s + t_f$

Delay Time

Initially, the transistor is in cutoff, since the input voltage is at V_2. Both junctions are reverse-biased, and the output voltage is equal to V_{CC}. When the input pulse is applied to the test circuit at $t = t_1$, there is a finite *delay time* t_d before any appreciable collector current starts to flow. The reason for this delay is that the voltages across the BE and BC transition capacitances cannot change instantaneously.

The total delay time can be considered to be the sum of three components:

$$t_d = t_{d1} + t_{d2} + t_{d3}$$

In a time t_{d1}, the BE transition capacitance is charged to the forward-bias cut-in voltage V_γ of approximately .5 V for Si BJTs. The second component t_{d2} is the time for the first minority carriers to traverse the base and reach the collector to form the collector current. Finally, at a time t_{d3} seconds later, the collector current rises from zero to 10% of its final value of $I_{C(eos)}$.

To find an expression for t_{d1}, we note that the transistor is in cutoff, so that we can make use of the cutoff mode charge-control equation for $i_B(t)$. We are concerned only with the depletion region charges Q_{BE} and Q_{BC}. During this time, the BE junction voltage is changing from V_2 to the cut-in voltage V_γ, and the BC junction voltage from $V_2 - V_{CC}$ to $V_\gamma - V_{CC}$. These changes in junction voltages imply changes in charge at these junctions, which are caused by the base current flowing through the resistor R_B. After the switching, the initial base current is

$$I_{Bi} = \frac{V_1}{R_B}$$

but t_{d1} seconds later it becomes

$$I_{Bf} = \frac{V_1 - V_\gamma}{R_B}$$

Therefore, the average base current during this time interval is given by

$$I_{B(av)} = \frac{2V_1 - V_\gamma}{2R_B}$$

This average current is equated to the changes in charge in both junctions as

$$I_{B(av)} = \frac{\Delta(Q_{BE} + Q_{BC})}{\Delta t}$$

Rearranging, we get

$$t_{d1} \equiv \Delta t = \frac{\Delta Q_{BE} + \Delta Q_{BC}}{I_{B(av)}} \tag{5.53}$$

Using the large-signal equivalent transition capacitance, we have

$$\Delta Q_{BE} = C_{tE(av)} \Delta V_{BE} = C_{tE(av)}(V_\gamma - V_2)$$

and

$$\Delta Q_{BC} = C_{tC(av)} \Delta V_{BC} = C_{tC(av)}(V_\gamma - V_2)$$

Therefore, from Equation (5.53) we get

$$t_{d1} = \frac{(C_{tE(av)} + C_{tC(av)})(V_\gamma - V_2)}{I_{B(av)}} \tag{5.54}$$

We could also find an expression for t_{d1} using the hybrid-pi circuit model of the BJT. Since t_{d1} is associated with the exponential charging of the BE transition capacitance, we use the following fundamental equation, which expresses the junction voltage as

$$v_{be}(t) = V_f + (V_i - V_f)e^{-t/\tau_d} \tag{5.55}$$

where V_f is the final value, V_i is the initial value, and τ_d is the *delay time constant*. In this particular case, although t_{d1} is the time necessary to change the voltage across the transistor's input capacitance from V_2 to V_γ, V_f is not the cut-in voltage but rather the final value to which the base voltage would head if allowed, that is, V_1.

The delay time constant is found from the equivalent circuit of Figure 5.12, which is valid as long as the transistor is in the cutoff region. The resistance r_π, being proportional to $1/I_C$, is large when the transistor is off. The voltage $v_{b'e}$ is zero; hence the current source in Figure 5.8 is open. Since the collector voltage does not change in cutoff, the right-hand side of C_μ in Figure 5.8 is at ac ground. Therefore,

$$\tau_d = (R_B + r_x)(C_\pi + C_\mu) \tag{5.56}$$

Note that the transition capacitances C_π and C_μ actually change as the BE junction reverse bias decreases. Thus, we must use the average values $C_{tE(av)}$ and $C_{tC(av)}$, respectively. Also note that since the BE junction is reverse-biased during this time, the associated diffusion capacitance is neglected.

Since $V_i = V_2$ and $V_f = V_1$, solving Equation (5.56) for $t = t_{d1}$ when $v_{be}(t) = V_\gamma$, we get

$$t_{d1} = \tau_d \ln\left(\frac{V_1 - V_2}{V_1 - V_\gamma}\right) \tag{5.57}$$

If $V_2 = V_\gamma$, that is, the transistor is initially at the edge of conduction instead of in cutoff, then $t_{d1} = 0$ because the BE junction was effectively never reverse-biased.

As far as the second delay time component is concerned, statistics are used to obtain an approximate expression:

$$t_{d2} = \frac{\tau_F}{3} \tag{5.58}$$

Figure 5.12 Equivalent circuit when the transistor is off.

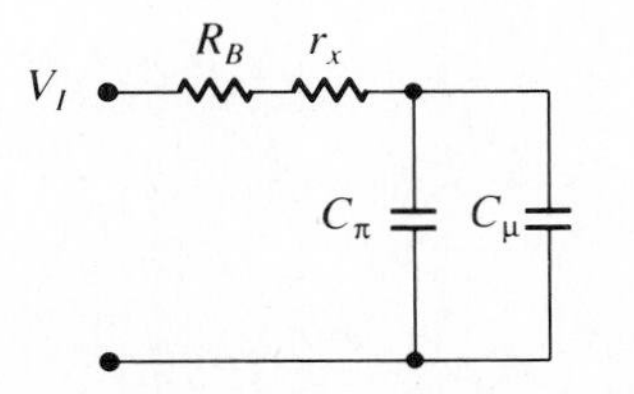

The final component t_{d3} is found by considering Equation (5.26) with $I_B = I_{B1}$:

$$i_C(t) = \beta_F I_{B1}(1 - e^{-t/\tau_{BF}}) \tag{5.59}$$

Note that the initial current I_i is zero, and the final value to which the collector current would lead if allowed is given by

$$I_f = \beta_F I_{B1} = \beta_F \frac{V_1 - V_{BE(sat)}}{R_B}$$

although it will actually reach

$$I_{C(eos)} = \beta_F I_{B(eos)} = \frac{V_{CC} - V_{CE(sat)}}{R_C}$$

because the external collector circuit can support only the latter value, even though $I_{B1} > I_{B(eos)}$. This distinction between the final value towards which the collector current heads and its actual final value is very important, as it greatly affects the transistor's switching time.

The collector current reaches 10% of its final value in t_{d3} seconds. Therefore, setting $i_C(t) = .1I_{C(eos)}$ and solving Equation (5.59) for $t = t_{d3}$, we get

$$t_{d3} = \tau_{BF}\left[\ln \frac{1}{1 - \dfrac{.1I_{B(eos)}}{I_{B1}}}\right]$$

$$= \tau_{BF} \ln\left[\frac{1}{1 - \dfrac{.1}{N_1}}\right] \tag{5.60}$$

where $N_1 \equiv I_{B1}/I_{B(eos)}$ is called the *forward overdrive factor*. It is a measure of the extra base current available above $I_{B(eos)}$. The larger N_1 is made, the smaller t_{d3} becomes, and the faster the device saturates. However, this does not mean that a large N_1 factor is necessarily desirable, as will be evident later.

Rise Time

Ignoring, for the time being, the effects of transition capacitances, we use Equation (5.59) to calculate the rise time t_r. Initially, assume that the transistor does not saturate, so that the actual final value will be $\beta_F I_{B1}$, with $I_{B1} < I_{C(eos)}$. Then

$$t_r \equiv t_3 - t_2 \tag{5.61}$$

where the times t_3 and t_2 are now defined by

$$.9\beta_F I_{B1} = \beta_F I_{B1}(1 - e^{-t_3/\tau_{BF}})$$

and

$$.1\beta_F I_{B1} = \beta_F I_{B1}(1 - e^{-t_2/\tau_{BF}})$$

respectively, to obtain

$$t_r = 2.2\tau_{BF} \tag{5.62}$$

Naturally, we want t_r to be as short as possible. Considering Equation (5.24), we see that t_r is directly proportional to $\beta_F \tau_F$, which implies an apparent conflict if both high-speed operation and high current gain are desired. Also note, however, that since the range of β_F, being at most a factor of 10, is relatively narrow, and the range of τ_F is usually at least two orders of magnitude, the control over rise time actually lies in τ_F. From Equation (5.15), we know that the mean forward transient time is expressed as $\tau_F = W_B^2/2D_B$. The diffusion coefficient is relatively constant, so the variation in τ_F is largely due to W_B. Consequently, a small W_B corresponds to a high-frequency transistor. High-power, low-frequency transistors, on the other hand, have longer rise times.

To decrease the rise time, the base current should be increased. A larger base current changes the stored base charge faster, thereby increasing the rate of change of the collector current and reducing the rise time. In other words, the speed with which the collector current can be changed depends on the speed with which the charge is pumped into the base by means of current in the base circuit.

As i_B is increased, the BC junction becomes forward-biased, and the transistor is driven into the edge of saturation. Recalling that the final value of i_C is $I_{C(eos)}$ and using Equations (5.59) and (5.61), we get

$$t_r = \tau_{BF} \ln \left[\frac{1 - .1/N_1}{1 - .9/N_1} \right] \tag{5.63}$$

It is evident from Equation (5.63) that a large overdrive factor N_1 reduces t_r and leads to a faster switching time, as before. Note that if $N_1 = 1$, Equation (5.63) reduces to (5.62), as it should.

In Equation (5.59), τ_{BF}, the effective minority carrier lifetime in the neutral base region, is used as the *rise time constant* τ_r, which determines the rate at which the collector current changes in the forward active mode. Now we use the hybrid-pi model to find an expression for τ_r in terms of the transistor parameters.

In Figure 5.8d, we can transform the capacitance C_μ into the input and the output using *Miller's theorem*. Consider the situation illustrated in Figure 5.13a. Two nodes and the ground terminal of a particular network are identified, and an admittance Y is connected between the nodes. Miller's theorem states that admittance Y can be replaced by two admittances, Y_1 and Y_2, as diagrammed in Figure 5.13b. Defining $V_2/V_1 \equiv K$, the current I_1 in Figure 5.13a is given by

$$I_1 = Y(V_1 - V_2) = Y(1 - K)V_1$$

From Figure 5.13b, we have

Figure 5.13 Illustration of Miller's theorem.

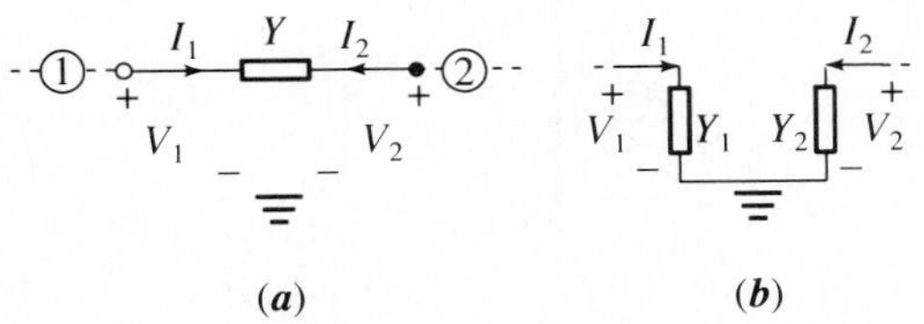

$$I_1 = Y_1 V_1$$

Equating these leads to

$$Y_1 = (1 - K)Y$$

A similar argument for I_2 would result in

$$Y_2 = \left(1 - \frac{1}{K}\right) Y$$

These expressions for Y_1 and Y_2 are both necessary and sufficient conditions for the network in Figure 5.13b to be equivalent to that of Figure 5.13a. Thus, applying Miller's theorem to the model of Figure 5.8d leads to the equivalent circuit of Figure 5.14. If $R_C \ll |\omega C_\mu|^{-1}$, we can neglect C_μ to get

$$K = \frac{v_{ce}}{v_{b'e}} = -g_m R_C$$

so that, at the input, the total capacitance is given by

$$C_T = C_\pi + C_\mu(1 - K) = C_\pi + C_\mu(1 + g_m R_C) \tag{5.64}$$

Assuming that $(R_B + r_x) \gg r_\pi$, we have

$$\tau_r = r_\pi C_T = r_\pi(C_\pi + C_\mu) + r_\pi g_m R_C C_\mu$$

$$= \beta_F \left(\frac{1}{\omega_T} + R_C C_\mu\right) \tag{5.65}$$

The second term on the right-hand side of Equation (5.65) is small and often neglected to simplify calculations. Hence, as expected, we get

$$\tau_r \approx \frac{\beta_F}{\omega_T} = \tau_{BF} \tag{5.66}$$

Another important contribution to rise time is due to the simultaneous changes in the depletion region charges Q_{BE} and Q_{BC}, the latter of which especially cannot be ignored. Integrating Equation (5.30c) from t_2 to t_3, we get

Figure 5.14 Equivalent circuit in the forward active mode obtained by using Miller's theorem.

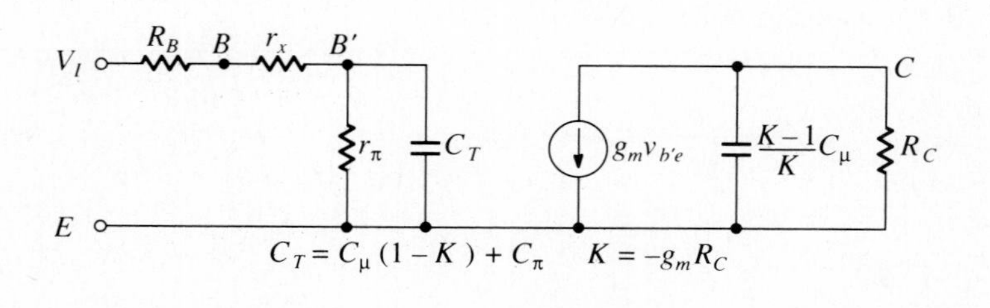

$$\int_{t2}^{t3} i_B(t)\,dt = \frac{1}{\tau_B}\int_{t2}^{t3} Q_F(t)\,dt + \Delta Q_F + \Delta Q_{BE} + \Delta Q_{CE} \qquad (5.67)$$

where

$$\Delta Q_F = Q_F(t_3) - Q_F(t_2)$$

and

$$Q_F(t_3) = .9\tau_F I_{C(eos)}$$

$$Q_F(t_2) = .1\tau_F I_{C(eos)}$$

Therefore, assuming a linear increase of Q_F during rise time, we have

$$\Delta Q_F = .8\tau_F I_{C(eos)}$$

At the beginning of the interval,

$$I_{Bi} = \frac{V_1 - V_{BE(on)}}{R_B}$$

and at the end,

$$I_{Bf} \approx I_{B1} = \frac{V_1 - V_{BE(sat)}}{R_B}$$

Therefore, the average base current during this interval is given by

$$I_{B(av)} \frac{2V_1 - (V_{BE(on)} + V_{BE(sat)})}{2R_B}$$

Furthermore, during rise time, the BE junction voltage changes from $V_{BE(on)}$ to $V_{BE(sat)}$ so that $\Delta V_{BE} \approx 0$. The BC junction voltage changes from about $V_{BE(on)} - V_{CC}$ to $V_{BE(sat)} - V_{CE(sat)}$; that is, $\Delta V_{BC} \approx V_{CC}$. Thus, Equation (5.67) becomes

$$I_{B(av)}(t_3 - t_2) = \left(\frac{1}{\tau_{BF}}\right) Q_{F(av)}(t_3 - t_2) + .8\tau_F I_{C(eos)} + C_{tC(av)} V_{CC} \qquad (5.68)$$

where $Q_{F(av)} = .5\tau_F I_{C(eos)}$. Rearranging yields a more accurate expression for the rise time than Equation (5.63):

$$t_r = t_3 - t_2 = \frac{.8\tau_F I_{C(eos)} + C_{tC(av)} V_{CC}}{I_{B(av)} - Q_{F(av)}/\tau_{BF}} \qquad (5.69)$$

Storage Time

Next, we consider the turn-off transient for the BJT. As the input voltage, V_I, returns to the low level, V_2, the collector current cannot respond immediately. It remains constant for time t_{s1} (the first component of the storage time, t_s), which is the time required to reduce the overdrive charge in the base to zero.

The change in the profile of the excess minority carriers in the base of a saturated transistor is shown in Figure 5.15. As the transistor is turned off, the extra stored

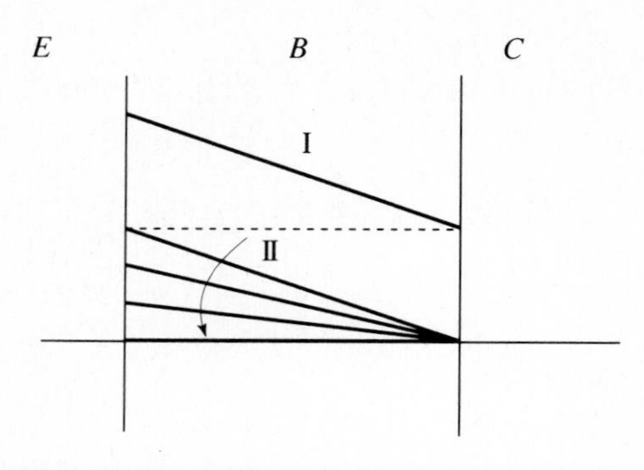

Figure 5.15 Profile of the excess minority carriers in the base during turn-off.

charge has to be removed first, during which time the profile changes from line I to line II. In the meantime, the base current reverses its direction because the BE junction voltage remains as $V_{BE(on)}$, while the input voltage is at the negative level V_2. This reverse current discharges the base and reduces the overdrive charge to zero.

To find t_{s1}, we make use of Equation (5.34). With the step change of input from V_1 to V_2, the current in the base becomes I_{B2}. Hence, we have

$$I_{B2} - I_{B(eos)} = \frac{dQ_S}{dt} + \frac{Q_S}{\tau_S}$$

whose solution is of the form

$$Q_S(t) = (I_{B2} - I_{B(eos)})\tau_S + Ke^{-t/\tau_S}$$

subject to an initial condition. We find the initial excess charge Q_{Si} over the critical charge Q_A needed to just saturate the transistor by assuming that the transistor had already reached the steady state in saturation before the switching occurred at the input. Hence, using Equation (5.34) again, we obtain

$$I_{B1} - I_{B(eos)} = \frac{Q_{Si}}{\tau_S}$$

Thus, from the preceding two equations, we find $K = (I_{B1} - I_{B2})\tau_S$ to get

$$Q_S(t) = \tau_S\left[(I_{B2} - I_{B(eos)}) + (I_{B1} - I_{B2})e^{-t/\tau_S}\right] \tag{5.70}$$

which is valid only up to the edge of saturation. Note that the final value of the excess charge is not zero but rather a negative value, $\tau_S(I_{B2} - I_{B(eos)})$, which is the value to which Q_S would go if left alone without the transistor limiting it.

After t_{s1} seconds, the overdrive charge is reduced to zero. Setting Q_S to zero in Equation (5.70), we solve for the first component of the storage time:

$$t_{s1} = \tau_S \ln\left(\frac{I_{B1} - I_{B2}}{I_{B(eos)} - I_{B2}}\right) \tag{5.71}$$

Note that $t_{s1} = 0$ only if $I_{B1} = I_{B(eos)}$, which means that the input pulse just saturates the transistor without producing any excess base charge.

Once the overdrive charge has been removed, the collector current begins to fall exponentially to zero, and hence the slope of the excess minority charge also decreases toward zero, as depicted in Figure 5.15. In the meantime, the reversed base current gradually decreases to zero as the BE transition capacitance charges up to the reverse bias voltage V_2. Therefore, t_{s2} seconds later, the collector current falls to 90% of its initial value of $I_{C(eos)}$. The calculation of the second component, t_{s2}, is similar to calculating t_{d3}. The result is

$$t_{s2} = \tau_{BF} \ln \left[\frac{1 - \dfrac{1}{N_2}}{1 - \dfrac{.9}{N_2}} \right] \tag{5.72}$$

where the negative quantity $N_2 \equiv I_{B2}/I_{B(eos)}$ is called the *reverse overdrive factor*. It is a measure of the amount of base current available to pull the transistor out of saturation.

Fall Time

The calculation of the fall time t_f is straightforward and similar to that of rise time because the transistor now operates again in the forward active mode. Referring to Figure 5.11, it is defined as

$$t_f \equiv t_6 - t_5$$

Proceeding as before, we obtain

$$t_f = \tau_{BF} \ln \left[\frac{1 - .9/N_2}{1 - .1/N_2} \right] \tag{5.73}$$

As in the case of the rise time, however, we have to take into account an important contribution due to changes in the depletion region charges Q_{BE} and Q_{BC}. The average base current during turn-off is obtained from the initial value,

$$I_{Bi} = I_{B2} = \frac{V_2 - V_{BE(sat)}}{R_B}$$

and final value,

$$I_{Bf} = \frac{V_2 - V_{BE(on)}}{R_B}$$

as

$$I_{B(av)} = \left| \frac{2V_2 - (V_{BE(on)} + V_{BE(sat)})}{2R_B} \right|$$

During fall time, the change in the BE junction voltage is given as $\Delta V_{BE} = V_{BE(on)} - V_{BE(sat)} \approx 0$. The BC junction voltage changes from $V_{BE(sat)} - V_{CE(sat)}$ to approximately $V_{BE(on)} - V_{CC}$, so that $\Delta V_{BC} = -V_{CC}$. Therefore, assuming a linear decrease of Q_F during fall time, we obtain a complete expression for the total fall time, similar to Equation (5.69), as

$$t_f = \frac{-(\Delta Q_F + \Delta Q_{BE} + \Delta Q_{BC})}{-(|I_{B(av)}| + Q_{F(av)}/\tau_{BF})}$$

$$= \frac{.8\tau_F I_{C(eos)} + C_{tC(av)} V_{CC}}{|I_{B(av)}| + Q_{F(av)}/\tau_{BF}} \tag{5.74}$$

Comparing Equations (5.69) and (5.74) reveals that the loss of carriers by recombination leads to an increase in the rise time, while it helps to remove charge from the base during the fall time. Moreover, as is obvious from the ongoing discussion of switching times, a large I_{B1} and N_1 reduce t_r and t_{d3}. Thus, to reduce the turn-on time, a large N_1 is desirable. However, this results in a longer t_s and hence a longer turn-off time because a large I_{B1} value drives the transistor deeper into saturation. On the other hand, a large $|N_2|$, achieved by a more negative V_2 value, reduces the fall time and the turn-off time. A large $|V_2|$ leads to a large BE reverse voltage, increasing the delay time and hence the turn-on time. The program listed in the appendix at the end of this chapter determines the inverter switching times, given the physical parameters of the transistor and the external parameters, along with the voltage input.

Speed-Up Capacitor

The rise time can be improved by shunting a so-called speed-up capacitor across the base resistor, as shown in Figure 5.16. The abrupt change at the input, which is considerably larger than the steady-state value, $V_{BE(on)}$, is transmitted to the BE junction by the capacitance. The additional voltage will inject the charge $Q_F = C(V - V_{BE(on)})$ into the base, reducing the rise time and producing a step change of collector current as

$$I_C = \frac{Q_F}{\tau_F} = \frac{C(V - V_{BE(on)})}{\tau_F}$$

Figure 5.16 The basic BJT inverter with speed-up capacitor.

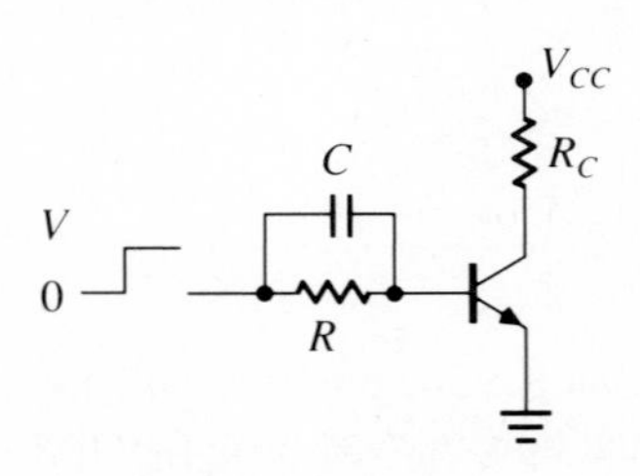

This is achieved if the available base current $(V - V_{BE(on)})/R$ is made equal to $Q_F/\tau_{BF} = C(V - V_{BE(on)})/\tau_{BF}$, the current needed to maintain this base charge. Equating them also provides the optimum value for the time constant,

$$RC = \tau_{BF}$$

Example 5.2

Consider the BJT inverter of Figure 5.11a. The transistor parameters and resistor values are given as follows: $\tau_{BF} = 15$ ns; $\tau_F = .15$ ns; $\tau_S \approx \tau_R = 15$ ns; $C_{tEo} = .4$ pF; $C_{tCo} = .2$ pF; $\phi_E = .9$ V; $\phi_C = .7$ V; $R_B = 10$ KΩ; $R_C = 1$ KΩ; $V_{CC} = 5$ V; and V_I switches from -5 V to $+5$ V and back to -5 V. To find the switching-time components, assume $V_\gamma = .5$ V; $V_{BE(on)} = .7$ V; $V_{BE(sat)} = .8$ V; and $V_{CE(sat)} = .1$ V.

Immediately after switching, the initial base current is found as

$$I_{Bi} = \frac{5\text{ V}}{10\text{ K}\Omega} = .5\text{ mA}$$

After t_{d1} seconds,

$$I_{Bf} = \frac{(5 - .5)\text{ V}}{10\text{ K}\Omega} = .45\text{ mA}$$

Therefore, the average base current during this time interval is $I_{B(av)} = .475$ mA. Now,

$$\Delta V_{BE} = \Delta V_{BC} = .5 - (-5) = 5.5\text{ V}$$

so that, using Equation (2.27), we find the large-signal equivalent capacitances across the BE and BC junctions as

$$C_{tE(av)} = .25\text{ pF}$$

and

$$C_{tC(av)} = .118\text{ pF}$$

Therefore, using Equation (5.54), we have

$$t_{d1} = \frac{(.25\text{ pF} + .118\text{ pF})(5.5\text{ V})}{.475\text{ mA}} = 4.26\text{ ns}$$

Using Equation (5.57), on the other hand, and neglecting r_x, would yield

$$t_{d1} = 2.95\text{ ns}$$

The second delay component is found using Equation (5.58):

$$t_{d2} = .05\text{ ns}$$

Next, we find the final component of the delay time. The dc current gain of the transistor is

$$\beta_F = \frac{15}{.15} = 100$$

The base current needed to saturate the transistor is then given by

$$I_{B(eos)} = \frac{(5 - .1)\text{V}}{1\ \text{K}\Omega} \times 100 = .049\ \text{mA}$$

while the actual value of the base current is

$$I_{B1} = \frac{(5 - .8)\text{V}}{10\ \text{K}\Omega} = .42\ \text{mA}$$

Therefore, the forward overdrive factor is determined as

$$N_1 = \frac{.42}{.049} = 8.57$$

Substituting this value into Equation (5.60), we have

$$t_{d3} = .18\ \text{ns}$$

To find t_r, we first note that

$$I_{C(eos)} = \frac{(5 - .1)\text{V}}{1\ \text{K}\Omega} = 4.9\ \text{mA}$$

so that

$$\Delta Q_F = .8(.15 \times 10^{-9}\ \text{s})(4.9 \times 10^{-3}\ \text{A}) = .588\ \text{pC}$$

On the other hand, during this interval, the initial value of the BC junction voltage is approximately given as $.7 - 5 = -4.3$ V while the final value can be approximated as $.8 - .1 = .7$ V. Thus, $\Delta V_{BC} = V_{CC} = 5$ V. The corresponding average value of the transition capacitance is found to be

$$C_{tC(av)} = .15\ \text{pF}$$

which in turn produces

$$\Delta Q_{BC} = (.15\ \text{pF})(5\ \text{V}) = .75\ \text{pC}$$

Also,

$$Q_{F(av)} = .5(.15\ \text{ns})(4.9\ \text{mA}) = .3675\ \text{pC}$$

Finally, the average value of the base current during rise time is found to be

$$I_{B(av)} = .425\ \text{mA}$$

Substituting these values into Equation (5.69) yields

$$t_r = 3.34\ \text{ns}$$

Next, we find the storage-time components. With the step change of input from 5 V back to -5 V, the base current becomes

$$I_{B2} = \frac{(-5 - .8)\text{V}}{10\ \text{K}\Omega} = -.58\ \text{mA}$$

Hence, using Equation (5.71), we have

$$t_{s1} = 6.95\ \text{ns}$$

During this time interval, the overdrive charge in the base is removed; then the collector current starts falling towards zero. It falls to 90% of its initial saturation value in t_{s2} seconds, which is obtained by using Equation (5.72), with $N_2 = -11.84$, as

$$t_{s2} = .13 \text{ ns}$$

Therefore, the total storage time of 7.08 ns is predominantly due to the time needed to remove the overdrive charge in the base.

Finally, during the fall time, the average value of the base current is found as

$$I_{B(av)} = -.575 \text{ mA}$$

Since the other quantities in Equation (5.74) are the same as in the case of the rise time, we find the fall time as

$$t_f = 2.23 \text{ ns}$$

Notice that the fall time is less than the rise time, due to large reverse current in the base.

5.3 The SPICE BJT Model

The earlier version of SPICE had two separate models for the bipolar junction transistor. The simpler one was based upon the *Ebers-Moll (E-M) model*, with the addition of *basewidth modulation*; the second was based upon the integral charge model as proposed by Gummel and Poon (the *G-P model*). Later versions merged both models into a *modified G-P model* that extends the original model to incorporate various effects at high-bias levels. This model automatically reduces to the E-M model when default values are used for certain parameters. This section presents these models in the context of the notation used in SPICE.

Basic Ebers-Moll Model

The Ebers-Moll model is capable of describing all four modes of operation with a single set of equations. To appreciate this, consider the minority carrier distributions in the base under different bias conditions, as sketched in Figure 5.17. Notice especially that the carrier distribution may be decomposed into two components in the saturation region. A saturated transistor can be represented by two transistors connected together with one operating in the forward active mode and the other in the reverse active mode.

Equation (5.17) expresses the current I_E that crosses the BE junction in the forward active mode. A large fraction of the minority carriers injected into the base region reach the reverse-biased BC junction, which can be represented by an $\alpha_F I_E$ current generator, as described in Equation (5.18).

In the reverse active mode of operation, the BC junction is forward-biased, while the BE junction is reverse-biased. This mode can then be described by a collector current, $I_C = I_{CS}(e^{V_{BC}/\phi_T} - 1)$, where I_{CS} is the reverse saturation current for the BC junction, and by a current generator $\alpha_R I_C$, representing the high impedance of the reverse-biased BE junction. Here, the small fraction α_R, as explained before, is due to the ratio of the minority carriers collected at the emitter to the number injected into the base at the forward-biased collector junction.

Merging these representations defines the classical bipolar transport model that applies for any arbitrary bias across the junctions. The Ebers-Moll model for the *npn* transistor, assuming both junctions injecting, is shown in Figure 5.18.

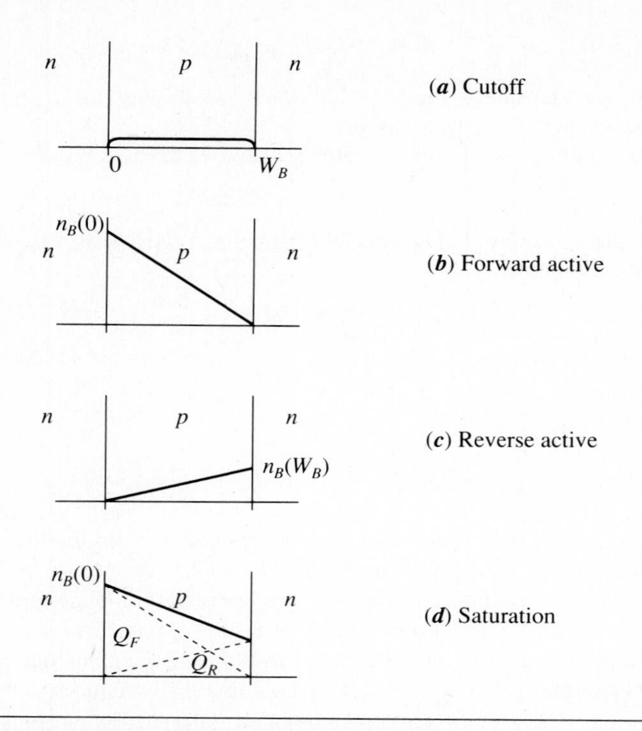

Figure 5.17 Minority carrier profiles in the base region of an *npn* transistor under all four modes of operation: (a) cutoff; (b) forward active; (c) reverse active; (d) saturation.

Figure 5.18 The Ebers-Moll model for the *npn* transistor.

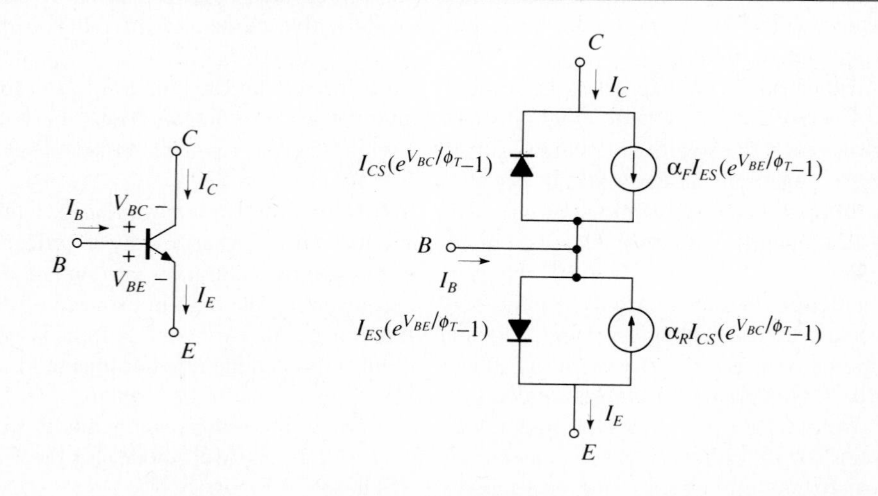

The general equations for the terminal currents I_E and I_C are then given by

$$I_E = I_{ES}(e^{V_{BE}/\phi_T} - 1) - \alpha_R I_{CS}(e^{V_{BC}/\phi_T} - 1)$$

$$= I_{ES}\lambda(V_{BE}) - \alpha_R I_{CS}\lambda(V_{BC}) \qquad (5.75a)$$

where

$$\lambda(V_{BE}) \equiv e^{V_{BE}/\phi_T} - 1$$

$$\lambda(V_{BC}) \equiv e^{V_{BC}/\phi_T} - 1$$

and

$$I_C = \alpha_F I_{ES}(e^{V_{BE}/\phi_T} - 1) - I_{CS}(e^{V_{BC}/\phi_T} - 1)$$

$$= \alpha_F I_{ES}\lambda(V_{BE}) - I_{CS}\lambda(V_{BC}) \qquad (5.75b)$$

Note that the currents are described by two variables, V_{BE} and V_{BC}, and four parameters: I_{ES}, I_{CS}, α_F, and α_R. For the ideal transistor, the parameters are related by the *reciprocity theorem* as

$$\alpha_F I_{ES} = \alpha_R I_{CS} \equiv I_S$$

where the reverse saturation current I_S is also referred to as the *transport current*. It is a function of area and integrated base doping. For an *npn* device,

$$I_S = \frac{qA_{min}n_i^2}{\int_0^{W_B} [N_{aB}(x)/D_B]\,dx}$$

where A_{min} is the minimum of the two junction areas; that is, $A_{min} \equiv \min[A_E, A_C]$. For uniform doping, and assuming $A_E = A_C = A$, we have

$$I_S = \frac{qAD_B n_i^2}{N_{aB}W_B}$$

Modified E-M Model

Figure 5.19 depicts the complete model for an *npn* transistor, as implemented in SPICE and used for both E-M and G-P representations. For a *pnp* device, the polarities of V_{BE}, V_{BC}, V_{CE}, I_B, and I_C must be reversed. The *modified* E-M equations for the terminal currents, as used by SPICE as a special case of the more complete Gummel-Poon model, are given by

$$I_C = \mathbf{IS}(e^{V_{BE}/\mathbf{NF}\phi_T} - e^{V_{BC}/\mathbf{NR}\phi_T})\left(1 - \frac{V_{BC}}{V_A}\right)$$

$$-(\mathbf{IS}/\mathbf{BR})(e^{V_{BC}/\mathbf{NR}\phi_T} - 1) \qquad (5.76a)$$

$$I_B = (\mathbf{IS}/\mathbf{BF})(e^{V_{BE}/\mathbf{NF}\phi_T} - 1) + (\mathbf{IS}/\mathbf{BR})(e^{V_{BC}/\mathbf{NR}\phi_T} - 1) \qquad (5.76b)$$

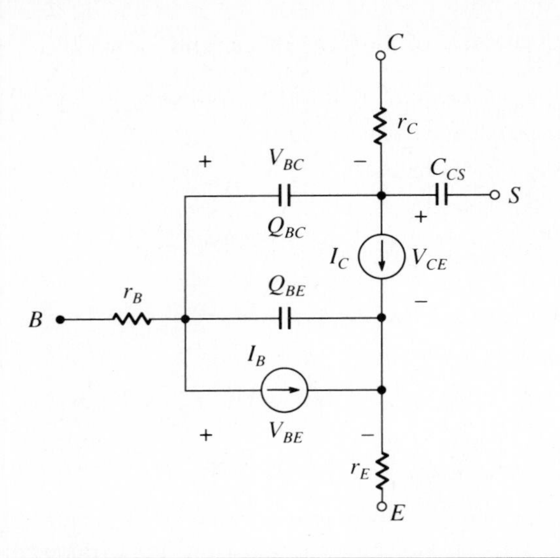

Figure 5.19 Complete model for an *npn* transistor as implemented in SPICE.

where **IS**, **NF**, **NR**, **BF**, **BR**, and **VA** are user-specified parameters, as defined in Table 5.3. The transport current **IS** is the extrapolated intercept current of $\log(I_C)$ vs. V_{BE} in the forward active mode, as plotted in Figure 5.20, and of $\log(I_C)$ vs. V_{BC} in the reverse active mode. It is related directly to the zero-bias majority carrier profile in the base. Experimentally, it is estimated from several (I_C, V_{BE}) points in the ideal region of operation. The parameters **BF** and **BR** are the maximum forward and reverse current gains, respectively. The parameter V_A (**VAF**), referred to as the *forward Early* voltage, produces a finite value of output conductance g_o due to the *basewidth modulation*, as explained below.

The basic E-M model embodies superposition of collector and emitter diode currents, while actual devices violate superposition due to certain nonlinear effects. For instance, the current gain is actually a function of the collector voltage. This so-called *Early effect* is especially important in narrow-base transistors, in which the increase of reverse bias on the BC junction decreases the neutral basewidth, thereby increasing the gain. The emitter current is approximately proportional to $1/W_B$ and will therefore increase as W_B is decreased. Furthermore, with a smaller W_B there is even less of a chance that electrons will recombine in the base, so that the number of electrons that reach the collector will increase. As depicted in Figure 5.21, the Early effect on the actual device output characteristics shows a finite slope throughout the forward active region. The collector current varies linearly with V_{CE} over a substantial range, and the curves, when projected by straight-line sections, intersect at a voltage point V_A. If the transistor really obeyed superposition, all straight-line sections of the curves would be parallel to one another.

The corresponding output conductance can be approximated as

Table 5.3 SPICE BJT Model Parameters

Name	Parameter name	Unit	Default
IS	Reverse saturation current	A	1.0E−16
ISE*	BE leakage saturation current	A	0
ISC*	BC leakage saturation current	A	0
IKF*	Corner for forward beta high current roll-off	A	∞
IKR*	Corner for reverse beta high current roll-off	A	∞
IRB	Current where RB falls halfway to its minimum value	A	∞
ITF	Coefficient for TF collector current dependence	A	0
VAF*	Forward Early voltage	V	∞
VAR*	Reverse Early voltage	V	∞
VTF	Coefficient for TF BC voltage dependence	V	∞
VJE	BE built-in potential	V	.75
VJC	BC built-in potential	V	.75
VJS	CS built-in potential	V	.75
RB	Zero-bias base resistance	Ω	0
RBM	Minimum base resistance at high currents	Ω	RB
RE	Emitter ohmic resistance	Ω	0
RC	Collector ohmic resistance	Ω	0
CJE	Zero-bias BE transition capacitance	F	0
CJC	Zero-bias BC transition capacitance	F	0
CJS	Zero-bias collector-substrate (CS) capacitance	F	0
TF	Ideal forward transit time	s	0
TR	Ideal reverse transit time	s	0
BF	Ideal maximum forward gain		100
BR	Ideal maximum reverse gain		1
NF	Forward current emission coefficient		1
NE*	BE leakage emission coefficient		1.5
NR	Reverse current emission coefficient		1
NC*	BC leakage emission coefficient		2
MJE	BE-junction grading factor		.33
MJC	BC-junction grading factor		.33
MJS	CS-junction grading factor		0
XTB	Forward and reverse beta temperature coefficient		0
XTF	Coefficient for TF bias dependence		0
XCJC	Fraction of CJC connected to internal base node		1
PTF	Excess phase at $1/2\pi$ TF Hz	°	0
EG	Energy gap	eV	1.1

(continued)

Table 5.3 *Continued*

Name	Parameter name	Unit	Default
KF	Flicker-noise coefficient		0
AF	Flicker-noise exponent		1
FC	Forward bias nonideal transition capacitance coefficient		.5

*G-P model parameter

$$g_o \equiv \frac{I_C}{V_{CE}} \approx \frac{I_C}{V_A}$$

It is estimated at a particular operating point either from small-signal or curve-tracer measurements. The forward Early voltage V_A is then determined from the above equation. For most digital applications, V_A can be defaulted to infinity.

It is also possible to estimate V_A if the neutral basewidth, W_B, can be represented as a function of V_{BC} by a Taylor series expansion:

$$W_B(V_{BC}) = W_B + V_{BC} \left.\frac{dW_B}{dV_{BC}}\right|_{V_{BC}=0}$$

Figure 5.20 Log current versus voltage plot, showing the forward active mode parameters in the Gummel-Poon model.

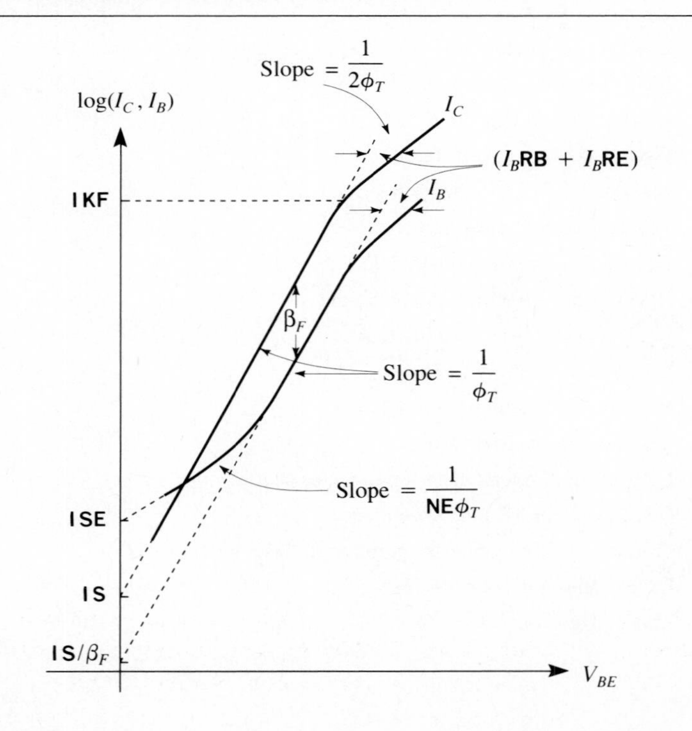

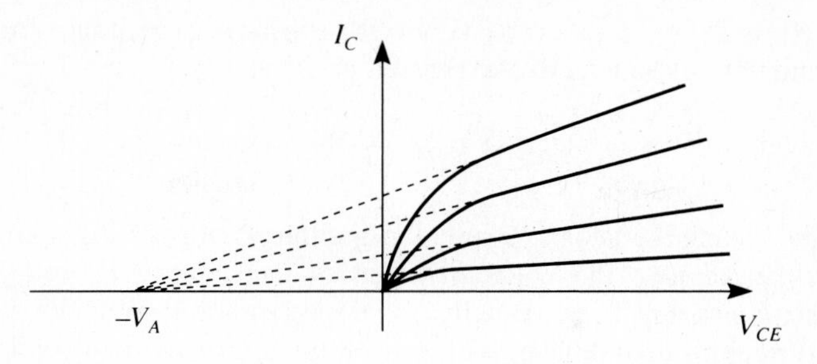

Figure 5.21 Plot of I_C vs. V_{CE} to show the effect of the basewidth modulation and the Early voltage, V_A.

Also assume that

$$\frac{W_B(V_{BC})}{W_B} = 1 + \frac{V_{BC}}{V_A}$$

Solving for V_A, we get

$$V_A = \frac{1}{\frac{1}{W_B}\left(\frac{dW_B}{dV_{BC}}\right)\Big|_{V_{BC}=0}} \tag{5.77}$$

Now, from Equation (2.23), we know that

$$W_B = \sqrt{\left(\frac{2\epsilon_s}{q}\right)\frac{N_{aB} + N_{dC}}{N_{aB}N_{dC}}\phi_C}$$

Thus,

$$W_B(V_{BC}) = W_B\sqrt{1 - \frac{V_{BC}}{\phi_C}}$$

Taking the derivative of the above equation with respect to V_{BC} at $V_{BC} = 0$, and substituting the result into Equation (5.77) yields

$$V_A = 2\left(1 + \frac{N_{aB}}{N_{dC}}\right)\phi_C \tag{5.78}$$

where ϕ_C itself is dependent on the impurity concentrations:

$$\phi_C = \phi_T \ln\left(\frac{N_{aB}N_{dC}}{n_i^2}\right)$$

As seen in Figure 5.19, the IC transistor has an additional component, C_{CS}(**CJS**), connected between the collector and substrate, representing the reverse-biased diode employed to isolate transistor collectors from each other and from the substrate. It is incorporated into the basic model as a collector-substrate capacitance for the transistor in the forward active mode and is expressed as

$$C_{CS} = \frac{C_{tSo}}{\left[1 - (V_{CS}/\phi_S)\right]^{m_S}}$$

where m_S and ϕ_S are the substrate junction grading factor and the built-in potential, respectively. In reality, the collector-substrate capacitance is voltage-dependent. Nevertheless, a constant C_{CS} is usually a valid assumption, in which case the BJT is treated as a three-terminal device, with C_{CS} returned directly to ground. Its value can be measured with a capacitance bridge. If not specified, it defaults to ground.

The Gummel-Poon (G-P) Model

Three major second-order effects are not represented by the basic E-M model:

1. collector-current-dependent output conductance due to *basewidth modulation*, that is, the Early effect, discussed above
2. *depletion layer recombination* at low-current levels
3. *high-level injection* effects such as *base stretching* and conductivity modulation in the base

The modified E-M model, incorporating only one second-order effect, lacks representation of other second-order effects that are present in actual devices. These effects are properly accounted for by the G-P model, which relates the electrical terminal characteristics to the base charge, thereby exhibiting a relationship to the charge-control model. It also takes into account the transport of minority carriers both by diffusion and drift in the base. Table 5.4 summarizes the second-order effects incorporated into the G-P model and the parameters that represent them. Even though the derivation of current equations for this model is beyond the scope of this book, we will briefly describe the effects themselves.

At low injection levels, the base current has *nonideal* components in the sense that its emission coefficient n deviates from 1. This is caused by the additional base current due to depletion-layer recombination, leading to current gain-degradation. It is particularly important for low-current I^2L circuits.

One effect of high injection is that, as the injected minority carrier density becomes comparable to the majority carrier density in the base, the hole current from the base to the emitter becomes a large portion of the total emitter current, reducing the emitter injection efficiency and leading to a decline in current gain. The *ideal*

Table 5.4 The G-P Representation of Second-Order Effects

Second-order effect	Parameters
Basewidth modulation	VAF VAR
Depletion-layer recombination	ISE NE ISC NC
High-level injection	IKF IKR

dependence of the collector current on the BE junction voltage is primarily due to the dependence of the minority carrier concentration in the base near the emitter as e^{V_{BE}/ϕ_T}. At a certain current density level, the low-level injection assumption is violated, and both the concentration and collector current vary as $e^{V_{BE}/n\phi_T}$ with $n \approx 2$. The intersection of the $n = 1$ and $n = 2$ asymptotes to I_C is called the *knee point*, as shown in Figure 5.20.

There are also additional complications due to *base stretching*, also known as the *Kirk effect*. Typically, an integrated transistor has a base doping much higher than that of the collector. As the current density crossing the BC junction increases, the electron concentration increases and approaches the density of the collector. The free electrons in this space-charge region influence the net charge density there by neutralizing the positively charged donor ions, so that the electric field can no longer be supported in the original BC transition region. Eventually, it collapses, and the base spreads over the collector epitaxial layer. The current gain degrades, the cutoff frequency is reduced, τ_F and τ_R increase, and hence the device slows down.

The charge storage in the BJT, in addition to the linear collector substrate capacitance C_{CS} discussed above, is modeled by two nonlinear storage elements, Q_{BE} and Q_{BC}, that are expressed as

$$Q_{BE} = \tau_F I_S (e^{V_{BE}/\phi_T} - 1) + C_{tEo} \int_0^{V_{BE}} \left(1 - \frac{V}{\phi_E}\right)^{-m_E} dV$$

$$Q_{CE} = \tau_R I_S (e^{V_{BC}/\phi_T} - 1) + C_{tCo} \int_0^{V_{BC}} \left(1 - \frac{V}{\phi_C}\right)^{-m_C} dV$$

where the two right-hand terms represent the injected minority-charge and space-charge contributions, respectively. Thus, the contribution due to the charge storage in the depletion regions is represented by the variable V_{BE} and the model parameters **CJE**, **VJE**, **MJE** or by the variable V_{BC} and the parameters **CJC**, **VJC**, **MJC**, depending on the junction. The contribution due to minority carrier injection across the junctions is modeled by **TF** for the BE junction and **TR** for the BC junction. To ensure a finite value for the transition capacitances for $V > \phi$, a linear approximation, controlled by the parameter **FC** as in the case of the diode model, is used in forward bias.

These elements can be equivalently represented by voltage-dependent capacitors as

$$C_{BE} = \frac{dQ_{BE}}{dV_{BE}} = \left(\frac{\tau_F I_S}{\phi_T}\right) e^{V_{BE}/\phi_T} + C_{tEo}\left[1 - \left(\frac{V_{BE}}{\phi_E}\right)\right]^{-m_E}$$

$$C_{BC} = \frac{dQ_{BC}}{dV_{BC}} = \left(\frac{\tau_R I_S}{\phi_T}\right) e^{V_{BC}/\phi_T} + C_{tCo}\left[1 - \left(\frac{V_{BC}}{\phi_C}\right)\right]^{-m_C}$$

In addition to the intrinsic elements discussed so far, the extrinsic elements shown in Figure 5.19 must be included. **RE**, **RB**, and **RC** are all constant parameters, representing passive components to terminal nodes, that is, bulk ohmic resistances of the three transistor regions. The former, having a typical value of approximately 1 Ω, is often neglected. When included, it reduces the BE junction voltage by **RE**I_E; so that it is equivalent to a base resistance of $(1 + \beta_F)$**RE**.

Values of **RB** range from about 10 Ω to several kilohms. Measuring this parameter is difficult because of its dependence on the operating point and also because of the error introduced by **RE**. Resistance **RC** decreases the slope of the curves on the I_C vs. V_{CE} characteristics in the saturation region for small V_{CE} values. Its value varies from a few ohms to hundreds of ohms. It is calculated from the slope of the characteristic, with β_F forced to 1. Thus, this technique yields the collector resistance in saturation mode that is appropriate for switching circuits. For linear applications, the higher value found by measuring the slope of the characteristic curve in the linear region should be used. SPICE considers it a constant.

The optional [(*area*) *value*] on the device card scales **RE**, **RB**, **RC**, **RBM**, **CJE**, **CJC**, **CJS**, **IS**, **ISC**, **ISE**, **IKF**, **IKR**, **IRB**, and **ITF**, and defaults to 1. When it is used, the capacitance and current values are multiplied by the area factor, while the resistance values are divided by it.

Example 5.3

The following is the SPICE input file for the basic BJT inverter of Example 5.2. The collector current and voltage waveforms as well as the voltage transfer characteristic are shown in Figure 5.22.

```
BASIC BJT INVERTER
VCC 4 0 5
VIN 1 0 PULSE -5 5 0 1P 1P 20N
RB  1 2 10K
RC  4 3 1K
Q1  3 2 0 Q1
.MODEL Q1 NPN BF 100 TF .15N CJE .4P CJC .2P PE .9 PC .7
+TR 15N MJE .5 MJC .5
.DC VIN -5 5 .05
.TRAN 1N 20N
.PROBE
.END
```

The following are the switching times in ns as obtained from SPICE and hand calculations:

	SPICE	Hand analysis
Delay time	3.40	4.49
Rise time	2.85	3.34
Storage time	2.06	7.08
Fall time	1.83	2.23

Example 5.4

The ring oscillator circuit shown in Figure 5.23 will be used to measure the average propagation delay time of the individual inverters. It is made up of a string of inverters, three in this case, in a series chain. Between successive nodes of the inverter chain, the

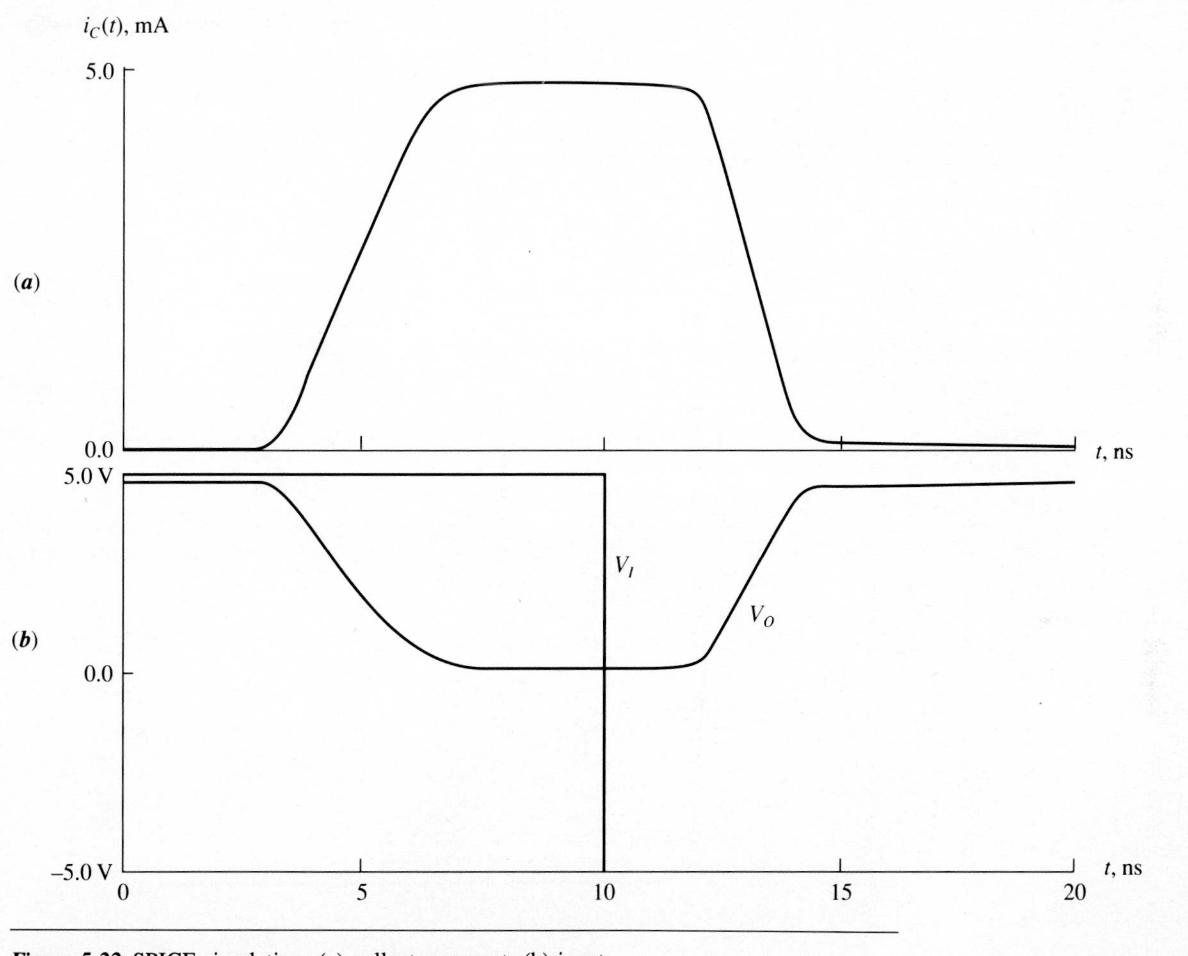

Figure 5.22 SPICE simulation: (a) collector current; (b) input and output voltage waveforms.

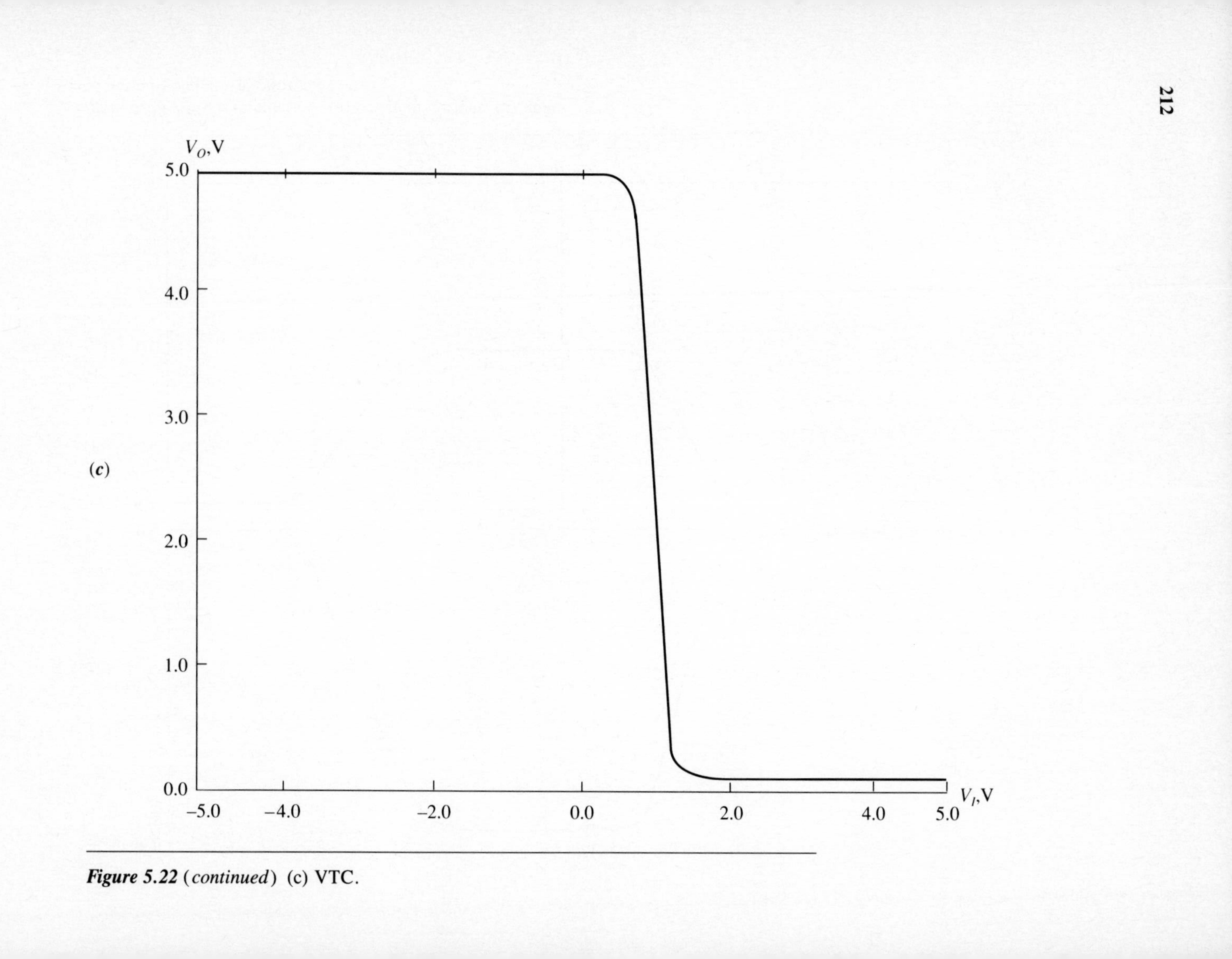

Figure 5.22 (*continued*) (c) VTC.

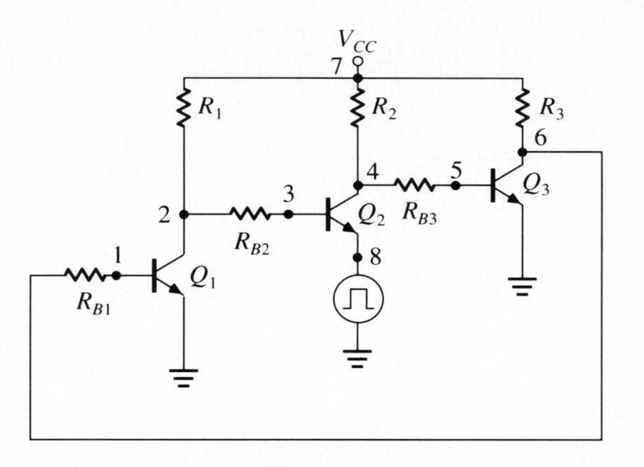

Figure 5.23 Three-stage ring oscillator circuit.

signal is opposite polarity under static conditions. With an odd number of gates, the output of the last stage is connected to the input to form the oscillator. The period of the oscillation frequency is equal to the total propagation delay time of the three inverters. Thus, the average propagation delay time is ideally the period of the oscillation divided by twice the number of gates within the ring. The following is the corresponding SPICE input listing used for simulation. A voltage pulse at the emitter of transistor Q_2 is employed to start the oscillator by offsetting the dc bias. This *trick* is not needed in practice as long as the number of inverters forming the ring is odd.

```
RING OSCILLATOR CIRCUIT
Q1   2 1 0 NPN
Q2   4 3 8 NPN
Q3   6 5 0 NPN
R1   7 2 240K
R2   7 4 240K
R3   7 6 240K
RB1 6 1 100
RB2 2 3 100
RB3 4 5 100
CS1 2 0 .5P
CS2 4 0 .5P
CS3 6 0 .5P
VEE 8 0 PULSE 0 .5 0 0 0 5N
VCC 7 0 5
.MODEL NPN NPN BF 70 IS 2E-16 RC 6 TF .2N TR 20N
+ CJE .3P CJC .15P VJE .9 VJC .7
.TRAN 10N 500N
.PROBE
.END
```

Figure 5.24 shows the resulting voltage waveforms at the collectors of Q_1 and Q_3. The oscillation period is 206 ns, so that $t_{PD} = 206/6 = 34.3$ ns. Also from Figure 5.24, $t_{PLH} = t_{PHL} = t_{PD} = 34.3$ ns.

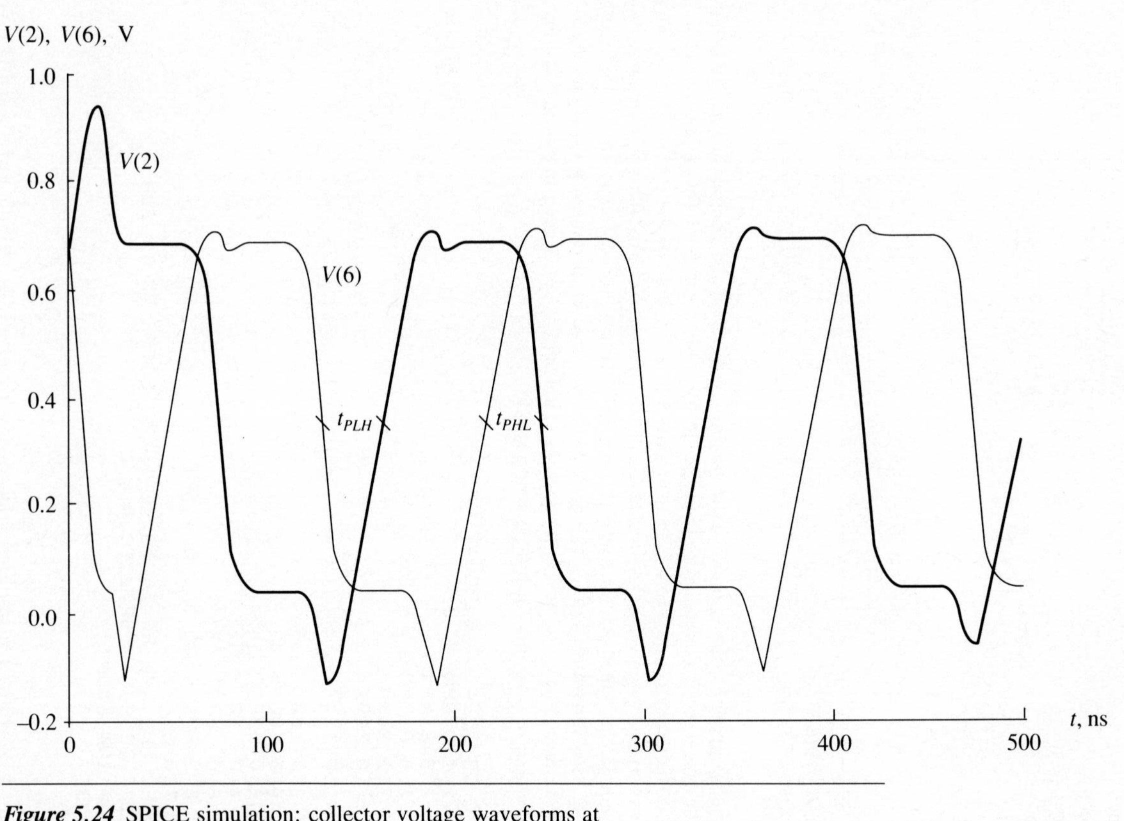

Figure 5.24 SPICE simulation: collector voltage waveforms at nodes 2 and 6 of the ring oscillator circuit.

Example 5.5

The circuit in Figure 5.25 will be used to examine the second-order effects in a BJT. Q_1 is a standard BJT while Q_2 includes the effects of depletion-layer recombination in the BE junction through SPICE model parameters **ISE** and **NE**, and of high-level injection by specifying **IKF**. V_{BC} is at -2 V reverse bias while V_{BE} is swept up to 1.1 V to secure the forward active mode of operation for both transistors. Figure 5.26 shows the plot generated by the following SPICE file. It illustrates the degradation of β_{F2} at low and high current levels.

```
SECOND-ORDER EFFECTS
VBC 1 0 2
VBE 0 4
Q1   1 0 4 Q1
Q2   1 0 4 Q2
.MODEL Q1 NPN BF 100 BR .1 IS 1E-14
.MODEL Q2 NPN BF 100 BR .1 IS 1E-14 ISE 1E-12 NE 2 IKF 5E-2
.DC VBE .4 1.1 .01
.PROBE
.END
```

5.4 BJT Design Considerations

Especially in linear operations and to some extent in digital circuits, one of the most important design objectives is to obtain a high current gain, β_F. To achieve this, the base current components must be minimized, which requires γ and α_T to approach unity. To see how basic semiconductor properties are associated with the design of a BJT to satisfy this objective, we need to develop sufficiently accurate expressions for I_E and I_C. For our purpose, it will be adequate to assume an ideal device with no recombination occurring in the base and having uniform doping densities in all regions. The former assumption can be considered to be valid in modern BJTs. Furthermore, such an ideal model greatly simplifies equations for emitter and collector currents. It enables us to avoid mathematical detail by not having to write minority carrier concentrations in terms of hyperbolic functions. With no base recombination

Figure 5.25 Circuit for the SPICE simulation to investigate the second-order effects on the BJT current gain.

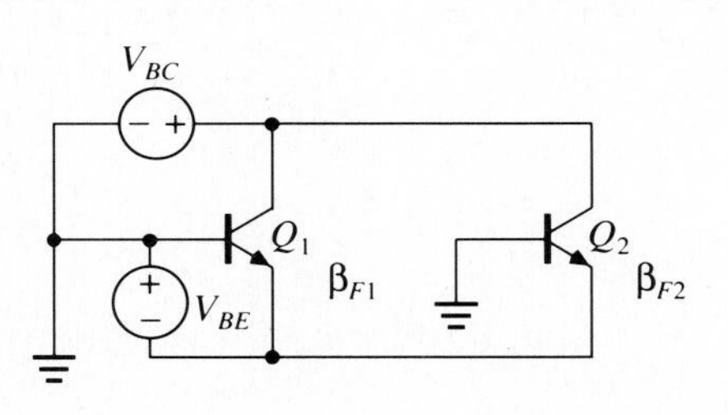

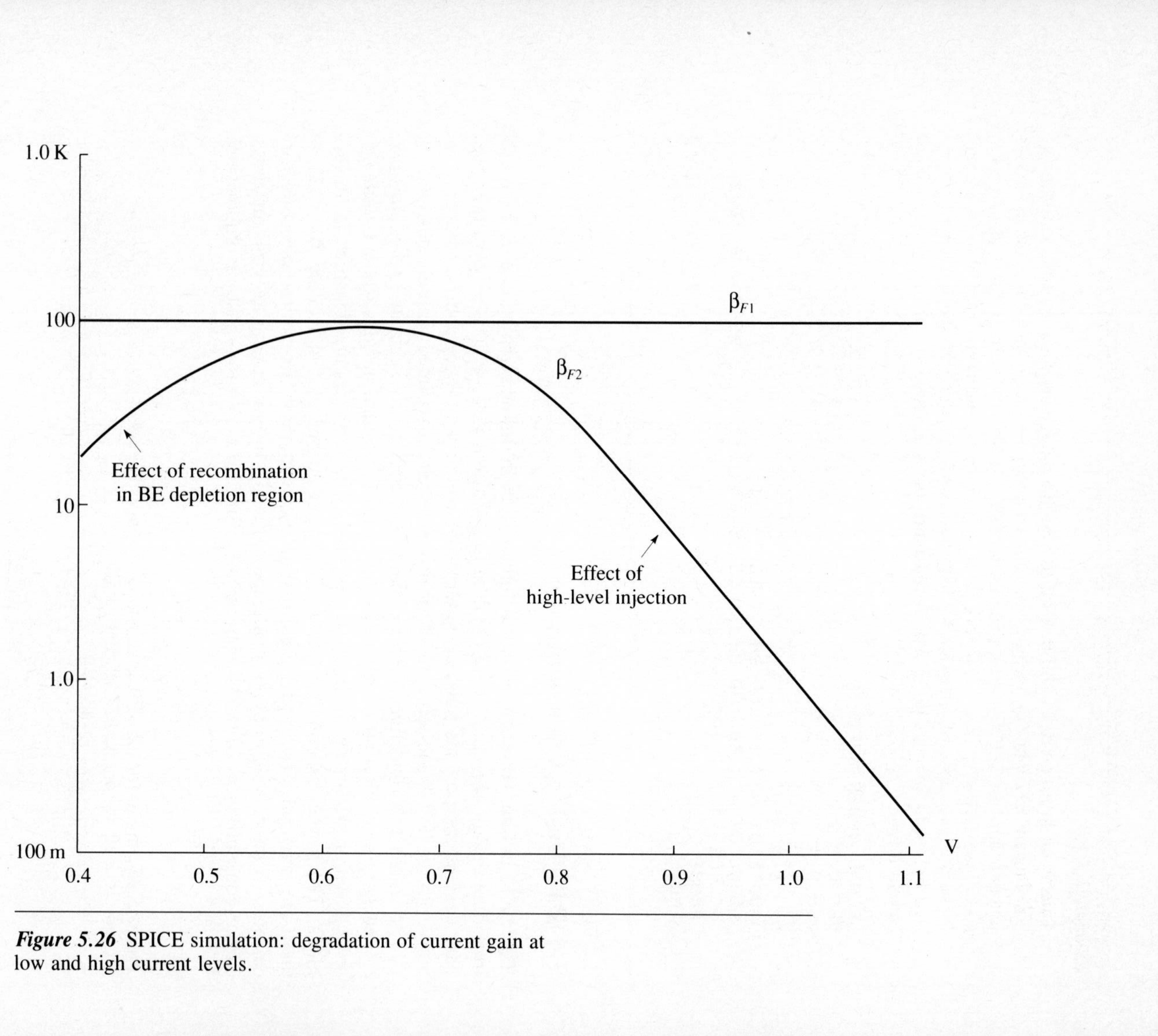

Figure 5.26 SPICE simulation: degradation of current gain at low and high current levels.

current (i.e., $I_{B2} = 0$), the lifetime, τ_{BF}, and L_B must approach infinity, or, in other words, $W_B \ll L_B$.

The emitter current crossing the BE depletion region is made up of two minority carrier diffusion currents:

$$I_E = I_{Ee}(0) + I_{Eh}(x_E) = qA_E D_B \left.\frac{d\Delta n_B(x)}{dx}\right|_{x=0} + qA_E D_E \left.\frac{d\Delta p_E(x)}{dx}\right|_{x=x_E}$$

where the first term on the right side is due to the electrons injected from the emitter into the base, while the second term is caused by holes injected from the base into the emitter. Notice that terms are evaluated at the edges of the depletion region, with x_E as defined in Figure 5.2c.

To solve for the first component, we express Equation (5.9) as

$$\frac{d^2\Delta n_B(x)}{dx^2} = \frac{\Delta n_B(x)}{L_B^2} = 0$$

whose solution is the equation of a straight line,

$$\Delta n_B(x) = K_1 x + K_2$$

Applying the boundary conditions yields

$$\Delta n_B(x) = \Delta n_B(0) - \left[\frac{\Delta n_B(0) - \Delta n_B(W_B)}{W_B}\right] x$$

Taking its derivative with respect to x and substituting Equations (5.8a) and (5.8b) produces the emitter current component:

$$I_{Ee}(0) = \left(\frac{qA_E D_B n_{Bo}}{W_B}\right)[\lambda(V_{BE}) - \lambda(V_{BC})]$$

Next, similar to Equation (2.6), we solve

$$\frac{d^2\Delta p_E(x)}{dx^2} = \frac{\Delta p_E(x)}{L_E^2}$$

for the excess hole concentration in the bulk emitter region. The solution is

$$\Delta p_E(x) = K_1 e^{-x/L_E} + K_2 e^{x/L_E}$$

Considering Equation (5.8c) and the fact that a hole cannot survive forever in the n-region when the region width is much greater than the minority diffusion length (i.e., $\Delta p_E(\infty) = 0$), we get

$$\Delta p_E(x) = p_{Eo}\lambda(V_{BE})e^{-(x-x_E)/L_E}$$

The hole current injected from the base to the emitter is obtained by taking the derivative of the above expression with respect to x at $x = x_E$:

$$I_{Eh}(x_E) = I_{B1} = \frac{qA_E D_E p_{Eo}\lambda(V_{BE})}{L_E}$$

The total emitter current is then

$$I_E = \left(\frac{qA_E D_B n_{Bo}}{W_B}\right)[\lambda(V_{BE}) - \lambda(V_{BC})] + \left(\frac{qA_E D_E p_{Eo}}{L_E}\right)\lambda(V_{BE})$$

$$= qA_E\left[\frac{D_B n_{Bo}}{W_B} + \frac{D_E p_{Eo}}{L_E}\right]\lambda(V_{BE}) - \left(\frac{qA_E D_B n_{Bo}}{W_B}\right)\lambda(V_{BC}) \qquad (5.79)$$

The collector current crossing the BC junction is determined by the electrons entering the depletion region from the base side and by the holes on the collector side. Evaluating each minority carrier diffusion current at the depletion edges, we get

$$I_C = I_{Ce}(W_B) + I_{Ch}(x_C) = qA_C D_B \left.\frac{d\Delta n_B(x)}{dx}\right|_{x=0} + qA_C D_C \left.\frac{d\Delta p_C(x)}{dx}\right|_{x=x_C}$$

A similar set of calculations carried out previously for the emitter current along with appropriate boundary conditions from Equation (5.8) and $\Delta p_C(\infty) = 0$ yield the collector current. For example, the second current component can be directly obtained from the expression for $I_{Eh}(x_E)$ with a change in sign and by exchanging the subscript E for C and V_{BE} for V_{BC}:

$$I_{Ch}(x_C) = -\frac{qA_C D_C p_{Co}\lambda(V_{BC})}{L_C}$$

With no recombination in the base, $I_{Ee}(0) = I_{Ce}(W_B)$. The total collector current is then

$$I_C = \left(\frac{qA_C D_B n_{Bo}}{W_B}\right)\lambda(V_{BE}) - qA_C\left[\frac{D_B p_{Bo}}{W_B} + \frac{D_C p_{Co}}{L_C}\right]\lambda(V_{BC}) \qquad (5.80)$$

If we now allow a finite I_{B2}, that is, holes furnished to the base for recombination with electrons, then $n_B(x)$ is changed from a straight-line to a hyperbolic function. However, with modern devices having $W_B/L_B < 1/10$, we can incorporate recombination into our calculations while still keeping the straight-line approximation to the minority carrier distribution in the base.

I_{B2} is due to the total excess charge in the base, which recombines on the average of τ_{BF} seconds so that, using Equations (5.23) and (5.12), we have

$$I_{B2} = \frac{Q_F}{\tau_{BF}}$$

$$= \left(\frac{qA_E}{2\tau_{BF}}\right)[\Delta n_B(0) - \Delta n_B(W_B)]W_B$$

$$= \left(\frac{qA_E W_B n_{Bo}}{2\tau_{BF}}\right)[\lambda(V_{BE}) - \lambda(V_{BC})] \qquad (5.81)$$

This expression is valid for all modes of operation.

Since I_{B2} is much smaller than I_{Ee}, and the latter is determined by the slope of $n_B(x)$ at $x = 0$, which in turn does not deviate from the straight-line approximation,

I_E does not change. On the other hand, with fewer electrons arriving at the collector, I_{Ce} will be reduced by an amount equal to I_{B2}.

Now, if we compare Equations (5.79) and (5.80) with Equation Set (5.75) and consider the reciprocity theorem, we obtain α_F and α_R in terms of physical parameters. Assuming $A_E = A_C$,

$$\alpha_F = \frac{1}{1 + \dfrac{D_E W_B p_{Eo}}{D_B L_E n_{Bo}}}$$

Since we assumed $\alpha_T = 1$, the emitter efficiency is given by $\gamma = \alpha_F$. Now, $p_{Eo}/n_{Bo} = N_{aB}/N_{dE}$, so that

$$\gamma = \frac{1}{1 + \dfrac{D_E W_B N_{aB}}{D_B L_E N_{dE}}} \tag{5.82}$$

Substituting Equation (5.82) into (5.7) yields

$$\beta_F = \frac{D_B L_E N_{dE}}{D_E W_B N_{aB}} \tag{5.83}$$

Similarly, we have

$$\alpha_R = \frac{1}{1 + \dfrac{D_C W_B N_{aB}}{D_B L_C N_{dC}}}$$

and from Equation (5.35),

$$\beta_R = \frac{D_B L_C N_{dC}}{D_C W_B N_{aB}} \tag{5.84}$$

To find a general expression for α_T, note that in the forward active mode, Equation (5.81) reduces to

$$I_{B2} = \left(\frac{qA_E W_B n_{Bo}}{2\tau_{BF}}\right)\lambda(V_{BE})$$

so that

$$\begin{aligned} I_{Ce} &= I_{Ee} - I_{B2} \\ &= \left[\frac{qA_E D_B n_{Bo}\lambda(V_{BE})}{W_B}\right]\left(1 - \frac{W_B^2}{2D_B\tau_{BF}}\right) \\ &= I_{Ee}\left(1 - \frac{W_B^2}{2D_B\tau_{BF}}\right) \end{aligned}$$

Hence, from Equation (5.1), we obtain

$$\alpha_T = 1 - \frac{(W_B/L_B)^2}{2} \tag{5.85}$$

Therefore, we have two important factors to consider in the design of a BJT:

1. From Equation (5.85), to improve α_T and hence β_F, we must have $W_B \ll L_B$; that is, the minority carrier diffusion length must be much greater than the basewidth.
2. From Equation (5.83), $N_{dE} \gg N_{aB}$; that is, doping the emitter much more strongly (n^+) than the base leads to a large current gain.

However, the emitter doping is limited by a process called *band-gap narrowing*, which causes the effective value of the energy gap to decrease. As doping levels increase, the doped atoms interact and cause the effective energy gap to decrease in value, resulting in an undesirable increase in the number of thermally generated minority carriers. Thus, there is only a certain range of values for impurity concentrations to balance the desired emitter efficiency with band-gap narrowing.

On the other hand, a very thin base increases α_T but also exposes the base either to *avalanche breakdown* due to the electric field created across the region or *punchthrough* due to the reduction in the effective width of the base region as the BC depletion region expands under reverse bias and touches the BE depletion region.

To avoid punchthrough while keeping the basewidth thin, the impurity concentration constraint cited above must be chosen such that most of BC depletion region will extend into the collector. Meanwhile, to minimize bulk resistance and hence excessive power dissipation, a lower limit on the collector doping level should be set by the maximum allowable bulk resistance, r_C, since from Equations (1.15) and (1.22),

$$r_C = \rho\ell/A_C \propto \frac{1}{N_{dC}}$$

To achieve both objectives, the collector may be doped at the lower level near the BC junction. Doping density is increased to an n^+ level away from the junction, as shown in Figure 5.27. In fact, it is common practice in BJT fabrication to employ a heavily doped thick n-type substrate for low collector resistance as well as for mechanical support. A lightly doped n-type thin epitaxial layer is then grown on top of this substrate. Care must be taken to separate the BC junction and n^+ region by a distance greater than L_B so that there will not be an interaction between the n^+ collector region and base.

For digital applications, a thinner base turns out to be an important design consideration, too, since this leads to an improvement in transit time and hence in the operating speed of the transistor. The transition between the logical states is governed by the time taken to establish a new minority carrier concentration profile in the base. In a long p^+n junction, the switching time is related directly to the minority carrier

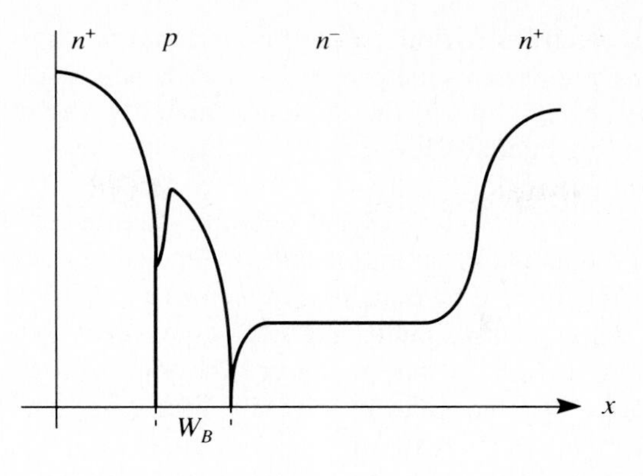

Figure 5.27 Typical impurity profile in an *npn* ($n^+pn^-n^+$) transistor.

lifetime because *all* the holes injected from the p^+ region recombine in the n material. However, in a BJT, the base region is intentionally made very narrow, as discussed above in detail, so that *most* of the electrons traverse the base and reach the collector. Therefore, a new minority carrier profile in the base is established by the transit time τ_F, which, as Equation (5.15) shows, is directly proportional to the square of the basewidth.

Summary

The BJT is made up of two back-to-back *pn* junctions sharing a common region, that is, the base. The base separates the collector from the emitter. The latter is more heavily doped than the base to improve the emitter injection efficiency. In a well-designed transistor, the basewidth is much less than the minority carrier diffusion length to improve the base transport factor. Two transistor structures, *npn* and *pnp*, are possible.

A transistor has four modes of operation. In the forward active mode, the BE junction is forward-biased, and the BC junction is reverse-biased. Exactly the opposite is true in the reverse active mode of operation. In saturation, both junctions are forward-biased, leading to a decreased value of the collector current below its value in the forward active mode. In cutoff, both junctions are reverse-biased, so that the terminal currents are negligible.

In the forward active mode, the collector current is almost equal to the emitter current, with the small base current, being the difference, controlling the collector current. The ratio of the collector current to the base current is the current gain. It is adversely affected at low values of the base-emitter voltage by recombination in the BE depletion region, and at high values of V_{BE} by high-level injection in the base.

In an *npn* transistor, a large portion, γ, of the emitter current is injected into the base in the form of electrons. Of this, a portion, α_T, makes it across the base into the collector. The transistor currents are proportional to the slopes of the minority carrier profile in the base at the edges of the depletion regions. The emitter current flows through the base at low injection levels by the diffusion mechanism because the electric field is small in this region.

The BJT is used as a switch in digital circuits, making use of the almost short-circuit behavior of the output at saturation and almost open-circuit behavior at cutoff. The switching speed is limited by transition capacitances and by charge accumulation in the base. The charge-control model provides convenient solutions for switching times. The basic parameters in this model are the total excess base charge in the forward active mode, Q_F; the minority electron lifetime in the base, τ_{BF}; and the transit time, τ_F, for these carriers to traverse the base region.

The Ebers-Moll transistor equations relate the emitter and collector currents to the BE and BC junction voltages. The Ebers-Moll circuit parameters can be expressed in terms of the physical device design parameters. The Gummel-Poon model incorporates certain second-order effects that were not included in the Ebers-Moll model.

Appendix 5A: Pascal Program for BJT Parameters and Inverter Transient Behavior

Given the impurity concentrations in all three regions of a BJT, the junction cross-sectional area, and the basewidth, the following program written in Pascal calculates the following parameters:

ϕ_E = built-in potential at BE junction
ϕ_C = built-in potential at BC junction
C_{tEo} = zero-bias transition capacitance at BE junction
C_{tCo} = zero-bias transition capacitance at BC junction
β_F = common-emitter forward current gain
β_R = common-emitter reverse current gain
I_S = transport current
V_A = forward Early voltage
τ_F = mean forward transit time in the base
τ_R = mean reverse transit time in the base
τ_{BF} = effective minority carrier lifetime in the base
α_F = common-base forward current gain
α_R = common-base reverse current gain
α_T = base transport factor
γ = emitter injection efficiency

In the second part, given the external parameters V_1, V_2, V_{CC}, R_B and R_C, switching times t_d, t_r, t_s, and t_f are determined for the basic BJT inverter. To avoid the infinite BC transition capacitance as predicted by Equation (2.27) for $V_{BC} > \phi_C$, its average value is modeled by a linear extrapolation beyond $V_{BC} \geq \mathbf{FC}\phi_C$, where **FC** is set to its SPICE default value of $1/2$. It is assumed that the low-level input voltage V_2 is less than ϕ_E.

```
program bjt_parameters;

uses crt;
var stat:char; phi_E,phi_C,C_tEo,C_tCo,b_f,b_r,Is,Va,a_t,a_f,a_r,gamma,t_f,t_r,t_bf:real;

procedure bjt_paras(var phi_E,phi_C,C_tCo,C_tEo,b_f,b_r,Is,Va,a_t,a_f,a_r,gamma,t_f,t_r,t_bf:real);

var N_dC,N_dE,N_aB,W_B,A,q,L_B,L_E,L_C,k,u_aB,u_dE,u_dC,D_aB,D_dE,D_dC,t_aB,t_dE,t_dC:real;

begin
  textbackground(red);
  clrscr;
  writeln('        **********   BJT PARAMETERS AND INVERTER TRANSIENT BEHAVIOR  *********');
  writeln('        *                                                                   *');
  writeln('        * Given doping densities, junction cross sectional area and the     *');
  writeln('        * basewidth, this program first determines the following parameters: *');
  writeln('        *********************************************************************');
  writeln;
  writeln('                φE   = built-in potential at BE junction');
  writeln('                φC   = built-in potential at BC junction');
  writeln('                CtEo = zero bias transition capacitance at BE junction');
  writeln('                CtCo = zero bias transition capacitance at BC junction');
  writeln('                βf   = common-emitter forward current gain');
  writeln('                βr   = common-emitter reverse current gain');
  writeln('                Is   = transport current');
  writeln('                Va   = forward Early voltage');
  writeln('                τf   = mean forward transit time in the base');
  writeln('                τr   = mean reverse transit time in the base');
  writeln('                τbf  = effective minority carrier lifetime in the base');
  writeln('                αf   = common-base forward current gain');
  writeln('                αr   = common-base reverse current gain');
  writeln('                αT   = base transport factor');
  writeln('                gamma= emitter injection efficiency');
  writeln;
  writeln('        *       and then calculates the switching times for the inverter    *');
  writeln('                Press any key to start....');stat:=readkey;
  clrscr;
  writeln('        Enter the concentrations in atoms per cm-3 in the form x(.x)Exx');
  writeln;
  write  ('        Enter N_dE = '); readln(N_dE);
```

```
write  ('        Enter N_aB = '); readln(N_aB);
write  ('        Enter N_dC = '); readln(N_dC);
writeln;
writeln('        Enter the junction area, A = Ae = Ac in cm² in the form x.(x)Exx');
writeln;
write  ('        Enter A = '); readln(A);
writeln;
write  ('        Enter the basewidth W_B in μm = '); readln(W_B);
writeln;

phi_E:=0.0259*ln(N_aB*N_dE/2.25E20);
phi_C:=0.0259*ln(N_aB*N_dC/2.25E20);
u_aB:=88+1252/(1+6.984E-18*N_aB);
u_dE:=54.3+407/(1+3.745E-18*N_dE);
u_dC:=54.3+407/(1+3.745E-18*N_dC);
D_aB:=0.0259*u_aB;
D_dE:=0.0259*u_dE;
D_dC:=0.0259*u_dC;
C_tEo:=A*sqrt((8.44E-32/phi_E)*N_aB*N_dE/(N_aB+N_dE));
C_tCo:=A*sqrt((8.44E-32/phi_C)*N_aB*N_dC/(N_aB+N_dC));
t_aB:=5E-7/(1+2E-17*N_aB);
t_dE:=5E-7/(1+2E-17*N_dE);
t_dC:=5E-7/(1+2E-17*N_dC);
L_B:=sqrt(D_aB*t_aB);
L_E:=sqrt(D_dE*t_dE);
L_C:=sqrt(D_dC*t_dC);
a_f:=1/(1+((D_dE*W_B*1E-4*N_aB)/(D_aB*L_E*N_dE)));
a_r:=1/(1+((D_dC*W_B*1E-4*N_aB)/(D_aB*L_C*N_dC)));
a_t:=1-0.5*sqr(W_B*1E-4/L_B);
b_f:=a_f/(1-a_f);
b_r:=a_r/(1-a_r);
gamma:=a_f/a_t;
t_f:=sqr(W_B*1E-4)/(2*D_aB);
t_r:=100*t_f;
Is:=1.602E-19*A*D_aB*sqr(1.5E10)/(N_aB*W_B*1E-4);
Va:=2*(1+(N_aB/N_dC))*phi_C;

writeln('τr will default to 100τf = ',t_r:9);
writeln;
write  ('Is this acceptable? (Y/N)');
```

```
  stat:=readkey;
  if not (stat='Y') and not (stat='y') then
    begin
      writeln;
      writeln;
      write('Enter the multiplication factor k (τ= k x τf) '); readln(k);
      t_r:=k*t_f;
    end;
  t_bf:=b_f*t_f;

  clrscr;
  writeln('          ***********************************************************');
  writeln('          * A Bipolar Junction Transistor with doping densities of *');
  writeln('          * Nab = ',N_aB:9,', Nde = ',N_dE:9,' and Ndc = ',N_dC:9,',  *');
  writeln('          * junction area of Ae=Ac =',A:8,' cm-2 and a basewidth  *');
  writeln('          * of Wb = ',W_B:4:2,' μm will have the following parameters:    *');
  writeln('          ***********************************************************');
  writeln;
  writeln('                      φC    = ',phi_C:5:3,' V');
  writeln('                      φE    = ',phi_E:5:3,' V');
  writeln('                      CtEo = ',C_tEo:8,' F');
  writeln('                      CtCo = ',C_tCo:8,' F');
  writeln('                      βf    = ',b_f:6:1);
  writeln('                      βr    = ',b_r:5:2);
  writeln('                      Is    = ',Is:9,' A');
  writeln('                      Va    = ',Va:5:1,' V');
  writeln('                      τf    = ',t_f*1E9:5:2,' ns');
  writeln('                      τr    = ',t_r*1E9:5:2,' ns');
  writeln('                      τbf   = ',t_bf*1E9:5:2,' ns');
  writeln('                      αf    = ',a_f:5:3);
  writeln('                      αr    = ',a_r:5:3);
  writeln('                      αT    = ',a_t:5:3);
  writeln('                      gamma= ',gamma:5:3);
  writeln;
end;

procedure bjt_times(var phi_E,phi_C,C_tEo,C_tCo,b_f,b_r,Is,Va,t_f,t_r,t_bf:real);

var td1,td2,td3,tr,ts1,ts2,tf,V1,V2,Vcc,Rb,Rc,Qfav,Cteav,Ctcav,Ibeos,Iceos,Ib1,Ib2,Ibav,N1,N2:real;
```

```
begin
  textbackground(red);
  clrscr;
  writeln;
  writeln('           **********   THE BJT INVERTER SWITCHING TIMES **********');
  writeln('           *                                                       *');
  writeln('           *        Given the following external parameters:     *');
  writeln('           *                                                       *');
  writeln('           *              V1,  high level input voltage           *');
  writeln('           *              V2,  low level input voltage            *');
  writeln('           *              Vcc, supply voltage                     *');
  writeln('           *              Rc,  collector series resistance        *');
  writeln('           *              Rb,  base series resistance             *');
  writeln('           *                                                       *');
  writeln('           * the program calculates the following switching times:*');
  writeln('           *                                                       *');
  writeln('           *                   td, tr, ts, and tf                  *');
  writeln('           *********************************************************');
  writeln;
  write('                         Enter V1  (V) = '); readln(V1);
  write('                         Enter V2  (V) = '); readln(V2);
  write('                         Enter Vcc (V) = '); readln(Vcc);
  write('                         Enter Rb (KΩ) = '); readln(Rb);
  write('                         Enter Rc (KΩ) = '); readln(Rc);

  Rb:=Rb*1000;
  Rc:=Rc*1000;
  Cteav:=(-2*C_tEo*phi_E/(0.5-V2))*(sqrt(1-(0.5/phi_E))-sqrt(1-(V2/phi_E)));
  Ctcav:=(-2*C_tCo*phi_C/(-0.5-V2))*(sqrt(1-((0.5-Vcc)/phi_C))-sqrt(1-((V2-Vcc)/phi_C)));
  Ibav:=(2*V1-0.5)/(2*Rb);
  td1:=(Cteav+Ctcav)*(0.5-V2)/Ibav;
  td2:=t_f/3;
  Ibeos:=(Vcc-0.1)/(b_f*Rc);
  Ib1:=(V1-0.7)/Rb;
  N1:=Ib1/Ibeos;
  td3:=t_bf*ln(1/(1-0.1/N1));
  Iceos:=(Vcc-0.1)/Rc;
  if (phi_C<0.7) then
   Ctcav:=(-4*C_tCo*phi_C/(phi_C-2*(0.7-Vcc)))*(0.71-sqrt(1-(0.7-Vcc)/phi_C))+(0.3535*C_tCo)*(3+2*0.7/phi_C)
   else
   Ctcav:=(-2*C_tCo*phi_C/(Vcc))*(sqrt(1-(0.7/phi_C))-sqrt(1-((0.7-Vcc)/phi_C)));
```

```
   Ibav:=(2*V1-1.5)/(2*Rb);
   Qfav:=0.5*t_f*Iceos;
   tr:=((0.8*t_f*Iceos)+(Ctcav*Vcc))/(Ibav-Qfav/t_bf);
   Ib2:=(V2-0.8)/Rb;
   ts1:=t_r*ln((Ib1-Ib2)/(Ibeos-Ib2));
   N2:=Ib2/Ibeos;
   ts2:=t_bf*ln((1-1/N2)/(1-0.9/N2));
   Ibav:=abs((2*V2-1.5)/(2*Rb));
   tf:=((0.8*t_f*Iceos)+(Ctcav*Vcc))/(Ibav+(Qfav/t_bf));

   clrscr;
   writeln;
   writeln('             **************************************************');
   writeln('             * The BJT inverter with V1 = ',V1:4:1,' V, V2 = ',V2:4:1,' V, *');
   writeln('             * Vcc = ' , Vcc:4:1,' V, Rb = ',Rb/1000:5:1,' KΩ and Rc = ',Rc/1000:5:1,' KΩ *');
   writeln('             *     will have the following switching times:    *');
   writeln('             **************************************************');
   writeln;
   writeln('                  td = ',td1+td2+td3:9,' s');
   writeln('                  tr = ',tr:9,' s');
   writeln('                  ts = ',ts1+ts2:9,' s');
   writeln('                  tf = ',tf:9,' s');
   writeln;
end;

begin {main program}
  stat := 'Y';
  while (stat='Y') or (stat='y') do
    begin
      bjt_paras(phi_E,phi_C,C_tEo,C_tCo,b_f,b_r,Is,Va,a_t,a_f,a_r,gamma,t_f,t_r,t_bf);
      write('            Do you wish to change data? (Y/N)');
      stat:=readkey;
    end;
  stat := 'Y';
  while (stat='Y') or (stat='y') do
    begin
      bjt_times(phi_E,phi_C,C_tEo,C_tCo,b_f,b_r,Is,Va,t_f,t_r,t_bf);
      write('            Do you wish to change data? (Y/N)');
      stat:=readkey;
    end;
end.{main program}
```

References

1. D. L. Pulfrey and G. Tarr. *Introduction to Microelectronic Devices*. Prentice-Hall, Englewood Cliffs, NJ: 1989.
2. M. Zambuto. *Semiconductor Devices*. McGraw-Hill, New York: 1989.
3. P. Antognetti and G. Massobrio, eds. *Semiconductor Device Modeling with SPICE*. McGraw-Hill, New York: 1988.
4. D. A. Hodges and H. G. Jackson. *Analysis and Design of Digital Integrated Circuits*. 2d ed. McGraw-Hill, New York: 1988.
5. E. S. Yang. *Microelectronic Devices*. McGraw-Hill, New York: 1988.
6. D. H. Navon. *Semiconductor Microdevices and Materials*. Holt, Rinehart & Winston, New York: 1986.
7. R. A. Colclaser and S. Diehl-Nagle. *Materials and Devices for Electrical Engineers and Physicists*. McGraw-Hill, New York: 1985.
8. S. M. Sze. *Semiconductor Devices, Physics and Technology*. John Wiley & Sons, New York: 1985.
9. M. I. Elmasry. *Digital Bipolar Integrated Circuits*. John Wiley & Sons, New York: 1983.
10. G. W. Neudeck. *The Bipolar Junction Transistor*. Addison-Wesley, Reading, MA: 1983.
11. H. E. Talley and D. G. Daugherty. *Physical Principles of Semiconductor Devices*. The Iowa State University Press, Ames, Iowa: 1976.
12. L. W. Nagel, "SPICE2, A Computer Program to Simulate Semiconductor Circuits." *ERL Memorandum ERL-M520*, University of California, Berkeley: May 1975.
13. H. K. Gummel and H. C. Poon. "An Integral Charge Control Model of Bipolar Transistors." *Bell Syst. Tech. J.* **49** pp. 827–852 (1970).
14. J. J. Ebers and J. L. Moll. "Large-Signal Behavior of Junction Transistors." *Proc. IRE* **42** pp. 1761–1772 (1954).

PROBLEMS

For the following problems assume a room temperature of 300 K.

5.1 Express γ in terms of conductivities in the emitter and base regions, and estimate the amount by which γ differs from unity.

5.2 Find an expression for $I_{B3} = I_{Ch}$ in terms of the physical parameters when the transistor is in the forward active mode.

5.3 For $W_B/L_B = 1, 1/2, 1/5, 1/10$, find β_F.

5.4 Given $W_B = L_E = 10^{-4}$ cm, base resistivity of .1 Ω-cm, and emitter resistivity of .005 Ω-cm, find the injection efficiency of electrons for the emitter.

5.5 Given an *npn* BJT operating with $I_C = 1$ mA, $h_{fe} = 60$, $f_\beta = 1$ MHz and $D_B = 20\ \text{cm}^2/\text{s}$, determine N_{aB}, C_π, W_B, and τ_{BF}.

5.6 Using the linear approximation of Figure 5.2c, show that

$$n_B(x) = n_{Bo}e^{V_{BE}/\phi_T}(1 - x/W_B)$$

5.7 a. Calculate C_π for a transistor characterized by $\beta_F = 50$, $C_\mu = 10$ pF, $f_T = 100$ MHz and $I_C = 2$ mA.
b. Given $\beta_F = 100$, $C_\mu = 3$ pF, $f_T = 700$ MHz and $I_C = 5$ mA, determine r_π and C_π.

5.8 In the text, the total base charge in the saturation mode, Q_B, is represented as the sum of Q_A and Q_S. Another way to describe it is by summing the forward and reverse components, as shown in Figure 5.17d. Therefore, we have

$$Q_B = Q_F + Q_R = Q_A + Q_S$$

Show that

$$Q_R = \frac{1}{1 + \tau_F/\alpha_R\tau_R} Q_S$$

Hint: First obtain the charge-control equation for i_C in the saturation mode in terms of Q_F and Q_R by noting that a slight change in either Q_{BE} or Q_{BC} can lead to respective components being neglected. Also, $i_C = I_{C(eos)} = Q_A/\tau_F$.

5.9 In the text, we assume a constant value for $V_{CE(sat)}$. This problem requires you to find an expression for $V_{CE(sat)}$ in terms of transistor parameters. The only assumption will be that both V_{BE} and V_{BC} are greater than $4\phi_T$.

Note that

$$I_E = I_B + I_C \tag{1}$$

$$V_{CE(sat)} = V_{BE(sat)} - V_{BC(sat)} \tag{2}$$

a. Multiply Equation (5.75b) by α_R, subtract from Equation (5.75a), use Equation (1) above to eliminate I_E, and solve for $V_{BE(sat)}$.
b. Multiply Equation (5.75a) by α_F, subtract Equation (5.75b) from the product, use Equation (1) above to eliminate I_E, and solve for $V_{BC(sat)}$.
c. Use Equation (2) above and the reciprocity theorem to eliminate I_{CS} and I_{ES}, and define $\beta_{(sat)} \equiv I_C/I_B$ in the saturation mode.
d. Finally, use Equations (5.7) and (5.35) to show that the *intrinsic* saturation voltage is given by

$$V_{CE(sat)} = \phi_T \ln\left[\frac{1/\alpha_R + \beta_{(sat)}/\beta_R}{1 - \beta_{(sat)}/\beta_F}\right]$$

Note that to obtain the *terminal* saturation voltage, we should also take into account the voltage across the bulk resistances r_C and r_E developed by terminal currents I_C and I_E.

5.10 Consider the basic inverter of Figure 5.11a with $R_B = 10$ KΩ, $R_C = 1$ KΩ, and $V_{CC} = 5$ V. Given $\phi_E = .9$ V, $\phi_C = .7$ V, $C_{tEo} = 1.5$ pF, $C_{tCo} = .7$ pF, $\beta_R = 1$, $\beta_F = 100$, $r_C = 30$ Ω, $r_B = 47$ Ω, and $I_{ES} = 10^{-14}$ A. Determine Q_{BE} and Q_{BC} when $V_I = -5$ V.

5.11 Repeat Problem 5.10 for $V_I = 5$ V.

5.12 Diodes used in ICs are formed from *npn* transistors connected as junction diodes. For the open-emitter diode-connected transistor shown in the diagram, use the E-M equations to find I_B as a function of V_{BC}.

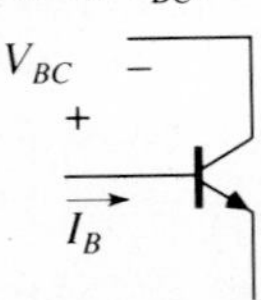

5.13 For the shorted base-collector diode-connected transistor shown in the diagram, find
a. I_E as a function of V_{BE},
b. V_{BE} if $I_E = 2$ mA and $I_{ES} = 10^{-14}$ A.

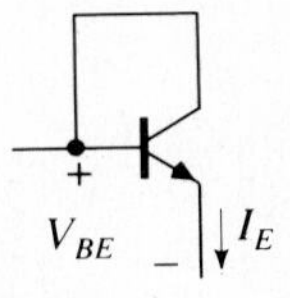

5.14 For the transistor shown in the diagram, $30 < \beta_F < 100$. Find the value of R_B that results in saturation with a minimum overdrive factor of $N1 = 10$. Assume $V_{CE(sat)} = .2$ V and $V_{BE(sat)} = .8$ V.

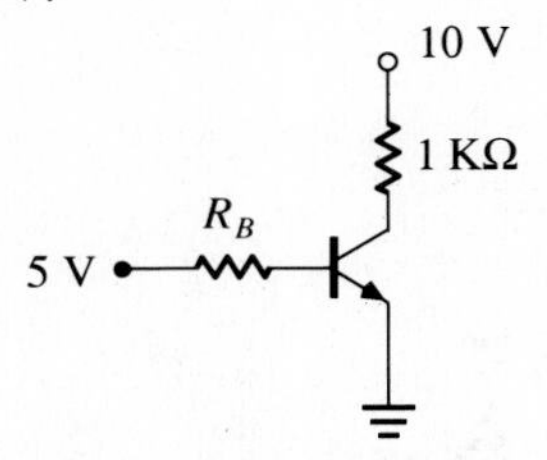

5.15 The following information is provided for an *npn* BJT: $\mu_{eE} = 450$ cm^2/V · s, $W_B = .7$ μm, $x_E = 1$ μm, $x_C = .9$ μm (so that the collector junction width $= .2$ μm), $A = 10^{-4}$ cm^2 and $\beta_F = 125$. Determine the hybrid-pi equivalent circuit parameters at $I_C = 5$ mA.

5.16 Given an *npn* transistor with a base width of 2×10^{-4} cm, $n_B(0) = 2 \times 10^{14}$ cm^{-3}, and $N_{aB} = 10^{15}$ cm^{-3}.
a. Calculate the diffusion current density of electrons $\mathbf{J}_{e(diffusion)}$ through the base region.
b. What should be the magnitude of the electric field in the base region for the electron drift current density to cancel the electron diffusion current density? Us : average electron density.
c. Find the corresponding voltage drop across the base.

5.17 The accompanying diagram shows the nonsaturating inverter circuit in which the SBD prevents the transistor from saturating, so that there is no storage time for the inverter circuit.
a. Explain the operation of the circuit qualitatively.
b. Compute the noise margins NM_H and NM_L.
c. Find the propagation delay times $t_{PHL} \equiv t_s + t_f/2$ and $t_{PLH} \equiv t_d + t_r/2$.
d. Verify (*b*) and (*c*) with SPICE.

Use the transistor data of Example 5.2. For the SBD, $C_{to} = .05$ pF and $\phi_o = .7$ V.

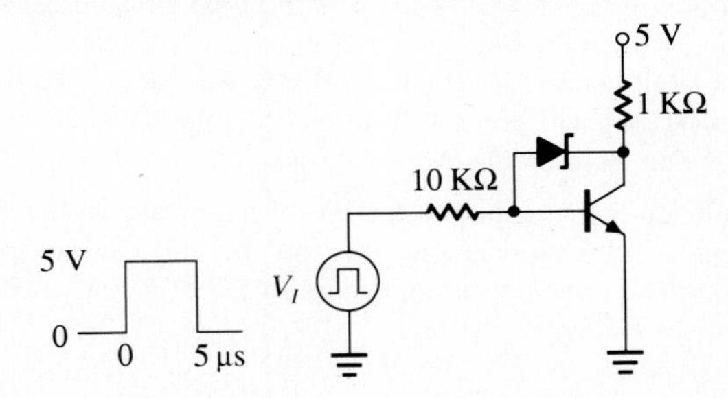

5.18 a. In the circuit shown in the diagram, determine the output rise time.
b. Explain the circuit operation after the active pull-up circuit is added to reduce the rise time.

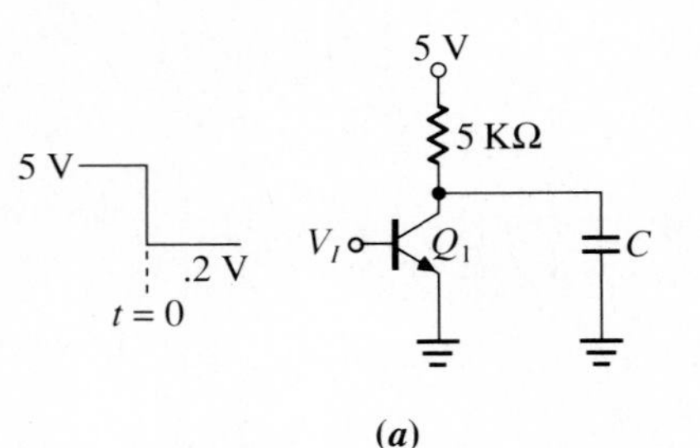

(*a*)

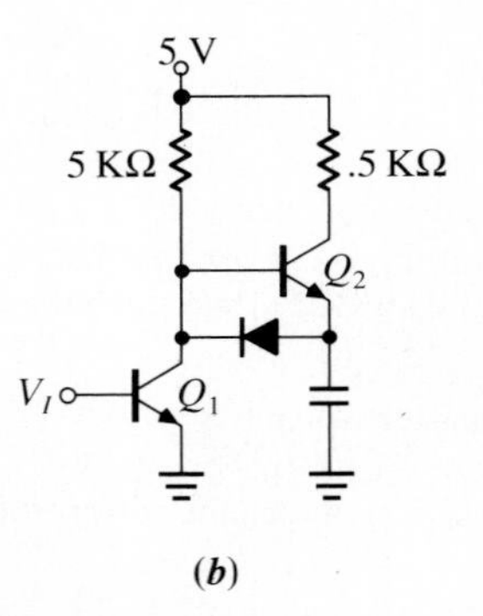

(*b*)

5.19 Design a simple circuit having five identical *npn* transistors connected in parallel that will allow you to examine the base current effect on the collector current. Each transistor is to be driven by a different base current, ranging from 0 to 12 μA in steps of 3 μA. Given the following physical properties for the transistors, prepare a SPICE input file to generate collector common-emitter characteristics over the range V_{CE} = 0 to 3 V. Specify **BF**, **BR** and **IS** in the model cards.

$$A = 5\ \mu\text{m} \times 5\ \mu\text{m}$$
$$W_B = .5\ \mu\text{m}$$
$$N_{dE} = 10^{19}\ \text{cm}^{-3} = 10^3 N_{aB} = 10^5 N_{dC}$$

5.20 Use the circuit of Problem 5.19 to investigate the effect of base impurity concentration on the BJT common-emitter characteristic at a fixed base current of 10 μA. Provide an input listing for SPICE specifying **BF**, **BR** and **IS** in the model cards for the transistors that will differ only in base doping densities over the range 10^{15} to 10^{19} cm^{-3}. Comment on the characteristics.

5.21 Repeat Problem 5.19 to show the effect of basewidth on the common-emitter collector characteristics by changing W_B over the range .5 μm to 2.5 μm in steps of .5 μm. Comment on the resulting change in the collector current.

5.22 Use SPICE to examine the base modulation effect by comparing the output characteristics of two transistors that are identical with the exception of the Early voltage. Specify the value of V_A for only one of the devices, given the following physical parameters:

$$A = 5\ \mu\text{m} \times 5\ \mu\text{m}$$
$$W_B = 1\ \mu\text{m}$$
$$N_{dE} = 10^{18}\ \text{cm}^{-3} = 10N_{aB} = 10^3 N_{dC}$$

5.23 a. Repeat Example 5.2 with V_I switching from 0 V to +5 V and back to 0 V after the circuit output settles to a stable logical state.

b. Verify with SPICE.

6. BIPOLAR LOGIC FAMILIES

Many different logic families can perform the basic logic operations of switching algebra. Of these, the most commonly used families are the *transistor-transistor logic* (*TTL*), the *emitter-coupled logic* (*ECL*), the *integrated-injection logic* (I^2L), and the *complementary metal-oxide semiconductor* (*CMOS*). The first three belong to the bipolar technology and are considered in this chapter. The CMOS is the topic of Chapter 4.

6.1 The Transistor-Transistor Logic

The first bipolar logic family to be discussed is the 54/74 line of TTL gates. TTL is the most widely used family for SSI and MSI circuits. The family parts for the standard subfamily are identified by a number of the form 54xx or 74xx where xx stands for a two- or three-digit number that signifies the function of the part. The military-grade 54xx series is functionally identical to the 74xx series but is manufactured to operate under stringent conditions. These devices are guaranteed to perform with a $\pm 10\%$ supply tolerance of ± 500 mV over an ambient temperature range of $-55°C$ to $+125°C$. On the other hand, the commercial grade 74xx series is guaranteed to perform with a $\pm 5\%$ supply tolerance of ± 250 mV over an ambient temperature range of 0°C to 70°C. The nominal supply voltage V_{CC} for all TTL circuits is +5 V. Currently, a number of different variations of the standard TTL circuitry are used to manufacture the family parts. The subfamily and the corresponding technology used to make a gate are identified by a letter or combination of letters embedded in its part number.

As the technology has advanced over the years since the standard subfamily first appeared on the commercial market in 1965, more efficient subfamilies, as measured by the speed-power product, evolved to satisfy the demanding performance requirements of more and more sophisticated applications. The available subfamilies of the 54/74 line of the TTL family are listed in Table 6.1. The last two, namely, 54AS/74AS and 54ALS/74ALS, were introduced in 1980, and are the most ad-

Table 6.1 Subfamilies of 54/74 TTL Family

Series	Name
54/74	Standard
54H/74H	High-speed
54L/74L	Low-power
54S/74S	Schottky
54LS/74LS	Low-power Schottky
54AS/74AS	Advanced Schottky
54ALS/74ALS	Advanced Low-power Schottky

vanced subfamilies so far. There is also the so-called 54F/74F *Fast* subfamily, introduced first by Fairchild, Inc., which is similar to the Advanced Schottky series.

The Basic NAND Gate

Since almost all the TTL products are derived from a common NAND logic structure, this circuit is also considered to be the basic TTL gate. A block diagram of its subcircuits appears in Figure 6.1b. The input circuit (*1*) is an AND gate fabricated with a *multi-emitter* transistor. The *phase-splitter* (*2*) provides the complementation and amplification in the circuit. It determines if the output is *active high* or *active low*. The *low-level driver* (*3*) is a saturated transistor designed to *sink*

Figure 6.1 The 7400 two-input TTL NAND gate: (a) the circuit diagram; (b) the functional block diagram; and (c) the truth table.

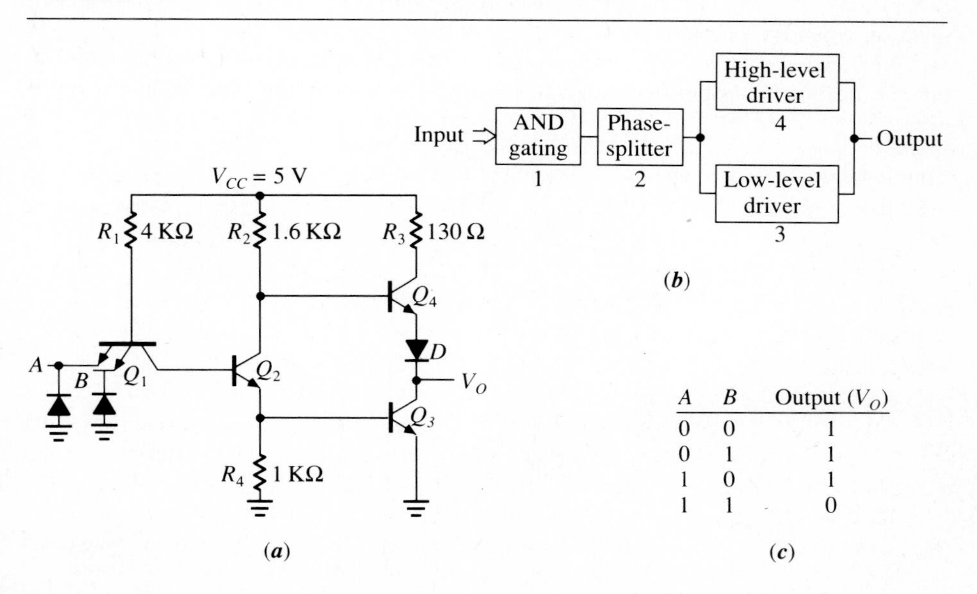

the maximum fan-out current, while the *high-level driver* (4) forms an emitter-follower circuit to *source* current to large capacitive loads.

Before discussing each of the stages of the basic NAND gate in more detail, we will briefly study an earlier form of bipolar logic known as the *diode-transistor logic* (*DTL*). The integrated circuit form of the DTL NAND gate, with only one input shown, is illustrated in Figure 6.2. Note that the input is a diode-connected transistor, Q_1, showing how the diodes are made in IC form. Q_2 and D are used for high fan-out and level-shifting purposes. The base resistor R_B provides a means of turning off the output transistor Q_3. The operation of this gate is easily understood qualitatively. If at least one of the inputs is low, the diode-connected transistor at that input conducts, and the voltage V_P at point P becomes low. Therefore, Q_2 and D are off, $I_B = 0$, Q_3 is off, and the output of Q_3, that is, V_O, will be logical **1** as expected. On the other hand, if all the inputs are at **1**, so that the input diodes are cut off, then V_P rises toward V_{CC}, and a base current I_B starts to flow. Eventually, Q_3 is driven into saturation, and the output V_O falls to its low state. Thus, the circuit satisfies the truth table for the NAND operation.

The disadvantage of the DTL gate is its slow transients both at the input and the output. When the input goes low, and Q_2 and D turn off, the charge stored in the base of the output transistor Q_3 leaks through R_B to ground. The initial reverse base current, $I_B \approx V_{BE(on)}/R_B$, is quite small in comparison to the forward base current. Therefore, the time required for the removal of the excess base charge is long.

Now, consider the operation of the common-emitter output stage when it is turned off. V_O cannot immediately rise to V_{CC} because at the output terminal there is a capacitive load, C_L, consisting of the capacitances of the reverse-biased diodes of the fan-out gates and any stray capacitance. When the output changes from the low to the high state, C_L charges up exponentially from $V_{CE(sat)}$ to V_{CC} through the collector resistor R_C, which is called the *passive pull-up*, resulting in a long turn-off time. The output delay may be reduced by decreasing R_C, but this will increase the power dissipation when the output is low.

Next, we see how these shortcomings of the DTL circuit are remedied with the advent of TTL. Consider the circuit of Figure 6.2. Remove the short circuit between

Figure 6.2 An integrated circuit DTL NAND gate driving a capacitive load (with only one input shown).

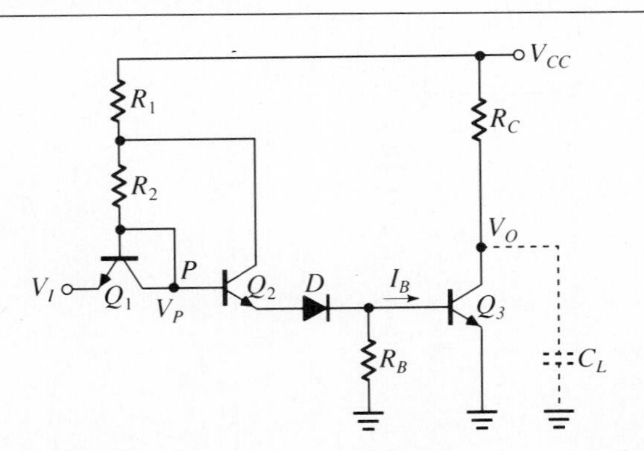

the base and the collector of Q_1, so that it now acts as a transistor. Also, eliminate Q_2, D, and R_B. The resulting circuit is shown in Figure 6.3a. Let the input V_I be logical **1**, forward-biasing the BC junction of Q_1 and reverse-biasing the BE junction. Therefore, Q_1 will be operating in the reverse active mode. The voltage polarities and current directions are shown in Figure 6.3b. The input base current I_1 is calculated as follows:

$$I_1 = \frac{V_{CC} - V_{BC1} - V_{BE3}}{R} = \frac{V_{CC} - 1.4}{R}$$

Figure 6.3 (a) The elementary form of the TTL NAND gate, (b) analysis of the circuit in (a) with a high input, and (c) analysis immediately after the input is brought down to low level.

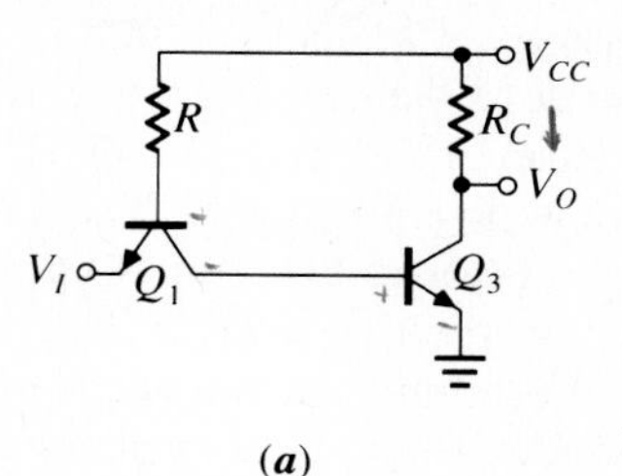

(*a*)

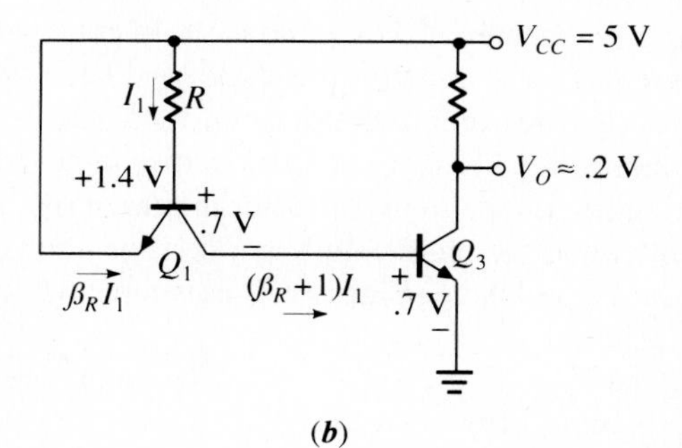

(*b*)

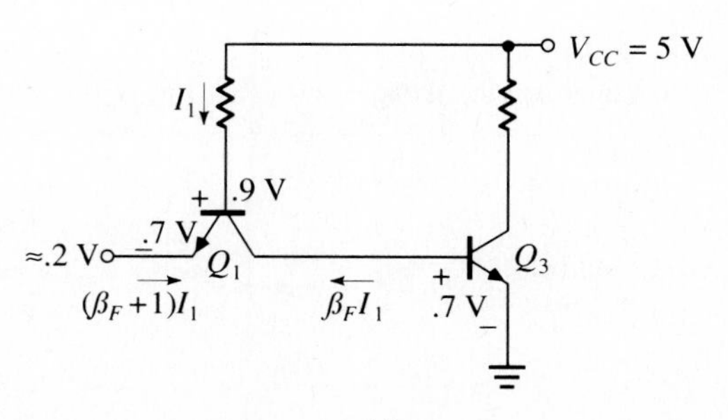

(*c*)

If the reverse current gain β_R is very low, as in the case of actual TTL circuits, the emitter current of the input transistor will be very small, and the base current of Q_3 will be approximately equal to I_1, which will be enough to drive the output transistor into saturation. Therefore, the output voltage will be around .2 V, corresponding to logical **0**.

Now, let the input voltage be brought down to the logical **0** level. This will cause I_1 to be diverted to the emitter of Q_1, as depicted in Figure 6.3c. The BE junction of Q_1 is forward-biased and the base voltage of Q_1 drops to .9 V. Since initially Q_3 was in saturation, its base voltage remains momentarily at .7 V until the overdrive charge stored in the base is removed. Meanwhile, Q_1 operates in the forward active mode, and its collector current becomes $\beta_F I_1$. This large current quickly discharges the base of Q_3. As the output transistor turns off, its base voltage is reduced, so that the BC junction of Q_1 becomes forward-biased, Q_1 enters the saturation mode, and its collector current, trying to draw base current out of Q_3, becomes negligibly small. Therefore, the base of Q_3 will be at about .4 V, keeping it in cutoff. Consequently, using a transistor at the input instead of a diode, as in a DTL circuit, speeds up the turn-off process and hence decreases the propagation delay time. It also makes efficient use of the silicon surface area.

The long rise time at the output of the DTL circuit is solved by modifying the output stage. When the common-emitter output stage of the DTL NAND gate in Figure 6.2 is turned on, the capacitive load C_L does not allow the collector voltage to change instantaneously. Therefore, Q_3 does not saturate but operates in the forward active mode for a short period of time, during which it sinks a relatively large current, $\beta_F I_B$, discharging C_L rapidly. However, it cannot provide fast charging, as already mentioned.

Faster turn-on time is accomplished if we have an emitter-follower output stage, as depicted in Figure 6.4, whose output resistance is characteristically low, leading to a fast charging of C_L when the transistor is on with a logical **1** input. The disadvantage with the emitter follower is that C_L has to discharge slowly through R_E when the input voltage V_I goes low and the transistor is turned off. Hence, a combination of the common-emitter and the emitter-follower configurations seems to be the optimum solution for the output stage. Figure 6.5 illustrates this combination as driven by two complementary signals, V_1 and V_2.

When V_1 = **0**, and V_2 = **1**, the common-emitter transistor, Q_3, will be saturated and supply fast discharging of the capacitive load, while Q_4 will be off. When V_1 = **1**, and V_2 = **0**, the emitter-follower Q_4 will be saturated and provide fast charging of

Figure 6.4 The emitter follower driving a capacitive load.

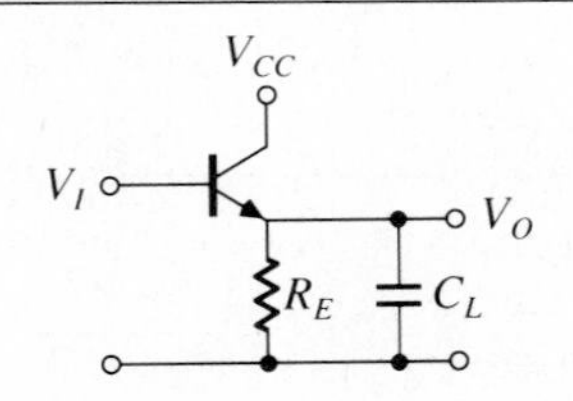

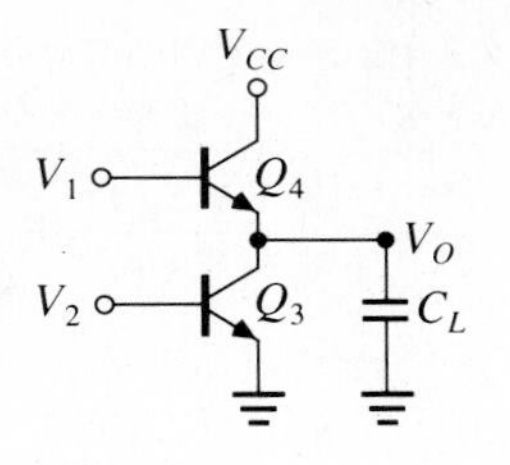

Figure 6.5 The totem-pole output stage with a capacitive load.

C_L. This way, the transition times at the turn-on and the turn-off will decrease. Therefore, the output stage features an *active pull-up* transistor, Q_4, in contrast to the simple resistor pull-up of the DTL gate. This output configuration is familiarly known as a *totem-pole* circuit because one transistor is stacked on top of the other.

Now, it is easier to proceed with the discussion of the complete NAND gate circuit of Figure 6.1a. The input circuit is basically an AND gate configuration designed with a *multi-emitter npn* transistor. Being operated in the reverse active mode, it shows a very high impedance in the high state, so that there is very little current flowing into the device. The low-level impedance of the input stage is determined by the 4 KΩ pull-up resistor. The inputs also have clamp diodes to minimize the negative *ringing* effects due to the lead inductance resonating with shunt capacitance. These diodes are reverse-biased for the steady-state conditions of logical **1** or **0** at the input. During transitions, any load inductance causes damped high-frequency oscillations because the wires connecting the inputs of the gate to the outputs of other gates behave as transmission lines that are not properly terminated. The input voltage can be greater than 5 V on the positive swing and can go below 0 V on the negative swing. The former reverse-biases the BE junction of Q_1. However, R_1 limits the current to prevent false switching at the output. On the negative swing, with Q_1 saturated, the collector voltage will at most be .2 V more positive than the emitter voltage, which can cause the normally reverse-biased collector-substrate isolation diode to be forward-biased, leading to an improper operation of the gate due to unexpected voltage spikes at other nodes of the circuit. The input diodes clamp the negative excursions of the input ringing signal at $-.7$ V.

The conduction of these clamping diodes also leads to a power loss in the transmission lines, resulting in the damping of the oscillating waveform; that is, when conducting, they lessen the energy being transferred between the distributed parameters.

The phase-splitter Q_2 generates the two complementary (i.e., out-of-phase) voltage signals required to drive the totem-pole output stage. Note that additional components, namely, a 130-Ω resistor and a diode, are added at the output stage.

In providing a detailed analysis of the standard two-input TTL NAND gate circuit, we will use the electrical characteristics given in manufacturers' data sheets, as listed in Table 6.2. We will also assume the following values for the junction voltages, where $V_{D\gamma}$ is the diode cut-in voltage:

Table 6.2 Electrical Characteristics for the 7400 Two-Input NAND Gate

	Min	Typ	Max
V_{CC}, V	4.75	5.0	5.25
V_{IH}, V	2.0		
V_{IL}, V			.8
V_{OH}, V	2.4	3.4	
V_{OL}, V		.2	.4
I_{IH}, mA			.04
I_{IL}, mA			−1.6
I_{OH}, mA			−.4
I_{OL}, mA			16.0

$$V_{CE(sat)} = .1 \text{ V},$$

$$V_{\gamma} = .5 \text{ V},$$

$$V_{D\gamma} = .6 \text{ V},$$

$$V_D = .7 \text{ V},$$

$$V_{BE(on)} = .7 \text{ V, and}$$

$$V_{BE(sat)} = .8 \text{ V}$$

Case 1: At least one input is low. The voltage and current values for this case are shown in Figure 6.6. With either input low, that is, $V_I = .1$ V, the BE junction of Q_1 will be forward-biased, Q_1 will saturate, and the base voltage will approximately be +.9 V. With $V_{CC} = 5$V, the base current is found to be

$$I_{B1} = \frac{5 - .9}{4 \text{ K}\Omega} = 1.025 \text{ mA}$$

Since .9 V is insufficient to forward-bias the series combination of the BE junctions of Q_2 and Q_3, the phase-splitter will be off. The collector current of Q_1 is thus limited to the negligible leakage current across the BC junction of Q_2. With $\beta_{F1} I_{B1} >> I_{C1}$, Q_1 must indeed be saturated, so that $V_{CE1} = V_{CE(sat)} = .1$ V. Therefore, $V_{C1} = V_{B2} = .2$ V, which implies that both Q_2 and Q_3 are off and the supply current through R_2 flows into the base of the pull-up transistor, Q_4, to turn both Q_4 and the diode D on. The high-level driver Q_4 acts as a source because it supplies the load current I_{OH} to the capacitive load.

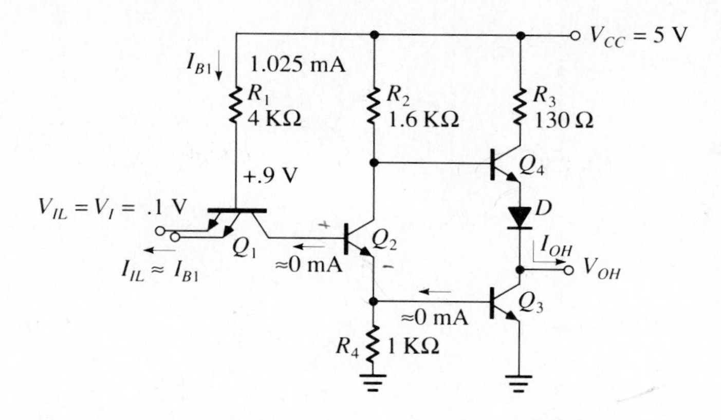

Figure 6.6 The analysis of the 7400 NAND gate when at least one input is low.

Depending on the value of I_{OH}, Q_4 is either in the forward active mode or in saturation, or even at only the cut-in condition. If the gate output terminal is open, I_{OH} is nothing but a leakage current through Q_3, so that Q_4 and D will both be at cut-in, with the BE junction of Q_4 and the diode barely conducting. In this case, we have

$$V_{OH} = V_{CC} - V_\gamma - V_{D\gamma} = 3.9 \text{ V}$$

If a single load is connected, then the load current will be increased, Q_4 and D will conduct more heavily, and Q_4 will remain in the forward active mode. The output voltage is then given by

$$V_{OH} = V_{CC} - \left[\frac{I_{OH}}{\beta_{F4} + 1}\right](1.6 \text{ K}\Omega) - V_{BE4(on)} - V_D$$

$$= 3.6 - \left[\frac{I_{OH}}{\beta_{F4} + 1}\right](1.6 \text{ K}\Omega)$$

where the second term on the right-hand side is the voltage drop across R_2. Thus, the output has a Thévenin representation consisting of a voltage of 3.6 V in series with an equivalent impedance $(1.6 \text{ K}\Omega)/(\beta_{F4} + 1)$.

As the fan-out is increased, Q_4 will eventually saturate, at which time the output voltage will be determined by the drop across R_3 as

$$V_{OH} \approx V_{CC} - 130 I_{OH} - V_{CE4(sat)} - V_D$$

$$= 4.2 - 130 I_{OH}$$

Figure 6.7 depicts the high-level output voltage V_{OH} as a function of the output current I_{OH}, as given in data sheets. When Q_4 is in saturation, the negative of the slope is found as

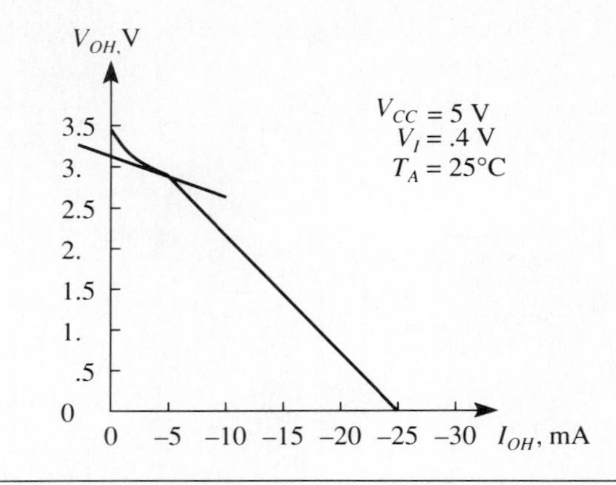

Figure 6.7 High-level output voltage V_{OH} vs. the output current I_{OH} for the 7400 NAND gate.

$$\frac{\Delta V_{OH}}{\Delta I_{OH}} \approx 150\ \Omega$$

This is the small output impedance as determined by the series combination of the 130-Ω resistance, the output impedance of the saturated transistor (about 10 Ω), and the forward resistance of the diode (about 10 Ω). Therefore, when switching from the low to the high state, the totem-pole output structure provides a low output impedance that is capable of rapidly charging the capacitive loads. We can also obtain the output impedance just before saturation, when Q_4 is still in the active region, by considering the tangent to the characteristic curve, as shown in Figure 6.7. The negative of the slope is then given as

$$\frac{\Delta V_{OH}}{\Delta I_{OH}} \approx 38\ \Omega$$

This corresponds to a dc-current gain of $\beta_{F4} \approx 41$.

Case 2: All inputs are high. The corresponding voltage and current values are indicated in Figure 6.8. The emitters of Q_1 are reverse-biased. Since the *p*-type base is connected to the +5 V supply, the BC junction is forward-biased. Therefore, Q_1 is operating in the reverse active mode. The high-level input current is given by

$$I_{IH} = \frac{\beta_R I_{B1}}{n} \tag{6.1}$$

where n is the number of emitters to the input transistor, which is two in this case.

The large collector current of Q_1 first turns Q_2 and Q_3 on and then drives them into saturation, so that the base voltage of Q_1 becomes 2.3 V. With $V_{CC} = 5$ V, we have

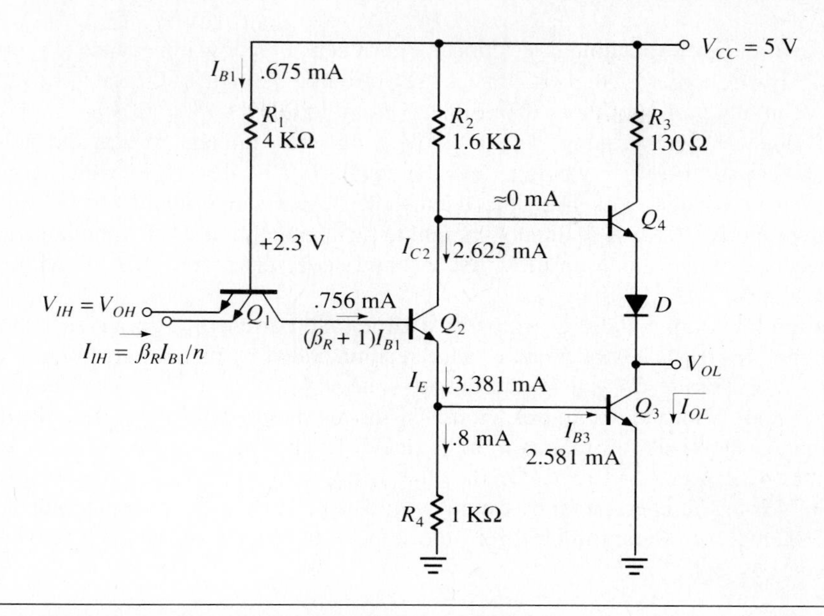

Figure 6.8 The analysis of the 7400 NAND gate when all inputs are high, and Q_2 and Q_3 have already saturated.

$$I_{B1} = \frac{5 - 2.3}{4 \text{ K}\Omega} = .675 \text{ mA}$$

Substituting this into Equation (6.1), using Table 6.2, and solving for the worst-case reverse current gain, $I_{IH} = 40\ \mu\text{A}$, we get

$$\beta_R = \frac{2(40\ \mu\text{A})}{.675 \text{ mA}} = .12$$

The collector current of Q_1 is obtained as

$$I_{C1} = (\beta_R + 1)I_{B1} = .756 \text{ mA}$$

Hence, with $V_{BE3} = .8$ V and $V_{CE2} = .1$ V, the collector voltage of Q_2 becomes .9 V and the collector current is found to be

$$I_{C2} = \frac{5 - .9}{1.6 \text{ K}\Omega} = 2.625 \text{ mA}$$

The emitter current of the phase-splitter is then given by

$$I_{E2} = 2.625 + .756 = 3.381 \text{ mA}$$

Since Q_3 is in saturation, the current through R_4 is given as $(.8 \text{ V}/1 \text{ K}\Omega) = .8$ mA, and the base current of the low-level driver becomes

$$I_{B3} = 3.381 - .8 = 2.581 \text{ mA}$$

Because of the corresponding large collector current and the low impedance to ground, Q_3 quickly discharges C_L and establishes the low output voltage of the gate as $V_{CE3(sat)} = .1$ V; in any case, not more than the maximum low-level value of $V_{OL} = .4$ V.

If diode D were missing, $V_{C2} = .9$ V would be sufficient to turn the pull-up transistor on because V_{BE4} would at least be equal to $.9 - .4 = .5$ V, which is equal to the cut-in voltage, V_γ. Then the totem-pole output stage would not be able to operate properly. Since .5 V are not enough to turn both Q_4 and D on simultaneously, the presence of the diode ensures that the high-level driver remains off while the output is at **0**.

In the low output state, Q_3 can sink a large load current I_{OL}. As given in Table 6.2, $I_{OL} = 16$ mA is a conservative value recommended by the manufacturers when $V_{OL} = .2$ V. Figure 6.9 shows the output voltage V_{OL} as a function of the output current I_{OL} at various temperatures. Actually, the maximum collector current the transistor can sustain and still maintain the logical **0** level, with V_{OL} still below .4 V, is more than 40 mA at $T_A = 25°$C. On the other hand, at high values of sinking current, especially at low temperatures, Q_3 cannot stay in saturation, and the output impedance rises. When Q_3 is in saturation at 25°C, the slope of the characteristic gives the output impedance as

$$\frac{\Delta V_{OL}}{\Delta I_{OL}} \approx 15\ \Omega$$

The Power Dissipation in the Standard NAND Gate

When all inputs are high, the current supplied per gate by the source is given by

$$I_{CCL} = I_{B1} + I_{C2} = .675 + 2.625 = 3.3 \text{ mA}$$

When at least one input is low, with zero fan-out, we can neglect the base and collector currents of the high-level driver, so that the supply current per gate is found as

Figure 6.9 Low-level output voltage V_{OL} vs. the output current I_{OL} for the 7400 NAND gate.

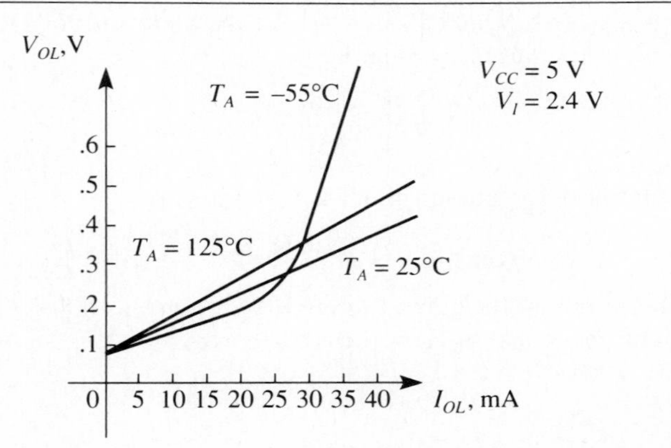

$$I_{CCH} = I_{B1} = 1.025 \text{ mA}$$

Therefore, the average static power dissipation per gate is given by

$$P_{(av)} = \frac{(3.3 + 1.025)5}{2} = 10.8 \text{ mW}$$

Note that I_{CCL} and I_{CCH} for 7400 quad two-input NAND gates are given in manufacturers' data sheets as 12 mA and 4 mA, respectively. This is because they are actually defined as the total current into the V_{CC} supply terminal of an IC when all (i.e., four for the 7400) of the outputs are at the low or high level.

Now, assume that, with $V_O = .1$ V, one of the inputs changes to $V_I = .1$ V to turn Q_2 and Q_3 off. The output remains momentarily at .1 V because the capacitive load does not allow it to change instantaneously. In the meantime, Q_4 goes into saturation and D conducts. Therefore,

$$V_{B4} = V_{BE4(sat)} + V_D + V_O = .8 + .7 + .1 = 1.6 \text{ V}$$

$$I_{B4} = \frac{V_{CC} - V_{B4}}{R_2} = \frac{5 - 1.6}{1.6 \text{ K}\Omega} = 2.125 \text{ mA}$$

$$I_{C4} = \frac{V_{CC} - V_{CE4(sat)} - V_D - V_O}{130 \ \Omega}$$

$$= \frac{5 - .1 - .7 - .1}{130 \ \Omega} = 31.5 \text{ mA}$$

Thus, as long as β_{F4} exceeds $31.5/2.125 = 14.82$, Q_4 will indeed be in saturation. Consequently, the dynamic power dissipation during turn-on is given by

$$P_{(dyn)} = (I_{B1} + I_{B4} + I_{C4})V_{CC} = (1.025 + 2.125 + 31.5)5 = 173.25 \text{ mW}$$

which is much higher than the steady-state value of 10.8 mW.

The 130-Ω Resistor

It is clear from the ongoing discussion that the omission of the 130-Ω resistor would result in faster switching during low-to-high transition. However, it is needed to limit the current spikes during the turn-on and turn-off transients.

When the input is brought down to the low level, Q_2 turns off relatively fast because Q_1 supplies a large reverse current to its base terminal. In the meantime, the base of Q_3 will have to discharge through the 1-KΩ resistor in order to turn off, while Q_4 will be turning on. Thus, with both totem-pole transistors conducting at the same time, a large current spike would flow, and the supply voltage would be short-circuited if the 130-Ω resistor were missing. While this current spike is generally harmful, it does, however, help Q_3 to come out of saturation quickly. With the resistor present, as calculated above, we have

$$I_{B4} + I_{C4} = 33.63 \text{ mA}$$

as the peak current to be drawn from the supply during the transient. Data sheets provide a *short-circuit output current* I_{OS}, which is defined as the current into an output when it is short-circuited to ground while the input conditions would otherwise establish a logical **1** at the output. This parameter is an important indicator of the ability of an output to charge a capacitive load. For the 7400, $I_{OS(max)}$ is specified as -55 mA.

Since the transient spikes also flow through the V_{CC} and the ground distribution system, the power and ground lines should be short and adequately decoupled by using *bypass capacitors* at frequent locations on the supply rail to lower the impedance of the voltage source, thereby reducing the magnitude of the spikes.

The VTC of the Standard NAND Gate

Figure 6.10 displays the voltage transfer characteristic of the standard TTL NAND gate in a piecewise-linear fashion, with no fan-out. With at least one input low, transistor Q_1 is saturated, Q_2 and Q_3 are off, and Q_4 is barely conducting. Therefore, we obtain segment AB with

$$V_{OH} = 3.9 \text{ V}$$

The first breakpoint in the characteristic occurs at point B, when the phase-splitter Q_2 starts conducting as the input voltage is increased. The base current to Q_2 is supplied by the forward-biased BC junction of the saturated Q_1. Since the cut-in voltage is .5 V, and $V_{CE1(sat)} = .1$ V, at the first breakpoint we have

Figure 6.10 The VTC of the 7400 NAND gate with no fan-out.

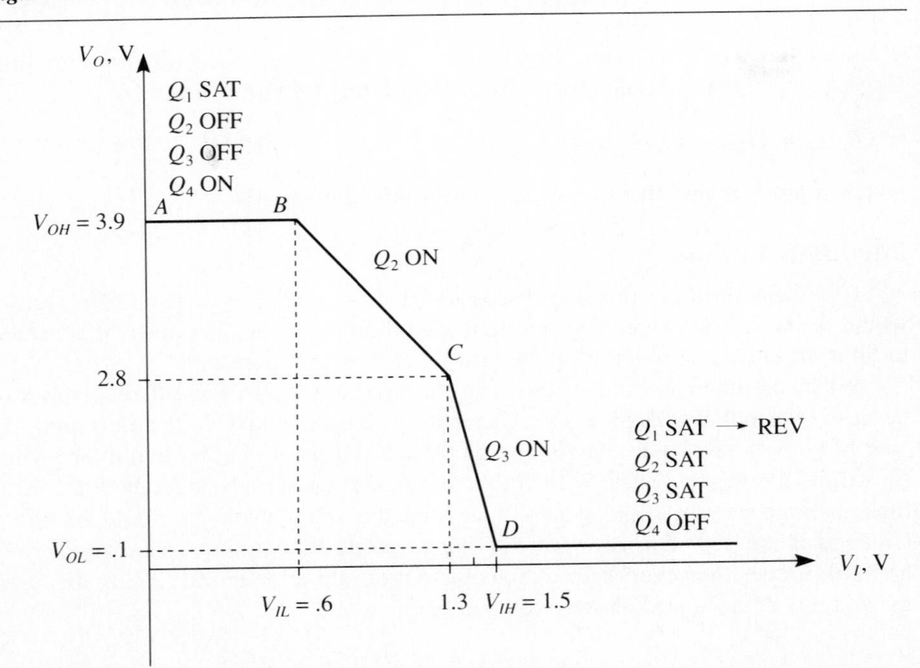

$$V_{IL} = .6 \text{ V}$$

Between points B and C, Q_1 remains saturated. However, its base current is increasingly diverted to its BC junction, and hence into the base of the phase-splitter. Q_3 does not turn on until the second breakpoint at C. The corresponding input voltage is found from

$$V_{I(C)} = V_{BE3(on)} + V_{BE2(on)} - V_{CE1(sat)} = .7 + .7 - .1 = 1.3 \text{ V}$$

Then, $I_{C2} \approx I_{E2} = (.7 \text{ V}/1 \text{ K}\Omega) = .7 \text{ mA}$. Thus, at point C, the voltage at the collector of Q_2 is $V_{C2} = V_{CC} - (1.6 \text{ K}\Omega)(.7 \text{ mA}) = 3.9 \text{ V}$ so that $V_{CE2} = 3.9 - .7 = 3.2 \text{ V}$ and Q_2 is operating in the forward active mode. The voltage at the gate output is then given by

$$V_{O(C)} = V_{C2} - (V_\gamma + V_{D\gamma}) = 2.8 \text{ V}$$

As the input is increased further, the low-level driver operates ultimately in saturation. Q_4 turns off, and the gate output voltage becomes

$$V_{OL} = .1 \text{ V}$$

Since Q_2 is also saturated, the base of Q_1 is clamped at 2.3 V, and Q_1 is still in saturation, so that

$$V_{IH} = 1.6 - .1 = 1.5 \text{ V}$$

As the input voltage is increased toward 2.3 V, the BE junction of the input transistor becomes reverse-biased, and Q_1 starts to operate in the reverse active mode.

54H/74H High-Speed and 54L/74L Low-Power Series

The speed of the standard TTL gate is limited by the finite storage times of transistors Q_1, Q_2, and Q_3 because they all saturate. The long time constants as the capacitive loads charge and discharge through the relatively large-valued resistors in the circuit also contribute to the transient delays.

One approach to speed up the operation of the TTL is to reduce the values of all resistances. The trade-off is a corresponding increase in power consumption. Figure 6.11 illustrates the high-speed 7400 two-input TTL NAND gate. In addition to the simple reduction in resistor values, a special feature of the 54H/74H series is its usage of a *Darlington pair* for the high-level driver in order to provide increased current gain, leading to an increase in the current sourcing capability. Together with the lower output impedance at logical **1**, this results in a reduction in the time required to charge the capacitive loads, allowing a higher speed of operation. Note that the additional transistor, Q_5, in the Darlington pair makes the diode D unnecessary. Resistor R_5 is provided to allow the emitter current to flow because Q_4, its BC junction being reverse-biased by the collector-emitter voltage drop of Q_5, is always in the forward active mode, and its base current is negligible.

Another subfamily of TTL, the low-power 54L/74L series, comes with increased circuit resistances, as compared to the values used in the standard TTL, to achieve reduced power dissipation. Table 6.3 shows the difference between the re-

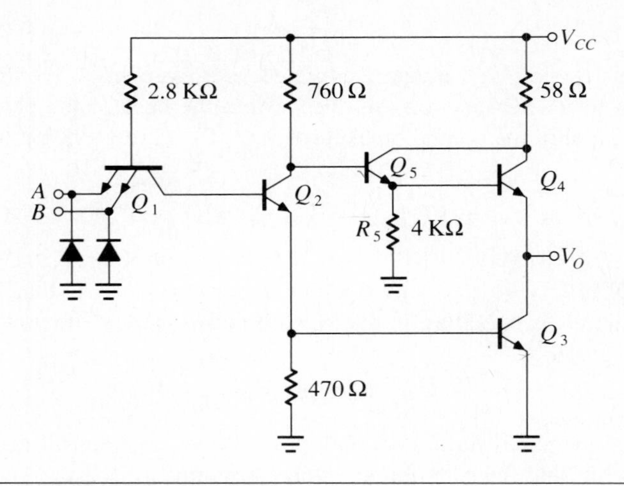

Figure 6.11 The 74H00 high-speed TTL NAND gate.

sistor values. Obviously, this subfamily is slower than the standard TTL. It was intended for use where a large number of units are drawing current from a limited power source. Both the 54H/74H and 54L/74L lines of TTL gates are now obsolete.

Schottky TTL

It should be obvious by now that a transistor must be prevented from entering the saturation region to obtain the fastest switching operation. There are mainly two circuit techniques that are employed to prevent the bipolar transistor from entering the saturation region. One way is to clamp V_{CE} to a value above $V_{CE(sat)}$ such that $V_{CE} >> V_{CE(sat)}$. This is achieved by using a Schottky-barrier diode between the base and collector, as indicated in Figure 6.12a. The combination is referred to as a *Schottky transistor*, and is represented by the symbol in Figure 6.12b. Another way is to limit the collector current, so that $\beta_F I_B << I_C$. This technique is utilized in ECL, where a current source I_o is used to limit the emitter current and hence the collector current to $\approx I_o$. These circuits are discussed in the next section.

The use of the *Baker clamp*, depicted in Figure 6.13, is a method to avoid the

Table 6.3 Comparison of Internal Resistors for 74 and 74L Gates (KΩ)

Resistor	54/74	54L/74L
R_1	4	40
R_2	1.6	20
R_3	.13	.5
R_4	1	12

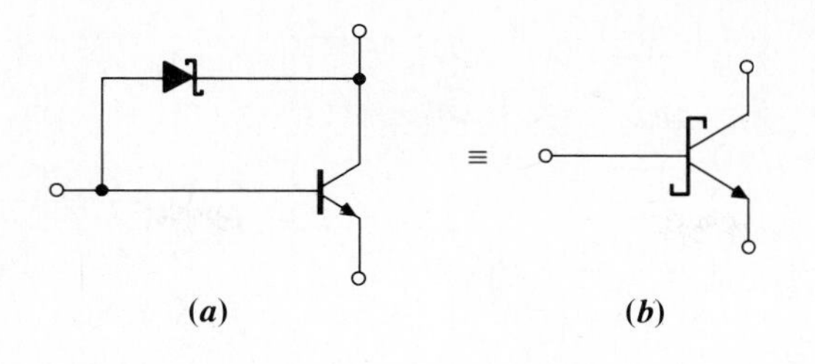

Figure 6.12 (a) A transistor with a Schottky clamp, and (b) the equivalent circuit symbol for the Schottky transistor.

saturation of a discrete transistor. The forward voltage drop of the germanium diode is .4 V as compared to $V_\gamma = .5$ V for the silicon transistor. As the input voltage is increased, the base current cannot drive the transistor into saturation because the drop in the collector voltage causes the diode to conduct, so that the excess base drive is diverted from the BC junction of the transistor. The BC junction voltage is clamped at .4 V, so that the transistor is at most held at the edge of saturation with no overdrive charge stored in the base, resulting in the elimination of the storage time and dramatic reduction in the turn-off time.

However, a Ge diode cannot be incorporated into a silicon IC. Thus, it has to be replaced with a silicon diode having a lower forward voltage drop than the BC junction of the transistor. A *pn* junction does not meet this requirement, so an SBD is used for this purpose.

Schottky transistors are used in the 54S/74S line of TTL gates, achieving a high speed of operation. Figure 6.14 shows the circuit for the two-input 74S00 Schottky TTL NAND gate. In this circuit, the Schottky-clamped transistors are at cut-in when the BE junction voltage is .7 V, and conduct fully with $V_{BE(on)} = .8$ V. Therefore, for a Schottky transistor, when the forward voltage drop of the SBD is at $V_{SBD} = .5$ V, we have $V_{CE(on)} = .8 - .5 = .3$ V.

In Figure 6.14, the high-level driver Darlington pair is similar to that of the high-speed TTL with the exception that Q_5 is now a Schottky transistor. Q_4 does not need a Schottky clamp because, when both transistors are conducting, $V_{CE4} = V_{BE4} + V_{CE5}$, so that Q_4 never saturates. Note that the input clamping diodes are also of the Schottky type.

Another added special feature is the replacement of R_4 between the base of Q_3

Figure 6.13 Baker clamp.

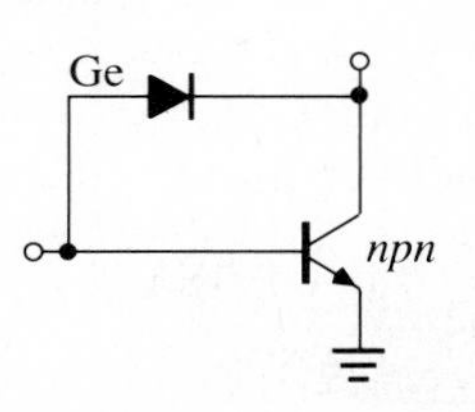

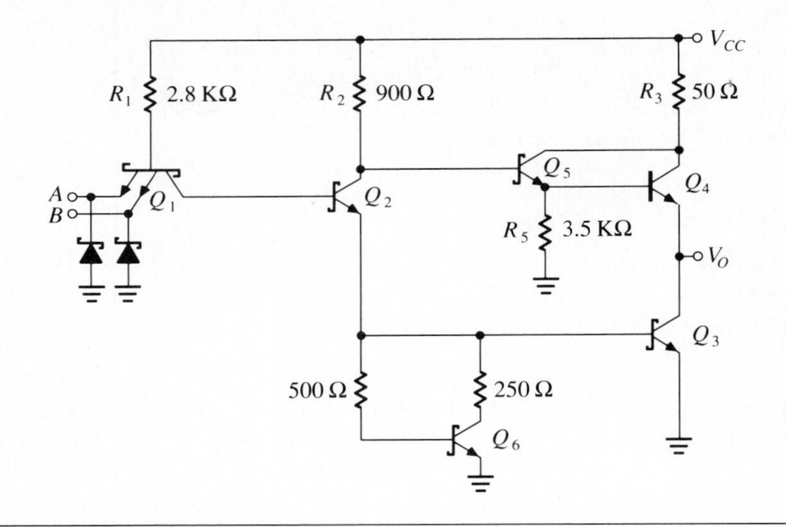

Figure 6.14 The 74S00 Schottky-clamped TTL NAND gate.

and ground with a *nonlinear resistance* implemented by Q_6 and two resistors. This subcircuit is known as an *active pull-down*. It conducts a very small current and behaves as a high impedance until the voltage across it reaches $V_\gamma = .5$ V and Q_3 starts conducting. In this way, the BC segment of the transfer characteristic in Figure 6.10 is eliminated, thereby improving the noise immunity. The BC segment exists because Q_2 turns on before Q_3 can turn on. Then V_O decreases steadily in the original circuit until the current into the base of Q_3 is sufficient to turn on and ultimately saturate the low-level driver. In the Schottky TTL, the active pull-down draws negligible current initially, so almost all the current supplied by the phase-splitter will be diverted into the base of Q_3, speeding up the turn-on of Q_3. Therefore, Q_2, Q_3, and Q_6 turn on at approximately the same input voltage, leading to a voltage transfer characteristic without the breakpoint C, and with a narrower transition width.

Since the active pull-down circuit *squares up* the characteristic curve, it is also known as the *squaring circuit*. The 500-Ω resistor determines the amount of the base currents I_{B3} and I_{B6}, while the 250-Ω resistor prevents the BE junction voltage of Q_3 from being clamped to

$$V_{BE(on)} - V_{SBD} = .3 \text{ V}$$

The active pull-down also helps to turn Q_3 off quickly, decreasing its turn-off time and hence t_{PLH}. With a logical **0** at the input, Q_1 becomes active, and the phase-splitter is cut off. As Q_3 begins to turn off, Q_6, which is now in the forward active mode, will initially draw a larger current than that drawn by a passive pull-down. This is because, with $V_{BE3} > V_{BE6}$, the turn-off base current of Q_3 flows both into the base and the collector of Q_6, which will lead to a rapid discharge of Q_3 because $I_{C6} = \beta_F I_{B6}$.

Finally, the resistor values R_1, R_2, and R_3 in the Schottky TTL are smaller than those of the corresponding resistors in the standard circuit. Consequently, 54S/74S power dissipation is larger than that of the standard TTL. However, the average propagation delay time is much smaller, so that the speed-power product has been improved.

Low-Power Schottky TTL

The 54LS/74LS series became the industry standard in the late seventies, replacing the original standard TTL with its low power and high speeds. The circuit for the two-input 74LS00 low-power Schottky TTL NAND gate is diagrammed in Figure 6.15. This circuit does not use the multi-emitter input structure that originally gave the TTL family its name; it uses a DTL-type input circuit with Schottky diodes D_1 and D_2 to perform the AND operation. This is because the Schottky-clamped phase-splitter Q_2 does not saturate, so that a transistor is not required at the input to remove the overdrive excess charge from the base of Q_2. Another reason is that the improvement of IC fabrication technology made it possible to reduce the silicon surface area of the input diodes, thereby reducing the parasitic capacitances.

All inputs are provided with usual clamping diodes. As discussed before, they are used to suppress transient currents. A clamp current exceeding 2 mA and with a duration greater than 500 ns can activate a substrate parasitic transistor, which in turn draws current from the internal nodes of the circuit to cause improper operation and logic errors.

Since the LS TTL also has the squaring circuit of the previous subfamily, the base of the low-level driver Q_3 is returned to ground through this circuit, which prevents conduction in the phase splitter Q_2 until the high-input voltage rises high enough

Figure 6.15 The 74LS00 low-power Schottky TTL NAND gate.

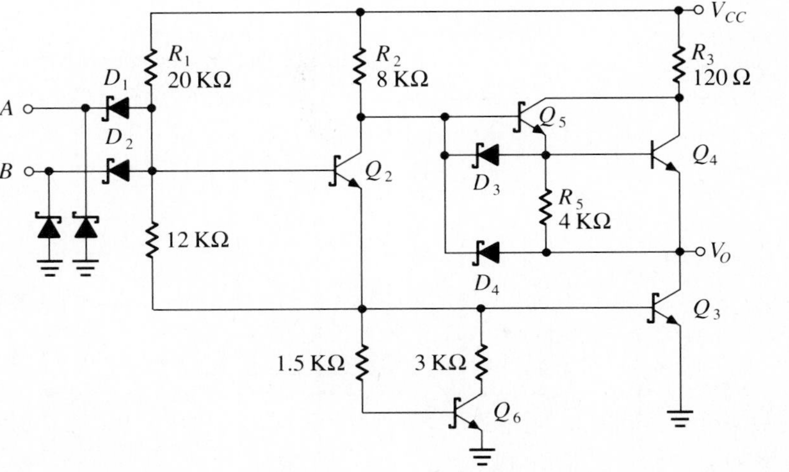

to allow Q_2 to supply the base current to Q_3. This circuit improves the propagation delay as well by providing a low-impedance path to the discharging transition capacitances of Q_3 during turn-off.

The active pull-up is again a Darlington pair. However, in contrast to the 74H and 74S, the base of the output transistor Q_4 is returned to the output terminal through a 4-KΩ resistor to reduce power dissipation. In addition, since the emitter current of Q_5 has a direct path to the output through R_5, it can charge the line capacitance before Q_4 is fully on, thus improving t_{PLH}.

The Schottky-barrier diodes, D_3 and D_4, are included to improve t_{PHL}. The reverse base current of Q_4, during its turn-off, switches them on, and a large current flows into the collector of the phase-splitter, thereby speeding up both the turn-off of Q_4, by discharging the capacitance at its base, and the turn-on of Q_3 by helping to discharge the load capacitance.

Finally, the 12-KΩ resistor across the BE junction of Q_3 supplies current into the base of the low-level driver when the inputs are pulled up to logical **1** level, even before the phase-splitter starts conducting.

The voltage transfer characteristic for this subfamily is depicted in Figure 6.16. The first breakpoint B occurs when Q_2 and Q_3 turn on simultaneously. At this time, we have $V_{B2} = 2V_\gamma = 1.4$ V at the base of Q_2 so that

$$V_{IL} = V_{B2} - V_{SBD} = .9 \text{ V}$$

With zero fan-out and Q_3 off, there will be no current flowing through R_5. However, Q_5 is on; thus, at point B we have

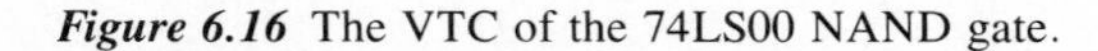

Figure 6.16 The VTC of the 74LS00 NAND gate.

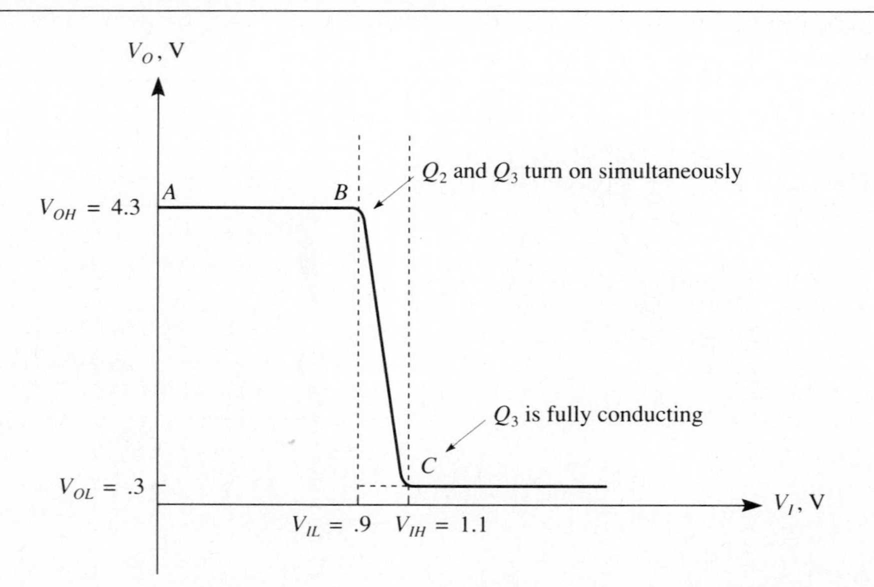

$$V_{OH} \approx V_{CC} - V_\gamma = 4.3 \text{ V}$$

neglecting the small voltage drop across R_2.

At the second breakpoint, C, Q_3 is fully conducting, even though it cannot saturate. Thus,

$$V_{OL} = V_{CE3(on)} = .3 \text{ V}$$

In the meantime, at the input, we have

$$V_{IH} = 2V_{BE(on)} - V_{SBD} = 1.1 \text{ V}$$

Note that the transition region and the logic swing have been narrowed, resulting in higher speed compared to the subfamilies studied so far.

Advanced Schottky Subfamilies

In the early eighties, continual semiconductor processing refinements led to the introduction of advanced versions of Schottky TTL subfamilies with smaller and shallower transistors. The latest developments in bipolar IC fabrication technology have allowed an appreciable reduction of internal feature size, down to 3 μm against about 5 μm in low-power Schottky chips. These new subfamilies improve the speed characteristics and decrease power dissipation. They are *advanced Schottky* (54AS/74AS) and *advanced low-power Schottky* (54ALS/74ALS), first developed by Texas Instruments, Inc., and the *Fast* (54F/74F) series, introduced by Fairchild, Inc. The latter can be considered an intermediate between the first two series. The ICs manufactured with new technologies are designed to be pin-for-pin replacements for older TTL parts. There are also high-density 20- to 24-pin chips available to perform more complicated functions.

New subfamilies make use of oxide isolation instead of the old junction-isolation technique to eliminate parasitic capacitance and leakage-current effects associated with *pn* junctions. The parasitic capacitance imposes frequency limitations on device operation. The new isolation method allows the internal circuit nodes to charge faster by eliminating many of the parasitic effects, resulting in a higher speed of operation. The high speeds of the AS and F series are able to push the system operating speeds into ranges previously reserved for the ECL logic family, but with a single, 5-volt power supply. The major shortcoming of these subfamilies is their increased susceptibility to damage from electrostatic discharge, as compared to the devices of earlier TTL subfamilies, due to the shallower diffusions and thinner oxides.

The two-input NAND gate circuits in the AS, F, ALS series are shown in Figures 6.17–6.19, respectively. These series have three stages of gain, consisting of Q_1, Q_2, and Q_3, instead of two stages as in older TTL subfamilies, to raise the input threshold voltage. The higher threshold voltage provides an improvement in noise immunity, leading to more reliable system design and operation. The input threshold voltage V_{th} of the devices at both logical levels are determined by the BE voltage drops of the transistors Q_1, Q_2, Q_3, and Q_5 as

$$V_{th} = V_{BE2} + V_{BE3} + V_{BE5} - V_{BE1}$$

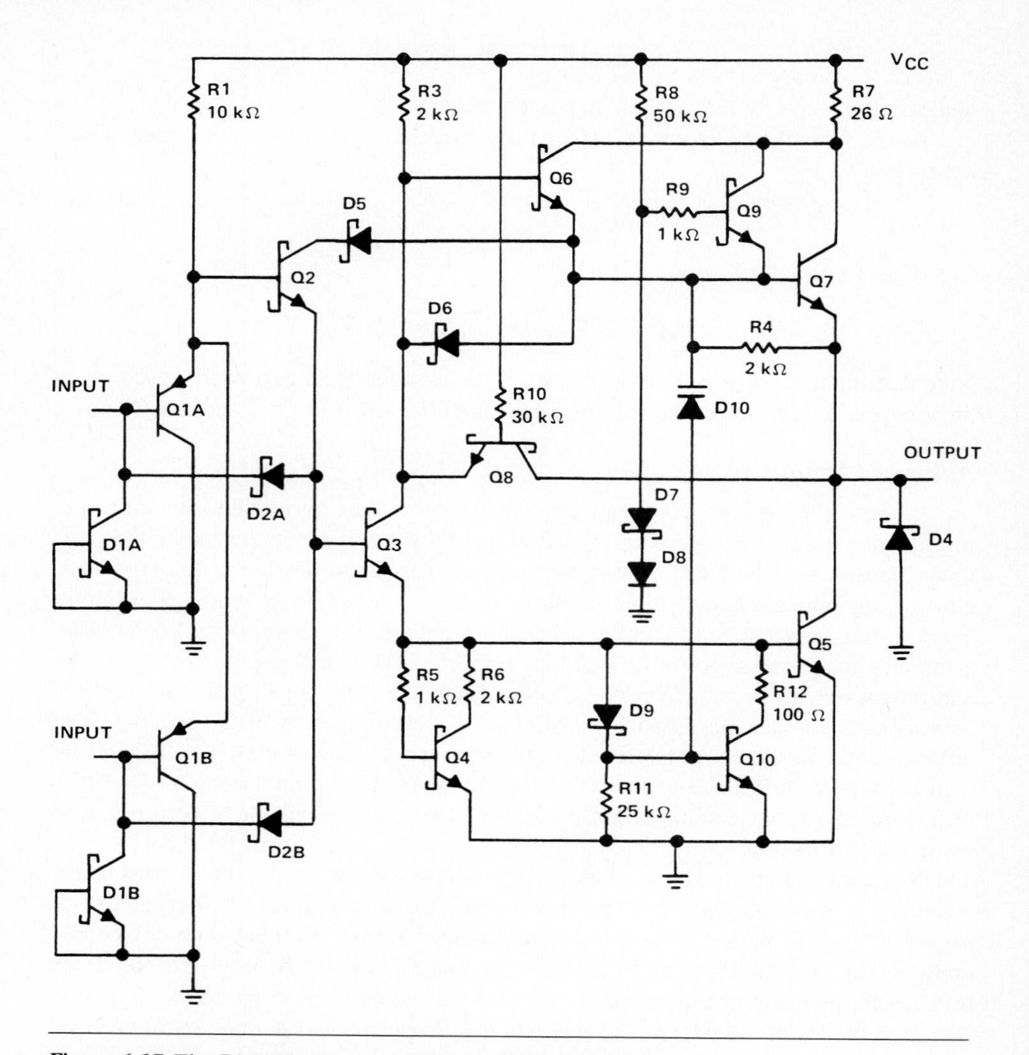

Figure 6.17 The 74AS00 advanced Schottky TTL NAND gate. (*Reproduced by permission of Texas Instruments. Copyright © 1986, Texas Instruments.*)

where V_{BE1} is across either Q_{1A} or Q_{1B} in Figures 6.17 and 6.19. In Figure 6.18, the voltage drop across D_1 or D_2 replaces V_{BE1} in the above equation. Thus, for the AS or ALS, the low-level input voltage is given as

$$V_{IL} = 3V_{\gamma} - V_{BE1(on)} = 1.4 \text{ V}$$

while the high-level voltage is

$$V_{IH} = 3V_{BE(on)} - V_{BE1(on)} = 1.7 \text{ V}$$

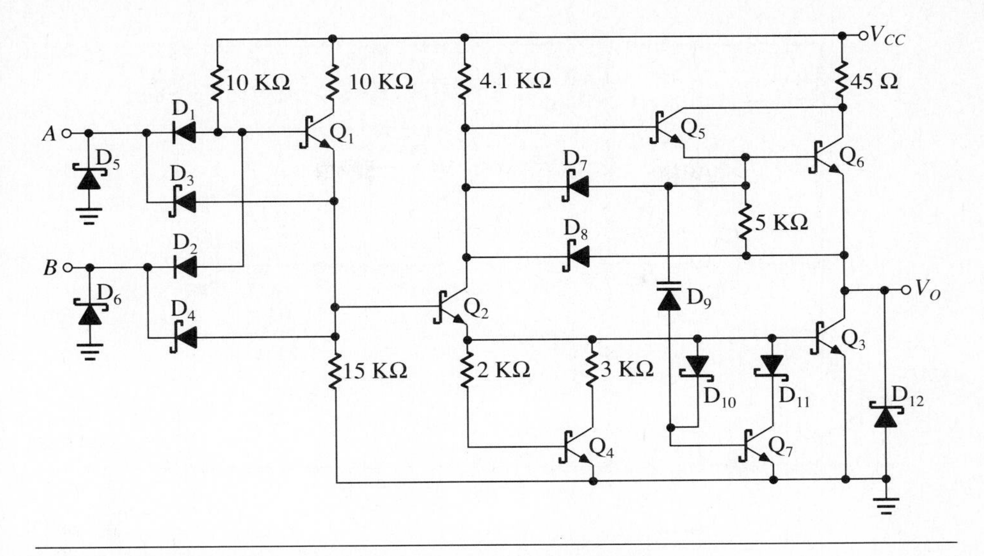

Figure 6.18 The 74F00 FAST NAND gate. (*Reprinted with permission of National Semiconductor Corporation.*)

The improvement in the input circuits of AS and ALS gates from using two emitter followers, Q_1 and Q_2, as buffers reduces the low-level input current I_{IL} and hence increases the fan-out. The utilization of Q_2 achieves high speeds internally, while the *pnp* emitter follower, Q_1, is needed to compensate for the voltage shift caused by the BE voltage drop of the former. Data sheets specify the maximum value of I_{IL} as -100 μA for the ALS, and -500 μA for the AS. In both cases, the specifications are one-fourth those of the LS and S subfamilies.

In Figures 6.17 and 6.19, diode D_2 provides a low-impedance path to ground during a high-to-low transition at the input to improve the switching time of the gate, while Q_1 delivers a faster transition, as compared to a Schottky-diode stage, when the input goes in the other direction.

Referring to the simpler ALS circuit of Figure 6.19, we see that the high-level output voltage V_{OH} is primarily determined by V_{CC}, the Q_6–Q_7 Darlington pair, and the resistors R_4 and R_7. With zero fan-out and no voltage drop across R_4, the output voltage is given, as in the case of LS, by

$$V_{OH} = V_{CC} - V_{\gamma} = 5 - .7 = 4.3 \text{ V}$$

As the number of loads is increased, the current drawn by them will be increased, and the Darlington pair will be fully conducting. For low values of current, the output voltage will be given as the difference between the supply voltage and the combined BE voltage drops of the Darlington pull-up transistor. Therefore, we will have

$$V_{OH} \approx V_{CC} - V_{BE6(on)} - V_{BE7(on)} = 5 - 1.6 = 3.6 \text{ V}$$

The actual base current to Q_6 through R_3 is typically less than 1 μA, so the voltage drop across R_3 is still negligible. As more current is drawn by the increasing number

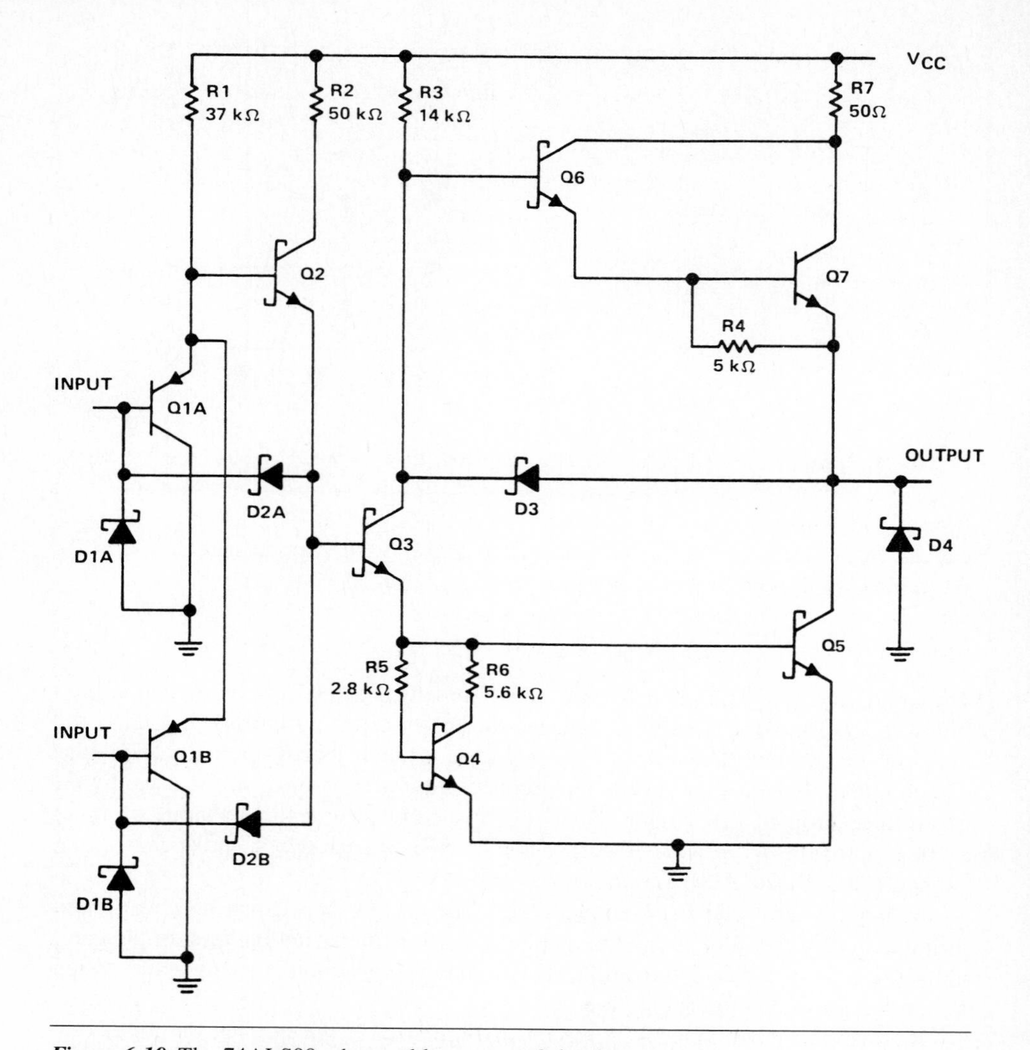

Figure 6.19 The 74ALS00 advanced low-power Schottky NAND gate. (*Reproduced by permission of Texas Instruments. Copyright © 1986, Texas Instruments.*)

of loads, the voltage across the limiting resistor R_7 at the Darlington collector will gradually increase until the Schottky clamping diode of Q_6 starts to become forward-biased, at which point the Darlington comes to the edge of saturation and the voltage across R_3 is no longer negligible. Thus, the high-level output voltage is expressed as

$$V_{OH} \approx V_{CC} - R_7 I_{OH} - V_{CE6(on)} - V_{BE7(on)} \approx 3.9 - 50 I_{OH}$$

Note that the pull-up current during rise time at the output is limited by the 50-Ω resistor as compared to being limited by 120 Ω for the LS TTL, which implies that the ALS is more capable of charging a load capacitance than the latter.

The operation of the active pull-down subcircuit, made up of Q_4, R_5, and R_6, has already been explained. In addition to producing a square transfer characteristic, it also reduces the turn-off time and hence the current transients caused by the momentary conduction overlap of the low-level driver and the Darlington pair high-level driver. On the other hand, the turn-off time of Q_7 is improved by diode D_3 and the phase-splitter Q_3, similar to the operation of the corresponding circuitry in the LS TTL. Moreover, during high-to-low transition at the output, the charge from the load capacitance is partially removed through D_3 and Q_3 to the base of Q_5 to increase the base drive of the latter and hence to lower the output voltage more rapidly.

The 74AS00 NAND gate of Figure 6.17 is a much more complex circuit, with subcircuits not found in the 74ALS00 gate. These are added to enhance the speed performance of this subfamily. The Q_8–R_{10} combination has the same effect as D_3 in the ALS. The transistor Q_{10} has been added to discharge the BC transition capacitance of the low-level driver. During low-to-high transition at the output terminal, the rising voltage injects current into the base of Q_5 through the BC transition capacitance to turn Q_5 on. However, the *varactor diode*, that is, D_{10} acting as a variable capacitor, causes Q_{10} to conduct, which in turn keeps Q_5 in cutoff. Diodes D_6 and D_9 provide a discharge path for D_{10}.

The 74F00 NAND gate is illustrated in Figure 6.18. The improved input circuitry is much simpler than its counterparts in the AS and ALS series. The higher threshold due to three stages of gain makes it possible to use *pn* junctions D_1 and D_2 for the input AND function.

During a high-to-low transition in at least one of the inputs, the corresponding Schottky diode, D_3 or D_4, acts as a low-impedance path to discharge the internal parasitic capacitances connected to the base of Q_2. In a similar way, as Q_2 turns on and its collector voltage falls, D_7 provides a discharge path for the capacitance at the base of Q_6. While D_3, D_4, and D_7 enhance the switching speed by helping to discharge the internal nodes, D_8 is utilized to discharge the capacitive load rapidly. Part of the charge stored in the load passes through D_8 and the phase-splitter, thereby increasing the base current of the low-level driver and hence its current-sinking capability during the high-to-low transition at the output.

The subcircuit consisting of D_9, D_{10}, D_{11}, and Q_7 improves the output rise time and minimizes the dynamic power dissipation during repetitive switching at high frequencies by providing a momentary low impedance at the base of Q_3 as the output rises. The rising voltage at the emitter of Q_5 causes a current to flow through the varactor diode D_9 and turn Q_7 on for a short period of time, which in turn pulls down the base of Q_3 and absorbs the current that flows through the BC transition capacitance of Q_3 during this time interval. Without this subcircuit, the current through the BC junction of the low-level driver acts as the base current, prolonging the turn-off of Q_3 and allowing a current to flow from the high-level driver Q_6 to ground through Q_3. Alluding to the Miller's theorem, which states that the BC transition capacitance C_μ of a transistor in the forward active mode is effectively multiplied by the voltage gain of the device, this subcircuit is familiarly known as the *Miller killer*! When Q_2 is turned on, the discharge path for the varactor diode is completed by D_{10} through D_7. Finally, D_{11} is used to limit the pull-down of the base of Q_3 to an adequate level, without sacrificing the turn-on speed.

Also seen in Figures 6.17–6.19 is a clamping diode at the output terminal. This diode serves the same purpose as the input clamping diodes of all TTL circuits by limiting the negative voltage excursions due to parasitic coupling in signal lines or unmatched transmission line effects.

The input of a 74F00 circuit can be represented by a small capacitance of around 5 pF in parallel with a V–I characteristic that displays different slopes over different ranges of input voltage, as shown in Figure 6.20. The flat vertical portion on the left gives the V_{IH} vs. I_{IH} characteristic, where the input diodes are reverse-biased, so that all of the current from the 10-KΩ input resistor flows into the base of Q_1, with only a leakage current, I_{IH}, flowing in the input diodes. At $V_I = 1.7$ V, the input diodes start to conduct, and some of the current from the 10-KΩ resistor is diverted from the base of Q_1, forming a knee on the curve. When the input voltage further declines to about 1.4 V, another knee is observed, at which point almost all of the current flows out of the input diode. The portion of the curve between these two values of the input voltage corresponds to the transition region of the voltage transfer characteristic. Below $V_I = 1.4$ V, the characteristic has the slope of the 10-KΩ input resistor. If the voltage at the input ever drops below −.3 V, the Schottky clamping diode conducts and the input current increases rapidly as V_I decreases further.

Figure 6.21a depicts the V–I characteristic of the 74F00 gate when the low-level driver is conducting. This curve illustrates the condition in discharging the load capacitance during a high-to-low transition at the output. When $V_{OL} = 3.5$ V, the circuit can absorb charge from the capacitive load at a rate of .5 A/s. As the output voltage decreases, this rate also decreases steadily down to about 100 mA/s at $V_{OL} = 1.5$ V. In this interval from 3.5 V to 1.5 V, part of the charge is fed back through D_8 and the phase-splitter to provide extra base current for Q_3, boosting its current-

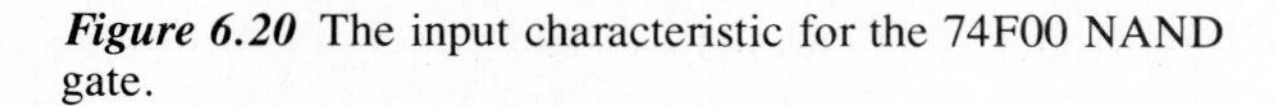
Figure 6.20 The input characteristic for the 74F00 NAND gate.

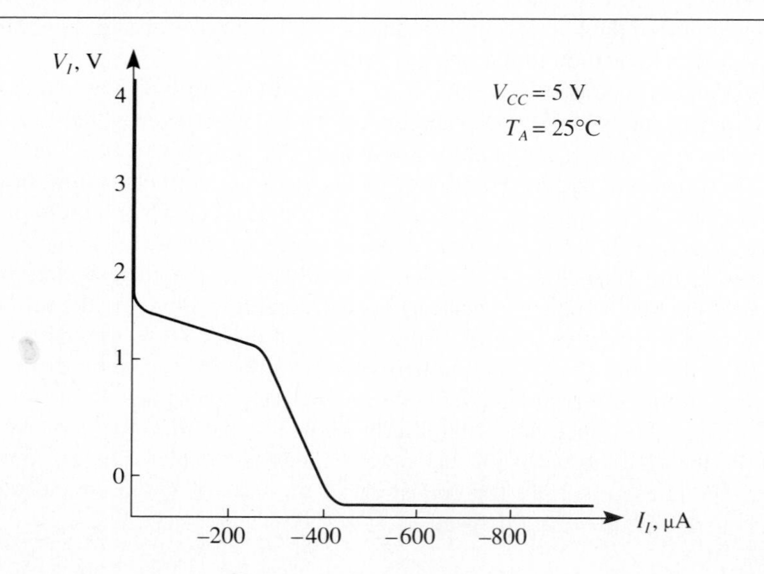

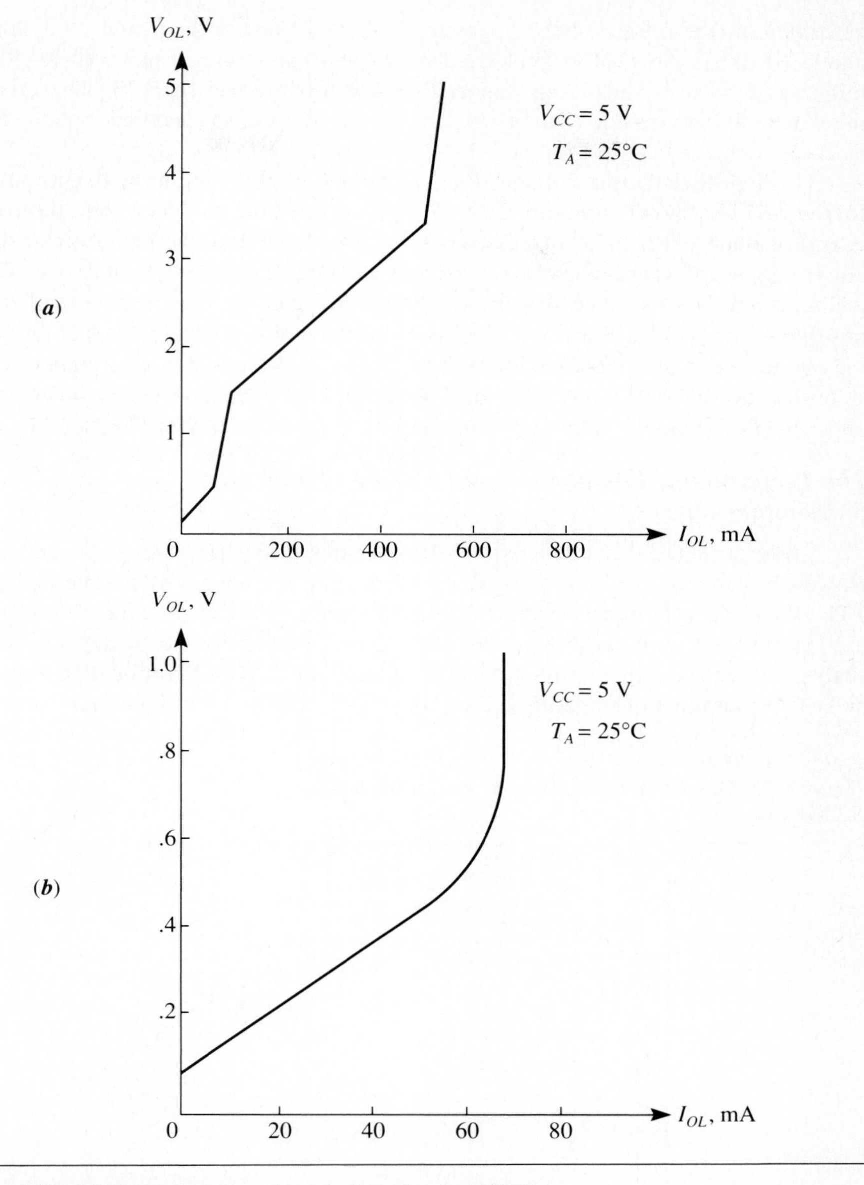

Figure 6.21 (a) The output low characteristic for the 74F00 NAND gate, and (b) characteristic on an expanded scale.

sinking capability and hence reducing the fall time. Below 1.5 V, Q_3 goes on to discharge the load without the extra base current from D_8. At .5 V, the integral Schottky-clamp diode across the BC junction starts to conduct and prevents Q_3 from going into deep saturation.

Figure 6.21b illustrates the same characteristic on a greatly expanded scale. With zero fan-out, the offset voltage is about .1 V, increasing with current on a slope of about 7.5 Ω. If the load current exceeds the current-sinking capability of Q_3, the output voltage rises steeply, as expected. It can be observed from Figure 6.21b that the worst-case maximum value of V_{OL} = .5 V at 20 mA, as specified in data sheets, is easily met.

The high-level output voltage, V_{OH}, as a function of the output high current, I_{OH}, for the 74F00 is shown in Figure 6.22. At low values of the output current, the voltage is approximately 3.6 V, as discussed above. We also obtain the short-circuit output current I_{OS} as the value where the characteristic intersects the horizontal axis. This is guaranteed to be at least 60 mA for a 74AS or 74F gate, as compared to a minimum of 40 mA for the 74S. Since it is a measure of the ability to charge capacitive loads, the output of the advanced Schottky subfamilies can obviously force a higher voltage step into the dynamic impedance of a long interconnection. The transmission-line effects in fast-switching digital circuits will be treated in detail in Chapter 11.

The Performance Comparison and Mixing of TTL Subfamilies

Typical electrical characteristics of each subfamily are given in Appendix A. However, Table 6.4 compares the key performance parameters of commercial-grade TTL subfamilies. For each series, the noise margins, propagation delay times, power dissipation, and speed-power products are listed. Figure 6.23 provides a graphical analysis of the speed-power relationships for all eight TTL subfamilies. Here we can make two important observations. First, the figure clearly shows how much improve-

Figure 6.22 The output high characteristic for the 74F00 NAND gate.

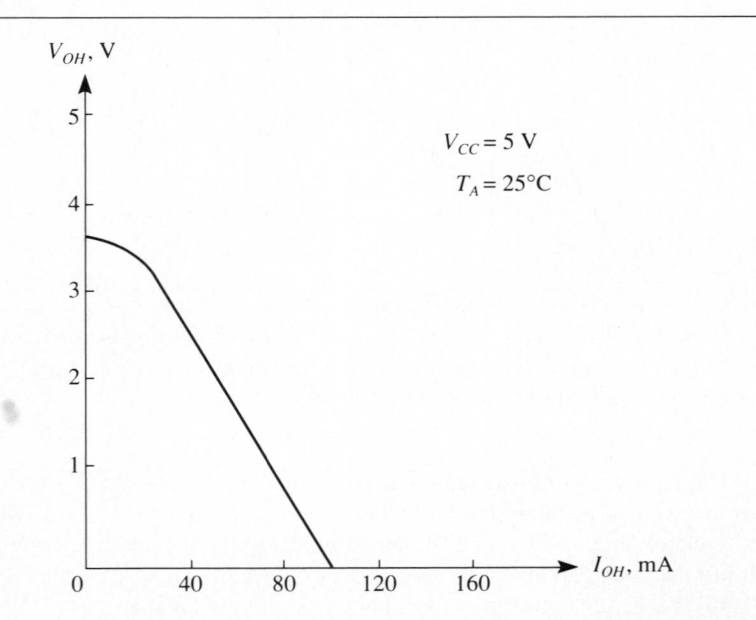

Table 6.4 Key Performance Parameters of 74 TTL Subfamilies

	74	74H	74L	74S	74LS	74AS	74F	74ALS
$V_{OH(min)}$, V	2.4	2.4	2.4	2.7	2.7	2.5*	2.7	2.5*
$V_{IH(min)}$, V	2	2	2	2	2	2	2	2
NM_H, V	.4	.4	.4	.7	.7	.5	.7	.5
$V_{OL(max)}$, V	.4	.4	.4	.5	.5	.5	.5	.5
$V_{IL(max)}$, V	.8	.8	.8	.8	.8	.8	.8	.8
NM_L, V	.4	.4	.4	.3	.3	.3	.3	.3
$I_{OL(max)}$, mA	16	20	3.6	20	8	20	20	8
$I_{IL(max)}$, mA	−1.6	−2	−.18	−2	−.4	−.5	−.6	−.1
$I_{OH(max)}$, mA	−.4	−.5	−.2	−1	−.4	−2	−1	−.4
$I_{IH(max)}$, μA	40	50	10	50	20	20	20	20
N_{max}	10	10	20	10	20	40	33	20
t_{PHL}†, ns	7	6.2	31	3	10	1.5	3.2	7
t_{PLH}†, ns	11	5.9	35	3	9	1.5	3.7	5
Power/gate, mW	10	22.5	1	19	2	8	5.5	1.2
SP, pJ	90	135	33	57	19	12	19	7

*In general, $V_{CC} - 2$.
†Typical values for the following capacitive loads:
C_L = 15 pF for 74, 74S, and 74LS,
C_L = 25 pF for 74H, and
C_L = 50 pF for 74L, 74AS, 74F, and 74ALS.

ment has been made on speeding up the low-power subfamilies. It also illustrates the achievement in reducing the power consumption of high-speed subfamilies over the years.

The 74H subfamily runs more than 30% faster than the standard TTL, but at the expense of over twice the power dissipation. Thus, the SP of 135 pJ is worse than that of standard TTL. As the manufacturers moved into Schottky-barrier diode bipolar technology, two fast and fairly low dissipation series, the 74S and 74LS, emerged. For high-speed applications, the 74S offers speed twice that of 74H, yet dissipates less power and, as a result, provides a superior SP of 57 pJ. On the other hand, the 74L series of low-power TTL offers an SP of 33 pJ, which was once well suited to both industrial and military applications. The power dissipation is one-tenth that of a standard TTL, whereas the average propagation delay is three times as bad as those of standard and 74LS gates.

Before the arrival of truly high-performance bipolar logic, the manufacturers of the high-speed CMOS logic, HCMOS, made comparisons against the 74LS subfamily. With an SP of 19 pJ, the 74LS was the ideal choice for the majority of system applications up until recently.

As shown in Table 6.4, the AS/F/ALS devices made significant strides over their S/LS counterparts. The 74ALS not only runs at better speeds than the 74LS, it

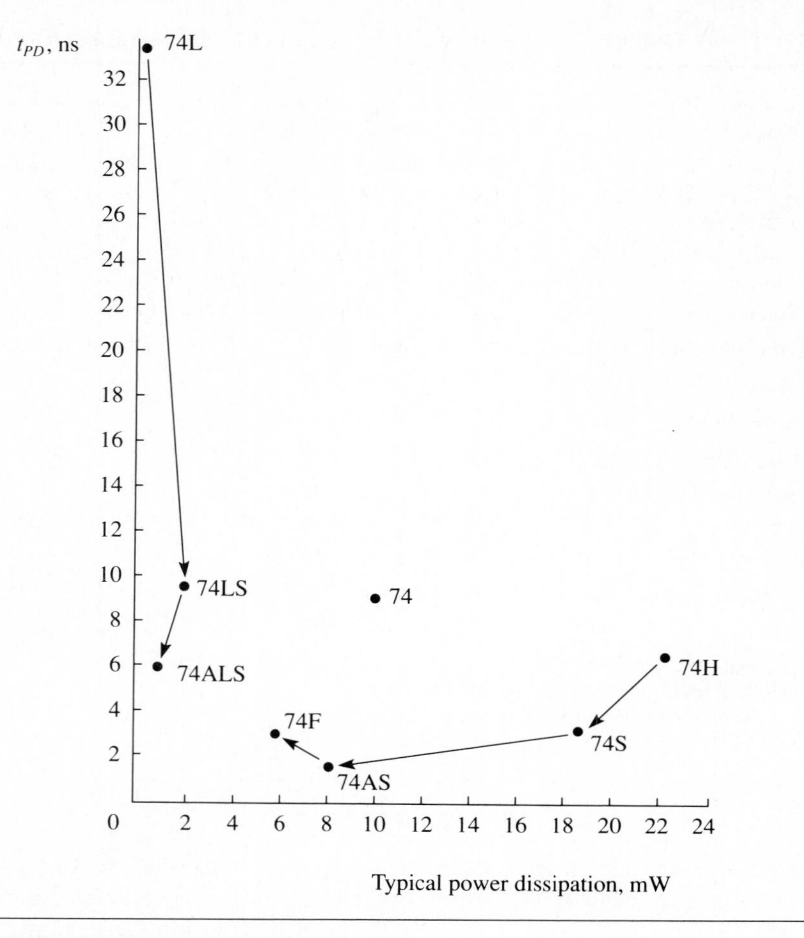

Figure 6.23 The speed-power relationships for TTL subfamilies.

achieves even less power consumption and hence has the best SP of 7 pJ above all the other series in the family. In the same way, the 74AS and 74F subfamilies are superior to the 74S. Running at a fast 1.5 ns delay time with just 8 mW of power consumption, the 74AS enjoys an SP of 12 pJ, as compared to 57 pJ for the 74S. In addition, the 74AS and 74F subfamilies have better driving capabilities than their 74S counterparts. The former, like a 74S gate, can sink 20 mA. A significant difference exists between the subfamilies, however, on the input side. Table 6.4 gives an I_{IL} value of 500 μA for the 74AS and 600 μA for the 74F. In both cases, the I_{IL} specifications are about one-fourth those of the 74S subfamily. Also, the sourcing capability of the 74AS is twice as much as that of the 74S. As a result, the maximum fan-out is 40 for the 74AS, 33 for the 74F, and only 20 for the 74S. Not much difference is observed, on the other hand, between the 74ALS and 74LS driving capabilities.

Figure 6.24 illustrates the propagation delay times as a function of the capacitive load for the advanced Schottky subfamilies.

Manufacturers' data sheets conservatively specify the maximum output low-level voltage V_{OL} as .5 V for all Schottky series. The maximum low-level input voltage V_{IL} that can be applied to a gate is given as .8 V, resulting in a low-level noise margin of NM_L = .3 V. This means that a signal can have .3 V of noise riding on top of it, but the driven gate will still recognize it as a **0**.

Noise margins in TTL circuits are related to the dc threshold voltages of individual transistors. V_{IL} for a typical 74LS gate was obtained above as .9 V. This was derived by adding the BE voltage drops of transistors Q_2 and Q_3 and subtracting the forward voltage drop across the input diode. If V_{OL} = .5 V, then the real NM_L is

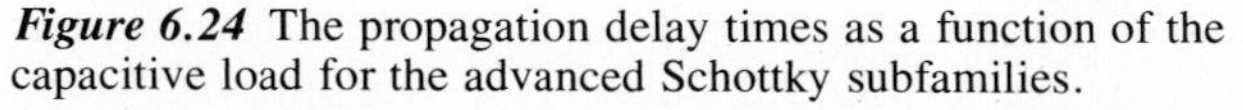

Figure 6.24 The propagation delay times as a function of the capacitive load for the advanced Schottky subfamilies.

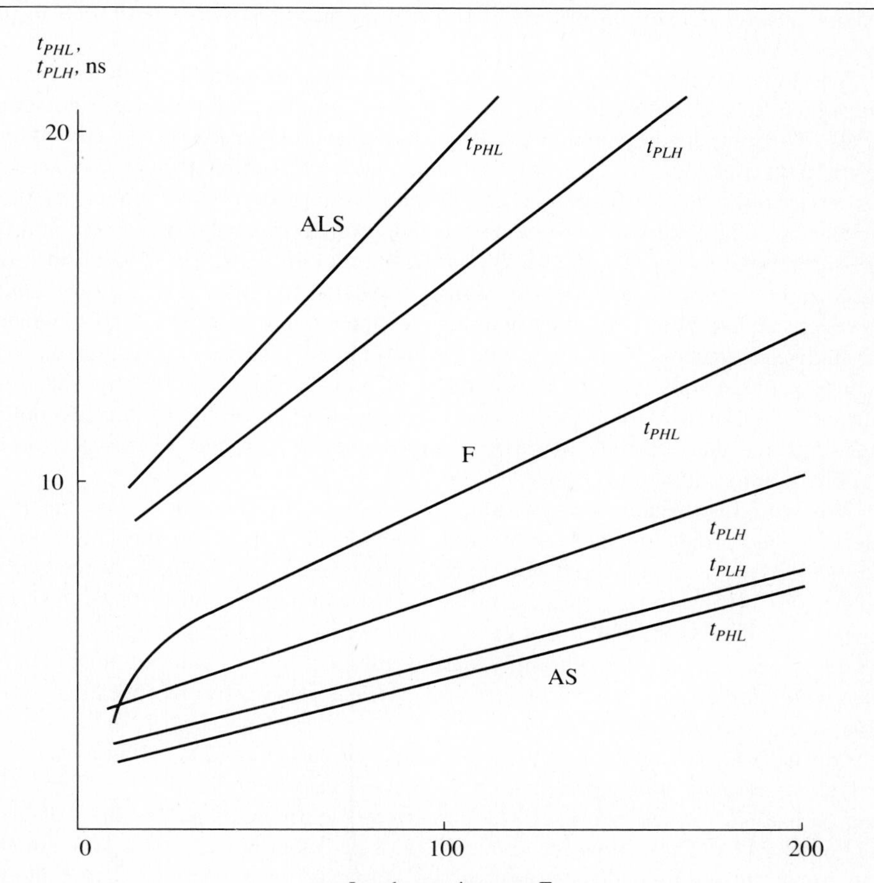

.4 V, which is 100 mV greater than the noise margin specification of the data sheets. However, .4 V is the noise margin only at 25°C. Since bipolar transistors have a negative temperature coefficient of V_{BE}, the V_{BE} values of Q_2 and Q_3 will be lower at the maximum operating temperature of 70°C for commercial grade ICs. In fact, the threshold voltage V_{IL} at 70°C turns out to be about .85 V. Therefore, the noise margin at this temperature is .35 V. Thus, on the data sheet, the low-level noise margin is specified conservatively as .3 V to account for processing and ambient temperature, among other factors.

The low-level input threshold for the AS/F/ASL logic is 1.4 V at 25°C, or .5 V greater than that of the S/LS. Thus, NM_L = .9 V. Naturally, the transistors in these circuits have the same negative temperature coefficient of V_{BE} as those of the S/LS circuits. At 70°C, V_{IL} will fall to about 1.1 V, giving NM_L = .6 V. Consequently, even though the data sheets still specify it as .3 V, we can easily see that the newer logic provides a noise margin improvement of roughly 2:1 over the older Schottky series, leading to more reliable operation. Moreover, the latest package styles, namely, *pin-grid arrays* and *surface-mount plastic chip carriers*, allow the designer greater density on printed circuit boards than is possible with *dual-in-line packages* (DIPs).

The logic levels of all the TTL products are fully compatible with each other. The subfamilies are intended to be used together, but this cannot be done indiscriminately. The input loading and output drive characteristics of each series in this family are different and must be taken into consideration when mixing them in a system. Fast devices like the 74AS and 74F are designed with relatively low input and output impedances. The speed of these devices is determined primarily by fast rise and fall times at internal nodes as well as at input and output terminals. These fast transitions cause various types of noise in the system (treated in Chapter 11 in more detail). However, we can briefly mention here the two most common types of noise generation in digital systems. The power- and ground-line noise is due to the large currents needed to charge and discharge the circuit and load capacitances during the switching transients. The signal-line noise, on the other hand, is generated by the fast output transitions and the relatively low output impedances, which tend to increase the reflections in long interconnections.

When mixing the slower speed subfamilies with the higher speed series, one must note that the former are more susceptible to induced noise than the latter simply because they have higher input and output impedances. For example, when wiring the 74S and 74LS together, separate or isolated power and ground systems should be utilized, and the 74LS input signal lines should not be run adjacent to lines driven by 74S devices since the 74LS subfamily is especially sensitive to induced noise. On the other hand, mixing standard TTL and the 74LS is less restrictive because slower signal transitions generate less noise.

Normalizing the input current requirements and output drive capabilities is useful for system designers working with more than one TTL subfamily. Table 6.5 provides a set of load factors as characterized by current requirements at the input of eight TTL subfamilies. The entries are obtained by using Table 6.4 and taking the smaller of the input current ratios $|I_{IHr}/I_{IHc}|$ and $|I_{ILr}/I_{ILc}|$ where subscripts r and c signify the row and column subfamilies, respectively. These values are then compared to the

Table 6.5 Normalized Input Currents for 74 TTL Subfamilies

	Replacing							
Replaced	74	74H	74L	74S	74LS	74AS	74F	74ALS
74	1	.8	4	.8	2	2	2	2
74H	1.25	1	5	1	2.5	2.5	2.5	2.5
74L	.11	.09	1	.09	.45	.36	.3	.5
74S	1.25	1	5	1	2.5	2.5	2.5	2.5
74LS	.25	.2	2	.2	1	.8	.67	1
74AS	.31	.25	2	.25	1	1	.83	1
74F	.38	.3	2	.3	1	1	1	1
74ALS	.06	.05	.56	.05	.25	.2	.17	1

maximum fan-out capability of the output being considered, as listed in Table 6.6. The former table is generally used in conjunction with the latter when replacing one series with another in an existing system; Table 6.6 can be used when designing an original system that employs several subfamilies to optimize performance. Every possible combination of the eight commercial-grade TTL subfamilies is included in these tables. Note that, as seen from the entries in the last row of Table 6.5, the ALS requires only a fraction of the input current demanded by other TTL series.

Example 6.1

Consider an existing system using 74LS series of logic gates. We want to replace some of it by series 74ALS logic. First, we use Table 6.5 to check if the 74ALS can be supplied with sufficient input current. The entry corresponding to the 74LS row and the 74ALS column is 1, which indicates that one 74ALS gate can be driven for each 74LS gate removed. However, if more 74LS gates are being driven by the replaced 74LS gate, then Table 6.6 is used to find the fan-out relationship between these two subfamilies. In this

Table 6.6 Maximum Fan-Out Capabilities of 74 TTL Subfamilies

	Driven							
Driving	74	74H	74L	74S	74LS	74AS	74F	74ALS
74	**10**	8	40	8	20	20	20	20
74H	12	**10**	50	10	25	25	25	25
74L	2	1	**20**	1	9	7	6	10
74S	12	10	100	**10**	50	40	33	50
74LS	5	4	40	4	**20**	16	13	20
74AS	12	10	111	10	50	**40**	33	100
74F	12	10	100	10	50	40	**33**	50
74ALS	5	4	40	4	20	16	13	**20**

table, the intersection of the 74ALS row and the 74LS column has an entry of 20. Thus, each 74ALS gate inserted will be able to drive 20 74LS gates.

Finally, one should note that electrically open inputs degrade the switching speed of a circuit. Therefore, in order to eliminate the distributed capacitance associated with the floating input and to ensure that no degradation occurs in the propagation delay times, the unused NAND and AND inputs should be maintained at a logical **1** voltage level, while the unused NOR and OR inputs must be returned to ground.

Example 6.2

The following is the SPICE input file for the two-input NAND gate of Figure 6.1 with all possible combinations of inputs:

```
TTL 7400 2-I/P NAND GATE
D1CLAMP 0  1     D
D2CLAMP 0  2     D
Q11IN   9  5 1   Q
Q12IN   9  5 2   Q
Q2      6  9 8   Q
Q3      3  8 0   Q
Q4      7  6 10  Q
D1      10 3     D
R1      4  5     4K
R2      4  6     1.6K
R3      4  7     130
R4      8  0     1K
VCC     4  0     5
V1IN    1  0     PULSE 0 5 0 10N 10N 10N 50N
V2IN    2  0     PULSE 0 5 0 10N 10N 20N 100N
.MODEL D D RS 10 CJO 3P VJ .7 TT 5N
.MODEL Q NPN BF 75 RB 100 CJE 1P CJC 3P
.TRAN 10N 150N
.PROBE
.END
```

The resulting output waveform for all combinations of inputs is plotted in Figure 6.25.

6.2 The Emitter-Coupled Logic

Another commercially available family of bipolar digital ICs is the *emitter-coupled logic* (*ECL*). Currently, it is the fastest form of logic, with typical propagation delay times of less than 1 ns. As in the case of the Schottky TTL, the active devices within the ECL are operated out of saturation to avoid overdrive charge storage. In addition,

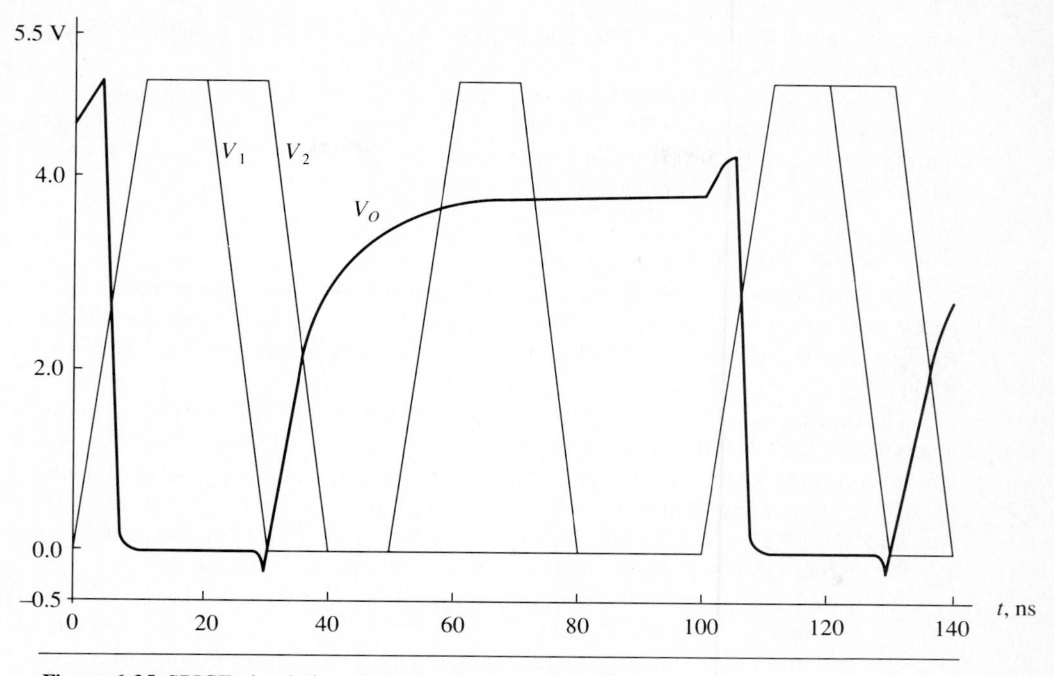

Figure 6.25 SPICE simulation: the output voltage waveform for all combinations of inputs for the 7400 NAND gate.

the logical signal swings are comparatively small; thus, the time needed to charge or discharge the load and parasitic capacitances is short.

ECL was first introduced in 1962 by Motorola, Inc. as the MECL I. This and a later version introduced in 1966, the MECL II, are now obsolete. They gave way to the currently available MECL III, 10K, 100K, and 10KH series.

The third subfamily, introduced by Motorola in 1968, was MECL III. It has the highest speed, with typical rise and fall times, also called *edge speeds*, of 1 ns that require a transmission-line environment. For this reason, all circuit outputs are designed to drive transmission lines and all output logic levels are specified when driving 50-Ω loads. The series finds applications in high-speed digital communications as well as in counter prescalers, VHF phase-locked-loops, digital signal processors, and in the critical timing delays of large systems. In 1971, the same company introduced the slower 10K series for general-purpose applications, such as large mainframe computers, and high-performance control and test systems. It has typical edge speeds of 2 ns. On the other hand, 10K ECL gates use less than one-half the power of MECL III. Although the circuits are specified to drive transmission lines, the slower edge speeds permit the use of wire-wrap and standard printed circuit lines.

In 1981, Motorola introduced the 10KH product subfamily, which features up to 100% improvements in propagation delay and edge speeds while maintaining the same power supply currents as the 10K subfamily. Noise margins have also been improved by 75% over the older series. With 1 ns propagation delay and a power consumption of 25 mW per gate, the 10KH enjoys the best SP of any subfamily available.

Finally, to minimize the variation with temperature and the dependence on power supply of the output levels and input threshold values, thereby making the voltage transfer characteristic independent of the supply voltage and temperature, Fairchild introduced in the early eighties a new ECL series, the 100K subfamily.

The Current Switch

The basic component of all ECL circuits is the nonsaturating *current switch*, also called the *voltage comparator* or *difference amplifier*. This circuit is also the basis of the operational amplifiers and comparators in analog electronics. It is illustrated in Figure 6.26.

The input signal is applied to the base of Q_1, and a constant reference signal V_R is applied to the base of Q_2. In Figure 6.26, V_R is at ground. Consider first the case where V_I is sufficiently lower than V_R, that is, a negative value. Q_1 will be cut off, Q_2 will be on and the current I_{EE} will flow through Q_2. The collector voltage of Q_1 will be $V_{O1} = V_{OH} = V_{CC}$, while $V_{O2} = V_{OL} \approx V_{CC} - R_C I_{EE}$ for $\beta_F >> 1$. R_C and V_{CC} are selected to assure that Q_2 operates in the forward active mode.

As V_I rises and Q_1 turns on, the emitter voltage V_E increases. Since $V_E = V_I - V_{BE1}$, Q_2 eventually turns off. Then Q_1 operates fully in the forward active mode. Hence, a variation at the input switches a nominally fixed emitter current from one transistor to the other. This switching is accomplished rapidly since the transistors are not saturated in the process.

Figure 6.26 Basic nonsaturating current switch.

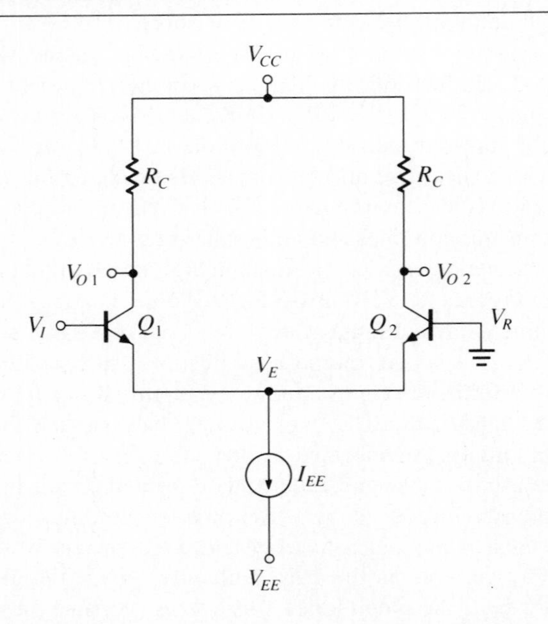

Important Features of the ECL

The important features and significant advantages of the ECL family over other logic families can be summarized as follows:

1. *Complementary outputs.* A majority of logic elements offer double-rail logic by having complementary outputs. Timing-differential problems arising from the time delays introduced by inverters are eliminated. This feature also reduces the package count by eliminating the need for the complementation operation and alleviates the system power requirements.
2. *Constant supply current.* The current drain is the same regardless of the state of the switches, so that the circuit presents constant current loads to the power supply, simplifying the power supply design and reducing costs.
3. *Minimal power-supply noise generation.* Unlike the totem-pole outputs of the TTL family, the differential amplifier design and the emitter-follower outputs almost totally eliminate the current spikes to which the power lines are subjected during switching. In addition, the small voltage swings lead to very small current spikes caused by charging and discharging the stray capacitance.
4. *Low crosstalk.* The increased susceptibility to crosstalk in high-speed circuits is the result of the very steep leading and trailing edges, that is, fast rise and fall times, of the signal. The current mode switching, as utilized by ECL, makes use of differential comparison techniques and avoids transistor storage delays. Therefore, switching times can be controlled by internal time constants. To minimize ringing and reflections on interconnection wiring, and hence to reduce the affinity for crosstalk without compromising the performance, the edge rates have been deliberately slowed in the 10K, 100K, and 10KH series. For example, the typical edge rate for 100K ECL is 1 μV/s, which is about 80% that of the Schottky TTL.

Finally, we can also mention that the high input impedance and low-impedance emitter-follower outputs allow large fan-out and versatile drive characteristics. One obvious trade-off in using the ECL family is the imposition of additional design rules and restrictions of a transmission-line environment. One can also mention the higher power consumption of the family compared to the advanced Schottky subfamilies.

The Basic ECL Gate

The basic ECL gate consists of three blocks: the *current switch*, the *output emitter-followers*, and the *bias network*, as illustrated in Figure 6.27. The output emitter-followers have high drive capabilities through impedance transformation. They also shift the level of the output signals by one V_{BE} drop to make the output logic levels compatible with the input logic levels. The current switch, as its name implies, contains the current-steering element that provides simultaneous outputs of both the OR operation and its complement, the NOR operation, as well as the voltage gain necessary for a narrow linear threshold region. The temperature- and voltage-compensated bias network generates a reference voltage V_{BB} whose value is made to change with temperature in a predetermined manner to keep the noise margins constant. It is also insensitive to variations in the supply voltage V_{EE}.

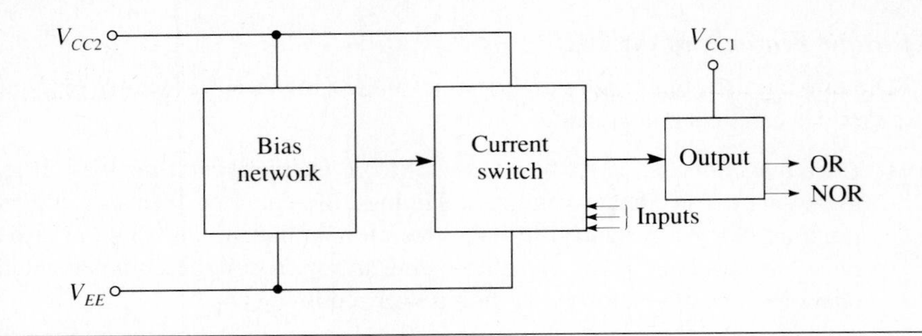

Figure 6.27 The functional block diagram of the basic ECL circuit.

The 10K Series

The schematics of the basic gate circuits for the first four subfamilies developed by Motorola are shown in Figures 6.28 through 6.31. As seen in these figures, the first variation employed as the technology advanced was the inclusion of the bias network in all subfamilies subsequent to MECL I, in which it was not an internal feature. The second change is in the resistor values. This difference in conjunction with transistor geometries, obviously not incorporated in the schematics, was necessary to achieve the varying speed and power improvements of the different lines.

Figure 6.28 The basic MECL I circuit, first introduced in 1962. (*Copyright of Motorola, Inc. Used by permission.*)

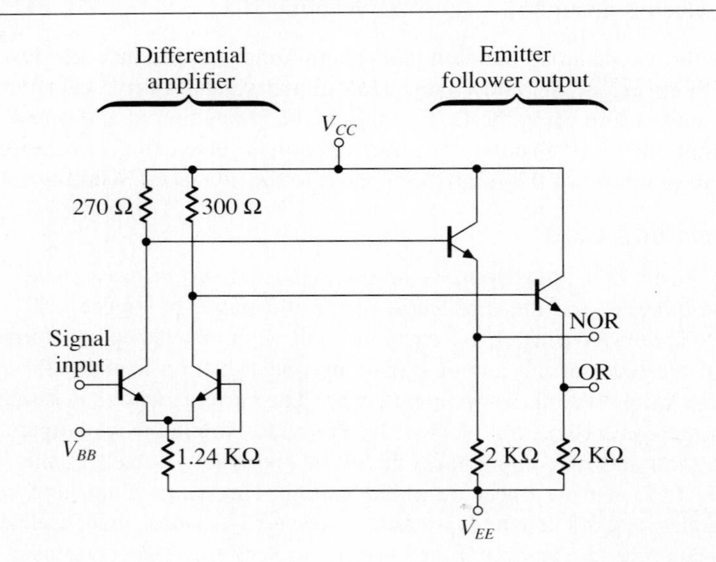

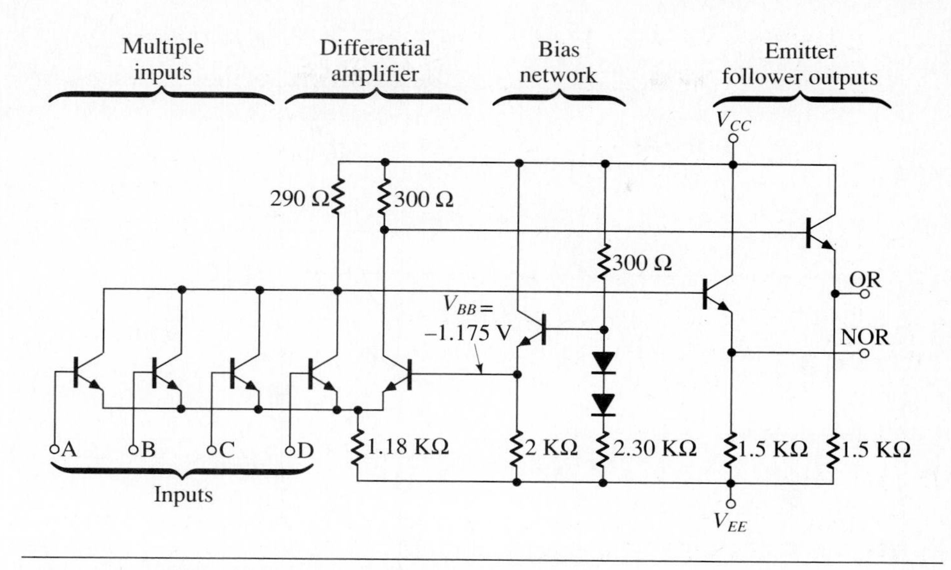

Figure 6.29 The basic MECL II circuit, first introduced in 1966. (*Copyright of Motorola, Inc. Used by permission.*)

Figure 6.30 The basic MECL III circuit, first introduced in 1968. (*Copyright of Motorola, Inc. Used by permission.*)

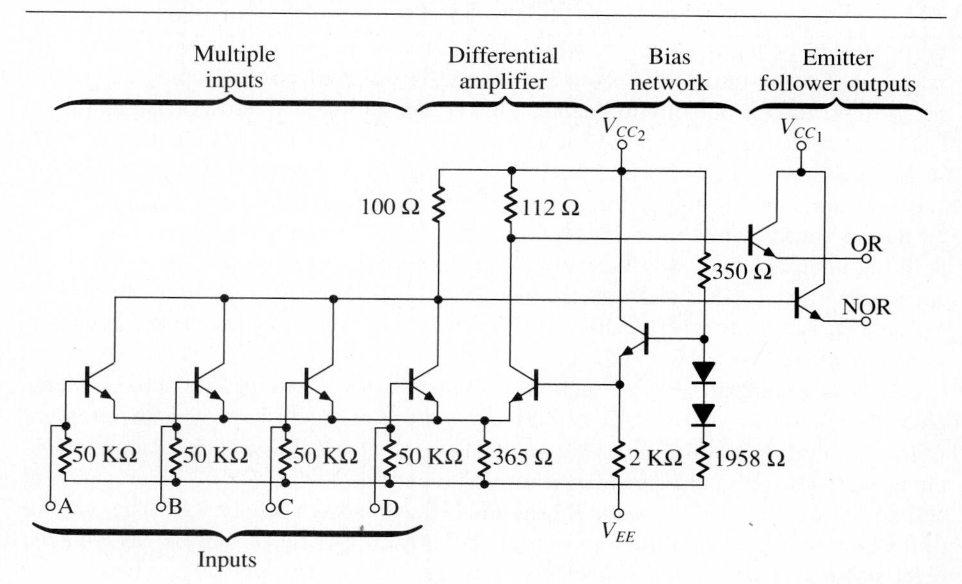

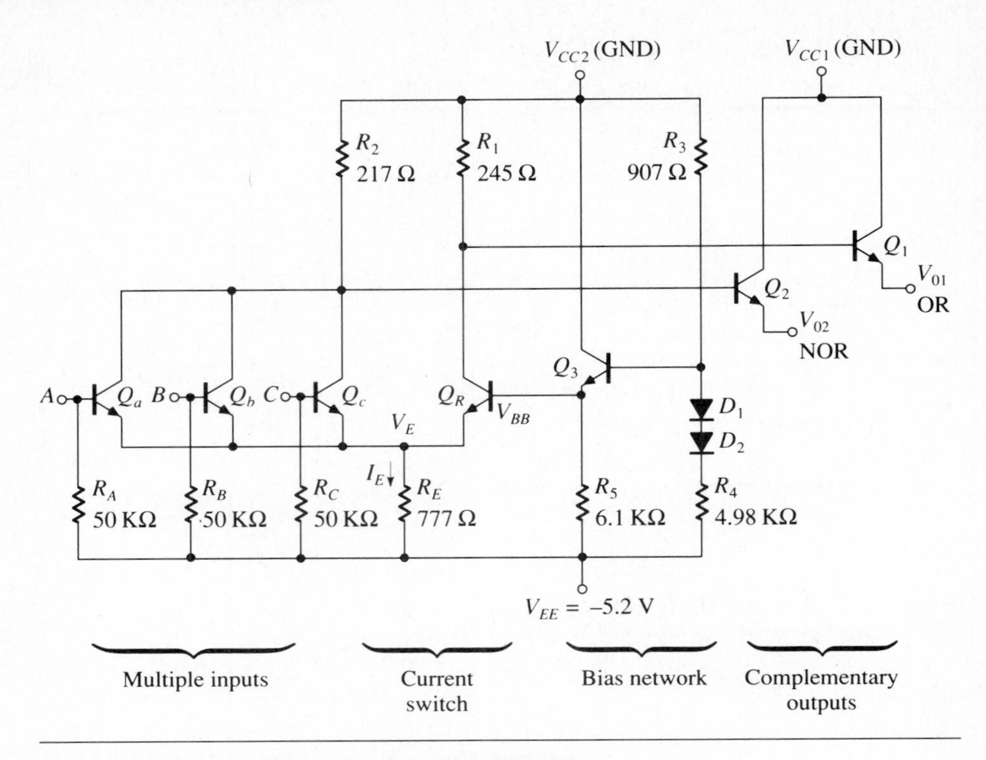

Figure 6.31 The basic two-input 10K ECL OR/NOR gate.

Another difference is in the output circuits. MECL I circuits are supplied with output pull-down resistors on the chip, while the later subfamilies do not have them. The reason is that the internal resistors are actually a waste of power in properly terminated transmission-line environments. This configuration also allows more complex LSI, and increased chip life and reliability. Finally, note that MECL III and 10K series circuits are supplied with 50-KΩ base pull-down resistors. These *pinch* resistors serve to drain off the input transistor leakage current by providing a path for the current to unused input bases, thereby causing them to be well turned off. Since they hold the unused inputs at a fixed low level, these inputs can be left open. For high speed operation, 50-KΩ resistors are usually paralleled by 50-Ω external resistors to −2 V. The reason for this practice will be clear later when we discuss the transmission-line effects.

Now, we concentrate our attention on the basic 10K circuit of Figure 6.31. Here, Q_R is the reference transistor. The reference voltage at the base of Q_R is the emitter-follower output V_{BB} of the bias network. Q_a through Q_c are the input transistors; they, along with Q_R, form the current switch. The current source I_{EE} of Figure 6.26 is realized using the 777-Ω resistor R_E and the −5.2 V power supply V_{EE}. The outputs of the current switch are through the emitter-followers Q_1 and Q_2, serving as voltage-level shifters as well as low-impedance voltage drivers. The first of the two power supplies, V_{CC1}, is connected to the collectors of the output emitter-followers, while

V_{CC2} is for the current switch and the bias network. As discussed above, the fast switching times of the output cause voltage transients, and hence current spikes, to appear in the collector circuits of Q_1 and Q_2, while the supply current to the current switch is essentially constant. Although both V_{CC1} and V_{CC2} are connected to ground, separate and short paths should be used to reduce the cross-coupling between the individual subcircuits and thereby minimize the effect of the output voltage transients on the current switch.

Therefore, the circuit is characterized with V_{CC1} and V_{CC2} at ground potential, while V_{EE} is at -5.2 V. This convention results in maximum noise immunity because any noise induced on the V_{EE} line is applied to the circuit as a common-mode signal that is rejected by the differential action of the current switch. However, noise induced into either V_{CC} line is not canceled out in this fashion. Thus, the V_{CC} bus requires a good system ground for best noise immunity. Manufacturers recommend a supply regulation of 10% or better. $V_{EE} = -5.2$ V results in the best switching times. A more negative value increases the noise margins at the expense of the power dissipation performance, while a less negative voltage leads to exactly the opposite effect.

The Voltage Transfer Characteristic

In the analysis to determine the voltage transfer characteristic of the 10K ECL OR/NOR circuit, we will assume that all transistors operate in the forward active mode with a minimum $V_{BE(on)} = .75$ V. This value is due to the small areas of the BJTs used in the circuit in order to have small capacitances and a high unity-gain cutoff frequency f_T. Actually, an even higher value of .8 V differential is a characteristic of the BE junctions of current-switch transistors. Also, sufficiently high current gain β_F will allow us to neglect the base currents.

The reference voltage is given by

$$V_{BB} = V_{B3} - V_{BE(on)}$$

where

$$V_{B3} = V_{CC2} - \frac{R_3}{R_3 + R_4}(V_{CC2} - 2V_D - V_{EE})$$

$$= 0 - \frac{.907}{.907 + 4.98}(0 - 1.5 + 5.2) = -.57 \text{ V}$$

Therefore,

$$V_{BB} = -.57 - .75 = -1.32 \text{ V}$$

In practice, the typical value of V_{BB} is given as -1.29 V by the manufacturers. Thus, we will use this value in our analysis.

Case 1: All the inputs are at logical **0**. Q_a, Q_b, and Q_c are off, and Q_R is on. Thus the common emitter will be one diode drop more negative than its base, or

$$V_E = V_{BB} - V_{BE(on)} = -1.29 - .8 = -2.09 \text{ V}$$

The current through R_E is obtained as

$$I_E = \frac{V_E - V_{EE}}{R_E} = \frac{-2.09 - (-5.2)}{.777} = 4 \text{ mA}$$

Neglecting the base current, $I_{CR} \approx I_E$ so that

$$V_{CR} = V_{CC2} - R_1 \times I_{CR} = 0 - 245 \times 4 \times 10^{-3} = -.98 \text{ V}$$

Thus, at the output we have

$$V_{O1} = V_{OL} = V_{CR} - V_{BE(on)} = -.98 - .75 = -1.73 \text{ V}$$

while (the input transistor being off) the drop across R_2 is zero. Since the base and collector of Q_2 are effectively tied together, it now behaves as a conducting diode. Hence, at the emitter we have

$$V_{O2} = V_{OH} = V_{CC1} - V_{BE(on)} = 0 - .75 = -.75 \text{ V}$$

The data sheets provide the minimum and maximum values for the logical **1** output voltage V_{OH} as $-.98$ V and $-.81$ V, respectively, while they are specified as -1.85 V and -1.63 V, respectively, for the logical **0** output voltage V_{OL}. The typical transfer characteristic curve varies from a low state of $V_{OL} = -1.75$ V to a high state of $V_{OH} = -.9$ V with respect to ground. The discrepancy between the calculated value of $V_{O2} = -.75$ V and the typical value of $-.9$ V is due to the finite base current flowing through R_2 when a load is applied to V_{O2}. Note that the typical output voltage levels are almost symmetrical about the reference voltage $V_{BB} = -1.29$ V, which implies that the outputs may be directly connected to the inputs of the loads.

With inputs at -1.75 V, Q_a, Q_b, and Q_c are indeed in the cutoff region, since

$$V_{BEa} = V_{BEb} = V_{BEc} = -1.75 - (-2.09) = .34 \text{ V}$$

This verifies the assumption that the input transistors are not conducting current: .34 V is less than their cut-in voltage.

Case 2: At least one input is high. When any one or all of the inputs are shifted upward from -1.75 V to the $-.9$ V logical **1** state, the BE junction voltage of the corresponding transistor increases beyond the cut-in voltage, and the current in the switch is diverted to that transistor, while Q_R is cut off.

With any input, say Q_a, conducting, V_E rises to $V_{Ba} - V_{BE(on)} = -.9 - .8 = -1.7$ V, and the emitter current becomes

$$I_E = \frac{V_E - V_{EE}}{R_E} = \frac{-1.7 - (-5.2)}{.777} = 4.5 \text{ mA}$$

Therefore, at the collector of Q_a we have

$$V_{Ca} = V_{CC2} - R_2 \times I_{Ca} \approx 0 - 217 \times 4.5 \times 10^{-3} = -.98 \text{ V}$$

Then, the output voltage is obtained as

$$V_{O2} = V_{OL} = V_{Ca} - V_{BE(on)} = -.98 - .75 = -1.73 \text{ V}$$

which corresponds to the logical **0** state, as expected. On the other hand, the base voltage of the fixed-bias transistor is held at -1.29 V so that $V_{BER} = -1.29 - (-1.7) = .41$ V, and Q_R is off.

To find a more realistic value of V_{OH} than that of the previous condition, consider the circuit of Figure 6.32, which illustrates the OR output of Figure 6.31 with a 50-Ω pull-down returned to -2 V, and assume that $\beta_{F1} = 100$. Then, writing KVL around the loop including the BE junction, we have

$$\frac{245}{101} I_{E1} + .75 + 50 I_{E1} - 2 = 0$$

or

$$I_{E1} = 22.3 \text{ mA}$$

Therefore, the high-level output is found as

$$V_{O2} = V_{OH} = -2 + 22.3 \times 50 = -.885 \text{ V}$$

which is practically equal to the typical value of $-.9$ V, as specified by the manufacturers.

The gate transfer curves for both outputs are shown in Figure 6.33. Note that the maximum V_{IL} and minimum V_{IH} values define the limits of the transition region. The study of the current switches shows that the transition region is centered on the reference voltage $V_{BB} = -1.29$ V. Indeed, data sheets provide $V_{IL(max)}$ and $V_{IH(min)}$ as -1.475 V and -1.105 V, respectively. Thus, the transition region width is 370 mV.

Figure 6.33 shows that the symmetry of the transfer characteristics is lost as the input is increased above $V_{IH(min)}$, and the output voltage decreases with a slope of about $-.24$ V. This is caused by the decrease at the collector input node due to the increasing collector current as V_I is increased further. With sufficient input voltage, Q_a will saturate. Assuming $V_{CE(sat)} \approx 0$, at this point, we will have

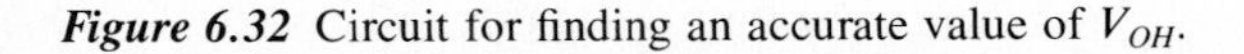

Figure 6.32 Circuit for finding an accurate value of V_{OH}.

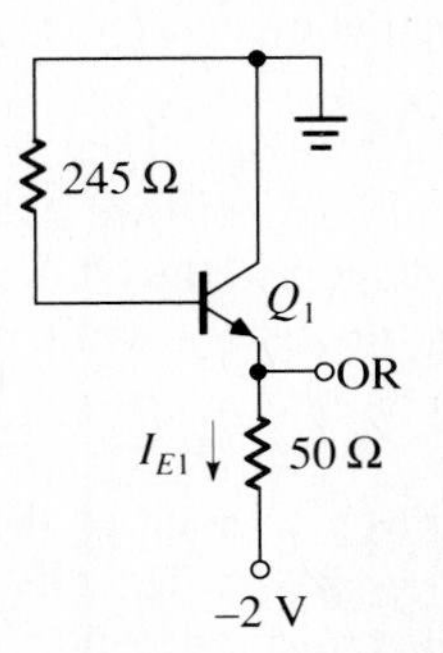

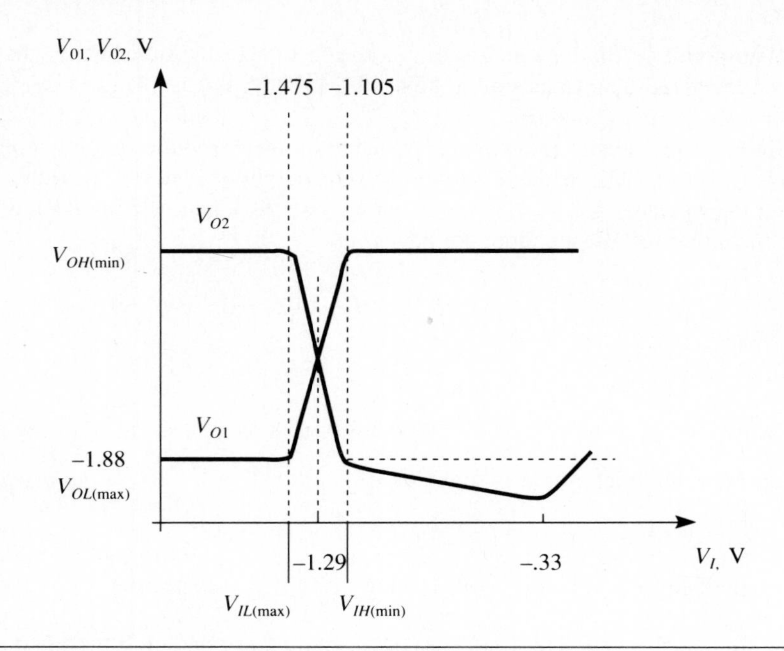

Figure 6.33 The VTC of the 10K ECL OR/NOR circuit.

$$V_E = V_{Ca} - V_{CE(sat)}$$

$$= V_{CC2} - \frac{R_2}{R_2 + R_E}(V_{CC2} - V_{CE(sat)} - V_{EE})$$

$$= 0 - .217\frac{0 + 5.2}{.217 + .779} = -1.13 \text{ V}$$

Therefore, at the input we will have

$$V_I = V_E + V_{BE(on)} = -1.13 + .8 = -.33 \text{ V}$$

The corresponding output voltage is given as

$$V_{O2} = V_{Ca} - V_{BE(on)} = -1.13 - .75 = -1.88 \text{ V}$$

Beyond that point, the BC junction is forward-biased to saturation, and $V_{BC(on)} = V_{BE(on)}$ for Q_a. The collector voltage and the NOR output will go more positive with the increasing input level. Thus, the output will be given as

$$V_{O2} = V_I - V_{BCa(on)} - V_{BE2(on)}$$

In normal system operation, however, none of the transistors saturate because the input voltages do not go above −.81 V. The OR output level is unaffected by the input voltage levels except in the transition region.

Instead of the unity-gain definition, to simplify the calculation of $V_{IL(max)}$ and $V_{IH(min)}$, let us arbitrarily define the threshold points as those points at which one transistor in the current switch carries 99.9% of the emitter current and the other carries the rest. Both transistors Q_a and Q_R are in the forward active mode over the transition region, so that we can use the exponential diode equations for their BC junctions. Thus, we have

$$I_{Ea} + I_{ER} = I_E$$

and

$$\frac{I_{Ea}}{I_{ER}} \approx e^{(V_{Ba} - V_{BB})/\phi_T}$$

where V_{Ba} is the base voltage of Q_a. Combining the above equations produces

$$I_{Ea} = \frac{I_E}{1 + e^{(V_{BB} - V_{Ba})/\phi_T}}$$

Thus, the total input voltage difference, corresponding to the transition width, is obtained as

$$V_{Ba}(I_{Ea} = .999I_E) - V_{Ba}(I_{Ea} = .001I_E) = \phi_T \ln (999{,}000) = 358 \text{ mV}$$

as compared to the actual width of 370 mV. This yields

$$V_{IL(max)} = -1.29 - \frac{.358}{2} = -1.469 \text{ V}$$

$$V_{IH(min)} = -1.29 + \frac{.358}{2} = -1.111 \text{ V}$$

which are very close to the values provided by the manufacturers.

The voltage gain of the basic 10K gate circuit is approximately 4 and essentially independent of the current gain. Therefore, the current gain can vary over a wide range, which facilitates the fabrication of the circuit. Another factor that eases the processing is that the output voltage depends on the resistor ratios, which can be held to within $\pm 5\%$ even though the absolute values can vary as much as $\pm 20\%$. Finally, since the transistors do not normally saturate, the gold doping commonly used to decrease the storage time in earlier TTL subfamilies or the Schottky-clamped transistor used in the Schottky series is not required in manufacturing ECL circuits.

Edge Speeds and Propagation Delays

Because of the emitter-follower outputs, the fall time of the output signal and t_{PHL} are longer than the rise time and t_{PLH}, especially for large capacitive loading. When the base voltage of an emitter follower changes abruptly to increase the emitter current, that is, on the leading edge of the output signal, the emitter follows the base, and any capacitance across it charges rapidly through its low output impedance, which is around 7 Ω in the 10K series. On the trailing edge, when the input change is in the reverse direction, the emitter follower cuts off. Thus, the emitter voltage cannot momentarily follow due to the coupled capacitor, and the load capacitance now has to discharge through the combination of load and pull-down resistors.

The recommended minimum and maximum values for the output pull-down resistors are 50 and 100 ohms if connected to −2 V, and 270 and 510 ohms when returned to −5.2 V, respectively. If the pull-down resistor R_T is connected to −2 V, the fall time is given by the manufacturers as

$$t_f \approx (1.1 R_T C_L + 2) \text{ ns} \tag{6.2}$$

where C_L is the load capacitance in pF. If, on the other hand, the pull-down is tied to V_{EE}, then the fall time is specified as

$$t_f \approx (.2 R_T C_L + 2) \text{ ns} \tag{6.3}$$

Equations (6.2) and (6.3) take into account the delays due to the parasitic capacitances within the IC. Figures 6.34 through 6.37 show the degradation in switching times and propagation delays with the 50-Ω load placed near the output pin. Here, since the input capacitance of a 10K basic gate averages about 2.9 pF, it is assumed that each added load contributes 5 pF, allowing for the capacitances associated with wired interconnections. Note that the edge speeds are measured between 20% and 80% values.

The Fan-Out

When the input is logical **0**, the input current is equal to the current flowing through the 50-KΩ pull-down resistor. Hence,

$$I_{IL} = \frac{-1.75 + 5.2}{50 \text{ K}\Omega} = 65 \ \mu\text{A}$$

Figure 6.34 The rise time versus capacitive loading at various temperatures for the 10K series with a 50-Ω pull-down resistor at the output. (*Copyright of Motorola, Inc. Used by permission.*)

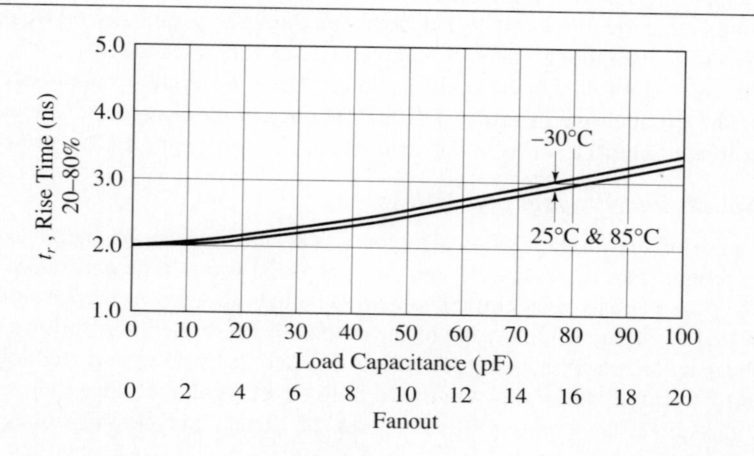

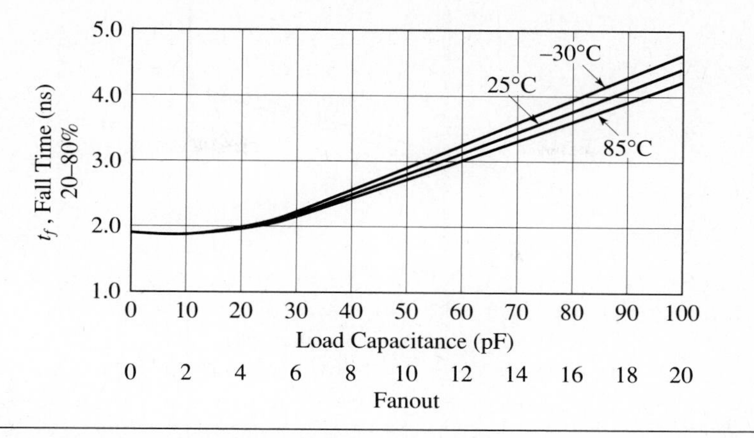

Figure 6.35 The fall time versus capacitive loading at various temperatures for the 10K series with a 50-Ω pull-down resistor at the output. (*Copyright of Motorola, Inc. Used by permission.*)

When the input is high, the input current has to feed the base of the input transistor, too. Thus, assuming $\beta_F = 100$,

$$I_{IH} = \frac{-.9 + 5.2}{50 \text{ K}\Omega} + \frac{4.5}{101} = 130.5 \ \mu\text{A}$$

So, the relatively high input impedance and hence small input current values at both logic levels make it possible for the circuit to drive a relatively large number of loads

Figure 6.36 The propagation delay t_{PLH} versus capacitive loading at various temperatures for the 10K series with a 50-Ω pull-down resistor at the output. (*Copyright of Motorola, Inc. Used by permission.*)

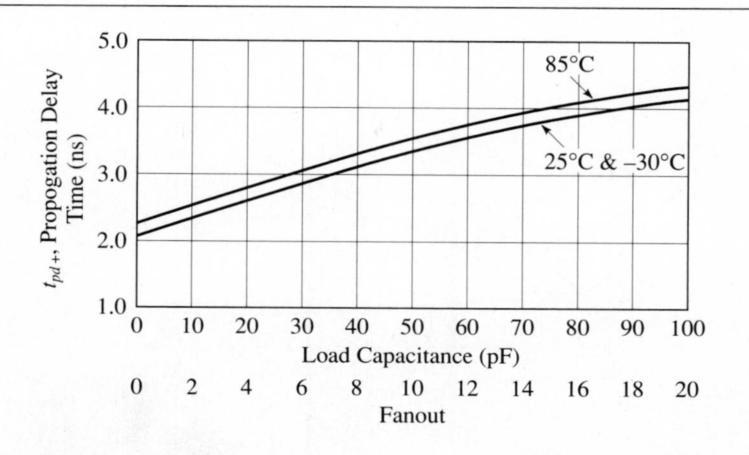

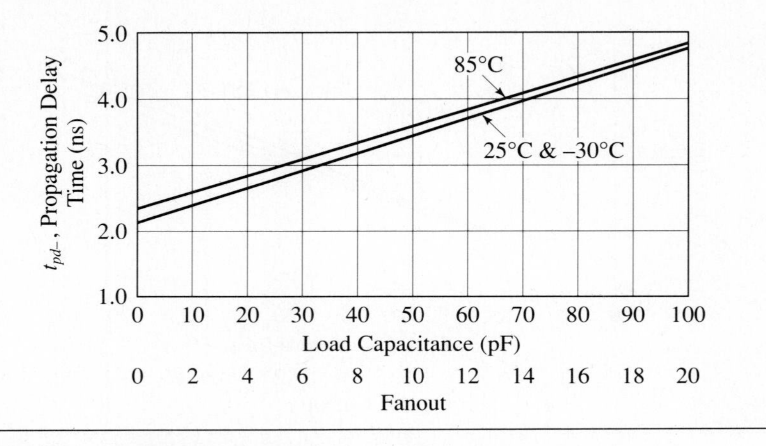

Figure 6.37 The propagation delay t_{PHL} versus capacitive loading at various temperatures for the 10K series with a 50-Ω pull-down resistor at the output. (*Copyright of Motorola, Inc. Used by permission.*)

without the deterioration of the guaranteed noise margins. Therefore, dc loading causes little change in the output voltage levels and does not present a design problem. For example, 10 gate loads, in addition to a 50-Ω pull-down to -2 V, will reduce the noise margin by less than 20 mV. Figure 6.38 provides output voltage levels of the 10K series as a function of dc loading with various pull-down resistor values. Maximum specified dc fan-out for this series is 90.

On the other hand, ac loading increases the capacitance associated with the circuit, thereby affecting the circuit speed. The input loading capacitance of a 10K device is 2.9 pF. Thus, the fan-out for the non-transmission-line environment should be limited to a maximum of 10 loads, due to the line delay increases.

Figure 6.38 Output voltage levels as a function of dc loading for the 10K series.

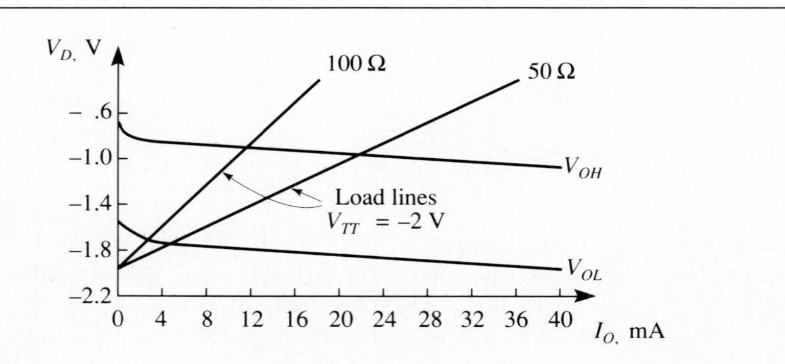

Noise Margins

Using the worst-case values, that is, the minimum guaranteed values for V_{OH} and V_{IH}, and the maximum guaranteed values for V_{IL} and V_{OL}, as specified by Motorola for the 10K series at room temperature, we determine the noise margins as

$$NM_L = V_{IL} - V_{OL} = -1.475 - (-1.630) = .155 \text{ V}$$

and

$$NM_H = V_{OH} - V_{IH} = -.98 - (-1.105) = .125 \text{ V}$$

Then, the lesser of the two, 125 mV, constitutes the guaranteed dc-noise margin. The typical noise margin is usually better than the guaranteed. It is specified as 210 mV by the manufacturer.

As found before, the transition region width is given as

$$V_{IH} - V_{IL} = -1.105 - (-1.475) = .37 \text{ V}$$

while the logic swing is obtained as

$$V_{OH} - V_{OL} = -.98 - (-1.630) = .65 \text{ V}$$

so that the midpoint of the logic swing, -1.305 V, is indeed very close to the reference voltage $V_{BB} = -1.29$ V.

The ECL family is sensitive to voltage differences between the V_{CC} buses of two given circuits. Thus, V_{CC} is made to be the ground bus to establish a stable reference level in the system. However, a more common problem for a system is with circuits that operate with different V_{EE} voltages. Any difference in the supply voltage causes a loss of noise immunity. The changes in output voltage levels and the reference voltage as a function of supply voltage for the 10K series are given as

$$\frac{\Delta V_{OH}}{\Delta V_{EE}} = .016$$

$$\frac{\Delta V_{OL}}{\Delta V_{EE}} = .25$$

and

$$\frac{\Delta V_{BB}}{\Delta V_{EE}} = .148$$

The change in the logical **1** output level is extremely small compared to the change in the low-level output value as a function of V_{EE}. The reference voltage, on the other hand, is designed to change at approximately one-half the low-level rate in order to stay at the center of the logic swing.

Example 6.3

Suppose that there is a 10% power supply difference between the driver and load gates such that the driving gate is at +5% of the nominal value (i.e., $V_{EE} = -5.45$ V), while $\Delta V_{EE} = -.26$ V for the driven gate, corresponding to −5% of the nominal value. Then the voltage values for both levels at the output of the driver are

$$V_{OH} = V_{OH(nom)} + \left(\frac{\Delta V_{OH}}{\Delta V_{EE}}\right) \Delta V_{EE}$$

$$= -.96 - .016 \times 5.2 \times .05 = -.964 \text{ V}$$

and

$$V_{OL} = V_{OL(nom)} + \left(\frac{\Delta V_{OL}}{\Delta V_{EE}}\right) \Delta V_{EE}$$

$$= -1.65 - .25 \times 5.2 \times .05 = -1.715 \text{ V}$$

On the other hand, the logic levels at the input of the load will be affected directly by the change in the reference voltage, so that

$$V_{IH} = V_{IH(nom)} + \left(\frac{\Delta V_{BB}}{\Delta V_{EE}}\right) \Delta V_{EE}$$

$$= -1.105 + .148 \times 5.2 \times .05 = -1.067 \text{ V}$$

and

$$V_{IL} = V_{IL(nom)} + \left(\frac{\Delta V_{BB}}{\Delta V_{EE}}\right) \Delta V_{EE}$$

$$= -1.475 + .148 \times 5.2 \times .05 = -1.437 \text{ V}$$

Therefore, the noise margins now become

$$NM_L = -1.437 - (-1.715) = .278 \text{ V}$$

and

$$NM_H = -.964 - (-1.067) = .103 \text{ V}$$

so that NM_H is reduced from .125 V to .103 V. NM_L would have been reduced, too, if the driven gate were also at +5% of the nominal value. Note that, since the worst-case conditions have been assumed here, the noise margins should typically be better.

The Temperature Dependence

The voltage transfer characteristics and hence the noise margins are also temperature dependent. The principal source of the dependence is the temperature variation of the forward-biased BE junction voltage drops. The bias network supplies a temperature-dependent reference voltage to ensure that the symmetry of the noise margins is maintained over a wide range of temperatures.

A temperature change ΔT causes a change in voltage across each forward-biased BE junction by the amount $\delta = -k\Delta T$, where k is a constant. This introduces to the bias network of Figure 6.31 two generators, δ and 2δ, representing the voltage increments due to Q_3, and D_1 and D_2, respectively. Therefore, the changes in transistor and diode voltage drops are considered as signals, while the power supply is regarded as a signal ground. Assuming that the emitter-follower voltage gain is unity, and neglecting the base current and incremental resistances of the diodes, we employ the principle of superposition to determine the change in V_{BB}.

The component of ΔV_{BB} due to the generator 2δ is equal to the voltage signal at the base of Q_3:

$$\Delta V_{BB1} = \frac{R_3}{R_3 + R_4} 2\delta = .31\delta$$

The component due to the generator associated with the transistor voltage drop is obtained by considering the reflection of the total resistance of the base circuit into the emitter circuit, that is, $(R_3 \| R_4)/(\beta_{F3} + 1)$. Thus, assuming $\beta_{F3} = 100$, we have

$$\Delta V_{BB2} = -\frac{R_5}{R_3R_4/(R_3 + R_4)(\beta_{F3} + 1) + R_5} \delta \approx -\delta$$

The total change in V_{BB} is then found as

$$\Delta V_{BB} = \Delta V_{BB1} + \Delta V_{BB2} = -.69\delta \tag{6.4}$$

The equivalent circuit to determine the change in the low-level output voltage is shown in Figure 6.39a. When Q_3 is on, the OR output is at **0**. Noting that we have two more generators corresponding to the voltage increments due to the conducting Q_R and Q_1, and using Equation (6.4), the change in the low-level output voltage is given as

$$\Delta V_{OL} = \frac{-R_1}{R_E} \Delta V_{BB} - \left(\frac{-R_1}{R_E}\right) \delta - \frac{R_T}{R_T + R_1/(\beta_F + 1)} \delta$$

Assuming $\beta_{F1} = 100$,

$$\Delta V_{OL} = \frac{R_1(\delta - \Delta V_{BB})}{R_E} - \delta = -.47\delta \tag{6.5}$$

In Equations (6.4) and (6.5), the changes in V_{BB} and V_{OL} depend on the ratios of resistors rather than their absolute values, as already discussed above.

The equivalent circuit to determine the change in the high-level output voltage is illustrated in Figure 6.39b. When Q_R is at cutoff, the OR output is at **1.** We then get the increment in the high-level output voltage as

$$\Delta V_{OH} = -\frac{R_T}{R_T + R_1(\beta_{F1} + 1)} \delta = -.95\delta \tag{6.6}$$

Motorola provides the typical *tracking rate* for the reference voltage in the 10K series as

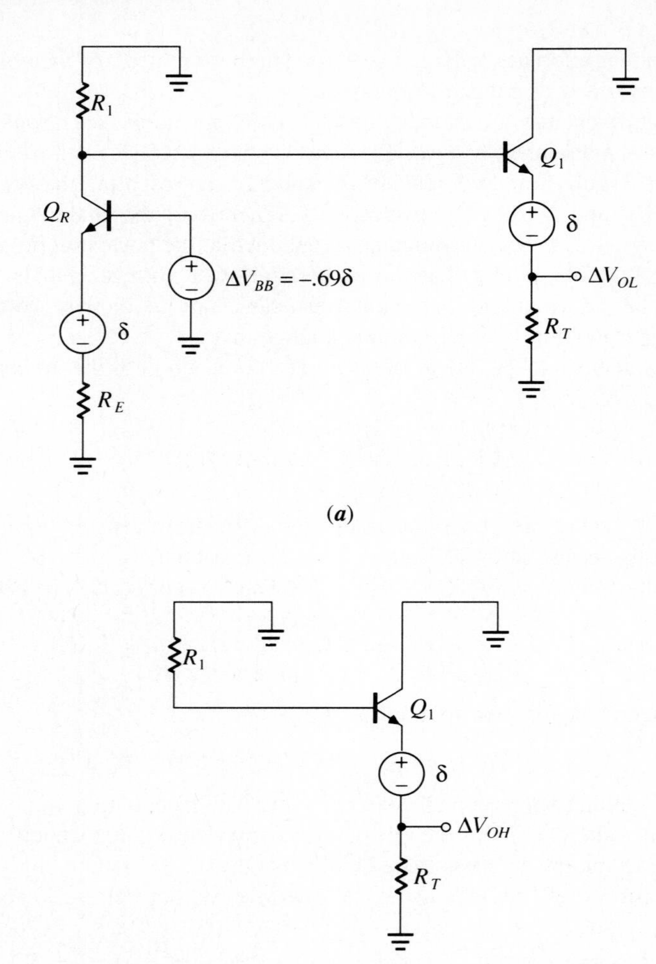

Figure 6.39 Equivalent circuits to determine the change in the (a) low-level output voltage, and (b) high-level output voltage.

$$\frac{\Delta V_{BB}}{\Delta T} = 1 \text{ mV}/^\circ\text{C}$$

Comparing this with Equation (6.4), we obtain $k = 1.45 \text{ mV}/^\circ\text{C}$, so that

$$\delta = -1.45 \, \Delta T \text{ mV}$$

Substituting this value into Equations (6.5) and (6.6), we have

$$\Delta V_{OL} = .68 \text{ mV}$$

and

$$\Delta V_{OH} = 1.38 \text{ mV}$$

Now, the corresponding typical values supplied by Motorola are a range of .5 mV to 1 mV per °C for ΔV_{OL}, and 1.3 mV per °C for ΔV_{OH}. Note that the average increment of the two logic levels, 1.03 mV, is close to the increment $\Delta V_{BB} = 1$ mV, which implies that, as the temperature changes, the midpoint of the logic swing follows the reference voltage V_{BB}. In other words, the bias network ensures that V_{BB} is centered between V_{OL} and V_{OH} to guarantee the noise immunity across the full operating temperature range from -30°C to $+85$°C.

The Power Dissipation

The total power consumption by the ECL circuits involves several components; the current switch, the reference voltage supply, the output emitter-follower transistor, and the terminating or pull-down resistor power dissipations. Since the last two depend on the method of termination, they are not included in the calculation of power dissipation by the basic gate itself.

For the bias driver, the current in Q_3 is given as $[-1.29 - (-5.2)]/(6.1 \text{ K}\Omega) = .64$ mA, while the current in the bias supply is $[0 - (-.57)]/(.907 \text{ K}\Omega) = .63$ mA. The corresponding power dissipation is obtained as $[(.64 + .63) \text{ mA}](5.2 \text{ V}) = 6.6$ mW. Since most ECL gates share a common bias driver, which is coupled through the emitter followers for isolation, the actual power dissipation per gate is less than this value. For example, a quad two-input OR/NOR gate has a typical per-gate bias driver dissipation of 1.65 mW.

The current-switch component of the total power consumption is calculated in a similar fashion. The average emitter current is found as $(4 + 4.5)/2$ mA $=$ 4.25 mA, so that the power required is obtained as $(4.25 \text{ mA})(5.2 \text{ V}) = 22.1$ mW.

Therefore, the total power dissipation per gate for a quad is found as

$$P_T = 22.1 + 1.65 = 23.75 \text{ mW}$$

The output power consumption, on the other hand, is a function of the load network, and it is computed for circuits operating at a 50% duty cycle.

Example 6.4

To calculate the typical power consumption at either output of an ECL circuit with an external 50-Ω resistor returned to -2 V, we first find the current through R_T, when the input is high, as

$$I_{OH} = \frac{-2 - (-.9)}{50} = -22 \text{ mA}$$

Thus, the power consumed by the output emitter follower is $(-22 \text{ mA})(-.9 \text{ V}) = 19.8$ mW while the 50-Ω resistor dissipates $(-22 \text{ mA})[\{-2 - (-.9)\} \text{ V}] = 24.2$ mW. Similarly, we find the low level current as

$$I_{OL} = \frac{-2 - (-1.75)}{50} = -5 \text{ mA}$$

which contributes to a power consumption of 8.75 mW at the output transistor. Moreover, the power dissipated in the external resistor is found to be (5 mA)[{−2 − (−1.75)} V] = 1.25 mW. Hence, averaging the logical **0** and **1** level powers yields

$$P_{R_T} = 12.73 \text{ mW}$$

for the pull-down resistor, and

$$P_{ef} = 14.3 \text{ mW}$$

for the output emitter followers. Thus, the total power required at the output, in addition to the power consumed by the gate, is given as

$$P_T = 12.73 + 14.3 = 27 \text{ mW}$$

The 100K Series

As in the case of advanced Schottky TTL gates, the 100K subfamily makes use of advanced fabrication technologies by utilizing oxide isolation and walled emitter structure, resulting in minimum-size transistors with very high and well-controlled switching speeds, extremely small parasitic capacitances, and an f_T in excess of 5 GHz.

The heart of the 100K subfamily is the compensation network, as shown in Figure 6.40. The vital element in this subcircuit is the *pnp* transistor Q_S, acting as a shunt regulator. Any change in the supply voltage is regulated by this transistor, so that the two output voltages V_{BB} and V_{CS} are insensitive to variations in the supply. One consequence is the matched output tracking rates with temperature for the gate.

Figure 6.40 The compensation network of the basic 100K ECL gate.

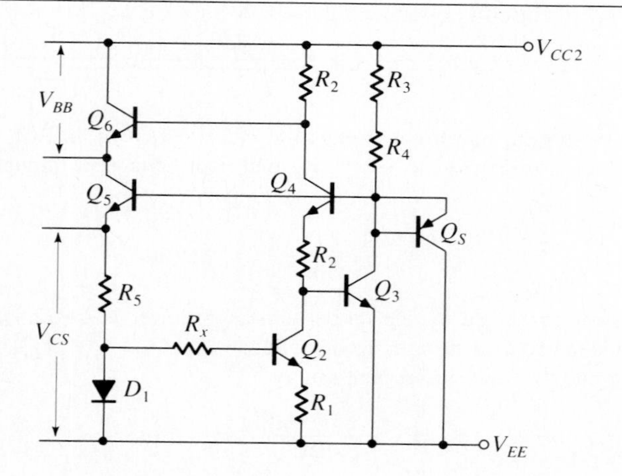

If V_{EE} is made more negative, for example, then I_{C3} tends to increase, causing an increased voltage drop across R_4, which in turn forces Q_S to conduct harder. This action diverts the extra current from the collector of Q_3 to the base of Q_S. Thus, the currents I_{R5}, I_{C2}, and I_{C3} are not affected by changes in V_{EE}. Now, since

$$V_{BB} = V_{BE6} + V_{R2} \tag{6.7}$$

and

$$V_{CS} = V_{BE3} + V_{R_2} + V_{BE4} - V_{BE5} = V_{BE3} + V_{R2} \tag{6.8}$$

and since no change in the aforementioned currents implies no change in the voltage drops of Equations (6.7) and (6.8), we can conclude that this network is indeed effective in compensating for the variations in the power supply.

D_1 is operated at a higher current density than is Q_2, resulting in $V_{R_1} = V_{D_1} - V_{BE2}$, and a positive temperature tracking coefficient for V_{R_1}. This voltage difference is amplified by the ratio R_2/R_1 to produce V_{R_2}. Note that R_2, being used twice, is instrumental in generating both V_{BB} and V_{CS}. The positive temperature coefficient of V_{R_2} cancels the negative temperature coefficient of V_{BE6} and V_{BE3} to make V_{BB} and V_{CS} insensitive to changes in temperature, respectively. R_x is used to compensate for process variations in β_F and V_{BE}.

The current switch of the basic 100K circuit is shown in Figure 6.41. Note that the supply voltage V_{EE} is changed to −4.5 V to reduce the power dissipation of the circuit. The gate-switch current is determined by the reference voltage V_{CS}, the emitter resistor R_S, and the BE voltage of the transistor current source Q_3. The ref-

Figure 6.41 The current switch of the 100K ECL OR/NOR gate with output emitter followers shown.

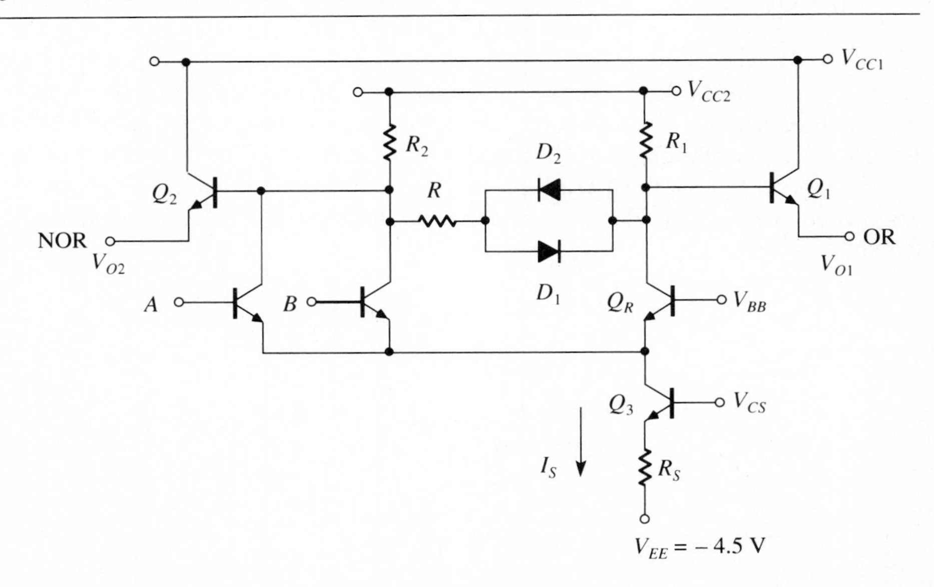

erence voltage is designed to remain fixed, as discussed above, with respect to the supply voltage, so that I_S is independent of V_{EE}. Regulating the source current this way simplifies the system design, since the output voltage levels V_{OH} and V_{OL} are primarily determined by the voltage drops across R_1 and R_2, respectively, which in turn result from the collector currents of Q_R and input transistors. Since the collector current of the conducting transistor is essentially equal to I_S, the voltage across the collector load resistor is not affected by variations in the supply. A 1 V change in V_{EE} causes the output level V_{OL} to vary by only 30 mV. Also, an I_S independent of V_{EE} makes the incremental changes in power consumption linear with changes in the supply, while an unregulated ECL circuit power dissipation changes quadratically with V_{EE}.

Since the compensation network holds V_{BB} fixed with respect to V_{CC2}, variations in the input threshold values with V_{EE} are minimized, too. For a change of 1 V in the supply voltage, V_{BB} changes by only 25 mV. Regulating V_{BB} also helps to make the input thresholds V_{IL} and V_{IH} insensitive to temperature.

Insensitivity of the output logic levels to temperature is achieved by a simple subcircuit, which is connected between the bases of the output emitter followers and consists of two back-to-back diodes, D_1 and D_2, in series with a resistor $R = R_1 = R_2$. With at least one of the input transistors conducting and Q_R off, most of the source current I_S flows through R_2; a small amount flows through R and R_1, since D_2 also conducts. As the junction temperature increases and the forward-biased BE junction voltage of the output emitter follower V_{BE2} decreases about 1.5 mV/°C, the base voltage must become more negative to offset the temperature dependence of V_{BE2}. This is accomplished by increasing the source current, causing a corresponding increase in the voltage drop of sufficient magnitude across R_2. Now, V_{CS} is independent of temperature, so the emitter current increases due to the temperature dependence of V_{BE3}. As the temperature increases, V_{BE3} decreases approximately 1.5 mV/°C, leading to an increase in the voltage across R_3, which causes I_S to increase.

In the meantime, the voltage across R_2 increases at the rate of about 1.5 mV/°C, too, while the voltage drop across D_2 decreases at the same rate. This results in a net voltage increase of 3 mV/°C across the series combination of R and R_1 that is equally divided between the two resistors. Thus, the base voltage of Q_1 decreases by 1.5 mV/°C, compensating for the temperature dependence of the BE junction of Q_1. When Q_R is on and the input transistors are off, D_1 is forward-biased, and a similar compensation occurs for V_{BE2}. The change rates for the 100K series are specified by Fairchild as

$$\frac{\Delta V_{BB}}{\Delta T} = .08 \text{ mV}/°\text{C}$$

$$\frac{\Delta V_{OH}}{\Delta T} = .06 \text{ mV}/°\text{C}$$

$$\frac{\Delta V_{OL}}{\Delta T} = .1 \text{ mV}/°\text{C}$$

The voltage transfer characteristics of the basic 100K circuit are given in Figure 6.42. The worst-case logic swing is seen to be

$$V_{OH} - V_{OL} = -1.025 - (-1.620) = .595 \text{ V}$$

The midpoint is at -1.32 V, which is identical to the value of the reference voltage V_{BB}, as provided by the manufacturer. We can also determine the worst-case noise margins as

$$NM_L = V_{IL} - V_{OL} = -1.475 - (-1.620) = .145 \text{ V}$$

and

$$NM_H = V_{OH} - V_{IH} = -1.025 - (-1.165) = .140 \text{ V}$$

The transition region width is found as

$$V_{IH} - V_{IL} = -1.165 - (-1.475) = .310 \text{ V}$$

Figure 6.43 compares the transfer characteristics of the uncompensated ECL circuit and the fully compensated 100K series. The latter exhibits relatively constant output levels and input thresholds over the 0°C to +85°C specified temperature range and -4.2 V to -4.8 V specified voltage range. The typical propagation delay of the basic circuit driving a 50-Ω transmission line is .75 ns. With a power dissipation of 40 mW, this leads to an SP of 30 pJ.

Figure 6.42 The VTC of the 100K ECL OR/NOR circuit.

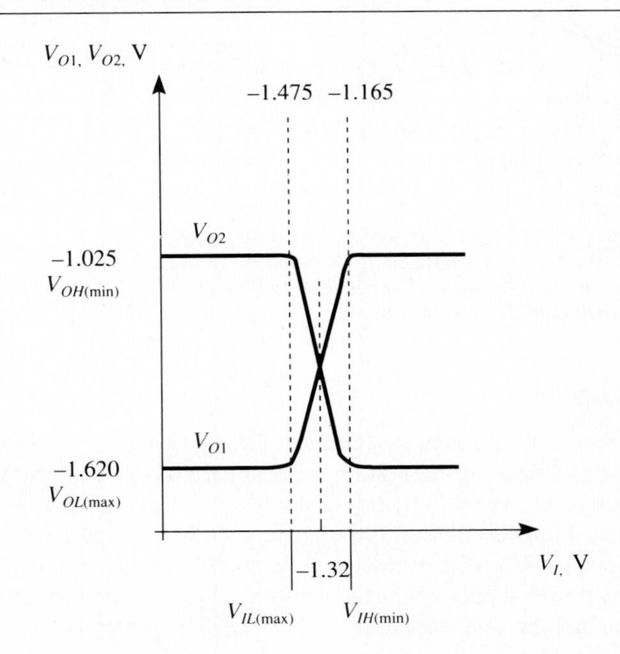

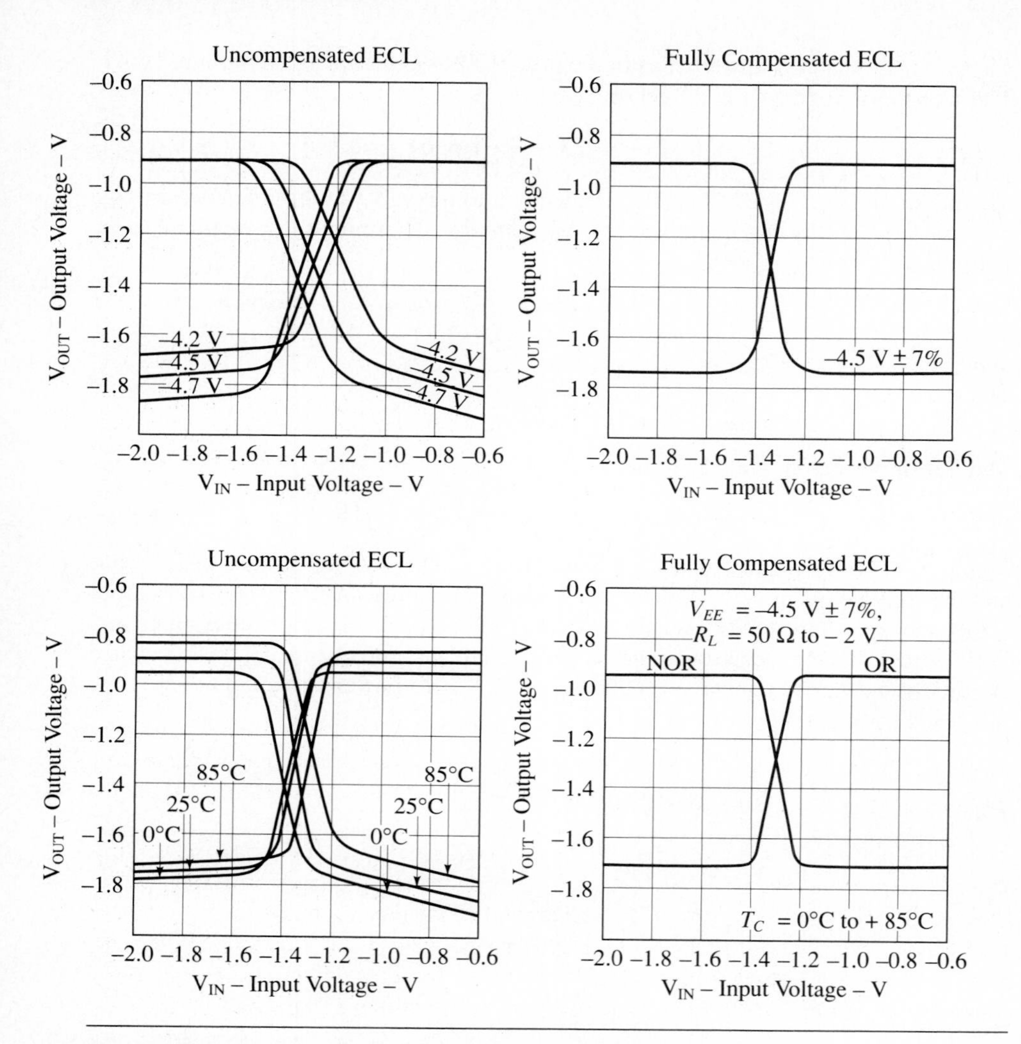

Figure 6.43 Comparison of the VTCs of the uncompensated ECL circuit and the fully compensated 100K series at various supply voltages and temperatures. (*Reprinted with permission of National Semiconductor Corporation.*)

The 10KH Series

10KH devices offer typical propagation delays of 1 ns at 25 mW per gate, resulting in an SP of 25 pJ, the best among available ECL subfamilies even though still inferior to those of advanced Schottky subfamilies of TTL. They also lack the excellent temperature compensation of 100K devices. When compared to the 10K series, the 10KH features 100% improvement in clock speeds and propagation delay while maintaining the power supply voltage and current values of the former. Noise margins are also better than the 10K subfamily over the ±5% power supply range. The operating temperature range, however, has been narrowed from the −30°C to +85°C range of the 10K to a range of 0°C to +75°C.

The 10KH transistor is fabricated using Motorola's so-called MOSAIC I process, which shows a 100% improvement in f_T, a reduction of more than 50% in parasitic capacitances, and a decrease in device area of more than 86%, as compared to the 10K. The transistors of both series are compared in Table 6.7.

In circuit level, the major difference between the two subfamilies is the replacement of the bias network by the compensation network, exactly as in the 100K series. In addition, the source current is established with a transistor current source in series with $R_E = 777\ \Omega$, and the collector resistors are matched as $R_1 = R_2 = 276\ \Omega$. Some comparative performance characteristics and ac specifications of both the 10K and 10KH appear in Table 6.8.

The improved performance in output voltage variations over the 10K is the result of the voltage regulator. In both logic levels, the voltage compensation has reduced the variations significantly. As far as the temperature compensation is concerned, the only improvement over 10K is in the tracking rate of V_{OL}. The 10KH lacks the cross-connected subcircuit of the 100K between the bases of the complementary outputs. The superiority of the 100K series in this respect is obvious.

6.3 The Integrated Injection Logic

The *integrated injection logic* (I^2L), also known as the *merged transistor logic* (*MTL*), is the latest form of bipolar transistor logic, developed in 1972 independently by two groups of engineers: one at the IBM Laboratories in Böblingen, West Germany, and the other at the Philips Research Laboratories in Eindhoven, The Netherlands.

The families described so far are predominantly used in SSI, MSI, and more recently in LSI. Because of their relatively large geometry and power consumption, their use in VLSI has been prohibitive. The I^2L, on the other hand, is mainly suitable for LSI and VLSI because it can blend the high speed of the BJT with the high density and low power of the MOSFET. Since it is based on what is called *direct-coupled transistor logic* (*DCTL*), we start our discussion with a brief description of DCTL.

Table 6.7 Comparison of 10K and 10KH Transistor Characteristics

	10K	10KH
Device area, μm^2	51 × 85 = 4323	16 × 37 = 592
Emitter, μm^2	4 × 20 = 80	3 × 8 = 24
C_{BC}, pF	.46	.16
C_{BE}, pF	.18	.07
C_{CS}, pF*	.83	.18
f_T, GHz	1.6	3.5

*Collector-substrate parasitic capacitance

Table 6.8 Performance Characteristics and ac Specifications of 10K and 10KH Series

	10K			10KH		
	Min	Typ	Max	Min	Typ	Max
Edge speed, ns		2		.7	1.5	2
t_{PD}, ns		2	2.9	.7	1	1.5
Power, mW		25			25	
SP, pJ		50			25	
Temp. range, °C	−30		85	0		75
$\Delta V_{BB}/\Delta T$, mV/°C	.8	1	1.2	.8	1	1.2
$\Delta V_{OL}/\Delta T$, mV/°C	.35	.5–1	1.55	0	.4	.6
$\Delta V_{OH}/\Delta T$, mV/°C	1.2	1.3	1.5	1.2	1.3	1.5
NM_H, mV	125	205		150	230	
NM_L, mV	155	275		150	270	
$\Delta V_{BB}/\Delta V_{EE}$, mV/V	110	150	190	0	10	25
$\Delta V_{OL}/\Delta V_{EE}$, mV/V	−30		0	−20		0
$\Delta V_{OH}/\Delta V_{EE}$, mV/V	200	250	320	0	20	50

Figure 6.44 shows the basic DCTL structure, which implements the NOR operation. Consider first that all inputs are in the **0** state. The input transistors are off, the output rises to V_{CC}, and the loads go into saturation. Therefore, the output is clamped at $V_O = V_{OH} = V_{BE(sat)} = .8$ V. That is, with all the inputs in the low state, the output goes high. Consider now that at least one input, say V_1, is in logical **1**. Then Q_1 goes into saturation, and the output becomes $V_O = V_{OL} = V_{CE(sat)} = .2$ V, corresponding to the logical **0** state. Hence, it indeed performs the NOR operation.

Figure 6.44 The basic DCTL structure implementing the three-input NOR operation.

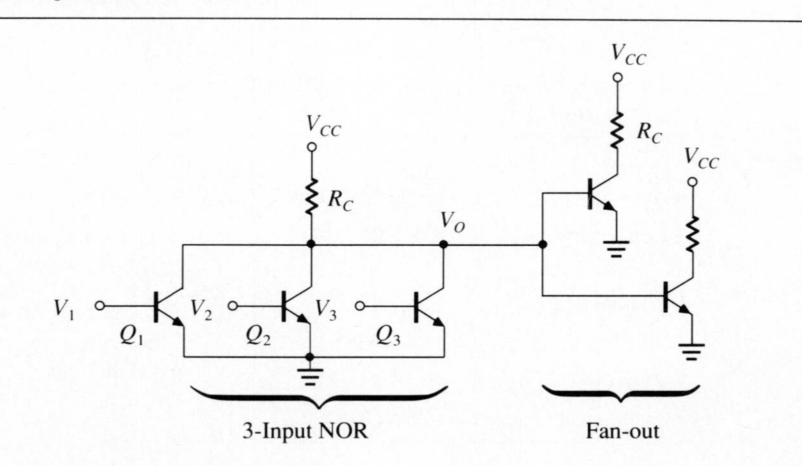

The disadvantages of DCTL can be summarized as follows:

1. Direct connection of the output collector to the input base, with $V_{CC} >> V_{CE(sat)}$ and $V_{CC} >> V_{BE(sat)}$, drives the load transistor into heavy saturation and results in a large overdrive base charge, corresponding to a slow switching speed.
2. Because of the small logic swing of about .6 V, current spikes can lead to improper operation.
3. At sufficiently high temperatures, the temperature dependence of the reverse collector saturation current can cause the output voltage to be too low to drive the loads into saturation.

Now, consider the development of I^2L from DCTL, as depicted in Figure 6.45. If the load resistance R_C of DCTL is associated with the input circuit, as illustrated

Figure 6.45 The development of the basic I^2L from DCTL: (a) transitional structure; (b) equivalent circuit; (c) circuit diagram; and (d) cross section.

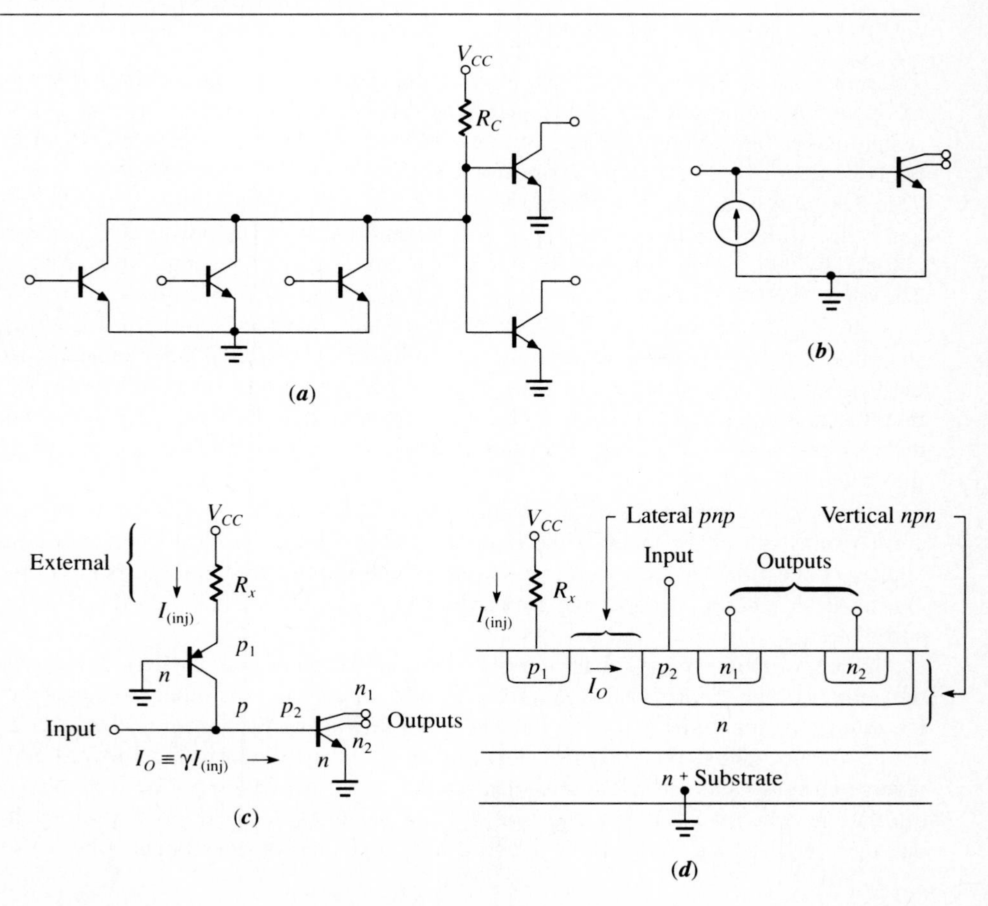

in Figure 6.45a, rather than the output, then the input circuit becomes a current source and the output a *multicollector npn* transistor, as shown in Figure 6.45b.

The diffused resistor wastes power and takes a lot of silicon area that cannot be afforded in LSI and VLSI fabrications. Note that in the original DCTL circuit, the effective load on the input transistors is equal to R_C in parallel with the input impedance of the load. This implies that an infinite R_C would be the ideal case because a large R_C improves the gain of the circuit. Thus, we would like to eliminate the collector resistor, so that the collectors of the input transistors are directly connected to the base of the fan-out transistors. However, the loads still need a biasing base current. Consequently, the reason we associate R_C with the input circuit is to emphasize that it is required only to provide the load base current. As a result, we omit R_C and replace it with a grounded-base *pnp* transistor acting as a current source, or *current injector*. This is diagrammed in Figure 6.45c. Note that the gate structure comprises a *lateral pnp* and a *vertical npn* transistor. The *pnp* is forward-biased using a power supply $+V_{CC}$, which, together with a biasing resistance R_x, determines the amount of the injected current as

$$I_{(inj)} = \frac{V_{CC} - V_{BE}}{R_x}$$

The range of operation extends from around 1 nA to 1 mA, so that the desired speed of operation can be selected by the user. The supply voltage can be as low as +1 V which makes the structure operate at low power levels. On the other hand, V_{CC} can be as high as 15 V, increasing the power consumption but also the operating speed. This flexibility is an advantage because I^2L can be operated at different current levels and hence with different speed and power-dissipation levels. The power consumption is found by taking the product of the injector current $I_{(inj)}$ and the supply voltage V_{CC}. The SP of the I^2L is essentially constant and in the range of .5 to 2 pJ.

The basic gate topology of Figure 6.45c reveals that the *pnp* and *npn* devices share a common ground n^+ region, and the collector of the *pnp* is interconnected to the base of the *npn*, another *p* region. Hence, it is possible to save silicon area by merging both transistors. The *pnp* collector and the *npn* base become one region, and the *pnp* base along with the *npn* emitter another. The cross section of the device is illustrated in Figure 6.45d.

The *npn* inverter transistor at the output of the I^2L circuit is similar to the multi-emitter transistor at the input of the standard TTL circuit but with the collector and emitter connections of the inverter reversed, which implies that the transistor is operating upside down. Hence, the forward current gain is smaller than the reverse current gain.

When a bipolar transistor is in the forward active mode, the carriers flow from the emitter to the collector, that is, *down* into the silicon. On the other hand, for the I^2L *npn* transistor in the active region, the main flow of carriers is *up* to the surface of the silicon. Thus, the subscript notation of F and R to signify the forward and reverse current gains, respectively, of a bipolar transistor have specific technology translations when we consider I^2L, for which F becomes *u*, and R becomes *d* such that $\beta_F = \beta_u$, and $\beta_R = \beta_d$. Here, *u* and *d* imply technology constraints, whereas F and R are the electrical circuit choices associated with modes of operation of the transistor.

To determine the voltage transfer characteristic of an I^2L inverter, it is best to consider it as a member of a chain of inverters, as shown in Figure 6.46. The outputs of the drivers are directly connected to the inputs of the loads. When V_{I1} is at **0**, Q_1 will be off, and the injected current $I_{(inj)}$ of the second gate will flow into the base of Q_2 and will saturate it. The collector of Q_2 will then sink the current $I_{(inj)}$ of the third gate, and Q_3 will be off.

When V_{I1} is high, the injected current will be diverted into the base of Q_1, and Q_1 will saturate. Then its collector will conduct the current $I_{(inj)}$ of the second gate, so that Q_2 will be off. Finally, the current $I_{(inj)}$ of the third gate will saturate Q_3. Thus, when any driver is off, $V_O = V_{BE(sat)}$; when it is on, $V_O = V_{CE(sat)}$. Hence, we have

$$V_{OH} = V_{IH} = V_{BE(sat)} = .8 \text{ V}$$

and

$$V_{OL} = V_{CE(sat)} = .1 \text{ V}$$

while

$$V_{IL} = V_\gamma \approx .6 \text{ V}$$

The voltage transfer characteristic is depicted in Figure 6.47. Note the rather small logic swing and noise margins. However, since a complete digital system can be designed on a single VLSI chip, even these margins are sufficient for proper logic operation. To interface to off-chip circuitry, translators and buffers are used at the input and output of the I^2L circuit.

The injector current $I_{(inj)}$ supplies the *npn* base with an input current $I_o = \gamma I_{(inj)}$, where γ is the emitter injection efficiency. Let us assume an ideal value of $\gamma = 1$. The input base current needed to turn on the *npn* transistor is nI_o/β_u, where n is the fan-out. The input current supplied from the previous stages when their outputs are logical **1** is the negligible leakage current. Thus, the *current noise margin* is the difference between the current supplied by the injector and the base current needed to turn the *npn* on, or

Figure 6.46 Cascade of three I^2L inverters.

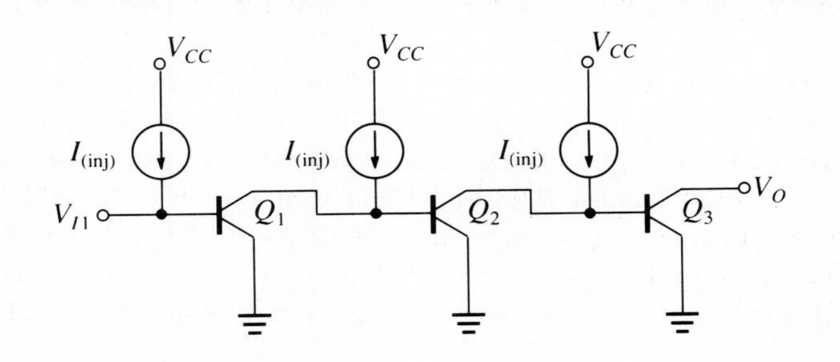

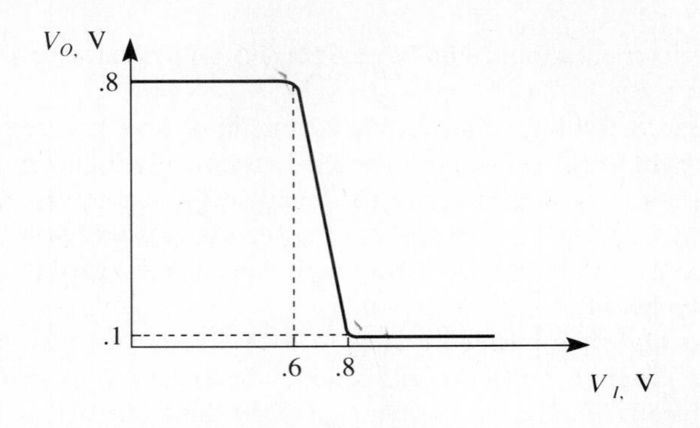

Figure 6.47 The VTC of the I²L inverter.

$$INM_H = I_o - \frac{nI_o}{\beta_u} = \left(1 - \frac{n}{\beta_u}\right) I_o \tag{6.9}$$

Similarly, the input sinking current supplied to the gate from the previous stages when their outputs are **0** is given as mI_o, where m is the number of saturated drivers. Thus, the net current supplied to the gate is $I_o - mI_o$ and the corresponding current noise margin is now the difference between the base current required to turn the *npn* on and the current supplied to the base, or

$$INM_L = \frac{nI_o}{\beta_u} - (I_o - mI_o) = \left(\frac{n}{\beta_u} - 1 + m\right) I_o \tag{6.10}$$

The worst-case occurs when $m = 1$ and Equation (6.10) reduces to $INM_L = nI_o/\beta_u$. Also note that increasing n improves INM_L while it degrades INM_H.

The collectors at the output can be tied together to form the so-called *wired-AND* operation, but the connection should be to only one input. This is because fanning out from the same collector to different bases will affect other inputs when that collector is at **0**.

Summary

The three bipolar logic families are TTL, ECL, and I²L. In standard TTL, the multi-emitter transistor input stage implements the AND operation, the phase-splitter generates a pair of complementary signals, and the totem-pole output stage provides the complementation. The basic TTL gate is NAND. To improve the switching speed of TTL, transistors are prevented from saturating by connecting Schottky-barrier diodes between the base and collector. The 54/74ALS subfamily has the smallest SP.

Very high speed of operation is accomplished in ECL by avoiding transistor saturation and by employing small logic swings. The basic ECL gate configuration is that of the current switch, with output stage emitter followers providing both OR and NOR operations. Because of the high switching speeds, transmission-line techniques are usually utilized.

Increased component density in bipolar fabrication is achieved by reducing and even eliminating the isolation islands that separate devices and also by eliminating area-consuming resistors. This is done in I^2L by merging devices to realize VLSI density without sacrificing the high speeds of bipolar transistors. The I^2L circuits can be operated over a wide range of speeds by simply changing the injector current through an external resistor. However, advances in CMOS technology in recent years have led to the decline in the use of I^2L.

References

1. A. S. Sedra. *Microelectronic Circuits*. 2d ed. Holt, Rinehart & Winston, New York: 1987.
2. Technical Engineering Staff. "Advanced Schottky Family (ALS/AS) Application." *Application Report*, Texas Instruments Inc., Dallas, TX: 1986.
3. Fairchild Memory & High Speed Logic Unit. *F100K ECL Data Book*. Fairchild Semiconductor Corporation, Puyallup, WA: 1986.
4. Fairchild Memory & High Speed Logic Unit. *F100K ECL Users Handbook*. Fairchild Camera and Instrument Corporation, Puyallup, WA: 1985.
5. Fairchild Digital Unit. *FAST Data Book*. Fairchild Camera and Instrument Corporation, So. Portland, ME: 1985.
6. M. I. Elmasry. *Digital Bipolar Integrated Circuits*. John Wiley & Sons, New York: 1983.
7. Motorola Computer Applications Engineering Dept. *MECL System Design Handbook*. 3d ed. Motorola Semiconductor Products, Inc., 1980.

PROBLEMS

For the following problems assume a room temperature of 300 K.

6.1 In the elementary TTL gate of Figure 6.3, let $V_{CC} = 5$ V, $R_C = 1$ KΩ, $\beta_F = 50$, and $V_{CE(sat)} = .2$ V.

a. Find the time necessary to charge a capacitive load of 5 pF from $V_{OL} = V_{CE(sat)}$ to $V_{OH} = 2.4$ V.

b. Find R and R_C such that $I_{C(sat)} = 20$ mA.

c. Verify (a) with SPICE.

6.2 From the standard TTL VTC of Figure 6.10 we see that the slope of the segment BC is $(3.9 - 2.8)/(.6 - 1.3) \approx -1.6$ Ω. Verify this by finding the expression for the output voltage in terms of the input voltage V_I during this interval.

In Problems 6.3–6.5, use the following transistor models:

Transistor:	Q_1	Q_2	Q_3	Q_4	Q_5	Q_6
Model:	C	A	B	B	A	A

	β_R	r_b, Ω	r_c, Ω
A	1	70	40
B	.2	20	20
C	.02	500	40

Also, for all models use: $\beta_F = 49$, $V_A = 50$ V, $I_S = 10^{-14}$ A. For the SBDs, $r_s =$ 15 Ω, $C_{to} = .2$ pF, $I_S = 5 \times 10^{-14}$ A. For the *pn*-junction diode, $r_s = 40$ Ω.

6.3 For the 74H00 high-speed two-input NAND gate of Figure 6.11

a. Sketch the VTC showing breakpoint values and determine the noise margins.
b. Use SPICE to find the VTC with a fan-out of 1.
c. Calculate the supply currents I_{CCH} and I_{CCL} and compute the power dissipation.
d. Calculate the maximum fanout.

6.4 Repeat Problem 6.3 for the 74L00 low-power two-input NAND gate.

6.5 Repeat Problem 6.3 for the 74S00 Schottky two-input NAND gate of Figure 6.14.

6.6 Use SPICE to compare the output characteristic of the active pull-down subcircuit in Figure 6.14 with that of a 500-Ω resistor that would be connected if the former were not used. Represent Q_3 with its BE junction diode. Use the parameters of Problem 6.5.

6.7 Within a panel of 10KH ECL logic, one card is operating near the inlet airflow duct at 25°C and another interconnected card, remote from the air inlet, is at 35°C. Devices on the former card have a typical V_{OH} of −.9 V and a V_{OL} of −1.7 V. Find the corresponding values for a device on the interconnected card.

6.8 Calculate the total power required at the output of a 10K ECL circuit with an external 510-Ω pull-down resistor connected to the supply voltage.

6.9 Explain the operation of the ECL circuit shown in the following diagram.

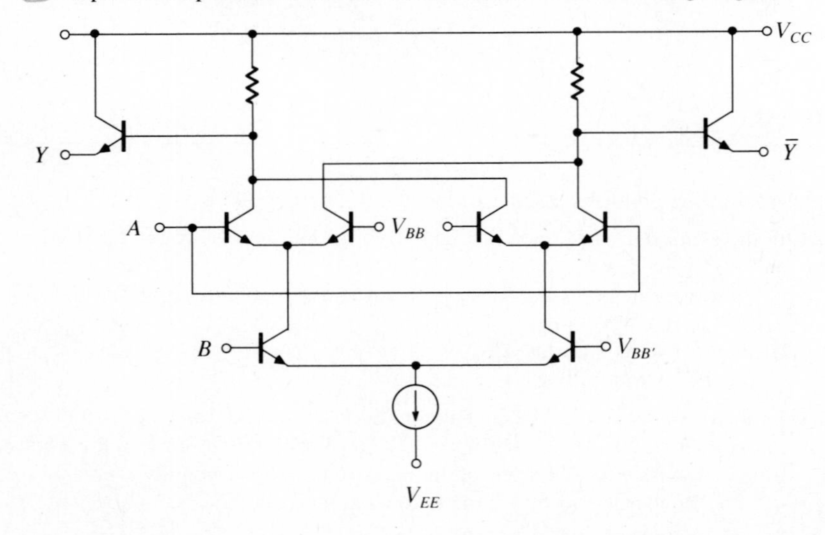

6.10 For the ECL inverter shown in the diagram, use SPICE to do the following:
a. Find the VTC.
b. Compute the average propagation delay, t_{PD}.

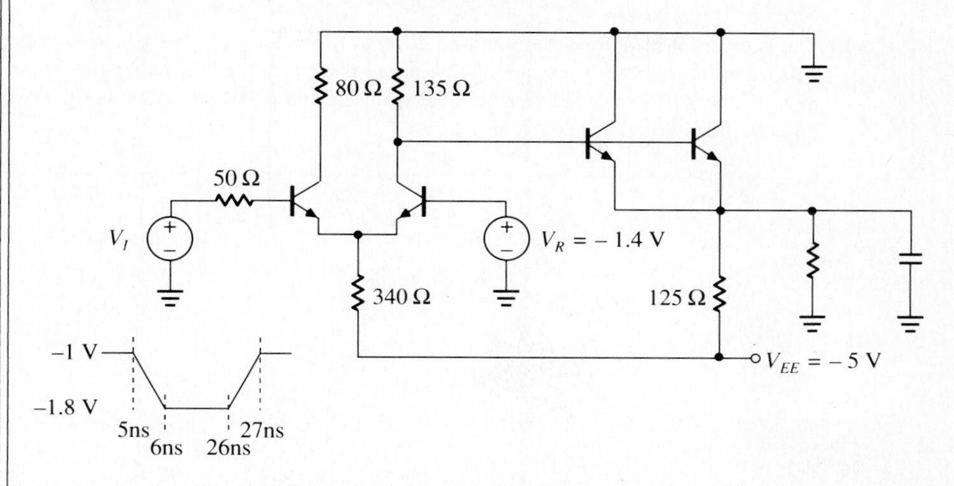

The transistor parameters are given as follows: $I_S = 10^{-16}$ A, $\beta_R = .1$, $\beta_F = 50$, $r_b = 50\ \Omega$, $r_c = 10\ \Omega$, $C_{tEo} = .4$ pF, $C_{tCo} = .5$ pF, $\phi_E = .8$ V $= \phi_C$, $C_{tSo} = 1$ pF, $\tau_F = 100$ ps, $\tau_R = 5$ ns, and $V_A = 50$ V.

6.11 Use SPICE to find the switching times for the ECL gate of Figure 6.31 when it drives a 20-pF load capacitance at its NOR output in addition to a 50-Ω pull-down to −2 V. Use the transistor parameters and the input pulse of Problem 6.9.

6.12 What is the maximum fan-out for an I²L inverter with $\beta_{(sat)} = 5$ for the output transistor?

6.13 The I²L wired-AND logic shown in the diagram results in faulty logic operation. Explain why. Show the circuit for correct logic operation.

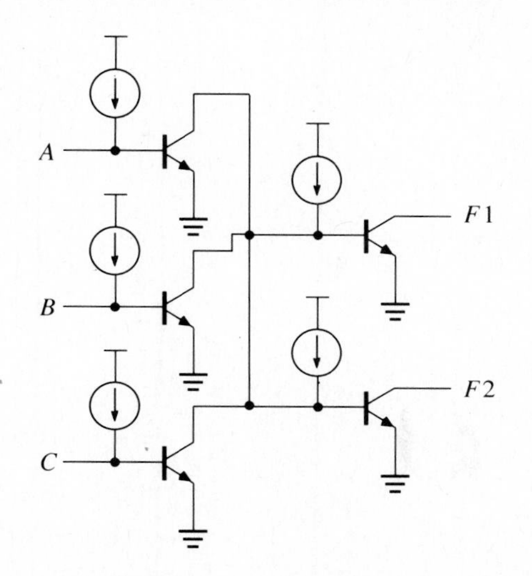

6.14 A system consisting of I^2L ICs, each incorporating 1000 injectors and using gates with an SP of 1 pJ, is designed in which a cascade of three stages must settle during one cycle of a 10 MHz clock. What is the total injector current that should be provided to each chip if $V_{CC} = 1$ V?

6.15 The standard TTL allows only around 20 gates/mm^2, while the I^2L allows about 200 gates/mm^2. Suppose that an IC requires 6000 gates with a maximum propagation delay of 10 ns/gate. Compare the chip size and power dissipation of the two logic families.

7. BUFFERS, DRIVERS, AND INTERFACING

Situations often arise where many components in a digital system must share a common path to be able to transfer data to one another. To reduce the number of interconnections, a small set of shared lines called a *bus* may be used. In general, outputs from different devices cannot be simultaneously present on the bus. Consequently, for proper operation, that is, to prevent *bus contention*, only one of the devices connected to a shared bus can place information on the bus at any time. Basically, there are two broad categories of devices that permit the simultaneous connection of the outputs of two or more devices to form a bus: (1) TTL open-collector devices, and (2) TTL and CMOS three-state devices.

In digital system design, it is frequently found that some line may have a maximum output current that is insufficient to drive all the output lines that must be connected to it. To remedy this problem, a current amplifier called a *driver* or a *buffer* may be used. It is a circuit that meets a particular electrical requirement without necessarily performing a specific logic function. Generally, a buffer furnishes drive current beyond the capacity of a gate.

Optimum performance of digital systems is usually achieved by employing more than one logic subfamily. Thus, it may often be necessary to interface a logic subfamily to other types of logic. When interfacing different logic series, one must consider both the voltage levels and the current requirements of each subfamily. This chapter examines different types of buffers/drivers as well as interfacing between TTL, CMOS, and ECL families.

7.1 Open-Collector Outputs

In many applications such as bus-organized digital systems where various outputs must be ANDed, using TTL gates with totem-pole outputs would require an AND gate with as many input lines as there are signals to be ANDed.

To overcome this problem, TTL gates are available with *open-collector* outputs. Additional logic is created when the outputs of two or more open-collector gates are tied together, as in Figure 7.1. This scheme is called the *wired-AND* and is used to save logic gates in comparison with other methods. The standard way to indicate the presence of a wired-AND gate in a logic diagram is to superimpose an AND symbol over the connector in question. Note that V_O will be logical **1** only when both outputs X and Y are high. Therefore, $V_O = X \cdot Y$; hence the name wired-AND.

It is not possible to interconnect TTL gates with totem-pole output stages in this configuration. Figure 7.2 depicts the high-level current path when the outputs of totem-pole gates are tied together. If gate A output is high, high-level driver Q_{4A} will be on and Q_{3A} will be turned off. If the output of gate B is low, then Q_{4B} will be off and Q_{3B} will be turned on. Under these conditions, gate A dissipates a large amount of power, and Q_{3B} is required to sink a current which may exceed its guaranteed 16 mA sink capacity.

Although the main application of the open-collector gate is to allow the formation of the wired-AND, it is also useful for driving an individual load like an LED or relay. Figure 7.3 illustrates the circuit of the two-input open-collector NAND gate 7401. Note that the emitter-follower transistor at the output is altogether deleted. The output stage consists solely of the common-emitter transistor Q_3 without even a collector resistance. This gate, when supplied with a proper load resistor R_L, may be paralleled with other similar TTL gates. At the same time, it will drive from one to

Figure 7.1 (a) Logic circuit containing the 7401 open-collector NAND gates, and (b) equivalent circuit with the 7400 totem-pole NAND gates.

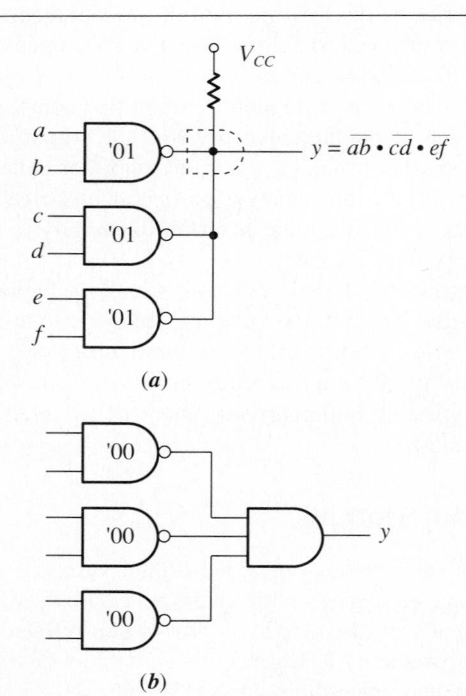

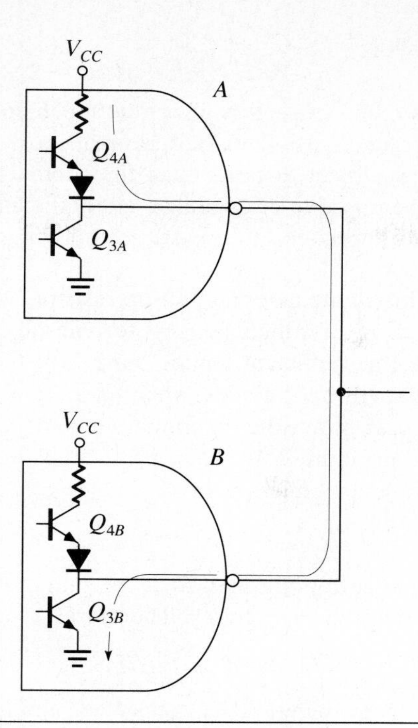

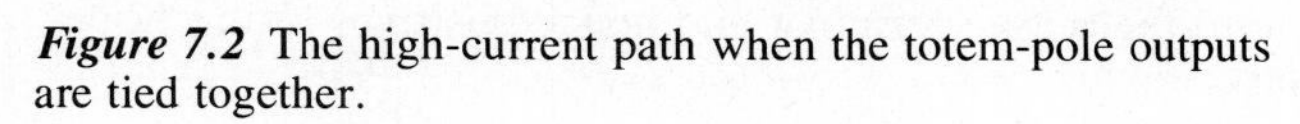

Figure 7.2 The high-current path when the totem-pole outputs are tied together.

Figure 7.3 The 7401 two-input, open-collector TTL NAND gate.

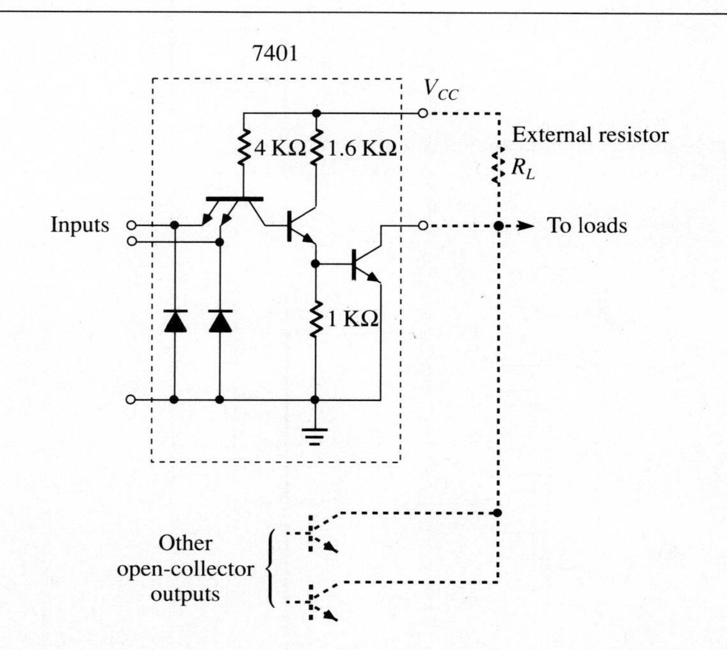

nine standard loads of its own series. When no other open-collector gates are tied, it may be used to drive ten loads. Their main disadvantage is that these gates are inherently slower and more subject to noise than their totem-pole counterparts. The pull-up resistor bias can be raised to any voltage within the breakdown voltage of the driver transistor to enable interfacing to a system employing voltage swings, like a CMOS.

To determine the value of the external pull-up resistor, we should find the upper and lower limits of the range of values the resistor can take. A maximum value is found which will ensure that sufficient source current to the loads and off current through paralleled outputs will be available when the output is logical **1**. Therefore, the total leakage current determines the maximum value of R_L when all driving transistors are off, as shown in Figure 7.4. When V_O is high so that the drivers are off, the voltage drop across R_L must be less than

$$V_{RL(max)} = V_{CC} - V_{OH(min)} \tag{7.1}$$

On the other hand, the total current through R_L is the sum of the load currents I_{IH} and the leakage currents I_{OH} through each driver. Therefore,

$$I_{RL} = \eta I_{OH} + NI_{IH} \tag{7.2}$$

where η is the number of gates wired-AND connected, and N is the number of standard loads. Note that I_{OH} is into the output terminal and hence positive. Using Equations (7.1) and (7.2), we find

$$R_{L(max)} = \frac{V_{RL(max)}}{I_{RL}}$$

$$= \frac{V_{CC} - V_{OH(min)}}{\eta I_{OH} + NI_{IH}} \tag{7.3}$$

Figure 7.4 High-level circuit conditions to determine $R_{L(max)}$.

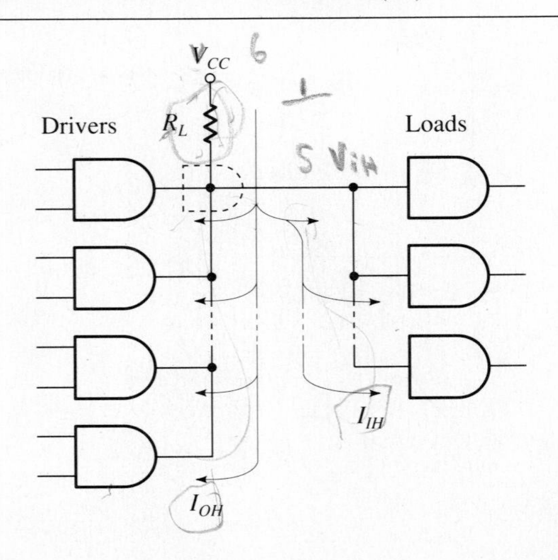

A minimum value for the pull-up resistor is established when V_O is logical **0**, so that the current through R_L and the total sinking current from the load gates do not cause the output voltage to rise above $V_{OL(max)}$ even if only one driving gate is sinking all the currents. Therefore, the current must be limited to the recommended maximum I_{OL}, which will ensure that the low-level output voltage will be below $V_{OL(max)}$. Since part of I_{OL} will be supplied from the loads, the amount of current that can be allowed through R_L will be reduced. Hence, neglecting the leakage currents of the turned-off drivers, we have

$$R_{L(min)} = \frac{V_{RL(min)}}{I_{RL}}$$

$$= \frac{V_{CC} - V_{OL(max)}}{I_{OL(max)} - N|I_{IL}|} \tag{7.4}$$

Logical **0** circuit conditions to calculate $R_{L(min)}$ are illustrated in Figure 7.5. Table 7.1 provides the electrical characteristics of the 7401 open-collector NAND gate.

Example 7.1

To determine the value of the pull-up resistor for optimum performance when $\eta = 4$ and $N = 3$, we use Equations (7.3) and (7.4) and Table 7.1 to get

$$R_{L(max)} = 2321\ \Omega$$

and

$$R_{L(min)} = 410\ \Omega$$

Therefore, any commercially available resistor between these two values would be an appropriate choice.

Figure 7.5 Low-level circuit conditions to determine $R_{L(min)}$.

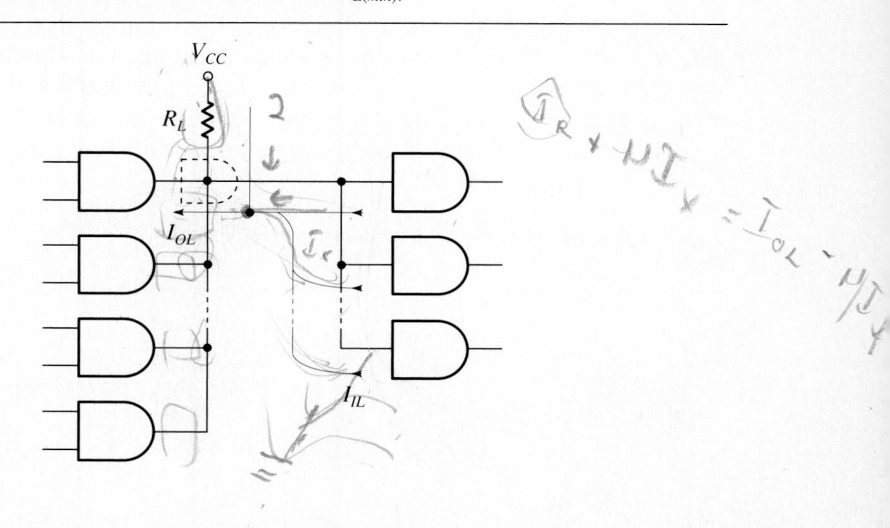

Table 7.1 Electrical Characteristics of the Two-Input NAND Gate 7401

$V_{OH(min)}$, V	2.4	$I_{OH(max)}$, μA	250	$t_{PLH(typ)}$, ns 35
$V_{IH(min)}$, V	2	$I_{IH(max)}$, μA	40	$t_{PHL(typ)}$, ns 8
$V_{OL(max)}$, V	.4	$I_{OL(max)}$, mA	16	$C_L = 15$ pF
$V_{IL(max)}$, V	.8	$I_{IL(max)}$, mA	−1.6	$R_L = 4$ KΩ for t_{PLH}
				= 400 Ω for t_{PHL}

7.2 Three-State Outputs

Another useful variant of TTL that can solve the problem of driving a common bus line by two or more logic circuits is the *three-state* output arrangement shown in Figure 7.6. The term *tri-state* is also used. However, it is a registered trademark of National Semiconductor Corporation, which introduced this design concept in 1970.

By combining the high-speed advantage of the totem-pole output with the advantages of an open-collector output, the three-state gates enable the connection of a number of gates to a common output line or bus. In addition to a totem-pole output, these gates have another terminal, called the *output enable*, which permits the device to function normally or the output signal to be disconnected from the rest of the circuit by going into a third state in which both output transistors are turned off, resulting in an extremely high output impedance. Therefore, a disabled gate can be assumed to have an open circuit in its output line, so that the high-impedance state may be equated to the voltage level of a conductor that has no sources connected to it, that is, it is floating.

The two most frequently used three-state ICs in the standard TTL logic subfamily are the 74125 and 74126 quadruple-bus buffers with independent output controls. Both are noninverting buffers, but the former's output is enabled by a logical **0** while the 74126 is enabled by an active high signal.

Operation of the 74125

Figure 7.7 shows the circuit of the 74125 buffer gate, which is basically an AND gate with one of the inputs connected to the output of an on-chip inverter. Note that, with Q_1 as the multi-emitter transistor, a high-speed 74H00 NAND gate would result if the collector of Q_1 were directly connected to the base of Q_2.

The totem-pole output of the inverter goes not only to the input transistor Q_1 but

Figure 7.6 Three-state buffer with control input G (output enable).

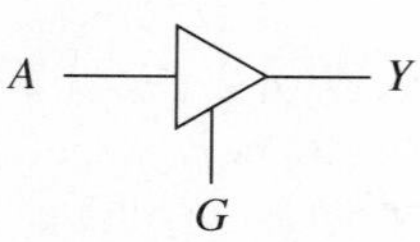

$$Y = \begin{cases} A & \text{when } G = \mathbf{1} \\ Z \text{ (Hi imp.)} & \text{when } G = \mathbf{0} \end{cases}$$

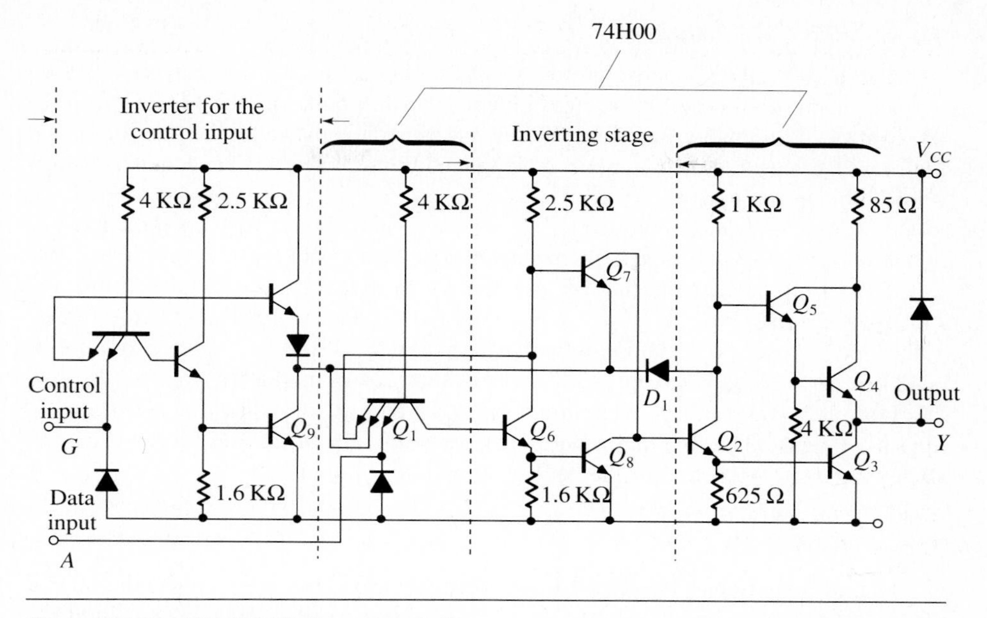

Figure 7.7 The 74125 TTL three-state buffer.

is also tied to the emitter of Q_7 and the cathode of the diode D_1. When this output is low, due to a high level on the control input G, the base of Q_1 goes low and Q_6 is cut off. Normally, under these conditions, the collector of Q_6 heads toward V_{CC}. However, the low voltage at the emitter of Q_7 causes Q_7 to turn on and hold the collector of Q_6 low. Therefore, instead of supplying base current for the phase-splitter Q_2, the current through the 2.5-KΩ collector resistor is diverted to and sunk by the low-level driver Q_9 of the inverter. With Q_2 off, Q_3 cannot turn on, either. To prevent the high-level Darlington pair from turning on, D_1 turns on and holds the collector of Q_2 at **0**. Thus, all the transistors in the AND gate are cut off. With the source and sink drivers both off, the gate cannot load a line activated by another driver.

When the control input G is low, the logical **1** inverter output reverse-biases both the BE junction of Q_7 and D_1, preventing them from interfering with the operation of the AND gate and hence allowing the gate to transmit the data input. In this case, when the data input is high, the base of Q_1 follows, and Q_6 is turned on as the base current of the input transistor is diverted to the base of Q_6. The collector voltage of the latter goes low and turns off Q_2 so that the current through the 1-KΩ resistor turns the Q_5-Q_4 Darlington pair on, driving the output high. On the other hand, if the data input is low, Q_6 goes off, its collector voltage rises to **1** to turn Q_2 on through the BC junction of Q_7, and the high-level driver is cut off while Q_3 is turned on, thereby pulling the output low.

Note that in the inverting stage, Q_8 is driven in phase with Q_6. By reinforcing the current through the 2.5-KΩ resistor, Q_8 boosts the voltage swing of the Q_6 collector, leading to an improvement in the switching times of Q_2.

Three-State CMOS Buffers

The CMOS three-state output buffer has logic elements in the gate connections to each of the transistors in the final inverter, so that both may be turned off under the control of an enable function. Figure 7.8 illustrates the logic diagram of such a buffer with active low-enable input. Note that additional inverters are added as buffers or to optimize timing.

A complete list of commercial-grade three-state buffers/drivers in advanced Schottky bipolar as well as in high-speed and advanced CMOS series is given in Table 7.2. For all bipolar parts in the list, the data sheets specify $I_{OH(max)}$ as -15 mA and $I_{IH(max)}$ as 20 μA. For ALS, $I_{IL(max)}$ is given as $-.1$ mA while it is -1 mA for the other bipolar parts. The maximum sinking capability of all AS/F parts is 64 mA. It is 48 mA for ALS parts except '1240, '1241, and '1244 for which $I_{OL(max)} = 24$ mA.

For all CMOS parts, the maximum input currents at both logical levels are specified as ± 1 μA. The maximum output currents are ± 6 mA for HCMOS (HC and HCT) parts and ± 24 mA for the ACL (AC and ACT) devices.

Transceivers

Digital signals can be placed on or received from the bus using bidirectional three-state buffers called transmitter-receivers, or *transceivers*. They have high drive-current outputs to enable high-speed operation when driving large bus capacitances. The principle of operation of a transceiver is depicted in Figure 7.9a. When the C line is active-high, Buffer 1 is enabled while Buffer 2 presents a high impedance to both the bus and the device so that the device can send data to the bus. When the C line is low, data is sent from the bus to the device via Buffer 2.

The TTL and the CMOS ICs '242 and '243 along with the '245 are the most commonly used transceivers. The former two are quad inverting and noninverting transceivers, respectively, designed for four-line, asynchronous, two-way data communications, while the '245 contains eight noninverting, bidirectional buffers with

Figure 7.8 The CMOS inverting three-state output buffer with active low enable: (a) the circuit, and (b) the truth table.

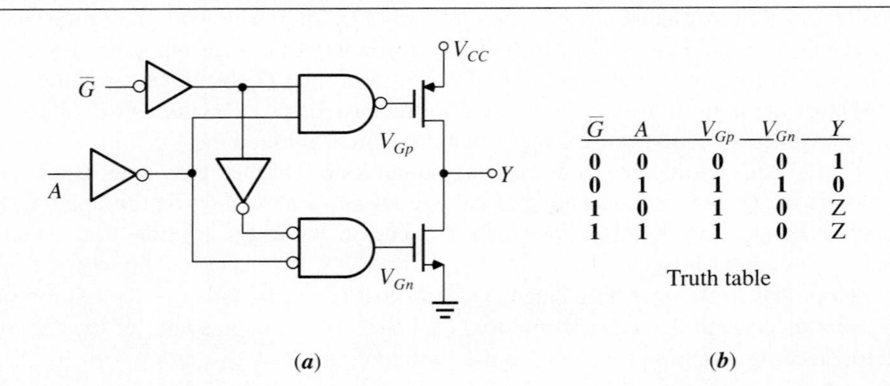

$\overline{G}$	A	V_{Gp}	V_{Gn}	Y
0	0	0	0	1
0	1	1	1	0
1	0	1	0	Z
1	1	1	0	Z

Table 7.2 Commercial-Grade Three-State Buffers/Drivers

Quad buffers/drivers			
Part	*Subfamilies*	*Output data*	*Control inputs*
'125	F/HCMOS	True	Individual active low enables
'126	F/HCMOS	True	Individual active high enables
Hex buffers/drivers			
Part	*Subfamilies*	*Output data*	*Control inputs*
'365	F/ALS/HCMOS	True	Two active low enables are NORed to control all gates.
'366	F/ALS/HCMOS	Inv.	Same as '365.
'367	F/ALS/HCMOS	True	One low enable controls four gates, while the other controls the remaining two.
'368	F/ALS/HCMOS	Inv.	Same as '367.
Octal buffers/drivers			
Part	*Subfamilies*	*Output data*	*Control inputs*
'230	AS/ALS	4 True/4 Inv.	Two active low enables independently control four gates.
'231	AS/ALS	Inv.	Each of the complementary enables controls four gates.
'240	AS/F/ALS/HCMOS/ACL	Inv.	Same as '230.
'241	AS/F/ALS/HCMOS/ACL	True	Same as '231.
'244	AS/F/ALS/HCMOS/ACL	True	Same as '230.
'465	ALS/ACL	True	Two low enables are NORed to control all eight gates.
'466	ALS/ACL	Inv.	Same as '465.
'467	ALS/ACL	True	Same as '231.
'468	ALS/ACL	Inv.	Same as '230.
'540	F/ALS/HCMOS/ACL	Inv.	Same as '465.
'541	F/ALS/HCMOS/ACL	True	Same as '465.
'1240	ALS	Inv.	Same as '230.
'1241	ALS	True	Same as '231.

(*continued*)

Table 7.2 Commercial-Grade Three-State Buffers/Drivers *Continued*

Octal buffers/drivers			
Part	*Subfamilies*	*Output data*	*Control inputs*
'1244	ALS	True	Same as '230.
'11240	ACL	Inv.	Same as '230.
'11241	ACL	True	Same as '231.
'11244	ACL	True	Same as '230.

three-state outputs. Table 7.3 summarizes the drive capabilities and speeds of these transceivers.

Figure 7.9b shows the logic diagram of the 'HC242, which has one active-high and one active-low enable. When C_1 is high, both transistors Q_1 and Q_2 are off, resulting in a high-impedance state of the output. Then, $C_2 = \mathbf{1}$ enables the B input to be seen by the gates of Q_3 and Q_4. If $B = \mathbf{1}$, Q_4 turns on, so A becomes low, while $B = \mathbf{0}$ turns Q_3 on, leading to the A output being pulled to V_{CC}, which corresponds to logical **1**. Therefore, A is the complement of B. When both control inputs C_1 and C_2 assume **0**, Q_3 and Q_4 turn off, and the direction of data flow reverses.

7.3 Peripheral Drivers

The peripheral drivers generally have open-collector output transistors that can switch hundreds of milliamperes at high voltages and are driven by standard logic gates. High-current and high-voltage peripheral drivers find many applications associated with digital systems, such as relays, printer solenoids, displays, lamps, and buses.

75451 Peripheral Driver

Figure 7.10a shows the circuit of the 75451, a typical peripheral driver.[1] The circuit is equivalent to a 7400 NAND gate driving an output transistor whose output current is rated as 300 mA at $V_{OL} = .7$ V, although, when on, it is capable of sinking more than 1 A. It can operate with voltages up to 30 V without breaking down. Figure 7.10b shows the I_C vs. V_{CE} characteristics of the transistor. BV_{CES} corresponds to the breakdown voltage when the output transistor is held off by the low-level driver Q_3, as would happen when $V_{CC} = 5$ V. BV_{CER}, on the other hand, is the breakdown voltage when the transistor is held off by the 500-Ω resistor, as would happen when the power is lost. LV_{CEO} is the value of the breakdown voltage if it could be measured with the base open. Note that all the breakdown voltages converge on LV_{CEO} at high currents. The destructive secondary breakdown voltage, shown as a dotted line in Figure 7.10b, occurs when the power dissipation of the device is exceeded. This value depends on the duration of the time the condition exists, the device temperature, and the voltage and current values.

[1] This subsection is based on Reference 4.

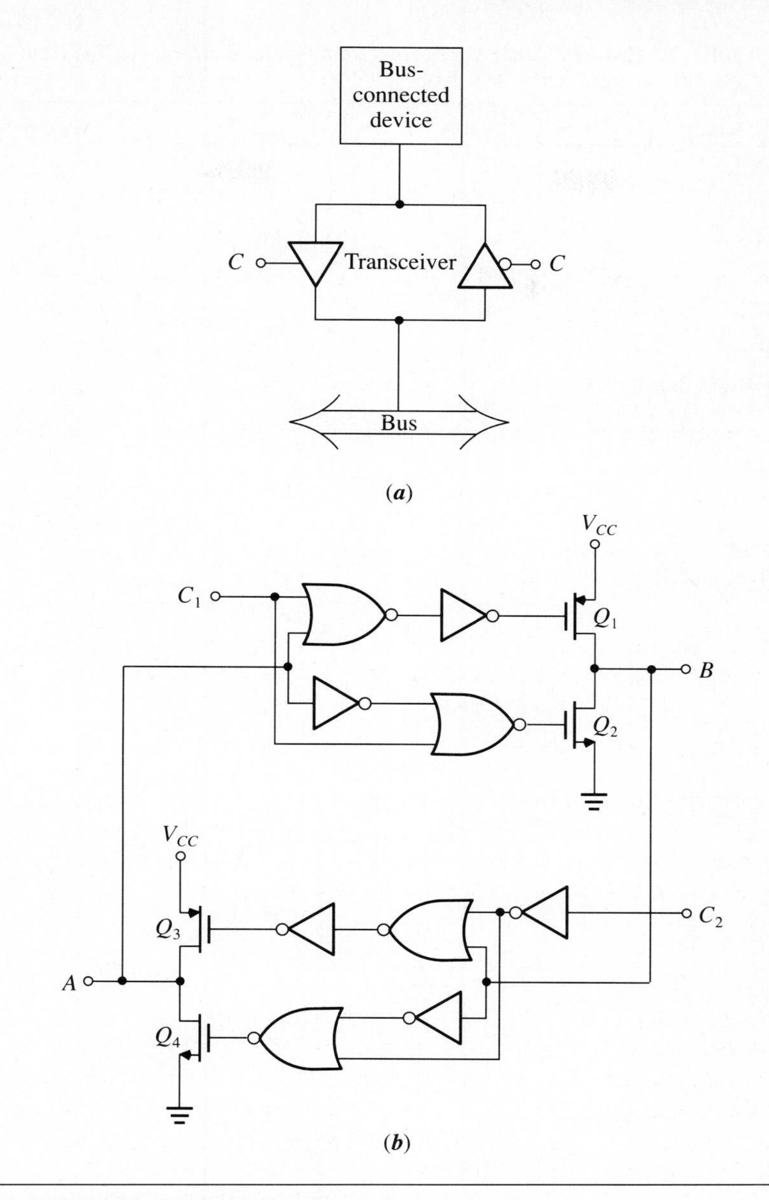

Figure 7.9 (a) Principle of operation of a transceiver, and (b) the logic diagram of 74HC242.

Consider the switching transfer characteristics superimposed on the dc characteristics of the output transistor with an inductive load, as illustrated in Figure 7.11. Note that the load voltage V_L exceeds LV_{CEO}. When the transistor turns on, the initial current through the inductor load is zero, and the transfer curve switches across to the left toward V_{OL} and slowly increases the inductor current. When the output tran-

Table 7.3 Electrical Characteristics of Commercial-Grade '242, '243, and '245 Transceivers at V_{CC} = 4.5 V

Subfamily	$I_{OH(max)}$, mA	$I_{OL(max)}$, mA	t_{PD}, ns*
AS	−15	64	6
F	−3	64†	4.5
ALS	−15	48	7
HC	−6	6	14
HCT	−6	6	16
AC	−24	24	6

*typical values for '245 at T_A = 25°C, C_L = 50 pF.
†only 20 mA at the A ports of 74F245.

Figure 7.10 (a) The 75451 peripheral driver, and (b) I_C vs. V_{CE} characteristics of the output stage.

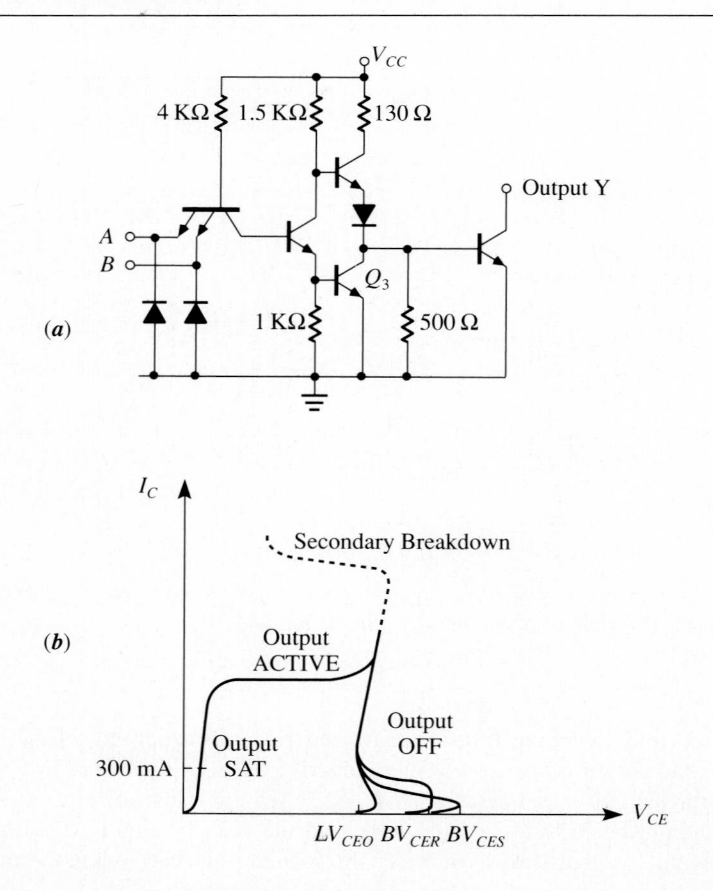

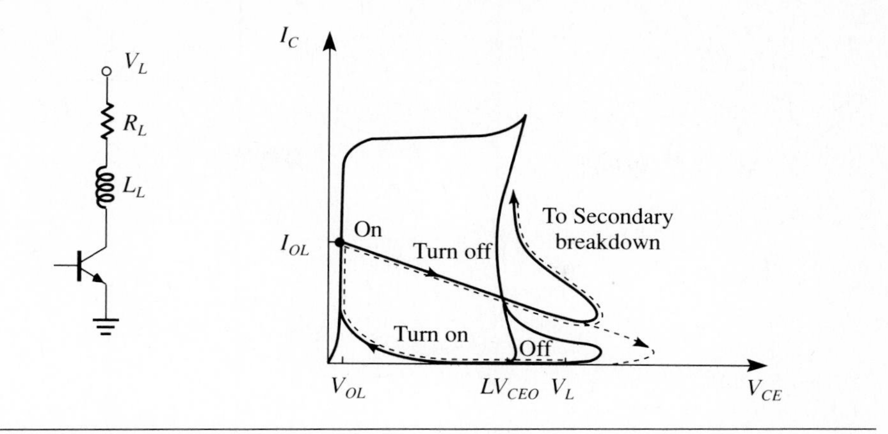

Figure 7.11 The inductive-load transfer characteristics of the output transistor with $LV_{CEO} > V_L$.

sistor turns off, the initial current is I_{OL}, which is sustained by the inductor. Then the curve switches across toward V_L through a high-current and high-voltage area that exceeds LV_{CEO}. Thus, instead of turning off completely by following the dotted line in Figure 7.11, the device goes into the secondary breakdown region. Consequently, it is not a good practice to let the output transistor's high-level voltage V_L exceed LV_{CEO}.

If the driver's load is located at a distance from the driver, there will be additional parasitic reactance, inductive and/or capacitive, which may cause ringing on the driver output, which in turn can surpass LV_{CEO} or the transient current that is larger than the sustaining current of the driver. Therefore, in order to reduce the noise associated with the transients and hence to prevent damage to the peripheral driver, the driver's output should be damped. Figure 7.12 shows an acceptable application with an inductive load. Not only is V_L less than LV_{CEO}, but the inductive voltage spike caused by the initial current is clamped by a diode returned to V_L.

Another method to quench the inductive voltage spike is illustrated in Figure 7.13. To critically damp the value of L_L, R_D and C_D are chosen such that

$$\left(\frac{L_L}{R_L + R_D}\right)\left(\frac{1}{\sqrt{L_L C_D}}\right) = \frac{1}{2} \tag{7.5}$$

In addition to the peak current and the breakdown voltage, another limitation that the design engineer should consider is the power dissipation, which in turn is limited by the IC package and the external thermal reactances, such as that of the *printed circuit board* (*PCB*) and the heat sink, as well as the maximum allowable junction temperature of the device. If the consumption is not properly calculated, the peripheral drivers can easily be used at power levels that inadvertently exceed the package rating. Nevertheless, a peripheral driver is able to dissipate peak power levels during switching that go well beyond the average dc power because the die and the package can consume the transient energy while still maintaining the junction temperature at a safe

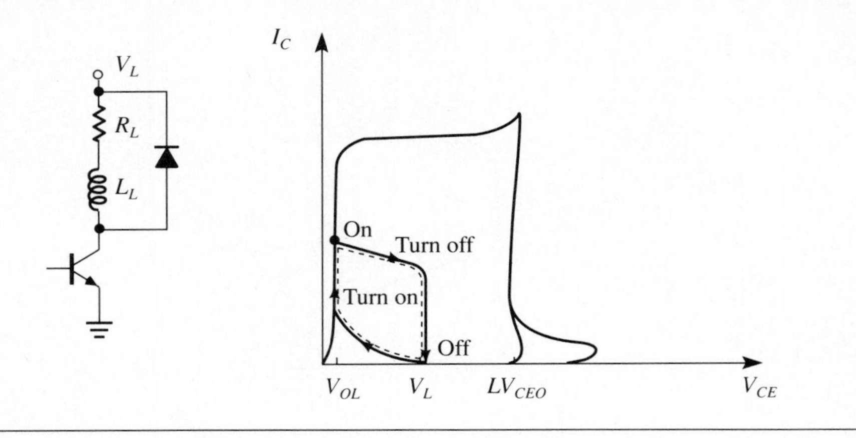

Figure 7.12 The inductive-load transfer characteristics of the output transistor with $LV_{CEO} < V_L$ and with a clamping diode.

level. In the laboratory under the microscope, it is possible to observe the device glow orange around the perimeter of the junction under excessive peak power without any damage to the device.

Consider, as an example, the 3686 NAND driver primarily used in telephone relay applications. The device package rating (i.e., the maximum power dissipation) is given as 787 mW at 25°C for the TO-5-type package. The driver IC has an internal clamp network to quench the inductive swing at 65 V. Yet, if the package rating is exceeded for a certain period of time, the driver succumbs to the thermal overload and becomes nonfunctional. However, since it is intended for telephone relays, in most applications it is not expected to switch more than about two dozen times a day.

The schematic diagram of 3686 is shown in Figure 7.14. The circuit incorporates a reference circuit made up of a Zener diode, which usually eliminates the need of an external clamping diode. When the output is turned off by input logic conditions, the resulting inductive voltage transient at the output is detected by the diode, which

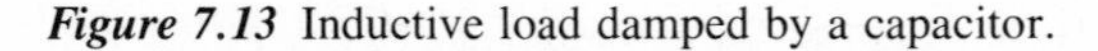

Figure 7.13 Inductive load damped by a capacitor.

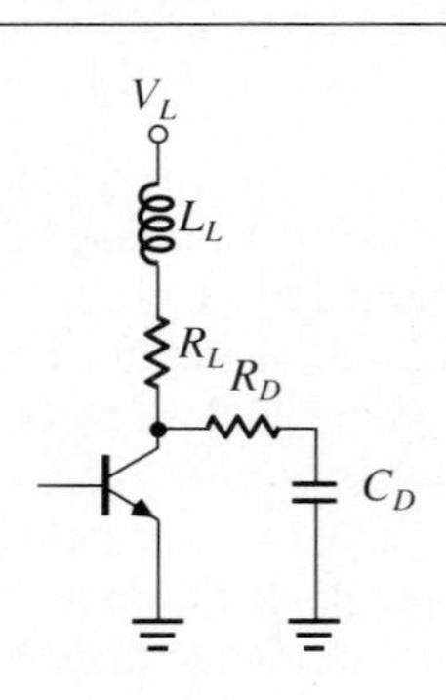

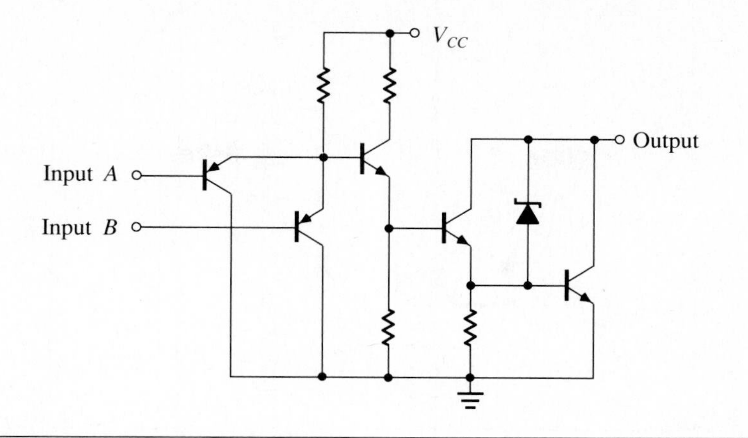

Figure 7.14 The 3686 NAND relay driver.

momentarily activates the output transistor long enough to discharge the relay energy. Note that the output stage is a Darlington pair to allow high-current operation even at low internal V_{CC} current levels. Therefore, the output requires very little drive from the logic circuit driving it and dissipates less power when the output is turned on and off compared to a single saturating transistor. If the power supply is lost, the output transistor is protected by being forced into the high-impedance state with the same breakdown levels as when V_{CC} was applied. The *pnp* transistors at the input provide both TTL compatibility and high impedance for low input loading.

Example 7.2

To calculate the output power dissipation of the driver of Figure 7.15 with an inductive load, we will use the following device and load characteristics:

Output on voltage $V_{OL} = 1.2$ V

Output clamp voltage $V_C = 65$ V

Load voltage $V_L = 30$ V

Load resistance $R_L = 120\ \Omega$

Load inductance $L_L = 5$ H

Switching times $t_{on} = t_{off} = 100$ ms

Case 1. The output transistor is turned on. The exponential characteristic equation for the instantaneous current through the inductive load is given by

$$i_L(t) = I_f + (I_i - I_f)e^{-t/\tau} \tag{7.6}$$

where I_f is the final value, I_i is the initial value, and τ is the inductive time constant given as

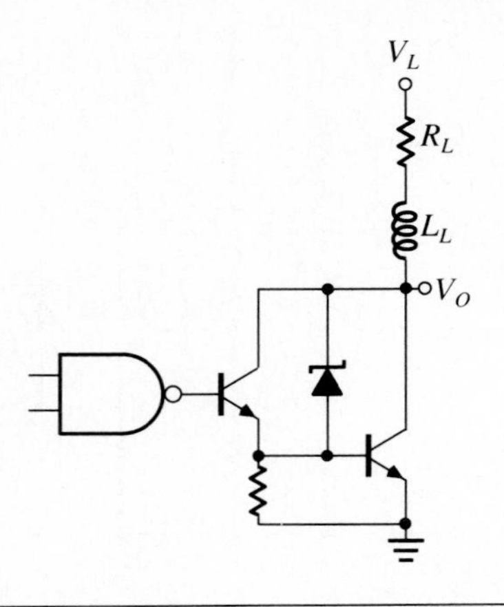

Figure 7.15 The 3686 peripheral driver with an inductive load.

$$\tau \equiv \frac{L_L}{R_L} = 41.67 \text{ ms}$$

The initial value of the inductive current is zero, and the final value I_f, to which it would head if allowed, is obtained as

$$I_f = I_L = \frac{V_L - V_{OL}}{R_L} = \frac{30 - 1.2}{120} = 240 \text{ mA}$$

so that Equation (7.6) becomes

$$i_L(t) = I_L(1 - e^{-t/\tau}) = 240(1 - e^{-t/41.67}) \text{ mA} \tag{7.7}$$

At $t = t_{on}$, the inductive current reaches its peak value, I_p. Substituting the values into Equation (7.7) and solving for I_p, we get

$$I_p = I_L(1 - e^{-t_{on}/\tau}) = 218.22 \text{ mA}$$

The average power dissipation in device output when it is on is expressed as

$$P_{on(av)} = V_{OL}I_{on(av)} = \left(\frac{V_{OL}}{T}\right)\int_0^{t_{on}} i_L(t)\,dt$$

$$= \left(\frac{V_{OL}I_L t_{on}}{T}\right)\left[1 - \int_0^{t_{on}} \left(\frac{e^{-t/\tau}}{t_{on}}\right) dt\right]$$

$$= \left(\frac{V_{OL}I_L t_{on}}{T}\right)\left[1 - \left(\frac{\tau}{t_{on}}\right)(1 - e^{-t_{on}/\tau})\right] \tag{7.8}$$

where $T \equiv t_{on} + t_{off}$ is the total period. Substituting the numerical values into Equation (7.8), we have

$$P_{on(av)} = 89.44 \text{ mW}$$

Case 2. The output transistor is turned off. When the device turns off, the output is clamped momentarily at 65 V before it becomes $V_L = 30$ V. During this time interval, the inductive current decays exponentially to zero with the same time constant τ. Now $I_i = I_p$, and the final value, if the current were left alone without the transistor limiting it, would be

$$I_f = I_R = \frac{V_L - V_C}{R_L} = \frac{30 - 65}{120} = -291.67 \text{ mA}$$

Therefore, using Equation (7.6), the instantaneous current during this period is expressed as

$$i_L(t) = I_R + (I_p - I_R)e^{-t/\tau} \tag{7.9}$$

The current decays to zero t_x seconds later. From Equation (7.9), t_x is found as

$$t_x = \tau \ln\left(\frac{I_p - I_R}{-I_R}\right) = 23.28 \text{ ms}$$

The average power dissipation during t_x, then, is given as

$$P_{off(av)} = V_C I_{off(av)} = \left(\frac{V_C}{T}\right)\int_0^{t_x} i_L(t)\,dt$$

$$= \frac{V_C t_x}{T}\left[(I_p - I_R)\int_0^{t_x} \frac{e^{-t/\tau}}{t_x}\,dt + I_R\right]$$

$$= \frac{V_C t_x}{T}\left[(I_p - I_R)\frac{\tau}{t_x}(1 - e^{-t_x/\tau}) + I_R\right] \tag{7.10}$$

Substituting the numerical values into Equation (7.10), we get

$$P_{off(av)} = 748.32 \text{ mW}$$

Therefore, the total average consumption in the device output is found to be

$$P_O = P_{on(av)} + P_{off(av)} = 89.44 + 748.32 = 837.76 \text{ mW}$$

The output voltage and current waveforms are shown in Figure 7.16. If the load were purely resistive, then the output power dissipation would simply be

$$P_O = \frac{V_{OL}(V_L - V_{OL})}{R_L}\frac{t_{on}}{T} = 144 \text{ mW}$$

Note that the total power dissipation in the IC must also include the consumption in the other output and the dissipation due to the power supply currents.

74C908 and 74C918 General Purpose Drivers

74C908 and 74C918 are dual, general-purpose, high-voltage drivers, each capable of sourcing a minimum of 250 mA at $V_O = V_{CC} - 3$ V and junction temperature $T_j = 65°$C.[2] As depicted in Figure 7.17, both ICs consist of two CMOS NAND gates driving an emitter-follower Darlington pair for high-current drive and high-voltage

[2] This subsection is based on Reference 5.

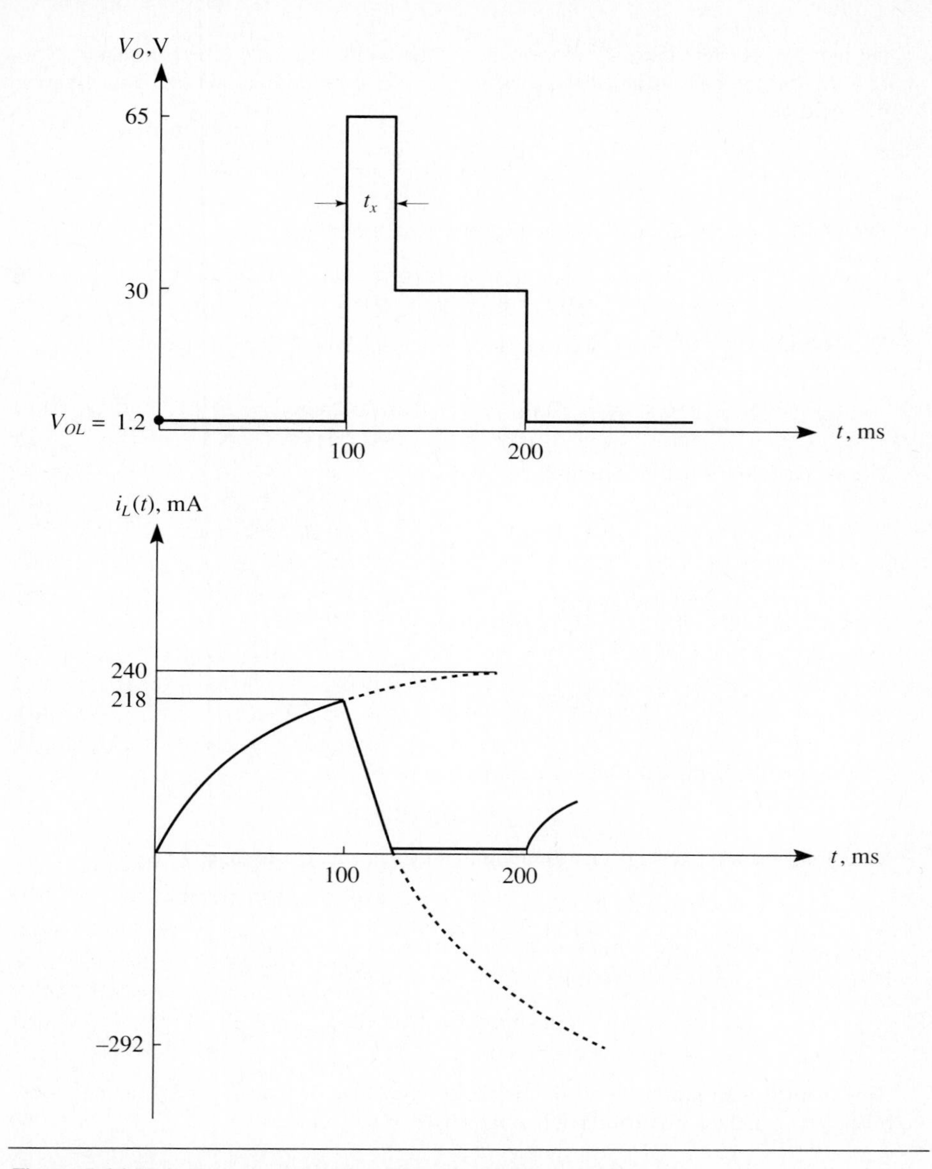

Figure 7.16 Voltage and current waveforms for the inductive load.

capabilities. Therefore, by combining the qualities of CMOS and bipolar technologies on a single silicon chip, these drivers provide a wide supply range from 3 V to 18 V, typical noise immunity of .45 V_{CC}, typical input impedance of 10^{12} Ω, and a typical low output on-resistance of 8 Ω. The high output current and the low on-resistance are achieved through the use of an *npn* Darlington pair at the output stage; the other properties are typical characteristics of CMOS technology. Another unique feature of

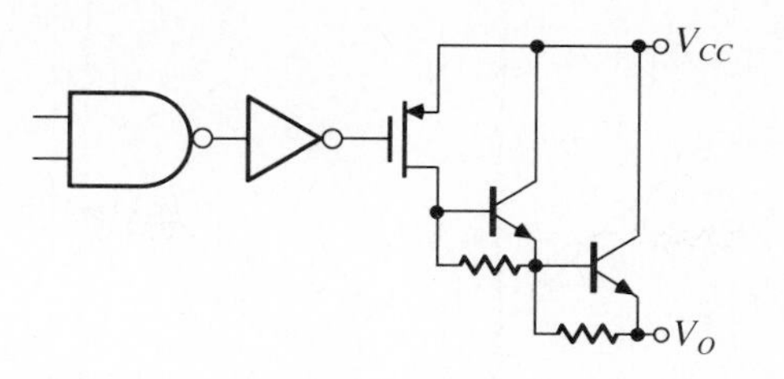

Figure 7.17 The schematic diagram of the 74C908/74C918 high-voltage, general-purpose drivers.

these drivers that is not available in other bipolar drivers is their exceptionally low quiescent power consumption per package of 750 nW at V_{CC} = 15 V. The 74C908 can dissipate at least 1.14 W, and the higher power version 74C918 has the power capability of a minimum 2.27 W.

When both inputs are high, the inverter output prevents the *p*-channel MOSFET from being turned on. Thus, the output is off, with only a small amount of leakage current flowing. When at least one input is low, the logical **0** output of the inverter turns on the PMOS and hence the Darlington pair.

To ensure the junction temperature of maximum 150°C or less, the on-chip power consumption must be limited to within the power rating of the package. Figure 7.18 depicts the maximum on-chip power dissipation as a function of the ambient temperature for both parts. The junction temperature T_j is related to the ambient temperature T_A by

$$T_j = T_A + P_D \Theta_{jA} \tag{7.11}$$

where P_D is the power dissipation and Θ_{jA} is the *thermal resistance* between the junction and ambience in °C/W. It gives the rise in junction temperature over the ambient temperature for each watt of dissipated power. Since it is desirable to dissipate a large amount of power without raising the junction temperature above $T_{j(max)}$, we need to have small values for Θ_{jA}. To find an expression for the on-chip power dissipation as a function of the output current, consider the general application circuit of Figure 7.19. The output current in either section is given as

$$I_{OA} = I_{OB} = \frac{V_{CC}}{R_{on} + R_L} \tag{7.12}$$

The device on-resistance as a function of T_j is expressed by

$$R_{on} = R_{on(max)}[1 + k_r(T_j - 25)] \tag{7.13}$$

where $R_{on(max)} = 9\ \Omega$ at $T_j = 25$°C, and the output resistance coefficient $k_r = .8\%/°C$ for the 74C908. Substituting these values, Equation (7.13) becomes

$$R_{on} = 9[1 + .008(T_j - 25)] \tag{7.14}$$

The power terms due to the leakage current, the internal capacitance, and switching are insignificant compared to the power dissipated at the output stages. The expression for the total output power consumption, then, is given by

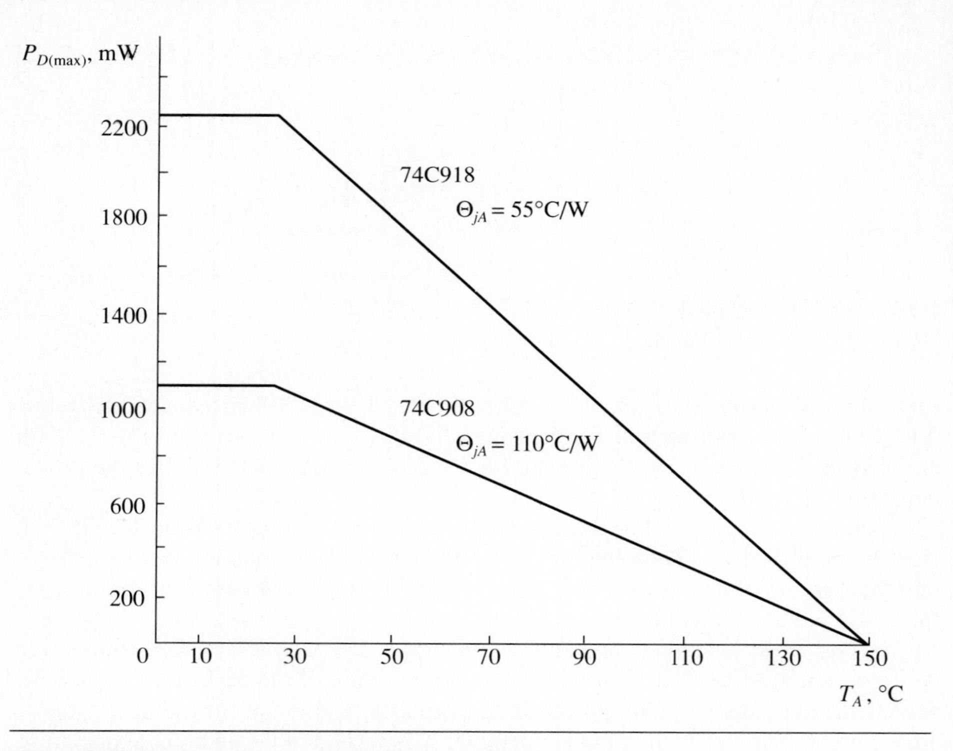

Figure 7.18 Maximum power dissipation vs. ambient temperature.

$$P_D = P_{DA} + P_{DB} = I_A^2 R_{on} + I_{OB}^2 R_{on} \tag{7.15}$$

The resistive load R_L is chosen to satisfy the load requirement, such as a minimum current to turn on a relay. At the same time, the power consumed in the driver package must be kept below its maximum power-handling capability. The following graphical technique may be used to obtain an optimal design and to minimize the design effort.

Figure 7.19 A general application circuit for the 74C908/74C918 drivers.

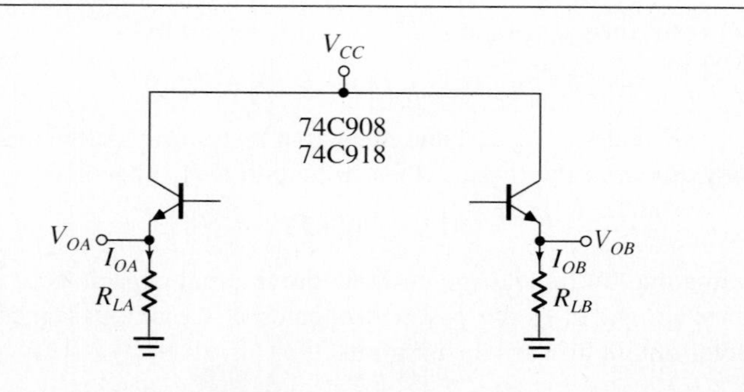

First, to simplify the computation, assume that both sections of the 74C908 are operating under identical conditions (i.e., $R_{LA} = R_{LB}$). Data sheets specify the typical thermal resistance, measured in free air with the device soldered into the PCB, as $\Theta_{jA} = 110°C/W$. The total worst-case power is dissipated when $T_j = 150°C$ so that, using Equation (7.11) at $T_A = 25°C$, we get

$$P_{D(max)} = \frac{150 - 25}{110} = 1.14 \text{ W}$$

Therefore, the maximum power allowed in each section is given as $P_{D(max)}/2 = .57$ W, from which a constant power curve

$$I_O(V_{CC} - V_O) = .57 \tag{7.16}$$

can be plotted, as shown in Figure 7.20. The region of operation is under this curve. For any given R_L, the load line

$$I_O = \frac{V_O}{R_L} = \frac{V_{CC}}{R_L} - \frac{V_{CC} - V_O}{R_L}$$

Figure 7.20 The constant power curve for the 74C908 at $T_A = 25°C$.

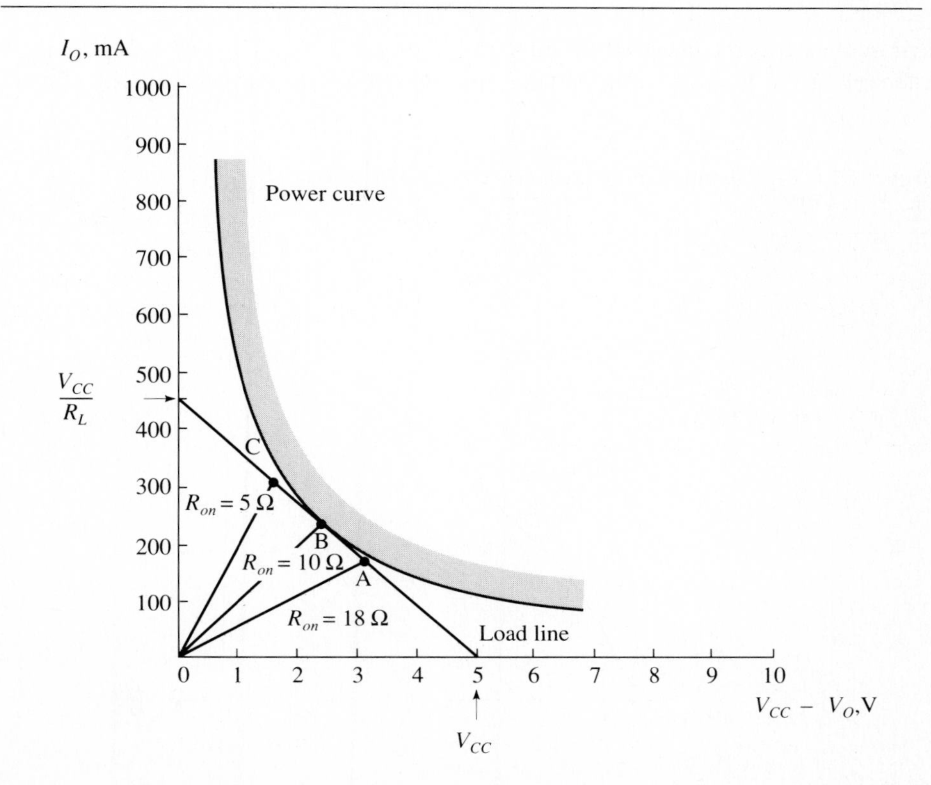

is superimposed on Figure 7.20. Note that this line intersects with the horizontal and vertical axes at V_{CC} and V_{CC}/R_L, respectively.

Given V_{CC}, $R_{L(min)}$ is obtained by drawing the load line tangent to the constant power curve. For example, in Figure 7.20, at $V_{CC} = 5$ V the load line intersects the I_O axis at $I_O = 450$ mA. Therefore, $R_{L(min)} = 5\text{ V}/450\text{ mA} = 11.1\ \Omega$. Any value below this will cause a section of the load line to extend into the shaded region, allowing the junction temperature to exceed $T_{j(max)} = 150°\text{C}$. Whether this situation will occur is determined by both the value of V_{CC} and the range of driver on-resistance.

R_{on} is actually a nonlinear function of the output current, as seen in Figure 7.21. The nonlinear characteristic at high values of V_O is due to the fact that the output *npn* transistor is not saturated. As the junction temperature is increased, the transistor reaches saturation at a lower value of the output current. As soon as it saturates (e.g., around $I_O = 150$ mA for $T_j = 150°\text{C}$), the curve becomes a straight line that can be extrapolated back to the origin. However, for all practical purposes, it is sufficient to consider R_{on} as a linear function of the output current. Therefore,

$$R_{on} = \frac{V_{CC} - V_O}{I_O}$$

so that

$$I_O = \frac{V_{CC} - V_O}{R_{on}} \tag{7.17}$$

represents a straight line passing through the origin with a slope of $1/R_{on}$ S, and intersecting the load line at a certain point depending on the value of R_{on}. From

Figure 7.21 Typical output transfer characteristic at $V_{CC} = 5$ V and $T_j = 150°\text{C}$.

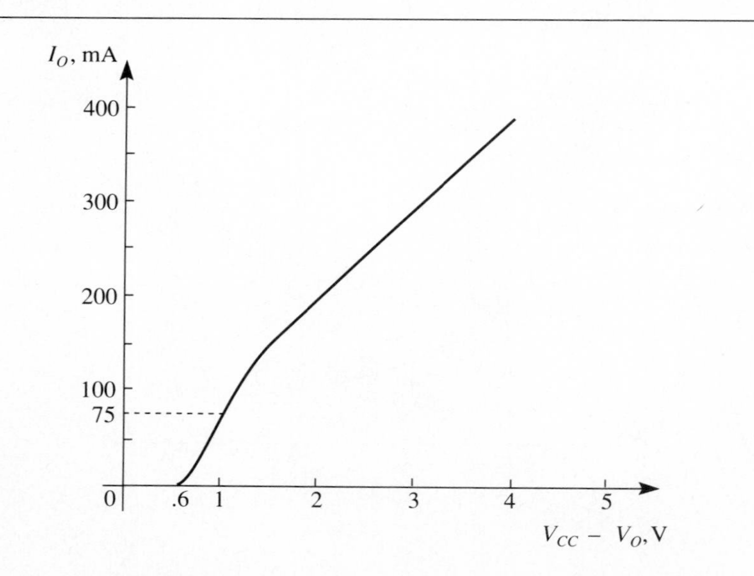

Equation (7.14), the maximum value of R_{on} at $T_j = 150°C$ is found to be 18 Ω. From Figure 7.21, we find an approximate value for $R_{on(min)}$ when the output transistor is not yet saturated as

$$R_{on(min)} \approx \frac{1 - .6}{75} \times 10^{-3} = 5\ \Omega$$

In the saturation region, the slope of the characteristic in Figure 7.21 yields the typical value of R_{on} as 10 Ω. To find a safe range of operation, we use Equation (7.17) to plot three straight lines corresponding to $R_{on(min)}$, $R_{on(typ)}$, and $R_{on(max)}$, passing through the origin with slopes of 1/5, 1/10, and 1/18, and intersecting the load line at points C, B, and A, respectively. For $V_{CC} = 5$ V, we observe that the tangent point falls between points A and B. Thus, unless $R_L >> 11.1\ \Omega$, part of the load line within the specified R_{on} range extends into the shaded region, jeopardizing the safe operation of the device by raising the possibility that $T_j \geq 150°C$ will occur. Consequently, the section of the load line between A and C should be the operating range for the circuit at $V_{CC} = 5$ V and $R_L \geq 11.1\ \Omega$. From Figure 7.20, we find the current and voltage ranges as

$$172\ \text{mA} \leq I_O \leq 310\ \text{mA}$$

and

$$1.9\ \text{V} \leq V_O \leq 3.4\ \text{V},$$

respectively. Note that as the operating point on the load line moves away from the tangent point, the actual junction temperature drops. At point A, for example, the device is actually running at a temperature cooler than $T_j = 150°C$, leading to a drop in the R_{on} value below 18 Ω and an operating point slightly different from A.

These drivers can be used to drive inductive loads such as relays, as depicted in Figure 7.22. A diode at the relay coil is connected to suppress transient spikes, as discussed above.

Figure 7.22 The 74C908/74C918 driving an inductive load.

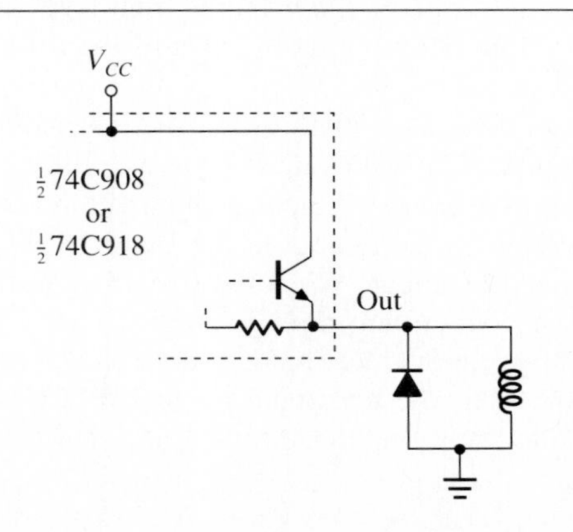

7.4 Interfacing

An interfacing problem arises when the output logic levels and/or the current requirements of the driving device are different from the input logic levels and/or the current requirements of the driven device.

CMOS and TTL

Since the high-speed CMOS is currently the most widely used CMOS subfamily, and various Schottky TTL are the most popular series, our coverage of interfacing will start with and mainly concentrate on these parts. Nevertheless, the ongoing discussion can easily be extended to other series in both families.

Let $V_{OH(nl)}$ designate the high-level output voltage expected when the output stage of the driver is unloaded. Similarly, $V_{OL(nl)}$ can signify the expected value of the low-level voltage when the output is unloaded. Usually, these two voltage values are not provided on the data sheets. For CMOS devices, the output can be considered to be switching between $V_{OH(nl)} = V_{CC}$ and $V_{OL(nl)} = 0$ V. For TTL devices, $V_{OL(nl)}$ is about $V_{CE(sat)} \approx .2$ V. Within the TTL family $V_{OH(nl)}$ varies such that standard TTL has a $V_{OH(nl)}$ within two V_{BE} drops of the supply voltage, while for LS TTL it is within one V_{BE} drop of V_{CC}. When the output stage is loaded, the values deteriorate even more.

When TTL is driving HC, there are some minor differences in TTL specifications for totem-pole outputs and HCMOS input specifications. The TTL output low level is completely compatible with HC input low, but TTL outputs are specified to have minimum high levels of 2.4 V to 2.7 V, while the HCMOS logical **1** input level is 3.5 V at $V_{CC} = 5$ V. Therefore, TTL is not guaranteed to pull a valid CMOS logical **1** level.

To see why standard TTL cannot pull up further, recall from Chapter 6 that, as the output of the standard TTL pulls up, it can go no higher than two diode voltage drops below V_{CC} due to Q_4 and D. Clearly, if the output is loaded, the output voltage will be lower. On the other hand, the outputs of low-power Schottky and advanced Schottky circuits (i.e., the LS and AS/F/ALS devices) can pull up to 4.3 V, which is more than adequate to drive a CMOS. However, when designing to the worst-case characteristics, greater compatibility is desired because the LS TTL specifications, for example, still guarantee only a 2.7 V output high level. One solution is to raise the output high level on the TTL output by placing a pull-up resistor from the TTL output to V_{CC}.

No additional interfacing is required for HCMOS devices to be able to drive TTL loads: the output voltages of HCMOS are compatible with the input voltage requirements of TTL devices. However, the input current requirements of the TTL devices do place a strict limitation on the fan-out of the HCMOS. Figure 7.23 illustrates an HCMOS gate driving a TTL input. When the CMOS gate drives the emitter of the input transistor Q_1 low, the current through R_1 flows into the HCMOS gate. The maximum guaranteed current that the HCMOS device can sink is 4 mA. Since $I_{IL} = -1.6$ mA for standard TTL, the maximum number of TTL loads that an HCMOS device can drive without exceeding the specified limit is two.

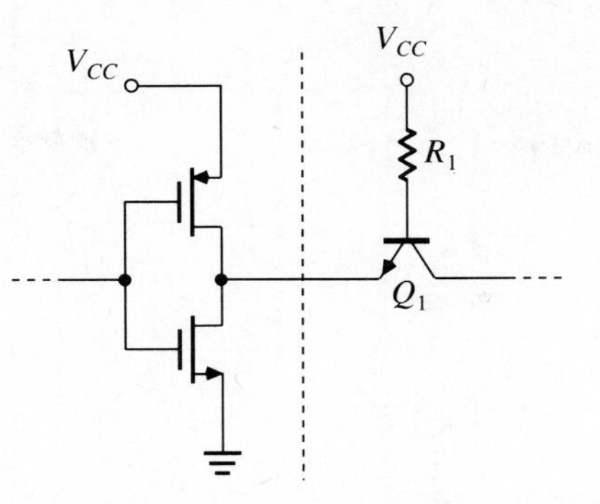

Figure 7.23 The HC-to-TTL interface.

Next, consider the simplified schematic of the input of an HCMOS gate, as depicted in Figure 7.24. The diode D_1 and the *npn* transistors Q_1 and Q_2 provide static discharge and input transient clamping for the device. Any input voltage value higher than V_{CC} + .5 V or lower than −.5 V will be clamped. The parasitic capacitances at the gate input are represented by C_1 and C_2. The data sheets specify the maximum value for $C_1 + C_2$ as 10 pF, while the typical value is about 5 pF. Note that the input capacitance is split between the supply and the input. Therefore, if the input is driven by a high-impedance source, then any transient noise on V_{CC} may be coupled back into the input.

Figure 7.25 shows the schematic of the TTL-to-HC interface. To eliminate the voltage incompatibility, a pull-up resistor R_p is used. The current-sinking capability of the driving device determines the lower limit of the pull-up resistor. When the TTL output goes low, the low-level driver Q_3 will be required to sink a current of $(V_{CC} - V_{OL(max)})/R_p$ in addition to the sum of the output currents of the driven devices. Therefore, we have

Figure 7.24 The HCMOS input stage.

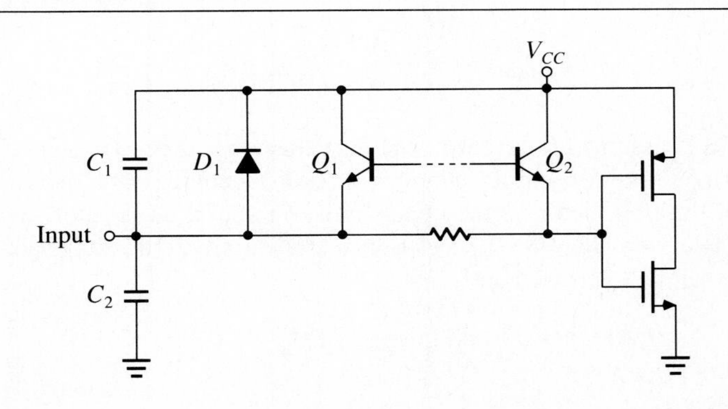

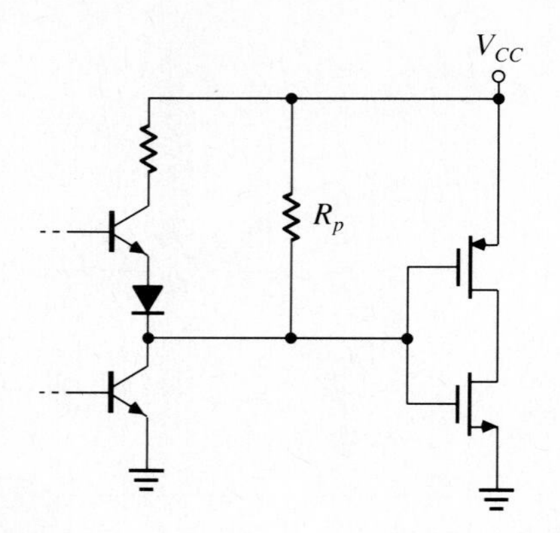

Figure 7.25 The TTL-to-HC interface.

$$R_{p(min)} = \frac{V_{Rp(min)}}{I_{Rp}}$$

$$= \frac{V_{CC} - V_{OL(max)}(\text{TTL})}{I_{OL(max)}(\text{TTL}) - NI_{IL}(\text{HCMOS})} \qquad (7.18)$$

where N is the number of standard loads being driven.

The upper limit of the pull-up resistor is determined by two factors; the total input capacitance of the loads, and the total high-level input currents of the loads. When the TTL output is logical **1**, not only Q_3 but also Q_4 is turned off due to the pull-up resistor. Therefore, almost all the current that flows into the loads flows through R_p. The input voltage of the driven devices rises exponentially with $\tau = R_p C_i$, where $C_{i(max)} = 10$ pF. On the other hand, the manufacturers' data sheets specify that the maximum rise time is 500 ns at $V_{CC} = 4.5$ V. Since the total input current must not cause the voltage drop across the pull-up resistor to exceed $V_{IH(min)}$ for HCMOS, we have

$$R_{p(max)} = \frac{V_{CC} - V_{IH(min)}(\text{HCMOS})}{NI_{IH}(\text{HCMOS})} \qquad (7.19)$$

Note that I_{OH} is assumed to be zero. The data sheets also specify the maximum value of the input rise time for various supply voltages. Generally, this rise time constraint is the actual limiting factor on the upper limit of the pull-up resistor. Since the input voltage of the driven HCMOS devices rises exponentially, the voltage across the input capacitance can be expressed as

$$V_{Ci}(t) = V_{CC}(1 - e^{-t/R_p C_i}) \qquad (7.20)$$

Solving Equation (7.20) for $R_{p(max)}$ we obtain

$$R_{p(max)} = \frac{t_r}{C_i} \frac{1}{\ln\left(\dfrac{V_{CC}}{V_{CC} - V_{IH(min)}}\right)} \tag{7.21}$$

Example 7.3

For a 7400 driving four 74HC00 NAND gates at $V_{CC} = 5$ V, we have $V_{OL(max)}(\text{TTL}) =$ .4 V, $I_{OL(max)}(\text{TTL}) = 16$ mA, and $I_{IL} = -1\ \mu$A so that

$$R_{p(min)} = 287\ \Omega$$

Since $V_{IH(min)} = 3.5$ V and $I_{IH} = 1\ \mu$A for HCMOS, using Equation (7.19), we obtain

$$R_{p(max)} = 375\ \text{K}\Omega$$

However, if the input rise time is calculated using the value and Equation (7.21), it will be seen that the recommended 500 ns at $V_{CC} = 4.5$ V is exceeded. The rise time cannot be slower than 500 ns because it may not trigger the device. Using Equation (7.21) with total $C_{i(max)} = 4 \times 10$ pF $= 40$ pF and $t_{r(max)} = 500$ ns, we get

$$R_{p(max)} = 8.3\ \text{K}\Omega$$

for $V_{CC} = 4.5$ V.

If open-collector TTL outputs with a pull-up resistor are driving the HC logic, then no interface circuitry is needed because the external pull-up will pull the output to a high level very close to V_{CC}.

The pull-up resistor interface should be used only if unavoidable because the propagation delay increases due to the combined effects of the time constant of the resistor, the stray capacitance, and the load capacitance. It also consumes extra power and reduces the low-level noise margin because of its active load. All of these conflict with the purpose of using HCMOS.

Another method to interface from any TTL subfamily to HCMOS is to use HCT devices. This is by far the easier method because the HCT device inputs are already TTL compatible, while the outputs are both TTL and HCMOS compatible.

The input configuration for HCT ICs is shown in Figure 7.26. It employs a level-shifting diode D_3 between the PMOS transistor Q_1 and the supply. The combined effect of D_3 and the large *n*-channel transistor Q_2, which has a higher gain than the former transistor, reduces the typical input switching level of HC from $V_{CC}/2$ to 1.3 V. The function of D_3 is to cut the power dissipation when a high input is applied from a TTL circuit. Even with $V_{IH(min)} = 2$ V, the *p*-channel conducts slightly in the absence of the diode, resulting in a current flow between V_{CC} and ground. However, the current becomes negligible with D_3 present and with the connection of the Q_1 substrate to V_{CC}. This combination also leads to lower power dissipation in the input stage, with switching levels still comparable with LS TTL.

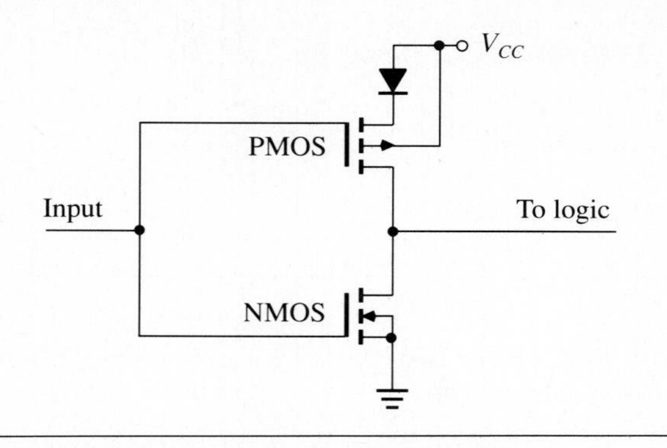

Figure 7.26 The input configuration for the HCT logic.

A third solution is to operate CMOS at $V_{CC} = 3$ V while TTL is connected to 5 V. This is particularly true when HCMOS or ACL is operated in a battery back-up application for a TTL system. Then CMOS can be directly connected to TTL because its input and output levels are already compatible with TTL, and the TTL output levels will now be compatible with CMOS inputs, as the dc specifications of CMOS devices at $V_{CC} = 3$ V in Table 7.4 show. Note from Figure 4.39, however, that the AC propagation delay at 3 V is increased by about 75% over the delay at 5 V.

Interfacing with ECL

Although all ECL subfamilies are usually directly compatible with each other, there may still be some difficulties in mixing them, as illustrated in Figures 7.27 and 7.28 for the 10K and 100K series. As discussed in Chapter 6, the 10K output levels and input thresholds vary with temperature, whereas 100K levels and thresholds remain essentially constant, resulting in variance of the noise margins with temperature even if the temperatures of the driver and the load track. Note from Figure 7.28 that the high-state noise margin is seen to be less than 100 mV above 40°C, which would not represent an acceptable dc margin in any real system.

ECL circuits normally operate with ground on V_{CC} and -5.2 V power supply on

Table 7.4 Worst-Case Voltage Values for the Advanced and High-Speed CMOS at $V_{CC} = 3$ V, $T_A = 25°C$

	AC	HC
$V_{IH(min)}$, V	2.1	2.1
$V_{IL(max)}$, V	.9	.6
$V_{OH(min)}$, V	2.9	2.9
$V_{OL(max)}$, V	.1	.1

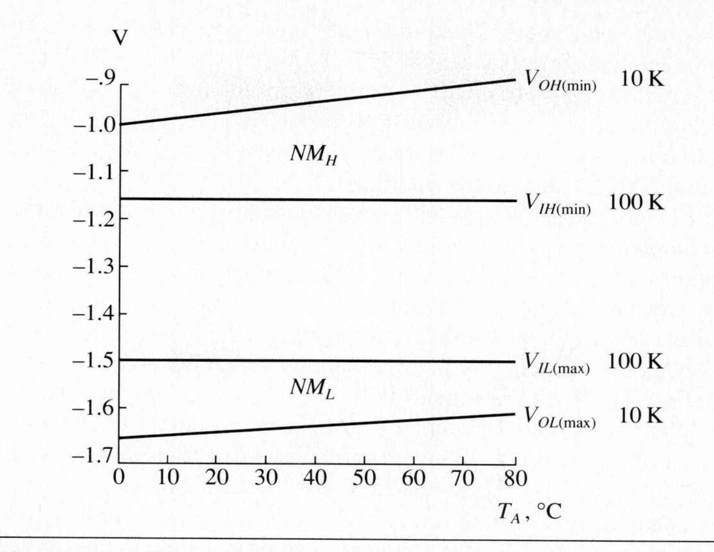

Figure 7.27 Noise margins at both levels when the 10K ECL is driving the 100K ECL.

V_{EE}. While ECL may be used with ground on V_{EE} and +5 V on V_{CC}, the negative supply operation leads to better noise immunity. In addition, ECL operates with a relatively small logic swing. Thus, it is not directly compatible with other logic families, such as TTL and CMOS. Translators must be used when interfacing these logic types with ECL.

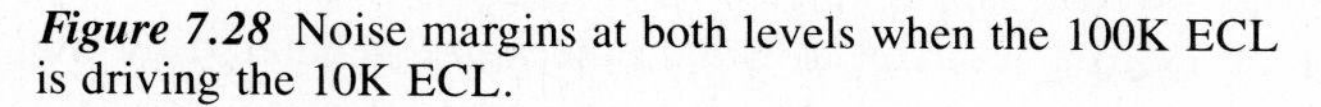

Figure 7.28 Noise margins at both levels when the 100K ECL is driving the 10K ECL.

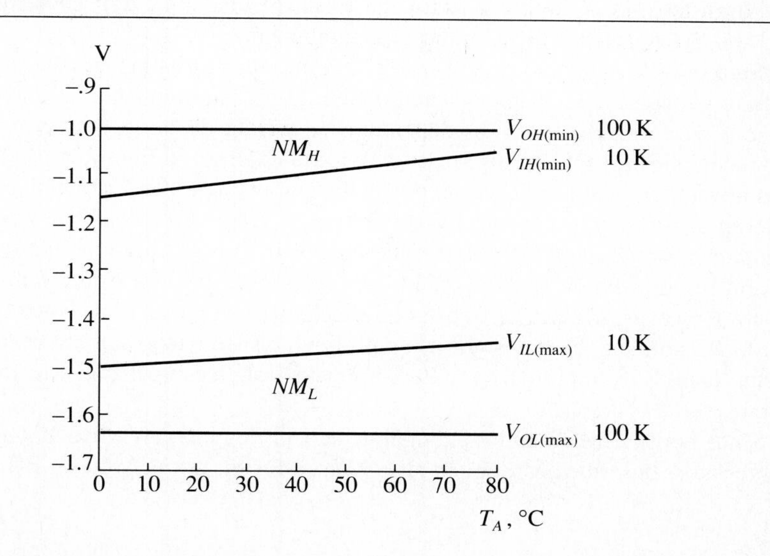

The most common interface requirement for ECL is with the TTL logic levels. Normally, the interface occurs when the ECL is powered with a −5.2 V power supply and the TTL with +5 V. The use of a common ground and separate power supplies helps to isolate the TTL-generated noise from the ECL supply lines. Depending on the ECL subfamily, the translator circuits 10124 and 10125, or 10H124 and 10H125, or 100124 and 100125 provide the interfaces.

The 10124 is a quad TTL-to-ECL level translator with a common TTL strobe input and complementary open-emitter ECL outputs that allow it to be used as an inverting/noninverting translator or as a differential line driver in which the TTL information can be transmitted differentially, via balanced twisted-pair lines, over long distances. The propagation delay through the circuit is typically 5 ns, and the maximum operating frequency is 85 MHz, above which the output fails to reach the specified levels. The 10125 is a quad ECL-to-TTL level translator with differential amplifier inputs and Schottky-clamped transistor totem-pole TTL outputs. The differential inputs allow for use as an inverting/noninverting translator or as a differential line receiver. In the latter capacity, the ECL-level information can be received, again through the balanced twisted-pair lines, in the TTL equipment, thereby isolating the ECL logic from the noisy TTL environment. The propagation delay for the circuit is a function of the fan-out. For example, it is typically 4.5 ns when there is no load and increases to 5.5 ns when driving four loads.

The typical propagation delay of the quad 10H124 translator has been improved to 1.5 ns, while it is specified as 2.5 ns for the quad 10H125 part driving a 25 pF load. The 100K counterparts of these two ICs, that is, 100124 and 100125, are hex translators. The minimum and the maximum propagation delays are given as .5 ns and 2.7 ns, respectively, for the 100124, and .9 ns and 3.5 ns, respectively, for the 100125.

In a system design that uses only a small number of ECL circuits, it may be desirable to operate both logic systems on a single +5 V supply. The ECL works properly in this mode as long as care is taken to isolate the noise generated by the TTL from the +5 V supply line for the ECL. In this case, translators to interface TTL and ECL can be built with discrete components.

Figure 7.29 illustrates a TTL-to-ECL translator consisting of three resistors in series to attenuate TTL output levels to ECL input requirements. The translation is normally performed under 1 ns depending on wiring delays and stray capacitance. Two techniques for interfacing ECL to TTL are depicted in Figure 7.30. The former takes advantage of the ECL complementary outputs to drive a differential amplifier made up of two *pnp* transistors. When driving a single TTL load, the speed of this translator is in excess of 100 MHz. The latter circuit uses only one *pnp* transistor to perform the translation at the expense of speed, as compared to the differential approach. However, the typical translation time is still less than 10 ns when driving one TTL load. Note that both designs use pull-down resistors to ground to be able to sink the low-level TTL input current. Resistor values should be changed to increase the fan-out.

Since both HCMOS and ACL operate at TTL logic levels when V_{CC} = 5 V, the IC translators described earlier can also be used for ECL-to-CMOS-to-ECL interface

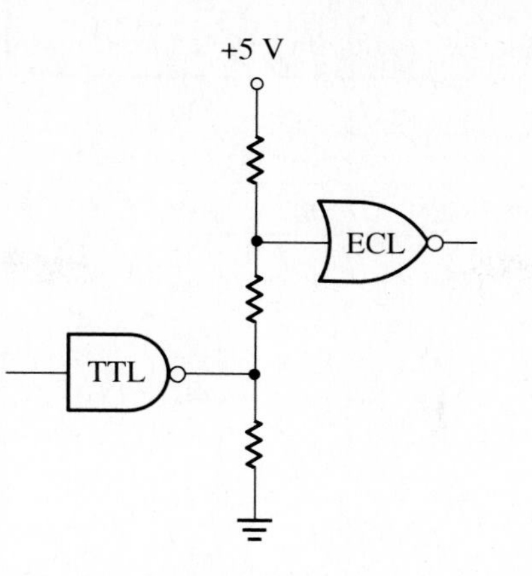

Figure 7.29 A common-supply TTL-to-ECL interface circuit.

requirements. Table 7.5 summarizes the interface compatibilities of various logic families.

7.5 Comparison of Logic Families

Currently, the most widely used logic families are TTL and CMOS. The HCMOS is a full-line subfamily developed to be designed into any application that has used the LS TTL, with substantial improvement in power consumption. In late 1985, several companies introduced new lines of advanced CMOS parts to challenge the advanced Schottky bipolar parts. Consequently, we concentrate on those subfamilies in this section.

Figure 7.30 Two techniques for interfacing ECL to TTL using a single +5 V supply.

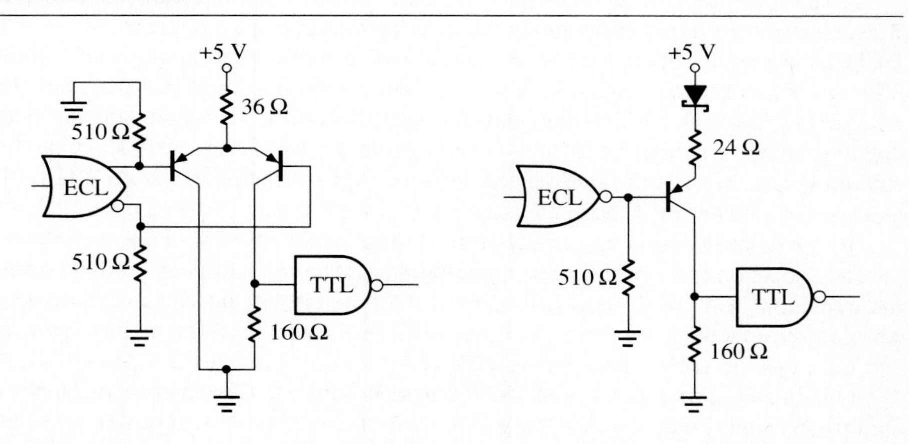

Table 7.5 Logic Family Interface Compatibility

	Driven								
	CMOS		TTL				ECL		
	HC	AC	LS	AS	F	ALS	10K	10KH	100K
Driver									
CMOS									
HC	D	D	D^1	D^1	D^1	D^1	L	L	L
AC	D	D	D^1	D^1	D^1	D^1	L	L	L
TTL									
LS	D^2	D^2	D	D	D	D	L	L	L
AS	D^2	D^2	D	D	D	D	L	L	L
F	D^2	D^2	D	D	D	D	L	L	L
ALS	D^2	D^2	D	D	D	D	L	L	L
ECL									
10K	L	L	L	L	L	L	D	D	D
10KH	L	L	L	L	L	L	D	D	D
100K	L	L	L	L	L	L	D	D	D

D = Direct interface.
D^1 = Direct when operating at the same supply voltage.
D^2 = Direct with the use of external pull-up resistor when operating at the same supply voltage, or using HCT or ACT parts.
L = Level translator required.

HCMOS vs. LS TTL

Table 7.6 compares the main characteristics of major commercial-grade subfamilies. One of the criteria in deciding which logic to use for a design is the speed. The HCMOS is intended to offer the same basic speed performance as low-power Schottky TTL while giving the designer the low power and high noise-immunity characteristics of CMOS. Figure 7.31 illustrates that the variation in the HC's propagation delay due to changes in capacitive loading is very similar to that of the LS TTL. While this subfamily has virtually the same speed and load-delay variation as the LS TTL, it is, as expected, slower than the ALS logic.

If input signals spend appreciable time during transition between the logic states, then the noise on the input or power supply may cause the output to oscillate during the transition. This could lead to logic errors in the circuit and could also dissipate extra power unnecessarily. For this reason, HC data sheets recommend that input rise and fall times be shorter than 500 ns at $V_{CC} = 4.5$ V.

One major advantage of all CMOS series over their TTL counterparts is that in the former family, only the switching gates contribute to the system power consump-

Table 7.6 Performance Comparison of Commercial-Grade Logic Families

	CMOS		TTL				ECL		
	HC	AC	LS	AS	F	ALS	10K	10KH	100K
t_{PD},* ns	8	5	9.5	1.5	3.5	6	2	1	.75
Power/gate, mW									
static	.0025	.0025	2	8	5.5	1.2	25	25	40
at 1 MHz	1.75	1.75	2	8	5.5	1.2	25	25	40
SP, pJ at 1 MHz	14	9	19	12	19	7	50	25	30
f_{max},† MHz	50	125	33	105	125	34	160	250	350

*C_L = 15 pF for LS
= 50 pF for others
†maximum clock frequency for the D-type flip-flop

Conditions: typical values at 25°C,
V_{CC} = 5 V for CMOS and TTL, V_{EE} = −4.2 V to −4.8 V for 100K
= 0 V for ECL = −5.2 V for 10KH and 10K

tion. This reduces the size of the power supply required and hence provides lower system cost and improved reliability through lower heat dissipation. On the other hand, the power consumption for an individual gate at higher speeds is comparable to that of the LS TTL, although in typical systems, since only a fraction of the gates are switching at the clock frequency, one can justifiably argue that significant power savings can still be realized by using the HCMOS. Even though power consumption is somewhat dependent on frequency in TTL devices, the LS, like other TTL parts, has essentially flat curves, except at very high frequencies, because the static currents

Figure 7.31 Comparison of propagation delays for NAND gates as a function of capacitive loads.

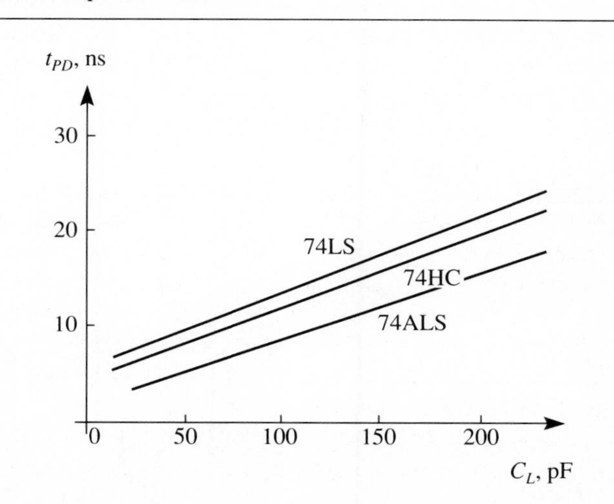

mask out internal capacitive effects. Therefore, the majority of power dissipated below 1 MHz is due to the quiescent supply current. The LS TTL contains many resistive paths from V_{CC} to ground, so that even when it is not switching, it draws a supply current that is several orders of magnitude greater than that of the HCMOS. The static power requirement per gate (i.e., $V_{CC}I_{CC}$) is around 2 mW for the LS, while it is only about 2.5 nW for HCMOS parts.

Note that in an HCMOS system, the typical capacitive load each gate output sees is $C_L = 50$ pF, whereas it is only 15 pF for LS TTL parts. From Equation (4.26), with $C_{PD} = 20$ pF, we obtain the power dissipation per gate for an HC part as $\approx$ 17.5 mW at 35 MHz. At the same frequency, an LS gate consumes only 10 mW for the same capacitive loading and a mere 5 mW for a more realistic C_L of 15 pF. Figure 7.32 compares the power consumption in the HCMOS and LS TTL as a function of frequency. The frequency at which the CMOS device draws as much power as the LS device is known as the *power crossover frequency* f_{co}. Since C_{PD} is specified in data sheets as 20 pF for the HC part, neglecting the static power, we obtain

$$f_{co} = \frac{2 \text{ mW}}{(20 \text{ pF})(5 \text{ V})^2} = 4 \text{ MHz}$$

for unloaded NAND gates, and

$$f_{co} = \frac{2 \text{ mW}}{(70 \text{ pF})(5 \text{ V})^2} = 1.1 \text{ MHz}$$

for a typical $C_L = 50$ pF for the CMOS. Note that when $C_L = 0$, the power dissipation per gate for the bipolar part stays constant until the operating frequency is over

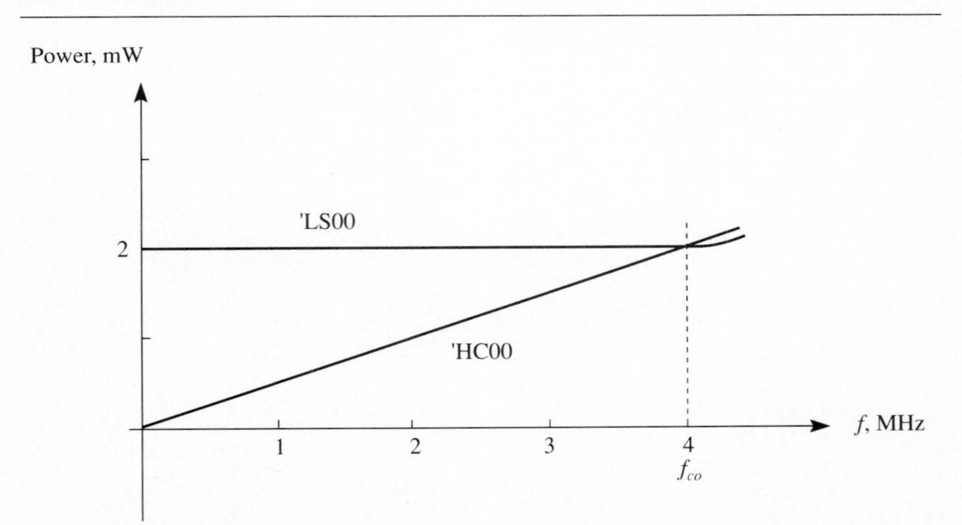

Figure 7.32 Power-consumption comparison in HCMOS and LS TTL NAND gates.

4 MHz. Therefore, above these frequencies, the total consumption for the CMOS device is actually more than that of its LS counterpart. This is also partly due to the lower average load current, for a given load, of the TTL because of its smaller logic swing. However, in a typical system with capacitive loads, as the frequency increases, the capacitive currents tend to dominate the bipolar power dissipation as well.

The power crossover frequency increases as the circuit complexity increases. There are two major reasons for this. For one, having more devices on an LS TTL IC means that more resistive paths between V_{CC} and the ground exist, and hence more static current will be required. In a CMOS IC, on the other hand, the increase in the supply leakage current is of such a small magnitude (in the order of nanoamperes per device) that the corresponding increase in the total power consumption is minimal. Secondly, as the system complexity increases, the percentage of the total system operating at the maximum frequency tends to decrease.

Assuming that on a system level the switching frequencies of individual gates are normally distributed between zero and the crossover frequency, the power saved with HCMOS at low frequencies can be significant, as illustrated in Figure 7.33, which is obtained by multiplying the individual gate characteristics of Figure 7.32 by the normal frequency distribution. The contribution to the total system power is the area under each curve. Since there is virtually no static power dissipation, it is obvious that CMOS is ideal for battery-operated systems or systems requiring battery back-up.

The improved noise immunity of the HCMOS over the bipolar series is due to the rail-to-rail, (i.e., V_{CC}-to-ground) output voltage swings. Figure 7.34 illustrates the noise immunity provided by the HCMOS subfamily compared to that of the LS TTL. This noise immunity makes it ideal for high-noise environments such as industrial and automotive applications. Minimum and maximum output voltages are guaranteed at ± 4 mA, and the noise immunity is not impaired as long as the output currents do not exceed these limits.

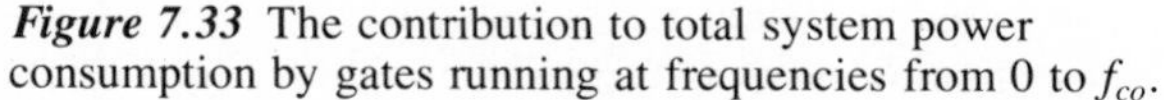

Figure 7.33 The contribution to total system power consumption by gates running at frequencies from 0 to f_{co}.

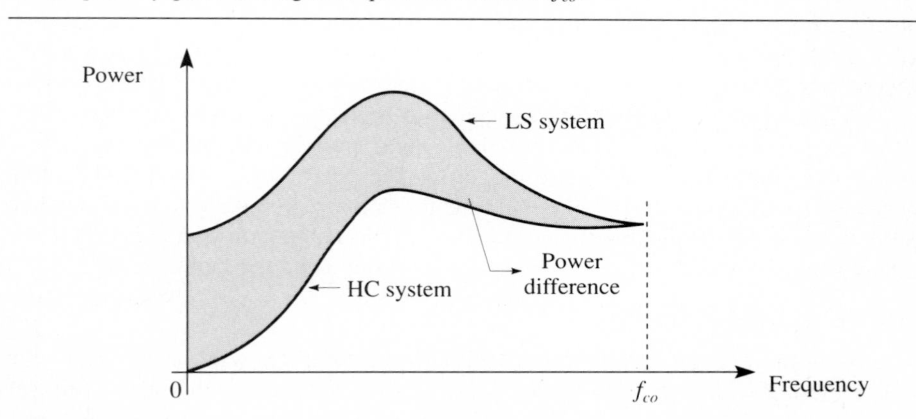

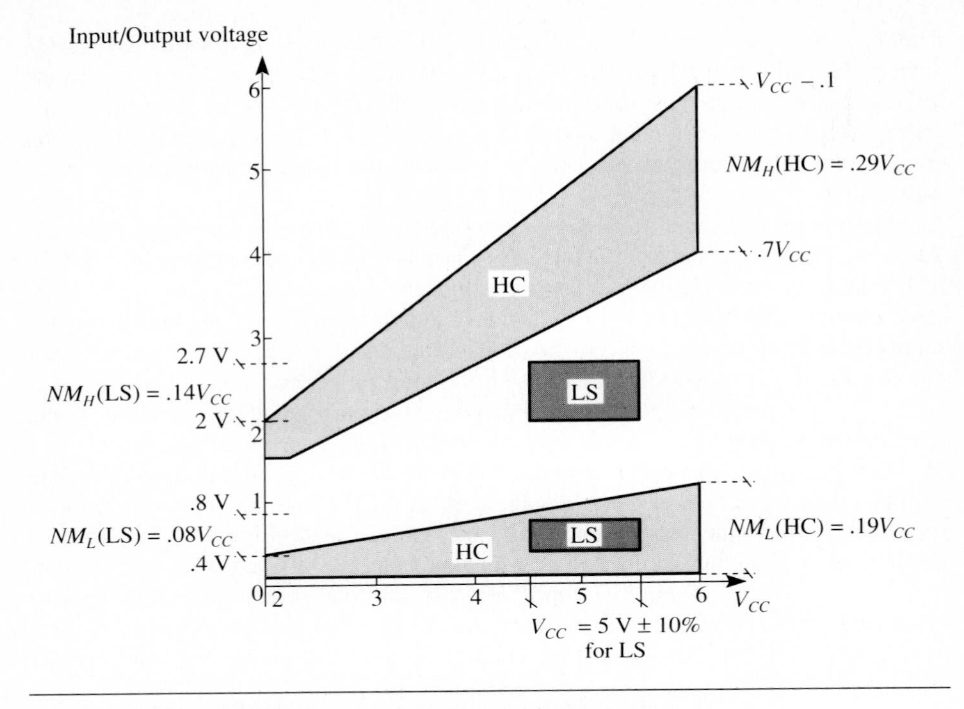

Figure 7.34 The comparison of noise margins for the HCMOS and LS TTL.

The HCT devices have input noise margins similar to those of the LS TTL because their inputs are TTL compatible. Although $V_{OH(min)}$ and $V_{OL(max)}$ are guaranteed for output currents up to ±4 mA, continuous currents up to ±25 mA can be obtained to drive LEDs or even relays. As the device sinks and sources more current, however, $V_{OH(min)}$ and $V_{OL(max)}$ levels will begin to fall and rise, respectively.

Figure 7.35 compares the transfer characteristics of the basic HC and LS NAND gates over the entire temperature excursion and emphasizes the stability as well as the centering of the HC transition region relative to the supply voltage. The HC part has V_{CC} and ground output levels and a very sharp transition at about 2.25 V, resulting in good noise immunity. The latter feature is due to the large circuit gains provided by triple buffering in the HCMOS compared to the single bipolar gain stage of the LS. The transition point is also very stable with temperature, drifting only about 50 mV over the entire temperature range. On the other hand, the LS TTL output makes the transition at about 1.1 V, and the transition region varies several hundred millivolts over the temperature range. In addition, since the transition region is closer to the logical **0** level, less ground noise can be tolerated on the input.

ACL vs. Advanced Schottky

Just as the HC/HCT high-speed CMOS logic competes with the LS TTL, the ACL represents a new CMOS line of logic rivaling the advanced Schottky bipolar subfamilies in a number of applications, including computers, peripherals, and tele-

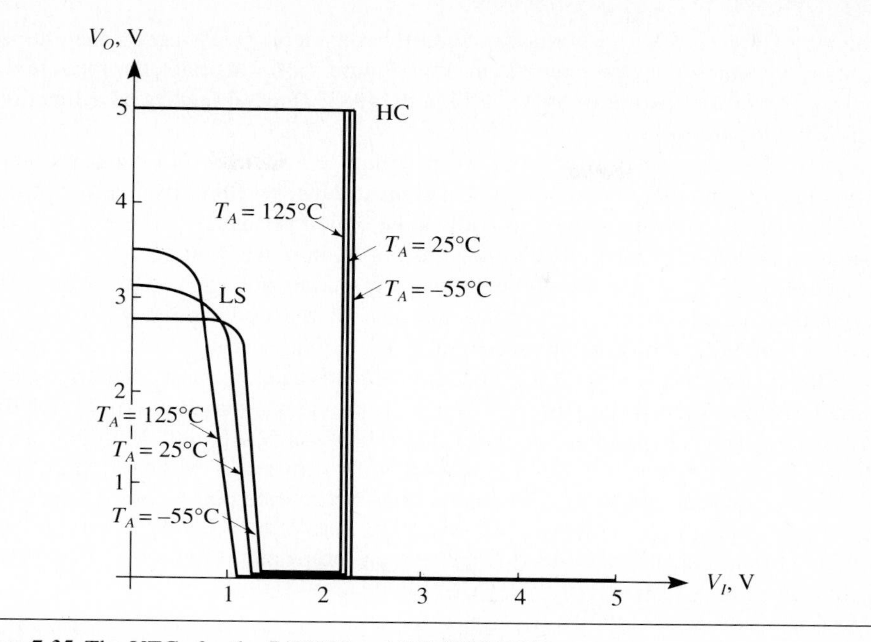

Figure 7.35 The VTCs for the 74HC00 and 74LS00 NAND gates at various temperatures.

communications. It was first introduced in 1985 to meet very high speed data processing requirements as an alternative to the AS/F/ALS TTL logic series with less power consumption.

The ALS is usually chosen over the LS because of its lower power consumption, not its higher speed. If, on the other hand, the ALS is selected solely because of its higher speed, then the ACL subfamily seems to be the better choice since its speed is higher than that of ALS parts operating at 3 V and above. However, there may be two penalties involved in using the AC/ACT to replace the LS TTL: (1) a high probability of logic race conditions and transient functional difficulty requiring extensive logic redesign, and (2) the greater cost of printed-circuit boards, decoupling, and the ICs used. Therefore, the HC subfamily should be preferred instead.

The advantages of the ACL subfamily compared to other logic series include the following:

1. Lower power dissipation at dc and lower frequencies
2. Balanced propagation delay
3. Larger noise margin due to the rail-to-rail output voltage swing
4. Wider operating supply voltage range, from 1.5 V to 6 V
5. Very stable switching-voltage levels over a wide temperature range

The power savings of ACL over bipolar logic are directly related to overall average logic switching rates. Although the quiescent power dissipation per gate is 2.5 μW for 74AC00, it becomes 1.75 mW at 1 MHz as compared to 1.2 mW for the ALS logic. Nevertheless, when the AC series is operated at its most efficient power supply voltage of $V_{CC} = 3$ V, it obviously consumes considerably less power than the ALS subfamily, which must be operated at 5 V. Figure 7.36 illustrates the measured operating power dissipation of 74ALS373 and 74AC373 octal latches as a function of the switching frequency.

A shift from bipolar to the ACL corresponds to a reduction in the power supply size and cost, and the elimination of heat sinks or fans, resulting in smaller space and size requirements for equipment as well as the option for battery power, low-voltage standby operation, or battery backup in case of main power outage.

Table 7.7 shows the worst-case input and output voltage values at both logic levels and the improvement in noise margins of the commercial-grade advanced CMOS subfamilies over the advanced Schottky bipolar series.

One major shortcoming of the ACL is that it still cannot provide the drive current often needed in high-speed systems. The ACL provides a maximum of 24 mA drive current, while such popular board-level products as VME and Multibus require 48 mA. However, limited CMOS parts, generally directed at particular applications such as bus drivers, are starting to emerge that incorporate both very high speed and high drive capabilities. For example, one ACT line of bus drivers has a maximum sinking current capability of 48 mA and typical propagation delays of 7 ns even at $C_L = 300$ pF.

Summary

Usually, many components in a digital system share a common path for transferring data to one another. To minimize the number of interconnections, buses are used. Outputs from different devices cannot be present on the bus at the same time. Thus,

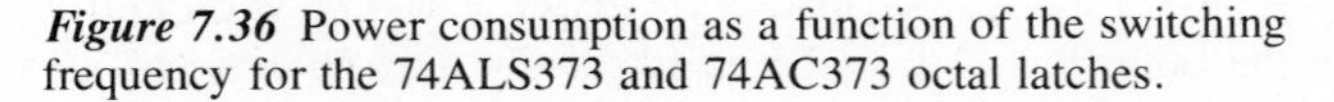

Figure 7.36 Power consumption as a function of the switching frequency for the 74ALS373 and 74AC373 octal latches.

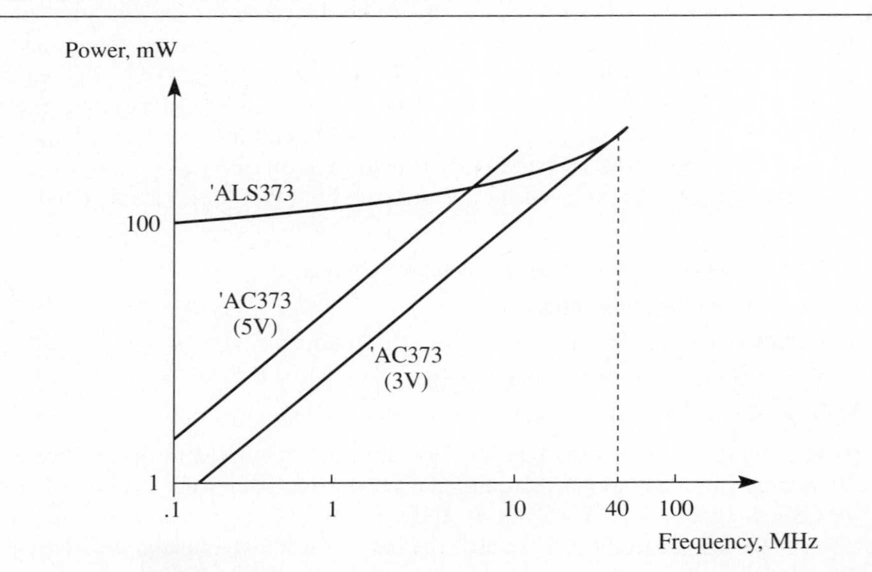

Table 7.7 Voltage Values and Noise Margins of ACL and Advanced Schottky Parts at $V_{CC} = 4.5$ V, $T_A = 25°C$

	74AS	74F	74ALS	74AC	74ACT
$V_{OH(min)}$, V	2.5	2.4	2.5	4.4	4.4
$V_{IH(min)}$, V	2	2	2	3.15	2
NM_H, V	.5	.4	.5	1.25	2.4
$V_{OL(max)}$, V	.5	.5	.5	.1	.1
$V_{IL(max)}$, V	.8	.8	.8	1.35	.8
NM_L, V	.3	.3	.3	1.25	.7

to avoid bus contention, only one of the devices sharing a bus can place information on the bus at any given time. Two classes of devices that permit the simultaneous connection of the outputs of two or more devices to form a bus are TTL open-collector devices and TTL and CMOS three-state devices. In digital systems, a certain line may not have sufficient output current to drive all the output lines connected to it. In that case, a current amplifier called a driver or a buffer may be employed.

Optimum performance of digital systems may be accomplished by utilizing more than one logic subfamily. Therefore, it may be necessary to interface a logic subfamily to other types of logic. When interfacing different logic series, both the voltage levels and the current requirements of each subfamily must be taken into account.

References

1. L. Wakeman. "An Introduction to and Comparison of 54HCT/74HCT TTL Compatible CMOS Logic." *Application Note 368.* National Semiconductor, Santa Clara, CA: March 1984.
2. L. Wakeman. "Interfacing to MM54HC/MM74HC High-Speed CMOS Logic." *Application Note 314.* National Semiconductor, Santa Clara, CA: June 1983.
3. L. Wakeman. "Comparison of MM54HC/MM74HC to 54LS/74LS, 54S/74S and 54ALS/74ALS Logic." *Application Note 319.* National Semiconductor, Santa Clara, CA: June 1983.
4. B. Fowler. "Safe Operating Areas for Peripheral Drivers." *Application Note 213.* National Semiconductor, Santa Clara, CA: October 1978.
5. J. Huang. "Designing with MM74C908, MM74C918 Dual High Voltage CMOS Drivers." *Application Note 177.* National Semiconductor, Santa Clara, CA: March 1977.
6. B. Blood. "Interfacing with MECL 10,000 Integrated Circuits." *Application Note 720.* Motorola Semiconductor Products, Inc., Phoenix, AZ: 1974.

PROBLEMS

7.1 In Figure 7.20, construct the load line for $V_{CC} = 10$ V and comment on the safety of the circuit operation.

7.2 We already know that $R_{L(min)} = 11.1\ \Omega$ for 74C908 at $T_A = 25°C$. Find $R_{L(min)}$ at $T_A = 45°C$, 65°C, and 85°C for $V_{CC} = 5$ V and $V_{CC} = 10$ V. Plot $R_{L(min)}$ as a function of T_A for both values of the power supply voltage.

7.3 Assume that in Figure 7.19 the drivers are not operating identically, so that the driver *A* has to deliver 200 mA to its load while driver *B* needs 100 mA. Examination of Figure 7.20 reveals that drivers with high R_{on} values cannot deliver 200 mA. However, we can reduce the power consumed in section *B* to compensate for the higher power requirement in section *A*. Using the graphical analysis, find R_{LA} and R_{LB} when $V_{CC} = 5$ V.

7.4 In Problem 7.3, although the direct use of the graphical method results in meeting the design requirements, there is not much margin left for tolerance in resistances and other circuit parameters by pushing at the power limits of the 74C908 package. To ease this restraint, we can either increase the power supply or use a higher power package such as 74C918, which is capable of delivering 2.27 W with $\Theta_{jA} = 55°C/W$. Repeat Problem 7.3 using this chip.

7.5 Repeat Problem 7.4 for $V_{CC} = 10$ V assuming that drivers *A* and *B* of 74C918 have to deliver 250 mA and 150 mA, respectively. What standard load-resistance values with 5% tolerance will guarantee satisfactory performance of the circuit?

7.6 Suppose that an output from a standard 7400-series TTL IC is driving two standard TTL loads. How many additional LS TTL loads can be connected to it?

7.7 Why in the standard TTL is the logical **1** range, which extends over 3 V, wider than the **0** range, which spans only .8 V?

7.8 An LED with a forward voltage of 1.6 V at 18 mA is connected for troubleshooting between the output of a 10K ECL NOR gate and the supply voltage.

a. Verify the proper operation of the circuit.
b. Find the value of the limiting resistor for proper operation. Note that gates in this series are specified as driving 50-Ω loads to −2 V.
c. Discuss the advisability of driving another ECL gate from an output driving an LED.

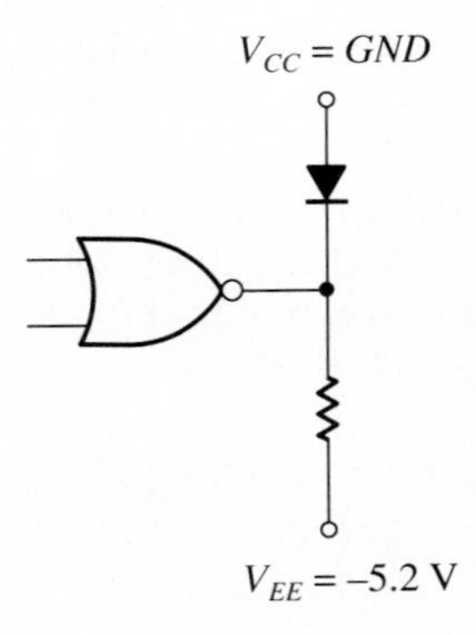

7.9 Select a pull-up resistor for a 7401 that will simulate the effect of a totem-pole source driver. How much fan-out current is available to drive other gates? How many 7401s can be driven?

7.10 Explain why so many of the buffers/drivers listed in Table 7.2 have control lines that are active low.

7.11 The 75451 is driving an incandescent lamp that is equivalent to a reactive load whose electrical model (a) and dc characteristic (b) are shown in the diagram. Note that the knee in the characteristic is at 2 V, where the power starts to be dissipated in the form of light.

a. Find an expression for the transient lamp current and sketch the corresponding waveform. Will the device be able to handle the peak current?
b. Calculate the energy dissipated by the driver in 100 ms.
c. To reduce the initial peak current, the lamp is biased to be partially on, so that the lamp filament will be warm, but the lamp will not emit light, as illustrated in the diagram (c). For the initial lamp voltage of 1 V, calculate R_B and the peak lamp current.

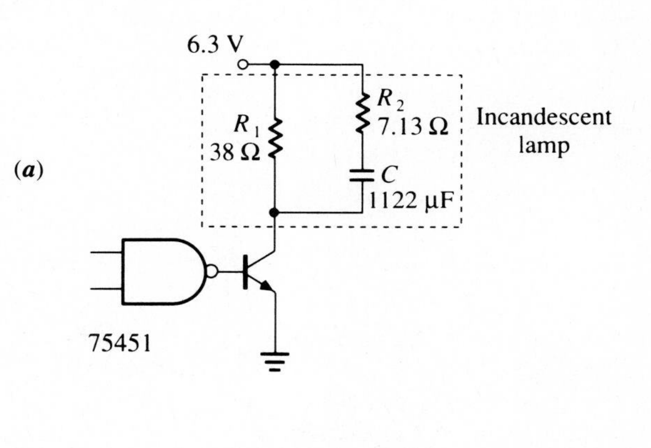

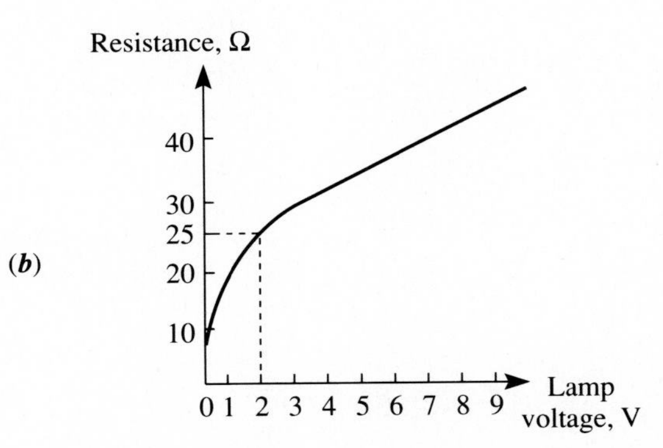

Problem 7.11 continued on p. 342

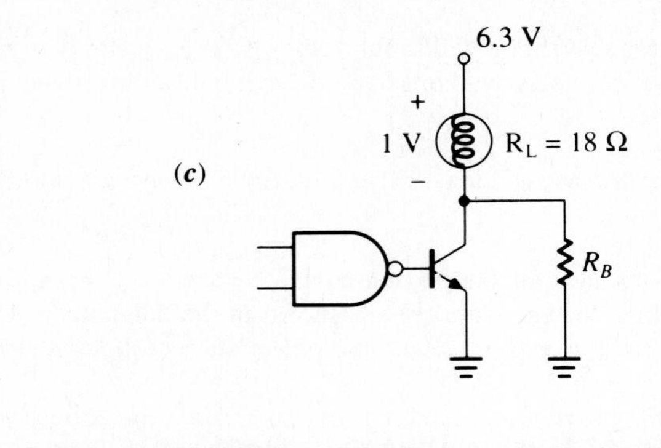

7.12 Using Table 7.4, determine an appropriate value for the pull-up resistor R_L that will allow three Schottky open-collector gates to drive eight similar gates.

8. DIGITAL REGENERATIVE CIRCUITS

Multivibrators are *regenerative* circuits that are used in timing applications. They are classified as

1. *Bistable* circuits
2. *Monostable* circuits
3. *Astable* circuits

Latches, *flip-flops*, and regenerative comparators, also called *Schmitt triggers*, are all bistable circuits. The most important characteristic of a bistable circuit is its maintenance of a given output state as long as there is no applied external signal. Application of an appropriate external trigger causes a change of state, and this level is maintained until a second signal is applied.

The monostable multivibrator, also referred to as a *one-shot*, generates a single pulse of specified duration in response to a trigger signal application. The monostable multivibrator remains in this *quasistable* state for a predetermined interval of time, after which it reverts to its original stable state without any external signal to induce this reverse transition. The duration of the pulse is not related to the properties of the triggering pulse.

The astable multivibrator has no stable states, and no external signal is required to produce changes in its state. It has two quasistable states, and the circuit remains in each state for predetermined intervals. Since it oscillates between states, it is used to generate periodic pulses such as the clock generators in digital systems. Astable and monostable multivibrators can be designed using operational amplifiers. However, here we concentrate mainly on the discussion of these circuits using logic gates.

Consider the common multivibrator configuration of Figure 8.1, made up of two inverting amplifiers *A1* and *A2* forming a positive feedback along with two couplers, *C1* and *C2*. From the voltage transfer characteristic of a practical inverter, as depicted in Figure 0.4, we observe that the magnitude of the slope of the characteristic is greater than unity in the *unstable* transition region. Since the slope is a measure of

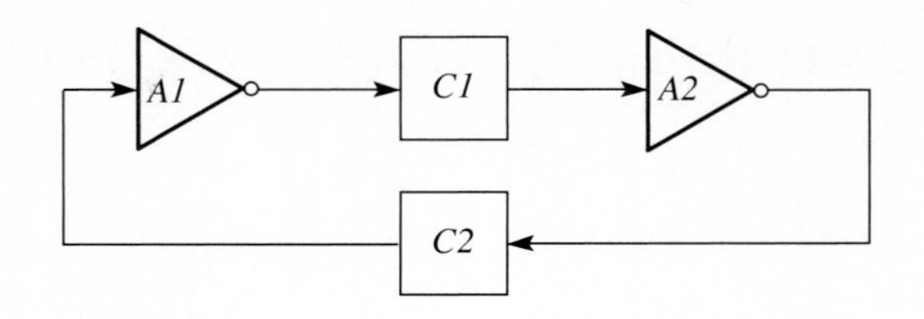

Figure 8.1 The basic multivibrator block diagram.

the voltage gain between the input and output, we can conclude that a voltage gain greater than 1 is needed to change states. Thus, in Figure 8.1, the two inverters provide a positive feedback, that is, a regenerative action, as they switch between the two logic levels. The two coupling networks, on the other hand, determine the type of the multivibrator. Resistive couplers correspond to bistable operation. A signal that is applied to the first inverter *A1* and that can produce a transition is transmitted via *C1* and causes *A2* to change state until another trigger signal is applied.

If either of the coupling networks contains a series capacitor, dc signals cannot be transmitted indefinitely. Since the voltage across the capacitor cannot change instantaneously, transition in *A1* or *A2* is transmitted for a short duration. The action of the capacitor charging as a result of the initial transition generates an internal trigger, leading to a circuit operation to revert to its initial state. Monostable circuits employ a single capacitive coupler; in an astable multivibrator, both coupling networks are capacitive.

8.1 Bistable Multivibrators

The Set-Reset (SR) Latch

In many digital systems, the need to keep track of any sequence of events requires circuits capable of storing binary signals. These circuits constitute the *memory*, and the logic circuits that incorporate memory are called *sequential circuits*. The basic digital memory element is obtained by cross-coupling two logic inverters, that is, NOT gates, as diagrammed in Figure 8.2a, and is called a *latch*. The most important property of the latch is that it is a bistable circuit: it can exist in one of two stable states. The inverters form a positive feedback loop.

To analyze the behavior of the latch, we break the loop at the input of *G1* and apply a signal V_I, as in Figure 8.2b. In order not to change the original loop voltage transfer characteristic, V_O vs. V_I, we assume that the input impedance of *G1* is large. The transfer characteristic, as shown in Figure 8.2c, is of two cascaded inverters. The superimposed straight line $V_O = V_I$ corresponds to the closing of the feedback loop. With the voltage gain being greater than unity in the unstable transition region and hence causing a positive feedback, the latch obviously cannot operate at point *B*. Thus, depending on the polarity of the initial voltage increment at the input, the operating point shifts either to point *A* or *C*, where no regenerative action can take place because the loop gain is close to zero at these points. As expected, the latch

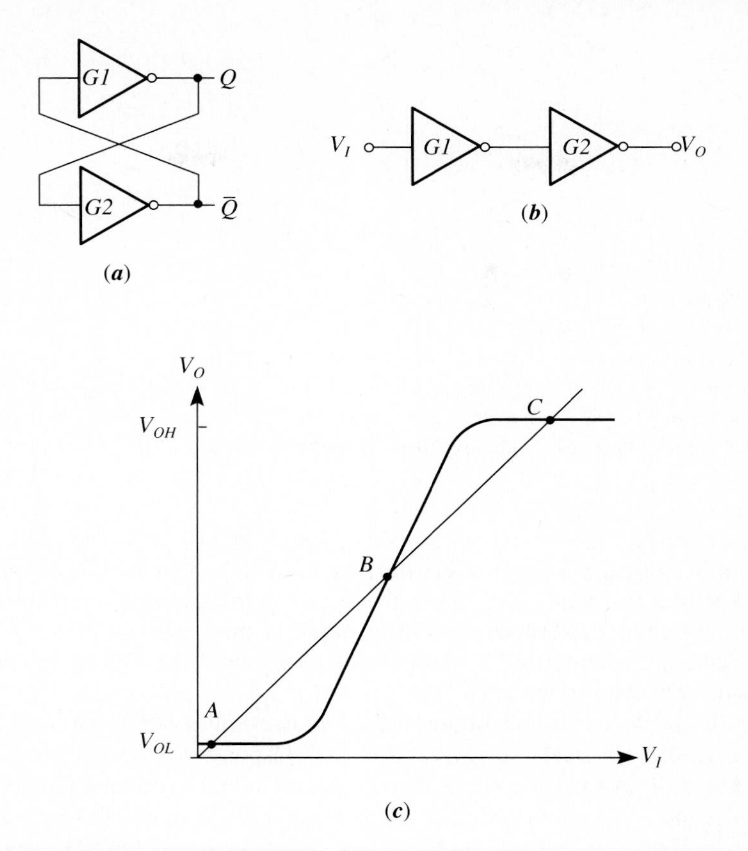

Figure 8.2 (a) Basic bistable circuit using two inverters;
(b) breaking the loop at the input of *G1* to apply a signal; and
(c) the VTC of the latch.

has two stable operating points. At point C, V_I and V_O are high, while the reverse is true at point A.

Note that the gate outputs are complements of each other in either stable state. The external excitation forces the latch to a particular state in which the latch stays indefinitely, thereby *remembering* or *memorizing* this external action. Thus, the latch stores one bit of information, corresponding to either $Q = \mathbf{1}$ or $Q = \mathbf{0}$.

In order to excite the latch, we need to incorporate a triggering circuitry. One simple way is to form the cross-coupling using two NOR gates. The second input of each NOR gate provides a means of triggering the latch from one stable state to another. This configuration is a simple bistable multivibrator, called a *set-reset* (*SR*) latch, and is shown in Figure 8.3a. It is the common building block of more complex bistable multivibrators such as flip-flops.

The SR latch is said to be *set* when the output is logical **1**, and *reset* when the same output is logical **0**. From Figure 8.3a, we see that a **1** at the S input sets the

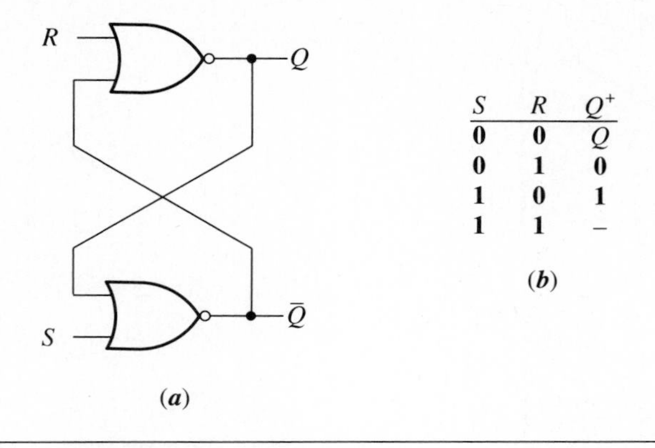

S	R	Q^+
0	0	Q
0	1	0
1	0	1
1	1	–

Figure 8.3 (a) The basic SR latch and (b) its characteristic table.

latch, while a **1** at the R input resets, or *clears*, it. Since the latch responds to high voltage levels at the inputs, they are called *active high* inputs. When both inputs are **1**, the outputs are not complements of each other, so this combination is not allowed. These results are summarized in the characteristic table of Figure 8.3b, where Q^+ denotes the next state of the latch.

The SR latch can also be designed using two two-input NAND gates, as in Figure 8.4a. The reader can easily verify that this latch responds to *active low* inputs. Note from Figure 8.4b that when both the $\overline{S}$ and $\overline{R}$ inputs are returned low, the output level is indeterminate. The circuit diagram of the low-power Schottky TTL version of the $\overline{\mathrm{S}}\ \overline{\mathrm{R}}$ latch is shown in Figure 8.5. It is made up of two-input NAND gates that are cross-coupled through the collector of the phase-splitter transistor. Hence, the regeneration is independent of any load at the output of the latch. The 74LS279 consists

Figure 8.4 (a) The NAND latch, and (b) its characteristic table.

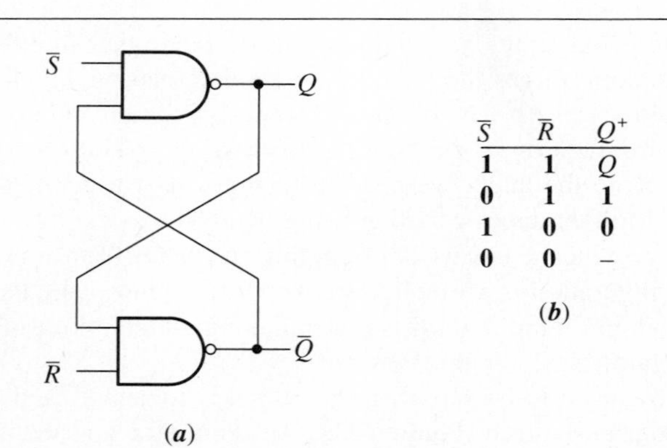

$\overline{S}$	$\overline{R}$	Q^+
1	1	Q
0	1	1
1	0	0
0	0	–

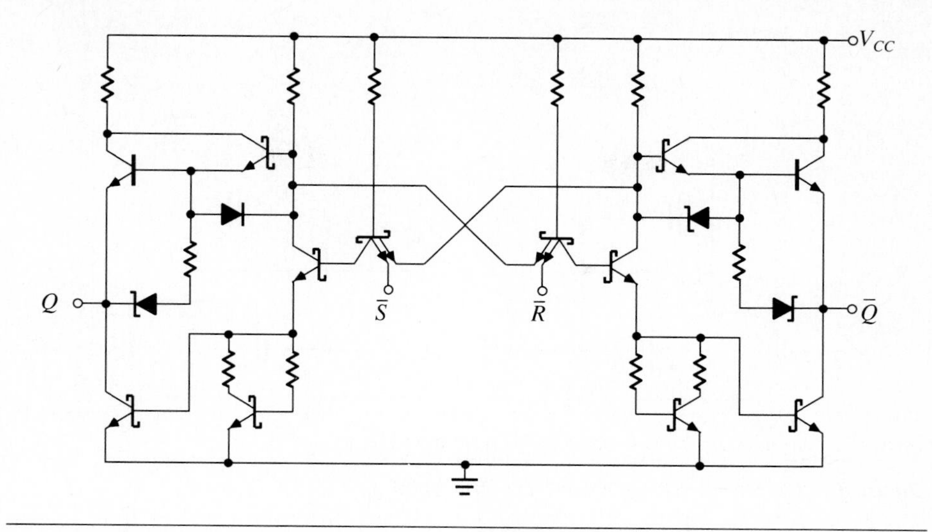

Figure 8.5 The circuit diagram of the LS TTL NAND latch.

of four NAND latches, two of which are identical to that of Figure 8.5. The other two have double $\overline{S}$ inputs, $\overline{S1}$ and $\overline{S2}$.

The ECL latch in Figure 8.6 uses the NOR outputs of the basic OR/NOR gate; the CMOS realization in Figure 8.7 employs a coupled NOR-gate topology. A simple I^2L NAND latch is shown in Figure 8.8. To avoid the connection of two input bases in parallel, which may result in increased collector current for the driver and hence fan-out problems, two collectors are used for each output inverter.

Figure 8.6 The ECL SR latch using the basic OR/NOR gates.

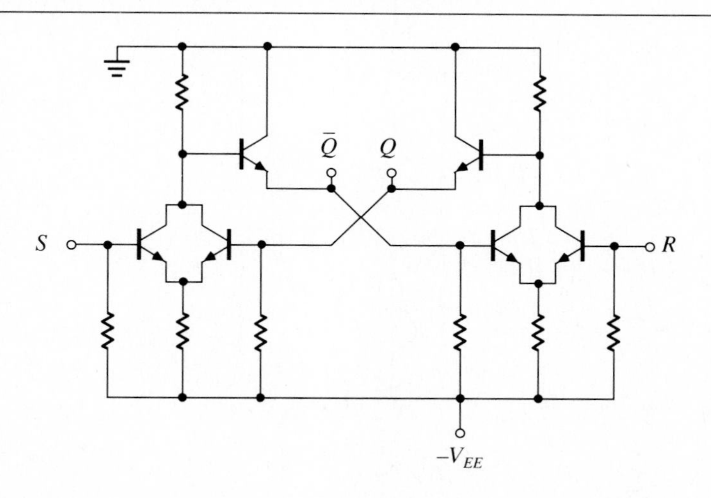

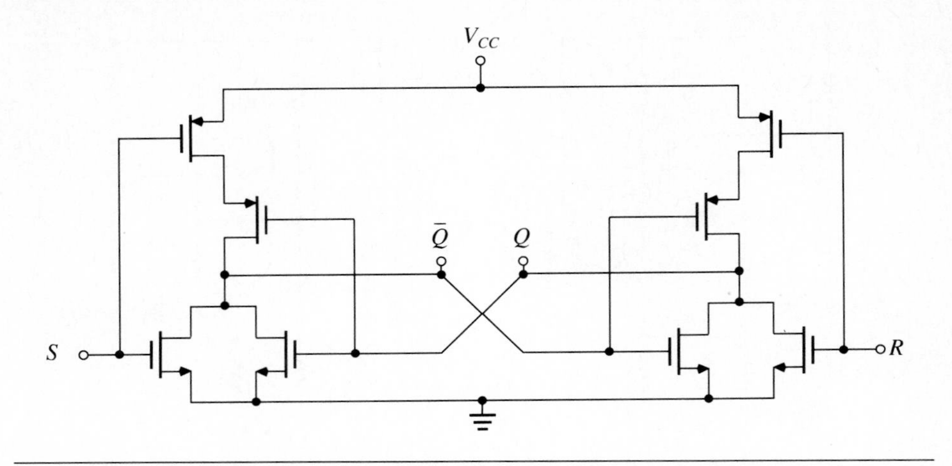

Figure 8.7 The CMOS SR latch using the basic NOR gates.

The JK Flip-Flop

By augmenting the basic SR latch with two AND gates and additional feedback lines, as depicted in Figure 8.9, the ambiguity in its characteristic table can be removed. This versatile and widely used circuit is known as a *JK flip-flop* (*JKFF*). Note the important addition of a clock input to synchronize the output logic states with a system clock. The inputs *J* and *K* are thus called the *synchronous inputs*. The

Figure 8.8 The I²L NAND latch.

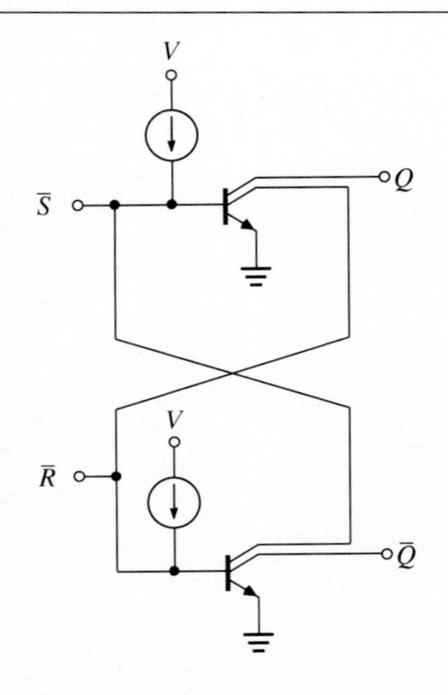

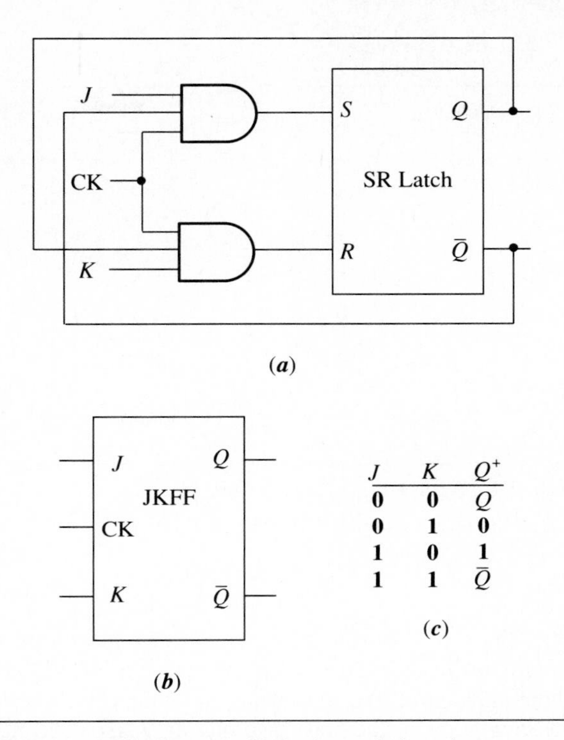

Figure 8.9 The JKFF: (a) logic diagram, (b) logic symbol, and (c) characteristic table.

first three rows of the characteristic table in Figure 8.9 are identical with the corresponding rows for an SR latch. However, if both J and K are high, the output is complemented by the clock pulse; that is, the unit *toggles* when clocked. When operated in this mode, the flip-flop is called a *T flip-flop* or TFF.

An all-NAND version of the JKFF is shown in Figure 8.10. There is an inherent problem with these constructions of the flip-flop. The operation will be unstable if the clock pulse is longer than the time required for the flip-flop to change state. For example, if $J = K = \mathbf{1}$ and a clock pulse is applied, then the output will oscillate back and forth between two logic levels unless the pulse width of the clock is less than the propagation delay time of the flip-flop. This is called a *race-around condition*. However, with today's IC technology the propagation delay is usually much less than the clock pulse width of the system; thus, the above restriction may not be satisfied, and the output may be indeterminate.

The Master-Slave JKFF

One way to circumvent this problem is to use the *master-slave* principle by simply cascading two JKFFs, with the master driving the slave, as illustrated in Figure 8.11a. Note the existence of the feedback from the output of the slave to the input of the master. A positive clock pulse is applied to the master and is complemented before

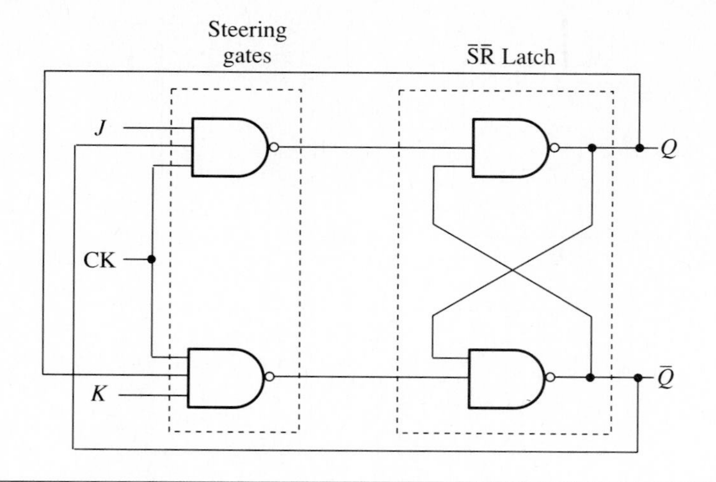

Figure 8.10 The logic diagram for the all-NAND version of the JKFF.

being used to excite the slave. Consider the clock pulse waveform of Figure 8.11b. As CK rises, $\overline{CK}$ falls to disable (at time t_a) the steering NAND gates of the slave, thereby isolating the slave from the master and maintaining its present state. At t_b, CK enables the steering gates of the master so that the master flip-flop can change its state, depending on the J, K and Q values. On the falling edge of the clock pulse, at time t_c, the master is inhibited. Finally, at t_d, the NAND gates to the slave are enabled so that the slave follows the state of the master $\overline{S}\ \overline{R}$ latch. Even though the clock pulse must be wider than the propagation delay time through the master flip-flop, there is no upper limit to its width, and the circuit functions properly for all combinations of input and output values as long as the data in the J and K inputs remains constant for the pulse duration. Otherwise, an improper operation may result. One way to overcome this restriction is to use *edge-triggered* flip-flops, as discussed below.

Clocked IC flip-flops often have so-called *asynchronous* inputs, which can be used to set or reset the flip-flop independent of the clock. The addition of the dashed PRESET and CLEAR inputs in Figure 8.11a allows the initial state of the flip-flop to be assigned. These inputs override the clock and synchronous inputs. The *bubble* on the CK line indicates that the change of the flip-flop state always occurs on the trailing edge of the clock pulse. Also note that the asynchronous inputs are active low signals.

JKFFs with Data Lockout

An early solution to the shortcoming of the master-slave operation was the JKFF with *data lockout*. In this type of flip-flop, the inputs are enabled only during a short period starting with and immediately following the rising edge of the clock pulse. After this, even if the inputs are changed while the clock is high, the state of the

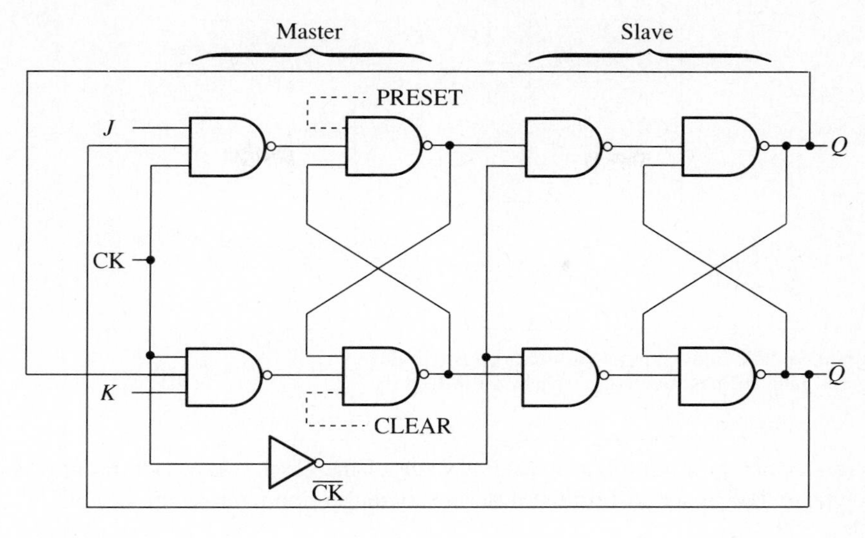

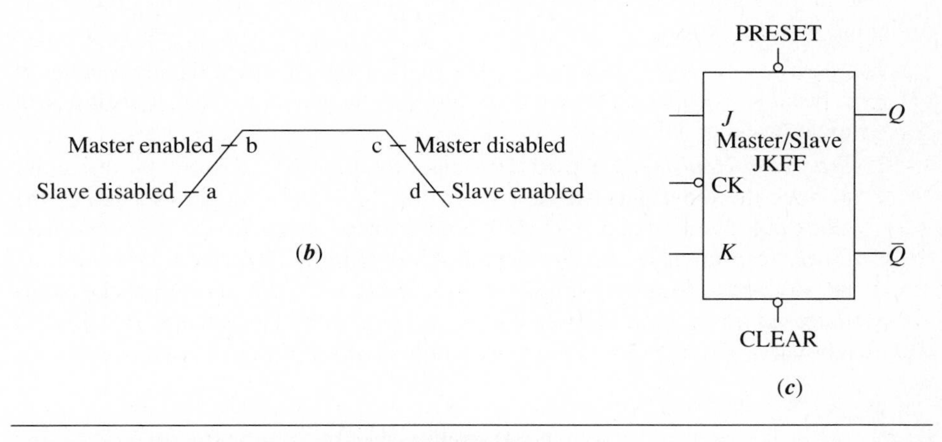

Figure 8.11 The master-slave JKFF: (a) logic diagram, (b) clock pulse waveform, and (c) logic symbol.

master is not affected. The flip-flop's and hence the whole system's susceptibility to noise are thereby effectively reduced. As before, at the threshold level of the falling edge of the clock pulse, the data stored in the master is transferred to the output.

Edge-Triggered Flip-Flops

The logic symbol for an edge-triggered JKFF is shown in Figure 8.12. It is differentiated from the master-slave by a small arrowhead (>) at the clock input. The presence or absence of the small complementary circle indicates whether the state

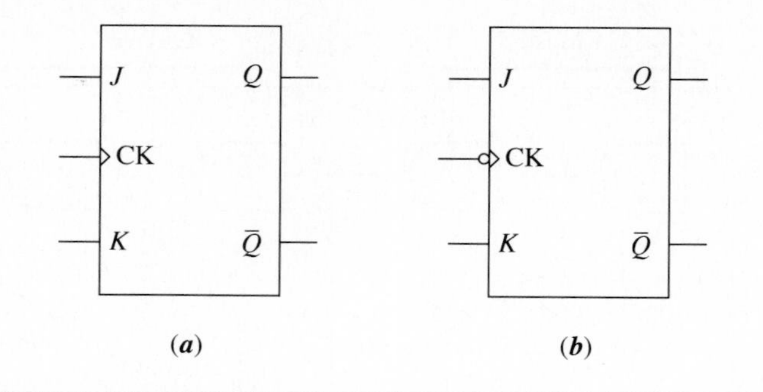

Figure 8.12 The logic symbol for the (a) positive-edge triggered, and (b) negative-edge triggered flip-flop.

changes occur on the trailing or leading edge of the clock pulse. Before discussing this type of JKFF, we will briefly describe some flip-flop parameters.

Flip-Flop Parameters

The following parameters specified by the manufacturers should be observed in designing with flip-flops:

1. *Maximum clock frequency*, f_{max}: The highest rate at which the clock input of a bistable circuit can be driven while still maintaining stable transitions of logic levels at the output.
2. *Setup and Hold times*, t_{su}, t_h: To trigger the flip-flop, it is usually necessary to have the input data arrive a short time before the triggering edge of the clock pulse and remain a short time after; these are called the *setup* and *hold* times, respectively, and are illustrated in Figure 8.13 for both positive- and negative-edge-triggered flip-flops. Manufacturers' data sheets list the minimum values in reference to a particular edge of the clock that is the shortest interval for which the correct operation of the flip-flop is guaranteed. For

Figure 8.13 Setup and hold times for the (a) positive-edge-triggered, and (b) negative-edge-triggered flip-flop.

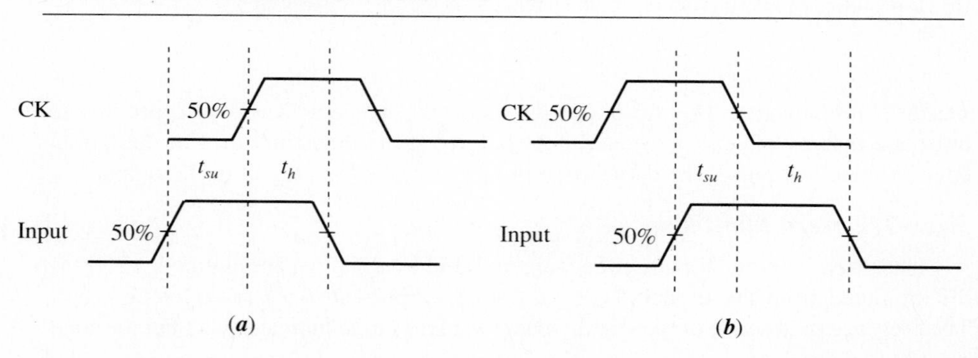

master-slave flip-flops with data lockout, the reference is to the triggering edge of the master. The input data must encompass the rising edge, even though it is transferred to the output at the falling edge of the clock. For other master-slave flip-flops, data can be set up any time prior to the positive transition of the clock and must be held until the negative transition to assure reliable triggering.

3. *Clock high and low pulse widths*, t_{wh}, t_{wl}: These are the minimum times the clock must remain in high and low states, respectively, for reliable clocking. Due to various delays within the chip, the clock has minimum pulse and space requirements.

The ac-Coupled Edge-Triggered Flip-Flop

Before considering the commercially-available edge-triggered flip-flops, we will discuss various approaches to achieve edge triggering for the basic SR latch. One way to accomplish this is by using the circuit of Figure 8.14, which is known as an *ac coupling circuit* because the capacitor C allows high-frequency sinusoidal signals to pass and blocks the direct current so that constant levels at the input have no effect on the output. As shown in Figure 8.14b, the circuit responds only to changes in the signal if the input is a pulse. Even if the transitions between the two levels at the input are exponentials with a rise time t_r, the circuit can still respond to the edges of the input pulse, and the output will attain a peak voltage approximately equal to that of the input as long as the circuit time constant $\tau = rC >> t_r$. However, as the rise

Figure 8.14 The effect of an ac coupling on a pulse input: (a) the circuit, and (b) input and output waveforms.

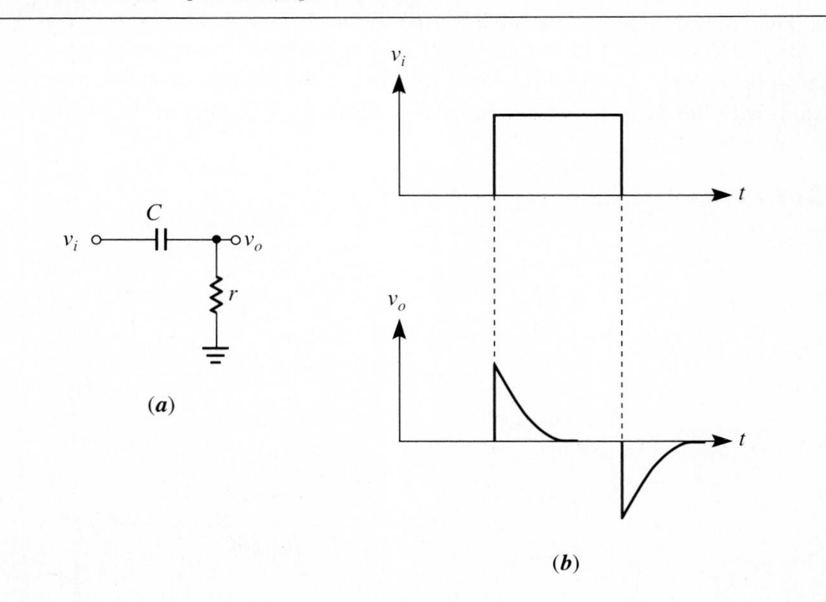

time of the input pulse increases, the pulse amplitude presented to S and R decreases. Eventually, when $t_r >> \tau$, the circuit acts like a differentiator.

Now consider the ac-coupled edge-triggered flip-flop of Figure 8.15, made up of the SR latch, the steering NAND gates, and the ac-coupling circuits in between them. Suppose that initially $S_D = \mathbf{0}$, $R_D = \mathbf{1}$, and clock is low. The output of *G1* is **1**, and hence $R = \mathbf{0}$ because the capacitor does not see a change at the input. The arrival of the clock pulse causes the output of *G1* to go low. Nevertheless, it has no effect on the flip-flop output because the negative-going transition, coupled by the capacitor to the R input of the latch, is still interpreted as logical **0**.

Next, the clock pulse makes its transition from high to low, forming a positive pulse at the NAND output that acts as a logical **1** and hence is coupled to the latch to reset it. Thus, the desired type of operation is obtained. The set input S_D operates in a similar way. One should note, however, that for the flip-flop to operate properly, the inputs to the latch should not be set to **1** at any time other than just after the clock pulse, which implies that the input signals cannot change during the clock pulse.

The Propagation Delay Edge-Triggered Flip-Flop

The technique employed with TTL parts to construct an edge-triggered flip-flop relies upon gate propagation delays instead of capacitors. The basic component of this type of flip-flop is shown in Figure 8.16. Independent of the A input, the output is **0** if $B = \mathbf{1}$. Assume that $B = \mathbf{0}$. If $A = \mathbf{0}$, the output will be a logical **0**. If $A = \mathbf{1}$, then Z is still **0**. Thus, it seems as if the output is always a **0**. Actually, this is not the case when propagation delays come into play.

Suppose that the A input switches on at time t_1, as depicted in Figure 8.16b. A propagation delay t_{PHL} seconds later, the NOR output goes low. Therefore, both inputs to the AND gate are **1** for a short period of time, approximately equal to t_{PHL}, during which the output will be **1** as long as the AND gate is faster than the NOR gate. The corresponding waveform is also shown in Figure 8.16b.

At t_2, A goes back to **0**. Since the NOR output does not switch to **1** before A has fallen to **0**, the circuit output does not respond to the change in A. Further, it can be verified that with $B = \mathbf{1}$, neither edge of A will have an effect at the output.

Figure 8.15 Ac-coupled edge-triggered flip-flop.

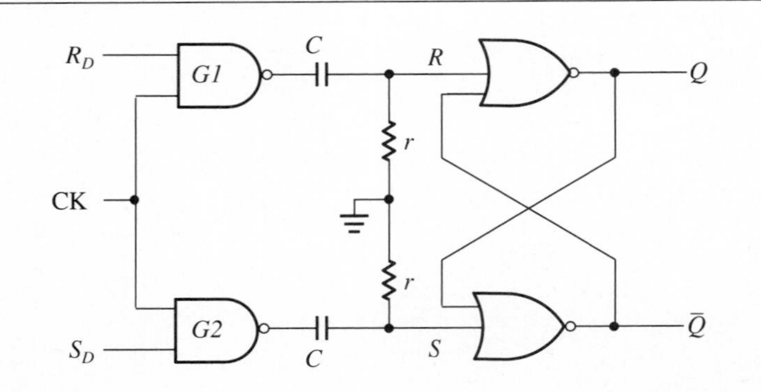

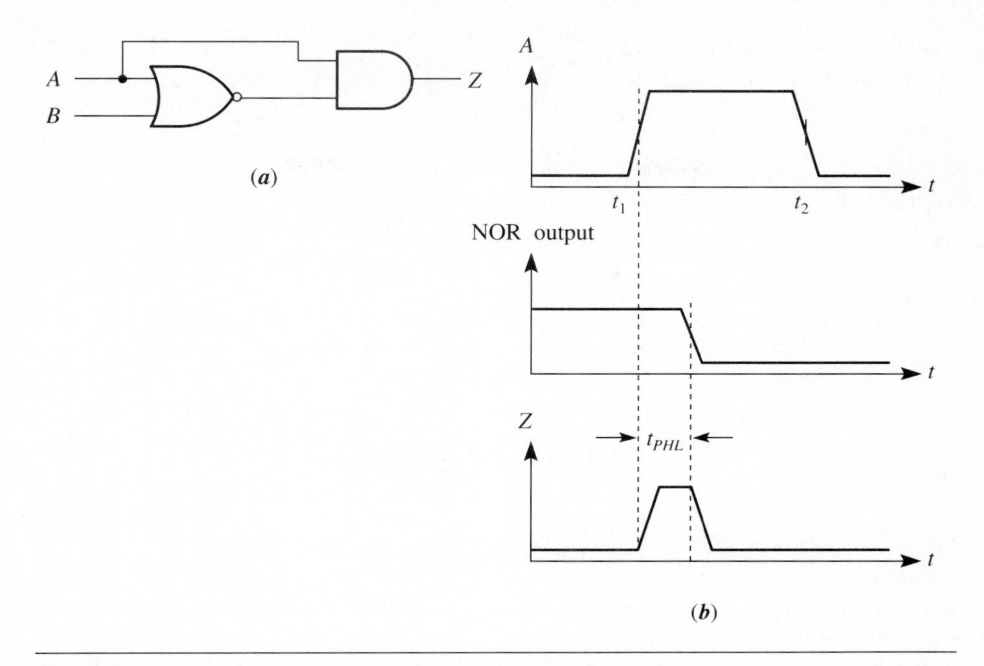

Figure 8.16 (a) The basic component of a propagation-delay edge-triggered flip-flop; (b) waveforms when $B = 0$.

In this circuit, A represents the clock pulse, which is subject to certain restrictions. Suppose that just before it switches on, B goes low. In order for the NOR gate output to switch to **1** and ensure proper operation, the setup time t_{su} should be sufficiently long. Also, if the rise time of A is too long, so that the NOR output falls to **0** before A rises to **1**, then both inputs to the AND gate cannot be **1** at the same time, and the output pulse does not result. Therefore, one should observe the limitations on the clock signal by the manufacturer.

A simple propagation delay edge-triggered flip-flop, based on the basic SR latch, is shown in Figure 8.17. Note that the primary inputs as well as the clock $\overline{\text{CK}}$ are all active-low.

Negative-Edge-Triggered JKFFs

The logic diagram of a negative-edge-triggered JKFF employed in LS/AS/F/ALS bipolar parts '112, '113, and '114 is shown in Figure 8.18. The only difference between these units is the presence or absence of the preset and/or clear inputs. The preset input, when available, is fed to gates *G2*, *G5*, and *G3*, while clear is tied to *G1*, *G6*, and *G4*. These asynchronous inputs are omitted for simplification. The NOR gates provide the complementary outputs, and only one of them is held high at a time due to the cross-coupling.

The clock signal CK is connected directly to *G5* and *G6*, and is complemented before being fed to *G3* or *G4* by the steering NAND gates *G1* and *G2*. Therefore,

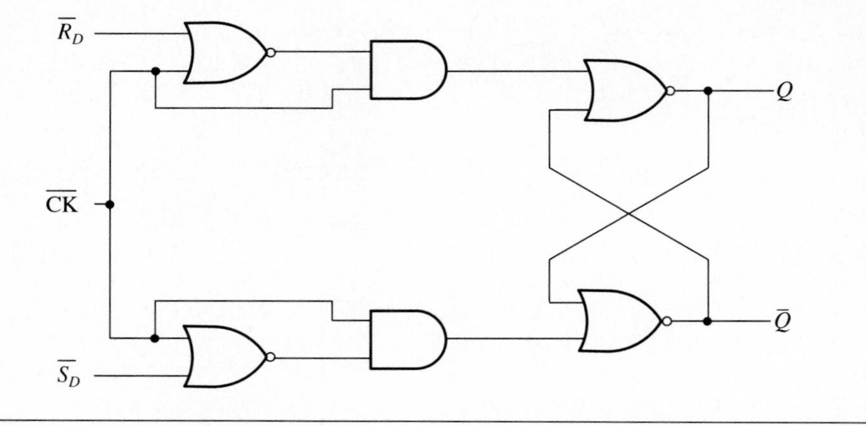

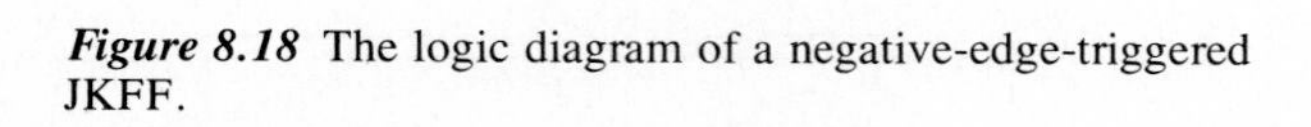

Figure 8.17 A simple propagation-delay edge-triggered flip-flop.

Figure 8.18 The logic diagram of a negative-edge-triggered JKFF.

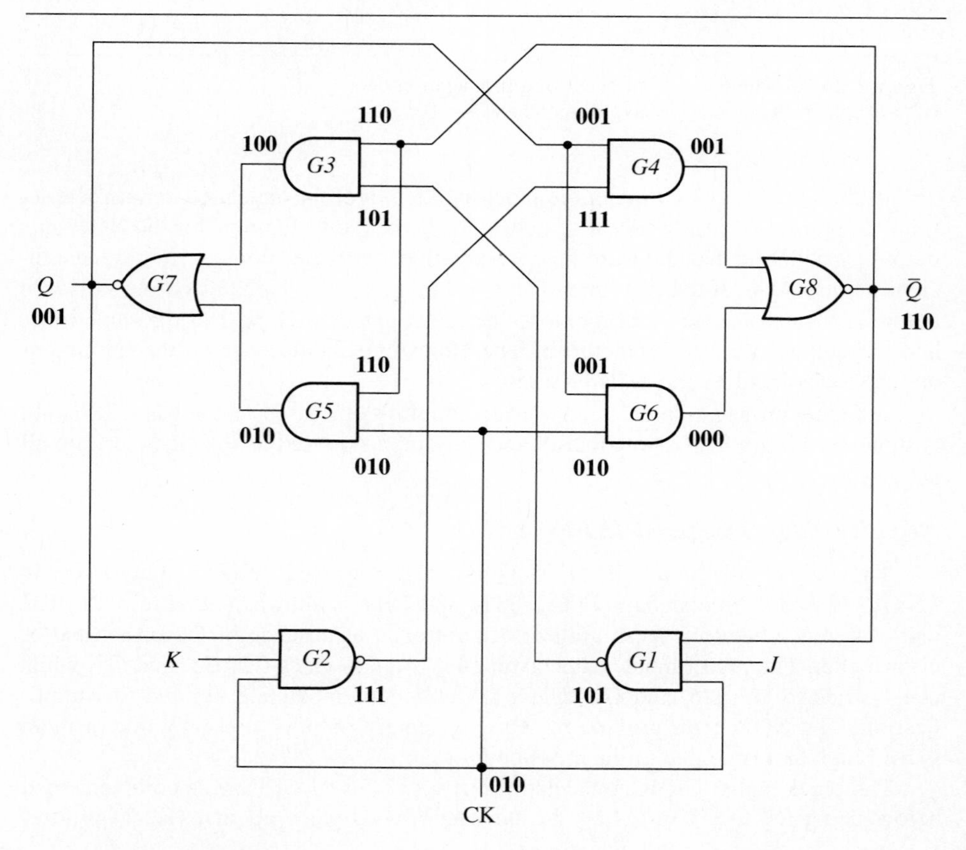

the complemented clock lags the original signal due to the propagation delay of the steering gates. These delays, coupled with the overlap of these clock pulses, inhibit any response until the trailing edge of CK, as the timing diagram of Figure 8.19 illustrates.

To analyze the behavior of the circuit, assume that the flip-flop is initially reset; that is, $Q = \mathbf{0}$ with $J = \mathbf{1}$ and $K = \mathbf{0}$. The values at various points in Figure 8.19 indicate the logical levels at three consecutive time intervals before, during, and after the clock pulse.

The appearance of the clock pulse causes the AND gate inputs to change. When the clock input to *G5* goes high, the complemented clock input to *G3* goes low, moving the logical **1** input from one leg of *G7* to the other. *G5* switches before *G3* because the clock is directly connected to the former. Consequently, *G7* has a high input during and after the transition, and Q remains low and hence continues to hold $\overline{Q}$ high through gates *G4*, *G6*, and *G8* while the clock is high.

When the clock returns to its low level during the third time interval, it disables

Figure 8.19 Waveforms for the flip-flop of Figure 8.18. The flip-flop is initially reset.

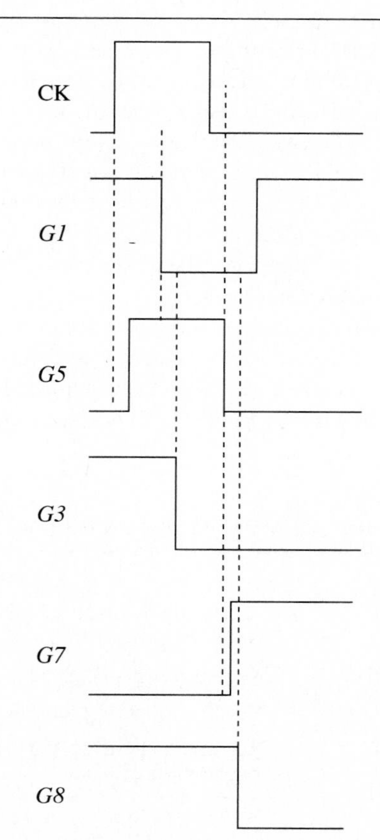

G5 while *G3* is still low due to the propagation delay of *G1* and the subsequent delay of the complemented clock. With both inputs low, Q goes high and pulls *G4* high, which has already been enabled by *G2*. Hence, $\overline{Q}$ goes low and imposes a low on *G3* before the complemented clock returns to high, thereby maintaining a high Q in conjunction with the low output from *G5*.

As is obvious from the ongoing discussion, the propagation delays in various parts of the circuit are crucial. Note that several gates must switch during the delay between the trailing edges of the direct and complemented clocks; thus, gates *G3* through *G7* must have very short delays as compared to those of the steering gates.

The flip-flop can be reset by reversing the function of the odd- and even-numbered gates. Note also that the connection between Q and *G2*, and $\overline{Q}$ and *G1* provide the cross-coupling necessary to toggle the flip-flop by opening one steering gate and closing the other when both J and K inputs are high.

A list of widely used TTL and CMOS dual JKFFs is given in Table 8.1. Table 8.2 shows the electrical characteristics of the commercial-grade '109 positive-edge-triggered J$\overline{\text{K}}$FF for various subfamilies.

Delay Flip-Flops

Another basic flip-flop available in the IC form is the clocked *delay flip-flop* (*DFF*). The state of the DFF after the clock pulse is equal to the input before the clock pulse. Therefore, it provides a delay such that the output does not change until a clock pulse is applied. One way to realize the clocked DFF is to modify the SR latch by adding an inverter and two AND gates, as in Figure 8.20.

The commercially available DFFs come in two types. They may be positive-edge-triggered, in which case the D input has no effect on the output when CK is either low or high. The input information is transferred to the output only on the rising edge of the clock pulse. Therefore, the next state of the flip-flop is equal to the D input value one setup time before clocking.

The second type of DFF is used in so-called *transparent* latches. The clock input is actually an enable input that allows the Q output to follow the D input only when it is high. When the enable is taken low, the output will be latched at the level of the latest data that is set up. The DFF of Figure 8.20 is an example for this type of usage.

Table 8.1 A List of TTL and CMOS Dual JKFFs

Part	Subfamilies	Description
'109	AS/F/ALS/HC/AC	Active low K input, positive-edge-triggered with clear and preset
'112	AS/F/ALS/HC	Negative-edge-triggered with clear and preset
'113	AS/F/ALS/HC	Negative-edge-triggered with preset
'114	AS/F/ALS/HC	Negative-edge-triggered with common clear, preset, and common clock

Table 8.2 Electrical Characteristics of '109 J$\overline{K}$FF

	AS	F	ALS	HC	AC
f_{max}, MHz	129	125	50	50	75
power/FF, mW	29	58.5	6	120	115
t_{su}, ns	5.5	3	15	20	4.5
t_h, ns	0	1	0	0	0
t_{wh}, ns	4	4	14.5	16	4.5
t_{wl}, ns	5.5	5	14.5	16	4.5

Example 8.1

Figure 8.21 illustrates the output waveforms of the two types of DFF for the same input D. When the enable of the latch is high, the output follows the level on the D input. When the enable goes low, Q will be whatever the value of D has been just before the negative transition of the enable. On the other hand, the positive-edge-triggered DFF takes a *snapshot* of the input at each rising clock edge.

The 7474 TTL DFF

The logic and circuit diagrams of the 7474, which contains two independent D-type positive-edge-triggered flip-flops with PRESET and CLEAR are illustrated in Figure 8.22. The asynchronous PRESET and CLEAR inputs are omitted for simplification from the logic diagram. A low level at either one of these inputs sets or resets the output regardless of the values at the other inputs. The output is a NAND latch, and gates *1A*, *1B*, and *2A*, *2B* are two interconnected steering latches.

When CLOCK = **0**, $\overline{S} = \overline{R} = \mathbf{1}$. Therefore, from the characteristic table of the $\overline{\text{S}}\,\overline{\text{R}}$ latch in Figure 8.3b, we see that the flip-flop does not change state no matter

Figure 8.20 The delay flip-flop that can be used as a *transparent* latch.

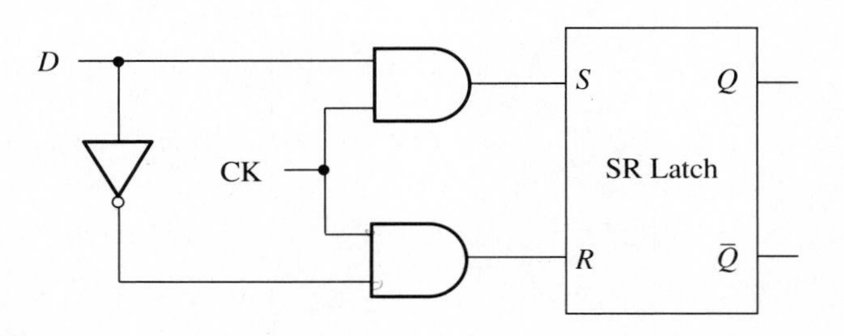

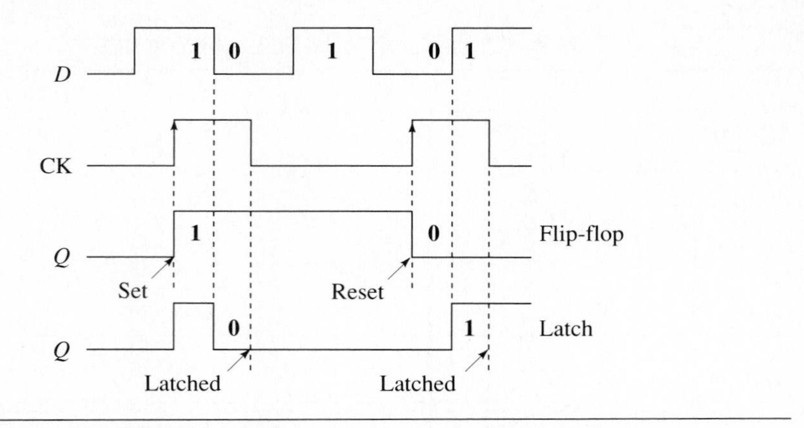

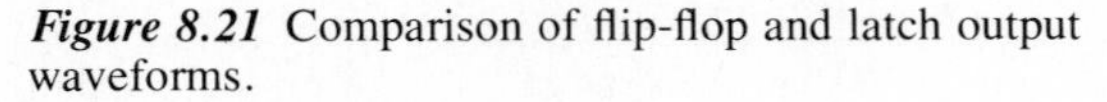
Figure 8.21 Comparison of flip-flop and latch output waveforms.

what the value of the D input is. If $D = \mathbf{0}$, both outputs of the lower left steering latch (gates $2A$, $2B$) are simultaneously high. This state is actually not allowed for the basic $\overline{\mathrm{S}}\,\overline{\mathrm{R}}$ latch. However, this situation does not lead to any indeterminate state because edge-triggering demands that the data at the D input be settled and waiting at the time the clock input goes high. Thus, as soon as the clock makes a positive-

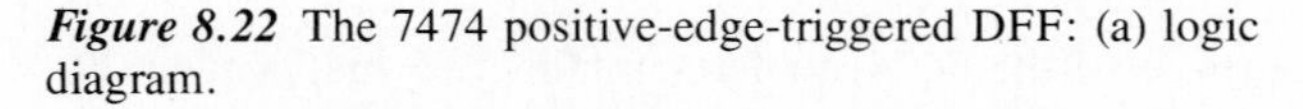
Figure 8.22 The 7474 positive-edge-triggered DFF: (a) logic diagram.

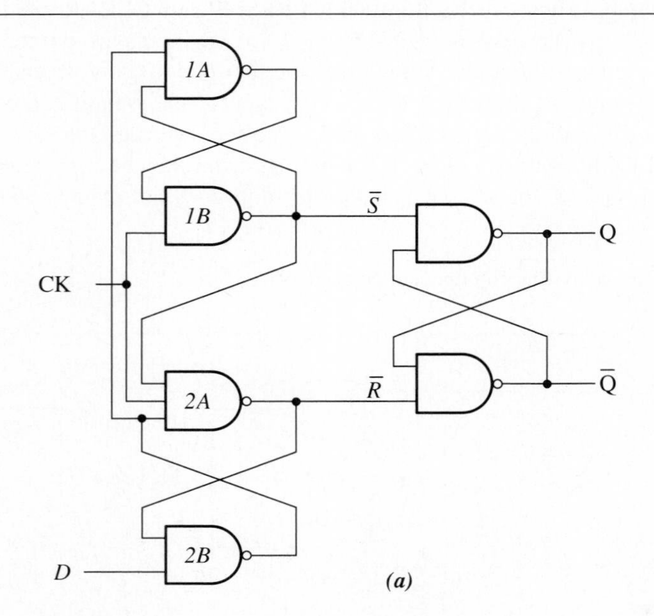

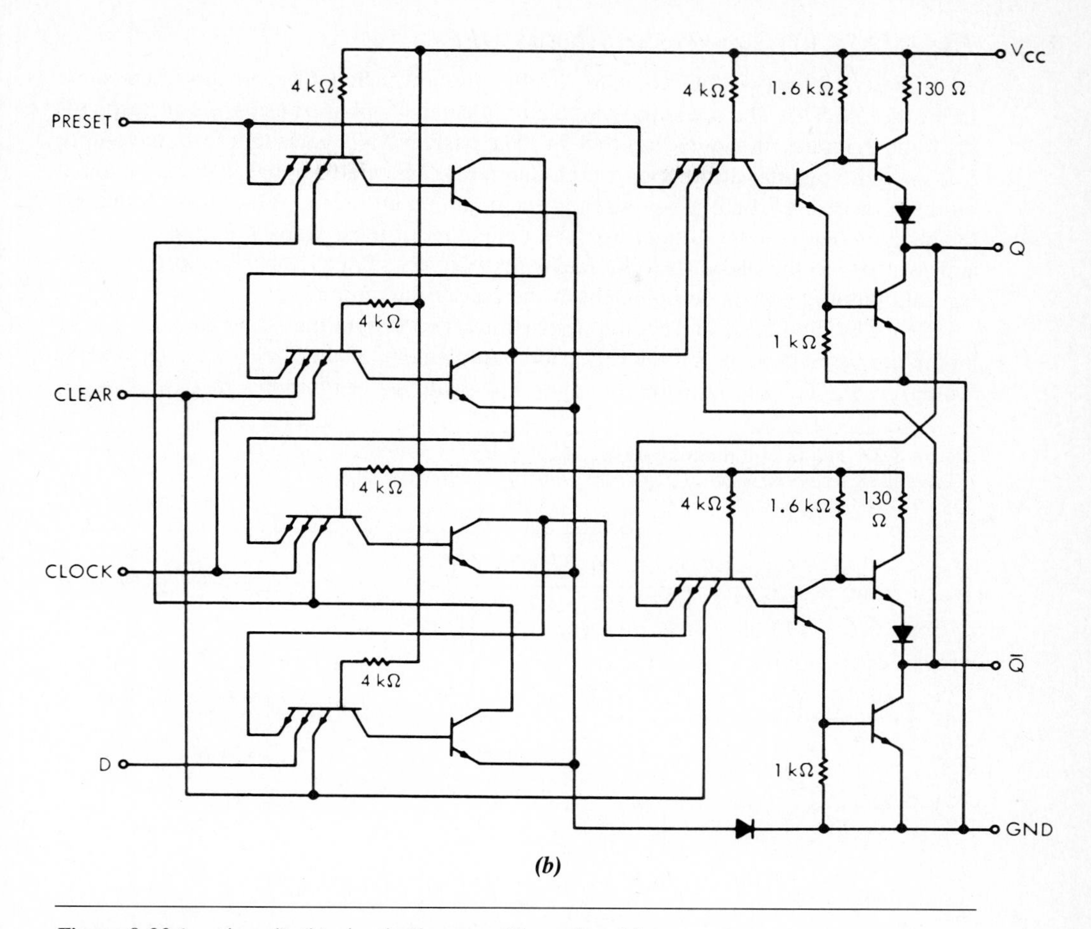

Figure 8.22 (continued) (b) circuit diagram. (*Reproduced by permission of Texas Instruments. Copyright © 1967, Texas Instruments.*)

going transition, gate *2A* flips to the **0** state so that the output latch is reset. Any subsequent change in *D* while the clock is high will not affect $\overline{R}$ because gate *2B* is now disabled by the **0** output from *2A*. Even if CLOCK returns to **0**, the output latch remembers its previous inputs.

A similar chain of events occurs when *D* is logical **1** at the time the clock goes high. In this case, the other steering latch is prepared for action with both outputs high, and it sets the output when the clock goes high.

In summary, when PRESET and CLEAR are high, data at the *D* input will be transferred to the output on the leading edge of the clock pulse as long as the input meets the setup time requirements. Following the hold time interval, the input data may be changed without affecting the levels at the output.

The 74LS74 TTL Low-Power Schottky DFF

Figure 8.23 shows the schematic of the low-power Schottky version of the same DFF, the 74LS74. The operation of the circuit can be explained using the corresponding logic diagram, shown in Figure 8.24. The basic NAND gates that form the output $\overline{S}\,\overline{R}$ latch are evident at the top of both diagrams. The clock signal is complemented by the transistor Q_3 before being fed to the common emitter of transistors Q_1 and Q_2 to achieve positive-edge-triggering. The collectors of these transistors provide the $\overline{S}$ and $\overline{R}$ inputs to the output NAND latch, respectively. Transistors Q_6 and Q_7 supply the data input D and its complement to the rest of the circuit.

Schottky diodes D_1 and D_2 and transistor Q_5 constitute the AND gate *G2* whose inputs are asynchronous preset $\overline{\text{PRE}}$, the complement of the input data, $\overline{D}$, and $\overline{S}$. Similarly, Q_4, D_3, and D_4 form the other AND gate *G1*, with inputs $\overline{R}$, $\overline{\text{CLR}}$, and D.

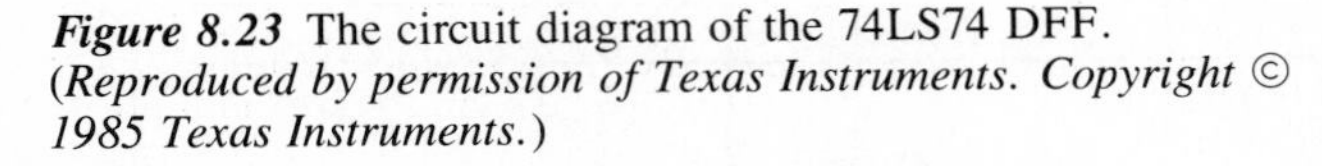

Figure 8.23 The circuit diagram of the 74LS74 DFF. *(Reproduced by permission of Texas Instruments. Copyright © 1985 Texas Instruments.)*

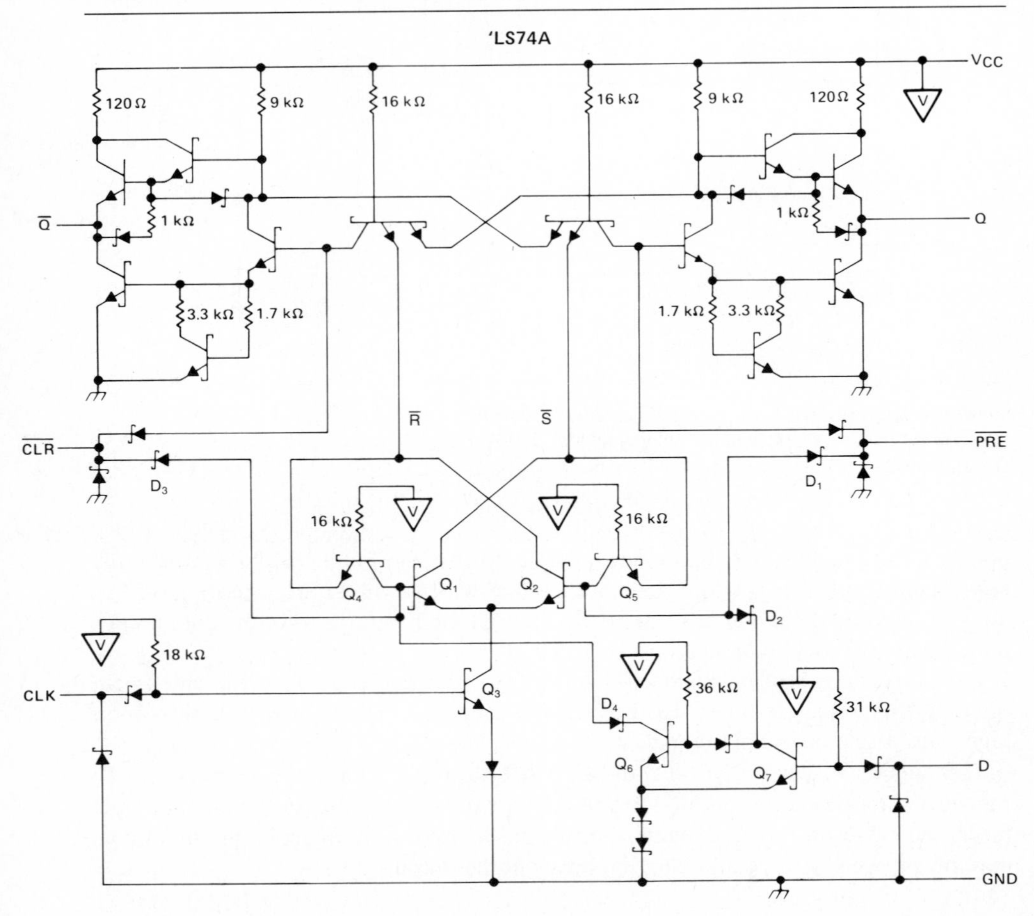

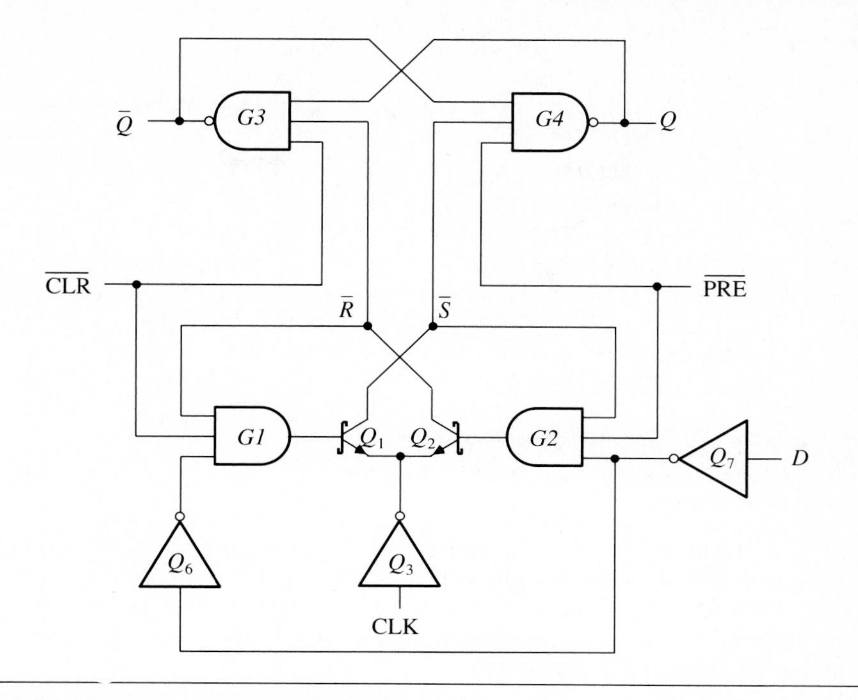

Figure 8.24 The logic diagram of the 74LS74 DFF.

Consider the synchronous operation with $\overline{\text{PRE}}$ and $\overline{\text{CLR}}$ high. As long as the clock line is low, the common emitter of Q_1 and Q_2 will be high, and both transistors will be off regardless of their base voltages. The state of the output NAND latch can only be changed by turning on Q_1 and Q_2, and this can only be brought about by the leading edge of the clock.

Let $D = \mathbf{0}$. As CLK makes a positive-going transition, Q_2 turns on because *G2* provides a logical **1** to its base while Q_1 remains cut off because of a disabled *G1* at its base. With Q_2 on, the current is pulled out of the $\bar{R}$ input to *G3*, and hence its output goes high. This causes the output of *G4* to go low due to the feedback. At the same time, the collector of Q_2 also conditions *G1* so that *D* is locked out at its input. Thus, any change in *D* while clock is high will not be able to alter the state of the flip-flop.

Similar action occurs when $D = \mathbf{1}$ and clock goes high. This time Q_1 turns on and $\bar{S}$ is pulled low to set the flip-flop. Note that the combination of *G1*, Q_1 or *G2*, Q_2 cannot be represented as a NAND gate.

CMOS Flip-Flops

The CMOS flip-flops are based not only on cross-coupled gates but also on *transmission gates* (*TGs*). A CMOS TG has four terminals and functions as a *single-pole, single-throw* (*SPST*) switch, as depicted in Figure 8.25. The opening and closing are controlled by the complementary gate voltages C and $\bar{C}$. If $C = \mathbf{1}$ and $A = \mathbf{1}$, then

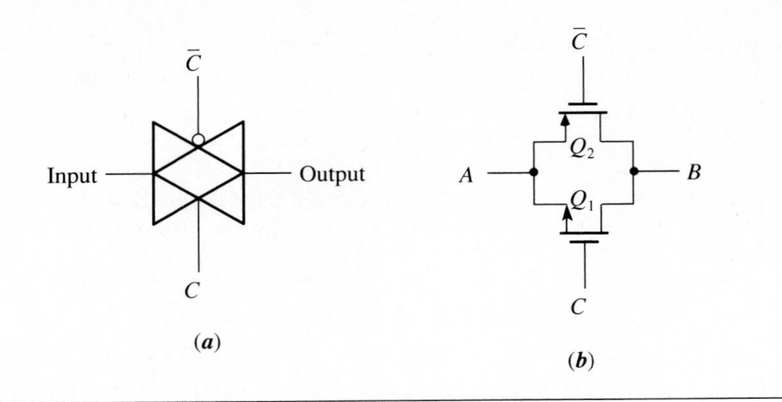

Figure 8.25 The CMOS transmission gate: (a) symbol, and (b) circuit.

$V_{GS1} = 0$ V and Q_1 is off while V_{GS2} is negative, and, since there is no drain voltage, Q_2 operates in the ohmic region and connects the output to the input. In a similar manner, if $C = \mathbf{1}$ and $A = \mathbf{0}$, then Q_2 is off, whereas Q_1 conducts and still $B = A$. On the other hand, if $C = \mathbf{0}$, independent of the value of the input, both transistors are off and no transmission is possible.

When $C = \mathbf{1}$, the total current through the TG is given by

$$I = I_{Dn} + I_{Sp}$$

while the voltage across it is

$$V_{DSn} = |V_{DSp}| = V_{CC} - V_O$$

so that the equivalent resistance is

$$R_{eq} \equiv \frac{V_{CC} - V_O}{I} = \frac{R_n R_p}{R_n + R_p} \tag{8.1}$$

where

$$R_n \equiv \frac{V_{CC} - V_O}{I_{Dn}} \tag{8.2a}$$

$$R_p \equiv \frac{V_{CC} - V_O}{I_{Sp}} \tag{8.2b}$$

are the resistances of the individual transistors. As long as $V_O < |V_{Tp}|$, both devices are saturated, and the equivalent resistances are

$$R_n = \frac{2(V_{CC} - V_O)}{k_n(V_{CC} - V_O - V_{Tn})^2}$$

$$R_p = \frac{2(V_{CC} - V_O)}{k_p(V_{CC} - |V_{Tp}|)^2}$$

The PMOS does not have any body-bias effect because $V_{BSp} = 0$; that is, $|V_{Tp}| = V_{TOp}$. However, this effect must be included for the NMOS transistor because $V_{SBn} = V_O$. Thus,

$$V_{Tn} = V_{TOn} + \gamma(\sqrt{2|\phi_f| + V_O} - \sqrt{2|\phi_f|})$$

When V_O reaches $|V_{Tp}|$, the p-channel device goes into the ohmic region and remains there until $V_O = V_{CC}$ is reached so that its equivalent resistance changes to

$$R_p = \frac{2}{k_p[2(V_{CC} - |V_{Tp}|) - (V_{CC} - V_O)]}$$

Finally, when $V_O > (V_{CC} - V_{Tn})$, the NMOS device will be in cutoff, so that $R_n \to \infty$ and $R_{eq} \approx R_p$. Meanwhile, the nonsaturated Q_2 controls the current flow.

Suppose that the output node of the TG in Figure 8.25b is connected to a capacitor C_L, which is at a voltage $V_O = 0$ V initially. Charging the capacitor through R_{eq} will yield

$$V_O(t) = V_{CC}(1 - e^{-t/R_{eq}C_L}) \tag{8.3}$$

with the time constant $\tau = R_{eq}C_L$. Note from the above discussion that the charging time and hence the transient performance is improved by increasing the aspect ratios for both devices, which in turn decrease the equivalent drain-source resistances. The trade-off, however, is the additional chip area.

Example 8.2

For a CMOS TG, characterized by $(W/L)_n = (W/L)_p = 2$, $\mu_e C_{ox} = 2.5\mu_h C_{ox} = 25\ \mu A/V^2$, $V_{TOn} = |V_{TOp}| = 1$ V, $\gamma_n = .4\ \sqrt{V}$, $\gamma_p = .5\ \sqrt{V}$, and $2|\phi_f| = .7$ V for both devices, the following SPICE simulation is performed to obtain the value of R_{eq} graphically:

```
CMOS TG
M1   1 3 4 0 MN
M2   4 0 1 3 MP
CL   4 0 .1P
VIN 1 0 PULSE 0 5 0 0 0 10N
VCC 3 0 5V
.MODEL MN NMOS L 5U W 10U VTO  1 GAMMA .4 KP 2.5E-5 PHI .7
.MODEL MP PMOS L 5U W 10U VTO -1 GAMMA .5 KP   1E-5 PHI .7
.TRAN .1N 15N
.PROBE
.END
```

Using Equation (8.3), $V_O = V_{CC}(1 - 1/e) = 3.16$ V when $t = \tau$ so that, from Figure 8.26, we get $\tau = 1.24$ ns. Therefore,

$$R_{eq} = \frac{1.24 \times 10^{-9}}{.1 \times 10^{-12}} = 12.4\ \text{K}\Omega$$

Using Equations (8.1) and (8.2) with $V_O = 0$ V would yield

$$R_{eq} = 13.9\ \text{K}\Omega$$

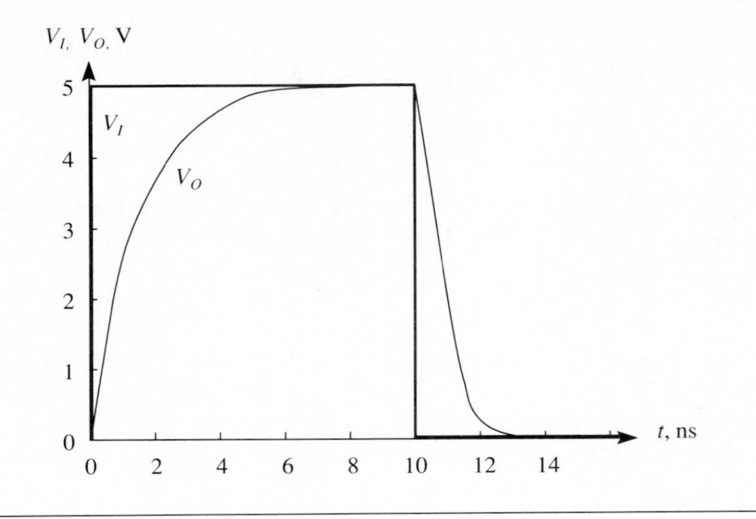

Figure 8.26 SPICE simulation: CMOS TG input and output waveforms.

Figure 8.27 shows a CMOS implementation of the positive-edge-triggered DFF with active high PRESET and CLEAR inputs. It is designed in a master-slave fashion, with two SR latches in cascade. Each latch has a TG inserted in its feedback loop. An open TG breaks the latch feedback loop, so that the NORs act as two inverters in series. A third transmission gate connects the D input to the master latch, while another connects the master to the slave. These are used to enter data into the input and output latches, respectively. The TGs provide a low-resistance path to charge or discharge the gate capacitances rapidly.

When the clock is low, *TG1* and *TG4* are short circuits, and *TG2* and *TG3* are open. As a result, the data from the D input is passed along to the master latch, which cannot yet store it. In the meantime, the slave is isolated from the master and retains whatever has previously been stored.

When the clock goes high, *TG1* and *TG4* turn off, and *TG2* and *TG3* turn on. Consequently, the master is disconnected from the input, and its feedback is closed. The value of the D input that is one setup time before the clock transition is transferred to both the slave, whose feedback loop is now open, and the output. When the clock falls again, the slave latches the master's state before *TG3* opens to disconnect the two latches. Note that only one latch is storing input data at any time.

8.2 Schmitt Triggers

An especially useful digital regenerative circuit is the Schmitt trigger, which converts a noisy or slowly varying input signal into a *clean* digital form. Its voltage transfer characteristic displays hysteresis by having two different input thresholds for positive- and negative-going voltage signals. Thus, the input voltage at which the output

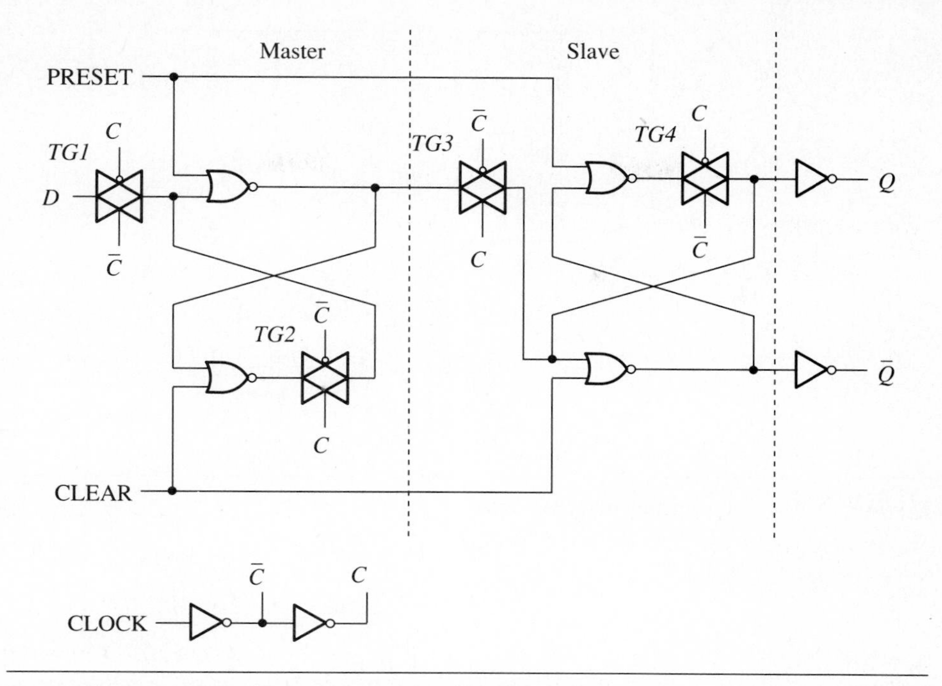

Figure 8.27 The CMOS implementation of the positive-edge-triggered DFF.

changes depends upon whether the output is currently high or low. The deadband separating the two thresholds does not respond to input signal changes; thus, after the first crossing of a threshold, later excursions due to noise have no effect as long as they do not exceed the deadband.

The 7414 TTL Schmitt Trigger

The operation of the basic emitter-coupled Schmitt-trigger circuit and its voltage transfer characteristic are best explained using the circuit in Figure 8.28 of the 7414 TTL hex inverter. Note that the totem-pole output stage is exactly that of a NAND gate. The basic Schmitt trigger, which is responsible for the hysteresis, is that part of the 7414 shown in Figure 8.29a.

Transistors Q_1 and Q_2 have different saturation currents because $R_1 > R_2$, which leads to the hysteresis as different input voltages are required to saturate and cut them off. Assume that V_I is initially at a low level to keep Q_1 off. Then the current through R_1 flows into the base of Q_2 and causes it to saturate with

$$V_O = V_{OL} = V_E + V_{CE2(sat)}$$

Noting that $V_E/R_E = I_{E2} = I_{B2} + I_{C2}$, and neglecting $V_{CE2(sat)}/R_2$ and $V_{BE2(sat)}/R_1$, the common-emitter voltage can be approximated by

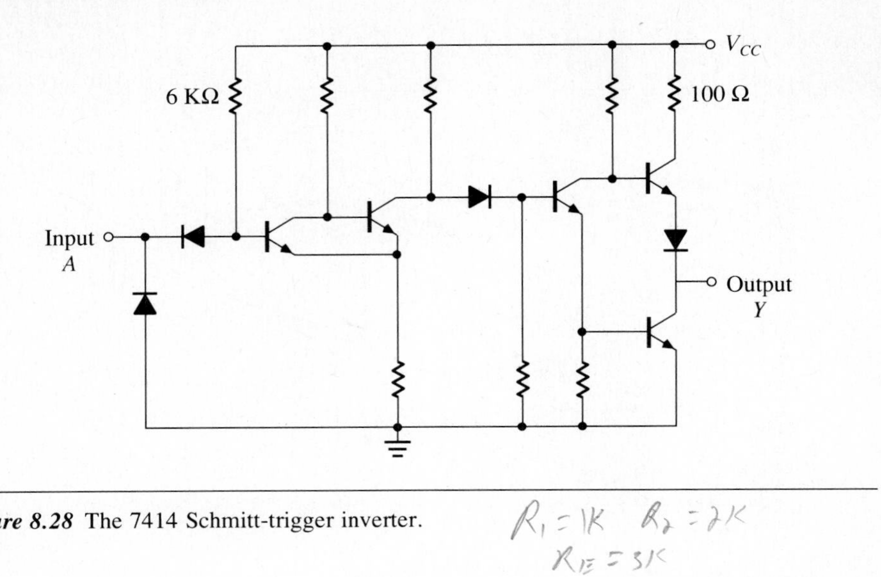

Figure 8.28 The 7414 Schmitt-trigger inverter.

$$V_E \approx \frac{1}{1 + \dfrac{R_1 R_2}{R_E(R_1 + R_2)}} V_{CC}$$

As the input voltage is increased, the current flowing through R_1 is diverted from the base of Q_2 to Q_1. When the input reaches

$$V_I = V_{T+} = V_E + V_{BE1(on)}$$

Figure 8.29 The basic Schmitt trigger proper: (a) circuit, and (b) VTC.

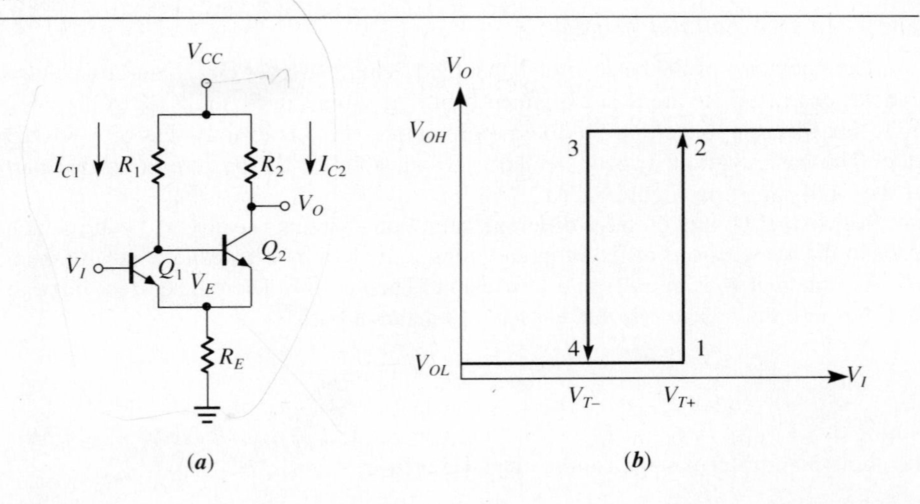

it diverts enough base current to cause Q_2 to drop out of saturation. Transistor Q_1 conducts, and the voltage $V_{C1} = V_{B2}$ decreases. Hence, as Q_1 heads toward saturation, Q_2 turns off and

$$V_O = V_{OH} = V_{CC}$$

Then the collector and emitter currents of Q_2 decrease, causing more emitter current to flow through Q_1 because V_E is being held constant within $V_{BE1(on)}$ of the input voltage. This in turn causes increased collector current through Q_1 and further diversion of current from the base of Q_2, resulting in a regenerative action. This process is rapid, and the transition at the output is a steep jump from point 1 to point 2, as seen in Figures 8.29b and 8.30a. Further increase at the input drives Q_1 into saturation.

With Q_2 off and Q_1 saturated, if the input heads negative, the output voltage remains high, as shown in Figure 8.30a, until V_I reaches the negative-going threshold

$$V_I = V_{T-} = V_E + V_{BE1(on)}$$

where now

$$V_E \approx R_E I_{C1} \approx \frac{R_E}{R_1 + R_E} V_{CC}$$

Figure 8.30 (a) The input and output waveforms of the basic Schmitt-trigger subcircuit, and (b) symbol for the Schmitt-trigger inverter.

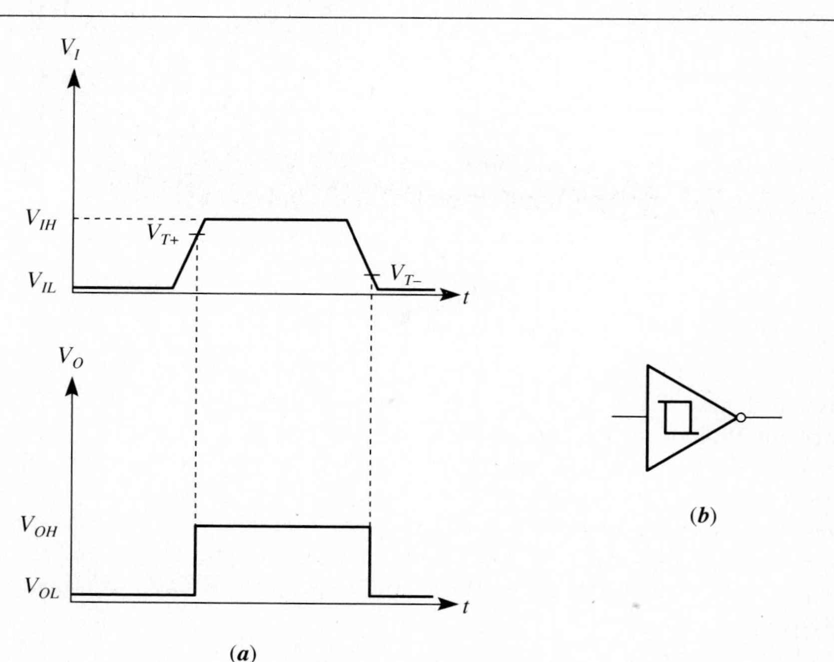

I_{C1} decreases as the current in R_1 diverts to the base of Q_2 and Q_1 goes out of saturation. Q_2 turns on, and its emitter current increases at the expense of the emitter current through Q_1 because V_E is clamped to V_I through $V_{BE1(on)}$. Thus, the regenerative process begins in the opposite direction. Q_1 turns off as Q_2 goes into saturation, leading to the steep change in the transfer characteristic from 3 to 4 where $V_O = V_{OL}$. Note that the hysteresis voltage $V_H \equiv V_{T+} - V_{T-}$ depends on the degree of the mismatch in R_1 and R_2. The typical threshold voltage values at $V_{CC} = 5$ V and $T_A = 25°\text{C}$ are given as $V_{T+} = 1.7$ V and $V_{T-} = .9$ V.

To indicate a Schmitt-trigger circuit at the input, a stylized form of the VTC is included in a logic symbol, as illustrated in Figure 8.30b.

Example 8.3

The following SPICE input file simulates the basic emitter-coupled Schmitt trigger of Figure 8.29a with $R_1 = 4$ KΩ, $R_2 = 2.5$ KΩ, and $R_E = 1$ KΩ. Note that the transfer characteristic of this circuit cannot be analyzed easily using a dc sweep. To measure the hysteresis, a piecewise-linear source of slowly varying ramp has been employed in one transient run rather than a dc analysis, which would need one run for each direction of the sweep.

```
EMITTER-COUPLED SCHMITT TRIGGER
Q1  2 1 5 Q
Q2  4 2 5 Q
R1  3 2 4K
R2  3 4 2.5K
RE  5 0 1K
VCC 3 0 5
VIN 1 0 PWL 0 0 5US 5V 10US 5V 15US 0V
.MODEL Q NPN IS 1E-16 BF 50 BR .1 RB 50 RC 10 TF 100P TR 5N
+ CJE .5P CJC .5P CCS 1P PE .8 PC .8 VA 50
.TRAN 1N 30U
.PROBE
.END
```

The resulting VTC is depicted in Figure 8.31, from which we obtain $V_{T-} = 1.65$ V and $V_{T+} = 2.57$ V; hand analysis would yield 1.56 V and 2.67 V, respectively.

The 74HC14 CMOS Schmitt Trigger

The circuit of 74HC14, the high-speed CMOS hex inverting Schmitt trigger, is shown in Figure 8.32. As in all CMOS parts, the input goes through the standard protection of diode clamps, not shown in Figure 8.32, to V_{CC} and ground to prevent damage due to the static discharge. Then it is tied to the gates of four stacked devices. The upper two are p-channel transistors, and the lower two are n-channel devices. Transistors P_3 and N_3 alternately operate in the source-follower mode and introduce hysteresis by feeding back the output voltage of the Schmitt trigger proper, V_x, to two different points in the stack. The output stage is made up of two inverters in cascade.

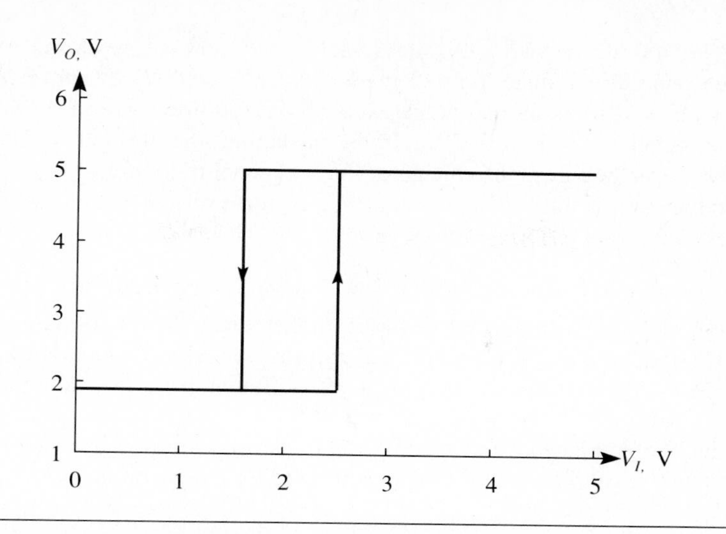

Figure 8.31 SPICE simulation: The emitter-coupled Schmitt-trigger VTC.

Figure 8.32 The 74HC14 HCMOS inverting Schmitt-trigger circuit.

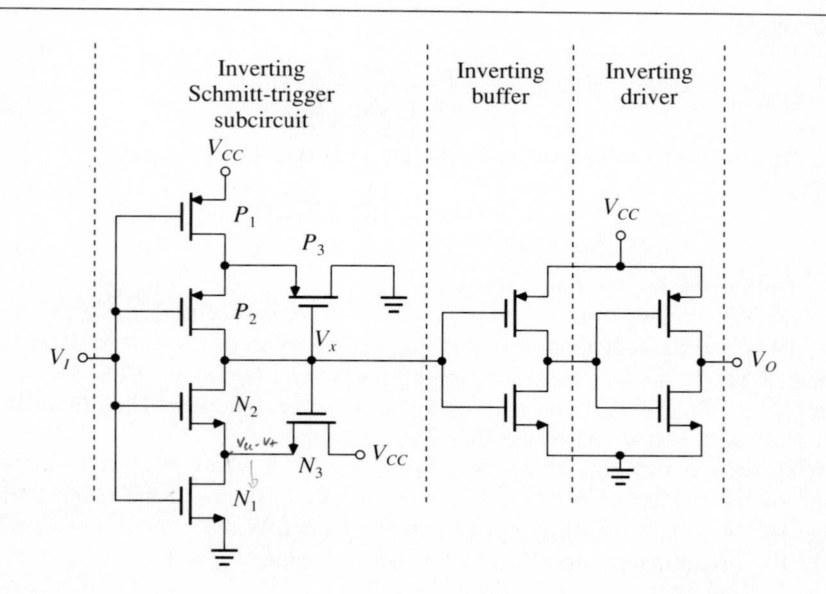

With $V_I = 0$ V, the two p-channel transistors P_1 and P_2 are on but conducting negligible drain current since the n-channel devices N_1 and N_2 are off. Hence $V_x = V_{CC}$, P_3 is off, and N_3 is on and acting as a source follower. After the two inverter stages, the output $V_O = V_{OH} = V_{CC}$. In the meantime, the drain of N_1 is at $V_{CC} - V_{Tn}$, where V_{Tn} is the threshold voltage of the n-channel transistors.

When the input voltage is increased to V_{Tn}, N_1 turns on and its drain voltage falls. As V_I rises to the forward trigger voltage

$$V_{T+} = V_{GS2} + V_{DS1} = V_{Tn} + V_{DS1}$$

N_2 turns on, and a rapid regenerative process takes over. With both N_1 and N_2 conducting, V_x drops to 0 V, N_3 turns off, and P_3 turns on. The conduction of P_3 brings the drain of P_2 low and aids in turning it off. Following the two inverting stages, $V_O = V_{OL} = 0$ V.

V_{T+} may be computed by evaluating the transistor currents. Ignoring body-bias effects to obtain a simple analytic expression, note that as N_2 turns on

$$V_{DS1} = V_{GS1} - V_{Tn}$$

so that N_1 is at the edge of saturation, with

$$I_{D1} \approx \frac{1}{2} k_{n1} (V_{T+} - V_{Tn})^2$$

The feedback transistor N_3 is also saturated, with

$$I_{S3} \approx \frac{1}{2} k_{n3} (V_{CC} - V_{Tn} - V_{DS1})^2 = \frac{1}{2} k_{n3} (V_{CC} - V_{T+})^2$$

Equating the currents to each other produces an approximation for the forward trigger voltage as

$$V_{T+} \approx \frac{V_{CC} + \sqrt{k_{n1}/k_{n3}}\, V_{Tn}}{1 + \sqrt{k_{n1}/k_{n3}}} \tag{8.4}$$

Thus, the driver-load ratio turns out to be the important design parameter:

$$\frac{k_{n1}}{k_{n3}} = \frac{(W/L)_{n1}}{(W/L)_{n3}} \approx \left[\frac{V_{CC} - V_{T+}}{V_{T+} - V_{Tn}}\right]^2$$

Notice that decreasing the ratio increases V_{T+}.

When V_I decreases from V_{CC} to 0 V, a similar process occurs in the upper portion of the stack, and the regenerative action takes place when the lower threshold V_{T-} is reached. When $V_I = V_{CC}$, P_1 and P_2 are off but N_1 and N_2 are on. Hence $V_x = 0$ V, so that N_3 is off, and P_3 is on and acting as a source follower. The circuit output $V_O = V_x = V_{OL} = 0$ V. Also, the source of P_1 is at V_{CC}.

When the input voltage is decreased to $V_{CC} - |V_{Tp}|$, where $|V_{Tp}|$ is the threshold voltage of the p-channel devices, P_1 turns on. As V_I drops to the reverse trigger voltage $V_{T-} = |V_{SD1}| - |V_{Tp}|$, P_2 will start to conduct. With P_1 and P_2 on, V_x rapidly rises to V_{CC}, N_3 turns on, and P_3 turns off. At the output $V_O = V_x = V_{OH} = V_{CC}$.

To approximate V_{T-}, suppose that body biases can be ignored. Then, both P_1 and P_3 are saturated, with

$$I_{D1} \approx \frac{1}{2} k_{p1}(V_{CC} - V_{T-} - |V_{Tp}|)^2$$

and

$$I_{S3} \approx \frac{1}{2} k_{p3}(|V_{DS1}| - |V_{Tp}|)^2 = \frac{1}{2} k_{p3}(V_{T-})^2$$

Setting $I_{D1} = I_{S3}$ yields

$$V_{T-} \approx \frac{\sqrt{k_{p1}/k_{p3}}\,(V_{CC} - |V_{Tp}|)}{1 + \sqrt{k_{p1}/k_{p3}}} \tag{8.5}$$

The reverse trigger voltage design equation is then

$$\frac{k_{p1}}{k_{p3}} = \frac{(W/L)_{p1}}{(W/L)_{p3}} \approx \left[\frac{V_{T-}}{V_{CC} - V_{T-}\ |V_{Tp}|}\right]^2$$

For a symmetric CMOS design, set

$$k_r \equiv \frac{k_{n1}}{k_{n3}} = \frac{k_{p1}}{k_{p3}}$$

and

$$V_{Tn} = |V_{Tp}| = V_T$$

Now, defining the *hysteresis voltage*—that is, the trigger voltage separation, for an ideal symmetric VTC as

$$V_H \equiv V_{T+} - V_{T-}$$

Equations (8.4) and (8.5) produce

$$V_H \approx \frac{(1 - \sqrt{k_r})V_{CC} + 2\sqrt{k_r}\,V_T}{1 + \sqrt{k_r}} > 0 \tag{8.6}$$

Rearranging yields a design equation

$$\sqrt{k_r} \approx \frac{V_{CC} - 2\Delta V}{V_{CC} + 2\Delta V - 2V_T}$$

that can be employed to obtain a symmetric hysteresis loop.

The typical threshold voltage values at $V_{CC} = 4.5$ V and $T_A = 25°$C are given as $V_{T+} = 2.7$ V and $V_{T-} = 1.8$ V. Therefore, with a typical hysteresis voltage of .9 V and the output pulling directly to the supply rails, the HCMOS Schmitt trigger has a high noise immunity. The variation of the threshold values as a function of the power supply voltage is shown in Figure 8.33.

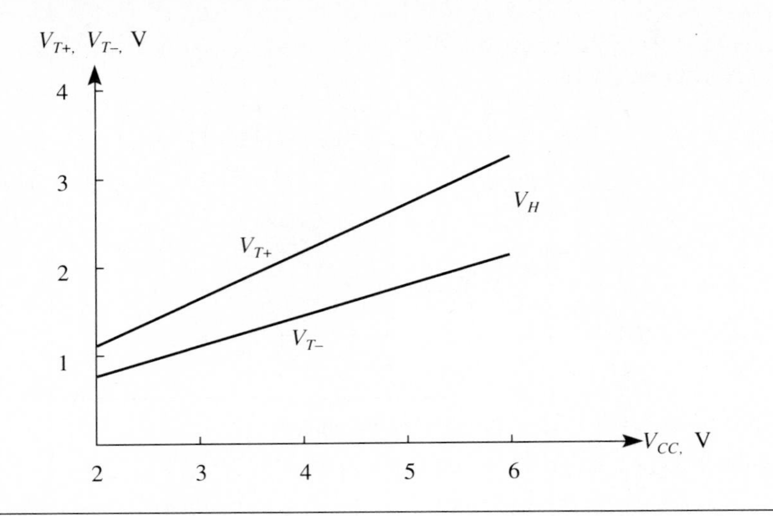

Figure 8.33 Input thresholds V_{T+} and V_{T-} vs. power supply voltage for the 74HC14.

8.3 Monostable Multivibrators

Monostable multivibrators are versatile ICs in digital circuits used in such applications as pulse generation, pulse shaping, time delay, envelope detection, pulse-width detection, noise discrimination, frequency discrimination, and bandpass filter design. A monostable, when triggered, produces a programmable output pulse whose width is independent of the input trigger pulse width. Normally, the output is reset until a trigger is applied, at which time it sets and remains at this quasistable state for a time period determined by a resistor and a capacitor.

Various device types in TTL and CMOS technologies feature single and dual one-shots, nonretriggerable and retriggerable parts, clearing inputs, and level- or pulse-triggering. Some devices also have Schmitt-trigger inputs, and may contain internal timing resistors.

Before discussing the commercially available monostable multivibrators, we first consider the simple one-shot of Figure 8.34 based on the SR latch. Note that the trigger is applied to the reset input, and the output is taken from the complemented output of the latch. Assume that the input impedance at S is large, and the output impedance at Q and $\overline{Q}$ is negligible.

When a positive trigger occurs at R, Q goes low and $\overline{Q}$ goes high. The drop in voltage at Q is coupled by C to the S input so that S is now at $-V_{OH}$. The current starts flowing through C and r, charging the capacitor at a rate of $\tau \approx rC$. The voltage at S continues to rise until S reaches logical **1**, at which time the latch sets and the output returns to low. The S input level then decays to $V_{IL} \approx 0$ V where it remains until the next trigger pulse. During the recovery time, the monostable should not be triggered because the logical level at S is indeterminate.

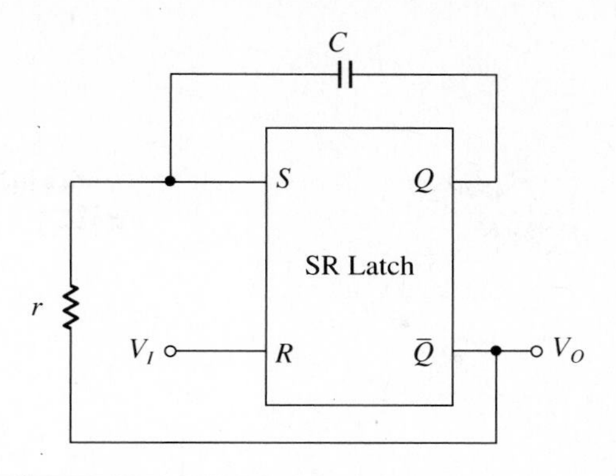

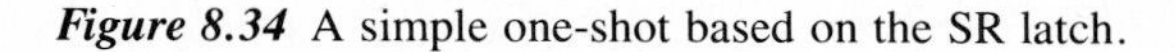

Figure 8.34 A simple one-shot based on the SR latch.

The *duty cycle* of a multivibrator relates the output pulse width t_w to the total period T and is expressed in percent as

$$\text{Duty cycle} \equiv \left(\frac{t_w}{T}\right) \times 100$$

The CMOS Monostable Multivibrator

A basic CMOS monostable multivibrator using two two-input NOR gates, a capacitor C, and a resistor R is depicted in Figure 8.35a. The coupler between the two NOR gates *G1* and *G2* is capacitive, and between *G2* and *G1* it is resistive. The input source V_I supplies the necessary triggering pulse for the one-shot. Recall from Chapter 4 the presence of the protective diodes at the input. For the two-input NOR gate of Figure 4.17b, these diodes are bridged between each input and the power supply terminals, and are used to prevent the input signal from rising above the supply voltage V_{CC} or falling below the ground voltage by more than one diode drop. This ensures that the gate voltages will never exceed the breakdown voltage of the oxide layer.

These clamping diodes affect especially the operation of *G2*, as will be clear later. Since this NOR gate is connected as an inverter, each pair of input diodes appear in parallel, leading to the equivalent circuit of Figure 8.35b. Even though the diodes provide a low-resistance path to V_{CC} for voltages exceeding the supply limits, the input current is essentially zero for intermediate voltage values.

For the sake of simplicity, we assume that the gates have V_{CC} and ground as logical **1** and **0** levels, respectively, and that the transition between the states is abrupt at the threshold voltage $V_{th} > 0$ V of the *n*-channel drivers. The latter assumption implies a small switching speed compared to the duration of the desired output pulse. With these assumptions, the waveforms of the monostable multivibrator appear as in Figure 8.36.

For $t < 0$, no current exists in R, and $v_x(t) = V_{CC}$. Thus, the output of *G2* is

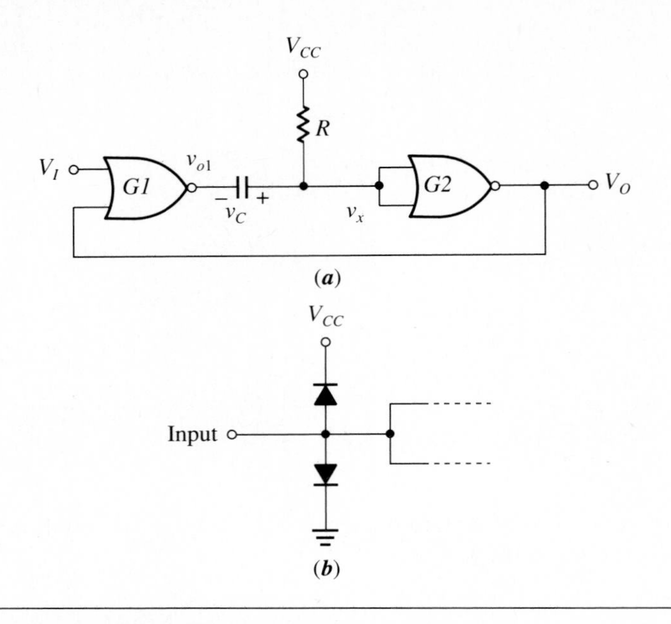

Figure 8.35 (a) A CMOS monostable multivibrator, and (b) equivalent circuit of the input clamping diodes.

low. With both inputs to the NOR gate *G1* at logical **0**, its output $v_{o1}(t) = V_{CC}$, which implies that the voltage across the capacitor is 0 V and it is initially discharged.

Now, the application of a short and positive triggering pulse $V_I > V_{th}$ at $t = 0$, as shown in Figure 8.36a, causes a transition in *G1*, and $v_{o1}(t)$ becomes logical **0**, as seen in Figure 8.36b. As a result of this, during the transient, at the beginning of the quasistable state, an instantaneous current flows from the supply through R and C, and sinks into the output of gate *G1*. Because of its finite on-resistance, R_{on}, *G1* output cannot go to 0 V. Instead, it drops abruptly, after a propagation delay t_{PD1}, by a value δ. A voltage divider is thus formed by R and R_{on}, from which the drop δ in $v_{o1}(t)$ is determined as

$$\delta = \frac{R}{R + R_{on}} V_{CC}$$

The voltage across the capacitor, $v_C(t)$, cannot change instantaneously, and C acts as a short circuit during the transient so that this abrupt drop is coupled through the capacitor to the input of *G2*, and $v_x(t)$ goes low, dropping by an identical amount δ. This causes *G2*'s output to go high, after t_{PD2}, to V_{CC}, which keeps the output of *G1* low even after the triggering pulse has disappeared. One should note here that propagation delays set a lower limit on the input pulse width t_w such that $t_w > t_{PD1} + t_{PD2}$.

During the quasistable state, the current through R, C, and R_{on} causes the capacitor to charge, and $v_x(t)$ rises from $V_{CC} - \delta$ toward its asymptotic value of V_{CC} as given by

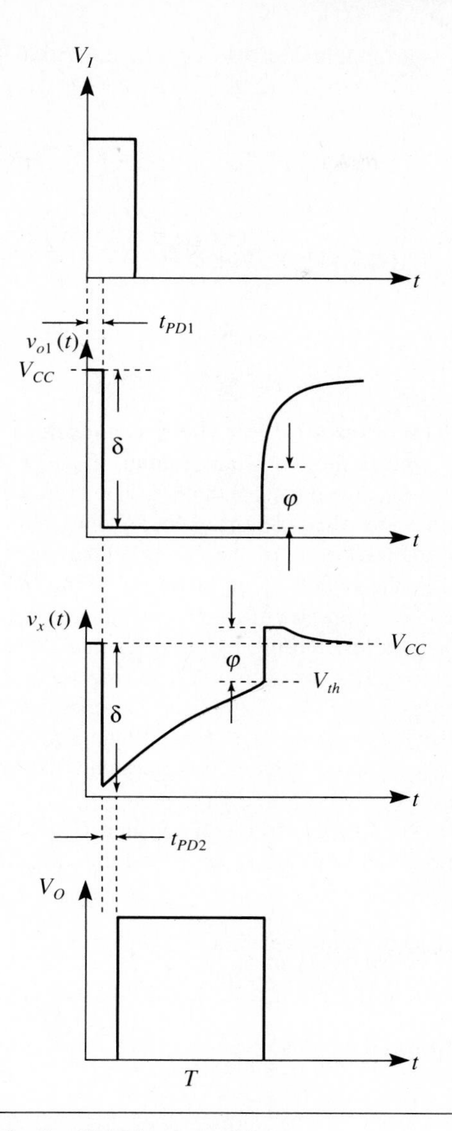

Figure 8.36 Waveforms for the CMOS monostable multivibrator.

$$v_x(t) = V_{CC} - \delta e^{-t/\tau} \tag{8.7}$$

where $\tau = (R + R_{on})C$. In the meantime, since the charging current decreases with time, the drop across R_{on} decays as well, and $v_{o1}(t)$ tilts downward with the same time constant τ.

When $v_x(t)$ reaches the threshold voltage V_{th}, $G2$ switches state, its output goes low, and the quasistable state is terminated. To find the interval T, we set $v_x(t) =$

V_{th} in Equation (8.7), rearrange the terms, take the logarithm of both sides, and solve for T to get

$$T = \tau \ln \left(\frac{\delta}{V_{CC} - V_{th}} \right)$$

$$= (R + R_{on})C \ln \left(\frac{R}{R + R_{on}} \frac{V_{CC}}{V_{CC} - V_{th}} \right) \tag{8.8}$$

In a typical case with $V_{th} = V_{CC}/2$ and $R >> R_{on}$, Equation (8.8) reduces to

$$T \approx RC \ln 2 = .69RC$$

The logical **0** at *G2* output causes *G1* to switch, as a result of which the output of *G1* attempts to rise to V_{CC} but is limited to an amount of φ. To analyze the behavior of the monostable at this point, we consider the equivalent circuit of Figure 8.37, which determines the charging and discharging cycles of the capacitor C. Here, R_p represents the volt-ampere characteristic of the series p-channel transistors, which themselves are symbolized by the switch S_p, at the output of the NOR gate of Figure 3.23b. It depends on the region of operation of the transistors and hence may not be a simple resistor. Likewise, the parallel combination of n-channel transistors is represented by the switch S_n. The diode D is open-circuited during most of the cycle. However, while *G2* is turned on at $t = T + t_{PD1} + t_{PD2}$, D conducts and ensures that $v_x(t)$ does not exceed V_{CC} by more than its cut-in voltage $V_{D\gamma}$. Actually, $v_x(t)$ may be slightly higher than $V_{CC} + V_{D\gamma}$ because of the small forward resistance r_d of the diode.

In the stable state, both inputs to the NOR gate *G1* are low. The series PMOS transistors are on, so S_p is closed, S_n is open, and the capacitor voltage is $v_C(0) =$

Figure 8.37 The equivalent circuit to determine the cycles of the timing capacitor.

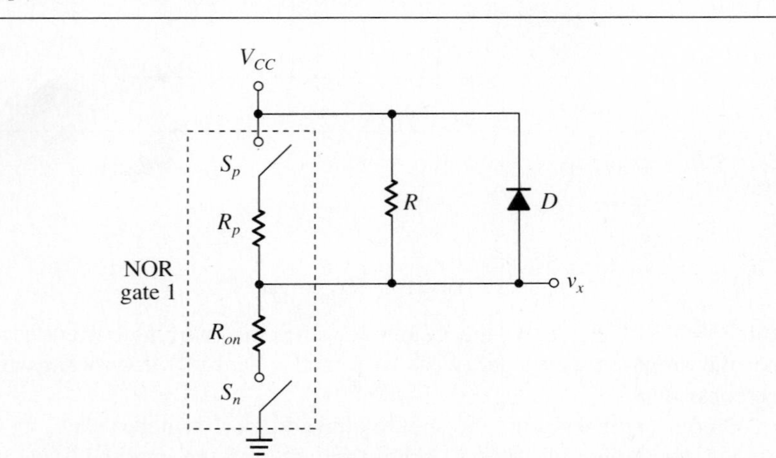

0 V. The quasistable state begins when S_n closes and S_p opens, and C charges from V_{CC} through R and R_{on}. If R is large enough, the voltage across the parallel n-channel transistors will be low enough so that the transistors will operate in the ohmic region, where they may be represented as a resistor, R_{on}. This state ends when p-channel devices turn on, that is, S_p closes and S_n opens.

During the quasistable state, the capacitor may charge up to V_{CC}. At the end of the quasistable state, C discharges through R_p and the diode because $r_d \ll R$, and their parallel combination is approximately equal to r_d. Depending on the initial value of $v_C(t)$, p-channel transistors may be in the saturation region, in which case the capacitor discharges at a constant rate. As $v_C(t)$ decreases, the transistors may be in the ohmic region, so that the discharge continues exponentially. When $v_C(t)$ drops to $V_{D\gamma}$, the diode turns off and the last part of the discharge takes place through R_p and R.

When S_p closes, S_n opens, and D starts to conduct, $v_x(t)$ becomes $V_{CC} + V_{D\gamma}$, jumping abruptly from V_{th} to $V_{CC} + V_{D\gamma}$, as seen in Figure 8.36c. Thus, the size of the jump is given by

$$\varphi = V_{CC} + V_{D\gamma} - V_{th}$$

This jump is transmitted through C to the output of *G1* and hence is also seen in the waveform of $v_{o1}(t)$. Thereafter, the capacitor discharges, as explained above, until $v_x(t)$ drops and $v_{o1}(t)$ rises to their steady-state values of V_{CC}.

The limiting effect of the clamping diode on the size of the increment φ prevents the output of *G1* from rising to V_{CC} instantaneously as a result of a change at the input. The monostable should not be retriggered until the capacitance has been discharged. Otherwise, it will not be able to provide a standard pulse at the output.

Although the input-output characteristic of a CMOS gate displays very little sensitivity to temperature, there may still be a spread in V_{th} from one production to another. Hence, the output pulse width may vary from circuit to circuit for given values of timing components.

The 9602 Bipolar Monostable Multivibrator

One popular bipolar IC is the 9602/8602 dual monostable multivibrator. Its block diagram and logic symbol are shown in Figure 8.38. The IC has two resettable, retriggerable one-shots with two inputs each, one active high and the other active low, so that both leading-edge and trailing-edge triggering can be utilized. Each one-shot employs a TTL NAND gate input and a conventional totem-pole output stage. The interior stages include a bistable latch, a monostable circuit, and an emitter-coupled Schmitt trigger. When the circuit is externally triggered, the bistable latch is set and reset immediately to produce a trigger pulse to the monostable, which in turn initiates a new cycle, and the external capacitor is allowed to rapidly discharge and charge again through different paths. The duration of the quasistable state depends on the values of the external components R_x and C_x, and on the threshold level V_{T+} of the Schmitt trigger, which is primarily used to produce sharp edge rates at the output. The minimum output width range is around 70 ns, but the retriggerable feature permits it to be extended infinitely to maintain a continuous true output. Any timing

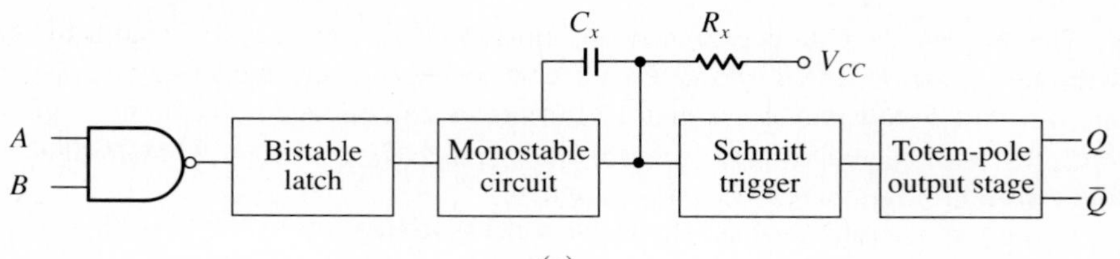

(*a*)

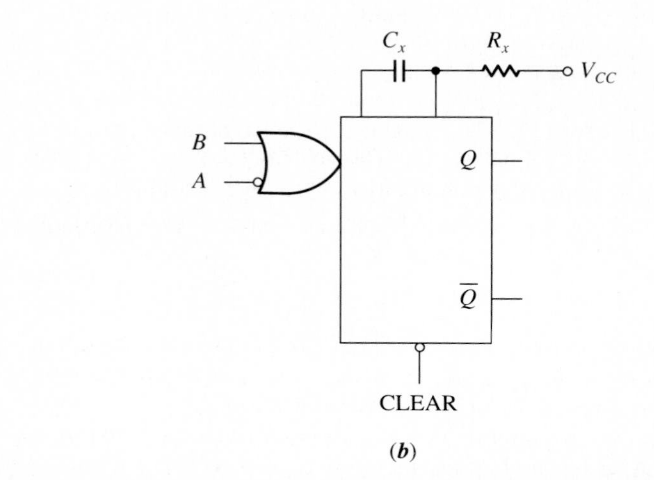

(*b*)

Figure 8.38 The 9602/8602 bipolar monostable multivibrator: (a) block diagram, and (b) logic symbol.

cycle at the output may be terminated by applying a low level to the overriding CLEAR input.

The output pulse width is defined for $C_x \geq 10^3$ pF as

$$t_w = KR_xC_x\left(1 + \frac{1}{R_x}\right) \text{ ns}$$

where R_x is in KΩ, C_x is in pF, and the multiplicative factor K depends on the timing capacitance. It is ≈ 34 for $C_x = 1$ nF and can be approximated as ≈ 31 as C_s is increased above this value.

Actually, the value of C_x may vary from zero to any positive value. However, if either the capacitor has forward leakages approaching 3 μA or the stray capacitance from either of its terminals to the ground is greater than 50 pF, then the timing equation may not represent the actual pulse width. For $C_x < 1$ nF, a graphical relation included in the data sheets must be used to determine the pulse width. The value of the external timing resistor is required to be between 5 KΩ and 25 KΩ. For the 8602 version, however, $R_{x(max)}$ is extended to 50 KΩ.

If electrolytic-type capacitors are to be used, the manufacturers recommend three different configurations for the external components.

1. The normal *RC* configuration of Figure 8.38 can be used as long as the forward capacitor leakage at 5 V is less than 3 μA, and the inverse leakage at 1 V is less than 5 μA over the operational temperature range.
2. If the electrolytic capacitor has high inverse leakage current, a switching diode should be connected, as in Figure 8.39a, to limit the leakage by preventing a reverse voltage from forming across the capacitor.
3. To obtain extended pulse widths at the output even with high inverse leakage capacitors, the configuration of Figure 8.39b can be used. However, the following relations should hold:
   a. $R_{x(max)} > R_y > R_{x(min)}$
   b. min $[.7h_{FE}R, 2.5\ \text{M}\Omega] > R$

   where h_{FE} is the current gain of Q_1. The pulse width is extended thanks to the larger timing resistor value achieved by beta multiplication.

For the last two configurations, the output pulse width is approximated as

$$t_w \approx .3RC_x$$

The triggering truth table for this monostable is shown in Table 8.3.

The *release time* is defined as the smallest time interval between one input going inactive and the other going active that is needed to guarantee the correct operation of the logic element. For the 9602, the setup and release times for either type of trigger input should be greater than 40 ns. Also, retriggering will not occur if the retrigger pulse comes within $\approx .3C_x$ ns after the initial trigger pulse, that is, during the discharge cycle.

Referenced to $T_A = 25°\text{C}$, the output pulse width for a 9602 varies $\pm 2.5\%$ with the ambient temperature between 0° to 70°C at $V_{CC} = 5$ V. It also varies $\pm 2\%$ as the supply voltage is varied about 5 V from 4.5 to 5.5 V at $T_A = 25°\text{C}$. The typical power dissipation is 125 mW.

Figure 8.39 Configuration (a) to limit leakage, and (b) to extend pulse width.

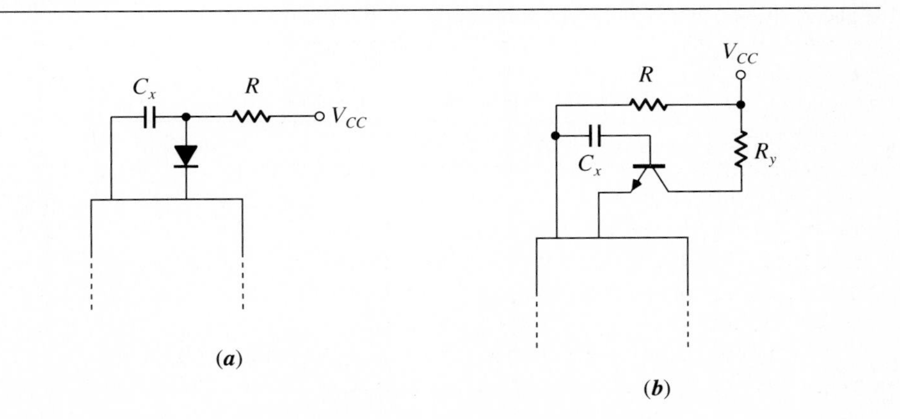

Table 8.3 Triggering Truth Table for 9602/8602

A	B	CLR	Operation
↓	**0**	**1**	Trigger
1	↑	**1**	Trigger
x	x	**0**	Reset

x = don't care

The 74LS221 Low-Power TTL Monostable Multivibrator

Another popular one-shot in IC form is the low-power Schottky 74LS221, a dual, non-retriggerable, monostable multivibrator chip with Schmitt-trigger inputs. Each one-shot has three inputs, permitting the choice of either leading-edge or trailing-edge triggering. The function table is shown in Table 8.4, and the logic diagram in Figure 8.40. It is evident from the last row of the function table that the clear input can also be used to trigger the device. $\overline{\text{SR}}$ latch should be conditioned to the logical **1** state prior to CLR going high by taking either A high or B low. Figure 8.41 illustrates this mode of triggering. In this example, the B input is set first, while the A and CLR inputs are maintained at logical **0** level. Then, with B high, the positive transition of the CLR input triggers an output pulse.

The hysteresis and hence a good noise immunity of typically 1.2 V are provided by the active high Schmitt-trigger input B that allows jitter-free triggering even for inputs with transition rates as slow as 1 V/s. The internal NAND latch at the input stage also provides a high noise immunity to voltage supply noise of typically 1.5 V.

As long as 1.4 KΩ $< R_x <$ 100 KΩ and $0 < C_x <$ 1 mF, the output pulse width is defined as

$$t_w = KR_xC_x \text{ ns}$$

where $K \approx \ln 2$, R_x is in KΩ, and C_x is in pF. Thus, the output pulse width ranges from 30 ns to 70 s. Once fired, the output is independent of further transitions of the A and B inputs, while it can be terminated by the overriding clear input.

Table 8.4 Triggering Truth Table for 74LS221

A	B	CLR	Operation
x	x	**0**	Reset
1	x	x	Reset
x	**0**	x	Reset
0	↑	**1**	Trigger
↓	**1**	**1**	Trigger
0	**1**	↑	Trigger

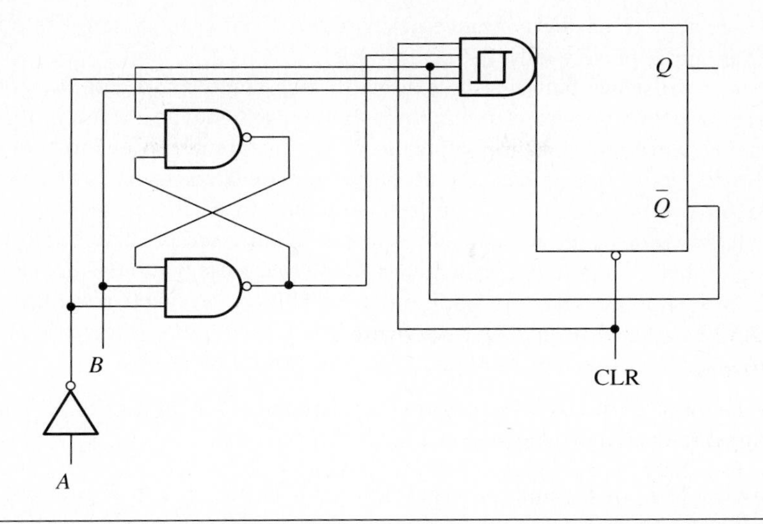

Figure 8.40 The logic diagram of the 74LS221.

If the timing capacitor has leakages approaching 100 nA, or if the stray capacitance from either terminal to ground is greater than 50 pF, then the timing equation given above may not represent the actual pulse width at the output. Unlike the 9602/8602, the 'LS221 does not need a switching diode to prevent high inverse leakage current when an electrolytic capacitor is used as the external timing component.

Pulse-width stability is achieved through internal compensation and is only limited by the accuracy of the external timing components. Referenced to $T_A = 25°C$ at $V_{CC} = 5$ V, the output pulse width varies $\pm 2\%$ with the ambient temperature between 0° to 70°C. Its variance with supply voltage at $T_A = 25°C$, however, is negligible. The power dissipation is typically 23 mW.

In general, when designing with monostables, a noninductive and low-capacitive

Figure 8.41 Clear mode of triggering.

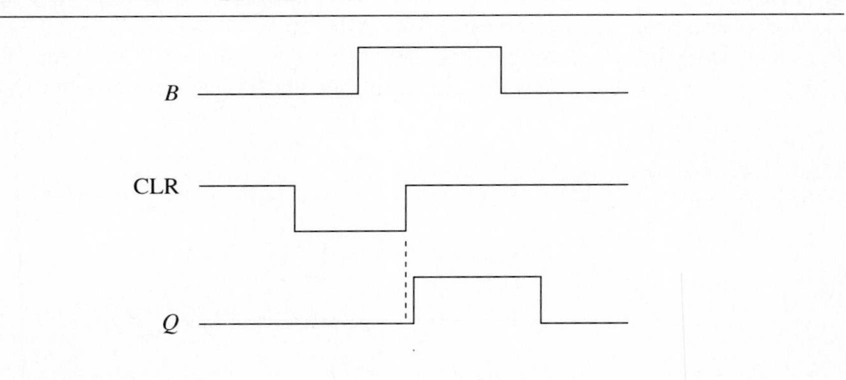

path is necessary to ensure complete discharge of C_x in each cycle of its operation, so that the output pulse width will be accurate.

Since any distance between the external timing components and the monostable IC causes so-called *time-out* errors in the output pulse width as a result of the voltage difference between the capacitor and the device, which in turn is due to both resistive and inductive series impedance, these components must be put as close to the IC as possible for precise timing. Having been designed to discharge the capacitor at a specific fixed voltage, the monostable is misled by any series voltage into releasing the capacitor before it is fully discharged, leading to a pulse width that appears shorter than the programmed value. In addition, for stability, precision resistors with low temperature coefficients, and capacitors with low leakage, good dielectric absorption characteristics, and low temperature coefficients should be used.

8.4 Astable Multivibrators

The simplest way of obtaining an astable circuit is depicted in Figure 8.42: three CMOS inverters, each having a propagation delay of t_{PD}, are cascaded in a loop. Since the output of *G3* is in phase with the *G1* input, the system will oscillate with a maximum frequency

$$f_{o(max)} = \frac{1}{2(3t_{PD})}$$

In general, any odd number of complementing logic gates will oscillate when they are tied together in this fashion. However, this configuration does not form a stable oscillator, Since CMOS propagation delay depends on the temperature, supply voltage, and external capacitive loading, the frequency will vary with these variables, too. Therefore, it is necessary to add external passive elements that not only determine the oscillation frequency but also reduce the effects of CMOS characteristics.

A simple and effective astable circuit can be formed using two NOR gates, a resistor, and a capacitor, as shown in Figure 8.43a. The waveforms are diagrammed in Figure 8.44. The gate output impedances are assumed to be negligible. Also, $t_{PD1} = t_{PD2} = t_{PD}$. For the time being, we will ignore the presence of the protective diodes at the gate inputs.

Since *G2* is used as an inverter, it is apparent that V_{O1} and V_{O2} are complementary. Assume that $v_x(t)$ is initially at the threshold voltage V_{th} of *G1*. Then, *G1* turns on, and after a time t_{PD}, V_{O1} drops to 0 V. After an additional delay t_{PD}, V_{O2} rises to V_{CC}. This step change of voltage is coupled through C so that $v_x(t)$ jumps from V_{th} to $V_{th} + V_{CC}$. Since $V_{O1} = 0$ V, the capacitor discharges through R toward 0 V,

Figure 8.42 A simple astable multivibrator.

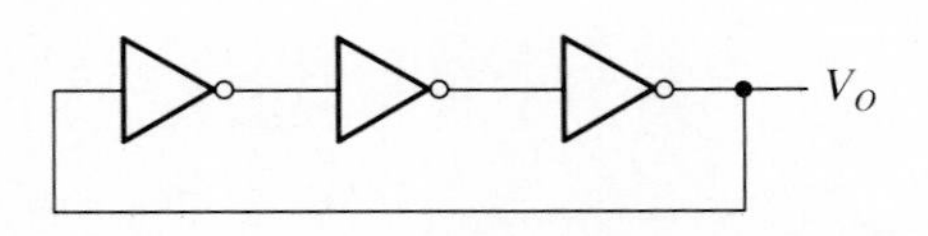

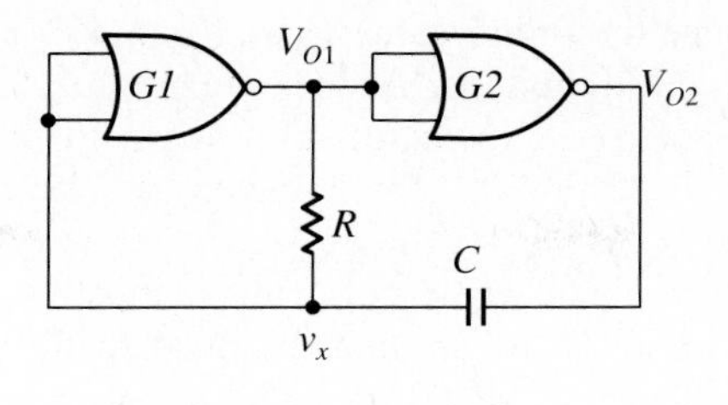

(*a*)

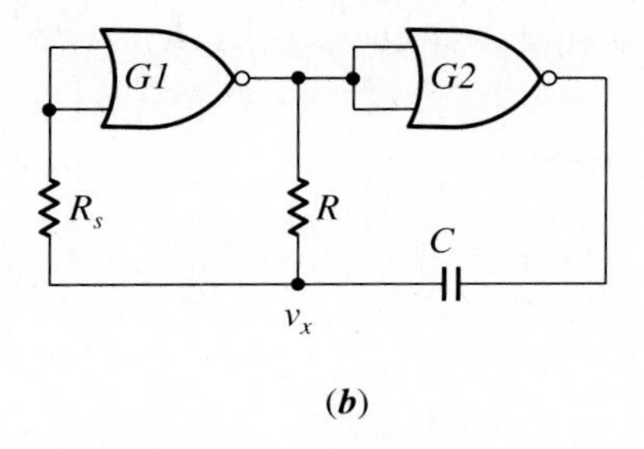

(*b*)

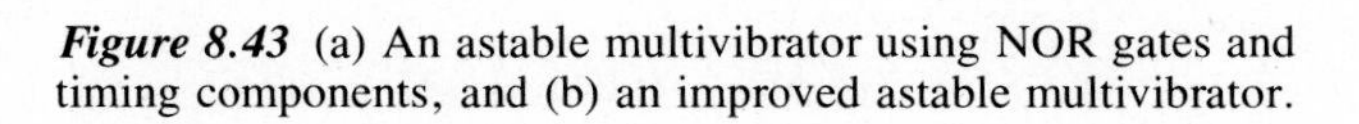
Figure 8.43 (a) An astable multivibrator using NOR gates and timing components, and (b) an improved astable multivibrator.

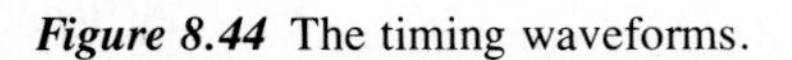
Figure 8.44 The timing waveforms.

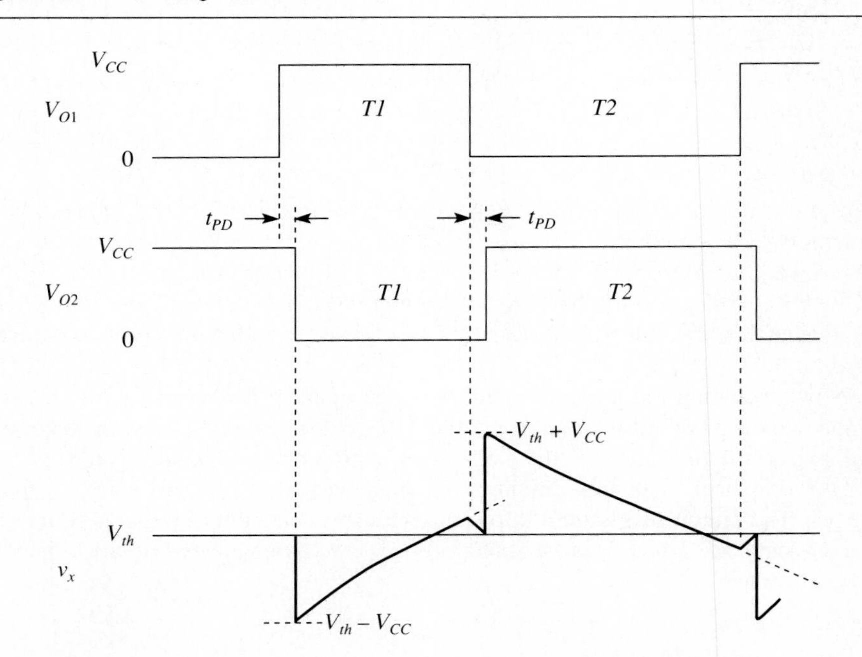

bringing $v_x(t)$ low. Since the initial value $V_i = V_{th} + V_{CC}$ and the final value $V_f = 0$, the voltage equation during the discharge is given by

$$v_x(t) = (V_{th} + V_{CC})e^{-t/RC}$$

When $v_x(t)$ crosses V_{th}, that is, at $t = T2$, where

$$T2 = RC \ln\left(\frac{V_{CC} + V_{th}}{V_{th}}\right)$$

n-channel transistors in $G1$ turn off, and t_{PD} seconds later the output V_{O1} rises to V_{CC}. After a time t_{PD}, V_{O2} falls to 0 V. Consequently, $v_x(t)$ goes from V_{th} to $V_{th} - V_{CC}$. The capacitor charges through R in an attempt to make $v_x(t) = V_{CC}$. Therefore, with $V_i = V_{th} - V_{CC}$ and $V_f = V_{CC}$, the charging equation is expressed as

$$v_x(t) = V_{CC} - (2V_{CC} - V_{th})e^{-t/RC} \tag{8.9}$$

The charging cycle is terminated when $v_x(t)$ reaches V_{th} again. Solving Equation (8.9) for $t = T1$ with $v_x(t) = V_{th}$, we obtain

$$T1 = RC \ln\left(\frac{2V_{CC} - V_{th}}{V_{CC} - V_{th}}\right)$$

at which time the whole cycle repeats itself and the astable operation continues indefinitely. The frequency of oscillation is therefore

$$f_o = (T1 + T2)^{-1} \tag{8.10}$$

If $V_{th} = \frac{1}{2} V_{CC}$, then $T1 = T2$, and the output waveform is symmetrical with a 50% duty cycle. Then Equation (8.10) reduces to

$$f_o = \frac{1}{2.2RC}$$

There is one important observation to be made. Note the behavior of $v_x(t)$ during the time interval $2t_{PD}$ after it reaches the threshold value. Since V_{O1} cannot change for t_{PD} seconds, the capacitor continues to charge or discharge, whatever the case may be. Then V_{O1} changes state, but now V_{O2} cannot do so until t_{PD} seconds later. Therefore, during this time interval the capacitor reverses its charging direction, as depicted in Figure 8.44.

Now, including the protective diodes at the input of both logic gates, $v_x(t)$ is clamped on the positive swing to one diode voltage drop above V_{CC}, while the downward swing will be limited to one diode voltage drop below ground voltage.

One drawback with this circuit is its frequency sensitivity to supply voltage changes. The circuit of Figure 8.43b minimizes this sensitivity by the addition of a second resistor R_s, which is large enough for $v_x(t)$ to be unaffected by the protective diodes at the $G1$ input.

Summary

Digital regenerative circuits are especially useful in timing applications. Three types of digital regenerative circuits are monostable, bistable, and astable multivibrators. The monostable multivibrator has one stable state that may be triggered into another quasistable state. When triggered, it generates an output pulse of constant width. The bistable multivibrator is a switching circuit with two stable states. The astable multivibrator has no stable state. The circuit oscillates between two quasistable states.

The simplest form of the bistable circuit is the SR latch, in which the trigger pulse is stored. However, a logical **1** at both inputs leads to an indeterminate state and hence is not allowed. Additional feedback lines of flip-flops overcome this ambiguity. Three basic types of flip-flops are JKFFs, TFFs, and DFFs.

Another regenerative circuit is the Schmitt trigger. An important feature of this circuit is that its VTC has different input thresholds for positive- and negative-going voltage signals. This hysteresis provides an effective means for rejecting interference. Schmitt triggers are also used for wave-shaping purposes. A sinusoidal voltage, for instance, can be converted into a square wave at the output. Finally, the circuit can respond to a slowly changing input waveform with fast transition times at its output.

References

1. J. P. Uyemura. *Fundamentals of MOS Digital Integrated Circuits*. Addison-Wesley, Reading, MA: 1988.
2. D. A. Hodges and H. G. Jackson. *Analysis and Design of Integrated Circuits*. 2d ed. McGraw-Hill, New York: 1988.

PROBLEMS

8.1 Assuming that the latch is initially reset, sketch the resulting Q waveform when the inputs of the following diagram are applied to the basic SR latch.

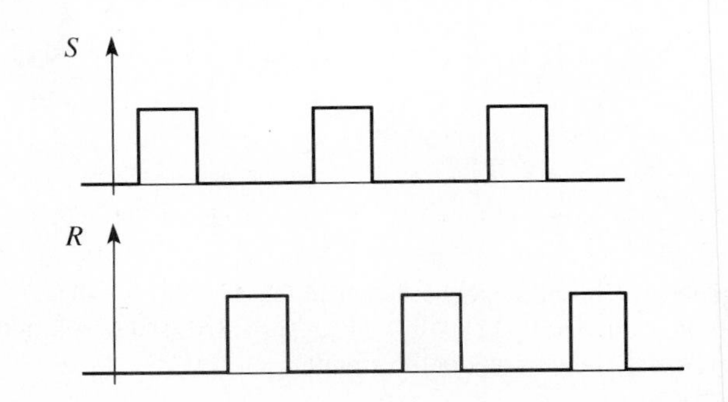

8.2 Assuming a master-slave flip-flop, find the J and K waveforms that will produce the output waveform illustrated in the diagram.

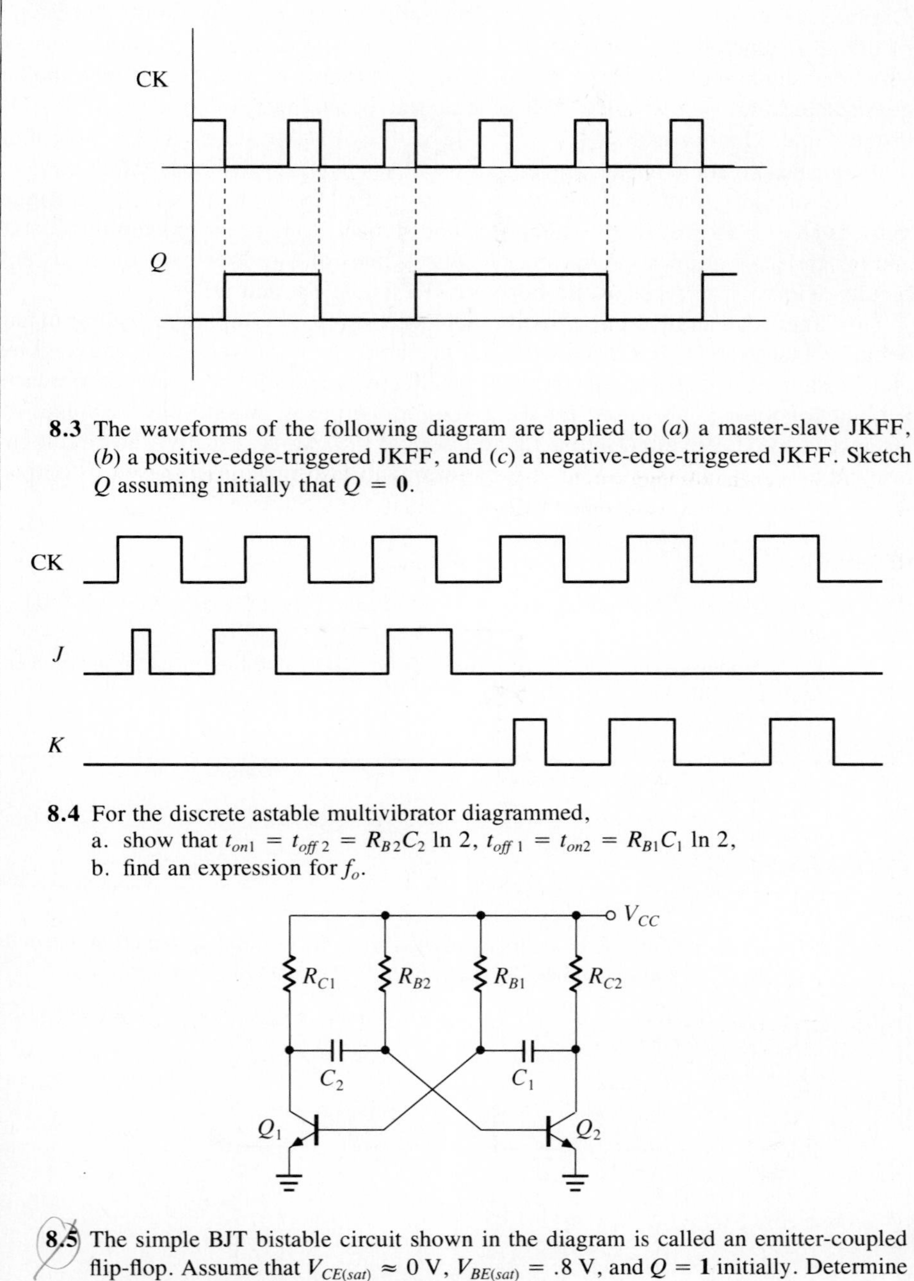

8.3 The waveforms of the following diagram are applied to (*a*) a master-slave JKFF, (*b*) a positive-edge-triggered JKFF, and (*c*) a negative-edge-triggered JKFF. Sketch Q assuming initially that $Q = \mathbf{0}$.

8.4 For the discrete astable multivibrator diagrammed,

a. show that $t_{on1} = t_{off\,2} = R_{B2}C_2 \ln 2$, $t_{off\,1} = t_{on2} = R_{B1}C_1 \ln 2$,

b. find an expression for f_o.

8.5 The simple BJT bistable circuit shown in the diagram is called an emitter-coupled flip-flop. Assume that $V_{CE(sat)} \approx 0$ V, $V_{BE(sat)} = .8$ V, and $Q = \mathbf{1}$ initially. Determine all the voltages and currents in the circuit.

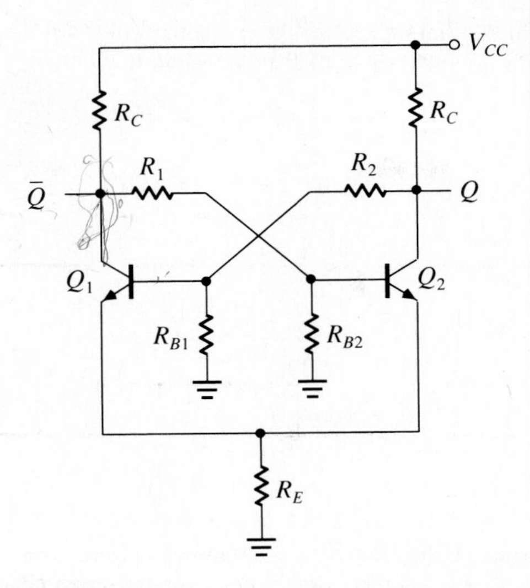

8.6 For the discrete emitter-coupled monostable illustrated, show that the output pulse width, that is, the time for Q_2 to reach turn on, is given by

$$t_w = -RC \ln \left(\frac{2V_{CC}}{V_{CC}} \right) = RC \ln 2$$

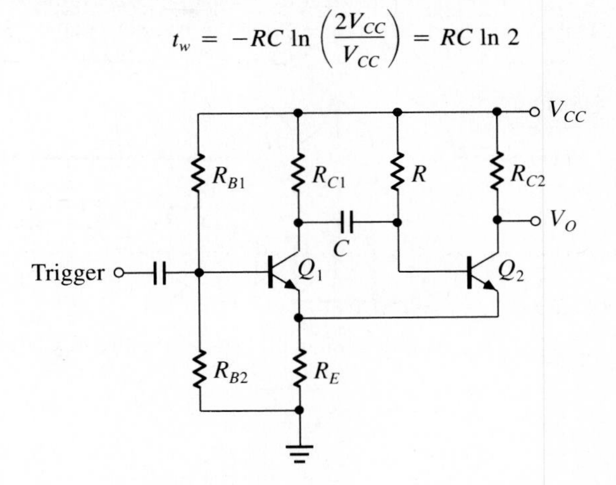

8.7 Determine the output pulse width of the CMOS monostable shown in the diagram. Assume $V_T = \frac{1}{2} V_{CC}$.

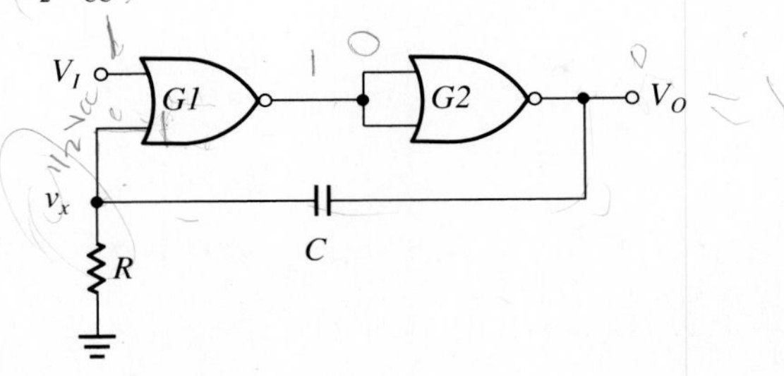

8.8 As depicted in the diagram, a Schmitt-trigger NAND gate is used as a monostable multivibrator. Show that the output pulse width is given by

$$t_w = RC \ln \left(\frac{V_{CC}}{V_{CC} - V_{T+}} \right)$$

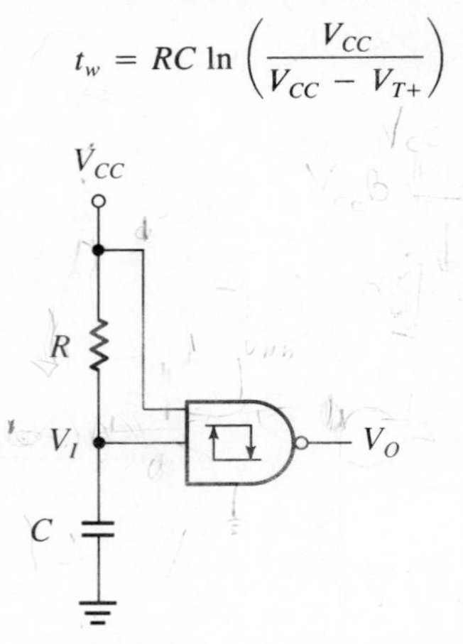

8.9 The following diagram shows a simple, astable multivibrator using a CMOS Schmitt trigger and its input and output waveforms. Find an expression for the period of the output waveform assuming that $t_1 + t_2 >> t_{PLH} + t_{PHL}$. If the Schmitt trigger is the TTL 7414, should there be an upper limit on the value of the resistor?

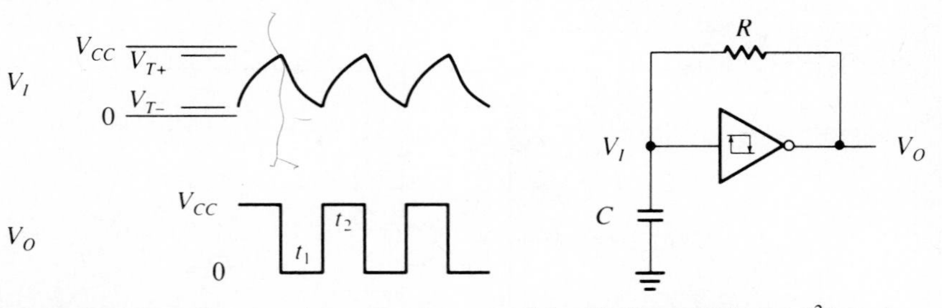

8.10 In the following diagram, determine the type of the flip-flop which uses I^2L logic.

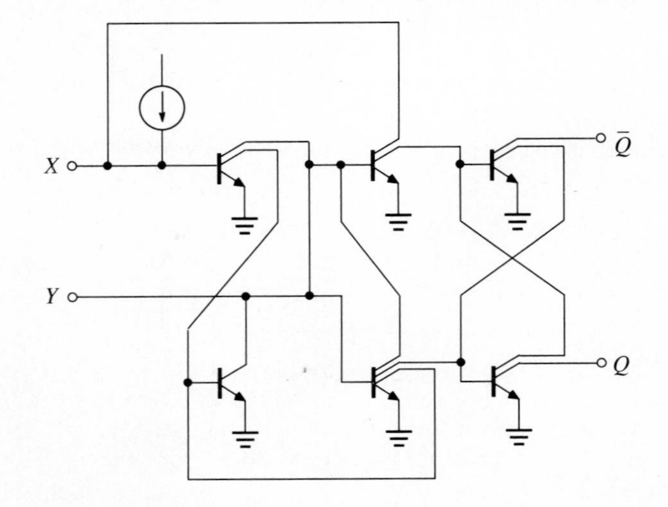

8.11 Using only CMOS TGs, design the following:

a. A two-input OR gate

b. A two-input AND gate

8.12 For the 7474 DFF, $t_h = 5$ ns, $t_{su} = 20$ ns, $t_{wh} = 30$ ns, and $t_{wl} = 37$ ns. Find the maximum clock frequency f_{max}. What should be the minimum input pulse width?

8.13 The following is the circuit diagram of the 74LS109 positive-edge-triggered J$\overline{\text{K}}$FF. Find the corresponding logic diagram.

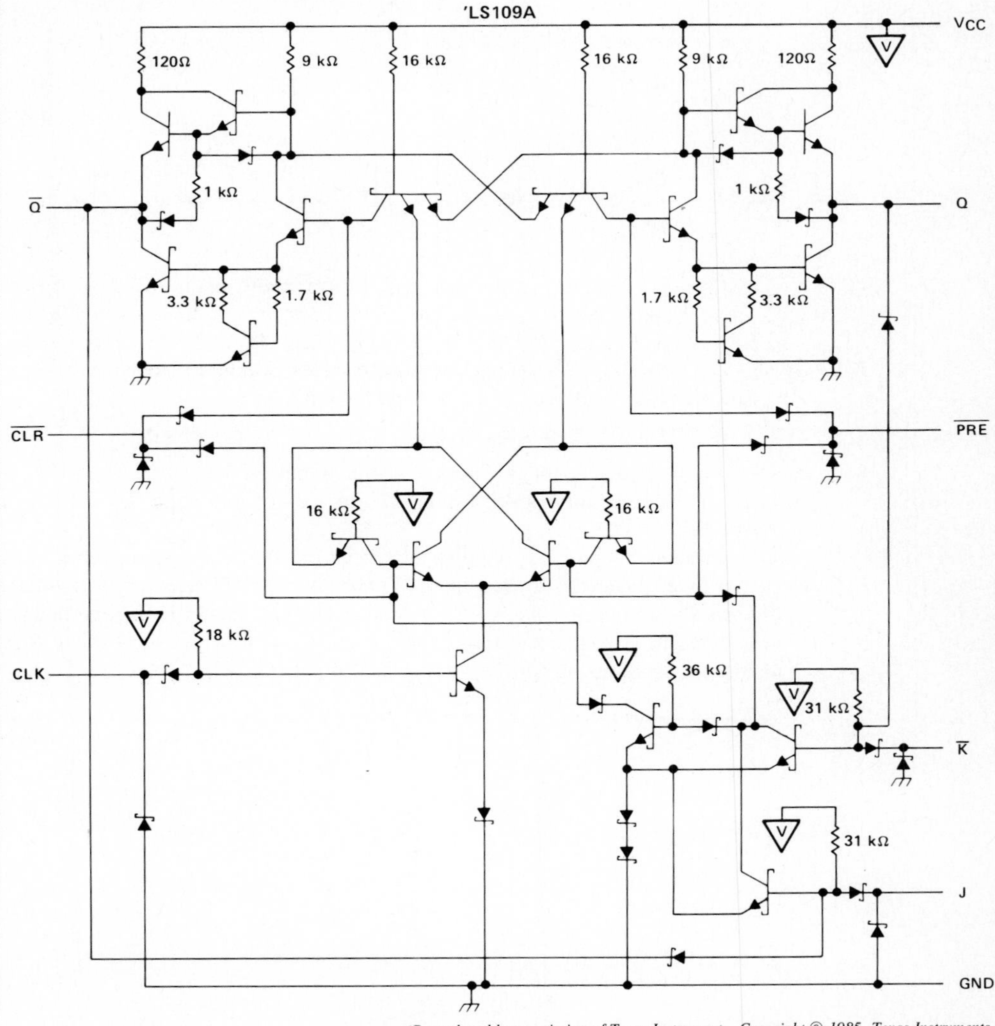

(Reproduced by permission of Texas Instruments. Copyright © 1985, Texas Instruments.)

8.14 A 2-V, 2-MHz sine wave is applied to a threshold detector that is to produce an output pulse if the input exceeds 1.7 V. What is the output pulse width for each of the following threshold detectors?

a. A NAND gate whose output changes when the input crosses 1.7 V.

b. A 7414 Schmitt trigger.

8.15 A special type of monostable multivibrator, called a *pulse stretcher*, is made up of an inverter, a Schmitt trigger, and external components, as depicted in the diagram. When CMOS chips are used and for very narrow output pulses under 100 ns, the capacitor can be omitted for the resistor to charge up the CMOS gate capacitance.

a. Describe the operation of the circuit.
b. For the inverter to be able to discharge the capacitor during the input pulse, what should be its minimum sinking capability?
c. Find an expression for the output pulse width.

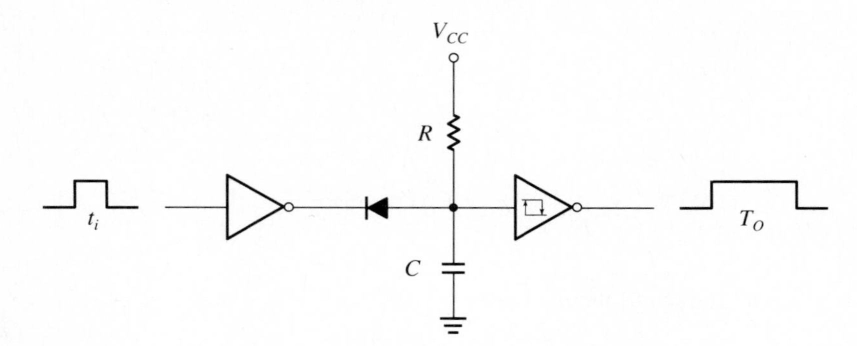

8.16 Given $V_{CC} = 5V_T = 5$ V, design a symmetric CMOS Schmitt trigger with $V_H =$ 1 V. Run a SPICE transient simulation on the circuit with

```
VIN 1 0 PULSE 0 5 0 1N 1N 5N
```

8.17 a. The astable multivibrator of Figure 8.43b may not oscillate if the transition regions of the gates differ even by only a few tens of millivolts. Use SPICE to verify this.

b. To overcome this problem, modify the circuit by adding an external element so as to provide hysteresis by delaying *G1*'s transition until *C* has sufficient voltage to ensure a smooth transition. Also, another element should be incorporated to provide positive feedback to expedite *G1*'s transition between logical states. Simulate your circuit using SPICE.

9. SAMPLE-AND-HOLD CIRCUITS AND DATA CONVERTERS

The inputs and outputs to the digital computers that are used to monitor and control processes are generally *analog* quantities. The input, then, must be converted to digital information before it can be used by the processor. This requires an *analog-to-digital* (A/D) conversion. Similarly, the purpose of *digital-to-analog* (D/A) conversion at the computer output is to produce a unique and consistent analog quantity (voltage or current) for a given digital input code. An *A/D converter* (*ADC*) can operate with great accuracy at high speeds with precise timing of samples only if a *sample-and-hold* (S/H) circuit is used between the input signal and the converter. Consequently, we will start this chapter by first discussing this circuit.

9.1 Sample-and-Hold Circuit

The sample-and-hold circuit is one of the key components in a data-acquisition system. In addition to being used to provide *quasi-dc* inputs for ADCs, they are also utilized in such applications as peak detectors, pulse stretchers, and *D/A-converter* (*DAC*) deglitchers. However, they are inherently troublesome since their operation may contribute errors to a signal, as will be clear below. They are often called *track-and-hold* (T/H) circuits because most of them can *track* a signal until the *hold* command is received, after which they hold. Therefore, an S/H circuit samples the input for a short time and stays in the hold mode for the duration of the duty cycle. A T/H circuit, on the other hand, spends most of the time tracking the input and switches to hold mode for only short intervals. Naturally, at very high speeds, the terms lose their distinction.

Sampling Theorem

The *Nyquist criterion* states that as long as the sampling rate f_s is greater than $2f_m$ samples per second, where f_m is the highest frequency component of the essentially band-limited signal $v(t)$, $v(t)$ can be determined uniquely by the sampled values. The frequency spectrum $V_s(f)$ of the sampled signal $v_s(t)$ will then consist of nonoverlapping repetitions of $V(f)$, the frequency spectrum of the original signal $v(t)$. This is illustrated in Figure 9.1a. However, if a signal is *undersampled*, that is, sampled at a rate below the *Nyquist rate* $f_s = 2f_m$, the frequency spectrum $V_s(f)$ will consist of overlapping repetitions of $V(f)$, as shown in Figure 9.1b. $V_s(f)$ will no longer have the complete information about $V(f)$ due to the overlapping tails; thus it will not be possible to recover $v(t)$ from $v_s(t)$. The passage of the latter through a low pass filter then leads to a distorted version of the original signal because (1) the tail of $V(f)$ is lost beyond $|f| > f_s/2$, and (2) the same tail appears folded onto the spectrum at the cutoff frequency. This *spectral folding* is called *aliasing*, as shown shaded in Figure 9.1b. The data will be ambiguous in the frequency band between $f_s - f_m$ and f_m. Higher sampling rates or the employment of sharp cutoff filters are two ways to eliminate the aliased portion. However, the latter technique will inevitably lead to a loss of some of the signal information.

Basic Open-Loop S/H Circuit

An idealized circuit to illustrate the operation of the S/H circuit is shown in Figure 9.2. The switch is closed at the start of each of the sample times and then is opened. The time between the switch closings is called the *sampling period*. The time that the switch is closed is short in comparison with the sampling period and much shorter than the period of the highest frequency in $v_i(t)$; thus, the input voltage will essentially be constant in this time interval, during which the capacitor charges to the instantaneous value of the input voltage. Then the switch opens, and the output voltage $v_o(t)$ remains at this value until the switch closes again at the next sampling period. One of the problems associated with this simple circuit is that the internal impedance of the input generator is assumed to be zero, as a result of which C is

Figure 9.1 (a) Continuous signal spectrum; (b) sampled signal spectrum with aliasing, $f_s < 2f_m$; and (c) sampled signal spectrum with $f_s > 2f_m$.

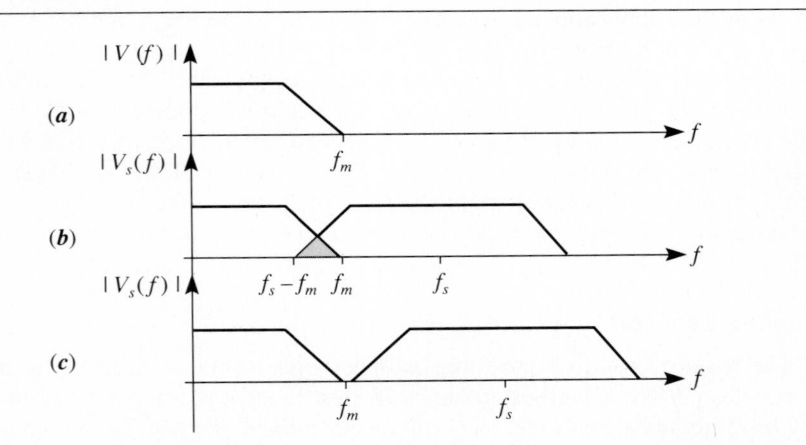

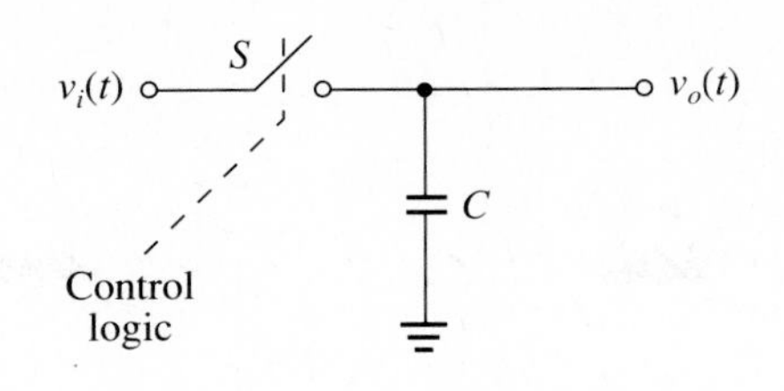

Figure 9.2 Ideal sample-and-hold circuit.

charged and discharged rapidly through the source. In addition, the voltage $v_o(t)$ does not actually remain constant when the switch is open because there will be a leakage current, however small, drawn from the capacitor.

A more practical form of the circuit in Figure 9.2 is shown in Figure 9.3. Here, R_i represents the source impedance in series with the switch resistance, and R_L represents the input impedance of the output buffer that is coupled across the capacitor to read the sampled value.

Figure 9.4 shows the basic open-loop S/H circuit, in which a MOSFET is used as a switch. When the clock pulse is present at the gate, the transistor switches on and shows a low resistance. The holding capacitor C then charges to the instantaneous value of the input voltage with a time constant R_aC, where $R_a = R_o + r_{ds(on)}$. Here R_o is the extremely low output impedance of the op amp *A1* connected as a voltage follower at the input, and $r_{ds(on)}$ is the on resistance of the MOSFET. In the absence of the clock pulse, the transistor is cut off and acts as a high impedance to isolate the capacitor. Thus, the sampled voltage value is held by the capacitor. The output voltage follower with its high input impedance is used as a buffer to isolate the capacitor from the external circuit. The individual gains of the op amps are ideally +1.

Another form of the switch used in S/H circuits is the diode bridge of Figure 9.5. It is a very fast and low-impedance circuit with low charge-transfer characteristics. The diode quad is available as an IC with very closely matched diodes having almost identical forward voltage drops V_f.

Suppose $|v_i| < V_1 - V_f$. When the first sampling pulse is applied to the V^+ input, D_1 becomes forward-biased and point a rises toward a value of $v_i + V_f$. Since D_3 is also forward-biased, it provides a path to charge the capacitor to v_i. The sampling pulse width t_s must be long enough to charge C through R. When v_i becomes

Figure 9.3 Practical S/H circuit.

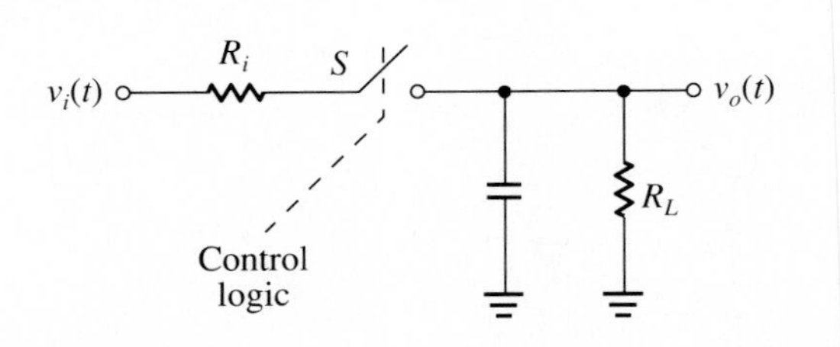

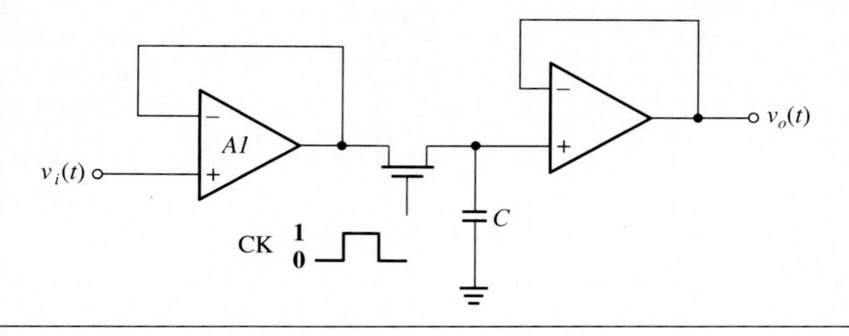

Figure 9.4 Basic open-loop S/H circuit.

more negative than its previous value, the capacitor is discharged through D_4 to v_i during the time V^- swings positive. Between sampling pulses, the diodes are reverse-biased, resulting in very little discharge of the capacitor.

The S/H circuit with its output voltage follower providing a low impedance to the ADC should be physically very close to the converter to minimize inductance and the ringing problems associated with it. Another factor affecting performance is the type of the holding capacitor. In discrete-component or hybrid circuits, a capacitor with polycarbonate, polyethylene, or polystyrene dielectric should be used; other types do not retain the stored voltage well due to the effective leakage resistance of the element. In addition, a phenomenon called *dielectric absorption* causes a capacitor to *remember* a fraction of its previous charge when there is a change in the capacitor voltage. This is due to the tendency of the charges within the capacitor to redistribute themselves over a period of time. This polarization changes the electric field intensity in the capacitor, leading to a long-term voltage error. It exhibits multiple exponential decays that can be modeled as series RC networks in parallel with the holding capacitor and with different time constants. In a polar dielectric such as a tantalum capac-

Figure 9.5 The diode bridge.

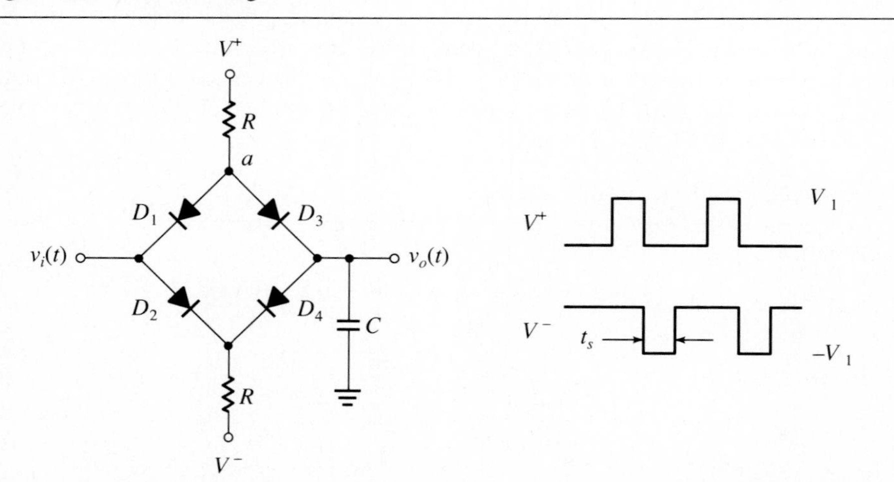

itor, the dielectric absorption can produce an error as large as 8%, while a nonpolar dielectric such as polystyrene produces errors of only about .04%. Dielectric absorption is usually negligible at resolutions up to 12 bits, but at 16-bit resolution it becomes a problem.

S/H Parameters

An S/H circuit is characterized by the following parameters:

Acquisition time: the time required for the capacitor to charge to a specified percentage of the analog signal value, including the time it takes the switch to close. In other words, it is the shortest time after a *sample* command has been issued that a *hold* command can be given and result in an output voltage that approximates the input voltage with the necessary accuracy. Therefore, it is not just the time required for the output to settle but also includes the time required for all internal nodes to settle so that the output assumes the proper value when switched to the hold mode. Figure 9.6 illustrates the acquisition time. Since the output voltage is an exponential, in order for it to be within, say, $\pm.01\%$ of the input, the time required will be approximately 9τ, where $\tau = R_aC$. If, on the other hand, R_o and $r_{ds(on)}$ are negligibly small, the acquisition time is limited by the *slew rate*, that is, by the maximum current I_{max} that the input operational amplifier can deliver. The capacitor voltage then changes at a peak rate of $dv_C(t)/dt = I_{max}/C$.

Aperture delay: the time interval between applying the hold control signal and the opening of the switch to hold the sample. Though this delay by itself does not really present a problem, its random sample-to-sample variation (i.e., its uncertainty), called the *aperture jitter*, coupled with the slew rate of the signal at the holding capacitor, may degrade system performance by causing an error in the held output voltage. Figure 9.6 illustrates the effect of this parameter.

Figure 9.6 Illustration of various S/H parameters.

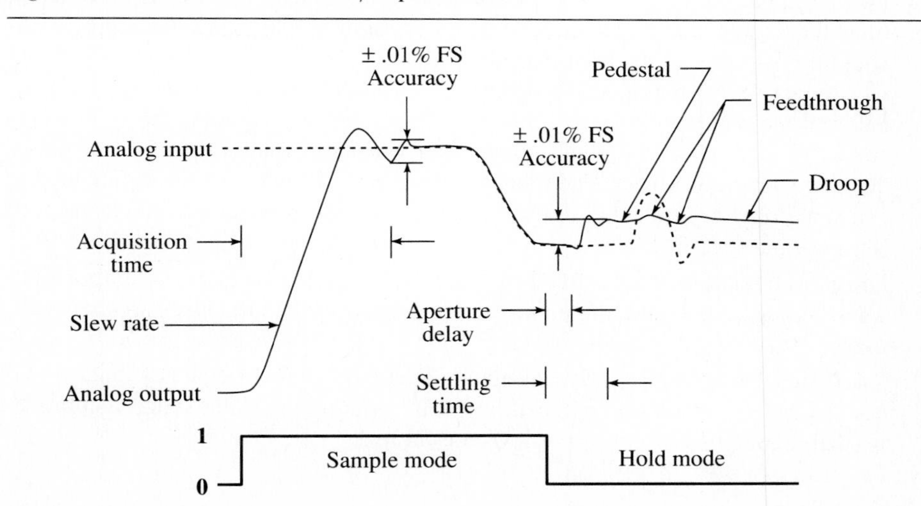

Droop rate: the rate of decay of the stored voltage, $dv_C(t)/dt$, caused by leakage during the hold time. It is inversely proportional to the capacitance because $dv_C(t)/dt = I_L/C$, where I_L is the net capacitor leakage current, also called the *drift current*. The drift current increases linearly with temperature.

Settling time: the time from the opening of the switch to the point when the output has settled to its final value within a specified percentage. In other words, it is the time required for the output to settle after the hold logic command. The A/D conversion should not begin until after the settling time.

Slew rate: the maximum possible rate of change of the voltage; therefore, it determines the maximum current I_{max} that the input op amp can deliver. It is usually limited by the rise and fall times of the op amp. The slew rate should exceed the maximum slope of the input signal to ensure that the circuit tracks the input signal.

Pedestal: the step error at the output caused by the charge transferred across the series switch onto the holding capacitor from the switch-control logic as the hold command is issued. In new-generation S/H circuits, its dependence on the input voltage has been eliminated. It may be calculated from the charge transfer, which is a specified parameter in data sheets, using the following relationship:

$$\text{pedestal (V)} = \frac{\text{charge transfer (pC)}}{\text{holding capacitance (pF)}}$$

Feedthrough: the error at the output caused by the fast-slewing input signals during the hold interval even though the series switch is open. The *feedthrough rejection*, the ratio of the signal that passes through the open switch to the analog input signal (expressed in dB), describes how well the switch keeps the input signals from feeding through to the output when the device is in the hold mode. It is caused primarily by the switch capacitance and varies with the signal frequency. Thus, the rejection is better for low-frequency inputs.

In addition, the feedthrough through the power supply can be improved if bypassing is properly done. The S/H ICs that have separate power supply pins for the input and output offer superior performance since the user is allowed to filter the supply line between the various sections to eliminate the supply as a coupling path during the hold period.

As a rule of thumb, the feedthrough rejection should be $(6n + 2)$ dB or better at the maximum analog input frequency, where n signifies the system accuracy in bits. This argument will be clear when we discuss the effect of noise due to quantization errors. The transients that feed through from the digital inputs and settle out very quickly are usually accounted for in the acquisition-time and settling-time specifications.

Full-power bandwidth: the frequency at which a full-scale input or output sine wave becomes slew-rate limited to -3 dB. It is also called the *large-signal bandwidth*.

Small-signal bandwidth: the maximum analog input signal frequency that can be tracked before the gain is reduced by 3 dB, assuming that the signal amplitude is small enough so as not to be slew-rate limited.

Closed-Loop S/H Circuit

The open-loop structure has the essential advantage of fast acquisition and settling time. If low-frequency tracking accuracy is more desirable than speed, then this is accomplished by closing the loop around the holding capacitor, as in Figure 9.7. In this configuration, the input follower is replaced by a high-gain difference amplifier to enforce the tracking accuracy.

When the switch is closed, the output, representing the charge on the capacitor, is forced to track the input within the difference amplifier's gain, bandwidth, and current-driving capabilities. Since both the input and output affect the charge on the capacitor, the acquisition and settling times are identical to each other. If the circuit is switched into the hold mode before the output has settled to its final value, the sample may be in error. In open-loop circuits, on the other hand, the opening of the switch once the charge has been acquired but before the output has settled does not affect the final output value.

When the circuit returns to its sample mode, the input stage must reacquire the input because the loop has been opened during the hold mode, resulting in a spike due to the high voltage gain of the input amplifier.

9.2 DAC Input and ADC Output Formats

Natural Binary

Even though the following discussion of binary codes is mainly based on DAC inputs, it applies to ADC outputs as well.

The most commonly used digital input code for an n-bit DAC is the *natural binary* number N represented as

$$N = a_{-1}2^{-1} + a_{-2}2^{-2} + \dots + a_{-n}2^{-n}$$
$$= \sum_{i=1}^{n} (a_{-i}2^{-i}) \tag{9.1}$$

where the coefficients a_{-i} assume the values of 0 or 1. Usually, the input code is written in the form of a binary integer with the fractional nature of the corresponding number understood. This data format is sometimes called *left-justified*, implying a

Figure 9.7 Closed-loop S/H circuit.

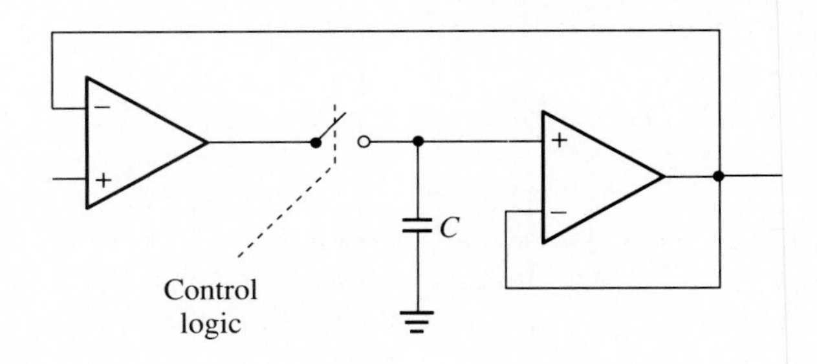

binary point to the left of the *most-significant-bit* (*MSB*). Thus, the MSB has a weight of 1/2, the next bit of 1/4, and so forth. The input binary digit carrying the lowest numerical weight (i.e., 2^{-n}) is called the *least-significant-bit* (*LSB*).

The analog output is related to its binary input as

$$V_O = NV_{FS} \tag{9.2}$$

where V_{FS} is defined as the nominal *full-scale* (*FS*) output of the DAC. The actual FS output, when all the input bits are 1, is given as

$$V_{O(FS)} = (1 - 2^{-n})V_{FS} \tag{9.3}$$

The term $V_{FS}/2^n$ is the smallest possible analog output step that the DAC can resolve. It is known as the 1 LSB output-level change.

The transfer function of an ideal 3-bit DAC is plotted in Figure 9.8. It has $2^3 = 8$ coded levels and hence eight corresponding outputs, ranging from 0 to 7/8 of the FS. Note that the FS is not available digitally. However, it represents the reference quantity to which the analog variable is normalized. The line that connects 0 and FS is called the *gain curve*. The output is *unipolar* because it swings in only one direction.

The transfer function of an ideal and unipolar 3-bit ADC, offset −LSB/2 at zero scale, is illustrated in Figure 9.9a. In many ADCs, an offset is intentionally introduced to adjust the positions of the transitions to suit the particular application.

The ideal analog values at which the transitions should occur in an n-bit ADC are calculated as

Figure 9.8 The ideal transfer function of a 3-bit DAC.

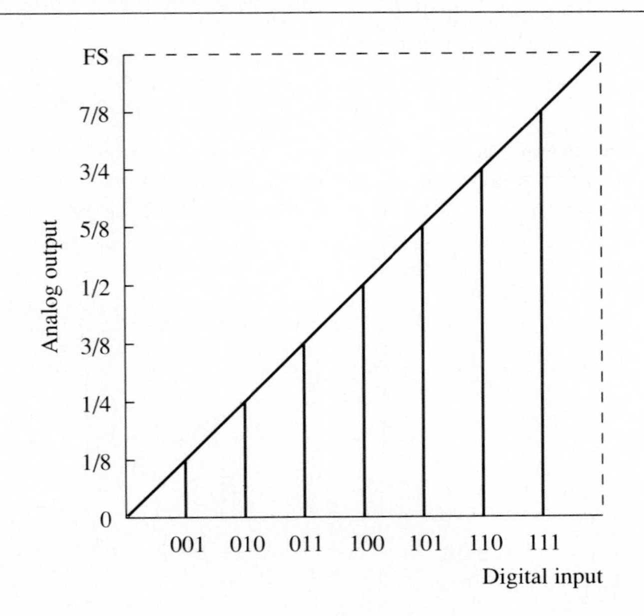

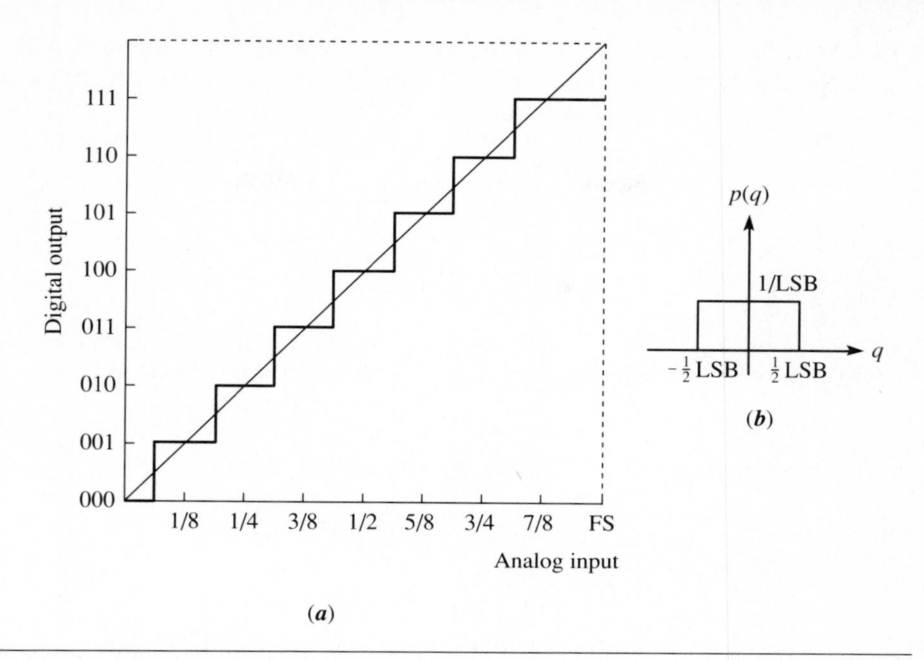

Figure 9.9 (a) The ideal transfer function of a 3-bit ADC; (b) the probability density function.

$$V_I = NV_{FS} + V_{off} \tag{9.4}$$

where V_{off} is the offset voltage.

The process of digitizing samples at the ADC input involves an approximation called *quantization*. In Figure 9.9a, the analog input values are quantized by partitioning the continuum into eight discrete ranges so that analog values within a given range are represented by the same digital code, generally corresponding to the mid-range value, and leading to an inevitable *quantization error* of $q = \pm\text{LSB}/2$ in addition to other conversion errors, as will be discussed below. Note that this value is the maximum error for an otherwise perfect ADC with a transfer function offset LSB/2 at zero scale. If the transfer function is not offset, the maximum quantization error will increase to 1 LSB because the sample values may differ at maximum by this amount.

To find an expression for the *signal-to-noise ratio* (*SNR*) for an ideal n-bit ADC whose error is solely due to the quantization, we first recall that the power S of a signal $s(t)$ that exists over the entire interval $(-\infty, \infty)$ is defined as the time average of the squared signal. Thus,

$$S = \lim_{T\to\infty} \frac{1}{T}\int_{-T/2}^{T/2} s^2(t)dt \tag{9.5}$$

where T is the period of the signal. Since a sample amplitude is approximated by the midpoint of the interval in which it lies, the quantization error has a range

$[-V_{FS}/2^{(n+1)}, +V_{FS}/2^{(n+1)}]$ for a full-scale range of V_{FS} volts. The probability density function of the error $p(q)$ is diagrammed in Figure 9.9b. Therefore, the mean square of the error, that is, the noise power, is given by

$$N = \int_{-\text{LSB}/2}^{\text{LSB}/2} q^2 p(q) dq = \int_{-\text{LSB}/2}^{\text{LSB}/2} \frac{q^2}{\text{LSB}} dq$$

$$= (\text{LSB})^2/12 = \frac{V_{FS}^2}{3 \times 2^{2(n+1)}} \tag{9.6}$$

Thus, once n and V_{FS} are known, the quantization noise is fixed. For a sinusoid having an amplitude of $V_{FS}/2$, using Equation (9.5), the signal power is found to be

$$S = V_{FS}^2/8 \tag{9.7}$$

Finally, using Equations (9.6) and (9.7), we obtain an expression for the SNR in dB as

$$\text{SNR} = 10 \log(S/N) = 10 \log(3 \times 2^{2n-1})$$

$$= [1.76 + 6.02n] \text{ dB} \tag{9.8}$$

Consequently, the SNR for an ideal ADC is dependent only on the resolution. Note that increasing the resolution by one bit quadruples S/N, corresponding to a 6.02 dB increase in the SNR.

Offset Binary and Two's Complement Binary

To convey the sign information of *bipolar* analog signals, either an extra bit is used or the MSB is reinterpreted as the sign bit. The extra bit doubles the analog range and halves the peak-to-peak resolution. If the existing MSB is used as the sign bit, then the analog range may still be doubled, but only at the expense of resolution.

Example 9.1

Consider an 8-bit code with a resolution of $1/2^8$ and representing a 10-V range. If it is extended to 9 bits to indicate the polarity, the peak-to-peak resolution will be $1/2^9$ and the analog range will double to ± 10 V.

If, on the other hand, the original 8-bit code is retained, the resolution will remain the same, but stretching the range to ± 10 V will double the magnitude of the LSB.

The *offset binary* code is obtained by offsetting the natural binary code by 1/2 scale, so that the half-scale code, 100 . . . 0, becomes zero. The maximum negative scale is represented by all zeros, while the maximum positive scale is represented as all ones. The disadvantage of this code is the major bit transition occurring at zero and involving all the bits. This can result in glitch problems due to possible difference in speed between bits turning on and off.

Another bipolar code widely used to represent negative values is the *two's complement binary* code, which is achieved by complementing the MSB of the offset binary code. Data converters using two's complement have the same disadvantage as those that respond to the offset binary. The ideal transfer function for a 4-bit DAC, 3 bits plus sign, using these bipolar codes is shown in Figure 9.10; Figure 9.11 depicts the ideal 4-bit bipolar A/D conversion relationship for the same codes.

9.3 Converter Parameters and Performance Specifications

Following is a list of important parameters and performance specifications for data converters:

Resolution applies to both types of data converters and specifies the smallest standard incremental change in the output voltage of a DAC or the amount of input voltage change required to increment the output of an ADC between adjacent codes. It may be expressed in percent of the FS or in binary bits. It is a design parameter rather than a performance specification. A 12-bit DAC will have $2^{12} = 4096$ possible output levels, including zero. An ADC with 12-bit

Figure 9.10 The ideal transfer-function for a 4-bit DAC with offset binary and two's complement codes.

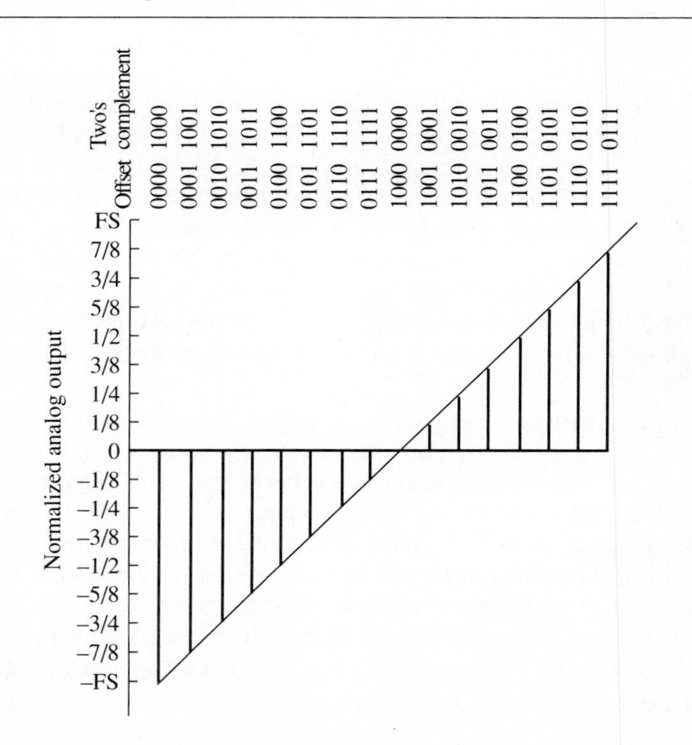

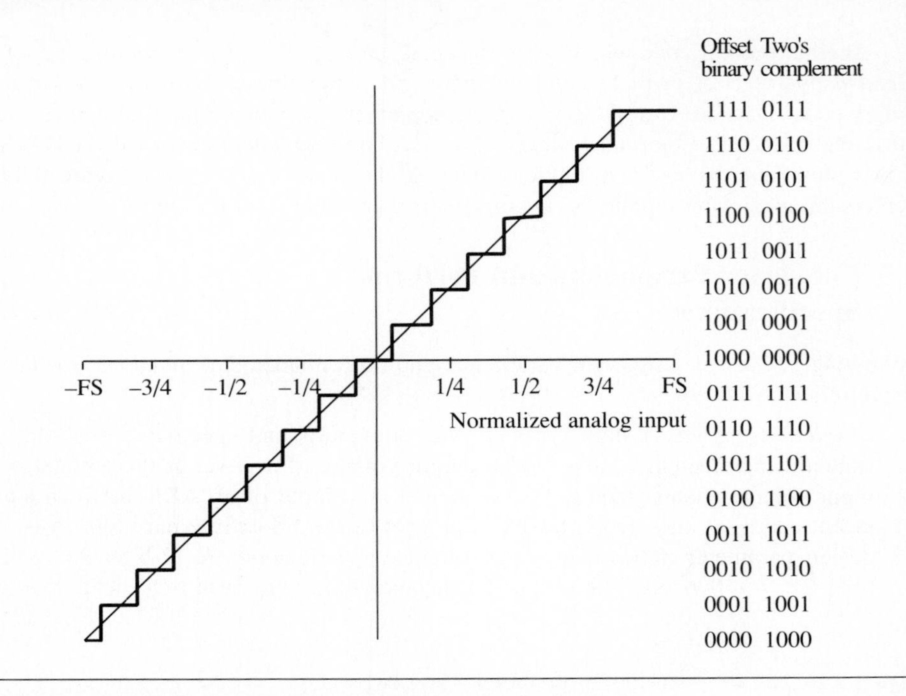

Figure 9.11 The ideal transfer function for a 4-bit ADC with offset binary and two's complement codes.

resolution can resolve one part in $2^{12} = 4096$ or .0244% of the full scale. Therefore, an ADC with 10 V FS will resolve a 2.44 mV change at the input, while a 12-bit DAC will exhibit an output voltage change of .0244% of FS when the input code is incremented by one. Table 9.1 lists the binary bit weights (i.e., the resolution) for numbers having up to 16 bits. The dB column represents the base-10 logarithm of the ratio of the LSB value to the FS, multiplied by 20. Thus, each successive power of 2 signifies a change of $20 \log_{10} 2 = 6.02$ dB. Table 9.2 lists, for 5 V full-scale and resolutions up to 16 bits, the LSB values, the maximum *all ones* values, and A/D converter transition points at LSB/2 and FS − 1½ LSB.

Accuracy is a measure of the worst-case deviation of the analog output level of the DAC from the ideal output value under any input combination. As applied to an ADC, it describes the difference between the actual input voltage and the FS-weighted equivalent of the binary output code. Accuracy covers all errors, including quantization error, system noise, and deviations from linearity. For a DAC with 12-bit resolution, an accuracy of ±LSB/2 corresponds to an error of magnitude $(1/2)(1/2^{12})$ or ±.0122% of the full scale. If a 12-bit ADC is specified to be ±1 LSB accurate, this is equivalent to ±.0244% of the FS.

Integral nonlinearity or *Linearity error* describes for both types of data converters the departure of the analog values from the ideal transfer curve, which passes

Table 9.1 Resolution up to 16 Bits

Bit	2^{-n}	Fraction	dB	Percentage
FS	2^{0}	1	0	100
MSB	2^{-1}	1/2	−6	50
2	2^{-2}	1/4	−12	25
3	2^{-3}	1/8	−18.1	12.5
4	2^{-4}	1/16	−24.1	6.25
5	2^{-5}	1/32	−30.1	3.125
6	2^{-6}	1/64	−36.1	1.563
7	2^{-7}	1/128	−42.1	.781
8	2^{-8}	1/256	−48.2	.391
9	2^{-9}	1/512	−54.2	.195
10	2^{-10}	1/1024	−60.2	.0977
11	2^{-11}	1/2048	−66.2	.0488
12	2^{-12}	1/4096	−72.2	.0244
13	2^{-13}	1/8192	−78.3	.0122
14	2^{-14}	1/16384	−84.3	.0061
15	2^{-15}	1/32768	−90.3	.0031
16	2^{-16}	1/65536	−96.3	.0015

through the *end points* 0 and FS. In a DAC with a nonlinear transfer function, the differences between the heights of adjacent bars are not equal, or they do not change uniformly. Similarly, an ADC has linearity error if the differences between the transition values are unequal or not uniformly changing.

Linearity error does not include quantization, gain drift or offset errors. It is measured after adjusting for zero and the full scale, and is a parameter intrinsic to the device that cannot be externally adjusted. It may be expressed in percent of FS or in fractional LSB. Figures 9.12 and 9.13 show the linearity error in DACs and ADCs, respectively.

Table 9.2 LSB and MSB Values and ADC Transitions for 5-V Full Scale

			ADC Transitions	
Bits	LSB	MSB, V	To ½LSB	To FS − 1½ LSB, V
8	19.55 mV	4.9805	9.75 mV	4.9705
10	4.885	4.995	2.44	4.9925
12	1.22	4.9988	.61	4.9982
16	76.5 μV	4.9999	38 μV	4.9999

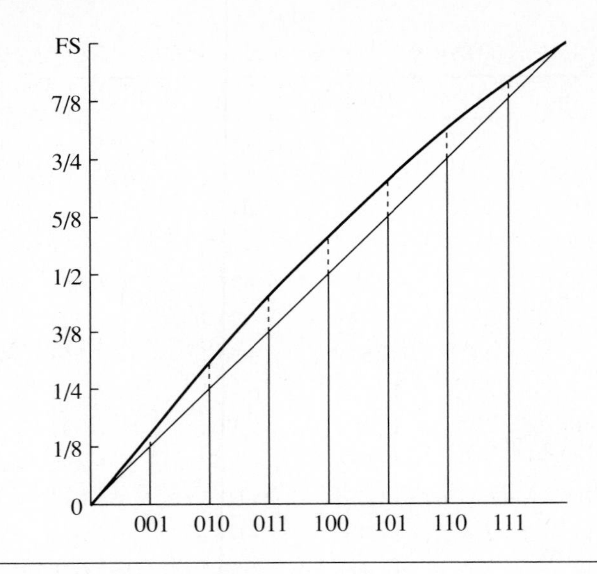

Figure 9.12 Nonlinearity in a 3-bit DAC.

Differential nonlinearity exists whenever there is an integral nonlinearity. It specifies the uniformity of a converter's bit width. That is, it indicates the difference between the actual analog voltage change and the ideal, $V_{FS}/2^n = 1$ LSB, voltage change for a 1-bit change in the input code of a DAC. For example, a DAC with a 1½ LSB step at a code change is said to exhibit LSB/2 differential nonlinearity.

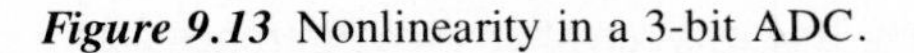

Figure 9.13 Nonlinearity in a 3-bit ADC.

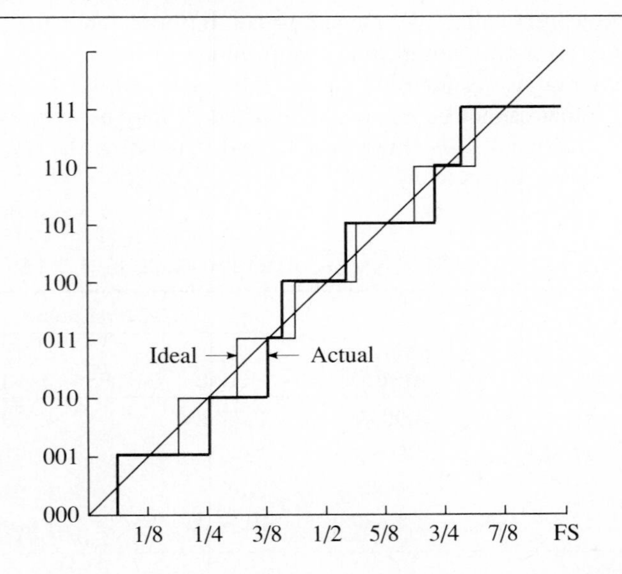

If the nonlinearity is sufficiently negative, the device may be *non-monotonic*; that is, one or more values of the analog output may actually be less than the values corresponding to the codes having similar weight.

Likewise, differential nonlinearity in an ADC is defined as the deviation in code width from the value of 1 LSB. If the nonlinearity is more negative than −1 LSB, the code width will vanish entirely, and the ADC will have a missing code analogous to the non-monotonic D/A conversion. Therefore, there will be no analog voltage value in the entire FS range that can generate that code at the output.

Gain drift or *FS error* is a measure of the change in the FS analog output of a DAC. In an ADC it is the departure of the actual input voltage from the ideal input voltage at the last transition, which should occur $1\frac{1}{2}$ LSB below the nominal full scale. This error appears as a change in the slope of the transfer function. Gain drift can be caused by errors in the reference voltage, amplifier gain or external resistor values. Figures 9.14 and 9.15 depict the gain drift in DACs and ADCs, respectively.

Offset error is a measure of the change in the analog DAC output with zero code input, or it is the required mean value of the input voltage of an ADC to set zero code out. Figures 9.16 and 9.17 show the effect of the offset error in DACs and ADCs, respectively.

Temperature coefficients for gain drift and offset errors specify the maximum change from the initial 25°C value to the value at T_{max} or T_{min}. They are expressed over the specified temperature range in parts per million per °C (ppm/°C) or in percent of FS per °C (%FS/°C).

Figure 9.14 Gain error in a 3-bit DAC.

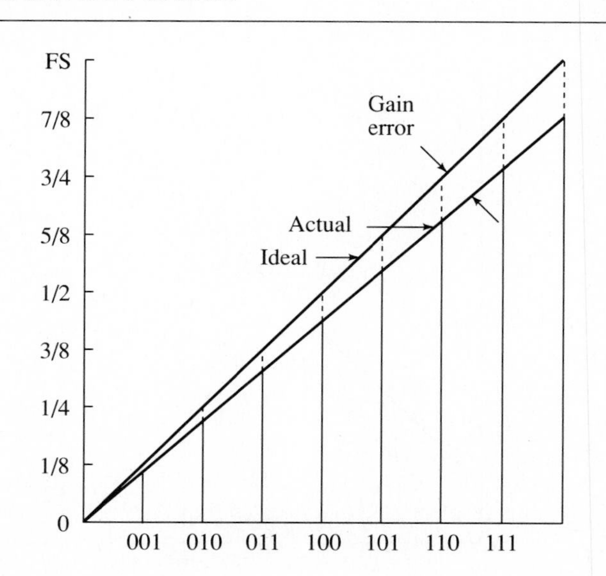

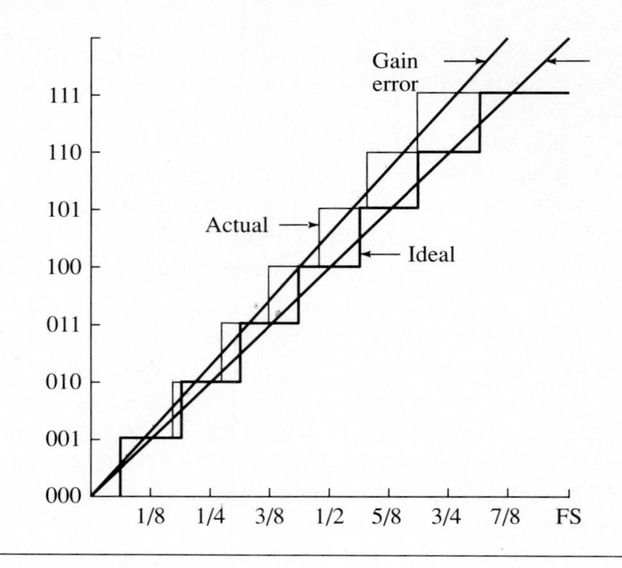

Figure 9.15 Gain error in a 3-bit ADC.

Settling time is the time elapsed for the DAC output to reach its final value within the limits of a defined error band, typically $\pm$LSB/2, after a code transition at the output.

Some DACs settle faster in one direction than the other. If bipolar operation is considered, voltage output in the negative-going direction can settle much more

Figure 9.16 Offset error in a 3-bit DAC.

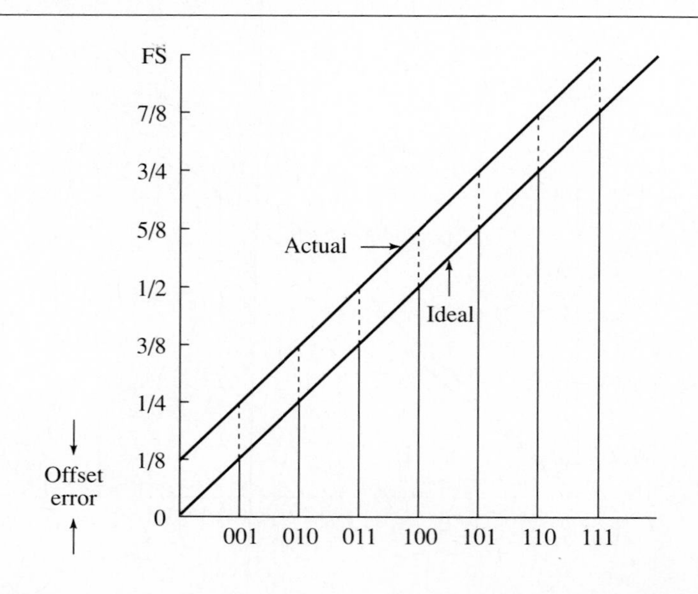

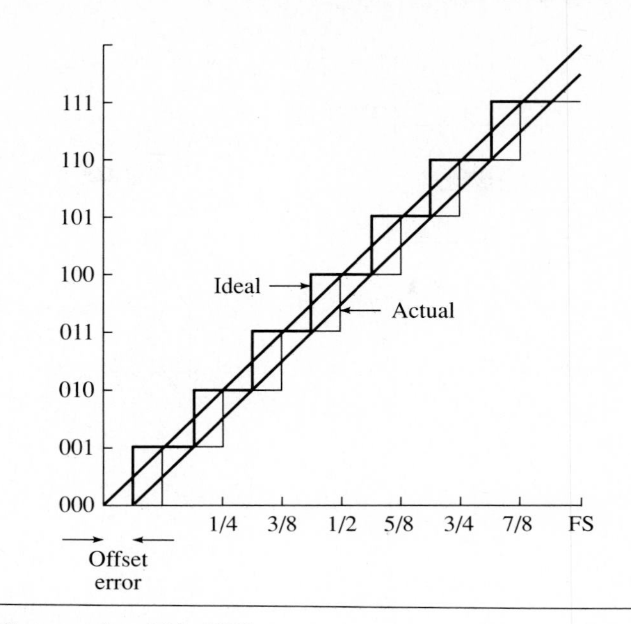

Figure 9.17 Offset error in a 3-bit ADC.

slowly than in the positive direction. The standard current mode for most IC DACs settles much faster than the voltage mode because the currents are simply steered one way or the other, as will be clear later, and there is no significant capacitive charging in the former, while the voltage mode requires an output op amp, resulting in a delay.

9.4 Digital-to-Analog Converters

Binary-Weighted DAC

The simplest approach to performing a D/A conversion uses *binary-weighted* precision resistors, a set of switches, and an op amp to realize Equation (9.2), as depicted in Figure 9.18. The parallel digital inputs a_{-i} are used to control the switches s_i. The value of the ith resistor is given by

$$R_i = 2^{i-1}R \tag{9.9}$$

where R is the one connected in the MSB position. The op amp holds one end of all resistors at virtual ground. Each switch that is closed adds a binary-weighted increment of the current $-V_R/R$ through the summing bus connected to the amplifier's inverting input. Therefore, the current I_f is obtained as

$$\begin{aligned} I_f &= \frac{-2V_R}{R}\left(a_{-1}2^{-1} + a_{-2}2^{-2} + \ldots + a_{-n}2^{-n}\right) \\ &= \frac{-2V_R N}{R} \end{aligned} \tag{9.10}$$

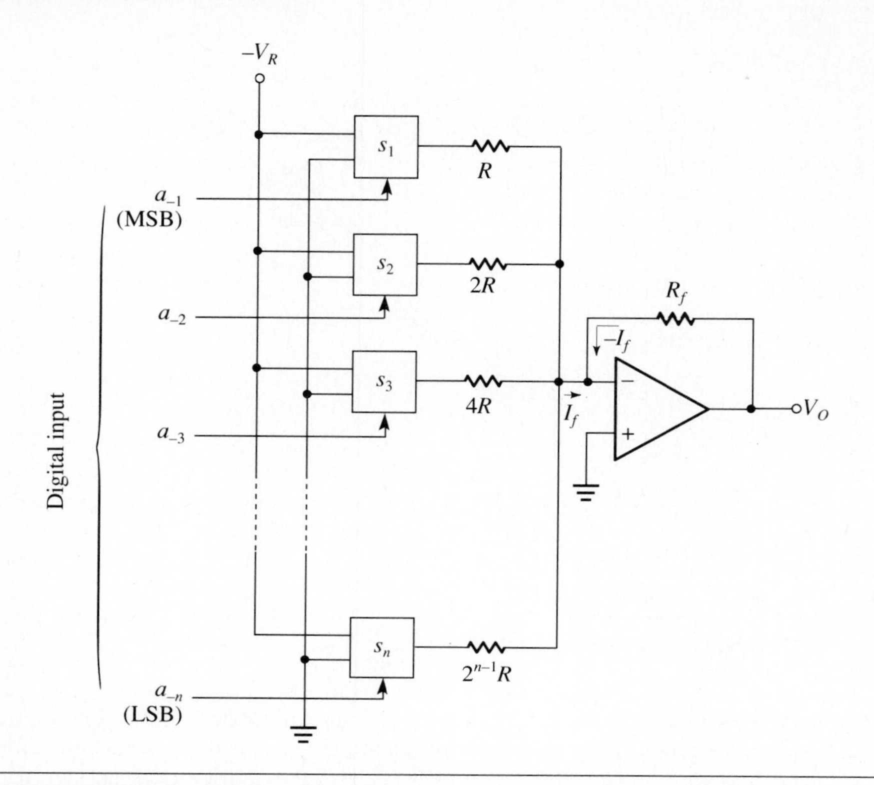

Figure 9.18 Binary-weighted DAC.

If all of the bits are set, Equation (9.10) becomes

$$I_{f(max)} = -\frac{V_R(1 - 2^{-n})}{R} \tag{9.11}$$

The op amp acts as a current-to-voltage converter, so that we get

$$V_O = -I_f R_f = \frac{2R_f V_R N}{R} \tag{9.12}$$

Thus, the resistors are weighted such that their resistances are inversely proportional to the numerical significance of the corresponding bit, and the analog output voltage is directly proportional to the numerical value of the digital input.

If the resistors are to be fabricated in IC form, the range of resistance values will be impractical even for an 8-bit DAC. If discrete resistors are used, the cost and size will be increased, and tracking advantages will be lost because the resistor temperature coefficients will not match. Consequently, this type of configuration is seldom used.

Ladder-Type DAC

One convenient approach to reduce the resistance range is shown in Figure 9.19. This popular form, called the *ladder-type*, utilizes twice as many resistors for the same number of bits as the binary-weighted resistor array but only two sizes R and $2R$, thereby simplifying the resistor trimming.

Binary-weighted currents flow continuously in the shunt arms of the resistive network. The switches steer the current to the appropriate output line in response to the digital input code. No matter what the switch settings are, the current finds a path either to a true ground or to a virtual ground at the op amp's negative input terminal. Both output buses I^+ and I^- are maintained at ground potential, either by the op amp feedback or by direct connection to the common ground.

Since the total resistance seen by the reference voltage is R, $I = V_R/R$ is constant, regardless of the digital input code. However, it is reduced by a factor of two at each node. The circuit has no voltage changes at the switch terminals, and there is no difference in propagation delay from left to right down the ladder associated with the switch closings.

Multiplying DAC

If V_R varies instead of being a fixed reference voltage, then the converter is called a *multiplying* DAC since the output is proportional to the product of the digital input and analog reference. The output polarity depends on the polarity of the analog signal and digital coding. If the DAC accepts reference signals of both polarities and the digital input is bipolar, then a *four-quadrant multiplication* is available. Single polarity of either the analog or the digital variable leads to a *two-quadrant multiplication*. Multiplying DACs may even be *fractional-quadrant* if the reference has a limited range of variation. The multiplying feature is useful for setting the gain of

Figure 9.19 Ladder-type DAC.

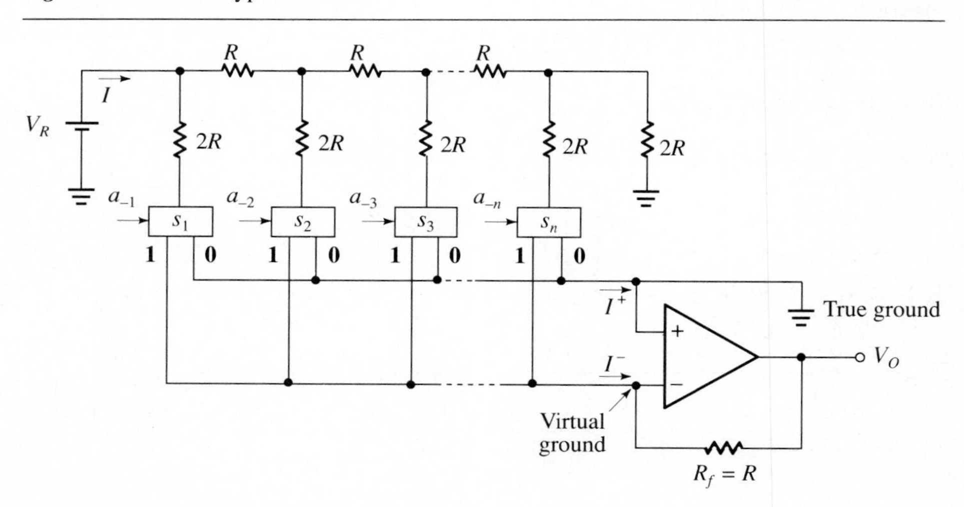

digitally programmable amplifiers and attenuators. They are also utilized to vary the output signals of control systems or audio equipment.

The internal configuration of a typical 10-bit multiplying DAC is diagrammed in Figure 9.20. An internal feedback resistor R_{fb} is included and must be used because an external resistor will not provide matching and temperature tracking.

For this particular converter, the reference voltage input can range from -10 V to $+10$ V. The digital input code controls the position of the MOS SPDT switches, which can switch currents of either polarity to achieve four-quadrant multiplication. To provide a voltage output, an external op amp is used as a current-to-voltage converter, as in Figure 9.21. The virtual ground at the inverting input due to the feedback action causes I_{out1} to be diverted to the feedback resistor to get

$$V_O = R_{bf}\,|I_{out1}|$$

The addition of a second op amp, as in Figure 9.22, gives, in effect, sign significance to the MSB of the input digital code to allow four-quadrant multiplication of the reference voltage. The applied digital word is in offset binary to get an output voltage

$$V_O = \frac{D - 512}{512} V_R \qquad 0 \le D \le 1023$$

where D is the decimal equivalent of the input with 1 LSB $= |V_R|/512$. Since the offset voltage error of the second op amp has no effect on linearity, only the offset voltage of the first amplifier needs to be nulled to ensure the linearity of the DAC. Another advantage of this configuration is that the external resistors do not have to match the value of the internal resistors but need only match and temperature-track each other. The operation is summarized in Table 9.3.

Figure 9.20 10-bit multiplying DAC.

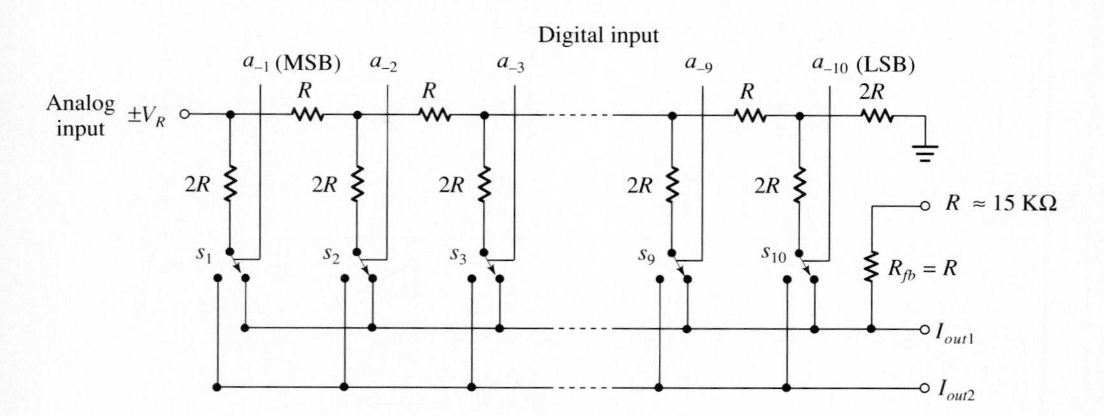

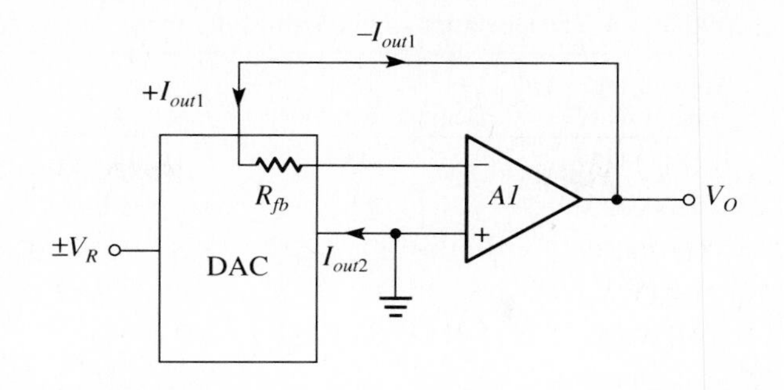

Figure 9.21 Multiplying DAC providing an output voltage (logic inputs omitted for clarity).

9.5 Analog-to-Digital Converters

Successive-Approximation ADC

Since they are capable of high speed as well as high resolution up to 16 bits, *successive-approximation* (*SA*) ADCs are by far the most popular type of converters and are widely used, especially for interfacing with computers. Only n clock cycles are needed to encode a number to n bits of resolution. Therefore, the conversion time is fixed and independent of the input signal magnitude. Also, since the internal logic is reset at the start of each conversion, conversions are independent of the results of previous conversions.

The SA technique compares the unknown analog input with a precise voltage or current output of a DAC, as depicted in Figure 9.23. At the heart of the circuit is the so-called *SA register* (*SAR*). After the conversion command is received and the converter is cleared, the conversion cycle starts with the SAR turning on its MSB, 1/2 of the full scale, to the DAC. This bit is compared with the analog input. If the input

Figure 9.22 Four-quadrant multiplication of the reference voltage.

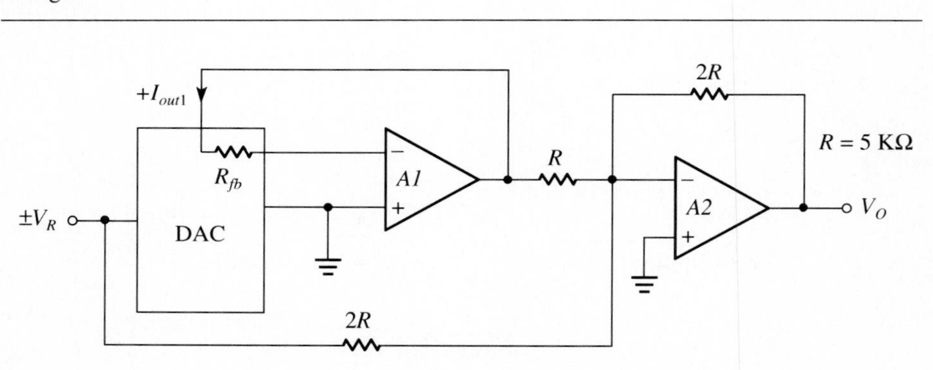

Table 9.3 The Operation of a 10-Bit Multiplying DAC

Applied input offset binary	Decimal equivalent	V_O
1111111111	1023	V_R − LSB
1100000000	768	½ V_R
1000000000	512	0
0111111111	511	−LSB
0100000000	256	−½ V_R
0000000000	0	$-V_R$

voltage is larger than FS/2, then the SAR keeps the MSB set. Otherwise, the MSB is turned off. In the next clock cycle, the next MSB is set, and this new approximation is compared with the input voltage. If the second MSB does not add enough weight to exceed the input, it is kept set, and the third bit is tested. The process continues until the LSB of the SAR is tried. At the end, the contents of the SAR form a digital code corresponding to the analog signal's magnitude, and a status line is raised to indicate that the encoding is completed and the output lines contain a valid binary word.

To obtain the desired accuracy in an SA ADC, one must ensure that the analog

Figure 9.23 Block diagram of a successive-approximation ADC.

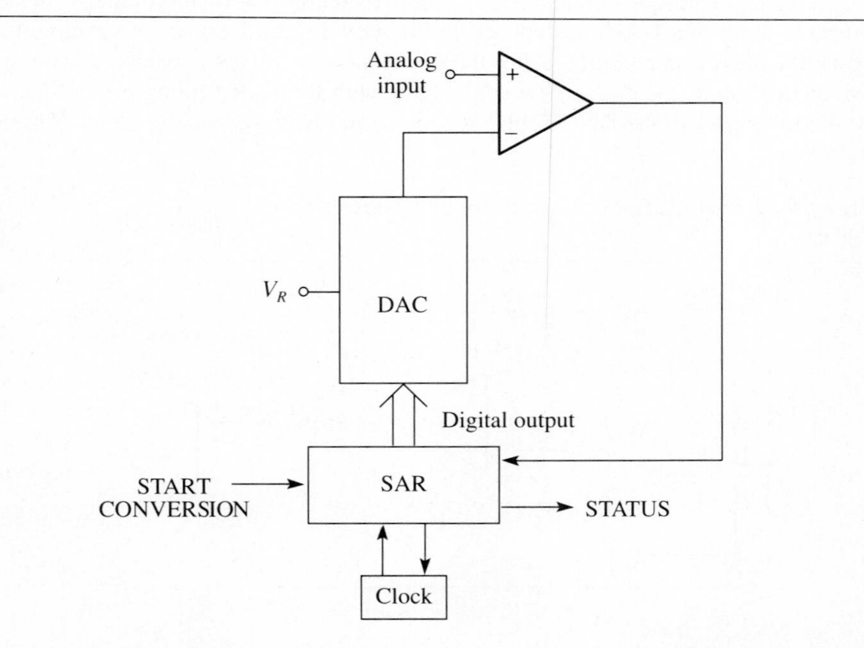

input signal does not change by more than ± LSB/2 during the conversion time. If the input signal is changing relatively fast during the conversion time interval, the algorithm may not be able to track changes in the signal because only smaller weights are being put on in succession. Thus, at high rates of change, the SA ADC generates linearity errors because it cannot tolerate the fast changes due to the nature of the weighing process. The error due to the time variation of the input signal is avoided by adding an S/H circuit before the converter and latching the sample in an aperture time t_{ap}.

Three types of limitations are imposed by an S/H circuit on the highest frequency component of the applied input. First, the analog channel full-power and small-signal bandwidths of the S/H circuit must be taken into account. These are usually more relaxed restrictions as compared to the other two.

The aperture jitter t_{aj} of the S/H circuit also imposes a limit on the input frequency. To allow the converter to resolve LSB/2 of the digital output, we must have an upper limit on the slew rate of the input signal as

$$\left.\frac{dv_i(t)}{dt}\right|_{max} = \frac{\text{LSB}/2}{t_{aj}}$$

$$= \frac{2^{-(n+1)}V_{FS}}{t_{aj}} \tag{9.13}$$

Suppose that the input is a sinusoid $v_i(t) = V_{FS} \sin 2\pi ft/2$. Then, the LHS of Equation (9.13) becomes

$$\left.\frac{dv_i(t)}{dt}\right|_{max} = V_{FS}\pi f \tag{9.14}$$

Equating Equations (9.13) and (9.14), we obtain an expression for the maximum frequency of a full-scale sinusoidal input and can be sampled to ±LSB/2 accuracy at n-bit resolution in terms of the aperture jitter as

$$f_m = [2^{(n+1)}t_{aj}\pi]^{-1} \tag{9.15}$$

While the presence of an S/H circuit permits the accurate digitizing of input signals with much higher frequencies, it also reduces the system *throughput* because its acquisition time t_{acq} and settling time t_{stl} now must be added to the ADC conversion time to determine how often new digital data can be obtained.

Therefore, t_{acq} and t_{stl} also place an upper limit on the maximum sampling rate and hence on the highest input frequency f_m because, according to the Nyquist criterion, the latter must not exceed one-half the sampling rate f_s to avoid aliasing errors. The maximum sampling frequency is then given as

$$f_s = [t_{acq} + t_{stl} + t_{conv}]^{-1} \tag{9.16}$$

where t_{conv} is the given ADC's conversion time.

Example 9.2

Consider a high-speed S/H circuit with 2 MHz small-signal bandwidth, an acquisition time of 1 μs, and a typical aperture jitter of 270 ps, and a 12-bit SA ADC with a conversion time of 25 μs. Neglecting the S/H's settling time of a few hundred nanoseconds and using (9.16), we see that the ADC can generate around 38,460 samples per second thereby allowing input frequencies of up to f_m = 19.23 KHz. In most applications, however, a low-pass filter is used to control the aliasing by attenuating the amplitude of the signal and the noise at and above the Nyquist frequency. This is usually accomplished by either increasing the sampling rate or the filter complexity. For instance, 12-bit accuracy requires a 5-pole filter and sampling at 11 times the highest frequency, limiting the input frequency to 38.46 KHz/11 $\approx$ 3.5 KHz.

On the other hand, Equation (9.15) leads to f_m = 143.9 KHz at 12 bits. Thus, the limit based on aliasing is a tighter restriction. However, without the S/H circuit, the allowable analog bandwidth can be much lower. For example, by replacing t_{aj} in Equation (9.15) with the ADC's conversion time ot 25 μs, the highest frequency signal that can be digitized is found to be only 1.55 Hz!

Since the S/H circuit is used to provide a constant input signal during the conversion interval of an ADC, it seems as if this circuit may not be needed if the original signal or its filtered version varies slowly enough to be considered effectively a dc signal. However, besides holding the input signal constant during the conversion, the S/H circuit also provides a low output impedance, which is usually a requirement for driving ADCs.

Figure 9.24 shows the interconnection of an S/H circuit and the input of a typical SA ADC. For the ADC of Example 9.2, the clock rate is found to be 12/25 μs = 480 KHz. At each test, the programmed current output from the internal DAC flows into the current summing junction and through R_{in} to develop a voltage drop across the resistor in opposite polarity to the analog input voltage. The net voltage is sensed by the comparator, which then makes its decision.

Therefore, the output circuit of the driver has to sink or source large high-frequency current transients (480,000 times a second for this particular circuit) and recover within one clock period. Since the driver will never be an ideal voltage source with zero dynamic impedance, there will be glitches at the ADC input that may lead to erroneous conversions. However, as long as the driver output impedance is sufficiently low at the switching frequency of the SAR so that the driver is capable of holding a constant output voltage even under dynamically changing load conditions, the small transient changes will not pose a problem, and the converter will be able to convert accurately. This can be accomplished by using an S/H circuit or a wideband high-slew-rate op amp configured as a voltage follower. Fortunately, there are S/H circuits available with typical output impedances of only .1 Ω.

Dual-Slope ADC

The block diagram of the *dual-slope* ADC is shown in Figure 9.25a. It first converts the analog input to a time function and then to a digital code, using a counter. The first part of the circuit is an integrator. Since the inverting input of the integrator

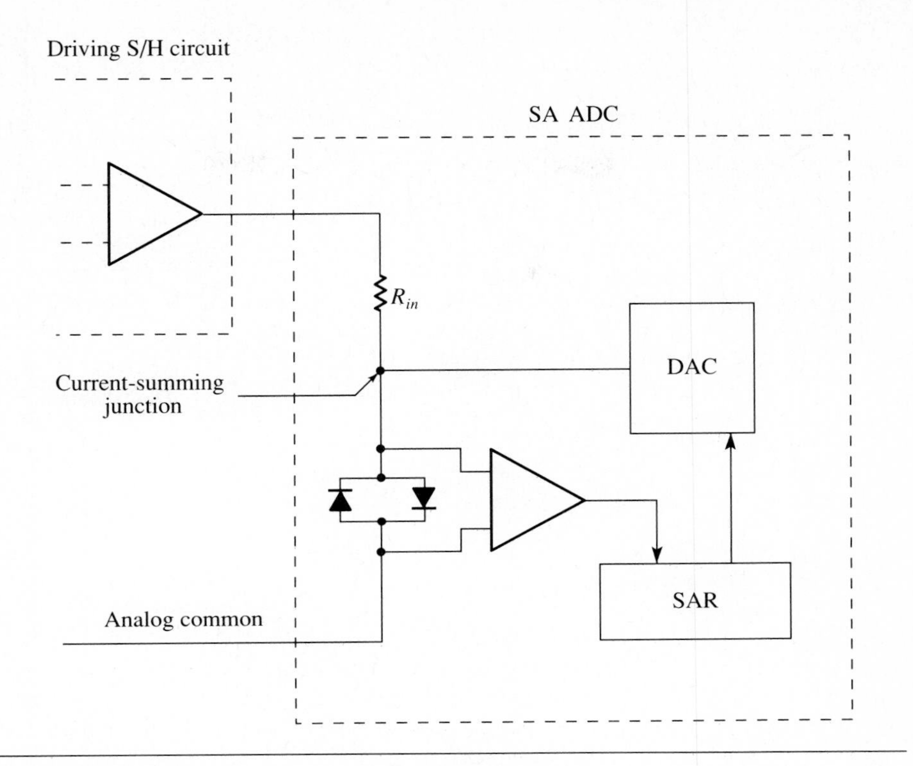

Figure 9.24 The interface between an S/H circuit and an SA ADC.

is held at virtual ground, the constant current caused by the input voltage flows through the resistor to the summing point and into the capacitor. The output of the op amp pulls this current from the other terminal of the capacitor to keep the inverting input at virtual ground. Therefore, the capacitor charges, and the output voltage of the integrator becomes more negative to keep the constant current flowing. The voltage across the capacitor is a linear ramp with a slope of $-V_I/RC$. A negative input voltage would cause a positive ramp.

A cycle starts with the counter reset to zero and the input switch connected to the input voltage. The actual value of the input that is converted by an integrating converter such as the dual-shape is represented by the average over the integration interval, which is a fraction of the total conversion cycle. If the input is positive, the integrator output goes negative to cause the comparator output to become positive. Thus, the AND gate enables the clock pulses, and the counter starts counting. After an interval of time

$$t_1 = \frac{N}{f_c} \tag{9.17}$$

where N is a fixed and predetermined number of counts, and f_c is the clock frequency, the counter is reset to count again, and the reference voltage of opposite polarity is

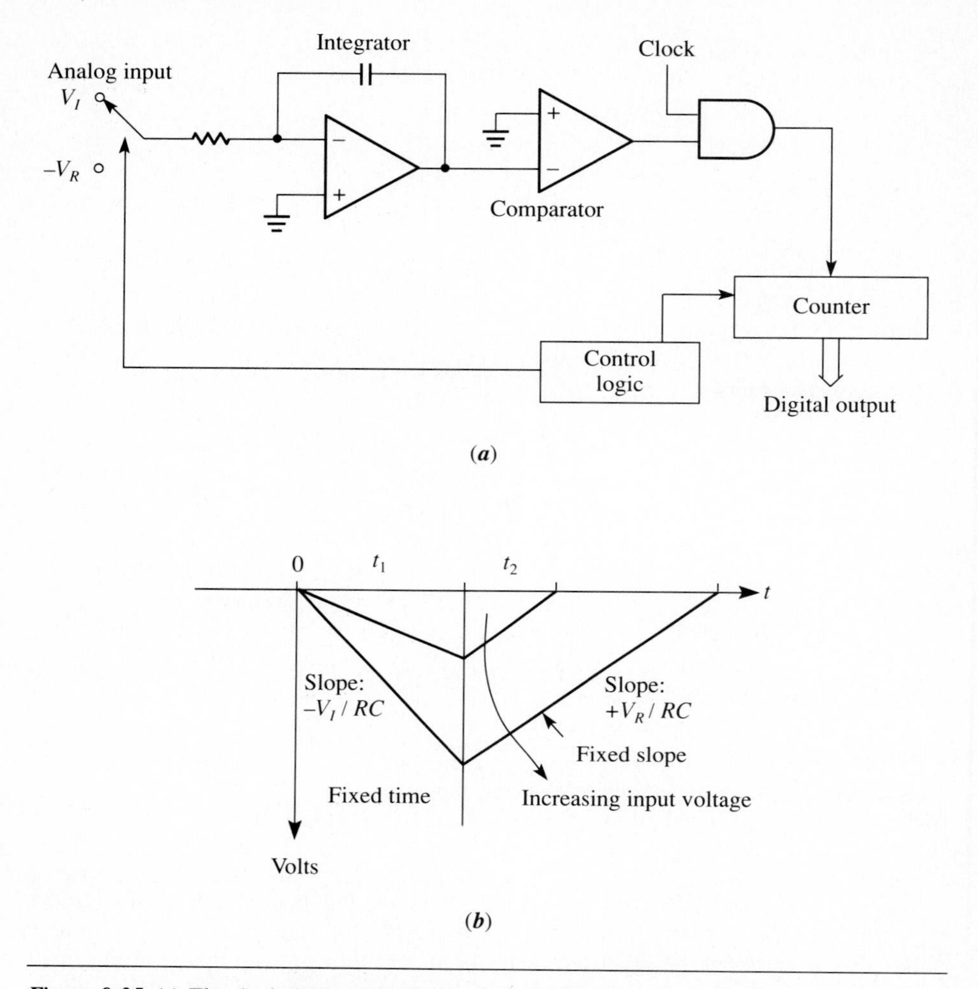

Figure 9.25 (a) The dual-slope ADC; (b) integrator output.

applied to the integrator. At this moment, the accumulated charge on the integrating capacitor is proportional to the average value of the input over the interval t_1.

The negative reference voltage causes the integrator output to ramp positive, as depicted in Figure 9.25b. Thus, the integral of the reference is an opposite-going ramp having a slope of V_R/RC. When the integrator output reaches zero again in t_2 seconds, the comparator output goes low, and the counter is stopped. The control circuitry strobes the counter outputs into the latches, resets the counter and switches the integrator input to the input voltage.

The charge gained during t_1 is proportional to $V_I t_1$, while the equal amount of charge lost during t_2 is proportional to $V_R t_2$. In fact, we have

$$t_2 = \frac{V_I}{V_R} t_1 \tag{9.18}$$

The number of counts stored at the end of t_2, n, is the binary representation of the input voltage. Using Equations (9.17) and (9.18), we get

$$n = t_2 f_c = \frac{N}{V_R} V_I \tag{9.19}$$

so that the counter output is directly proportional only to V_I since V_R and N are constants.

The conversion accuracy is independent of R, C, and the clock frequency because these parameters affect both integration periods in the same ratio. Differential linearity is very good because the input is continuous and the output codes are generated by a clock and a counter. Another advantage of the dual-slope type ADC is that it is inherently a low-pass filter. The integration rejects high-frequency noise and averages possible changes at the input during the conversion period. If a low-frequency or dc quantity is to be converted in the presence of a high-frequency ripple, a successive-approximation ADC will convert the instantaneous values of the noisy signal, producing an erroneous digital output even if preceded by an S/H circuit. On the other hand, an integrator attenuates high-frequency components to null out the integral multiples of $1/t_1$ completely during the fixed averaging period. Therefore, the sampling time t_1 is often determined by the fundamental frequency to be rejected.

The dual-slope ADC is too slow for fast data acquisition but suitable in applications such as digital voltmeters, battery discharge, and thermocouplers, in which a relatively lengthy time may be taken for the conversion to obtain noise reduction.

Counting ADC

The *counting* ADC, as depicted in Figure 9.26a, compares the analog input V_a with the DAC output V_d. The DAC input is fed by a binary counter. At the start of the conversion cycle the counter is reset. As long as the analog input is greater than V_d, the comparator output is high and the AND gate permits the transmission of the clock pulses to the counter. Since the number of clock pulses counted increases linearly with time, the binary number representing this count is used as the input to the DAC, whose output is the staircase waveform shown in Figure 9.26b.

When V_d exceeds the analog input, the comparator output goes low to disable the AND gate, which in turn causes the counting to stop. Therefore, the counter counts until the DAC output V_d reaches V_a. At this point the conversion stops, and the contents of the counter are stored before the counter is reset for the next conversion cycle.

For a given sampling rate and number of output bits, the counting ADC is much slower than the SA ADC. A conversion operation needs up to 2^n clock cycles, but it needs only n in the successive-approximation method. Note that the clock frequency increases exponentially with n in the counting converter and linearly in the SA ADC.

Tracking ADC

An improved variation of the counting ADC, called a *tracking* or *servo* converter, is obtained by using an up-down counter. Neither a clear pulse nor an AND gate is needed. The comparator output is directly tied to the up-down control of the counter. If the DAC output V_d is less than V_a, then the positive comparator output causes the

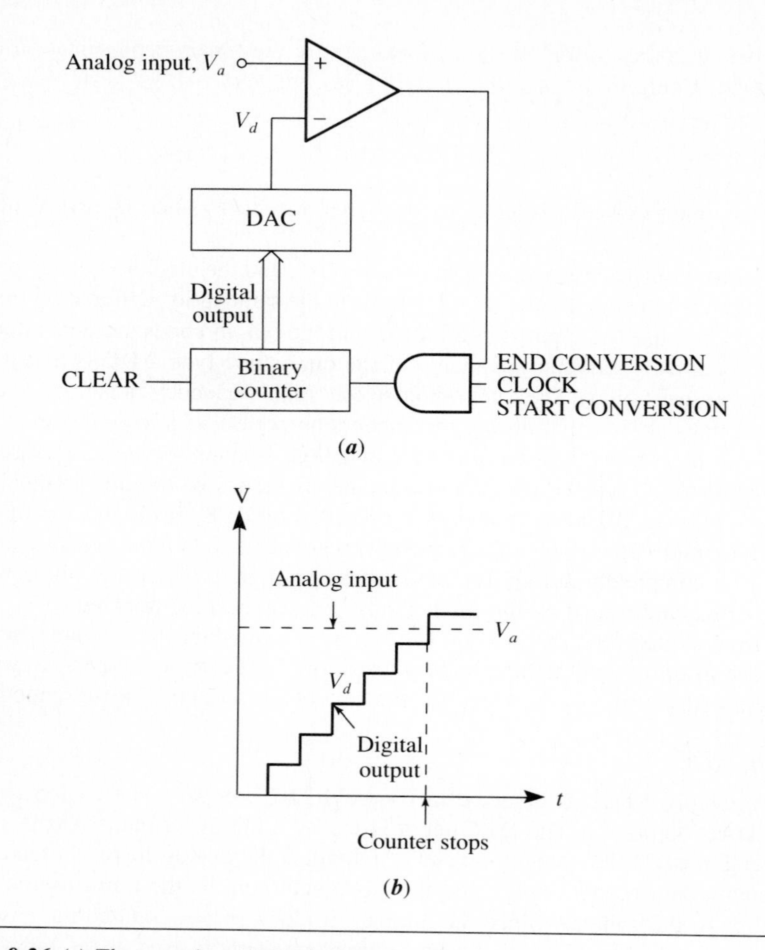

Figure 9.26 (a) The counting ADC; (b) the staircase waveform.

counter to count up. The DAC output increases with each clock pulse until it reaches and exceeds V_a, at which time the counter control line changes state so the counter will count down by only 1 LSB. This in turn causes the control to change again, so the count increases by 1 LSB. This process continues, and the digital output V_d hunts back and forth around the correct value V_a by ± 1 LSB. Compared with the counting DAC, a tracking converter can operate at twice the speed, requiring only half as many counts on the average to complete a conversion.

Flash Converter

The simplest and the fastest type of ADC is the parallel n-bit *flash* converter diagrammed in Figure 9.27. The single reference voltage V_R is applied across an array of 2^n identical resistors that divide the analog operating range into equal quantization levels. In this way, the converter has $2^n - 1$ latched comparators biased 1 LSB apart.

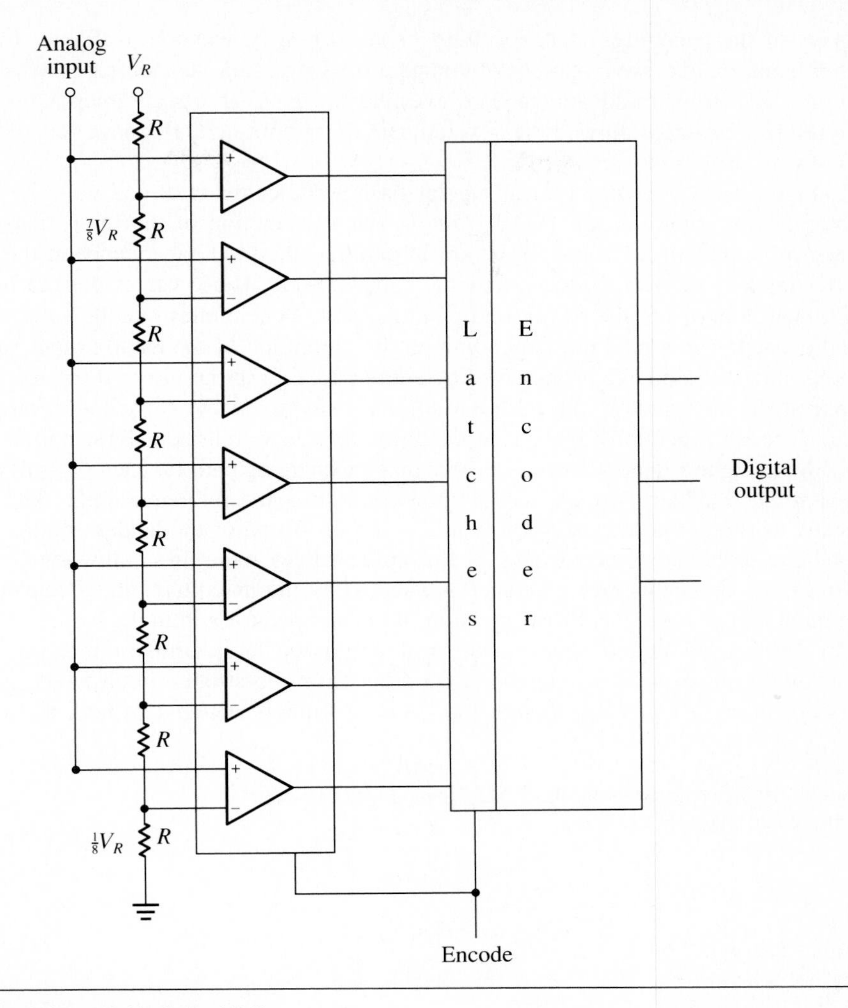

Figure 9.27 A 3-bit flash ADC.

When the encode command is logical **0**, the latches are transparent, and the converter is in the sample mode. When it goes high, the latches go into a hold condition to freeze the comparator outputs.

For 0 input, all comparators are off. The output of each comparator goes high if the input voltage on its positive input is greater than the reference voltage on its negative terminal. As the input increases, it causes an increasing number of comparators to switch state. The number of comparators tripped to a high output indicates the amplitude of the analog input voltage. The outputs of the comparators are not in binary, so the latched data are applied to an encoder that provides a set of outputs in the desired form.

The major advantage of this type is that since the conversion occurs in parallel, the speed is limited only by the switching time of the comparators and the propagation delays of the encoding gates, resulting in the fastest conversion available. On the other hand, an excessive number of comparators increasing geometrically with resolution is the major disadvantage. However, advances in large-scale integration techniques have recently allowed the development of monolithic flash converters of up to 10 bits of resolution.

Figure 9.28 depicts a typical bipolar flash ADC comparator cell with the associated analog input capacitance C_{ai} due to the BE junction of the input transistor. Since the inputs of all comparators are in parallel, the total capacitance at the converter input is the sum of the individual capacitances, whose values depend on the BE junction biases of the corresponding transistors. This implies that the value of the total capacitance is random, depending on the amplitude of the analog input signal. The problem could have been solved by adding a buffer at the input, if the reference input of the comparator cell were not similarly subject to the capacitance variation due to the BE transition capacitance C_{ri} of the reference transistor in the cell.

Referring to Figure 9.29, we observe that the charging path for each C_{ri} is through the reference ladder, forming an RC time constant at the reference input. The time needed to charge these capacitors results in the distortion of the ladder voltage division because the reference voltages lag the rapid changes in the current flowing through the ladder. This capacitive reactance is affected by the magnitude and frequency of the input signal as well as by the position of the comparator within the ladder. Therefore, a buffer cannot solve capacitance problems associated with the reference input.

Another error source is parasitic coupling from the strobe circuit to the ladder network through C_{ri}. Even though the strobe is equally coupled to both sides, the

Figure 9.28 A typical bipolar flash ADC comparator cell with parasitic capacitances.

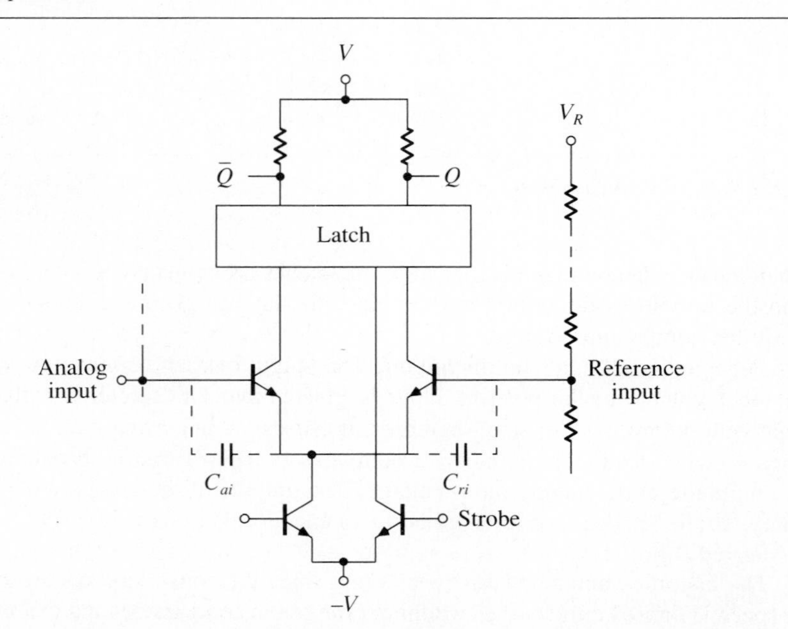

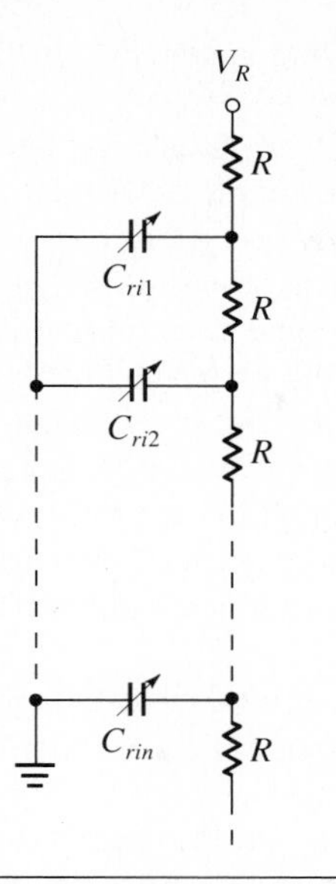

Figure 9.29 The equivalent reference resistor network.

lower source impedance at the analog side reduces its amplitude, leading to an imbalance between the inputs. The simplest and most effective way to equalize the differential input impedances is to use an external resistor in series with the analog input, thereby improving the integral linearity of the comparator array. Its value should be equal to the parallel combination of the upper and lower halves of the ladder network, that is, about one-fourth of the total ladder resistance.

A final potential source of conversion error is the sample delay variations. Each comparator can be visualized as having a variable delay line in series with the latch input that depends on the chip layout and the strobe frequency. Therefore, latch-time disparities can result in missing codes and differential nonlinearities.

Thus, even though the internal latches essentially perform a T/H function, high-speed, high-resolution systems still require an external T/H circuit because the inherent capacitance characteristics of flash converters affect their ability to handle high-frequency analog signals. At high frequencies, these converters retain only a fraction of their low-speed accuracy.

The performance of a flash ADC is improved if a fast T/H circuit is used ahead of the converter. Besides maintaining a constant input while the comparator outputs are being latched, it can also be timed to allow the parasitic input capacitances of the comparator to charge and settle before the conversion takes place.

Subranging ADC

Whenever very high resolution is required for the conversion function, a multi-stage converter, called a *subranging* ADC, is utilized. In two-stage subranging converters, flash converters are used in pairs to provide high resolution with much fewer comparators and simpler logic but at somewhat slower speed than the pure flash conversion.

Figure 9.30 illustrates a 12-bit, two-stage subranging ADC constructed with two flash ADCs having a total resolution of 13 bits. The analog signal from the T/H circuit is applied simultaneously to a 6-bit flash and a delay line. The converter converts the analog signal to a binary code, producing the 6 most significant bits, which are stored in a holding register and also applied to a 6-bit DAC having an accuracy of at least 12 bits. The *coarse* output of the DAC is subtracted from the delayed T/H output in a summation network, and the *residue* is amplified before being fed to a 7-bit flash that converts it to a *fine* code to represent the least-significant portion of the information. Finally, the outputs of the holding register and the 7-bit flash are fed to a combinational correction logic, where the coarse and fine codes are combined to yield a 12-bit parallel binary output in a manner that corrects for the error of the LSB of the most-significant portion.

Figure 9.31 shows the functional block diagram of a simpler realization of the subranging algorithm that utilizes one flash converter and only three timing intervals to encode the analog input signal.

During the first timing interval, the multiplexer connects the input signal to the 7-bit flash ADC, which encodes it and stores the coarse code in the MSB latch. Next, the DAC uses this code to generate a coarse signal that is subtracted from the analog input signal and amplified back up to the correct full scale by the differential amplifier. Finally, the multiplexer connects the flash ADC to the op amp output, and the converter encodes the residue signal, that is, the scaled difference voltage, to a fine code and stores this in the LSB latch. The combinational error-correction logic at the out-

Figure 9.30 The block diagram of a two-stage subranging ADC.

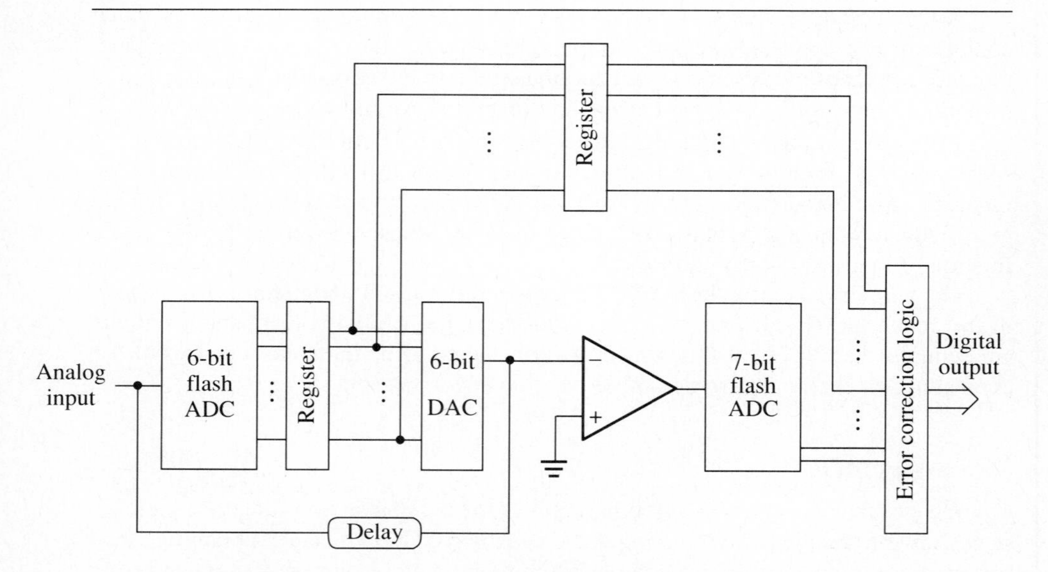

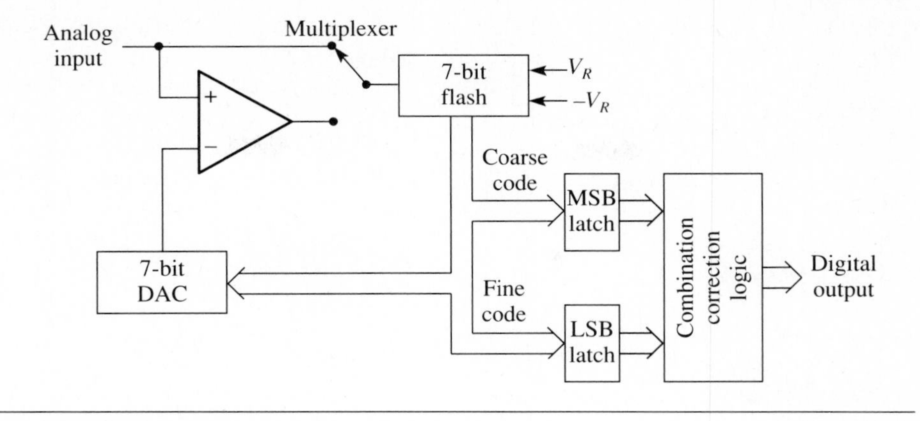

Figure 9.31 The block diagram of a 12-bit, two-stage subranging ADC with one flash converter.

put, made up of an adder, combines the contents of the latches to yield the final 12-bit digital output representation of the analog signal.

The most critical component in this two-step subranging ADC is the 7-bit DAC, which must have a linearity error equivalent to that of a 13-bit DAC. Using 14 bits, 7 for each encoding, to achieve 12-bit performance results in an overlap of coarse and fine encoded words. Therefore, particular attention must also be paid to the correcting technique at the output. Since an adder is used for this purpose, the coarse code must always produce a result that is less than the actual digital value of the analog input. The fine code, then, will always have to be added to the former. To accomplish this, both positive and negative reference voltages of the flash ADC must be offset by 2 LSBs during the first time interval, so that the analog input will always produce a coarse code output which is 2 LSBs lower than its actual value. Therefore, the DAC output must be corrected for the enforced offset error, so that the signal subtraction during the second time interval will be accurate. This is realized by adding 2 LSBs to the DAC output. Also, 2 LSBs are added to the op amp output during the fine encoding to remove the effect of the reference offset.

In this way, a very accurate representation of the erroneously low MSB code will be subtracted from the analog input. Suppose that the coarse code is low by 1 LSB due to some error. An accurate representation of this code will be subtracted from the analog input signal to yield a residue signal that is 1 LSB high. Therefore, the fine code will subsequently be high by the same amount, and the adder will correct the initial error when both codes are combined, as shown in Figure 9.32.

9.6 Bipolar vs. CMOS Data Converters

Digital ICs are designed for the highest possible speed and density and lowest power, leading to small geometries and low breakdown voltages, thus limiting the power supply and the logic swing. These features, however, restrict the dynamic range and resolution of the data-converter circuits.

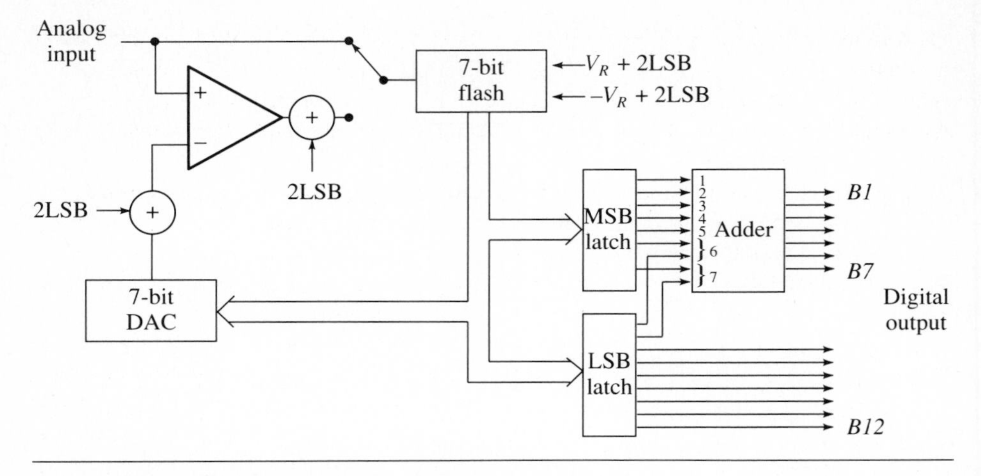

Figure 9.32 Illustration of the error compensation and correction.

Today, specific application areas require very different parameters for the data converters they utilize. Precision instrumentation and audio applications need 16 bits of accuracy, with conversion rates of 100 KHz or less; video and graphics applications require conversion rates of 20 MHz and higher but with lower resolutions of 8 bits or less.

With a growing demand for high circuit density and low power consumption, the CMOS technology is beginning to be preferred for data-converter designs. In the past, it was mainly used in low- and medium-speed applications such as industrial control, medical analyzers, and automatic test equipment. However, the reduced-geometry devices, capable of speeds comparable to bipolar logic, along with the introduction of the CMOS flash converters in recent years have enabled the CMOS technology to be an attractive alternative to bipolar converters, which are still commonly used in high-resolution, high-speed applications such as video and graphics, military radar, and medical imaging.

In the linear world of op amps and data converters, being a current-controlled device with a forward-biased BE junction is a disadvantage for a bipolar transistor when compared with a voltage-controlled MOSFET having a capacitive gate input. In order to operate, the BJT requires dc power at the input in the form of the base current and results in a relatively low input impedance, loading the signal source and making the input offset voltage dependent on the signal impedance. This leads to a complex bias compensation scheme to track the input bias current, as well as to voltage and temperature variations. The MOSFET, on the other hand, has virtually infinite low-frequency impedance, so that it draws negligible input current and hence needs little dc power, leading to a theoretically infinite power gain. However, since the parasitic capacitances increase with frequency, its power gain decreases with the increasing frequency.

The advantage of bipolar circuits over the CMOS is the inherently high transconductance of the bipolar transistor, which, being proportional to the collector current, can be increased effectively with bias current, resulting in greater gain per device and hence in high driving capability even at high frequencies. One drawback here is that current gain in the BJT is difficult to control in manufacturing and dependent on the input current in application, leading to highly variable input impedance specifications.

The CMOS technology has much more to gain from the continuous shrinkage of device geometries than the bipolar technology. First, the transconductance parameter of a MOSFET varies with the transistor's aspect ratio, while the transconductance of a BJT is independent of its dimensions. Furthermore, the parasitic capacitance of a MOSFET decreases more rapidly with the lateral dimensions of the device.

In CMOS devices, carriers flow at the surface region of the silicon substrate. Thus, the traps that exist at the gate-oxide and silicon interface and are associated with contamination ions and defect sites affect the current flow in the channel and contribute directly to the *flicker noise* that can only be reduced either by improvements in process technologies or circuit-design techniques such as the use of chopper stabilizers. In bipolar devices, on the other hand, the vertical flow of carriers in the bulk of the silicon wafer where the number of traps is small causes little flicker noise. However, thermal runaway in bipolar linear ICs remains a problem, as the output current doubles every 10°C.

When the system requirements call for both digital and precision linear functions to be incorporated on one chip, the silicon-gate CMOS has the edge over the bipolar because the latter's wafer fabrication process is not the same for both linear and digital functions.

One key advantage of the CMOS, related to DACs, is that almost all of them operate as multiplying DACs to perform four-quadrant multiplication. The CMOS technology is extremely suitable for this type of application because the analog signal path within the CMOS DAC, which is a function of resistors and switch settings, is essentially passive, so that it does not distort the input reference, thereby allowing the usage of a wide variety of signals as reference. In contrast, bipolar DACs generate their analog output from transistor switched-current sources.

Besides simply improving conventional specifications such as power consumption and density, the CMOS technology has also changed the design emphasis by influencing the techniques used to perform data conversions. The bipolar SA A/D conversion that relies on resistor matching and voltage levels is being replaced by what are called *charge redistribution* techniques. In this architecture, capacitors replace some or all of the resistors in the converter's resistive ladder. Since they depend only on device geometry, the capacitances can be set with high accuracy and are ideal with high-speed, high-resolution ADCs. During the conversion process, packets of charge rather than precise voltages are manipulated.

Figure 9.33 shows an n-bit ADC that implements the successive-approximation algorithm using the charge redistribution architecture. Instead of utilizing the traditional resistor array, the internal DAC is made up of binary-weighted capacitors that share a common node at the comparator's output. Other terminals can either be connected to the analog input V_I, reference voltage V_R, or the analog ground. The dummy capacitor serves the purpose of terminating the array and making the total capacitance equal to $2C$.

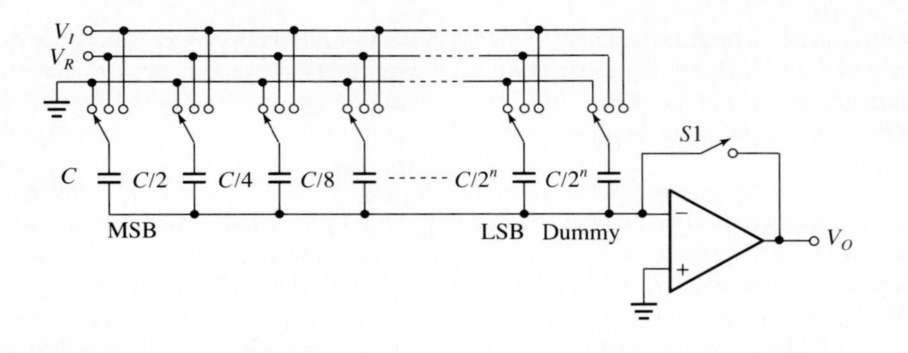

Figure 9.33 The internal DAC of the SA ADC, using charge redistribution.

Initially, before the conversion process takes place, all capacitors are tied to V_I, forming a total capacitance of

$$C_t = C + \frac{C}{2} + \cdots + \frac{C}{2^n} + \frac{C}{2^n} = 2C$$

where n is the resolution of the DAC. In addition, the switch $S1$ is closed and the total stored charge on the array, $Q_I = 2CV_I$, tracks the input signal V_I. This is shown in Figure 9.34a. Thus, during the tracking mode, a sample of V_I is taken, and a proportional amount of charge is stored on the capacitor array.

The conversion command opens the switch $S1$. As a result, Q_I is trapped on the comparator side of the capacitor array, leading to a floating node at the op amp input and causing the analog input to be ignored. Therefore, the DAC capacitor array functions much like a holding capacitor in an S/H circuit. During conversion, the free plates of the capacitor array to ground and V_R are manipulated to form a capacitive divider. The equivalent circuit is seen in Figure 9.34b, where V_{fn} signifies the floating-node voltage. The charge at this node stays fixed, so that V_{fn} depends on the amount of capacitance tied to V_R as opposed to that tied to the ground. This propor-

Figure 9.34 Equivalent circuit in (a) tracking mode, and (b) convert mode.

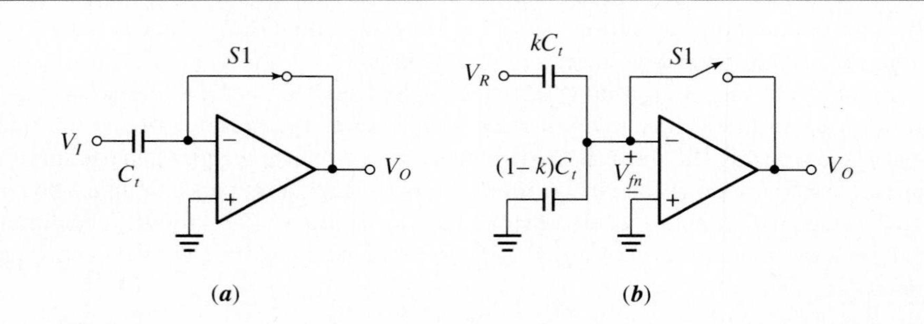

tion, k, is determined by the SA algorithm in such a way that when connected to the reference, the binary fraction kC_t, representing the ADC's digital output, reduces V_{fn} to zero. Thus,

$$k = \frac{V_I}{V_R} \qquad \text{for } V_{fn} = 0$$

The connection of the bit switches at the end of the charge redistribution phase produces the output digital word. An input switch connected to V_R indicates a **1**, whereas connection to ground shows a **0** value for the corresponding bit. At the end of the conversion mode, all the charge is stored in capacitors corresponding to **1** bits, while the capacitors of the **0** bits have been discharged.

A relatively recent process technology is the BiCMOS, which merges the best features of bipolar and CMOS technologies. It combines low-power CMOS logic with high-speed and precision bipolar circuitry in one chip. The converter manufacturers have already begun to focus their design and process development efforts toward BiCMOS. Data converters using this technology are already available.

Another emerging technology utilizes *gallium arsenide* (*GaAs*) instead of silicon. GaAs data converters promise speed and bandwidth capability beyond what is available with the current technologies. However, since the technology is still in its developmental stage, converters based on it are not expected to gain wide usage until the mid-1990s. This technology will be especially useful in fiber-optic and very high frequency communications, pattern recognition, and very low noise applications. Besides their exceptional speed and bandwidth advantages, GaAs converters also have low power consumption.

Summary

Analog-to-digital conversion translates continuous physical quantities into digital form to make these signals suitable as inputs for digital computer processing. Depending upon the conversion speed, accuracy, cost, and desired noise immunity, there are a variety of ways to achieve A/D conversion. The integral nonlinearity, differential nonlinearity, offset error, and gain drift of an A/D converter are temperature-dependent.

Digital-to-analog conversion is employed whenever discrete information from a digital computer needs to be translated into a physical quantity such as voltage or current. The information may be in any coded form. It may represent positive and/or negative quantities. One advantage of utilizing DACs in analog applications is the ability to keep the information in digital form, making it more immune to noise than the analog information. The resolution of a DAC depends on the number of bits representing the analog output quantity; its accuracy is governed by the precision and stability of all components used in the circuit.

A sample-and-hold circuit is required in A/D conversion when the input voltage variations during the conversion time exceed the resolution of the ADC. Samples of the analog signal are taken at regular intervals and are held until the next sample is taken. During the hold time, the signal sample is converted to digital form by the ADC.

References

1. "Track and Hold Amplifiers Improve Flash A/D Accuracy." *Application Note TH-06*. Comlinear Corporation, Fort Collins, CO: May 1987.
2. "Selecting and Using High-Speed Track and Hold Amplifiers." *Application Note TH-05*. Comlinear Corporation, Fort Collins, CO: March 1987.
3. The Engineering Staff of Analog Devices, Inc., *Analog-Digital Conversion Handbook*. 3d ed. Prentice-Hall, Englewood Cliffs, NJ: 1986.
4. "Flash Analog to Digital Converters: Optimizing Performance." *Application Note 103-1*. Comlinear Corporation, Fort Collins, CO: August 1983.
5. J. Williams, "Circuit Applications of Multiplying CMOS D to A Converters." *Application Note-269*. National Semiconductor, Santa Clara, CA: September 1981.
6. J. Sherwin, "Specifying A/D and D/A Converters." *Application Note 156*. National Semiconductor, Santa Clara, CA: February 1976.

PROBLEMS

9.1 Assuming $f_m = 3$ KHz and 128 levels of quantization are adequate, what is the number of bits required to store 1 minute of digitized speech signal?

9.2 Given $V_{FS} = 10$ V. If the digital code is 110001000001 using natural binary, find the corresponding V_I for an ADC.

9.3 Given $V_{FS} = 10$ V. Using bipolar offset, the digital code is given as 011000000001. What is the corresponding analog value? What does the same code correspond to if the input span is 20 V?

9.4 Is the gain nonlinearity of $\pm.0005\%$ for an S/H circuit less than LSB/2 at 16 bits?

9.5 Is a 20-μV rms noise level acceptable for 16-bit accuracy if the full-scale range is 20 V?

9.6 The maximum droop rate for a certain S/H circuit is specified at .05 μV/μs. How long can it hold a signal for a 16-bit converter with a $\pm$10-V output range without losing accuracy?

9.7 Does a feedthrough rejection of 98 dB compromise $\pm$LSB/2 accuracy at 16 bits for a $\pm$10-V input?

9.8 In a *compact disc* (*CD*) recording system, each of the two stereo signals is sampled with a 16-bit ADC at 44,100 samples/s.

a. Determine the output SNR for a full-scale sinusoid.

b. The music heard on a CD actually accounts for only 50% of the bits on the disc. The number of bits of data is increased by the addition of synchronization, display, control, and error-correcting bits, so that 2 bits are stored for each bit of digitized music. Determine the output bit rate of the system.

c. If the CD can record 75 minutes' worth of music, determine the number of bits recorded on a CD.

d. An unabridged dictionary contains 2100 pages, 3 columns/page, 100 lines/column, 7 words/line, 6 letters/word, and 6 bits/letter. Find the number of bits needed to represent the dictionary and estimate the number of similar books that can be stored on a CD.

9.9 What is the maximum frequency of a ± 10-V input sine wave that can be accurately digitized by a 12-bit SA ADC with a conversion time of 2 μs?

9.10 a. To what frequency will the full-power bandwidth of the SA ADC of Problem 9.9 be increased for the same input if an S/H circuit with an acquisition time of .2 μs, settling time of .1 μs, and an aperture uncertainty of 50 ps is used ahead of it?

b. What will be the highest frequency component of the input sinusoid based on the Nyquist criterion?

9.11 Assuming that the input is not slew-rate limited, derive an expression for the error in the output of an S/H circuit as the percentage of the maximum input signal amplitude V_i caused by an aperture jitter t_{aj} when a sinusoidal signal of frequency f and amplitude A is applied. What is the maximum percentage error if $t_{aj} = 500$ ps and $f = .5$ MHz?

9.12 The error caused by the aperture jitter of an S/H circuit may be obtained from the accompanying graph when the input is a sinusoid and not slew-rate limited. Any straight line drawn on the graph from top to bottom relates the corresponding t_{aj}, percentage error, and the input frequency.

a. Repeat Problem 9.11 using this graph.

b. What is the maximum error for an ADC with a conversion time of 2.2 μs and an input with a maximum frequency of 20 KHz?

c. To reduce the maximum error to .1% in (b), what should be the aperture jitter of the S/H circuit that is placed at the ADC's input?

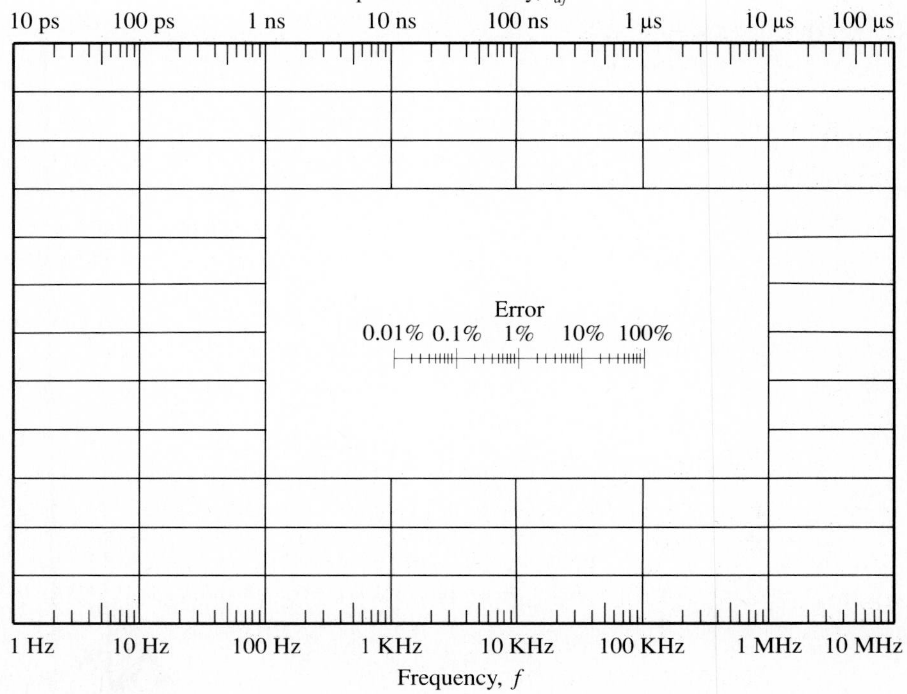

9.13 The timing diagram shown is for a 12-bit SA ADC. The conversion starts on receipt of a CONVERT START command. At $t = t_o$, the MSB is set, and the others are reset unconditionally. At $t = t_1$, the MSB decision is made, and *B2* is unconditionally set. This sequence continues until the LSB decision is made at $t = t_{12}$. After a short delay, the STATUS flag is reset to inhibit the gated clock and signify the completion of the conversion. What is the binary code at the end of the conversion? What is the value of the analog input if $V_{FS} = 10$ V and natural binary coding is used?

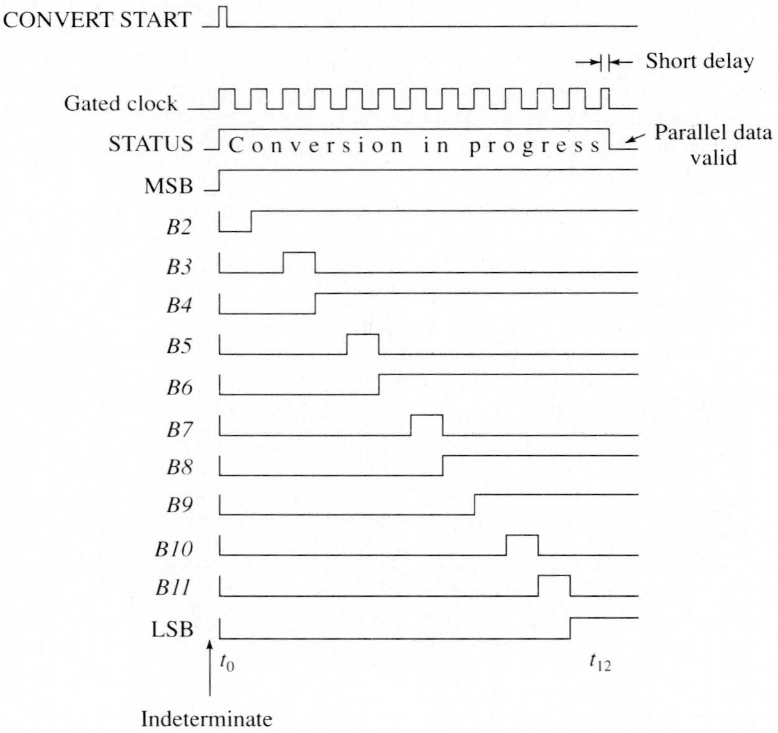

CONVERT START

Short delay

Gated clock

STATUS Conversion in progress

Parallel data valid

MSB

B2

B3

B4

B5

B6

B7

B8

B9

B10

B11

LSB

t_0 t_{12}

Indeterminate

9.14 Complete the following timing diagram for the SA ADC of the previous problem if the output code is 011001110110.

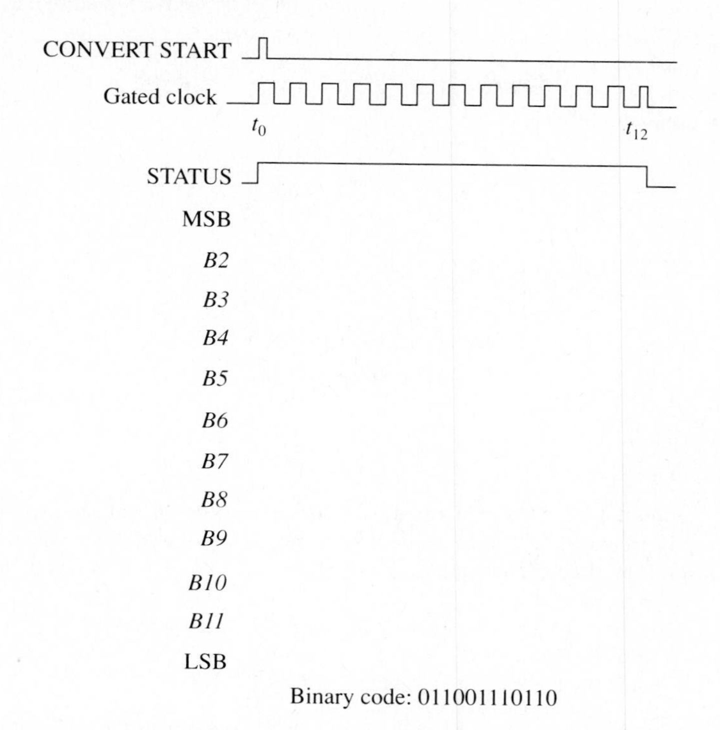

Binary code: 011001110110

9.15 One of the CMOS current switches of a multiplying DAC similar to that of Figure 9.19 is shown in the diagram. Explain the operation of the circuit. Comment on the relative values of the on-resistances of the CMOS switches. For a 10-V reference input and R_{on} = 20 Ω for the MSB switch, what is the voltage drop across each switch in the resistor array?

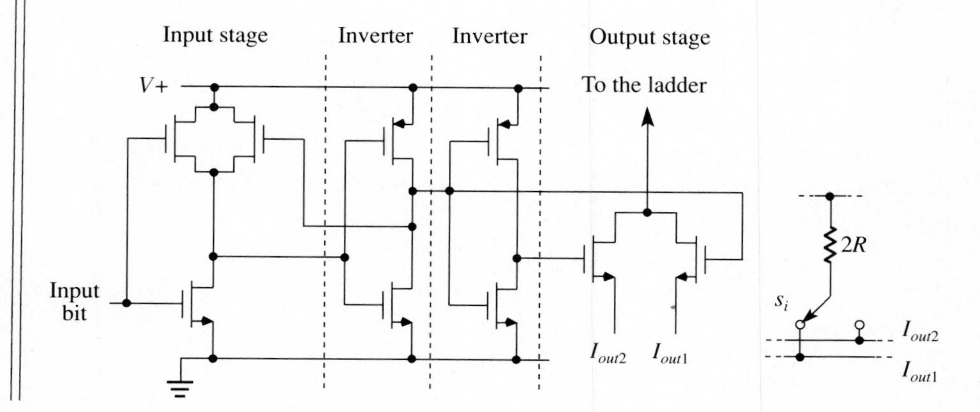

9.16 An analog/digital divider is obtained by connecting the DAC of Figure 9.19 in the feedback of an op amp, as shown in the diagram. Derive the transfer function of the circuit to verify the division of the analog variable V_I by a digital word. What happens when all bits are zero? What is the gain when

a. only the LSB is on?
b. all bits are on?
c. only the MSB is on?

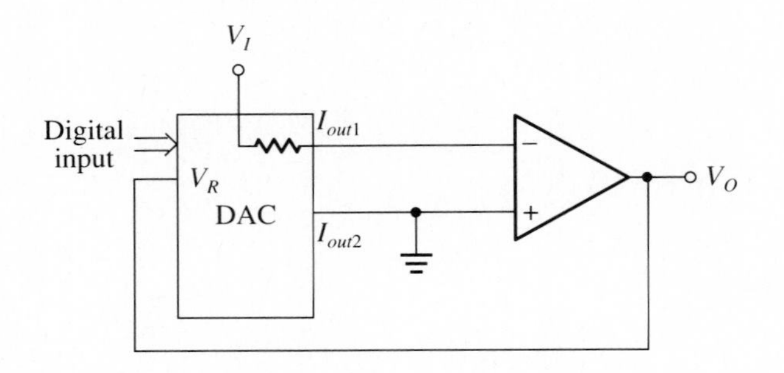

9.17 The multiplying DAC of Figure 9.19 can be used in the voltage mode if the I_{out2} bus is grounded, a voltage supply V_s is connected to the I_{out1} bus, and the output is taken from the reference input. A bipolar output voltage is obtained by using an op amp, as illustrated.

a. Find an expression for the output voltage V_o in terms of V_R.
b. What is the value of the output voltage for a digital input of all zeros?
c. What is the output voltage swing for $V_s = 2.5$ V?

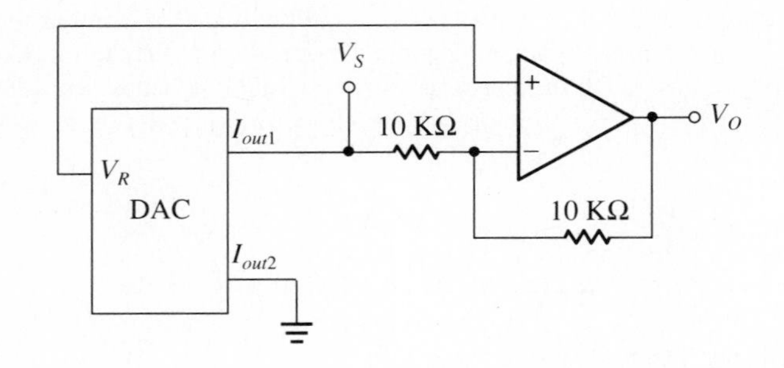

9.18 Verify that the circuit shown in the diagram generates the powers of 2 by finding an expression for V_n in terms of the reference voltage V_R and the digital input D.

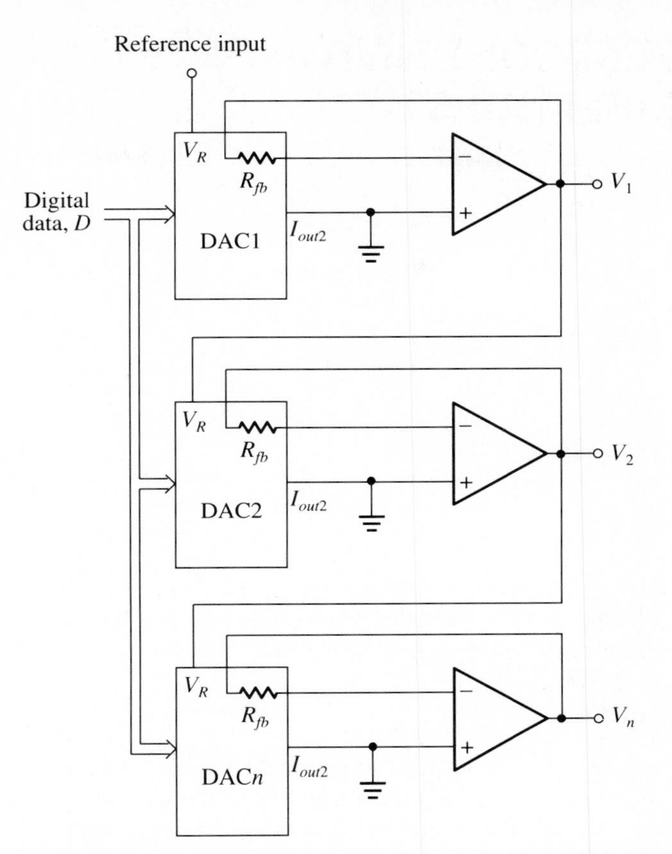

9.19 Design a 10-bit binary-weighted DAC with $V_R = +10$ V and $R = 2$ KΩ such that the error due to resistance tolerances should not exceed ±LSB/2 of FS. Determine the tolerances of the resistors in the MSB and LSB positions.

9.20 A DAC can be designed for any weighted code using appropriate resistors. Design a 2-bit *binary-coded decimal* (*BCD*) DAC whose block diagram is shown at right. The output is obtained across the finite load resistor R_L. Find V_o corresponding to the *BCD* input 1000 1000 = 88_{10}.

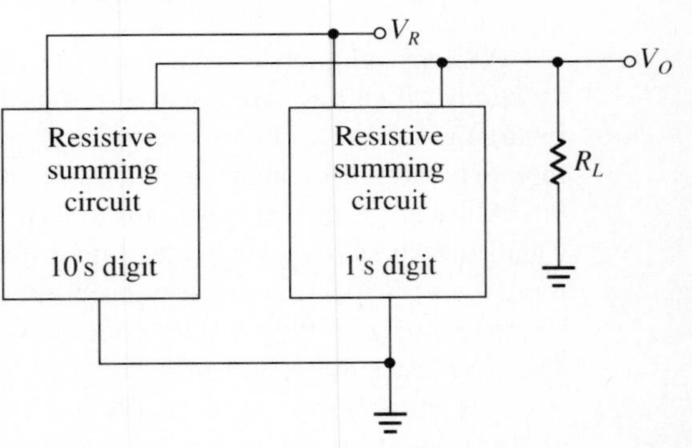

9.21 Consider the computerized control of the speed of a motor using a DAC whose analog output current is amplified to produce speeds from 0 to 2000 revolutions per minute (rpm).

a. Determine the number of bits for the computer to produce a motor speed that is within 2 rpm of the desired speed.
b. Using the result of (a), how close to 824 rpm can the speed be adjusted?

10. SEMICONDUCTOR MEMORIES

The main topic of this chapter is semiconductor memories. Our primary concern will be the CMOS technology that reflects current trends. We will also briefly consider other high-density circuits such as programmable logic devices and gate array chips.

10.1 Definitions

Memories are devices that respond to operational orders, usually from the *central processing unit* (*CPU*) of a digital computer; they are used to store large quantities of digital information. The memory of a computer is not a single entity. It is rather a collection of units dispersed in space and function with a variety of speeds, capacities, and costs. Even though our emphasis will be on the physical properties of the devices, we will first examine briefly their functional characteristics and logical organization.

A block diagram of a memory unit is shown in Figure 10.1. Its basic unit is a *location*, which has two essential properties: an *address* and contents. The address is invariant, but the contents may or may not be changed depending on the type of the memory, as will be clear later. The basic information element is a *bi*nary digi*t*, the *bit*. Although a bit organization is used for certain purposes such as flags, bits are usually organized together into larger entities called *words*, each occupying one or more memory locations and representing a number, an instruction code, alphanumeric characters or any other binary-coded information. A word of 8 bits is called a *byte*. The smallest subdivision of a memory unit into which a bit of information can be stored is called a *memory cell*. The size of the memory unit is specified by the number of locations it contains and the number of bits in each location. The address lines select one particular memory location out of the m locations available. Thus, a k-bit address can select any one of $2^k = m$ locations. Memory storage capacities are integral powers of 2 and units of K for *kilo*, M for *mega* or G for *giga* are used when referring to the number of bits or bytes in a memory unit. Thus, 1 Kbit equals 1024

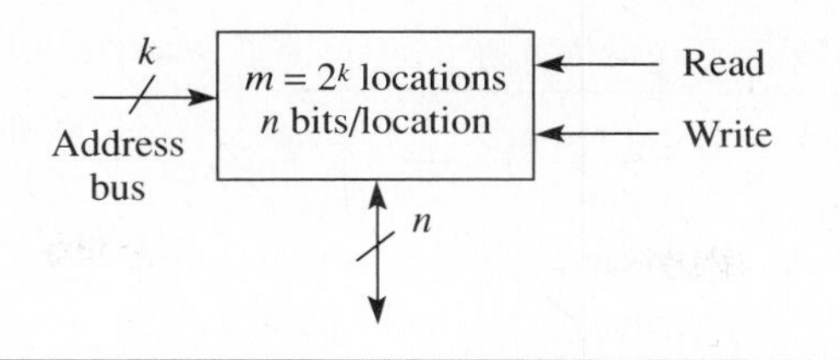

Figure 10.1 The block diagram of a memory unit.

or 2^{10} bits, 1 Mbit equals 1,048,576 or 2^{20} bits, and 1 Gbyte equals 1,073,741,824 or 2^{30} bytes.

The two control signals in Figure 10.1 are called *read* and *write*. A read signal specifies a transfer-out operation; a write signal specifies a transfer-in operation. Upon receiving a write signal, the internal control circuitry of the memory unit transfers the n data input bits from the data bus into the location pointed to by the contents of the address bus. A read control signal causes the contents of the memory location selected by the address lines to appear on the bidirectional data bus.

In general, there are three main modes of memory access:

1. *Sequential access* memories are those organizations where the next destination may only be the previous or the subsequent location. To reach address M from N, all the locations between them must be accessed. Thus, the next address is accessible in a time period that is a *linear* function of the address space between it and the current address. Examples of this organization are the magnetic tape, magnetic bubble memory, and the charge-coupled device, or CCD. Sequential memories use much less hardware and are slower than the other organizations. They are also inexpensive.
2. *Direct access* memory organization has the characteristic that the next address to be accessed may be any of those available and is accessible in a time interval that is a *nonlinear* function of the address space between it and the current address. The three most common direct access memory devices are drums, and fixed-head and moveable-arm disks.
3. *Random access memory* (*RAM*) is the fastest form of memory organization. The location at any address can be the next destination, which is reached in a fixed and known time. Solid-state semiconductor memories and the obsolete core memories are the prime examples.

The total storage capacity of a computer should be visualized as being a hierarchy of memory units. It consists of slow but high-capacity auxiliary memory devices, faster main memory, and smaller and very fast buffer memory accessible to the high-speed processing logic. The information can be moved from the cheaper and slower devices up a chain of faster and more expensive devices eventually to the *scratch pad* memory within the CPU and back down to the auxiliary devices as is convenient and cost-effective. Naturally, in terms of cost per unit of storage, the faster memories are more expensive and usually smaller; the slower memories are less expensive and larger. Table 10.1 summarizes the major characteristics of each type of memory.

Table 10.1 Summary of Memory Characteristics

Type	Primary use	Access	Volatility	t_a
Magnetic tape	Auxiliary	Sequential	Nonvolatile	300 ms
Magnetic bubble	Auxiliary	Sequential	Nonvolatile	2 μs
CCD	Auxiliary	Sequential	Volatile	10 μs
Magnetic disk	Auxiliary	Direct	Nonvolatile	100 ms
Magnetic drum	Auxiliary	Direct	Nonvolatile	100 ms
Core	Main	Random	Nonvolatile	2 μs
Semiconductor RAM	Main, cache	Random	Volatile	50 ns
Semiconductor ROM	Main	Random	Nonvolatile	50 ns

Figure 10.2 depicts the components in a typical memory hierarchy system. At the bottom are the relatively slow magnetic tapes used to store removable back-up files. Above it are the magnetic drums and disks used as secondary storage. At the central position is the main memory directly communicating with the CPU and the auxiliary devices via an *input/output* (I/O) processor. The main memory is made up of semiconductor memory modules with typical response times of around 100 nanoseconds or less, and stores the data and instructions of the currently executing application programs. A special ultrafast memory unit is used to increase the speed of the processing by making segments of the programs that are currently being executed and the associated data available to the CPU at a rapid rate. This type of buffer memory is called a *cache* memory. The data movement between these units is under the control of the *operating system*. The part of the operating system software that supervises, with hardware assistance, the flow of information between these storage devices is called the *memory management unit*.

In this chapter, we will discuss LSI and VLSI memories, known as *semiconductor memories*. Magnetic core, magnetic bubble, and moving-surface memories will not be covered.

Those types of semiconductor memories that lose their stored information when the power is turned off are called *volatile* memories. The others retain their stored information even when there is no power, and they are known as *nonvolatile* memories. The *read-only memory* or *ROM* is designed to allow its stored data to be read but not to be changed during normal system operation. The *read-and-write memory* (*RWM*), which is misleadingly called RAM, can be read from and written into during normal operation. Even though both ROM and RWM are random access memories, the term RAM is often used instead of RWM. To be consistent with the current literature, we will also have to use in this text the abbreviation RAM for RWM when no misunderstanding will result. On the other hand, to define ROM as a type of *memory* is somewhat inappropriate, too, because it is not a sequential but a combinational circuit. ROMs are nonvolatile, while RWMs are volatile memories.

A measure of the operating speed of a memory is its *access time* t_a, which is the maximum time required to read a word from the memory. It is measured from the application of a new address on the address bus to the appearance of the valid data

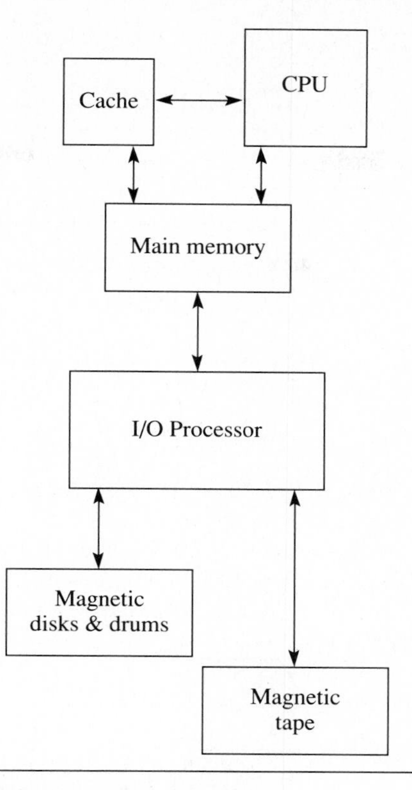

Figure 10.2 Memory hierarchy in a computer system.

on the data bus. The process of reading or writing a word requires that various signals be applied to the address, data, and control buses, and is called a read or write *memory cycle*. The *memory cycle time* t_c is the minimum time that must elapse between the initiation of two successive memory operations. Typically, $t_c \geq t_a$. These parameters are depicted in Figure 10.3 for both read and write cycles.

Consider a memory technology that can deliver or absorb an addressed data item in 50 ns and a data bus that is wide enough to carry all bits of a word simultaneously. If it is a *nibble* (that is, it has 4 bits), then this memory can deliver a maximum of 80 million *bits per second* (*bps*). If the word length is a byte, then the maximum delivery rate will be 160 million bps. Therefore, increasing the size of the data path will allow relatively slow, inexpensive memories to have acceptable data transfer rates. Table 10.2 illustrates the relationship between the memory delivery size, access speed, and the delivery rate.

10.2 Semiconductor Read-Write Memories

Semiconductor RAMs have been developed at a very rapid pace for the last 15 years. They are divided into two major types: bipolar RAMs, utilizing BJTs and manufactured using TTL, ECL or I^2L technologies, and MOS RAMs using NMOS or CMOS.

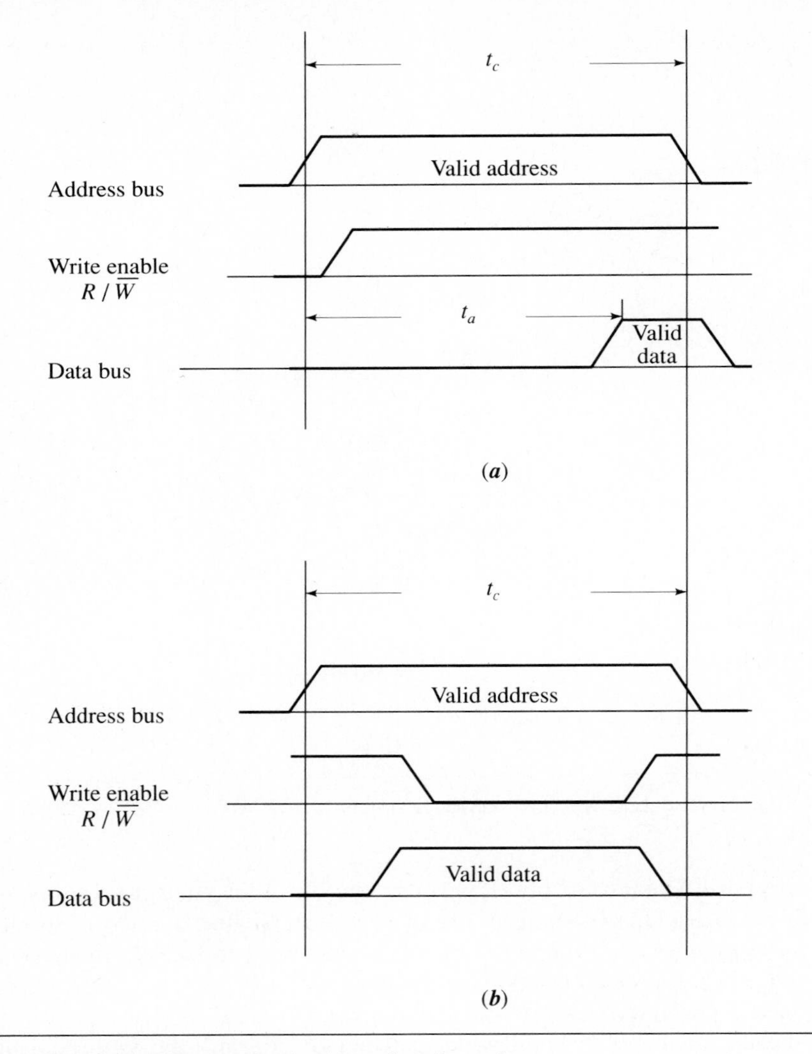

Figure 10.3 Timing diagrams for the (a) read cycle, and (b) write cycle.

In bipolar RAMs each memory cell is a flip-flop, and the typical access times are on the order of 10 ns, at the expense of large power dissipation. Therefore, they are especially useful in high-speed applications. On the other hand, the recent dominance of MOS devices in the RAM market comes about as a result of high packing density, low power consumption, increasingly high operating speed, and low cost per bit. The current level of integration of up to 4 Mbit in commercially available IC memories is made possible specifically by the use of *dynamic* data storage. Currently, the cycle times of the fastest CMOS memories are around 100 ns, and access times are below 50 ns.

Table 10.2 Effect of Increasing the Size of the Data Bus and the Access Time on the Transfer Rate

Size of each transfer (bits)	Memory access time (ns)			
	50	100	200	400
	(million bps)			
4	80	40	20	10
8	160	80	40	20
16	320	160	80	40
32	640	320	160	80
64	1280	640	320	160

When first introduced in the late sixties, semiconductor memories were much more expensive than the magnetic core memories they eventually replaced. However, since continuous advances in VLSI technology have led to a dramatic drop in their cost, they are now used exclusively in the implementation of cache and main memories.

10.3 Dynamic RAMs

Initially, MOS RAMs used cross-coupled flip-flops as storage cells containing up to eight transistors. The first breakthrough was the development of the concept of dynamic data storage, which represented the logical levels by a low or high voltage across a capacitor in a three-transistor (3-T) cell. However, in order to compensate for the leakage of charge off the capacitor and hence to retain the data, the data had to be sensed and restored to its original voltage level periodically by a *refresh* operation. Thus, the term *dynamic* implies the continual refreshing of the data in memory cells to ensure the validity of the stored data.

The second major breakthrough was the development of the single transistor (1-T) cell, which occupied less than half the area of the earlier configuration and permitted the integration of 4096 cells into a chip, compared with only 1-Kbit RAMs using the 3-T cell. The delay in the introduction of the 1-T cell was due to the difficulty in sensing the small signal from the cell. For the first time, there was no amplifier built into every cell. Instead, *sense amplifiers* were developed, as will be discussed later in detail. Moreover, improvements in the internal peripheral circuits made the new generation of ICs much easier to use. The 1-Kbit circuits required multiple and critically timed clock signals, whereas in the 4-Kbit chips they were replaced by a single clock for the 22-pin versions and by two TTL-level clocks for the 16-pin versions. The 1-Kbit chips required high voltages for address and data inputs that were replaced by TTL-level inputs in the 4-Kbit chips. Also, the high-impedance output of the former generation, requiring an external sense amplifier, was replaced by a low-impedance output capable of driving TTL loads. Finally, slower PMOS technology was displaced by the *n*-channel technology. These advancements made the MOS cost-competitive with magnetic cores for the first time.

While using the 1-T cell increased the bit density on a chip, it also degraded the access time by about 25% due to the delays through the sense amplifiers in detecting and amplifying the very small signals from the memory cells. However, these delays made the *multiplexing* of addresses an attractive means of reducing package pin-count for increased density on a printed circuit board.

Even though the internal organization of RAM chips varies, they share some common features that can be discussed in general terms. A memory chip is physically arranged as a two dimensional array of cells. Certain address inputs are used for row selection, and the remaining address lines are used for column selection. Figure 10.4 shows the organization of a 1K × 1-bit chip that is logically organized as 1024 words by 1 bit, implying that each bit is individually addressable, and physically organized in a square 32 × 32 array, meaning that the array has 32 rows and 32 columns. Each cell is connected to one of the row lines, called *word lines*, and one of the column lines, known as *bit lines*. A particular cell is selected by activating its word and bit

Figure 10.4 Physical organization of a 1K × 1-bit memory IC.

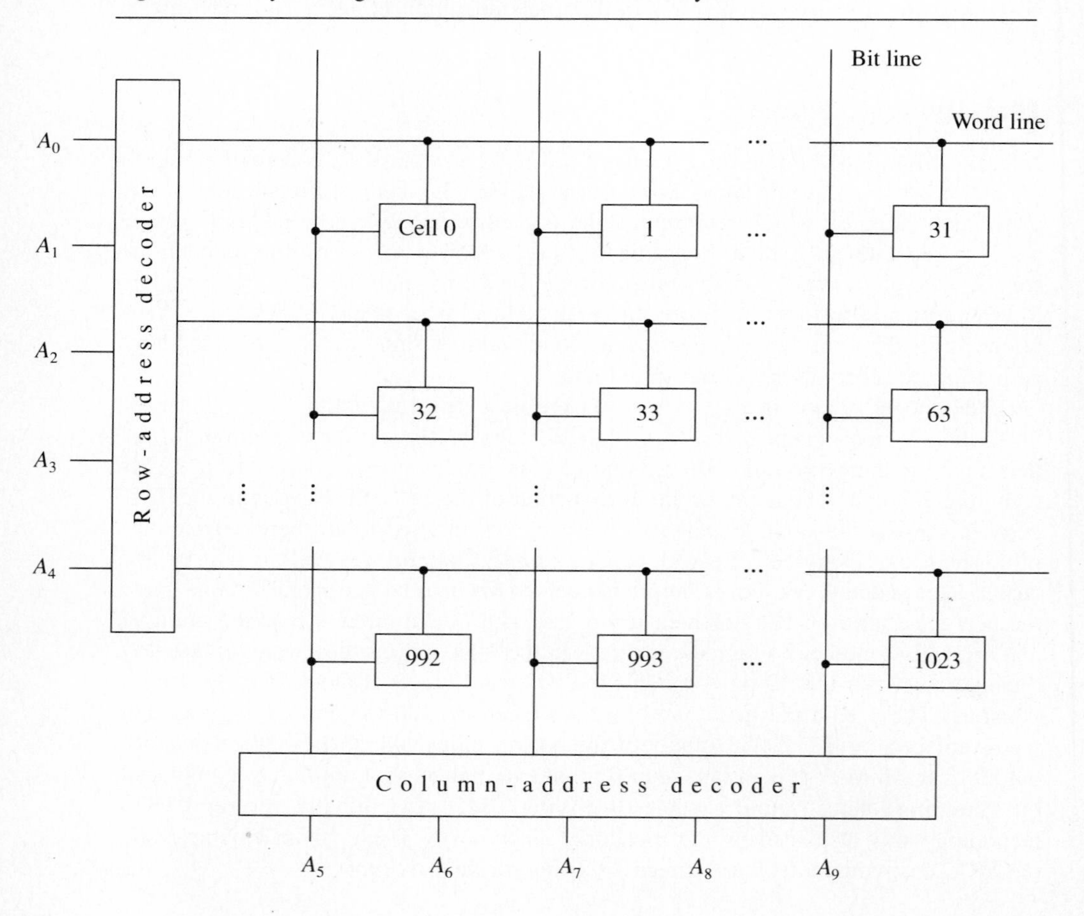

lines, which in turn is accomplished using the row-address decoder and the column-address decoder. Five of the 10 address bits needed form the row-address bits, labeled A_0 to A_4, and are fed to the row address decoder, which selects 1 out of the 32 word lines. Similarly, the other 5 address bits, labeled A_5 to A_9, are fed to the column address decoder, which chooses 1 out of the 32 bit lines.

The 3-T Cell

A 3-T memory cell, made up of three *n*-channel devices, is shown in Figure 10.5. There are two address leads, *w* and *r*, for write and read, respectively. A logical **1** on the *w* line causes Q_1 to conduct so that the circuit is connected to the input data line *D*. A logical **0** on this line, on the other hand, turns Q_1 off to disconnect the circuit from the data line. Similarly, the signal *r* and Q_3 perform the same function with the output $\overline{D}$ line. The information is stored by the capacitor *C*, which represents the stray capacitance between the gate and source of Q_2.

Suppose that both *w* and the data line are high, effectively connecting *C* to the supply through the low resistance of the data line. If *w* and data-line signals persist for a sufficiently long time, then *C* charges to the supply voltage that represents logical **1** in positive logic. Now suppose that *w* goes low, turning Q_1 off. Since the gate of Q_2 acts as an open circuit, assuming that the charge cannot leak off the capacitor, the information is stored in the cell, completing the writing of a **1**.

To see how a **0** is written, consider that *w* is high and the data line is low, so that Q_1 conducts. Since *C* is effectively connected to ground, it discharges, corresponding to storing a logical **0** in the cell.

To read the content of the memory cell, *r* is raised to logical **1**. Suppose that $\overline{D}$ is also **1**. If *C* is initially charged, then since gate of Q_2 will be positive, this transistor will turn on, resulting in a current flow through Q_3. If *C* is already discharged, then Q_2 will be cut off, and there will be no current flow. Thus, the state of the memory cell can be determined using the $\overline{D}$ line. Note that since the capacitor is isolated from the output by the gate-substrate insulation of Q_2, this circuit action does not affect the stored data.

Actually, the gate-channel resistance of Q_2 and the resistance of the cutoff Q_1 are not infinite, so the stored charge slowly leaks off, and the memory eventually loses its information. To prevent this, the memory is refreshed by periodically recharging the capacitors that store charge. A nonzero current through Q_3 implies a **1** in the

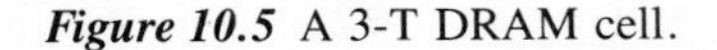

Figure 10.5 A 3-T DRAM cell.

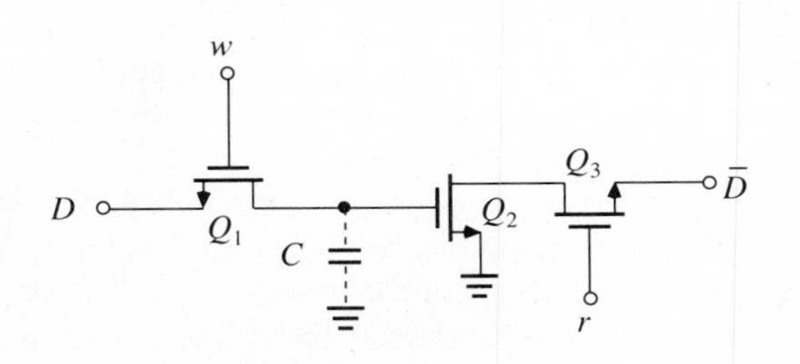

memory cell, and a positive signal placed on D and w lines replaces any charge leaked off the capacitor. Obviously, the capacitor must be recharged frequently enough to secure sufficient charge on and hence voltage across it. Thus, Q_2 will be on and the cell will be refreshed. During the refresh cycle, the entire memory is refreshed and no other operation can be performed, adversely affecting the speed of the memory. While some older 16-Kbit DRAMs must be refreshed every 2 ms, newer 1-Mbit DRAMs specify a 10-ms refresh rate.

The 1-T Cell

The simplest of all DRAM cells is the 1-T cell, illustrated in Figure 10.6a, utilizing an enhancement-mode n-channel MOSFET as a switch to charge the capacitor or remove the charge already stored. The gate of the transistor is connected to the word line and its drain tied to the bit line. It was first used in the 4-Kbit DRAMs that were available in 1973. As in the case of 3-T cell, the stored charge will be removed typically in a few milliseconds by the leakage currents of the capacitor unless refreshed.

Figure 10.6b shows the cross section of a 1-T cell using a *single polysilicon* gate process. The capacitor uses the channel region as one plate and the polysilicon gate as the other, while the gate oxide acts as the dielectric. To eliminate the concern over propagation delays down the long lines, a metal such as aluminum track is employed as the word line. The bit line is formed by n^+-diffusion. The source of the MOSFET serves as a conductive link between the inversion layers under the storage and transfer gates.

Figure 10.6c illustrates the cross section of a 1-T cell using a *double polysilicon* approach, in which the employment of two levels of polysilicon eliminates the need for a layout space to separate the two components. Therefore, the transistor and the capacitor should be regarded as one component only. Manufacturers resorted to this process to reduce the area of the memory cell.

The second polysilicon electrode, that is, the transistor gate, is separated from the first polysilicon capacitor plate by an oxide layer. The charge from the diffused bit line can therefore be transmitted directly to the area beneath the storage gate by the continuity of the inversion layers under the transfer and storage gates. Thus, the information is stored as an inversion charge in the poly MOS capacitor. The metal word line contacts the second poly level, isolating the storage cell from the bit line. The cell is insensitive to variations in the doping level of either polysilicon. In fact, its performance is primarily affected by the junction depth, oxide thickness, and the mask geometry, all of which remain constant.

Accessing an individual cell requires the selection of a word line by raising it from its *precharged* level of 0 V to the supply voltage value to turn the transfer gate transistor on, and the selection of a bit line that has been precharged to V_{CC}, as have all the other bit lines on which information is placed. When a row is enabled by the row decoder, all transistors in that row become conductive, transferring charge from their respective capacitors to their respective bit lines and thereby destructively reading the data. Each bit line or column has its own sense amplifier, whose function is to detect this charge and to amplify the signal caused by it. The amplified signal is a full logical level, ideally either V_{CC} or ground. The cell transistors remain on through-

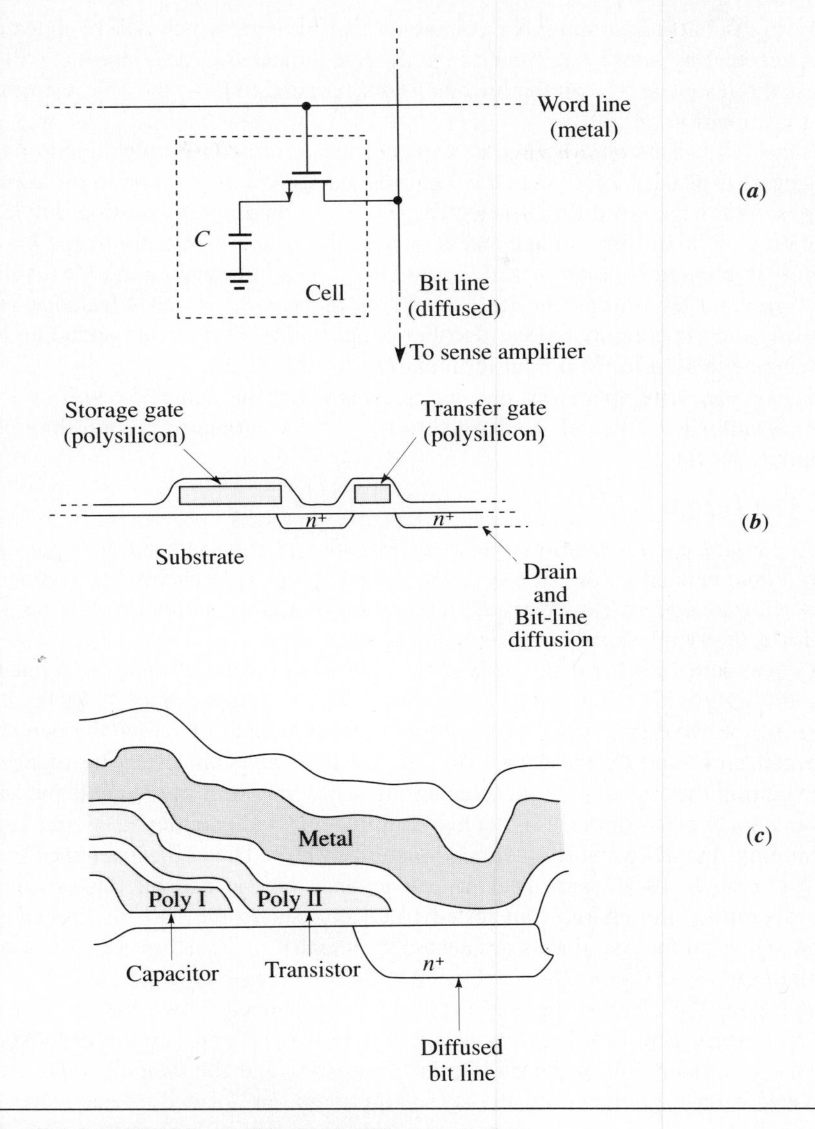

Figure 10.6 1-T DRAM cell: (a) circuit, (b) cross-section, single poly, and (b) cross-section, double poly.

out this period so that the amplified signals from the sense amplifiers can be fed back into their respective cells to refresh the stored charges.

Note that during a read cycle, the storage capacitor of each of the cells in the selected row is shorted to the corresponding bit line, and hence the storage capacitance shares charge with the bit-line capacitance. Now, shorting a storage capacitor C with an initial voltage of 0 V to the bit-line capacitance C_b will cause C to charge and C_b to discharge, dropping the voltage on that bit line, which will be detected by the corresponding sense amplifier to produce a logical **0**. The amplifier will then impress the resulting 0 V on the bit line to discharge C to 0 V, thereby restoring the stored information.

If the cell has a stored **1**, then its capacitor has originally been charged to a high voltage that is usually lower than the ideal logical **1** level of V_{CC} due to the capacitor leakages. When the word line is selected, C charges up slightly, causing only a negligible drop in the bit-line voltage that is interpreted by the sense amplifier as **1**, which in turn is impressed back on the bit line to charge C all the way to the ideal value of V_{CC}. Consequently, during the read operation all the cells in the selected word are refreshed, and the column-address decoder connects the bit line corresponding to the cell being addressed to the output terminal D_{out} of the chip.

During the write operation, depending on whether the data to be written is **1** or **0**, the capacitor C of the cell is charged to V_{CC} or discharged to ground through the appropriate bit line.

A 16-Kbit DRAM

To maximize the signal into the sense amplifier, that is, to keep the signal attenuation due to capacitive division to an acceptable level, a large cell capacitance and a small C_b are desired. However, the integration of a large number of bits on a chip calls for a physically small cell size and hence a small cell capacitance. The high density also demands that many cells share a common bit line, causing this line to be physically long with a high stray capacitance. The cell capacitance is increased by using a double-layer polysilicon fabrication process, which increases the percentage of the cell area used by the capacitor. The bit-line capacitance can be reduced by simply cutting the line in half and placing the sense amplifier in the center of the bit line, as depicted in Figure 10.7. This simplified block diagram is of the earliest 16-Kbit chip, the 4116, built by various manufacturers. The cell capacitance for this particular chip is 40 fF, and the stray capacitance of one half-bit line is typically 1 pF. Therefore, the charge transferred from any cell to the half-bit line causes a voltage signal on the line that is attenuated by a factor of 25 before being sensed by the amplifier.

As Figure 10.7 shows, the chip is physically organized as two 8-Kbit subarrays, forming a single 128 × 128 balanced array. The two 1-of-64 column decoders and 128 sense amplifiers are in the middle of the matrix, and the *dummy cells*, used to establish a voltage reference for the sense amplifiers, are located on each side. One of the subarrays complements the data and stores an input of logical **1** as a low level in the storage cell. A second complementation is performed by the output circuitry to recover the actual data.

The control circuitry surrounding the array is controlled by a network of clock generators which are activated by the external *row-address strobe* ($\overline{\text{RAS}}$) and *column-address strobe* ($\overline{\text{CAS}}$) signals. The address input buffers are multiplexed between row and column addresses, and the corresponding decoder circuits are independent of each other, reducing the input capacitance at these terminals as compared with other multiplexed RAM organizations, in which each address pin is connected to two input buffer circuits.

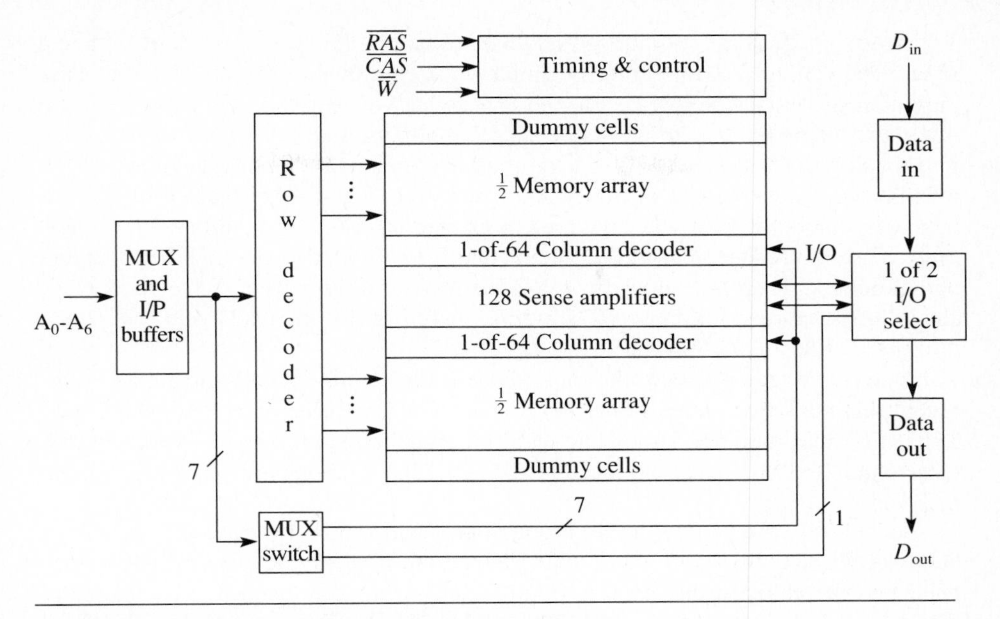

Figure 10.7 Functional block diagram of the 4116 16-Kbit DRAM.

Although this particular chip requires three power supplies, this inconvenience was eliminated in later DRAMs of 4-Kbit and higher densities. Currently, all DRAMs need only one 5-V power supply and generate other required voltages using on-chip circuitry.

First, the row selection is needed before the sense amplifiers can begin their relatively slow detection process. The column selection is not required until the outputs of the sense amplifiers are valid because its role is to gate data from the selected sense amplifier to the data output circuitry. The multiplexed addressing scheme takes advantage of this delayed need for the column address. Instead of using 14 address pins to select one of 16,384 memory cells, only 7 address pins are used to first select one of 128 rows, and subsequently the same pins are used to select one of 128 columns, resulting in a 16-pin package rather than a 22-pin package.

Although the address multiplexing provides system benefits, it complicates the timing, too, by requiring two timing signals. The sequence of events to address the chip is as follows:

1. Present the 7-bit row address to the address lines labeled A_0 to A_6.
2. After the input address bits have settled, activate the signal $\overline{\text{RAS}}$ by pulling it low to initiate a memory cycle and latch the row address.
3. Maintain row addresses for some minimum hold time.
4. Apply the 7-bit column address to the same address terminals.
5. Activate $\overline{\text{CAS}}$ by bringing it low.
6. Hold column addresses for some minimum time.

Note that all addresses must be stable on or before the trailing edges of $\overline{RAS}$ and $\overline{CAS}$. The former control signal is similar to a chip enable; it activates the sense amplifiers as well as the row decoder. It is necessary to keep this signal low for some minimum length of time to allow the sense amplifiers time to restore the data back into the destructively-read cells. These amplifiers maintain this data as long as $\overline{RAS}$ remains active. At the end of the cycle, $\overline{RAS}$ is taken high, the selected row is immediately turned off, and the precharge of all internal circuitry is initiated for a new cycle. $\overline{CAS}$ is used as a chip select, activating the column decoder and the input and output buffers. Therefore, it also controls the transfer of data from the selected sense amplifier to the output circuitry. The timing waveform for the read cycle is depicted in Figure 10.8.

During a write operation, the same sequence of events occurs as in a read cycle, except that the write-enable signal, ($\overline{W}$), is activated by bringing it low. This signal causes the input data to be strobed into the chip, buffered, and written into the selected sense amplifier and hence into the selected cell. The write cycle is shown in Figure 10.9.

To simplify the system timing problem and compensate for timing uncertainties that may be encountered in multiplexing operation, a *gated* $\overline{CAS}$ feature is incorporated into the chip, so that even if $\overline{CAS}$ occurs earlier than needed, it is internally delayed until the row-address hold-time specification has been satisfied, and the address inputs have been changed from row-address to column-address information.

Figure 10.8 Read-cycle timing.

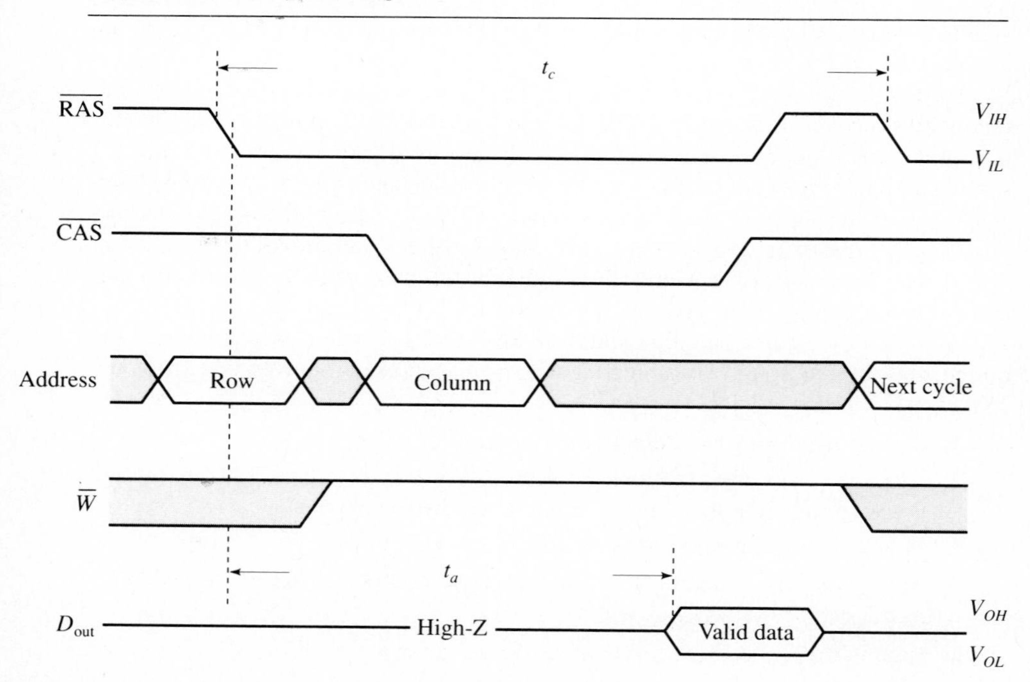

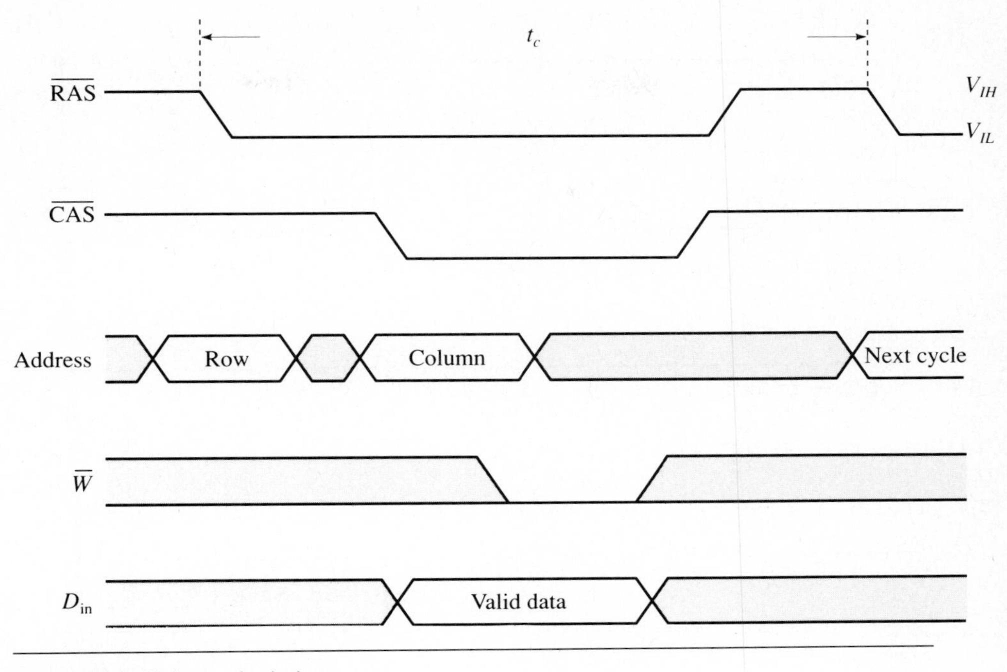

Figure 10.9 Write-cycle timing.

Page-Mode Addressing

Since $\overline{RAS}$ enables an entire row of memory cells and $\overline{CAS}$ transfers a single bit of data from the selected bit line to the output buffer, data can be transferred into or out of multiple column locations of the same row by having several column cycles during a single active row cycle. This permits a faster operation than is possible in the normal operating mode. When a row has been selected, the contents of all cells in that row are available in their respective sense amplifiers, so that the delay through the sense amplifier adds only to the access time of the first column. This mode of operation is called the *page mode* and is well suited to long, uninterrupted data transfers. It reduces the power consumption while typically doubling the maximum operating frequency. The page-mode read cycle timing is shown in Figure 10.10. The maximum pulse duration of $\overline{RAS}$ is the limiting factor for the maximum number of cycles that can occur in this mode because it must be taken high to satisfy the precharge requirements of the DRAM.

The Sense Amplifier

Since the sense amplifier is called upon to detect severely attenuated signals, it is the most important circuit in a DRAM. It is a differential amplifier with a high common-mode rejection in order to be able to reject picked-up interference due to crosstalk from other parts of the chip. There are two types of sense amplifiers that are utilized in commercially available ICs. These are variations of the *static* amplifier

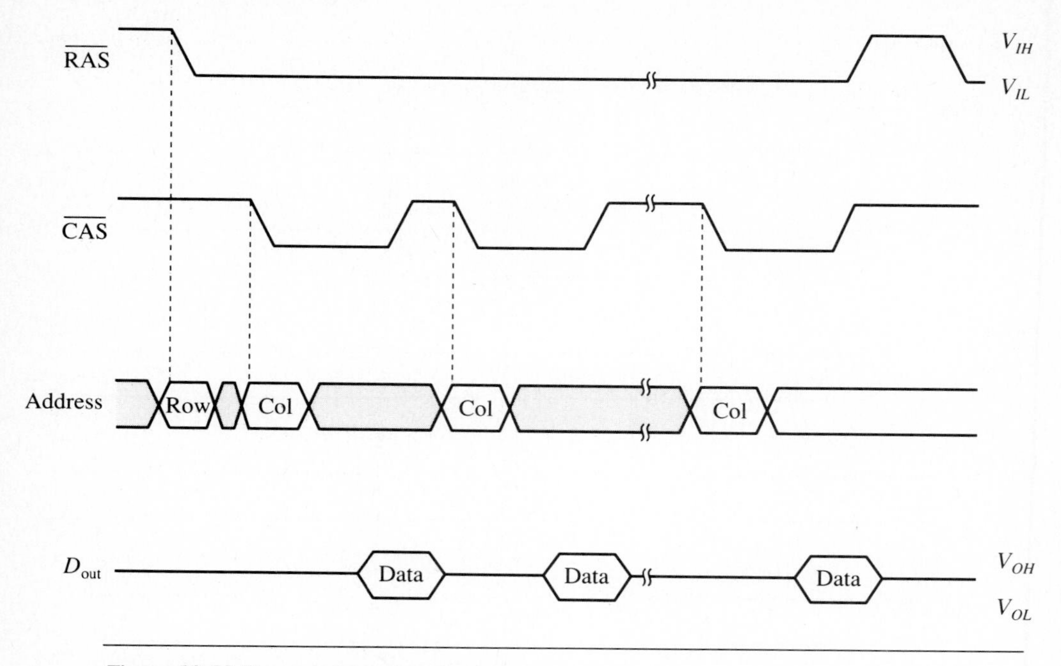

Figure 10.10 Page-mode read-cycle timing.

of Figure 10.11 and of the *dynamic* amplifier of Figure 10.12. Even though they are about equal in their ability to detect and amplify signals, the load resistors R_1 and R_2 in static amplifiers consume a considerable amount of power. Since dynamic amplifiers have no load resistors and are activated only when required to do the sensing, their power consumption is much less than that of the former. There are, however, design and layout problems associated with their use.

Static Sense Amplifier

The static sense amplifier is essentially a flip-flop formed by transistors Q_1 and Q_2, with resistors R_1 and R_2 functioning as loads. As depicted in Figure 10.11, the bit line is divided into two halves, with each half connected to memory cells as well as a dummy cell. The top half-bit line is connected to the bidirectional true data bus by means of Q_3, which is turned on by the output of the column decoder when this particular column is selected. Note that, to conserve power, the flip-flop is gated by Q_5, which is normally off and is turned on by the internal clock circuitry to do the sensing. Notice also that the capacitor is returned to the supply instead of the ground.

Between the cycles, the two halves of each bit line are precharged to precisely the same voltage. When a cycle is initiated by $\overline{\text{RAS}}$ going low, these lines are allowed to float briefly. Then, as a row is selected, charge is transferred from the enabled cell in each column to its respective half-bit line. Thus, the half-bit line will eventually be at the same logical state as the original state of the selected cell. In the meantime, the half-bit line that does not contain the addressed cell is adjusted to a voltage some-

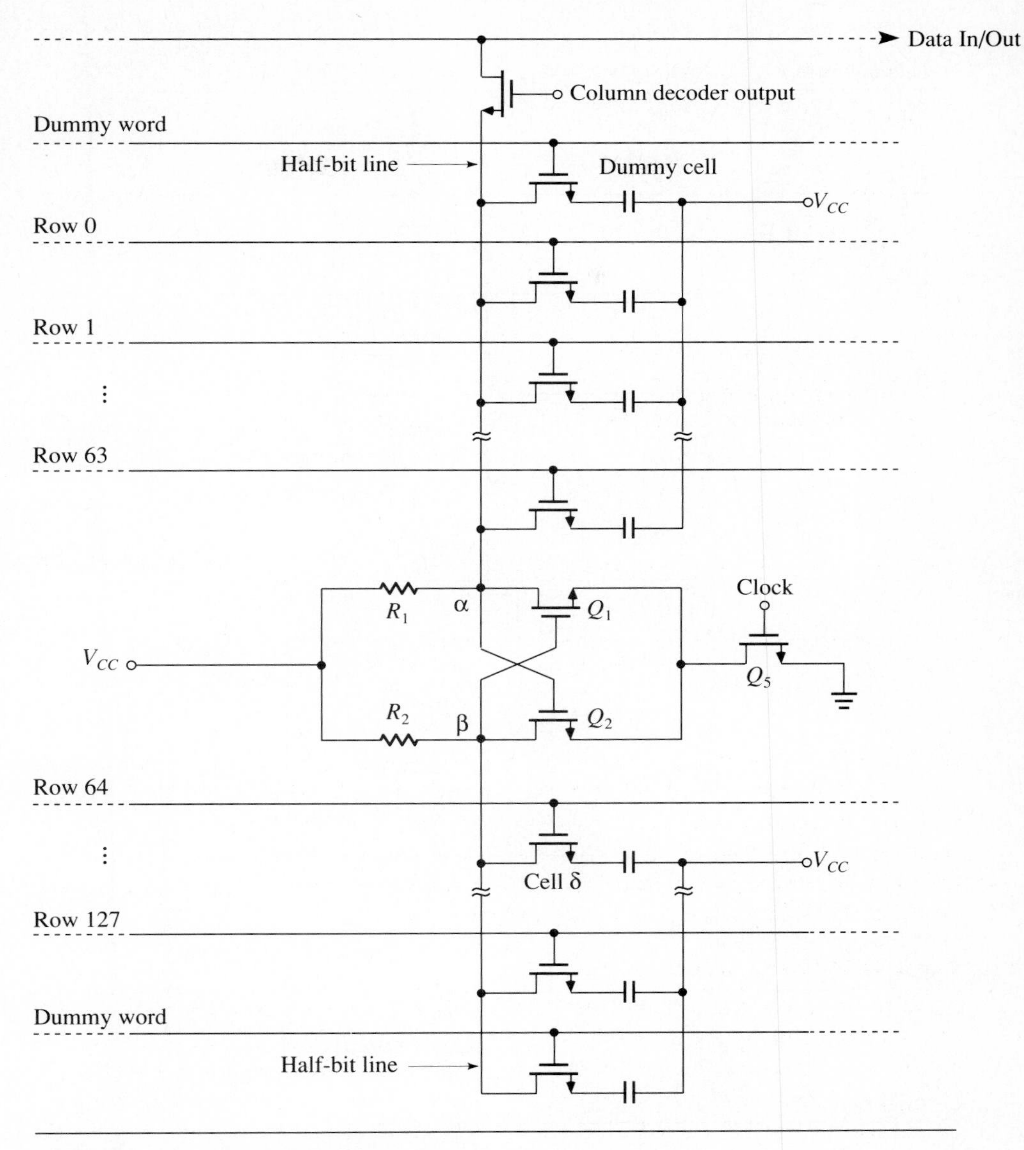

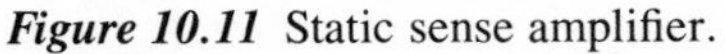
Figure 10.11 Static sense amplifier.

where around .25 V, a value in the middle of the low-to-high range of around .5 V. This is accomplished by simultaneously selecting the corresponding dummy row and connecting the proper dummy cell to this half-bit line. Therefore, if the selected cell originally contained a high voltage, the voltage of its half-bit line would be approximately .25 V above the adjusted intermediate voltage of the other half-bit line. Similarly, if it contained a low voltage, the voltage of its half-bit line would be around

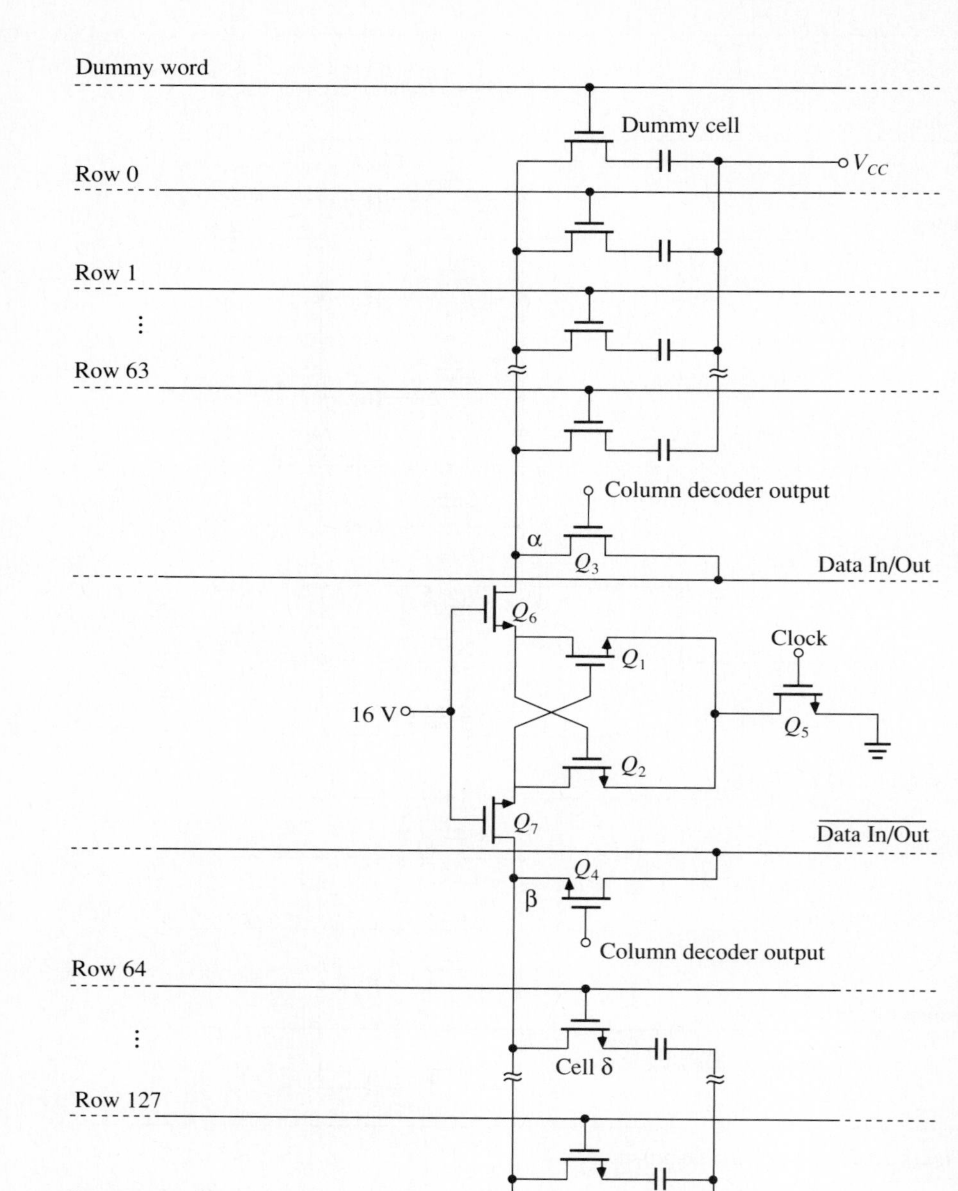

Figure 10.12 Dynamic sense amplifier of the 4116.

.25 V below the intermediate voltage of the other half-bit line. Consequently, the addressed cell together with the dummy cell secure an initial voltage imbalance for the regenerative action of the flip-flop to latch it up. The half-bit line that has the lower initial voltage is forced to ground, while the other line rises to logical **1**.

Suppose, as an example, that the cell δ in Figure 10.11 has stored a low voltage and has been selected. Then, node α will be higher than node β by about .25 V. This difference will be amplified by the positive feedback of the flip-flop. The transistor Q_1 will then turn off and Q_2 will turn on; as a result, node β will be pulled to ground through Q_5 to restore a **0** in cell δ.

Meanwhile, node α will be pulled to **1** through R_1, and Q_3 will be turned on so that the high voltage of node α, that is, the complement of the stored data, will be impressed on the internal data bus. Thus, the control logic will complement the data before it is presented to the D_{out} terminal.

Now, assume that we want to write a **1** in cell δ. Since this cell is connected to the lower half-bit line, the data input buffer must drive the data bus to ground. The transistor Q_3 will then force node α to ground to turn Q_2 off, which in turn will allow R_2 to pull node β to high as required to write the high level into the storage cell. Therefore, with these load resistors, data can be written into a cell in either half of the matrix using a single data bus.

The disadvantage of the static sense amplifier is its utilization of load resistors. Since either R_1 or R_2 dissipates power at any given time, a low value leads to a high power consumption. Increasing their values, on the other hand, results in excessively slow write times because the large bit-line capacitance produces a long time constant.

Dynamic Sense Amplifier

Referring to the dynamic sense amplifier of Figure 10.12 that is used in the 4116, we note that the flip-flop load resistors are replaced by two MOS transistors, Q_6 and Q_7, whose gates are connected to an internally generated 16-V power supply. Moreover, it requires both a true and a complement data bus. This means that the column-decoder outputs should be available in both halves of the matrix, causing a layout problem. Placing a single column decoder below or above the memory was initially not possible because it was not practical to run its outputs through the matrix to the other side.

One solution is to use two separate decoders, one above the top half of the array to service the true data bus, and the other below the bottom half of the array to service the complement data bus. However, this duplication consumes silicon area, leading to an increase in the cost of the chip.

Another solution is to use a single column decoder situated in the center of the memory array along with the sense amplifiers. This time, however, it becomes topologically necessary for the bit lines to cross the buffered address signals. Even only one address signal, going from ground to supply voltage, capacitively couples more signal onto a bit line than that provided by the memory cell. There will be no problem if all lines crossing the bit line are kept inactive until the sense amplifier detects and amplifies its signal. Using a multiplexed design alleviates the problem considerably: it is easy to assure that the buffered column-address lines remain static during this time because multiplexing inherently causes the column address to be processed after the row addresses have been processed.

Consider again the process of writing a **1** in cell δ. When this particular column is selected, its top half-bit line is connected to the true data bus via Q_3, and its bottom half-bit line is connected to the complement data bus through Q_4. Now the input buffer drives the true data bus to ground, with Q_3 causing the top half-bit line to follow, and forces the complement data bus to **1**, with Q_4 causing the bottom half-bit line to follow. Therefore, the complement data bus achieves what was previously performed by the load resistor. The high voltage on this half-bit line is transferred into the selected cell, and the write operation is completed. Note that Q_3 and Q_4 act only as switches and can have extremely low resistances to expedite the write time. In addition, since the half-bit line is driven directly by the data bus, the delay encountered in the static amplifier is significantly reduced. In short, memory designs using dynamic sense amplifiers dissipate much less power and write faster than do designs using static sense amplifiers.

Stacked-Capacitor Cell

The cell area of a 1-T cell is limited by the storage capacitor area that is needed to supply a detectable signal. An early attempt to increase the storage capacity of a high-density DRAM was to increase the effective dielectric constant of the insulator used to form the storage capacitor. One viable choice is silicon nitride, whose dielectric constant is higher than that of the silicon dioxide film used as the storage-capacitor insulator in conventional cells.

In the *stacked-capacitor* (*STC*) cell, which has a *triple polysilicon* structure, the capacitor is stacked above the transistor, bit lines, or field oxides. This type of cell was first utilized on a 5V-only 16-Kbit DRAM. Three types of STC cell structure that differed in the position of the stacked capacitor were originally proposed, but only one type was chosen for the memory design by evaluating and comparing their *charge-transfer ratios*

$$T = \frac{C}{C + nC_b}$$

where n is the number of cells on the bit line, to those of the conventional cell. Naturally, the larger the charge-transfer ratio, the larger the signal voltage.

In the particular STC cell chosen, each capacitor is stacked only on the field oxide, as depicted in Figure 10.13a, and is composed of a first polysilicon electrode, an Si_3N_4 film with at least twice the dielectric constant compared to that of SiO_2, and a second polysilicon electrode. Even though the transistor gate and word lines are the third polysilicon layer, and bit lines are aluminum in the original structure, the chip was designed using the STC cell with an aluminum word line and a diffused bit line to utilize the peripheral circuits of a conventional DRAM. The cross-sectional view of the modified cell is shown in Figure 10.13b.

Alpha-Induced Soft Errors

As the cell size shrinks to accommodate higher densities, the data are stored in much smaller capacitances, on the order of 30 to 100 fF with fewer electrons differentiating between logic states, resulting in a susceptibility to error-producing physical

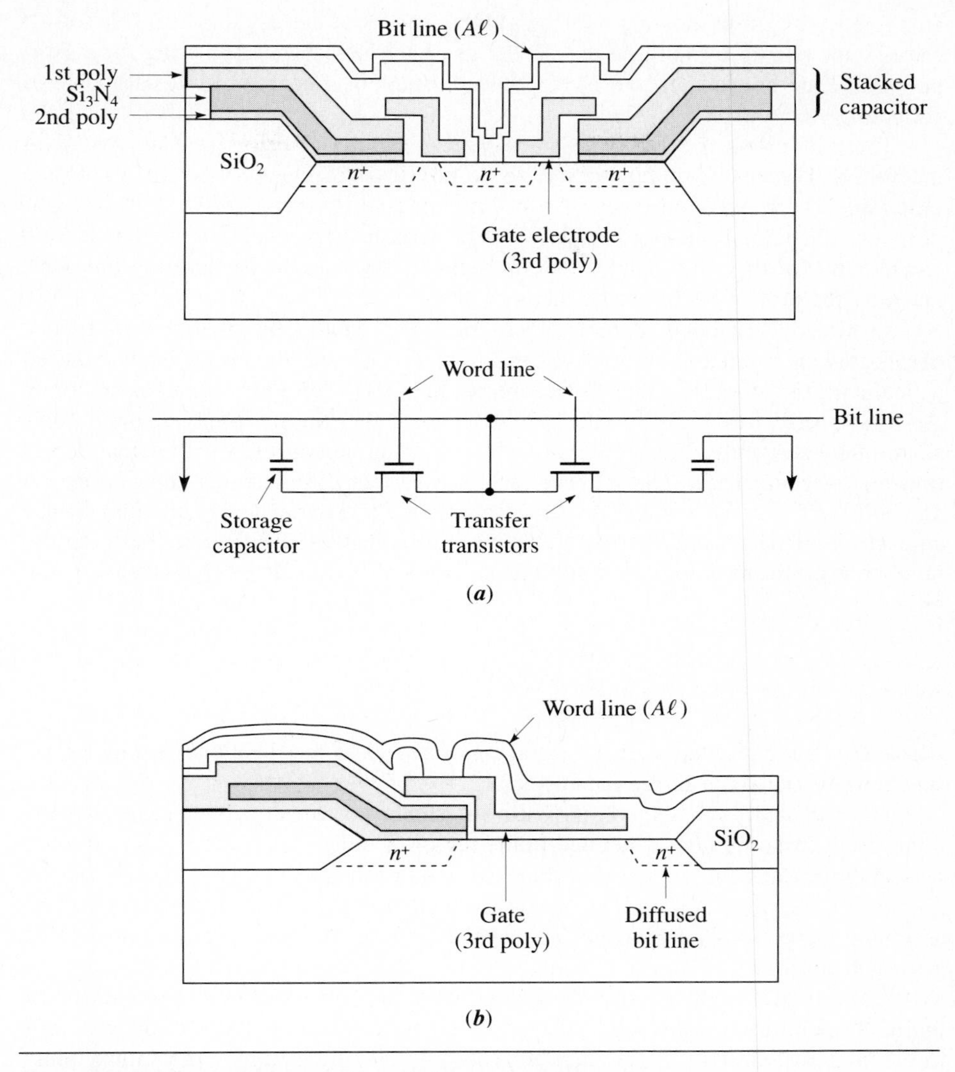

Figure 10.13 The stacked-capacitor cell: (a) cross-section and equivalent circuit of the triple-poly stacked-capacitor cell, and (b) the cell as implemented.

mechanism, which in turn may provide the ultimate limit to the trend of smaller stored charge packets as they are currently implemented. The *alpha* (α) *particles* that are emitted from natural radioactivity in the IC such as from the glass used to seal the package or the gold-plated metal lid used in side-brazed packages produce electron-hole (e-h) pairs as they penetrate the depletion layer of the MOS capacitor, leading to electrons being collected at the surface. Electrons from e-h pairs generated outside the depletion region can also diffuse toward the surface and get collected, giving rise to an increase in the error rate. This type of error is known as a *soft error*. In addition, electrons collected by floating nodes in other portions of the circuitry can lead to soft errors. One example would be the precharged bit line that is left floating for a short period of time prior to readout of the cell information, during which it is capable of collecting electrons.

The soft errors are random, nonrecurring, single-bit errors in semiconductor memories. The errors are temporary, so no physical defects are associated with the failed bit. There are other types of soft errors caused by system noise, sense amplifiers, or voltage marginality, but these can be eliminated by standard noise-reduction techniques. On the other hand, soft errors caused by α-particles can only be eliminated by the proper design of the memory device.

As already discussed, dynamic memories store data as the presence or absence of minority carrier charge on storage capacitors, so that periodic refreshing is required to maintain the stored charge. For example, in NMOS DRAMs, the capacitor is in the form of a *potential well* in the p-type silicon under the positively charged polysilicon gate electrode. This potential well should not be confused with actual depletion region geometries. The number of electrons that differentiates between empty and full wells can be reduced by an incomplete charge transfer to bit lines, sense amplifier sensitivity, and thermal generation. We define the difference between the number of electrons at logical **1** and **0** levels as the *critical charge* Q_c, which is given as

$$Q_c = \frac{1}{2} C(V_{CC} - V_T)$$

where C is the capacitance of the storage cell and V_T is the threshold voltage of the transfer-gate transistor of the memory cell.

The *collection efficiency* is the ratio of α-induced carriers collected by storage nodes to the total number generated in the region. Limited only by the small amount of recombination, the collection efficiency is almost unity under the gates. If the number of electrons that end up in storage wells exceeds the critical charge, then, assuming that the other error sources have already been eliminated, an α-induced soft error will result.

Alpha particles are doubly charged helium nuclei ejected from the nucleus of high-Z atoms during radioactive decay. They are emitted at discrete energies and travel in almost straight lines. Their emissions are not affected by temperature, pressure, and so forth. It has been observed that up to 2.5×10^6 e-h pairs can be generated in the exceptionally short time of a few picoseconds by the passage of an ultra-high energy, naturally occurring α-particle. In addition, an exact solution of the diffusion equation shows the collection process is complete within microseconds. The flux and energy of ionizing radiation, target area, cell geometry, and critical charge, which are functions of technology parameters, and the package type and composition, all contribute to the soft-error rate of a device.

The stages of the soft-error generation by α-particles are shown in Figure 10.14. The **0** state corresponds to a potential well filled with electrons; the well is empty for the logical **1**. As depicted in Figure 10.14b, a 5-MeV α-particle typically generates 1.4×10^6 e-h pairs, corresponding to 3.5 eV/e-h pair, and penetrates 25 μm in

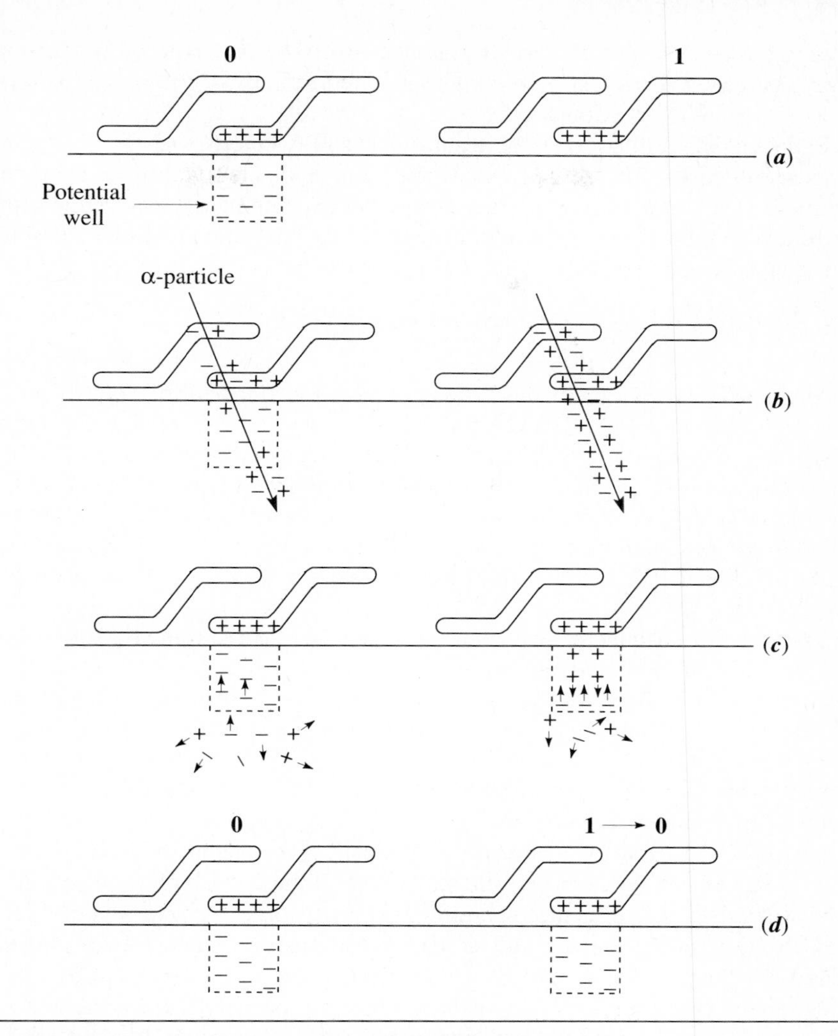

Figure 10.14 The stages of soft-error generation by α-particles: (a) steady-state before the radiation, (b) α-particles hit to generate e-h pairs, (c) transient-state after the hit, and (d) steady-state resulting in a soft error.

silicon. Figure 10.14c illustrates the resulting diffusion of the e-h pairs to the surrounding regions. The electrons reaching the depletion region are swept by the electric field into the potential well, while the holes are repelled. A soft error results when the collection efficiency times the total number of electrons generated is greater than the critical charge. Note that even a single α can cause an error.

It has also been reported that α-particles with sufficiently high energy can induce physical defects. These defects lead to an increase in the generation current of storage cells, which in turn degrades the refresh time, which decreases monotonically with increasing α-particle irradiation. In practice, however, this type of degradation is unlikely because the number of α-particles emitted from usual packaging materials is not large enough to cause such defects.

Reducing the α flux levels of natural package materials such as glass and ceramics seems to be difficult because of the costs associated with their purification. Moreover, it is unlikely that all soft errors can be eliminated, especially with the current level of integration. Therefore, internal error detection and correction is becoming increasingly desirable.

High-Density DRAMs and Other Second-Order Effects

Semiconductor memories have seen a continuing trend toward higher levels of integration since the 1-Kbit DRAM with a 3-T cell was introduced in 1971, followed by the 1-T cell 4-Kbit DRAMs in 1973. These circuits have evolved in the past few years to densities of 1-Mbit per chip, which were first sampled in 1985 and were available in large quantities the following year. Even denser 4-Mbit DRAMs were offered in sample quantities by a number of manufacturers as early as 1988. 1989 saw the 16-Mbit DRAMs going from the experimental laboratory to production prototypes.

The push for raising chip density has resulted in a quadrupling of memory density and a doubling of logic integration for every product generation to come along since the early seventies. Increased levels of integration have been achieved through improvement in sense-amplifier sensitivity and continuously refined lithography techniques. Recent advances in optical lithography will allow reductions in device geometries down to .3 to .5 μm in the mid 1990s. The X-ray techniques are expected to allow even smaller geometries.

As MOS transistors are scaled down to smaller dimensions, however, numerous second-order effects become prominent. For one thing, the threshold voltage of a scaled circuit becomes difficult to control, decreasing sharply as channel-gate lengths approach 1 μm and increasing just as sharply as transistor channel widths decrease. In addition, there is the so-called *drain-induced barrier lowering* (*DIBL*), which causes subthreshold conduction—a serious problem especially at the very short channel lengths needed for high-density memories.

Hot carriers, as covered in Chapter 3, cause a shift in the threshold voltage of a submicron MOSFET as the channel electric field in the device becomes higher with reduced feature size, which degrades the reliability of the device. One effective way of improving this situation is to reduce the channel electric field by using a *lightly doped drain* (*LDD*) structure, which has been demonstrated to reduce, at low gate bias, the substrate and gate currents greatly while increasing the source-drain breakdown voltage. It also offers smaller C_{gd} and C_{gs} for better frequency response, and helps threshold and punchthrough control. The disadvantages of the LDD structure are the transconductance reduction for a given device size and hence the reduced current drive capability, and added process complexity.

With a sufficiently high electric field in a semiconductor such as a MOSFET, an electron in the conduction band can gain kinetic energy from the field before it collides with the lattice, upon which the electron gives most of its kinetic energy to break a bond, thereby ionizing a valence electron from the valence band to the conduction band to create an e-h pair. This is referred to as *impact ionization*. Similarly, the generated pair creates additional e-h pairs by accelerating in the field, thus acquiring kinetic energy and finally colliding with the lattice. This process is called *avalanche multiplication*, which increases the reverse current very rapidly, eventually leading to the nondestructive mechanism called *avalanche breakdown* that imposes an upper limit to the drain voltage of a MOSFET.

As the channel length is reduced, the drain and source depletion regions extend into the channel, the channel length is reduced, and the drain current, displaying a significant dependence on the drain voltage, rises rapidly as the device enters the punchthrough state.

Therefore, the maximum usable voltage in a MOSFET is determined by the lowest punchthrough, junction breakdown, and hot electrons. The latter can limit the maximum voltage in two ways. They can lead to substrate currents, which induce a latch-up phenomenon that limits the maximum voltage to a value much lower than the breakdown voltage. They can also produce secondary carriers that penetrate the gate oxide to produce drift in the threshold voltage, as already mentioned.

A reverse bias on one diffused junction creates an electric field pattern that can lower the potential barrier separating it from an adjacent diffused junction. When this barrier lowering is large enough, the adjacent diffusion acts as a source, giving rise to an unwanted current path. The DIBL prevails if the potential distribution that causes the barrier lowering at the source cannot attain the gradient at the drain that is necessary to start the avalanche breakdown. It may take place when the source/drain diffusions are deep, the substrate resistivity is high, or the diffusion-to-diffusion spacing is narrow. The DIBL increases the threshold variation in a transistor and can cause interaction between otherwise isolated devices. Consequently, by determining the ultimate proximity of surface diffusions, it can be considered as one of the main electrical limitations for memory-chip density.

There are two effects that bring about the DIBL. When $V_{DS} = 0$ V, even though the potential distribution under the gate of a short-channel MOSFET is symmetric, the distance between the source and drain diffusions may not be sufficient to accommodate their depletion regions, resulting in DIBL because the potential peak is lower than for the long-channel case. When $V_{DS} > 0$ V, electric field lines penetrate from drain to source, leading to further lowering of the barrier.

The DIBL effect can be likened to the shielded dipole model, as illustrated in Figure 10.15. The electric field pattern in the silicon can be roughly approximated by a two-dimensional dipole in free space with the line charges situated at the source and drain. Then the effect of the gate is simulated by adding above the dipole a conducting plane whose charge imaging reduces the drain-to-source coupling. Finally, a movable conducting plane whose position represents the changeable gate and drain voltages is added below the dipole. The additional imaging lowers the coupling further. Evidently, the coupling is minimized when the distance to one of the conducting planes is small compared to the spacing of the line charges. As the fixed conducting plane is moved further from the silicon, more field lines can penetrate

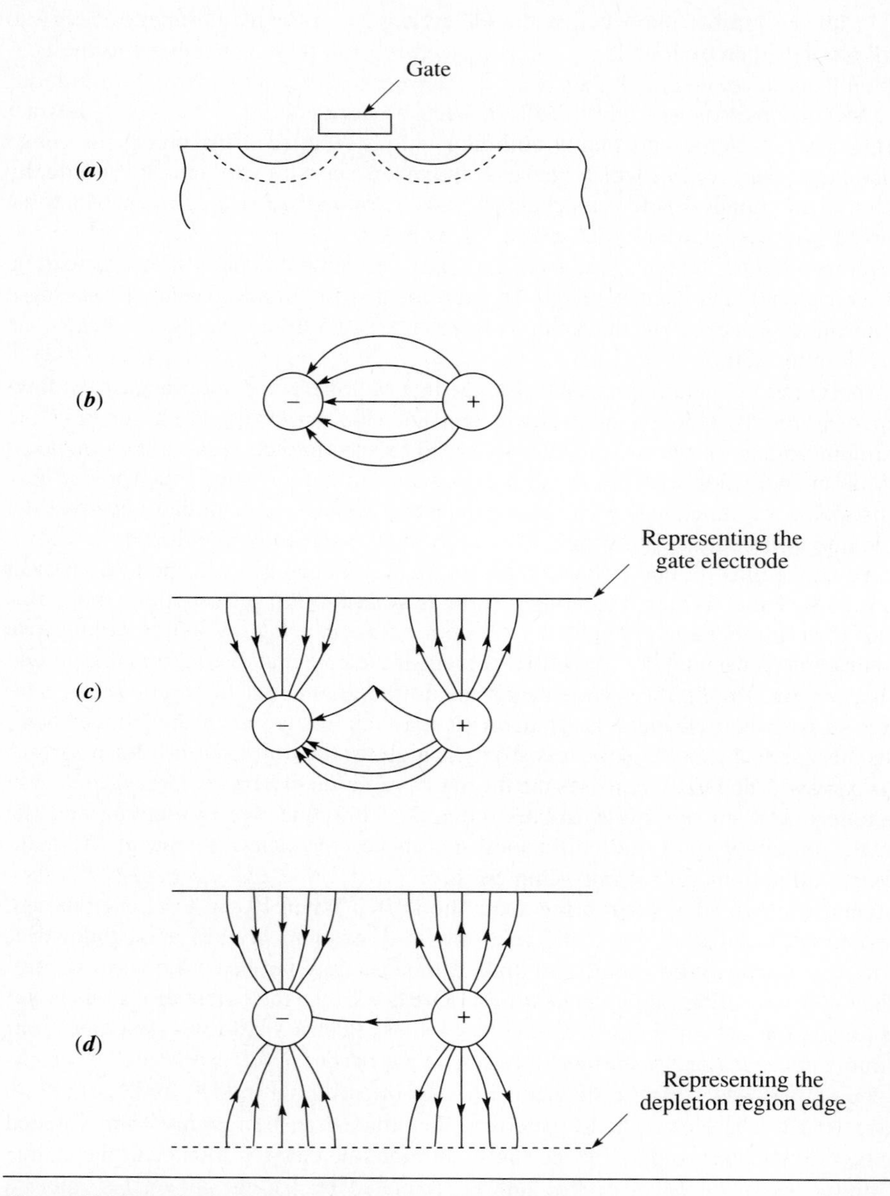

Figure 10.15 A dipole representation of the electric field in a MOSFET: (a) MOSFET structure with gate and depletion region, (b) dipole in free space, (c) dipole adjacent to the conducting plane, and (d) dipole between fixed and movable conducting planes.

from drain to source. Therefore, the potential peak is more effectively reduced with increasing drain voltage.

New Trends in High-Density DRAMs

The traditional transistor structures and bulk-silicon fabrication techniques imply a certain maximum in the density of DRAMs that can only be achieved if the die size is allowed to expand beyond its present limits. Thus, in recent years, it has been observed that semiconductor manufacturers have been moving to vertical structures to squeeze the most in the least amount of space. A variety of DRAM cell structures such as the *trenched-capacitor* cell and *trenched-transistor* designs have been employed to move densities toward 4- and 16-Mbit ULSI levels.

In addition, the *silicon-on-insulator* (*SOI*) technology, whose characteristics make it an attractive technology to pursue for scaling devices, is emerging to replace the bulk-silicon techniques. SOI refers to a class of materials in which a crystalline silicon film is grown epitaxially on an electrically insulating substrate, which provides flexibility in the design and fabrication of passive components such as resistors and capacitors. Moreover, it significantly reduces the parasitic capacitances, leading to further cuts in power dissipation and increases in the speed of MOS devices to levels near those of bipolar ECL. Since the silicon portion of the IC is limited to the thin film on the insulating substrate, the silicon volume available for minority carrier generation is small, appreciably reducing the soft error rates. The SOI does not require the channel-stop isolations or deep *n*- or *p*-type wells that are necessary to eliminate latch-ups. They are also inherently radiation-hardened.

As the leap from 16-Kbit chips to 64-Kbit DRAMs was made in the early eighties, laser programming of redundant cells was first utilized to increase yields. In a redundantly designed DRAM, spare circuitry consists of extra rows and columns that can be substituted for defective rows or columns in the regular memory array. The redundant circuitry imposes only a slight area increase (e.g., approximately 1.5% for a 1-Mbit device). Once a memory-test system identifies defective rows or columns on a memory chip, the laser-repair system produces a pulsed laser beam that vaporizes polysilicon links on the chip to disconnect the defective circuitry. Cutting links also logically connects a good redundant row or column and programs the chip's appropriate decoder to recognize it. This process is totally transparent to the device user, who does not notice a change in either performance or reliability of the part.

Initial megabit products resulted from enhanced 256-Kbit processing methods with minimum features that shrank from 2 to 1.2 μm. The CMOS, with its low power dissipation and high density, replaced NMOS in more than 50% of the new designs. Top speeds of the latest 1-Mbit models have already pushed below 50-ns access times. The *static-column decoding* architecture, in which column-address buffers, column decoders, and output control are all static circuits, even offers access speeds of around 25 ns.

As shown in Figure 10.16, operation in the static-column mode is similar to that in the page mode, but the DRAM does not require the toggling of the $\overline{\text{CAS}}$ clock. The access and cycle times are almost equal, ensuring a high data-transfer rate. A chip-select signal $\overline{\text{CS}}$ that has no timing restriction is used instead of $\overline{\text{CAS}}$ to enable the output.

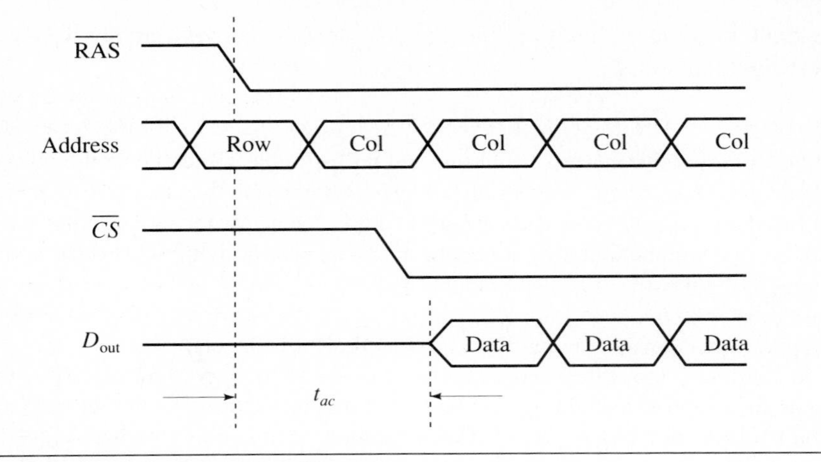

Figure 10.16 Static column-mode timing diagram.

The three-state column-address buffer, used instead of a column-address latch, is in the high-impedance state when the $\overline{\text{CAS}}$ line is high. Asserting the $\overline{\text{RAS}}$ and then the $\overline{\text{CAS}}$ lines initiates the first static-column access. Holding both lines low throughout the access period, subsequent columns are accessed by altering the column address. This method eliminates the setup, hold, transition, and precharge times associated with toggling the $\overline{\text{CAS}}$ line. Therefore, static-column mode is much faster than page mode. Essentially, though, both provide the central processing unit with a low-cost cache memory by virtue of their fast access to a small block of the system's main memory. The drawback of static-column access is that more power is consumed as compared to page-mode access because the output buffers are continuously enabled during the access. Moreover, the access is slower than direct read and write cycles when random rows are accessed.

Using conventional design methods, the cell sizes for 1- and 4-Mbit DRAMs are about 30 and 10 μm^2, implying capacitor areas of 10 and 3.3 μm^2, respectively, because the storage capacitor area is usually less than $1/3$ of the cell size. On the other hand, the amount of storage charge needed for sufficient soft-error immunity was calculated to be at least 200 fC, which in turn requires a capacitor oxide thickness of about 10 and 3 nm, respectively, for 1- and 4-Mbit DRAMs. This requirement is impossible to implement with a 5-V supply voltage.

One approach to overcome this problem employs trenched capacitors and on-chip error detection and correction. Internal error correction serves as an alternative to the 200 fC requirement by permitting the cell capacitance to be reduced to a value limited only by the sense-amplifier input-signal level, which stays constant provided that the operating voltage and the capacitor oxide thickness are reduced with the same scaling factor. The value of the capacitor can be as low as 30 fF and still allow a sufficient signal level to be attained if the operating voltage is reduced to 3.3 V by using an internal voltage converter. The low voltage value is also useful to prevent short-channel MOSFET degradation due to hot electron injection. This capacitance value is achievable when a trenched capacitor is utilized with a cell area of 10 μm^2 and a trenched depth of 3 μm.

In trench technology, the capacitor is carved down into the silicon substrate to decrease the surface area it consumes without reducing its overall size, as depicted in Figure 10.17. The difficulty with this approach is its demand of high-level precision from the production equipment to control the profile of the etched groove. Its advantage is that deeper trenches can be dug all the way down through the lightly doped epitaxial layer where active devices are formed and into the heavily doped substrate for more active area if more capacitance is needed. The capacitor plate holding a charge is the single-crystal silicon substrate. Trench capacitors of up to 60 fF have been designed, while some megabit planar capacitors are as small as 30 fF.

Further scaling can be accomplished by also dropping the memory-cell transistor into the trench. One 4-Mbit DRAM design incorporates a 9-μm^2 1-T memory cell in which the capacitor and transistor are built into the sidewalls of an 8-μm deep trench that is etched into a 1.3 $\times$ 1.5 μm aperture, with the 2-μm transistor region being stacked on top of the 6-μm capacitor region. Since the source voltage must be supplied through the substrate, polysilicon plate inside the trench holds the charge. This is diagrammed in Figure 10.18. At this level of density, the combined depth of all trenches totals 30 meters!

At higher memory densities, interconnections occupy a greater percentage of the circuit. To reduce the die size, the lines are made narrower and thinner, and more interconnection levels are stacked. As the lines get thinner and longer, however, interconnection resistances increase and time delays stretch out.

The capacitive loading of a metal bit line is significantly less than for a diffused line. Thus, for bit lines, aluminum is used down to 1 μm instead of the diffused bit lines of the previous generations. Below 1 μm, tungsten is the material of choice. For word lines, polysilicon silicide composites, called *polycides*, are preferred. A polycide is a polysilicon material that has been sputtered with a metal to enhance signal propagation. Its resistance is less than that of polysilicon by a factor of ten, allowing

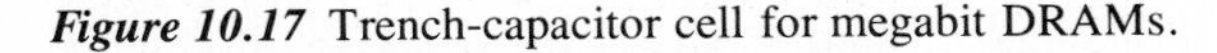

Figure 10.17 Trench-capacitor cell for megabit DRAMs.

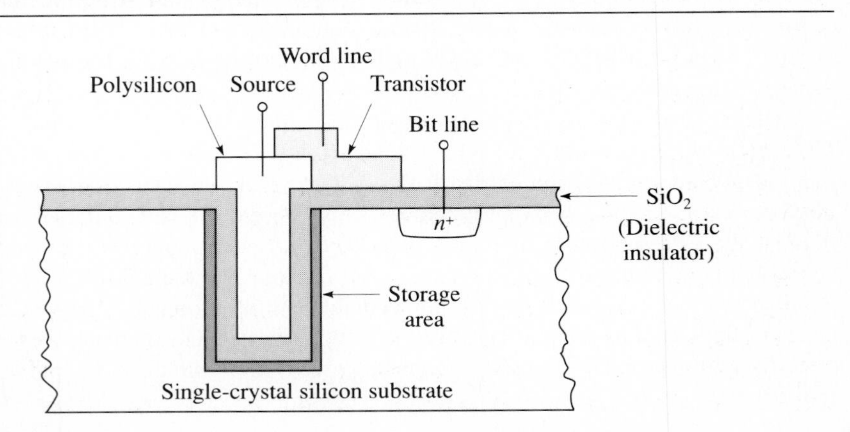

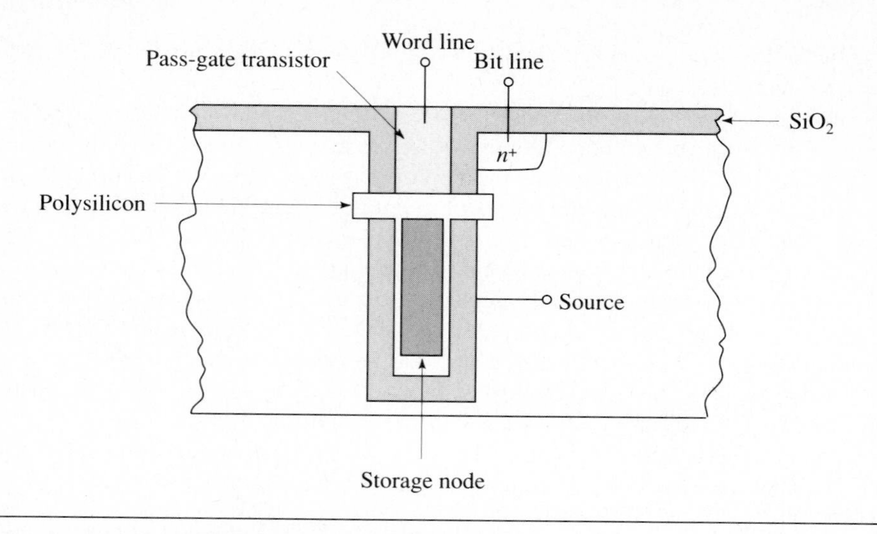

Figure 10.18 Trenched-transistor cell of a 4-Mbit DRAM.

reduced signal delays and hence fast access times that are necessary to accommodate demanding system performance.

Another way to accomplish fast access times is to utilize metal interconnects for both horizontal and vertical lines in the array. In a *double-level* interconnect system, which is becoming the standard in 16-Mbit DRAMs, both the bit and word lines are available in metal to reduce the worst-case time constant of any signal in the matrix.

NMOS vs. CMOS

The NMOS technology has traditionally been used in DRAMs mainly because of the cost benefit but at the expense of complex peripheral circuits and of power consumption. As compared to NMOS, the CMOS devices require less power, generate less heat on the average, and greatly simplify the design of peripheral circuits within the chip, while still matching and even exceeding the NMOS in processing speed. Therefore, CMOS designs will be dominant as the density, and the design and process complexity continue to grow.

The demand for CMOS also results from the fact that other components, such as VLSI gate arrays and microprocessors, have already shifted to CMOS; thus, a CMOS memory would simplify the design at the system level. Besides, in order to achieve maximum performance from an n-channel-only process, extensive use of *bootstrapping*, that is, driving word lines to a value above the V_{CC} supply to restore the full supply-voltage value in the memory cell, is required. At the 1-Mbit range and beyond, where very thin gate oxides are needed, the higher bootstrapping voltages are a severe limitation to scalability and reliability. With CMOS, not only is such extra circuitry eliminated, but a single p-channel device can often replace several n-channel transistors.

Theoretically, the CMOS technology has a 2 : 1 speed advantage over an equivalent NMOS cell with the same geometry due to its use of both active pull-up and pull-down in its gates. The price paid is higher dynamic power consumption. However, this active power is on the average much lower than that of the NMOS, since all devices are rarely active at the same time.

The CMOS circuits are fabricated by placing either n-channel devices in p-type wells or p-channel devices in n-type wells, leading to high-density memories much more resistant to alpha-particle bombardment because any e-h pair formed by these particles striking the surface will be formed primarily in the substrate below the well.

At 1 Mbit, the short-channel effects are barely manageable in NMOS, but beyond 1 Mbit the situation becomes completely intolerable. With scaled down NMOS, the leakage currents that occur as a result of these unwanted second-order effects can result in the discharge of the positively charged circuit nodes. On the other hand, with CMOS it is possible to clamp all the sensitive nodes without either increasing the standby power dissipation or complicating the design of the circuit.

The stored-charge problem is less critical in CMOS memories. For the same die area, CMOS DRAMs have significantly more supply-to-ground capacitance than their NMOS counterparts due to the presence of the well-to-substrate capacitance. In addition to its use as an anti-latch-up mechanism, the use of an epitaxial layer enhances this capacitance as the depth of the well approaches the thickness of the layer.

Power Reduction in High-Density DRAMs

Through the advances in fabrication technologies and the resulting evolution in DRAMs, it has become clear that one of the major problems of megabit DRAM design is the value of the power supply voltage because it is closely related to power consumption, reliability of small-sized transistors, and the operating voltage margin of the cell.

The cell voltage margin is limited by various noise-producing factors, some of which are directly related to the bit lines. Electrical imbalances in the lines and capacitive coupling between them contribute to obscuring the signal from the cell. The cell itself is affected by leakage-current noise and interference from stray α-particles, as already discussed in detail. These effects become especially more disruptive when the total amount of charge in the cells is small.

The curvature and convoluted surfaces of the three-dimensional trenched capacitors and transistors that are increasingly being used in megabit DRAMs concentrate charge to create hot electrons. Packing in enough electrons to allow sensing from the edge of these chips energizes them to the point where the delicate convolutions of the cells erode, especially if the current 5-V level is maintained. One response to the hot-electron degradation of capacitor cells has been to dope the capacitors with hardening agents such as tungsten silicide, which creates added process complications. In addition, having higher conductance, the doped structures make it more difficult for the cell to retain charge.

A consideration of power dissipation is also important in choosing the next power supply standard that is expected by DRAM designers to replace the existing 5-V supplies as the dimensions shrink further. Although a 3.3-V power supply voltage was proposed with earlier megabit designs, this could not be implemented because of the concern that chips at that level could not interface directly with the huge number of digital electronics lines that were designed on the basis of the 5-V standard. Furthermore, analog circuits suffer a loss in their dynamic range when operating at lower voltages. However, with minimum dimensions shrinking below the 1-μm mark, a lower supply standard seems inevitable in near future. Already, most of the 4- and 16-Mbit DRAMs designed so far have utilized on-chip voltage limiters to lower the current supplied to the memory arrays by reducing the incoming 5 V to 3.3 V. This method provides highly reliable performance in the small-sized transistors used in DRAMs.

While reducing the voltage level could preserve the delicate cell structure, it also creates problems. The bit-line capacitance becomes more significant as the cells become smaller. A minimum voltage must be maintained to ensure that the cell has enough charge to be able to drive the bit lines.

Another approach to lower the memory array current is to employ a $V_{CC}/2$ precharge of the bit lines. DRAMs using a $V_{CC}/2$ bit-line precharge voltage have been commercially available for several years, beginning with the 5-V single supply 16-Kbit DRAMs introduced in 1980. This half-precharge method offers a number of circuit advantages over the previous full-precharge bit-line scheme. Since the bit-line voltage swing is halved, coupling noise from bit lines into the substrate is reduced by a factor of two. Supply current spikes are reduced, too, because high-going bit lines are charged from $V_{CC}/2$ to V_{CC} rather than from ground to V_{CC}, also cutting the matrix power in half. The cell transistor leakage is also decreased. When a **0** is stored, V_{DS} is decreased to $V_{CC}/2$ which lessens the effect of the drain-induced barrier lowering. It has been shown that the most promising approach for reducing power dissipation in megabit DRAMs is to use the CMOS technology with a $V_{CC}/2$ precharge.

In terms of the bit-cost of DRAMs, the trend line descends from about 30 millicents per bit in 1980 to about 1.2 millicents by 1990, putting the newcomer 4-Mbit DRAMs around $50 each that year. The bits per unit area of semiconductor memories are rapidly increasing and will be about the same as magnetic memories with the 4-Mbit DRAMs so that they are expected to replace the magnetic media eventually.

10.3 Static RAMs (SRAMs)

Memories are said to be *static* if no periodic clock signals are needed to hold stored data indefinitely. RWMs that are based on latches are static. Both bipolar and MOS technologies are used in *static RAMs* (*SRAMs*), with the latter gradually dominating the market. The memory capacity of CMOS SRAMs has quadrupled every two years during the last five generations, resulting in a rate of increase that is faster than that of DRAMs. In 1982, 64-Kbit CMOS SRAMs were announced, and the 256-Kbit SRAMs were introduced two years later. In 1986, the design of the first 1-Mbit SRAMs was reported.

Once overlooked by those who were not willing to sacrifice density for high speed and ease of integration, static RAMs now offer features and levels of performance that make them an alternative to DRAMs. Although SRAMs remain a generation behind DRAMs with respect to density, 1-Mbit densities or access times under 10 ns make SRAMs attractive for a variety of applications. Their advantages over DRAMs lie in their high speed, low power, and highly reliable operation without having complex refresh control circuits. Furthermore, since SRAMs are easy to customize, many companies offer their products with a number of *application specific* features.

Bipolar memories using ECL produce the fastest access times available in SRAMs, but they consume at least twice the power of slower CMOS RAMs. Bipolar SRAMs also call for larger die sizes than CMOS memories. The latter, however, are often associated with quality and reliability problems stemming from their high susceptibility to electrostatic discharge and latch-up. The gate-oxide breakdown characteristics of the CMOS along with their high input impedance make them susceptible to ESD damage. Nevertheless, the design of an increasing number of battery-backed systems adds to the demand for low-power CMOS components.

Today's large on-line databases require far more extensive memory than has traditionally been used in microprocessor-based systems where banks of DRAMs were generally sufficient. In addition, the DRAMs' relatively slow access speeds make it difficult to take full advantage of a CPU's power, particularly that of the 32-bit microprocessors (μPs), whose speed can dramatically outpace the capabilities of large DRAM arrays, making memory the constraining factor in determining system speed and performance. An SRAM cache memory, on the other hand, can compensate for the slow cycle times and other speed-limiting factors inherent in such arrays.

When used alone, DRAMs pose their own problems. Their timing, address multiplexing, and refresh requirements add to circuit complexity, increase overall system delays, and use up valuable board space. Top-of-the-line 32-bit μPs, such as Intel's 80386, operating at 16-MHz and faster clock speeds require at least 70-ns memories to eliminate wait states. Even though DRAMs can have access times in the sub-100-ns range, some applications need parity and/or error correction, so that the associated delays demand DRAMs with even faster access times that are not generally available. Also, a DRAM with an access time of 100 ns probably has a memory cycle time close to 200 ns.

On the other hand, an SRAM can be used to gain the necessary speed. In addition, SRAMs do not need refresh, clock, or complex timing circuitry, simplifying the cache design, which is used to provide memory-access times compatible with the operating speed of the CPU. The cache operates as a high-speed buffer memory between the CPU and the main memory. In fact, its speed allows the processor to run without waiting for the main memory.

SRAM Memory Cells

There are mainly two types of cells employed in MOS SRAMs. The 4-T cell has four *n*-channel transistors and two polysilicon pull-up resistors; the 6-T cell replaces the two resistors with two *p*-channel transistors. Since transistors are more resistant to α-particle radiation and temperature variations than resistors, the 6-T cell reduces the possibility of soft errors. An added bonus may be the reduced power consumption if CMOS technology is employed because its loads need extremely low standby current. Therefore, the 6-T cell CMOS designs are preferred by military markets as high-performance, low-power, and reliable SRAMs with high immunity to single-event soft errors and radiation. Most SRAM suppliers use easily-built 4-T cells having smaller die size for the price-competitive commercial markets.

Figure 10.19 shows typical 6-T cells in CMOS and NMOS. Both cells utilize a pair of cross-coupled inverters, Q_1 through Q_4, as the storage latch, and two access

Figure 10.19 6-T SRAM cells: (a) NMOS, and (b) CMOS.

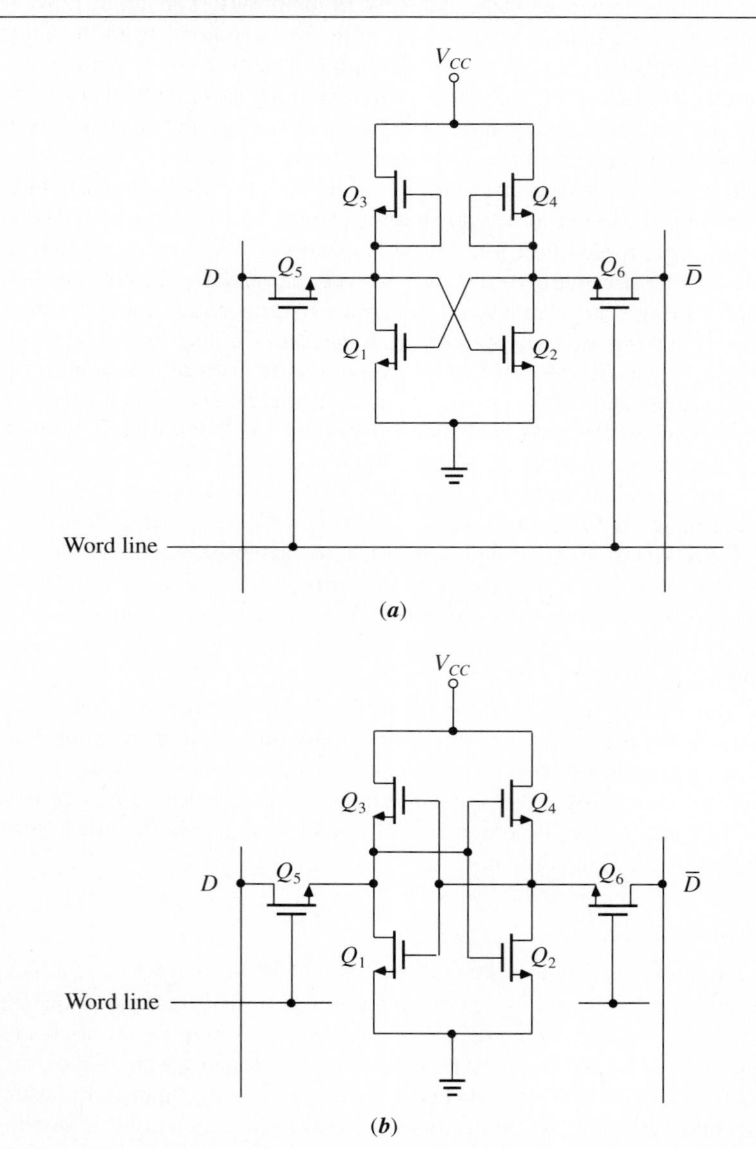

transistors, Q_5 and Q_6. The CMOS provides the advantage of negligible power consumption because the static power is determined only by junction leakage currents through the channels of the transistors that are due to thermally generated carriers. These currents increase as the temperature increases. On the other hand, in the NMOS circuit, one inverter is always on, drawing current from the supply.

The 4-T cell of Figure 10.20 employs two undoped polysilicon resistors as inverter loads, replacing the depletion or enhancement loads Q_3 and Q_4 of the 6-T cells to minimize the chip area at the expense of extra processing complexity. The selection of the resistance values is crucial. It cannot be too low because the standby supply current will then be unacceptably high. It cannot be too high, either, because the cell, which may behave like a DRAM cell, will be marginally stable and may lose data since there are no refresh cycles. If the bit lines are not precharged properly, the data may change when reading a marginally stable cell. In addition, due to the inverse relation between the temperature and the polysilicon resistor's resistance, the resistance decreases several orders of magnitude as the temperature rises toward the high end of the military temperature range. Finally, in the high-impedance 4-T cell, a transient current pulse due to large numbers of e-h pairs generated by high-energy α-particles in the vicinity of a reverse-biased junction results in a change in the internal voltage level that in turn may cause a change of state, resulting in data loss.

On the other hand, CMOS 6-T cells are inherently more stable than 4-T cells because the p-channel pull-up transistors have a lower impedance when turned on than that of the polysilicon resistors, even under worst-case conditions. Consequently, it is more difficult to pull the high side of the latch low even if the bit lines

Figure 10.20 4-T SRAM cell.

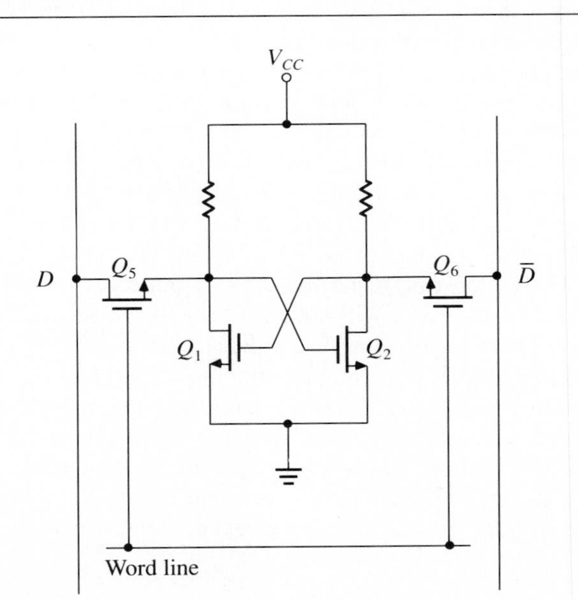

are not *precharged* and *equalized* prior to reading the cell. Moreover, tests have verified that the PMOS load transistors can source sufficient current to ensure that data is not lost by a soft error. Finally, it has also been shown that 6-T cells have tolerance to *total doses* of *gamma radiation* in excess of 10 Krads while 4-T cells fail at total doses of less than 6 Krads.

Gamma radiation is short-wave electromagnetic energy originating from the fragments of heavy atoms split in a nuclear explosion. It is difficult to shield against gamma rays because of their high energy, in excess of 1 MeV. The rate at which transient ionizing radiation arrives at a point is measured in units of rad/s where 1 rad is the amount of radiation that deposits 100 ergs in 1 g of a material. The integral of all ionizing radiation accumulated by a part is known as the total dose and is measured in units of rad. While bulk MOS parts are sensitive to transient ionizing radiation due to their large *pn*-junction volume, CMOS circuits are less sensitive because of their much smaller junction volume. A reduction in gate-oxide thickness also improves the radiation hardness of MOS circuits.

The reason complementary data paths D and $\overline{D}$ are needed is because it is difficult, due to variations in device parameters and operating conditions, to achieve reliable operation with a single access line at high speeds. The word line is held low unless the cells tied to it are to be accessed for reading or writing, at which time its voltage is raised to turn the access transistors on. In this way the two bit lines D and $\overline{D}$ provide the data path.

Writing is accomplished by transferring the bit to be written and its complement to the D and $\overline{D}$ lines, respectively. For example, if it is a logical **1**, then the D line is raised to V_{CC} while the $\overline{D}$ line is lowered to ground. The conducting access transistors cause the high and low voltages to appear at the gates of Q_2 and Q_1, respectively. The cell is designed in such a way that the conductance of access transistors is sufficiently larger than that of loads so that the drain of Q_2 and hence the gate of Q_1 may be brought below the threshold voltage V_T for Q_1 to turn off. Therefore, its drain voltage rises due to the currents from Q_3 and Q_5 to turn Q_2 on, thereby forcing the latch to set. The word line may now return to its low level and maintain this state until another write operation.

If a **0** is stored in the cell, then Q_1 is on and Q_2 is off. When the cell is selected to read this value, current flows from the D line through Q_5 and Q_1 to ground, and from V_{CC} through Q_4 and Q_6 to the $\overline{D}$ line, resulting in the voltage of D being pulled down to ground and the voltage of $\overline{D}$ rising toward the supply. Note that the conductance of Q_1 and Q_2 must be larger than that of the access transistors, so that the drain voltage of the on transistor will not rise above V_T; this avoids altering the state of the latch during a read operation. The voltage difference between the bit lines is detected by the sense amplifier to complete the operation.

The currents through Q_1 and Q_4 along with the bit-line capacitances determine the signal transient times on these lines, which in turn affect the access time of the SRAM. Another contribution to access time is from the finite rise time of the signal, especially at the end of the polysilicon word line due to the relatively high sheet resistance of the polysilicon, the line capacitance, and the resulting large RC time constant.

The Address-Transition Detection Circuit

One way to reduce the access time is to precharge the bit lines to a value around $V_{CC}/2$, as discussed above for DRAMs, so that as the access transistors turn on, these lines will not have to charge and discharge from the extreme values of 0 V and V_{CC}. This approach was first used in 1980 on the IMS1400, a 16-Kbit NMOS SRAM by Inmos Corporation, that employed an *address-transition detection circuit* (*ATDC*) to initiate the equalization and precharging of the bit lines during the word-line delay time.

Figure 10.21 illustrates the block diagram of this SRAM. The row-address lines are fed to both the row decoders and the ATDC so that a transition on any row-address input is sensed by the latter to trigger a pulse generator, whose output pulse drives the *equilibration* circuitry, which in turn equalizes and precharges the bit lines. Since column-access time is much faster than row-access time, this method is not utilized for column-address transitions in this chip.

Figure 10.22 shows the simple circuit of the ATDC. For every row-address line at the input, the row-address buffer generates two signals and their complements. The signal A_{new} and its complement $\overline{A}_{new}$ are fed to the drains of transistors Q_1 and Q_2, which form an *exclusive-or* arrangement. Meanwhile, two delay networks at the out-

Figure 10.21 The IMS1400 16-Kbit NMOS SRAM block diagram.

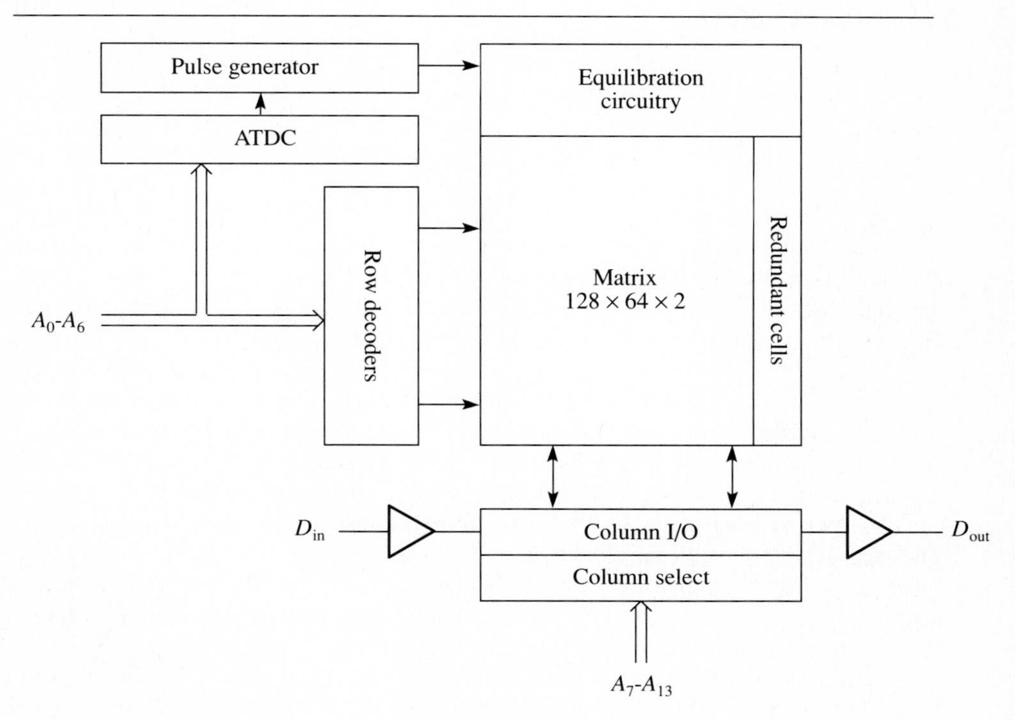

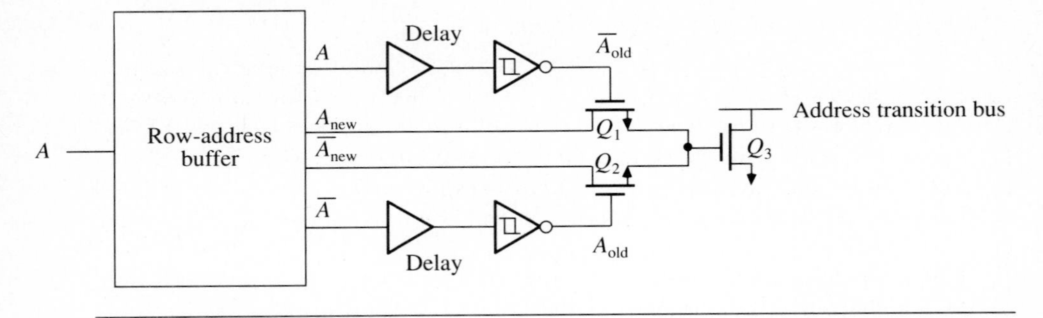

Figure 10.22 The address-transition detection circuit.

put of the row-address buffer feed delayed versions of the address signal and its complement to two Schmitt triggers to generate A_{old} and $\overline{A}_{old}$ that drive the gates of Q_1 and Q_2.

When an address transition occurs at the input, A_{new} and $\overline{A}_{new}$ can respond rapidly to drive the common source of Q_1 and Q_2 high momentarily until A_{old} and $\overline{A}_{old}$ switch, at which time it is again brought to low. Therefore, a short pulse on the gate of Q_3 turns it on to pull the address-transition bus, which is common to all seven row-address buffers, low to initiate the equilibration of the bit lines.

High-Density SRAMs

Currently, the most dense commercially available SRAMs use the CMOS technology and offer a maximum storage capacity of 1-Mbit with access times as low as 25 ns.

In 1984, Toshiba introduced the *dynamic double word line* (*DDWL*) scheme on its 256-Kbit CMOS SRAM, which was a combination of a double word-line structure and an automatic power-down feature. An ATDC was also incorporated to generate the trigger pulse for the power-down function as well as to prompt the equalization of the bit lines. In this way, it was possible to reduce the access time to 46 ns, about one-half that of a conventional SRAM of the same density. The device offered 30 μW standby and 10 mW operating power at 1 MHz.

Using 1.2 μm design rules with a .8-μm effective channel length, a *p*-well CMOS process, in which all 4-T cells are embedded in the well, is preferred to alleviate the soft-error problems. To speed the peripheral circuitry, 5-V internal operation is used. Since a conventional submicron *n*-channel MOSFET suffers from hot carriers at this voltage level, an LDD process has been employed for reliable operation.

To improve both the device performance, by decreasing the parasitic capacitance, and the packing density, the double-level polysilicon process is employed. The first layer is used for the transistor gates; the second layer serves as the polysilicon loads with sheet resistivity of over 100 G$\Omega/\square$. The second layer is also used for supply lines to memory cells by lowering the sheet resistivity to less than 50 $\Omega/\square$.

As the block diagram of Figure 10.23 illustrates, there are two levels of word lines. The *main* word line is an *Aℓ* layer to reduce the delay. Also, it is not directly connected to memory cells, so that its capacitance is relatively small, resulting in a

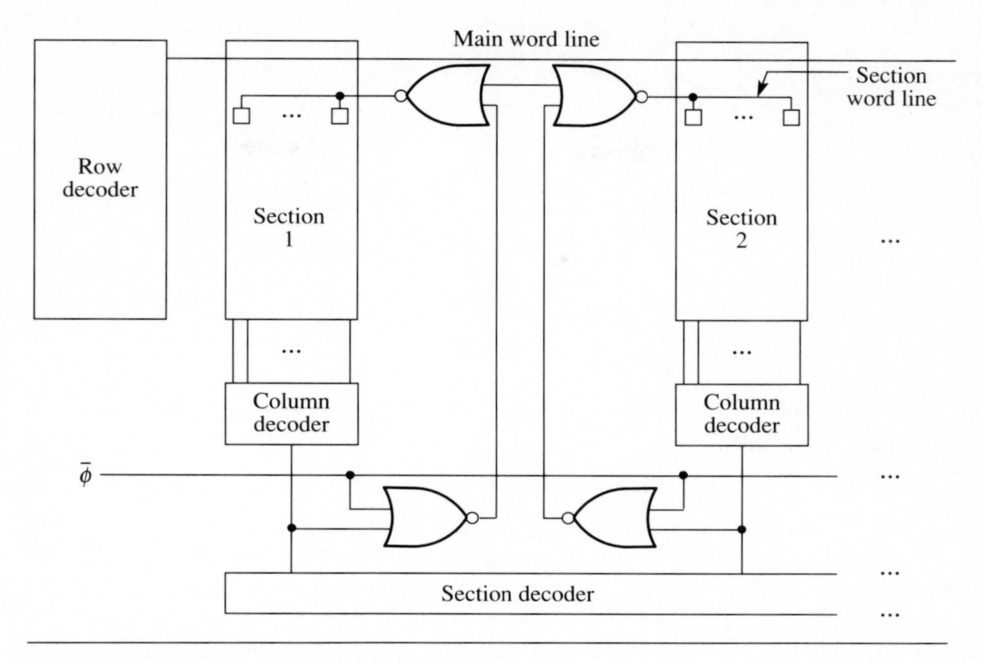

Figure 10.23 The block diagram of the DDWL scheme.

further reduction of the word-line delay. On the other hand, since one *section* word line is selected at a time, and hence only a few cells are activated, the power dissipation is also reduced. As soon as an address transition is detected by the ATDC, it generates an active low pulse $\overline{\phi}$. This clock pulse enables a simple NOR circuit, which in turn selects a section word line, depending on the input address.

In 1985, to achieve fast access times, NEC Corporation replaced the conventional polysilicon word lines with titanium polycide gate process in the design of their 256-Kbit *p*-well CMOS SRAM with a 32K $\times$ 8-bit organization. This process not only decreases the propagation delay of the word lines to about 4 ns, a value $1/10$ that of the polysilicon gate process, but also contributes to the reduction of the memory-cell size. This is because of the reduction in the width of the polycide ground line to about 1 μm due to the low resistivity of the titanium polycide. The ground line is connected to the n^+ diffusion layers of the memory cells using buried contacts.

One way to decrease the standby power is to increase the length of the polysilicon load resistor of a 4-T cell. On the other hand, a long resistor is not acceptable from the cell-size point of view. In order to reduce the standby power dissipation without compromising the cell size, the loads for the cells of this particular SRAM are designed to be 5 μm long and 2.3 μm wide at $T_A = 25°C$, corresponding to a resistance of around 680 GΩ. The result is a standby power of 10 μW, an active power of 175 mW, an access time of 55 ns, and a cell size of just below 90 μm^2.

This design also employs the ATDC for internally synchronous operation. However, instead of the conventional design discussed above, it uses a simplified ATDC with only one delay circuit. Its logic diagram is shown in Figure 10.24.

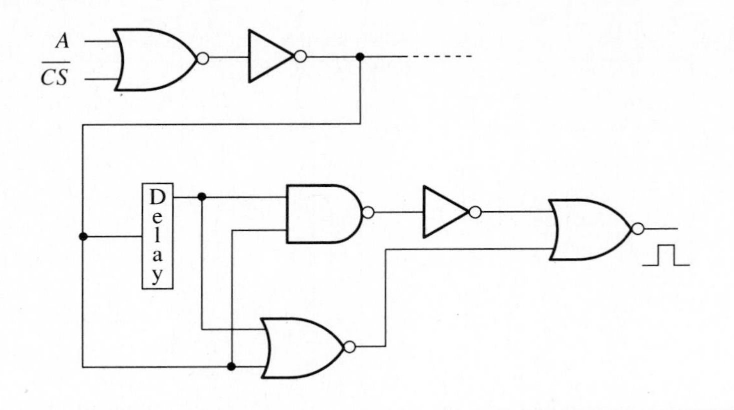

Figure 10.24 A simple ATDC.

As the devices gain speed, noise problems caused by fast-switching output pins are encountered in system design. The solutions may be new pinout schemes and *self-timing* architectures. With access times for 256-Kbit SRAMs already around 20-ns levels, the fast-switching output pins cause noise. For CMOS SRAMs in conventional DIPs, the inductance along the leads amplifies ground-bounce spikes. Some suppliers suggest new standards to increase and move to the center the ground and power pins.

There is also the problem of system synchronization with fast SRAMs. To keep the data valid with current SRAMs, either the cycles should be lengthened or external latches need to be added. Therefore, some companies have started to add internal latches with fast SRAMs that have a *synchronous-clocked architecture*. It seems that synchronous SRAMs will become standard for high speed within the next few years.

In one such architecture from Motorola, Inc., which has been incorporated into a family of 16K × 4-bit SRAMs with access times ranging from 10 to 25 ns, the traditional self-clocked ADTC is replaced and internal input and output latches are added to eliminate as much as 10 ns of interconnection delay on both input and output. It also eliminates circuitry that was required to make asynchronous devices appear synchronous in high-performance cache memory systems, which depend heavily on the synchronization of critical timing parameters.

Also incorporated on the chip are drive transistors capable of driving buses with capacitive loads of up to 130 pF without additional circuitry. The geometries are also enlarged without increasing the chip size to increase the inherent drive capability of the devices. Standard 30-pF devices with 400-to-600-μm widths have been replaced with transistors on the order of 1500 μm wide. This compares with 6-μm geometries in the memory array. Moreover, to achieve higher speed in spite of the high-drive currents, n-channel transistors are used rather than the slower PMOS, and these output devices are speeded up by incorporating a separate ground-supply pin for the output drivers, allowing more current in them without corrupting the operation of the rest of the circuit.

In 1987, Philips was the first to announce the availability of working samples of a 1-Mbit SRAM, produced in full CMOS submicron technology with 6-T cells. Its minimum features are .7 μm, packing 128-Kbytes onto a 10.7 $\times$ 12.2 mm chip with a cell size of 5 $\times$ 12 μm. It has an access time of 25 ns and consumes a typical 150 mW at 20 MHz and a mere .5 μW in standby. The low standby power is suitable for battery backup systems in data processing and communications.

Pressured recently by a faster generation of μPs with clock speeds up to 33 MHz, high-density, high-performance SRAMs have resorted to employing the BiCMOS technology. Purely bipolar SRAMs seem to be stuck at the 64-Kbit density level primarily because of their poor power dissipation. The operating speed of the CMOS SRAMs, on the other hand, can improve only if the minimum feature size goes well below 1 μm which requires aggressive scaling. For example, in 1989, Hitachi announced a 1-Mbit CMOS SRAM with a 9-ns access time using .5-μm technology. Meanwhile, Toshiba's 1-Mbit BiCMOS SRAM with ECL I/O buffers is not only slightly faster, with an 8-ns access time, but was produced with a .8-μm process. Thus, with TTL and/or ECL, which are used as I/O buffers and internal logic for driving, restoring, and sensing bit or word lines, surrounding the dense and low-power CMOS memory-cell array, fast BiCMOS SRAMs have already been manufactured with more conservative design rules. However, the extendability of digital BiCMOS to below .5 μm and beyond seems to require major circuit innovations.

BiCMOS offers on one chip what bipolar and CMOS can only provide separately. High-speed ECL-based systems such as mainframes need sophisticated cooling equipment as well as efficient IC packaging and printed circuit boards to channel the heat away. Due to its simpler gate construction, CMOS is more easily scalable than the bipolar technologies, consumes less power, costs less, and offers better noise margins. Bipolar has greater current drive and better immunity to latch-up. One concern, though, is the possible introduction of the 3.3-V operating voltage to facilitate the CMOS scaling and reduce the field stress. The lower voltage will degrade the ECL performance by limiting the viability of some circuit constructions for this bipolar family. In addition, the growth of an epitaxial layer over the substrate that is required by bipolar circuitry makes the BiCMOS process more complex and expensive than a CMOS-only process.

10.4 Semiconductor Read-Only Memories

A ROM is a combinational circuit that consists of an array of semiconductor devices, such as diodes, bipolar transistors, or MOSFETs, which are interconnected to store an array of binary data, thereby realizing a prescribed set of functions of the same set of inputs. The data stored can be read out whenever desired but cannot be altered under normal operating conditions except in the case of E^2PROMs, as will be clear later. A ROM with m input lines and n output lines contains an array of 2^m memory locations, with each location being n-bit long. A $2^m \times n$ ROM can realize n functions of m variables.

A ROM is basically made up of a decoder and an encoder memory array, as diagrammed in Figure 10.25. When an m-bit address is applied to the decoder inputs, only one of the memory locations in the array is selected, and its contents are transferred to the data bus.

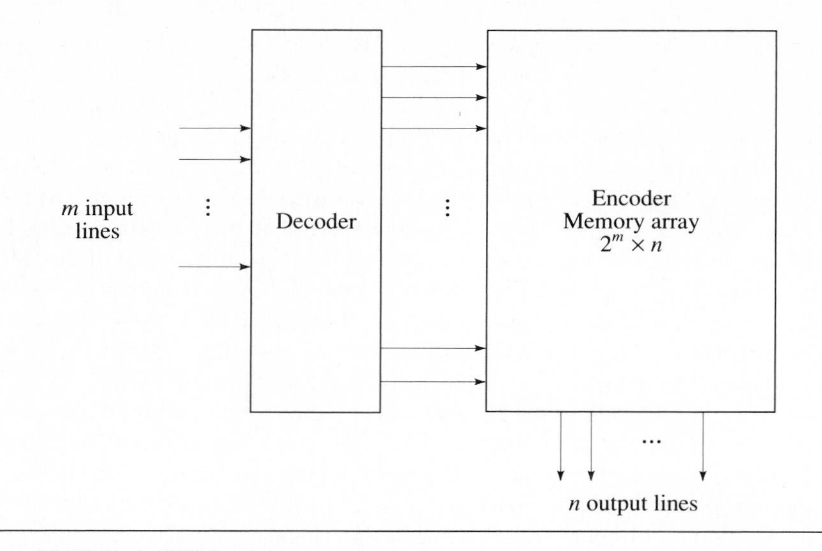

Figure 10.25 Basic ROM structure.

Four basic types of ROMs are *mask-programmable* ROMs, *field-programmable* ROMs that are usually called PROMs, *erasable programmable* ROMs (that is, EPROMs), and *electrically erasable programmable* ROMs or E^2PROMs. In the former, the data array is permanently stored at the time of manufacture by selectively including or omitting the switching elements at the row-column intersections of the matrix, requiring the preparation of a special *mask* to be used during the fabrication. PROMs, on the other hand, are typically manufactured using bipolar technology; all switching elements are present in the matrix, with the connection at each intersection made by means of a *fusible* link. These links are selectively *blown* by the user by means of a PROM programmer in order to store data permanently. Although very fast, bipolar PROMs are also the most power-hungry nonvolatile memories developed. The relatively small memory in each package leads to higher system chip count, increasing system size and power consumption. Figure 10.26 shows a typical cell used in bipolar ROMs. Note that, in a PROM, the emitter is connected to the bit line through a polysilicon fuse that is irreversibly blown by applying a large current, thereby opening the connection between the emitter and the bit line to store a **0** in the cell.

Both mask- and field-programmable ROMs have one serious limitation: a single incorrect bit of information renders the entire IC useless. Moreover, it is often necessary to modify the stored data during the design cycle of a digital system. To overcome these limitations and to avoid the expense of using a new ROM each time the data must be changed, EPROMs, which can be reprogrammed many times, may be used. Instead of fusible links, EPROMs employ a special charge storage mechanism to enable or disable the switching elements in the memory matrix. Any stored data can be erased by exposing the chip to intense *ultraviolet* (*UV*) radiation through a transparent lid at the top of the IC.

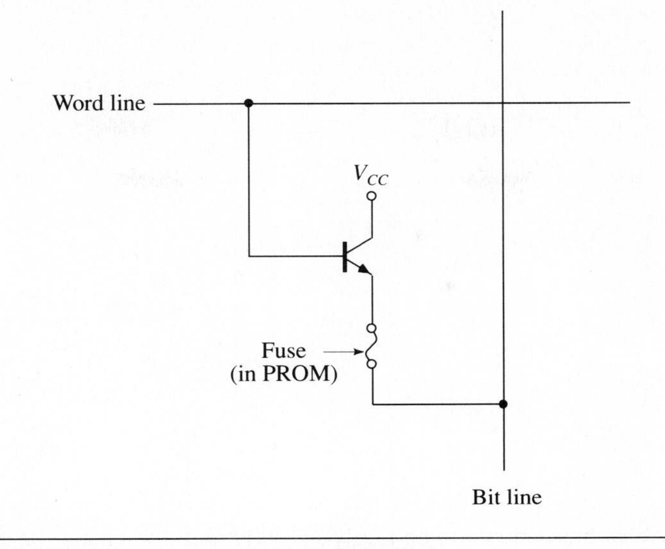

Figure 10.26 Typical fuse cell of a bipolar ROM/PROM.

Even better than the EPROM is the E^2PROM, which offers several significant advantages over the former. The erasure and reprogramming are accomplished in circuit and in a very short period of time by applying electrical pulses instead of exposure to UV light. Other advantages include the ability to erase and reprogram individual bytes or pages, and a very low programming-current requirement.

The densest mask-programmable ROM so far is the 16-Mbit device announced by Toshiba in 1989. Using .7-μm design rules, it has an access time of 120 ns with a cell size of $2.16 \times 1.44 = 3.11\ \mu m^2$ and a chip size of $7.1 \times 16.55 = 117.5\ mm^2$.

10.5 EPROMs

The basic principle of operation of an EPROM involves the storage of charge by trapping it at the gate of a special type of MOSFET known as the *floating-gate avalanche-injection MOS*, or FAMOS, whose basic structure is shown in Figure 10.27. A logical **0** or **1** is indicated by the presence or absence of this trapped charge on the floating gate and is read out nondestructively, as explained below.

In floating-gate MOS transistors, an additional metallic gate is embedded in the excellent insulator SiO_2 above the substrate. The transistor is programmed by injecting high-energy hot electrons through the SiO_2 onto this isolated floating gate, where they remain indefinitely, thereby biasing the transistor permanently because there are no electrical connections to the gate. As a result, a stored charge can be retained on the gate for an extremely long period of time even with the frequent removal of small amounts of charge each time the stored charge is read. To erase the bit pattern previously established, the chip is removed from the circuit and exposed to intense UV light of the correct wavelength for a specified duration, causing the insulation around

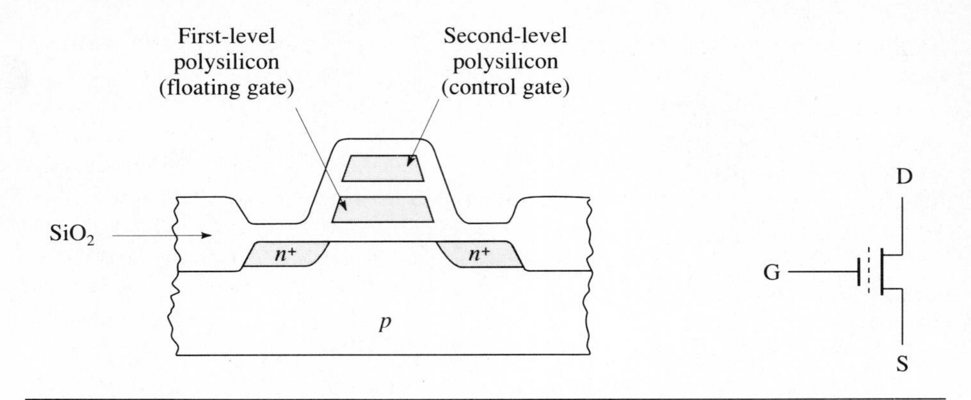

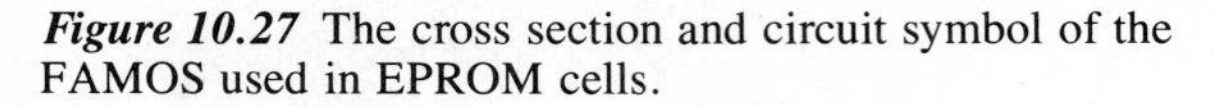
Figure 10.27 The cross section and circuit symbol of the FAMOS used in EPROM cells.

the buried gate to break down temporarily and allowing the charge on the isolated gate to leak off.

As depicted in Figure 10.27, the composite gate has a regular control or select gate and a floating gate. The former is connected to row decoders; the latter is isolated in the surrounding silicon dioxide. Before the cell is programmed, there is no charge on the floating gate, and the device operates as a regular n-channel enhancement MOSFET, exhibiting the transfer characteristic of Figure 10.28a, in which case the

Figure 10.28 Shift in the characteristic curve of the FAMOS due to programming.

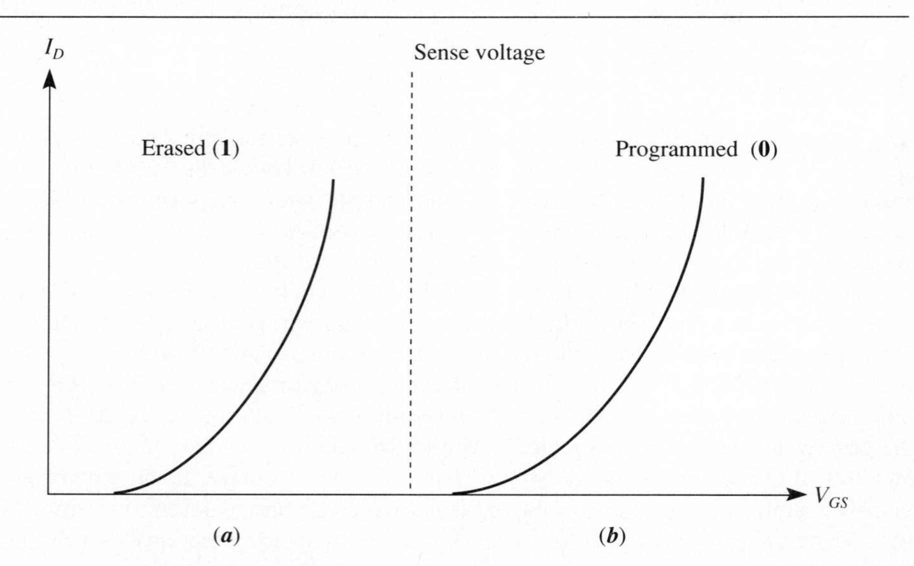

threshold voltage V_T is comparatively low. This state is known as the *erased* state and represents a stored **1**.

Programming or writing is done by avalanche injection of hot electrons from the substrate through the isolating oxide. This can occur with oxides as thick as 1000 Å, making the device relatively easy to fabricate, and is accomplished by simultaneously applying a large positive voltage to the control gate, and between the drain and source. As illustrated in Figure 10.29, an n-type inversion layer is created at the substrate surface as a result of the voltage applied to the control gate. It has a tapered shape because of the large positive drain voltage, which accelerates electrons through the channel. As these electrons reach the drain end of the inversion layer, they gain sufficient kinetic energy to be referred to as hot electrons.

The electric field established in the insulating oxide due to the large positive voltage on the control gate attracts the hot electrons and accelerates them toward the floating gate. Therefore, the electrons jump over the potential energy barrier between the silicon substrate and the silicon oxide, penetrating the oxide, flowing to the floating gate, and causing an electric charge to be accumulated and trapped on the floating gate. The process is self-limiting: the negative charge collected on the floating gate reduces the strength of the electric field in the oxide; thus, the modified field eventually becomes incapable of accelerating any more of the hot electrons.

As the gate becomes more charged, the trapped negative charge causes electrons in the oxide field to be repelled. Hence, a larger gate bias is required to overcome this repulsion. As V_T shifts to the right, the characteristic curve of the programmed transistor is altered as shown in Figure 10.28b. In this *programmed state*, the cell stores a **0**. Once programmed, the trapped charge can be stored in the floating gate even if the applied voltage is removed or the power supply is turned off because the oxide conductivity is sufficiently low, so that the decay time is years long. To read the content of the cell, a voltage value between the low and high threshold values is applied to the control gate. A programmed device does not conduct, whereas an erased one does.

An EPROM is erased optically by shining UV light through a window on the

Figure 10.29 Creation of an n-type inversion layer and hot electrons.

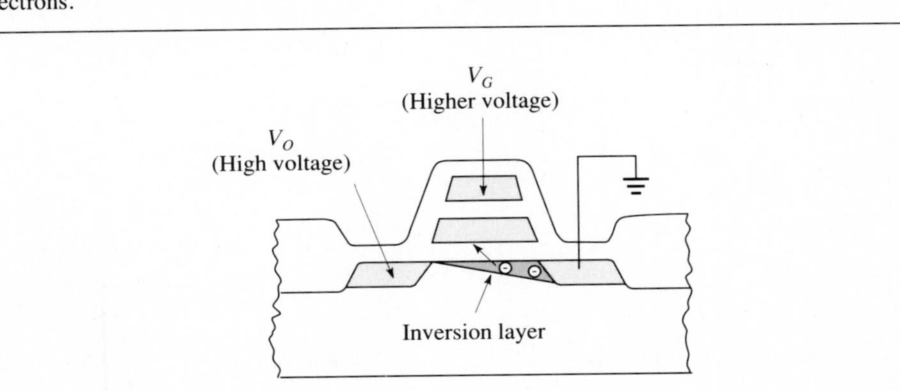

device. This temporarily increases the energy of the floating-gate electrons to a level at which they jump the inherent potential energy barrier between that gate and the field oxide, and hence the electrons leak away from the floating gate to the substrate.

There is a trade-off for the EPROM's low cost and small cell size, which is about 20 μm^2 at the 1-Mbit level. Designers have to settle for low *endurance*, with a maximum of 1000 erase/program cycles. In 1987, Toshiba was the first to announce a prototype 4-Mbit EPROM. The chip, which uses a .8-μm *n*-well CMOS process, has an access time of 120 ns.

10.6 E^2PROMs

The Fowler-Nordheim Tunneling

The nonvolatile storage and erasure in E^2PROMs occur through a *tunneling* mechanism that was first described by Fowler and Nordheim in 1928 for the case of electrons being emitted from metals into vacuum; hence it was named after them. The basic idea behind the *Fowler-Nordheim tunneling* is illustrated by the energy-band diagrams of the Si/SiO_2 system of Figure 10.30. We know from Chapter 1 that the energy gap in Si is about 1.1 eV, while it is approximately 9 eV for the excellent dielectric silicon dioxide. When the two materials are joined together in thermal equilibrium, as shown in Figure 10.30a, the difference in conduction-band energies becomes 3.25 eV. Furthermore, the valence band in Si is observed to be 4.27 eV above that in SiO_2. Therefore, the probability that an electron in silicon can gain enough thermal energy to overcome the barrier and enter the conduction band in the oxide is exceptionally small because its average thermal energy at room temperature is only .025 eV.

In the presence of a sufficiently high electric field, however, the energy bands will change, as diagrammed in Figure 10.30b, in which case there will be a finite

Figure 10.30 Energy-band diagrams of a Si/SiO_2 interface in (a) thermal equilibrium, and (b) during programming.

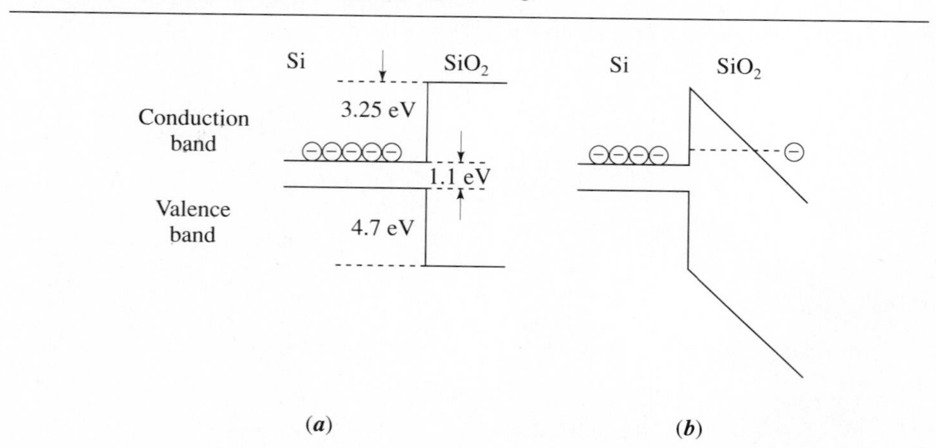

probability that a conduction-band electron will tunnel through the energy barrier and end up in the conduction band of the silicon dioxide. The Fowler-Nordheim current exponentially increases with the applied field and becomes observable for the Si/SiO_2 system when the Si surface is *smooth* for fields on the order of 10 MV/cm.

E^2PROM Cell Structures

The *metal-nitride-oxide-semiconductor* (*MNOS*) structures yielded the first non-volatile memories that were called *electrically alterable* ROMs or EAROMs. In an MNOS device, as depicted in Figure 10.31, a nitride layer is placed on top of the gate oxide, and the tunneled electrons are retained at the oxide-nitride interface.

When a positive gate voltage is applied, electrons tunnel through the thin oxide layer and are captured by the traps at the aforementioned interface to become stored charges. The charge stored causes a shift in the threshold voltage, and the device remains at a higher threshold voltage state corresponding to a logical **1** until a reverse gate voltage is applied to erase the memory and return the device to a lower threshold voltage to represent a **0**.

Besides encountering problems such as data disturbance during the read operation and the loss of data over time, EAROMs required multiple power supplies, one of which was negative, and signal swings beyond TTL levels. Further complicating their use was the fact that the addresses and data had to be stable for the entire cycle lasting up to 40 ms. Moreover, they were reprogrammable only after an entire memory array or at least one page was electrically erased.

Second-generation *thin-oxide* E^2PROMs stored data by trapping charge on floating gates using a transistor structure that resembled the FAMOS used in EPROMs. The difference was the addition of a small *tunnel-oxide* region above the drain of the floating gate transistor that allowed the charge to move bidirectionally under the influence of an electric field.

Figure 10.31 The cross section of the MNOS device used in first-generation EAROMs.

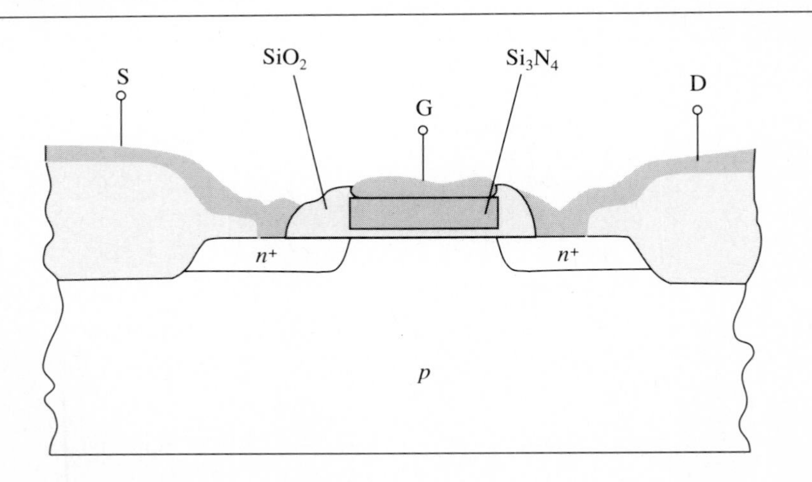

This structure, shown in Figure 10.32, improves the data integrity as compared with MNOS parts. The programming or erasure of the E^2PROM cell requires the application of a large reversible gate field of around 10^7 V/cm on a thin gate oxide of about 100 Å separating the floating gate from the substrate. This field must be strong enough to produce a measurable current through the thin oxide. At the same time, the field in the interpolysilicon oxide has to be maintained at a relatively low value to prevent unwanted transport of electrons between the floating and control gates.

Programming is accomplished by tunneling electrons from the floating gate to the n^+ drain diffusion by grounding the control gate and applying the programming voltage to the drain diffusion, hence producing a logical **0** state in the cell. Erasure is performed by charging the floating gate, resulting in a logical **1** state in the cell. This is achieved when the source, drain, and substrate are all grounded, and the control gate is raised to a high voltage.

Unlike an EPROM cell, in which the control gate acts as the select transistor, an E^2PROM cell needs a separate select transistor connected in series with the floating-gate transistor in order to read the device. During a read operation, the cell's state is determined by current sensing through the select transistor. Therefore, a typical E^2PROM cell, having two transistors (one used for reading and the other for programming and erasure), are larger than EPROM cells. In addition, the tunneling mechanism needs gate areas several times larger than those required for the avalanche-breakdown method used to charge EPROM cells.

Although second-generation E^2PROM chips used TTL levels, they still needed an externally generated high-voltage pulse to alter data as well as latches to hold the address and data signals.

Third-generation E^2PROMs eliminate the external components by incorporating all of the required supporting hardware for performing such functions as voltage generating and pulse shaping on the chip. They operate from a single 5-V power supply, thereby removing the high currents typically used to alter data by utilizing instead on-

Figure 10.32 Thin-oxide double polysilicon E^2PROM device.

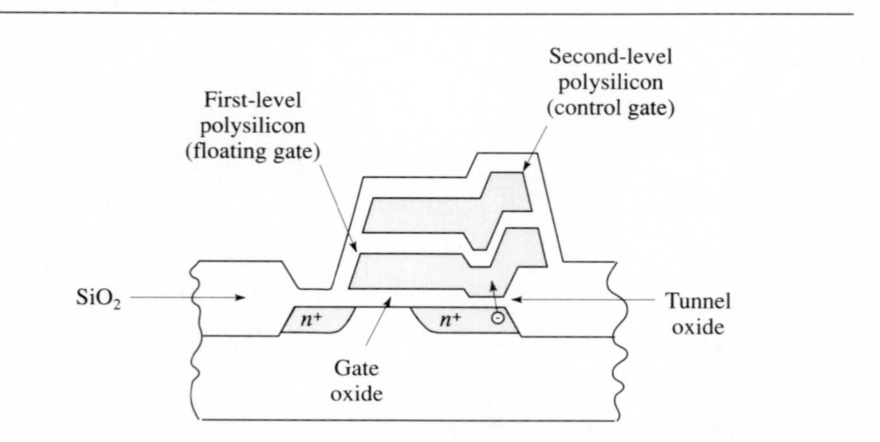

chip *charge-pump* circuits. They automatically perform a byte-erase cycle prior to each byte-write. Latches are also added on the address and data inputs to hold the information during the write cycle.

Textured Surfaces

The voltage magnitude required for programming a floating gate is related to the intensity of the electric field generated by that voltage at the oxide-polysilicon interface. Naturally, it is desirable to reduce the on-chip voltage level presented to the cell to initiate electron tunneling. The electric field strength at the interface can be increased by using an extremely thin oxide, as already explained, on the order of 100 Å. A second technique uses *textured* polysilicon to locally enhance the field at the surface and achieve electron tunneling.

It has been observed that the electron emission can be enhanced if the silicon surface has a *texture*, implying *bump* and *dimple* features on the order of a few hundred angstroms. Due to a well-controlled oxidation step in the fabrication process, the properties of the textured emitting surface are regular.

Initially considered to be undesirable side effects of MOS processing, these bumps and dimples develop because the oxidation grows faster along some crystal directions than others. Due to the random nature of crystal orientation in a deposited polysilicon, oxide growth is enforced in some parts of an integrated circuit surface. Even single-crystal polished silicon wafers have surface features on the order of 5 Å. Ordinary polysilicon has even larger variations. Thus, textured structures can be considered as having features that are deliberately accentuated.

A textured surface emits more electrons than a smooth one for a given voltage and oxide thickness. This enhanced emission allows the use of thick oxides instead of thin layers that are much harder to produce reliably. The topology of a textured polysilicon surface causes the electric field lines to diverge and converge instead of being parallel as in the case of parallel emitting and collecting surfaces. As depicted in Figure 10.33, the field lines converge near surfaces of positive curvature (that is, bumps) and diverge near surfaces of negative curvature (i.e., dimples). Thus, the

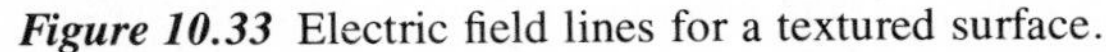

Figure 10.33 Electric field lines for a textured surface.

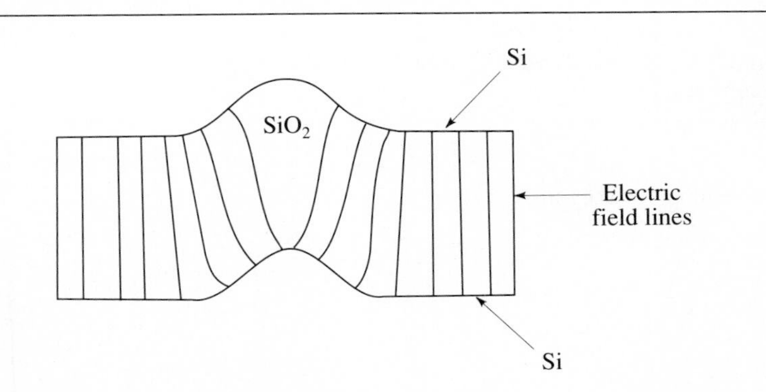

magnitude of the tunnel current emitted from textured polysilicon surfaces through SiO_2 layers depends on the surface topology. The shape of the bumps, serving as elecron emitters, tends to increase the electric field at the crest of the bumps, enhancing the electron emission substantially and allowing the use of thick oxide layers. Increasing the voltage not only increases the emission but also enlarges the area from which it occurs, as shown in Figure 10.34.

More important, the tunnel current decreases more rapidly with decreasing voltage than does the emission through the thin oxide grown on conventionally used smooth surfaces. E^2PROM cells that make use of the asymmetric nature of textured surfaces have lower leakage currents from the floating gate during storage and read operations of the device. Thus, textured structures have a significant advantage in *data retention*.

The schematic of an E^2PROM cell employing a textured surface triple poly n-channel process is shown in Figure 10.35. The cross-section of the floating-gate transistor is diagrammed in Figure 10.36. The data is stored as the presence or absence of charge on a piece of second-level polysilicon that acts as a gate for a readout transistor. This polysilicon is surrounded by approximately 500 to 800 Å of thermally grown SiO_2 layers. The charge is transferred into and out of the storage gate by means of the quantum mechanical phenomenon of Fowler-Nordheim tunneling, as described above.

The programming tunnel mechanism occurs between the first and second floating-gate poly layers; the erase tunneling action takes place between the second and third. A selection transistor isolates the selected cell on a column while a coupling capacitor develops enough voltage across a tunneling device to make electrons tunnel on and off the floating gate. This voltage is sensed by a MOSFET whose gate is formed by the second poly layer.

The coupling capacitor's size contributes to the smaller cell area. To induce tunneling, the floating-gate voltage is raised or lowered through capacitive coupling to a bias voltage supply. To avoid excessively high bias voltages, efficient coupling to the floating gate must be achieved by making the coupling capacitance much higher than all other floating-gate capacitances combined.

When the third polysilicon word-select line is brought to low, the select transistor is off. When the select line is raised to high, the conduction through the select transistor depends on the charge on the floating storage gate. If this gate is initially charged

Figure 10.34 An increase in voltage leads to the enlargement of the emission area.

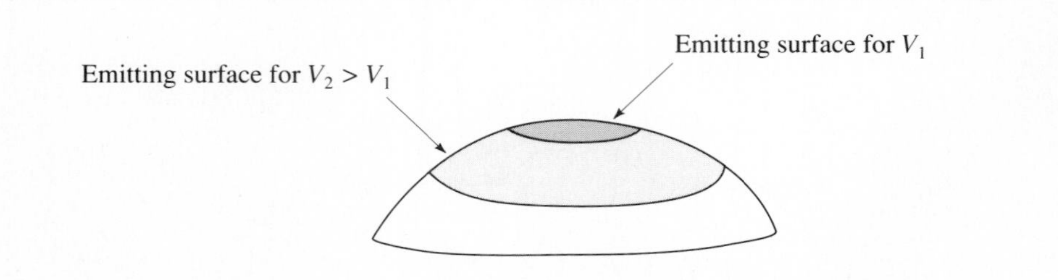

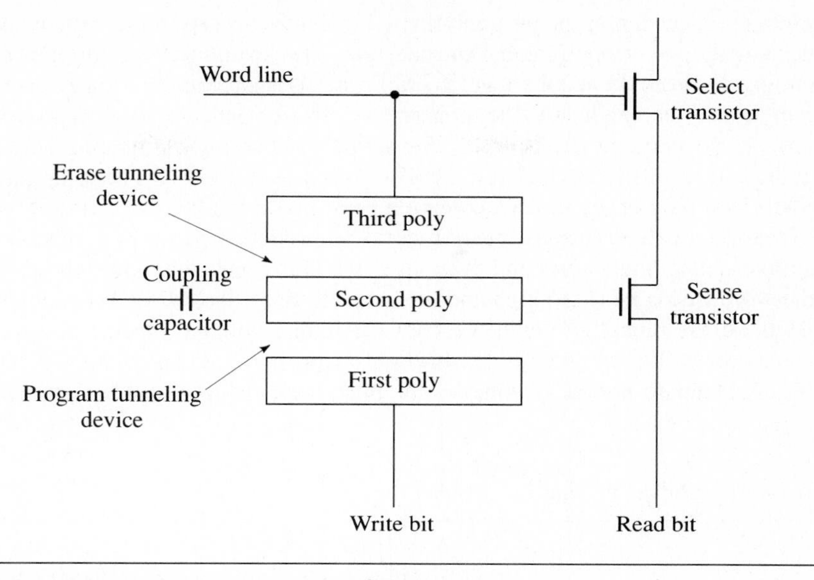

Figure 10.35 Textured-surface triple-poly E^2PROM cell schematic.

negatively, then no conduction occurs, corresponding to reading a logical **0**. If the floating gate is charged positively, then the channel below it is in depletion, and the channel conducts; thus, the cell is read as a **1**.

To program the floating gate, the word-select line is taken to high while the first poly write-bit line is held low. For the electrons to tunnel onto the floating gate, a bias voltage is applied to the coupling capacitor to pull the floating gate high and

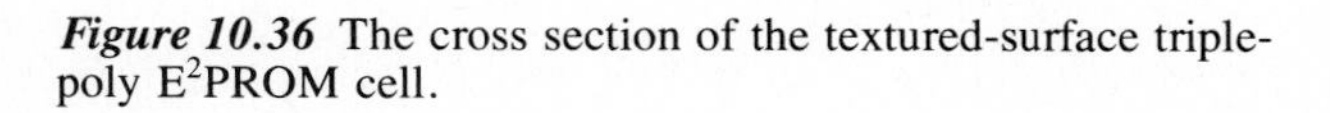

Figure 10.36 The cross section of the textured-surface triple-poly E^2PROM cell.

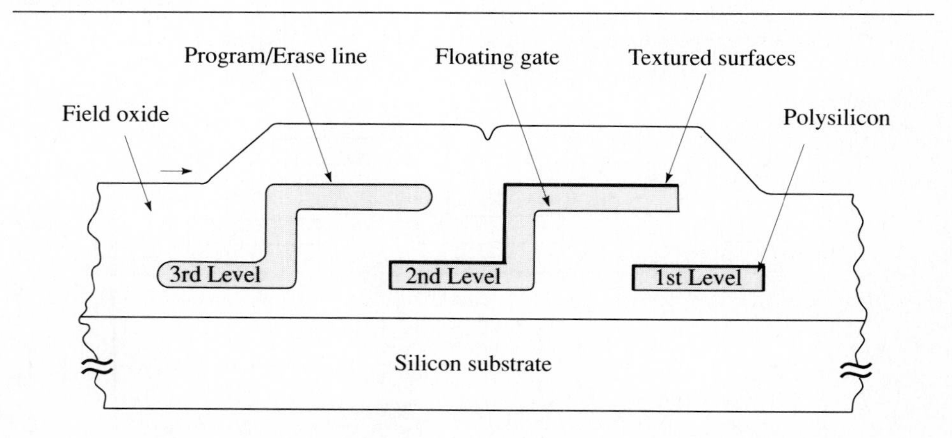

develop a voltage across the program tunneling device. When this voltage reaches the tunnel voltage, electrons tunnel from the first poly level's surface through the programming device to the second-level floating gate. When the applied voltage is brought back to normal reading levels, the programmed floating gate carries a negative voltage because of the extra electrons on it. The action of charging the floating gate to program the cell is illustrated in Figure 10.37a. When read, the MOS floating-gate sense transistor is turned off by the negative voltage to produce a **0** at the E^2PROM output.

To erase a cell, electrons must tunnel off the floating gate. This is achieved by capacitively coupling the second poly level's floating gate low while the third poly level's word line is raised to high and the write-bit line is held low. When the voltage across the erase tunneling device reaches the tunnel voltage, electrons tunnel from the second poly floating gate to the third poly word line. When the applied voltages are brought back to normal reading levels, this erased floating gate has a net positive

Figure 10.37 (a) Charging the floating gate to program the cell, and (b) discharging the floating gate to erase the cell.

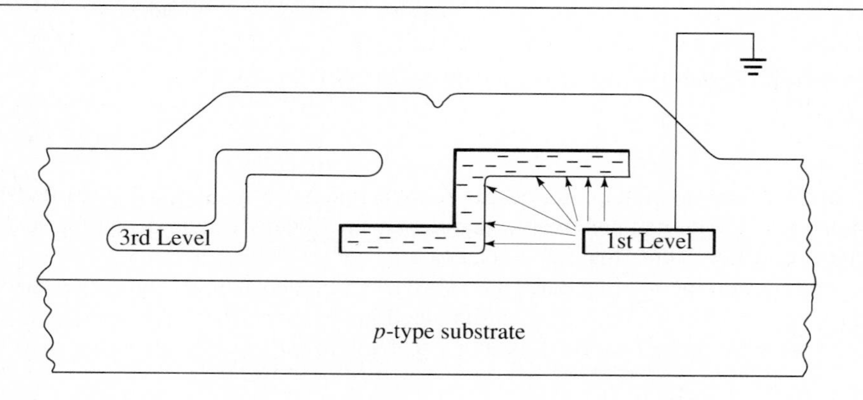

(a)

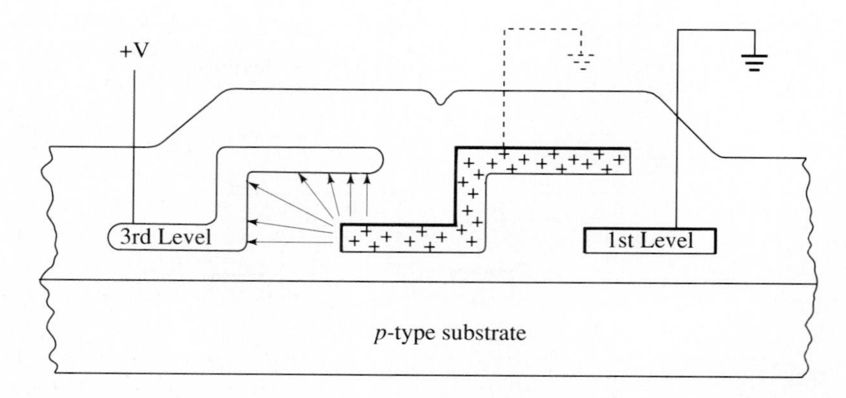

(b)

voltage because of the lack of electrons on its surface. The cell operation during erasure is shown in Figure 10.37b. When read, the MOS floating-gate sense transistor is turned on by this positive voltage to produce a **1** at the E^2PROM output.

Since the floating gates are completely surrounded by thick thermal oxides, similar to an EPROM, data retention is excellent even at very high temperatures, with typical retention being more than 2 million years at 125°C even though the manufacturer guarantees only 100 years conservatively. This performance is retained as the devices are scaled because, for a given maximum read voltage, a textured surface requires a lower programming voltage than a conventional planar thin-oxide structure.

The first 256-Kbit E^2PROMs that utilized textured surfaces and were fabricated with 2-μm design rules were the 32K × 8-bit NMOS X28256 and its CMOS version, the X28C256, both of which were introduced by Xicor, Inc. in 1986. The 256-Kbit E^2PROMs with conventional thin-oxide tunneling structure in both NMOS and CMOS were also available the same year. They are still the densest E^2PROMs available commercially.

The chip area of thick-oxide 256-Kbit E^2PROMs is equal in size to thin-oxide 64-Kbit parts using the 1.5-μm process and one-half the size of thin-oxide 256-Kbit E^2PROMs designed with 1.2-μm geometries. Both the X28256 and X28C256 feature 150-ns access times and support 64-byte page-write operations. A write cycle takes 31 μs/byte, enabling the entire memory to be written in less than 1 second. Each cell measures only 68 μm^2 compared with at least 80-μm^2 cells of conventional 1-μm thin-oxide designs.

In 1989, Mitsubishi announced a prototype 1-Mbit thin-oxide E^2PROM designed with a 1-μm double-metal process. It has a 120-ns access time, a cell size of 30 μm^2 and a chip size of 91.5 mm^2.

Endurance

Endurance describes the ability of a nonvolatile memory cell to sustain repeated data changes without failure. Such a data change occurs when a stored **0** is changed into a **1** and vice versa. Continual changes between two logic states are called *cycles*.

The limit of endurance, expressed in terms of cycles, is reached when the very first bit on the chip is found to be in error due to a permanent change in the cell's characteristics after a required data change under normal operating conditions as specified by the data sheet. Therefore, the worst-case endurance on any cell in a memory matrix, that is, the minimum number of data changes that is sustained by all cells until a cell produces an erroneous output, defines the endurance of the entire chip.

Regardless of the materials used or the approach to the cell design, there is a finite upper limit to the number of cycles that a nonvolatile memory cell can sustain since it depends on nonlinear conduction properties of the solid-state dielectrics utilized to form the cell. As treated in detail above, at high electric fields, these dielectrics permit a predictable current to pass from one electrode to another that is used to program or erase such a cell.

At low intensities, no electron is transmitted, so that transferred charges will remain in these isolated locations indefinitely. However, during the transfer of charge while programming or erasing, a very small fraction of the electrons passing through are trapped in the dielectric. Thousands of such occurrences contribute to the amount of trapped charge, thereby creating an *electric retarding potential* in the dielectric until its endurance limit is eventually reached.

We have already seen that textured surfaces are particularly effective as nonlinear conduction elements. As a result of the combination of surface curvature and thick oxide, they need relatively low voltages for programming and hence provide excellent retention of data. The small variations in the intricate fabrication steps of the silicon chip, however, lead to a range of endurance values for the memory cells on the same chip. Moreover, there will be small variations from chip to chip on the same wafer, and from wafer to wafer.

There are also external parameters that affect the endurance of the nonvolatile memory cells. For example, both the charge-transfer process within the cell and the operation of the internal high-voltage generating circuit are affected by temperature. The logarithm of the endurance is linearly related to temperature, doubling for every 50°C increase.

Another parameter is the cycling frequency. At the end of a storage cycle, the newly trapped electrons are in a relatively high free-energy state. Longer periods between cycles provide these charges and their surroundings more time to relax to a lower energy state. This should also affect the measured endurance.

It has been empirically found that the endurance statistically fits into the so-called *logarithmic extreme value distribution*, in which the cumulative probability density function is defined as

$$\Phi(x) \equiv \exp\left(-e^{-y}\right) \tag{10.1}$$

where the *reduced* variate y is related to the distributed property x by

$$y = \sigma(x - \mu) \tag{10.2}$$

with σ and μ being constants.

One important observation is that it is not the number of cycles but its logarithm that fits into this statistical model, so that x represents the logarithm of the observed cycles of individual chip endurances, σ is the *standard deviation* of the distribution of these endurances, and μ is the *mode*, that is, the observed or interpolated value of the logarithm of the endurance at $\Phi = .366$, where the maximum of the distribution is located. This means that statistically 36.6% of all devices will fail to reach this endurance level.

Since it involves a complex computational procedure to determine σ and μ from a given set of data, it is easier to obtain, using Equations (10.1) and (10.2), a linearized relationship between the logarithm of endurance cycles and the variate

$$y = -\ln\left[-\ln \Phi\right]$$

as

$$x = \mu + \left(\frac{1}{\sigma}\right) y \tag{10.3a}$$

or

$$\log c_i = \log c_m - \left(\frac{1}{\sigma}\right) \ln\left[-\ln \Phi_i\right] \tag{10.3b}$$

where c_i and c_m are the endurance of the ith part and the maximum value of the distribution of cycles, respectively. The maximum value occurs at $\Phi = .366$, while $1/\sigma$ signifies the slope of the straight line. To associate the correct value of Φ_i with a given observed endurance c_i, all the endurances from a group of devices are *ranked* in ascending order of cycles by assigning numbers from 1 through n, where n is the total number of devices tested. Then Φ_i can be found for each c_i from

$$\Phi_i = \frac{i - 1/2}{n} \tag{10.4}$$

A typical plot is shown in Figure 10.38. The lower horizontal axis is for the linear extreme value variate, and the upper axis corresponds to the cumulative failure percentage. Note that endurance-life tests are performed on a sample basis because the screening process is destructive.

Firm Errors in Nonvolatile Memories

Ionizing radiation incident on floating-gate nonvolatile memories can lead to three types of undesirable effects. The radiation may induce damage in the peripheral circuitry, giving rise to hard errors. It may cause upset of the sense circuitry, resulting in soft errors as in volatile RAMs, or it may bring about data loss by transfer of charge from the floating gate, leading to *firm errors*.

Figure 10.38 Plot of the logarithm of endurance cycles as a function of the reduced voltage.

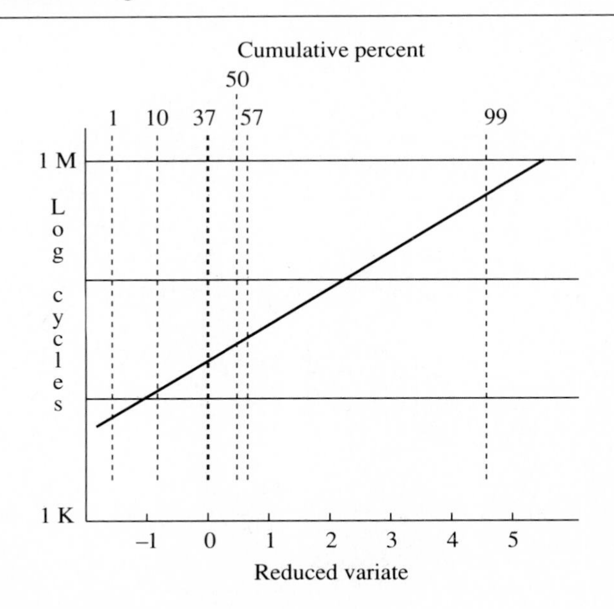

The carriers collected on the floating gate may come from two sources. They are created by the ionizing radiation in the SiO_2 that lies between the floating gate and another electrode or the substrate when they are at a different potential. Electrons are also excited in the floating gate if they have enough kinetic energy to surmount the potential barrier between the conduction bands in Si and SiO_2. Unlike soft errors, which are caused by a single ionizing particle, charge transfer to a floating gate is cumulative, so that the effects of many ionizing events occurring over an extended period of time must be considered. While soft errors occurring in floating-gate memories can be corrected by rereading the floating gate, firm errors result in read errors that cannot be corrected by rereading but can be corrected by rewriting the cell in question.

Figure 10.39 shows the cross-section through a floating-gate memory structure with the e-h pairs generated along the track of an α-particle and the corresponding energy-band diagram. Those pairs generated in the oxides on both sides of the floating gate drift apart to discharge the gate. Electrons injected from the floating gate into either oxide will also discharge the gate.

Flash E²PROMs

Until recently, there has been a gap in the programmable semiconductor memory spectrum where neither EPROMs nor E²PROMs were cost-effective. E²PROMs cannot match the memory density of the former. However, they are easily programmable

Figure 10.39 The cross section and the corresponding energy-band diagram of a floating-gate transistor with the floating gate programmed, and the e-h pairs generated due to ionizing radiation.

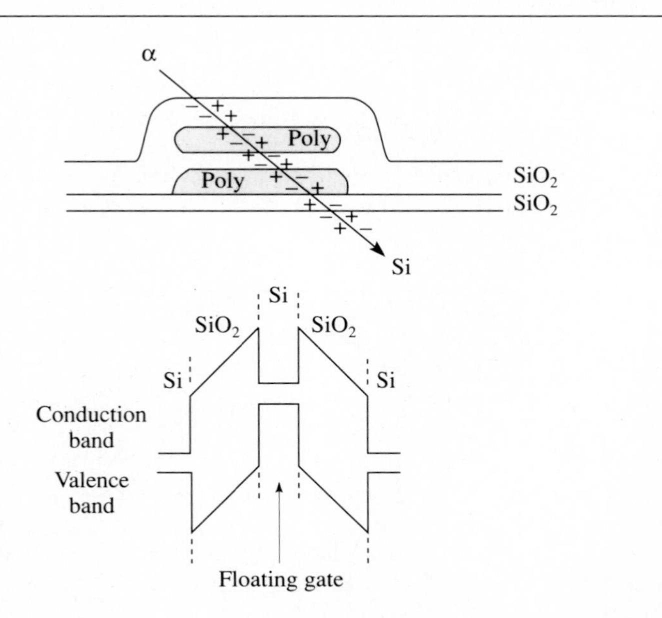

on circuit, even though only a byte at a time. Certain applications need more memory capacity than E^2PROMs can furnish, and reprogramming must be done faster and more often than can be achieved with EPROMs.

In 1988, Intel, Seeq Technology, and Toshiba independently introduced so-called *flash* E^2PROMs that can not only be reprogrammed as quickly and easily as conventional E^2PROMs but offer full-chip erasure like EPROMs as well. Furthermore, their single-transistor memory cells are more easily scalable to provide high density.

Flash E^2PROMs do not have the full nonvolatile functions that E^2PROMs possess, and their endurance is limited to 100 to 1000 programming cycles. Nevertheless, for many applications they cost much less than E^2PROMs do.

The 512-Kbit flash E^2PROM from Seeq supports a minimum of 100 cycles, with a typical access time of 200 ns and a maximum erasure time of 7.5 s. Toshiba offers a 256-Kbit flash memory with an access time of 170 ns and erasure time of 1 s that also supports a minimum of 100 write cycles. Intel's 256-Kbit flash E^2PROM has an access time of 150 ns and erasure time of 1 s, with a minimum of 100 cycles. All of these devices are dual-power-supply flash E^2PROMs that need 12 V for programming and erase, and 5 V for read.

The first experimental single-power-supply flash E^2PROM was announced by Texas Instruments, Inc. in 1989. All operating voltages, that is, +18 V for programming, +7 V for program inhibit, and −11 V for erasure, are internally generated from the 5-V supply through charge-pump circuits. The access time is reported to be 170 ns with an erasure time of only 10 ms and an endurance of 10,000 cycles. With a minimum feature of 1.5 μm, a cell size of 40 μm^2 and a chip size of 30 mm^2 have been achieved.

10.8 Programmable Logic Devices

Even though the term *programmable logic* alludes to a broad class of ICs, in practice it almost always refers to a narrower group of devices, as explained below. In this text, the term *programmable logic device* (*PLD*) is used in its narrower sense and will not refer to various types of PROMs that have already been discussed in previous sections. With their highly regular geometric structures, PLDs can be designed more compactly than gate logic.

Bipolar PLAs and PALs

PLDs were first developed in the mid-seventies by Signetics Corporation in the form of *field-programmable logic arrays* (*FPLAs*), or simply PLAs, as an extension of bipolar PROMs, in which only the OR array is programmable and the input address decoder acts as a fixed AND plane, whereas in the PLA both AND and OR matrices are programmable. Their internal organization differs from that of the PROM, in that the address decoder is replaced by an AND array that implements product terms of the input expression, and the OR array takes the logical *sum of the products* (*SOP*) to form the output functions.

The *programmable array logic* device, or PAL, also introduced in the seventies by Monolithic Memories, Inc. (MMI), now a subsidiary of Advanced Micro Devices,

Inc. (AMD), has a programmable AND matrix and a fixed OR matrix. Therefore, the basic difference between the PLA and PAL is the removal of the programmability of the OR array from the latter and fixing the number of product terms fed to each of the output sum terms. The schematics of all three architectures are shown in Figure 10.40, using programmable logic symbology to simplify the graphical depiction of PLDs. In this convention, the multiple input lines to the AND and OR gates in the arrays are replaced by a single line, called the *product line*, coming into the gate, and the actual number of inputs that the line represents is given by the number of input lines crossing it. The cross mark (×) indicates an intact fusible link that can be removed to represent a blown fuse when no connection is desired at that particular point of intersection. A dot (•) at the intersection of any line represents a hard-wired connection. Note that buffers are added at the inputs to drive AND gates and to provide double-rail logic.

In addition to specifying the number of inputs and functions, the PLA manufacturer also cites the number of product terms (p-terms) available because, with n as the number of input variables, there are less than 2^n terms, in contrast to the PROM,

Figure 10.40 The basic architectures of (a) PROM.

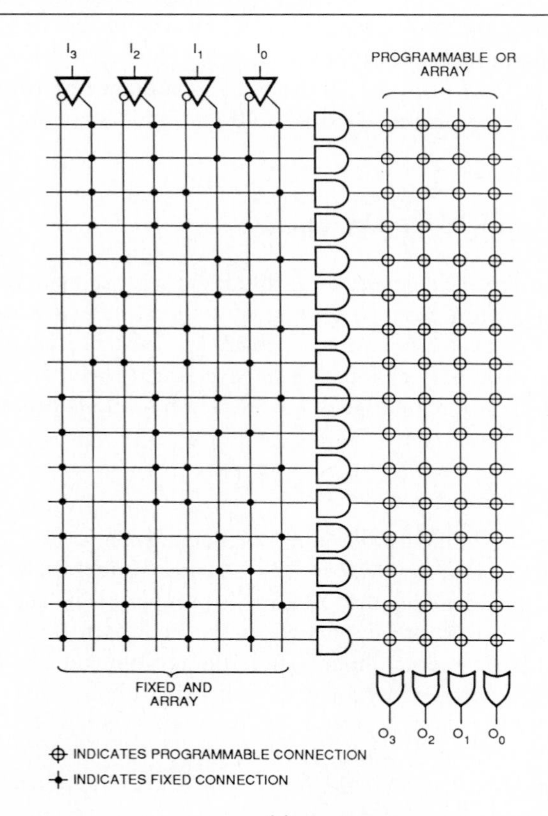

(*a*)

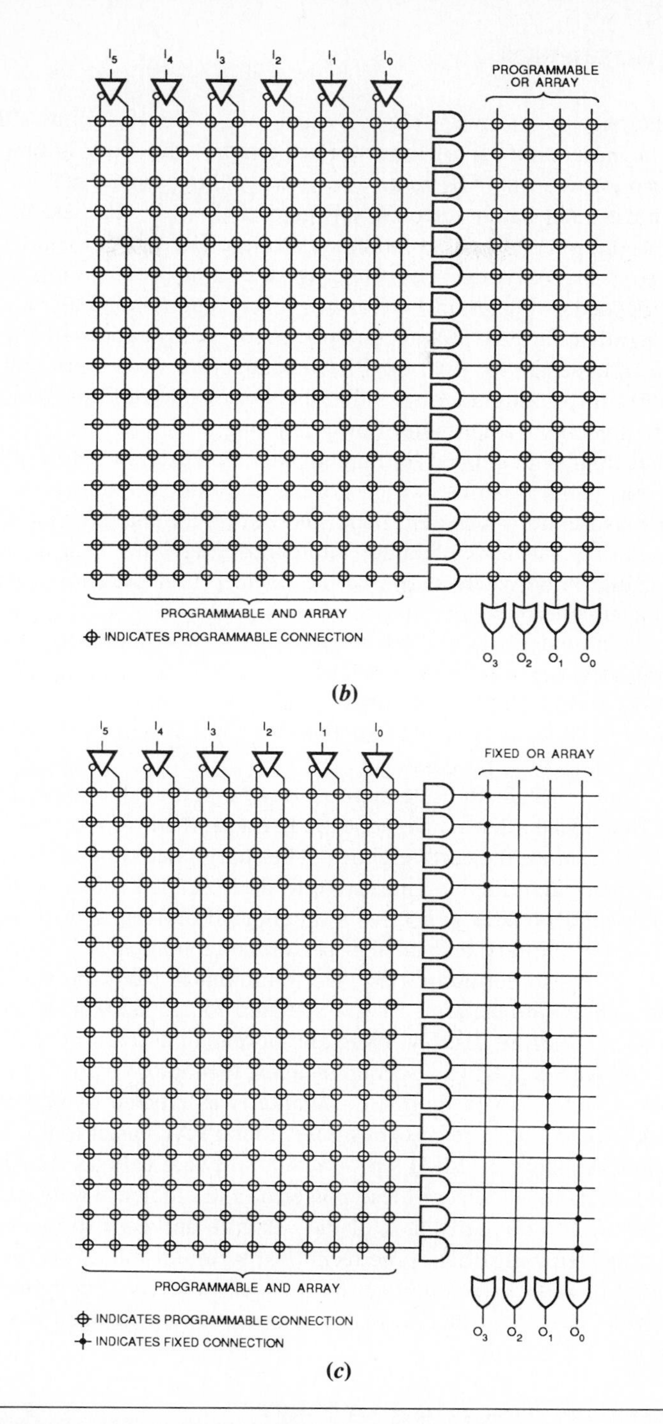

Figure 10.40 *(continued)* (b) PLA, and (c) PAL. (Copyright © Advanced Micro Devices, Inc., 1988. Reprinted with permission of copyright owner. All rights reserved. PAL and PALASM are registered trademarks.)

in which the number of p-terms is always equal to 2^n. This fact allows the PLA to accommodate a larger number of inputs. Furthermore, the PLA is unrestricted in combining the p-terms in the OR matrix, thereby adding considerable flexibility to the device. On the other hand, since the OR gates in PAL devices are prewired, the degree to which the p-terms can be combined at these OR gates is restricted. Therefore, in the case of the PLA, the user can assign the same product term to more than one output as needed, whereas in PAL it has to be programmed as many times as needed. It is clear that the PAL architecture places severe limitations on the utilization of p-terms. To compensate for this, the PAL manufacturers offer different part types that vary the OR array configuration, so that the specification of this matrix becomes a task of device selection rather than of programming.

The oldest technique employed for implementing the user programmable switches in an IC provides paths or links, which are narrow ribbons of metal or any other conductor, such as titanium-tungsten, deposited during the manufacture of the chip, between two crossing surfaces of either interconnecting lines or a device and the interconnecting line in a switch matrix of the device, as depicted in Figure 10.41. Application of a sub and/or supervoltage to one of the lines for a short period of time while the other is grounded would cause a temporary flow of high current through the fuse, generating local heat dissipation to melt the crosspoint link and open the connection. PLDs featuring this kind of fuse are commonly referred to as *fuse-link* or *fuse-programmable* PLDs and include all PROMs, PALs, and PLAs manufactured using bipolar TTL and ECL processes.

At their advent in 1978, the selection of user-programmable logic devices consisted only of TTL PLDs, that is, Signetics' PLA and MMI's PAL, and the bipolar PROM outsold PLDs until the early eighties. The sharp price drop of PLDs in recent years has made them an attractive alternative to the so-called standard *glue* logic of SSI and MSI ICs on a broader variety of systems than those designed for the traditional markets in the military and in high-performance computers. Still, the distribution of digital IC consumption in 1987, worth around $21 billion, was as follows: 29.6% memory, 28.1% processors, 21.9% standard logic, 9.5% gate arrays, 8.5% full-custom, and 2.3% PLDs. Today, there are more than 200 different PLD devices with close to 1000 parts available if second sources, technology variations, and package outlines are counted. Even though the market was created by Signetics, MMI won out with its PAL not only because designers found PAL easier to use but because it was also supported better by MMI's proprietary software called PALASM.

The initial reluctance to accept these powerful and efficient *semicustom* devices can also be attributed to the early lack of the versatile and easy to use CAD/CAE tools required to design with PLDs. The recent explosion of microcomputer technology and the flood of associated software design tools, dedicated computer workstations, and hardware development systems have considerably changed the ways logic designers approach logic design.

Enhancements to the Basic PAL Architecture

To make design easier, various features have been added to the newer generation of PAL devices, making the originally highly-structured PAL architecture increasingly more flexible. In many designs, there is a need to feed the outputs of the SOP

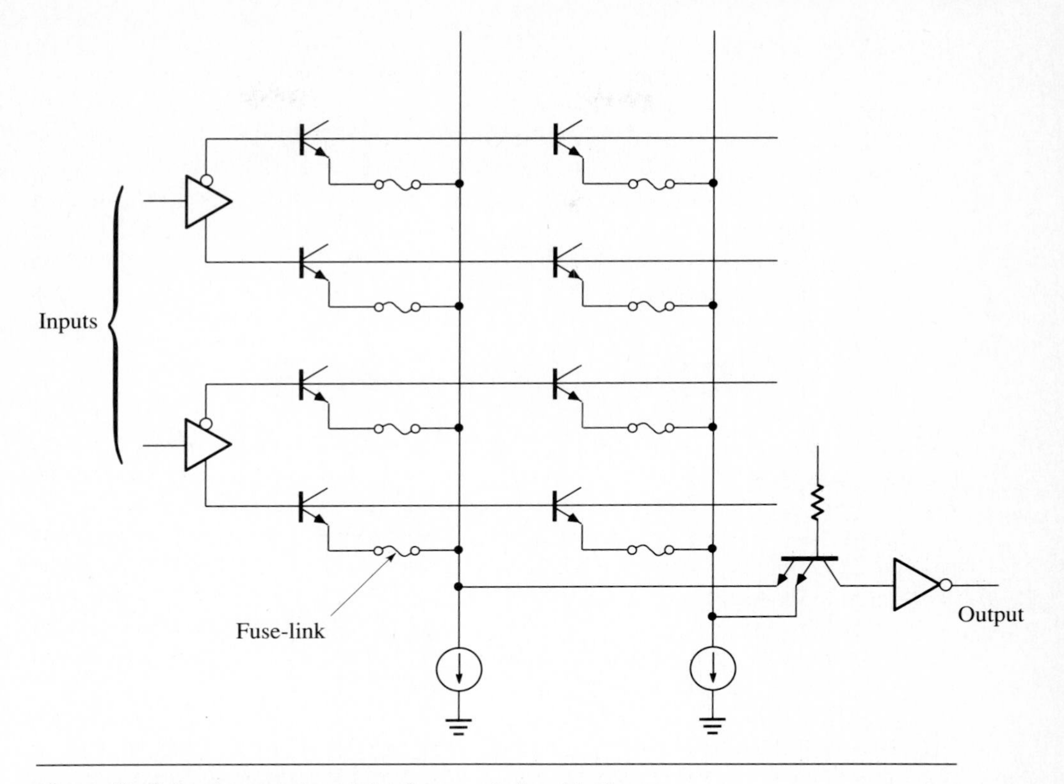

Figure 10.41 Implementation of the fuse matrix in a bipolar PLD.

terms back to the AND array to make them available as regular inputs for other product lines. In addition, the inclusion of three-state buffers at the output of a PLD under a single p-term control serves two purposes. First, it allows *programmable I/O*, in which the PLD's output also functions as an input because the user now has the option of switching the output pin to a high-impedance state, so that this pin can subsequently be controlled by an external signal and, through the feedback buffer, delivered to the array. The same pin can be defined as an input by simply leaving all the fuses in the three-state control p-term intact to enable the high-impedance state permanently. Second, the three-state buffer can also be employed in bus-oriented applications.

Figure 10.42 shows the logic diagram of the popular PAL16L8 whose architecture incorporates the special circuit features just discussed. Note that it has ten simple inputs and six of the outputs operate as I/O ports to allow feedbacks into the programmable AND array, which, with 64 product lines by 32 input lines, contains 2048 fuses. One AND gate in each SOP realization controls each of the three-state outputs. An important feature in this diagram and all other logic diagrams supplied in data sheets is that there are no ×'s marked at every fuse location in order to make the diagram useful for a user who wants to generate specific functions by inserting ×'s

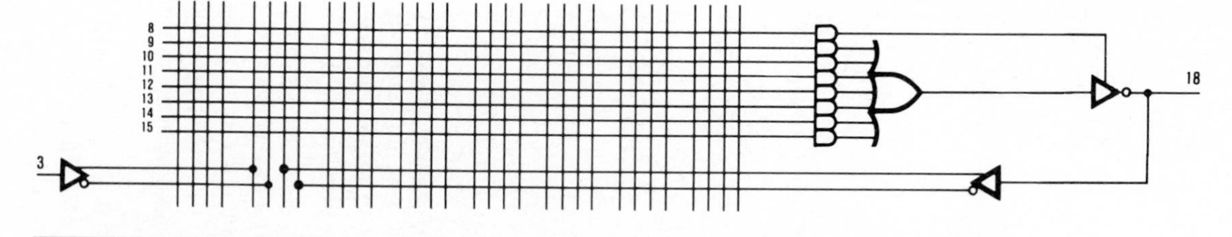

Figure 10.42 The simplified logic diagram of the PAL16L8. (Copyright © Advanced Micro Devices, Inc., 1988. Reprinted with permission of copyright owner. All rights reserved. PAL and PALASM are registered trademarks.)

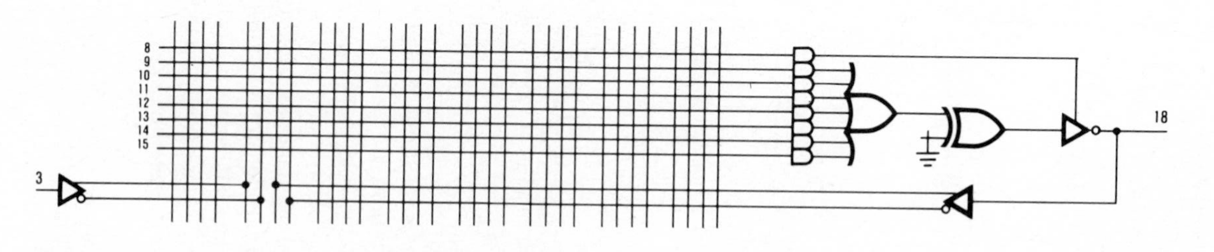

Figure 10.43 The simplified logic diagram of the PAL16P8. (Copyright © Advanced Micro Devices, Inc., 1988. Reprinted with permission of copyright owner. All rights reserved. PAL and PALASM are registered trademarks.)

wherever an intact fuse is desired. For performance-driven applications, this bipolar TTL logic device designed in MMI's PAL architecture has emerged as the industry standard and is widely second-sourced in both bipolar and CMOS technologies. It is typically packaged in a 20-pin plastic dual-in-line case with a power dissipation of 1 W. Now that 10-ns versions are available by using advanced Schottky technology, its utilization in support of the 20-MHz μPs has become possible.

The next enhancement of the combinational PLD architecture was to add individually *programmable output polarity* to overcome p-term limitations, especially in PALs, and to provide both active-high and active-low outputs. In 1982, Harris Corp. was the first to introduce a bipolar part, the PAL16P8, whose logic diagram is illustrated in Figure 10.43, that included an extra *exclusive-OR* (*XOR*) gate placed at the output of each SOP expression so that, depending on the state of the polarity fuse, the final polarity of the output could be programmed as either **0** or **1**. Blowing the polarity fuse ties the input of the XOR gate to high so that the output is complemented.

Sequential designs require that input data be processed along with the data generated during the previous state of the circuit and stored in the storage elements. The class of PLDs suitable for emulation of sequential logic circuits is called *registered* PLDs. The noninverting outputs of the storage elements, the delay flip-flops used as registers, are employed, through three-state buffers, as the device outputs, while the complemented outputs are routed back to the AND array to allow the implementation of the previous state sequential feedback inputs. The logical diagram of the registered version of the PAL16L8, denoted as PAL16R8, is shown in Figure 10.44. Since all large sequential circuits are built as synchronous designs, a way of clocking the internal registers must be provided. The PLDs developed for sequential designs feature a single clock line that is connected to all output flip-flops in parallel. Note also from Figure 10.44 that the three-state buffers are no longer a p-term control but are all connected to a single input because the output Q_i can no longer be used as a bidirectional port. One important disadvantage of this registered architecture is that no means to preset the output registers to a desired initial state is provided unless, of course, one of the p-terms is used as a *preset* term.

Recently, power-up reset of registered devices has been included to ease system initialization requirements. Also, a new *preload* feature allows the user to initialize the registered devices to a known state by loading the data placed on the output pins prior to testing the device, thereby simplifying and shortening the testing procedure of sequential designs. When testing sequential circuits, not only those needed in the normal machine operation, but all possible state transitions must be verified, because certain events such as line voltage glitches, brown-outs, and even power-up may throw the logic into an unspecified state. To test a design for proper treatment of these conditions, a way must be furnished to break the feedback paths and force any desired unspecified state into the registers, so that the machine can then be sequenced to test the next state of the outputs. Otherwise, a state sequencer would be required for test coverage.

A majority of first generation PLDs in use today are packaged in 20-pin DIPs, which impose a limitation for the number of input and output pins. Some PAL devices, housed in 24-pin DIP packages, meet the demand for a greater number of

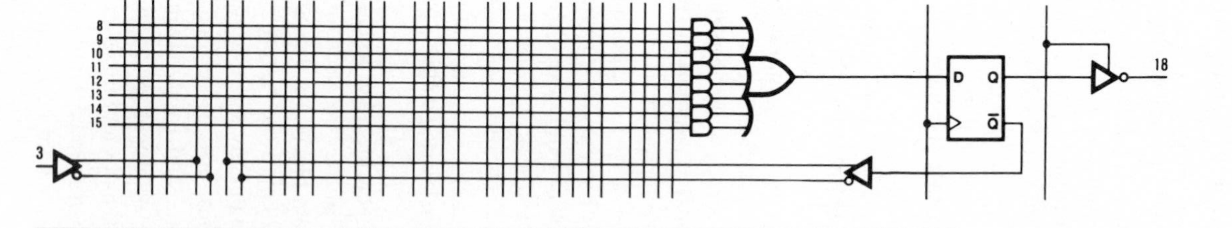

Figure 10.44 The simplified logic diagram of the PAL16R8. (Copyright © Advanced Micro Devices, Inc., 1988. Reprinted with permission of copyright owner. All rights reserved. PAL and PALASM are registered trademarks.)

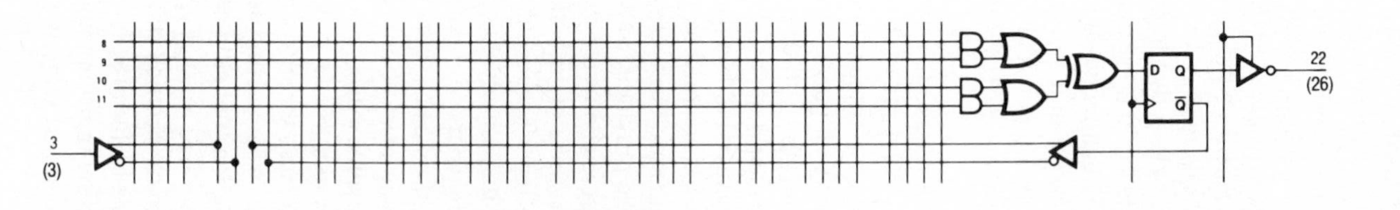

Figure 10.45 The simplified logic diagram of the PAL20X10. (Copyright © Advanced Micro Devices, Inc., 1988. Reprinted with permission of copyright owner. All rights reserved. PAL and PALASM are registered trademarks.)

outputs and wider p-terms in the array by duplicating the architectures of the 20-pin PLDs and simply expanding their I/O widths and array sizes. Certain devices such as PAL20X10, as illustrated in Figure 10.45, include XOR gates at the outputs, each fed by two adjacent p-terms. The outputs of the XORs, in turn, are fed to the output registers. This configuration provides an easy realization of fast counters and other state sequencers that progress through a binary-encoded counting sequence in such applications as accessing memories, counting events or sequential control in high-speed processor systems. Another device, the PAL20S10, incorporates *product-term sharing*, as diagrammed in Figure 10.46, in which 16 p-terms are shared between two output cells. However, a p-term can be used by only one output at a time. A registered version, the PAL20RS10, is also available, and is shown in Figure 10.47. Table 10.3 shows how to read the common TTL PAL part numbers.

PAL22V10

PLDs with traditional PAL architecture and registered outputs remain the mainstay of PLD-based state-machine design. Today, the most popular universal device is AMD's PAL22V10, which permits the development of custom LSI functions of 500 to 800 equivalent gate complexity with its 22 inputs, 132 p-terms, and 10 outputs. First introduced in 1983, it is widely second-sourced in bipolar TTL and CMOS technologies. The device has the familiar PAL architecture, as illustrated in Figure 10.48, but, unlike its predecessors, it permits the user to configure the characteristics of each output through the use of a fuse-programmable logic circuit, called the *macrocell*, whose logic diagram is shown in Figure 10.49.

The 22V10 incorporates the capability to define and program the architecture of each macrocell on an individual basis. Depending on the state of the two fuse selects $S0$ and $S1$, the ten macrocells can take on one of four configurations; registered output or combinational I/O pin with active high or active low polarity, as depicted in Figure 10.50. $S1$ controls whether the outputs are sequential or combinational, whereas $S0$ determines whether they are active high or active low. With no fuses blown, the output is sequential and active low; blowing both changes the output to combinational and active high. Blowing only $S1$ gives a combinational active low output, and blowing only $S0$ realizes a sequential active high output.

Therefore, since each of the ten output pins may be individually configured as inputs, expressions requiring up to 21 inputs and a single output or down to 12 inputs and 10 outputs are possible. The macrocell logic is programmed into the device at the same time the AND array is programmed.

Another novelty is the introduction of the *variable p-term distribution*, in which from 8 to 16 p-terms can be allocated to each output to allow more complex functions to be realized than in previously available devices. The 22V10 also has a power-up reset function, synchronous set and asynchronous reset p-terms common to all registers, and a preload feature for structured testing.

However, the device suffers from one basic problem that involves the architecture of its output macrocell. Note that the input feedback multiplexer selects between the output pin when used as an input and the register's complemented output, so that once the output pin operates as an input, the register cannot feed its state back to the

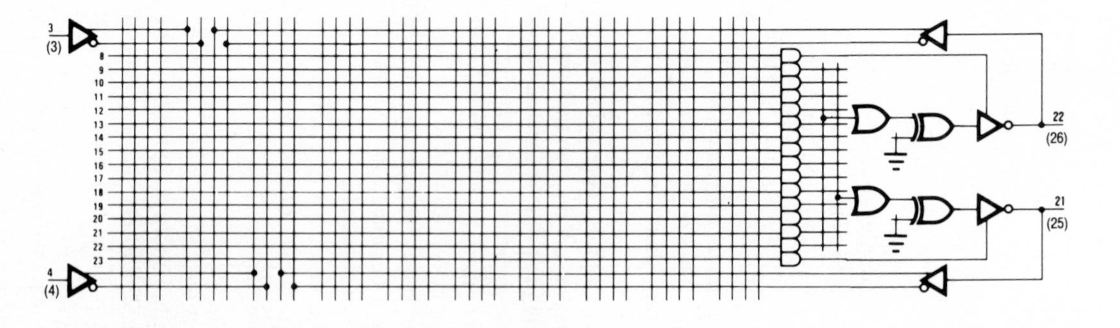

Figure 10.46 The simplified logic diagram of the PAL20S10. (Copyright © Advanced Micro Devices, Inc., 1988. Reprinted with permission of copyright owner. All rights reserved. PAL and PALASM are registered trademarks.)

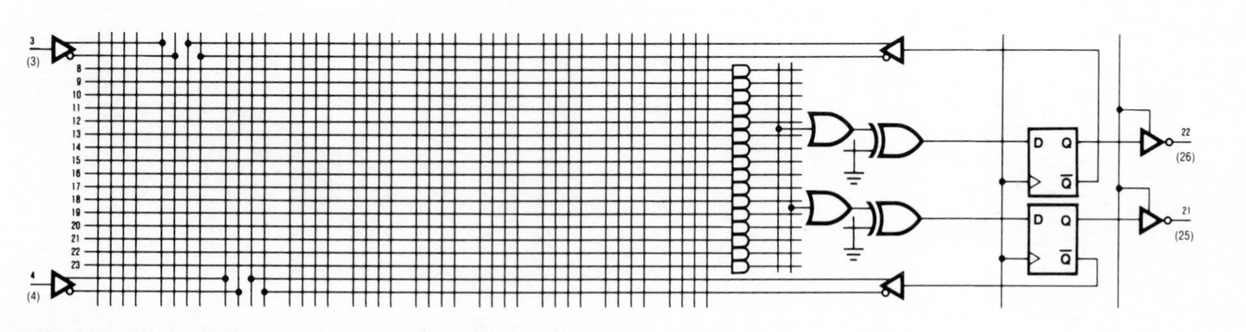

Figure 10.47 The simplified logic diagram of the PAL20RS10. (Copyright © Advanced Micro Devices, Inc., 1988. Reprinted with permission of copyright owner. All rights reserved. PAL and PALASM are registered trademarks.)

Table 10.3 Part-Number Interpretation for TTL PAL Devices

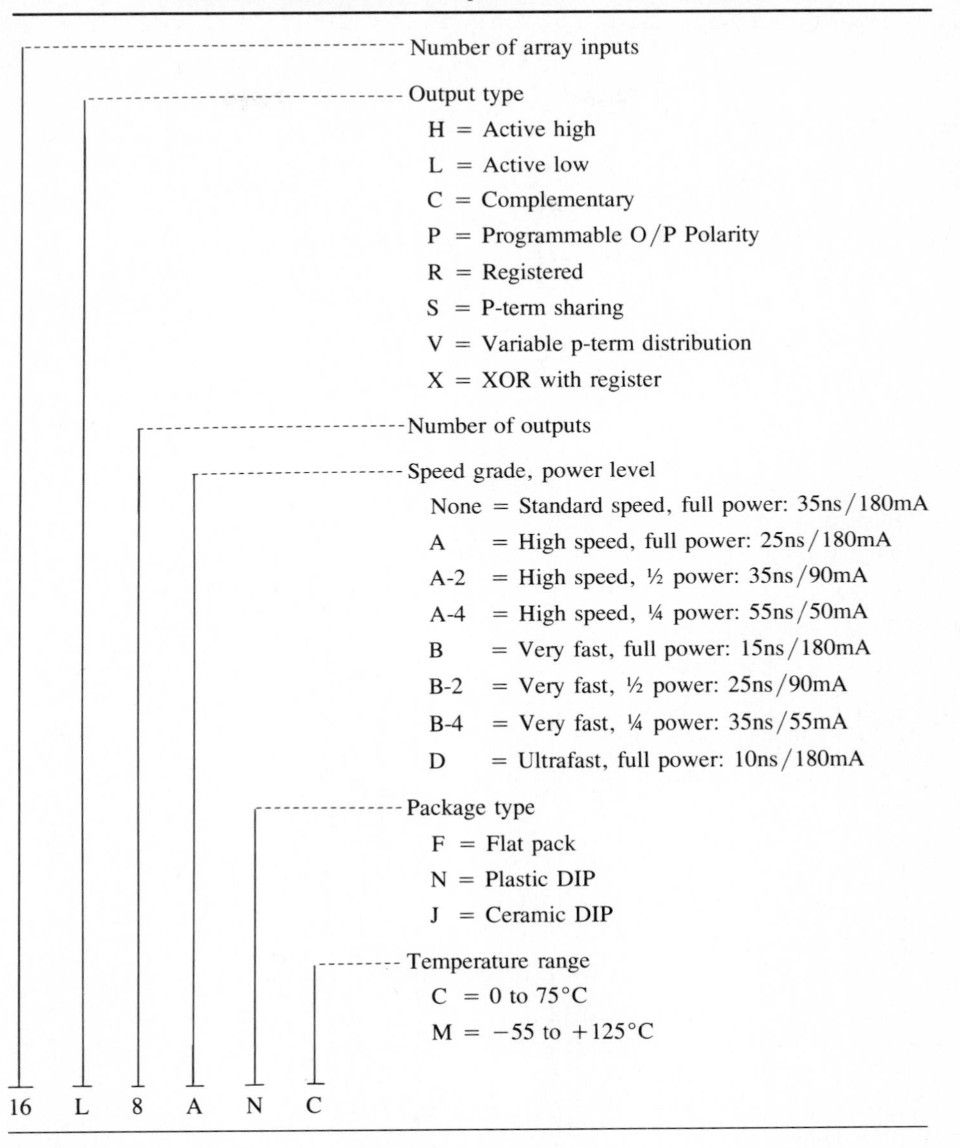

input of the AND/OR array anymore. Thus, the register cannot be *buried* or *embedded*, limiting either the number of I/O pins or registers available to the user.

To counter this problem, a number of PLDs provide *buried registers* to create storage elements for state-machine implementations without sacrificing the employment of the corresponding I/O pin as an input. For example, AMD introduced the

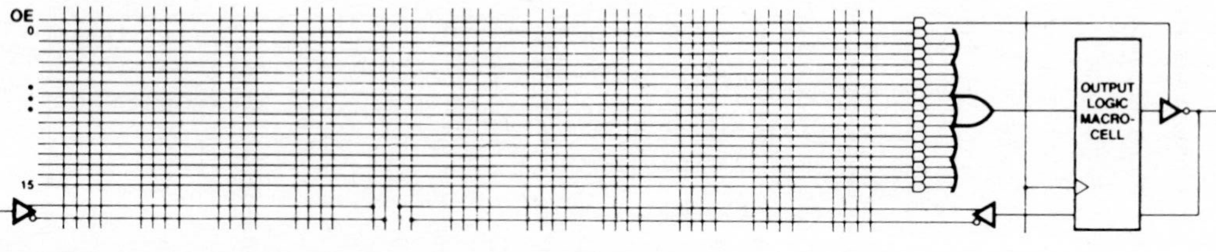
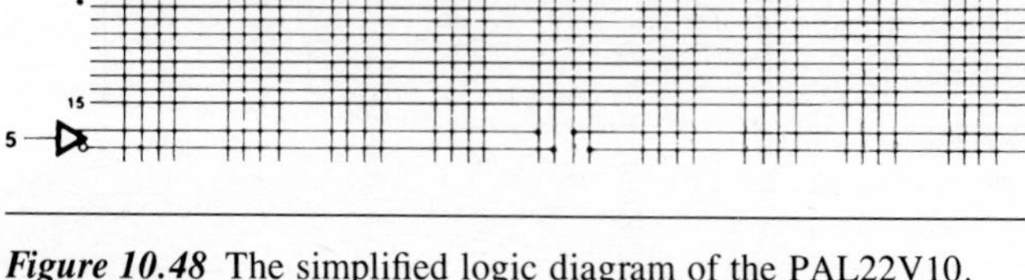

Figure 10.48 The simplified logic diagram of the PAL22V10. (Copyright © Advanced Micro Devices, Inc., 1988. Reprinted with permission of copyright owner. All rights reserved. PAL and PALASM are registered trademarks.)

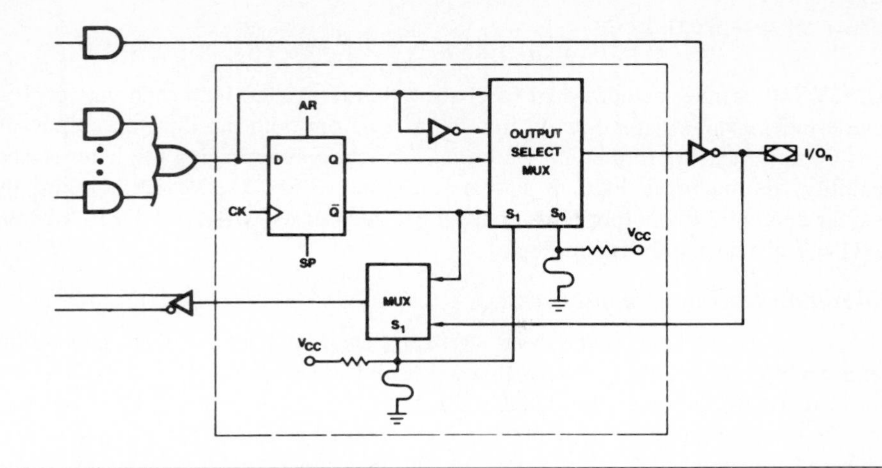

Figure 10.49 The logic diagram of the PAL22V10 output macrocell. (Copyright © Advanced Micro Devices, Inc., 1988. Reprinted with permission of copyright owner. All rights reserved. PAL and PALASM are registered trademarks.)

Figure 10.50 Four possible configurations of the PAL22V10 macrocell: (a) registered, active-low output, (b) registered, active-high output, (c) combinational, active-low output, and (d) combinational, active-low output. (Copyright © Advanced Micro Devices, Inc., 1988. Reprinted with permission of copyright owner. All rights reserved. PAL and PALASM are registered trademarks.)

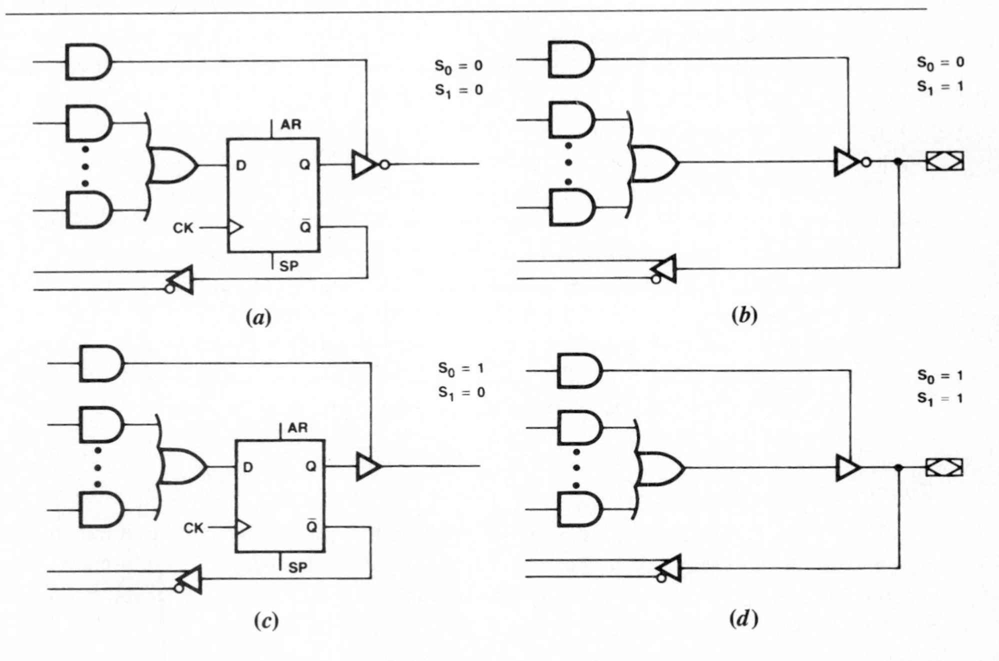

PAL32VX10, which includes dual feedback paths associated with each macrocell to provide independent paths directly into the array from both the flip-flop output and the I/O pin; thus, the former can be buried without compromising the latter's input capability. Its macrocell logic is shown in Figure 10.51. The device also lets the designer configure the output register as an SR latch or as a JK, T or DFF, whereas the 22V10 provides only D flip-flops.

Programmable Logic Sequencers

Programmable logic sequencers, or PLSs, are PLAs with storage capabilities similar to those of registered PALs that are useful in sequential circuit implementations. Figure 10.52 shows the details of a 16 × 45 × 12 bipolar PLS, Signetics' PLS155. A registered logic element with three-state outputs, it features four dedicated inputs and four registered I/O outputs, made up of positive-edge-triggered JKFFs, as well as eight bidirectional I/O lines. This yields variable I/O gate and register configurations through control gates D_0 through D_7, L_A and L_B, ranging from 16 inputs to 12 outputs with 45 p-terms. One group of AND gates drives bidirectional I/O lines, denoted as B_i, $0 \leq i \leq 7$, in Figure 10.52, whose output polarities are individually programmable via a set of XOR gates to implement AND/OR or AND/NOR logic functions. Another group, as shown in Figure 10.53, drives the J and K inputs of the flip-flops. The two p-terms L_A and L_B in the control matrix control the loading of the flip-flops in pairs that are synchronous with the clock CK.

The OR matrix is made up of three distinct parts. The first part consists of eight 32-input gates coupled to an output-polarity-controlling XOR array. The second part has 12 additional OR gates that control four flip-flops. The third part is a simple complement array: a single OR gate with its output complemented and fed back into the product matrix to enable a particular p-term to become a common factor of any or all of the remaining p-terms thereby making it possible to realize factored SOP expressions. It can also be used in conjunction with a p-term to provide escape vectors in case the state machine gets into one of the undefined states during power-up or a timing violation due to asynchronous inputs. Without the complement array, an alternate way would be to define all possible states by adding redundant p-terms.

The flip-flops have overriding asynchronous preset and clear inputs that are controlled by terms in the OR matrix. Their three-state output buffers are controlled from the enable pin $\overline{OE}$. They can also be enabled or disabled permanently by blowing fuses or leaving them intact in the enable array. They can be dynamically changed to T- or D-type flip-flops by using the three-state inverter between the J and K inputs that is under the control of the p-term F. When the inverter is in the high-impedance state, the flip-flop is a JKFF or a TFF if $J = K$. When the inverter is enabled, $K = \overline{J}$ so that the flip-flop becomes a DFF. The K input must then be disconnected from the OR matrix.

Bipolar vs. CMOS

Since a PLD usually replaces multilevel random logic, its performance must match that of the elements replaced. The TTL gate delays have been reduced to around 3 ns, so the same technology enhancements are being used to speed up programmable devices. Refinements such as trench isolation, implant technology, and

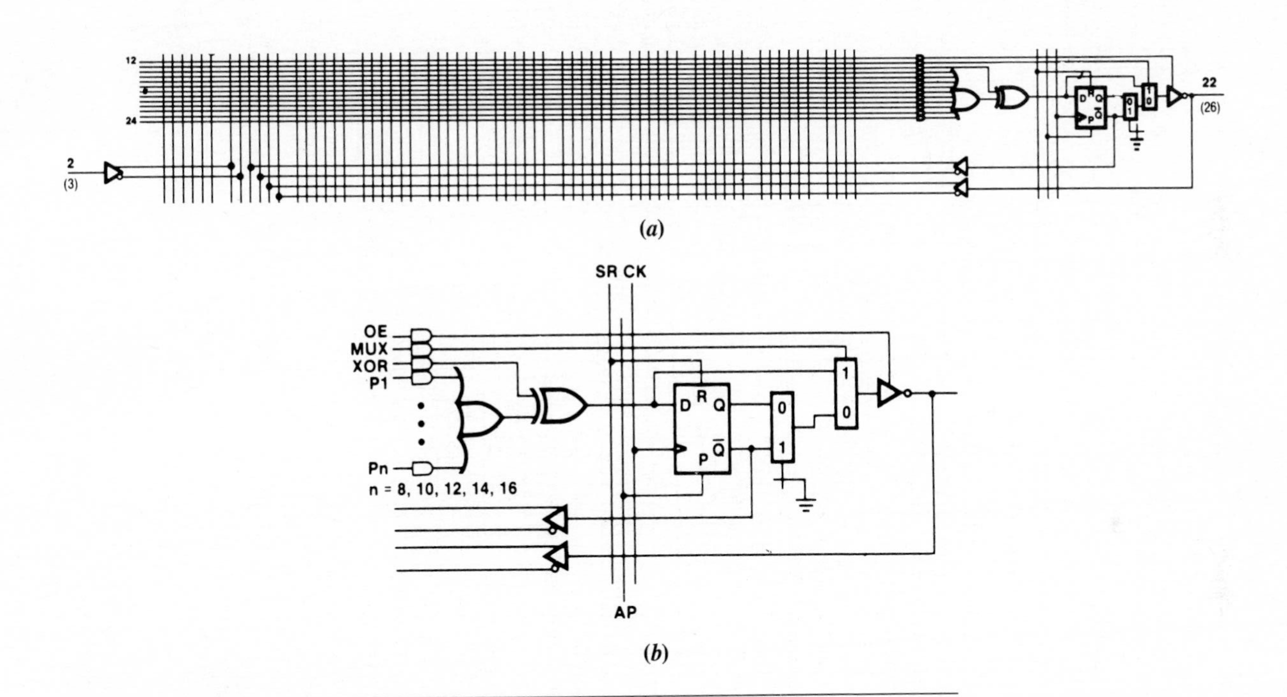

Figure 10.51 (a) The simplified logic diagram of the PAL32VX10, and (b) one of its macrocells. (Copyright © Advanced Micro Devices, Inc., 1988. Reprinted with permission of copyright owner. All rights reserved. PAL and PALASM are registered trademarks.)

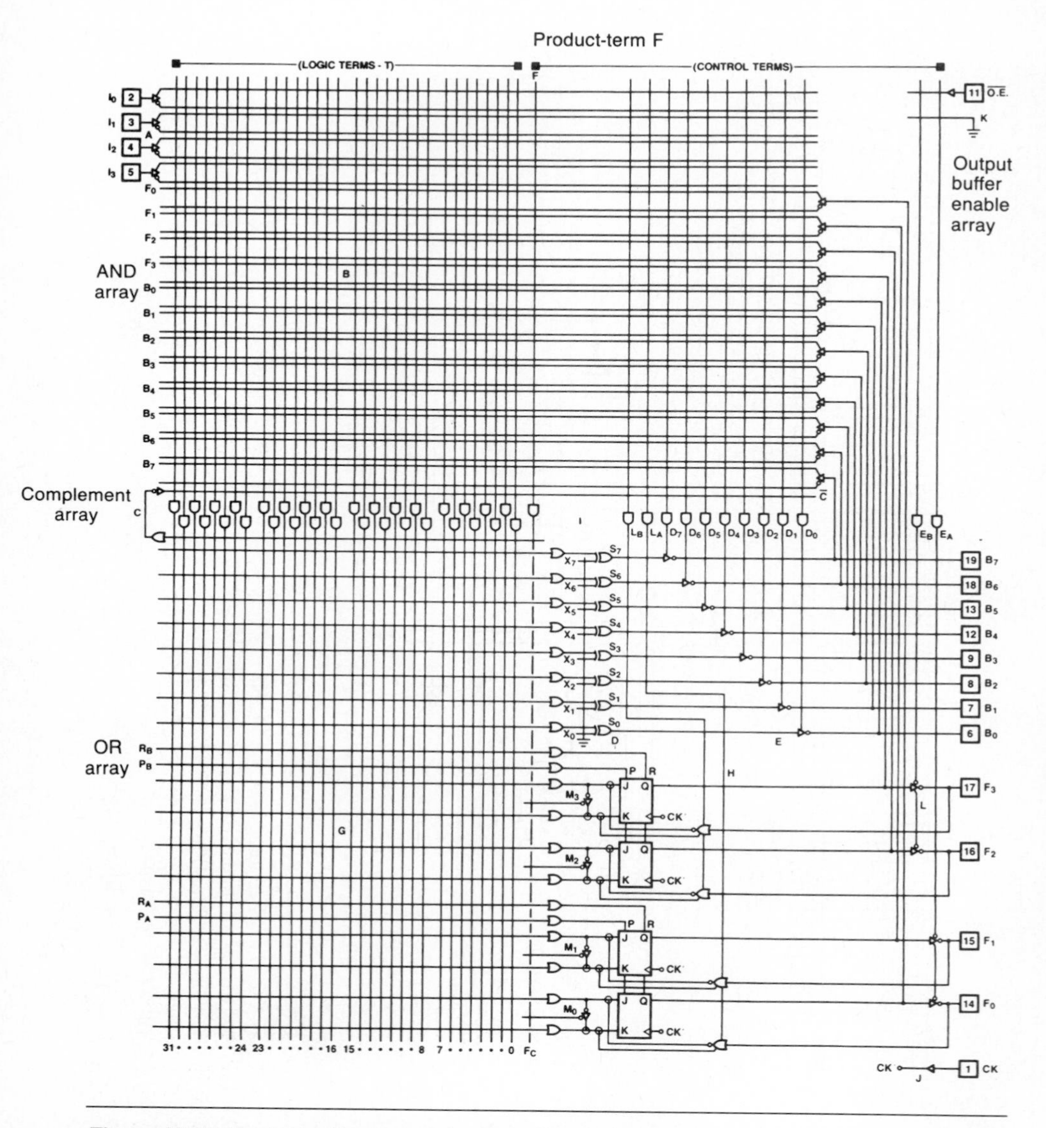

Figure 10.52 The logic diagram of the PLS155. (Courtesy of the copyright owner, North American Philips Corp.)

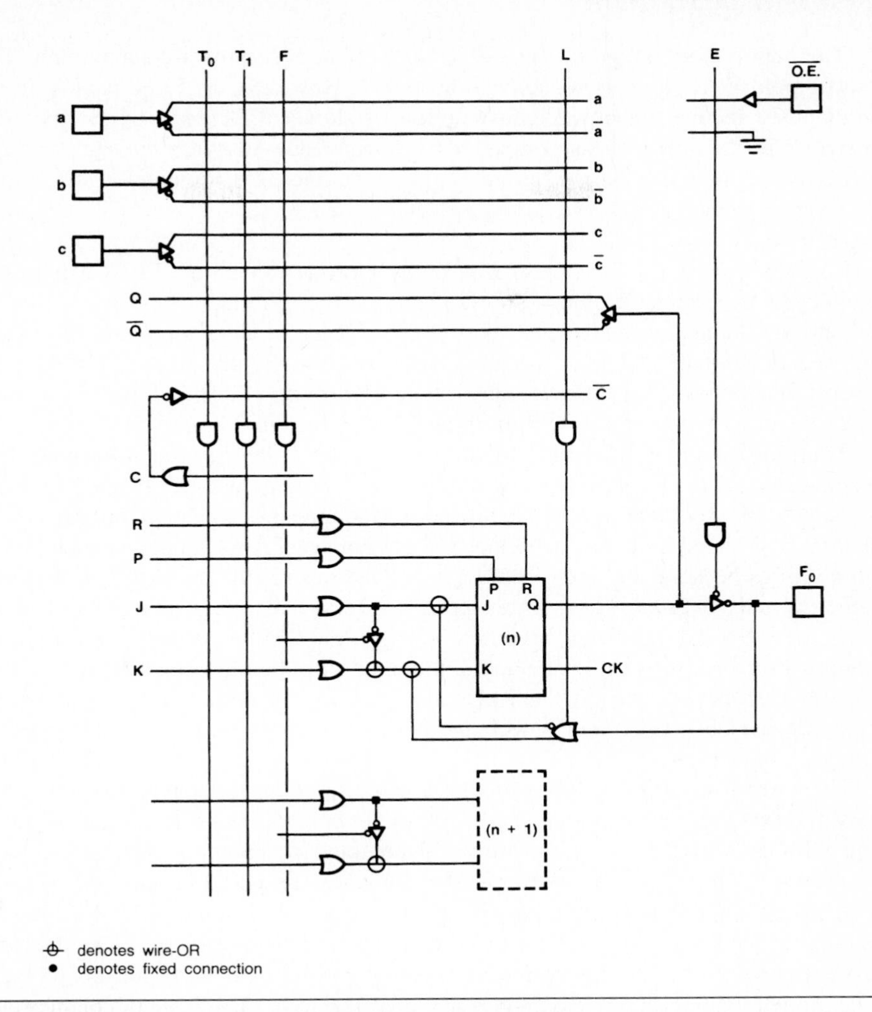

Figure 10.53 The architecture of the PLS155 flip-flop circuitry. (Courtesy of the copyright owner, North American Philips Corp.)

photolithography improvements have significantly reduced device sizes, improved switching speeds, and added significant functionality. As a result, propagation delays for the bipolar TTL PDL devices have been reduced to 10 ns, accommodating the demands of fast 32-bit μPs such as Intel Corp.'s 80386 and Motorola, Inc.'s 68030 that operate at 16-MHz or higher clock rates. If high speed and moderate functionality are the key requirements, then TTL devices are a good choice. Moreover, they offer high output drive capability making them useful in bus driving applications. With 90% of the PLDs in volume production utilizing this technology, they are also inexpensive. However, TTL is under increasing pressure to keep pace with advances that are driving propagation delay times under 10 ns.

The major shortcoming of the TTL process is its high power consumption. The fastest bipolar PAL devices consume almost 1 W. However, the PLD manufacturers continuously reduce the power consumption of slower TTL parts of the previous generation as the new and faster devices are introduced. Another drawback of these devices is that they can be programmed only once because they employ fuse links. The unprogrammed AND/OR array cannot be used to test dc, ac, and functional characteristics.

To pick up where TTL leaves off as far as speed is concerned, PLDs using ECL technologies have been developed for ultrafast computer architectures. The ECL PLDs facilitate the highest possible speed with 3-ns propagation delays but still have only moderate functionality. The power dissipation is only slightly higher than the full-power TTL options. These devices are available in both 10KH and 100K versions, so that the designer can pick the appropriate device for the design.

Traditional bipolar PALs are difficult to scale up in density much beyond 2000 gates because of their committed logic structures, such as macrocells and registers, associated with specific pins, so that when one function and one pin are utilized, other functions associated with that particular pin are not used. This cannot be tolerated at high densities because the size of the registers and macrocells increases, as does the number of unused gates. Thus, when speed can be sacrificed for lower power dissipation, high density, and hence high functionality, the CMOS, as usual, becomes the preferred process. However, one should note that with the high clock rates at which these devices operate, the CMOS PLDs also end up consuming as much power as bipolar devices because their power dissipation increases with the operation frequency.

There are four different types of PLDs using the CMOS technology. Developments in recent years have produced *erasable programmable logic devices*, or *EPLDs*, using CMOS EPROM switching arrays in contrast to bipolar parts that are programmed predominantly with fuses. We have already discussed in previous sections a way of implementing an array of switches using special versions of MOSFETs. As diagrammed in Figure 10.54, due to certain inherent properties of MOS devices, quasi-permanent states of device interconnection can be induced by means of local application of voltages to selected nodes of the switch array. This state of conductivity is called *floating-gate* generation. Such states can be maintained for a prolonged period of time depending on specific processing and device geometry, with no detectable degradation. Thus, floating local charges can be used as a means of opening the otherwise closed switches at the crosspoints of the switch arrays in the EPLDs as well as in EPROMs. In addition to windowed ceramic packages, plastic EPLDs, which are windowless *one-time-programmables* (*OTPs*), are also available.

The newest class of devices using a similar fusing mechanism consists of the so-called *electrically erasable PLDs* (*E^2PLDs*), again by analogy to E^2PROMs. These are actually a natural extension of EPROMs, in that they too utilize the electrically removed, induced floating charge as the means of shorting or opening the connections in the matrix intersections. Besides allowing the user to program the settings of the switches, they can be electrically erased and reprogrammed in a selective and convenient way. However, floating-gate switches are not as reliable or as radiation-tolerant as melted fuses.

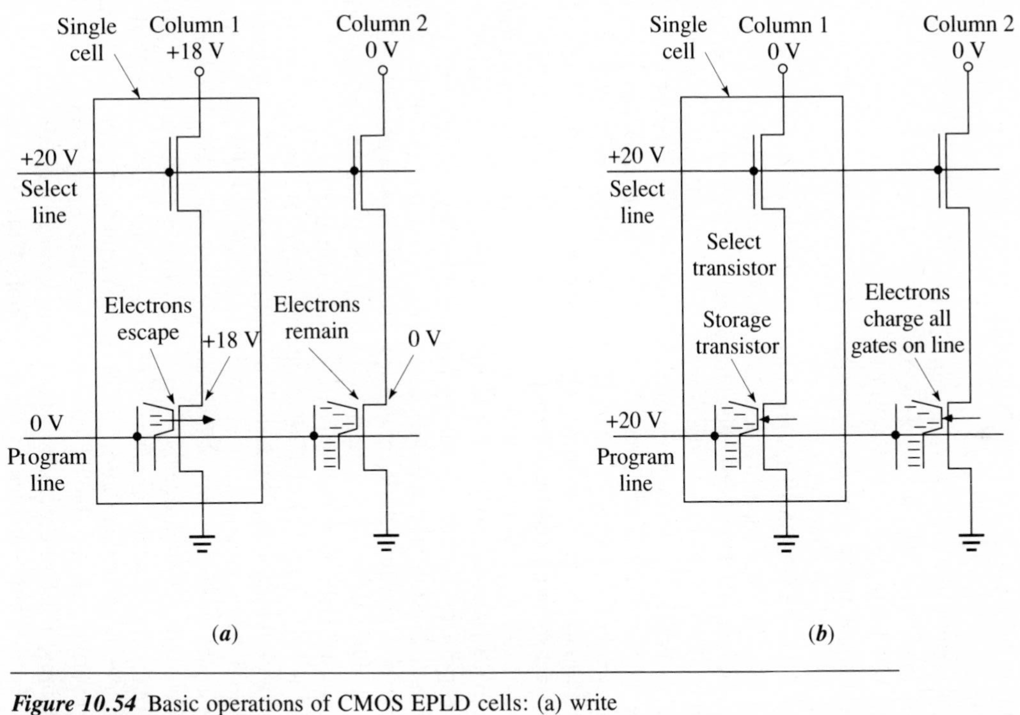

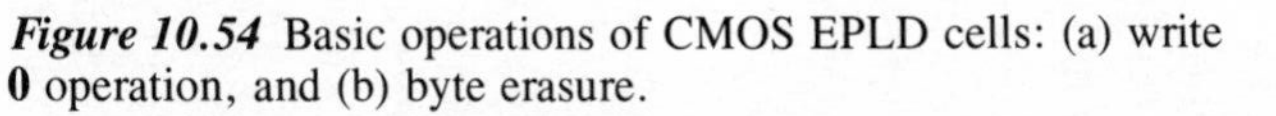
Figure 10.54 Basic operations of CMOS EPLD cells: (a) write **0** operation, and (b) byte erasure.

While fuses have the advantage of offering unlimited data retention, EPROM and E^2PROM cells retain data only a little more than ten years. Lack of permanence with these switches over the life of the device may be considered both an advantage and a disadvantage. Unlike the fuse-programmable PLDs, EPLDs can be erased by exposing them to a given amount of UV radiation. The energy of UV radiation applied for a certain period of time is sufficient to give the electrons in the floating-gate region enough energy to disperse and thus close the open cross-point connection. Obviously, this erasability becomes a drawback when the device must be operated in adverse environments.

While EPROM cells are smaller than E^2PROM cells, this fact does not affect significantly the size of the chip because the AND and OR arrays are small percentages of the IC. Furthermore, since EPLDs are not easy to reprogram, they must contain special test circuits that take up space. On the other hand, these circuits are not needed in E^2PLDs, which can be erased and reprogrammed easily and quickly.

The third type of CMOS PLDs are those that are *SRAM-based*, which are easily reprogrammed, but their volatility does not allow a logic pattern to be saved when the power is turned off. Therefore, the application must be stored in a separate device and must be *booted-up* every time the power is turned on. Finally, *fuse-based* CMOS PLDs are also available, but they are manufactured only by Harris Corp.

For top-performance circuits, bipolar PLAs still have the speed advantage with propagation delays of around 10 ns for TTL PLDs and 3 ns for ECL devices. The fastest CMOS PLDs have 15-ns delays. On the other hand, complete testing of bipolar PLAs at the factory is impossible without programming all the device fuses. With the introduction of EPLDs, this problem has been eliminated; the manufacturer can now guarantee device functionality by programming, testing, and then erasing all switching elements prior to shipment, resulting in higher quality and reliability.

CMOS and EPROM technologies also make much higher levels of integration and sophistication possible. Using much less stand-by power than bipolar parts, the former allows much more logic to be put on a single chip, thereby reducing the power consumption; the latter permits higher integration because its cell size is about one-tenth that of a fuse. Thus, most CMOS manufacturers are focusing on developing new devices with high complexity instead of replacing existing bipolar PLDs because of the cost and power advantages that CMOS provides as the chips become denser. Nevertheless, almost 90% of all PLDs today are still made with bipolar TTL technology, which is the dominant technology of the market-share leaders AMD/MMI, Signetics, Texas Instruments, and National Semiconductor, even though these companies have also been moving into CMOS PLDs recently. On the other hand, Intel Corp., and relatively small, start-up companies such as Altera Corp., Cypress Semiconductor Corp., Exel Microelectronics, Harris Corp., International CMOS Technology, and Lattice Semiconductor Corp. offer only CMOS parts.

Improvements in PLD Testing

In the past, bipolar PLDs earned a bad reputation; a variety of problems would surface because complete testing of the PLD at the factory was impossible without all the device fuses being programmed. Due to process variations, not all fuses in every device would program successfully, and not every device would function after programming because its internal circuitry had not been fully exercised.

Later, test fuses that are invisible to the user were incorporated. Still, manufacturers are unable to guarantee that a device will perform the user's logic function, leading to the failure of a small percentage of devices for program/verify, basic logic function or ac/dc performance. Most bipolar PLD manufacturers specify the failure rates that a user can expect for programming and basic function fall-out. They range from 1 to 3% for program/verify failures, and .1 to 3% for basic function and ac/dc parametric failures.

Considerable progress has been made in eliminating these problems, especially by vendors of erasable CMOS devices. While EPLDs allow greater testability than their bipolar counterparts, their long UV erase cycles or windowless OTP packaging limit the number of tests that manufacturers can perform. Most factory testing of EPLDs is done during wafer-sort testing of unpackaged dies at room temperature. If packaged as OTP EPLDs, the devices can no longer be erased, so that the actual memory cells to be used cannot be programmed to allow high-temperature testing. Instead, *phantom* arrays are used to correlate the function and performance of the rest of the chip.

E^2PLDs, on the other hand, provide instant erasability and unlimited reprogrammability for thorough testing at both wafer sort and final test. Thus, manufacturers of E^2PLDs guarantee device functionality by programming, testing and then erasing all fuses prior to shipment. In the same fashion, ac performance is verified by programming and testing worst-case patterns at the factory.

Bipolar PLD producers are also making good progress in this area. A new fuse technology, called *vertical fuse*, which relies on a simple transistor aligned vertically in the silicon as the programmable element instead of the metallic fuse, has been adopted to introduce the smallest and fastest array structure, increase programming yields, and eliminate reliability problems. The detailed electrical data, which can be obtained from each transistor prior to programming, and worst-case fusing paths, which are checked by programming test arrays of fuses, provide the highest possible fusing yield.

The vertical fuse cross-section and its equivalent circuit when unprogrammed and programmed are illustrated in Figure 10.55. It takes advantage of the properties of silicon and aluminum. The unprogrammed fuse is a three-layer device with a shallow layer of *n*-type Si on top, a *p*-type silicon layer in the middle, and a layer of *n*-type Si on the bottom, forming a pair of *pn* junctions connected back-to-back. Being excellent blocking elements with current leakages in the order of nanoamps, these act as an open circuit under normal conditions.

During programming, high-current conditions are induced by avalanche breakdown of the reverse-biased diode so that the cap of aluminum on top of the structure spikes through the shallow *n*-type layer, and the top diode is shorted out. After programming, the whole vertical fuse is set as a well-defined *pn* junction.

Programmable Logic vs. Gate Arrays

Another type of semicustom VLSI circuit, the *gate array*, has been around for a decade longer than PLDs, but early devices suffered from a number of limitations, including those of bipolar technology upon integration density and power. The gate array approach was boosted in the early eighties by the evolution of the CMOS technology to the point that allowed gate delays in the range of a few nanoseconds; com-

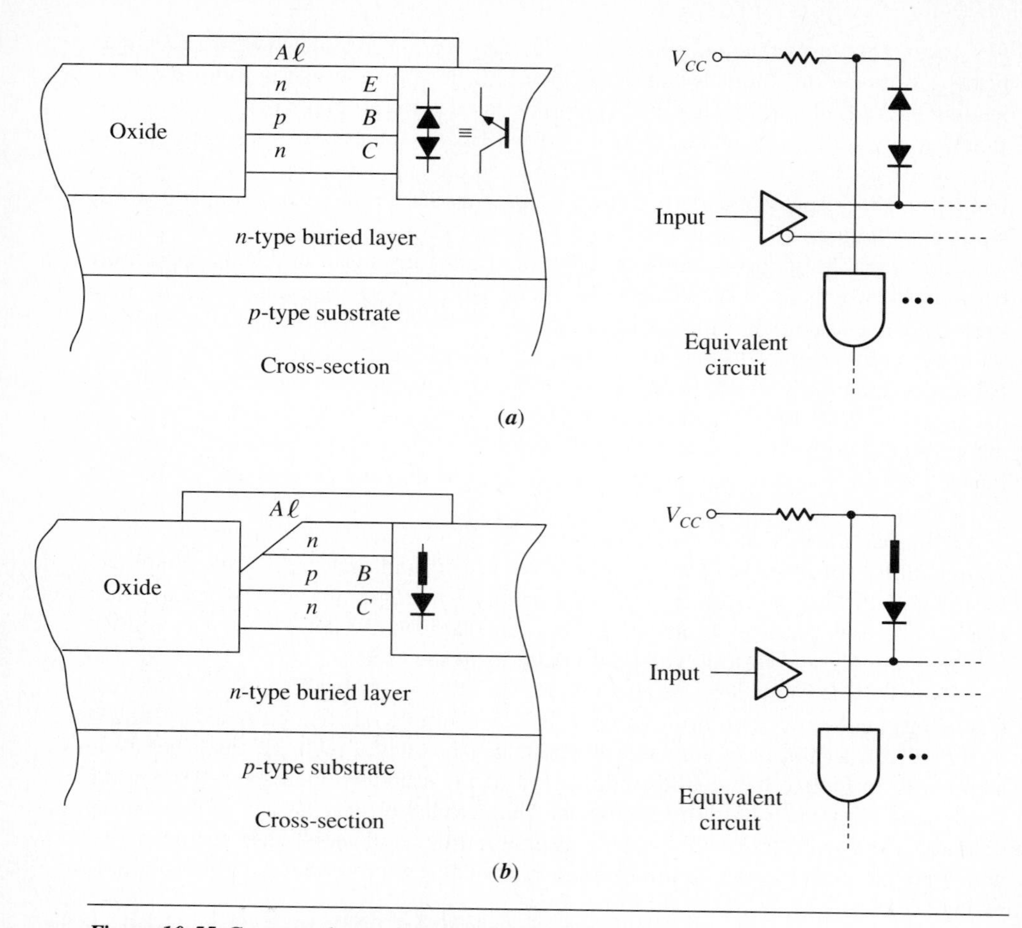

Figure 10.55 Cross section and equivalent circuit of the (a) unprogrammed vertical fuse, and (b) programmed vertical fuse.

bined with low power consumption, this led to the creation of arrays with complexities on the order of several thousand gates. Today, they are approximately a generation ahead of EPLDs in both complexity and cost per gate; gate arrays having over 20,000 gates are available. However, the gap is narrowing rapidly, and just as EPROMs have made ROMs virtually obsolete in the memory sector, the user-configurable EPLDs now are replacing masked gate arrays in the majority of general-purpose applications where they meet the performance and functional requirements.

Gate arrays are circuits that are made up of prediffused arrays of logic and are generally 70% to 80% complete in wafer form as stocked by the manufacturer. However, unlike programmable logic, they require additional masking and diffusion steps for completion and dedication that can only be performed at the factory.

The customer pays a *nonrecurring engineering* (*NRE*) charge for computer-aided mask design in order to dedicate each partially completed gate array. The additional diffusion steps, dicing, and packaging involve delays of weeks, depending on the complexity of the part and the manufacturer's work load.

By contrast, PLDs are totally completed at the factory and are sold as *off-the-shelf* devices that can be programmed in the field with the same equipment used to program PROMs but modified with plug-in *personality cards* to provide the required voltages for the devices to be programmed. Therefore, PLDs have very fast turn-around time, whereas it takes weeks for gate arrays to go from design to prototype. Also, the NRE charges for a large gate array, including design costs, masks, a possible design fix, and test programs, can be tens of thousands of dollars, while a PLD has no NRE changes at all.

Second-Generation CMOS PLDs

In order to address applications requiring high densities and provide an alternative to gate arrays, at least at lower gate counts, PLD manufacturers have recently been looking to new architectures that could combine user programmability with higher density. Some suppliers have elected to concentrate on incremental improvements of existing designs; others are actively engaged in increasing the equivalent gate density or adding functions. Such companies are expanding their user-programmable arrays to useful gate densities of 2000 gates or more, thereby entering the density territory that had once been the province of factory-programmed gate arrays.

Makers of CMOS PLDs have lately introduced a second generation of products, as listed in Table 10.4 and briefly explained below, that break records for density, gate utilization, flexibility, and speed by using a variety of architectural and process innovations. However, by moving out from the conventional PLD architectures that still dominate at low densities, the manufacturers must now invest substantially in software to ease the transition to higher densities and new architectures.

The Multiple-Array Matrix

The most recent approach, using an architecture developed by Altera Corp., is the *multiple-array matrix* (*MAX*), manufactured by Cypress Semiconductor Corp. employing a .8-μm CMOS process. This radical new architecture, with internal clock rates as high as 60 MHz and arrays with up to 5000 gates, consists of multiple *logic array blocks*, or *LABs*, linked by a dedicated programmable interconnect array. The

Table 10.4 Second-Generation PLD Architectures

Manufacturer	Architecture	Technology
Altera Corp.	MAX	EPLD
Exel Microelectronics, Inc.	ERASIC	E^2PLD
ICT, Inc.	PEEL	E^2PLD
Lattice Semiconductor Corp.	GAL	E^2PLD
Signetics Corp.	PML	Bipolar
Xilinx, Inc.	LCA	SRAM

LABs consist of three elements: the macrocell array, the *logic expander*, and the I/O block, as shown in Figure 10.56. The various members of the MAX family incorporate from one to eight LABs with macrocells per LAB ranging from sixteen to thirty-two.

The logic diagram of a macrocell appears in Figure 10.57. Each macrocell contains three p-terms and a programmable flip-flop in addition to multiple control inputs, so that the cell can be controlled independently of the other cells. Each flip-flop can be configured as an SR latch or as a D, T or JKFF, or bypassed entirely for purely combinational functions. Furthermore, each of them has both asynchronous preset

Figure 10.56 The MAX logic array block. (Reprinted with permission from Altera Corp.)

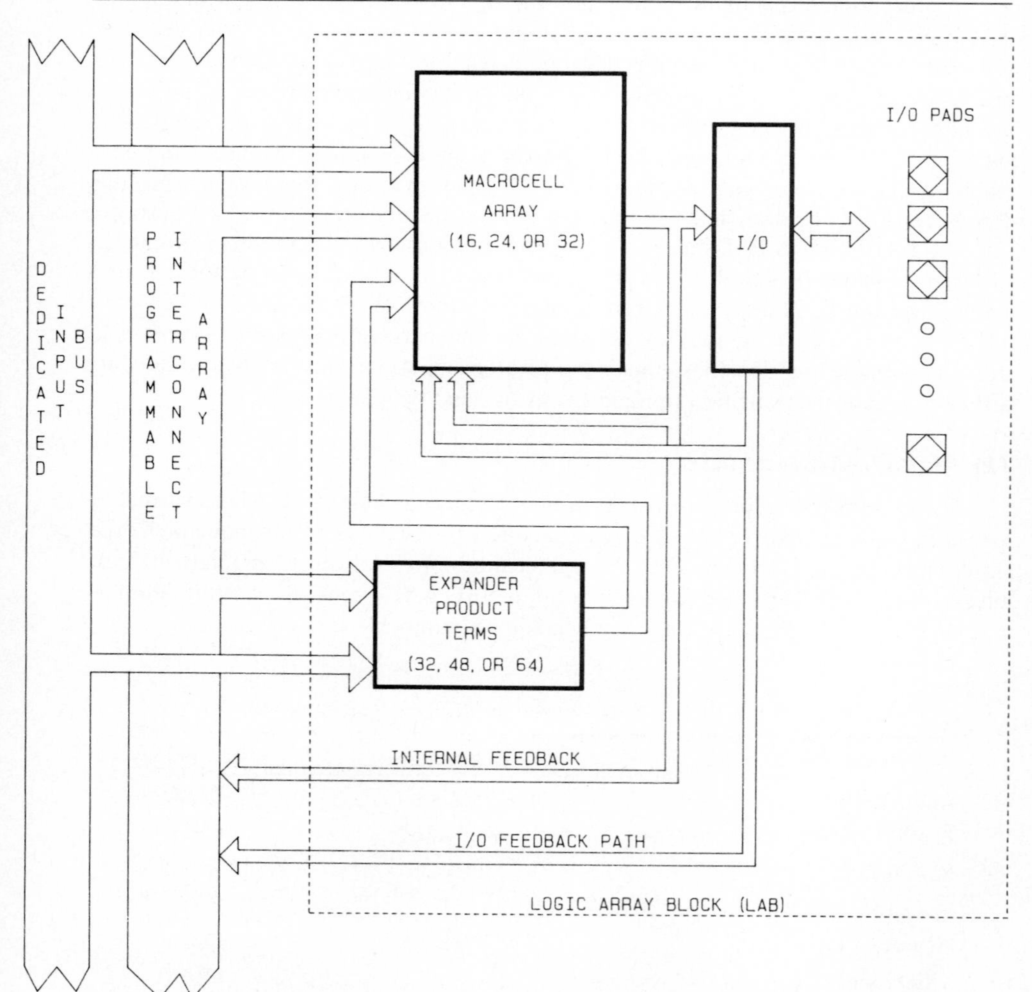

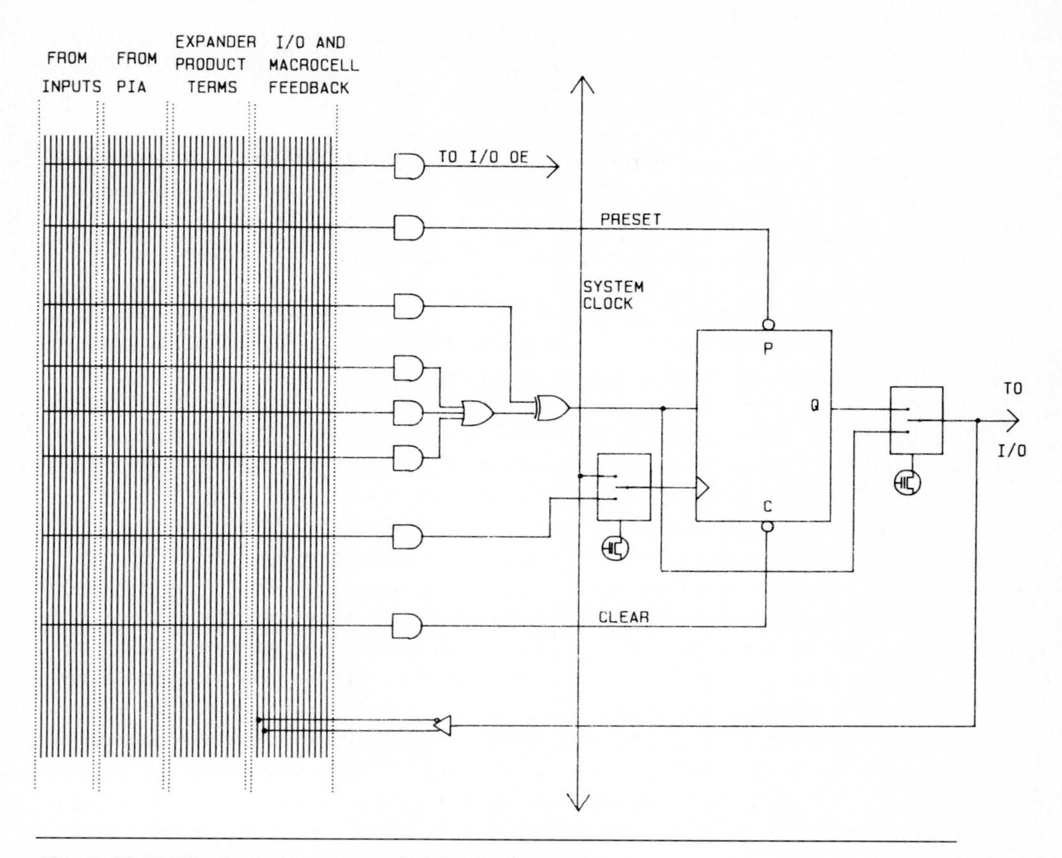

Figure 10.57 The logic diagram of the MAX macrocell.
(Reprinted with permission from Altera Corp.)

and clear inputs to allow loading of counters or shift registers. If the logic function to be implemented requires many p-terms, expander blocks can be distributed to any macrocell requiring more than three product terms.

A flip-flop can be programmed to operate as either a flow-through latch or an edge-triggered storage element. The former provides minimum propagation delays for high-speed applications such as chip-select decoding; the latter guarantees glitch-free outputs for sequential applications.

To address two problems common to most PLD structures, that is, undesired skews among logic signals due to the varying length of interconnections between gates as well as the often long delay paths, the MAX devices incorporate a special *programmable interconnect array*, or *PIA*, that links each LAB to another, thereby providing a single, uniform delay between two points to eliminate glitches in internal or external logic.

The first device in the family, the EPM5032, introduced in 1988, is a 32-macrocell device containing 240 product terms and the equivalent of 1200 gates in a 28-pin DIP. It is intended for sequential functions, featuring eight dedicated input signals and 16 dual feedback I/O pins with a total of 32 user-definable flip-flops.

The ERASIC

A second approach is the use of a *folded-array multiple-logic* structure, favored by companies such as Signetics Corp., with its NAND-based bipolar *programmable macro logic* (*PML*), discussed below in more detail, and Exel Microelectronics Corp. with its NOR-based CMOS *erasable reprogrammable application-specific IC* (*ERASIC*).

Besides their limited densities, another drawback to traditional PAL and PLA structures is that they essentially implement only two-level logic circuits, thereby reducing the chip count only as far as the glue logic of SSI and MSI ICs is concerned. In order to realize an LSI or VLSI system function, it becomes necessary to use multiple PLDs. Both Exel and Signetics have introduced PLDs that abandon the AND/OR concept and instead use folded logic planes. This architecture comprises a single NOR plane in Exel's ERASIC, or a single NAND plane in the case of Signetics' PML that feeds back to its inputs. As any Boolean expression can be implemented with NOR or NAND gates, the respective devices are *functionally complete*. The advantage of this approach lies in its avoidance of the two-level logic restriction of conventional PLDs. For instance, the ERASIC offers the equivalent of forty-two levels of internal logic. Since the signals remain on-chip, the device does not compromise performance and still provides increased functionality, allowing it to implement complex Boolean expressions.

For building multilevel logic, the foldback architecture provides an additional set of gates that have their inputs folded back to a switch matrix with bipolar or E^2 fuses to provide the user with signal paths through exactly the number of gates the expressions require. Once the desired function has been defined, it can be directed through the output macrocell to a pin. It can also be routed through the same macrocell back into the circuit, where it may be combined with other functions or input signals to realize even higher level expressions.

Foldback PLDs also have an advantage over conventional architectures for a single-level logic. For a design requiring only a single AND gate, for instance, an OR gate and the other AND gates connected to its inputs are inevitably wasted with the conventional architecture. With the foldback design, on the other hand, the function is not trapped and can go directly to an output pin.

The logic diagram of the XL78C800 is shown in Figure 10.58. Packed into a PAL-compatible 24-pin DIP device, it is equivalent to 800 gates, and provides up to 42 levels of NOR gates. The chip's I/O structure consists of 10 dedicated inputs as well as 10 I/O macrocells that include buried registers in the form of positive-edge-triggered JKFFs, whose feedback paths are independent of the I/O pin. The flip-flops can emulate both D- and T-type flip-flops. The logic diagram of a macrocell is depicted in Figure 10.59.

The Programmable Electrically Erasable Logic

International CMOS Technology, Inc. (ICT) opted to extend traditional PALs to higher functional densities by modifying them to increase gate utilization to 60% to 80% from the present 30% to 50%. Until recently, their main emphasis was the development of direct replacements for current 20- and 24-pin devices with their *programmable electrically erasable logic* (*PEEL*) devices. In 1987, however, they went into production with what was the first in a family the company called *superset replacement* PEEL devices that incorporated not only the functions of traditional PAL devices but a range of new functions as well.

Depending on the particular IC, between 18 and 22 inputs and 8 to 10 macrocells are available. The key to the flexibility of the approach is that each logical sum on the device is directly associated with a multifunction I/O macrocell and its I/O pin. Each of these macrocells contains a DFF, an output multiplexer, and a feedback multiplexer as well as four E^2PROM memory cells that can be programmed to configure each macrocell in 12 different ways, as diagrammed in Figure 10.60. This enables the user to control the output polarity, feedback path, and the output type—registered or combinational, dedicated input or output, or bidirectional I/O—so that one can emulate over 30 conventional PLDs with a single PEEL device.

The Generic Array Logic

Lattice Semiconductor Corp. turned completely away from the simpler but more limited PAL structure to the more flexible PLA by adding features and options to increase the functionality and flexibility of this architecture. In the *generic array logic* (*GAL*) series, each programmable output macrocell gives the designer five configuration options: output polarity, feedback, combinational logic, registered outputs, and input selection. The outputs can be connected to one Output-Enable signal or separate product terms can provide individual enabling controls. In addition to emulating many PAL series devices, the GAL offers some combinations of options not previously available.

The GAL16V8 and GAL20V8, which were originally introduced in 1985, are pin-for-pin replacements for all 20- and 24-pin bipolar PAL devices, respectively, and offer 15-ns speeds over the entire temperature range at half-power levels of less than 90 mA, with 24 mA output drives. The major drawback of these E^2PLDs is that, uncharacteristically, they are not on-board programmable because they require *supervoltages*, that is, programming voltages identical to those of fusible-link parts. The logic diagrams of the GAL20V8 and one of its output macrocells are depicted in Figures 10.61 and 10.62, respectively.

The Programmable Macro Logic

An alternative foldback approach, undertaken by Signetics Corp., is the PML architecture. It combines a NAND-based network to allow the direct interconnection of any number of logic nodes within the single matrix, with macrocells formed and interconnected to the I/O structure.

Instead of NOR gates and an E^2 matrix, these parts use fuse links and NAND gates, which are the fastest gates that can be built in bipolar technology. Thus, the parts cannot be reprogrammed. However, due to its bipolar construction, it can pass a signal through a single level of logic in 18 ns compared to the ERASIC's 35-ns propagation delay. Internal foldback terms cost an additional 8-ns delay in the PML; in the ERASIC they add up to 20 ns. On the other hand, the PML draws 250 mA on the average, hence dissipating 1.25 W typically. In contrast, the CMOS ERASIC is

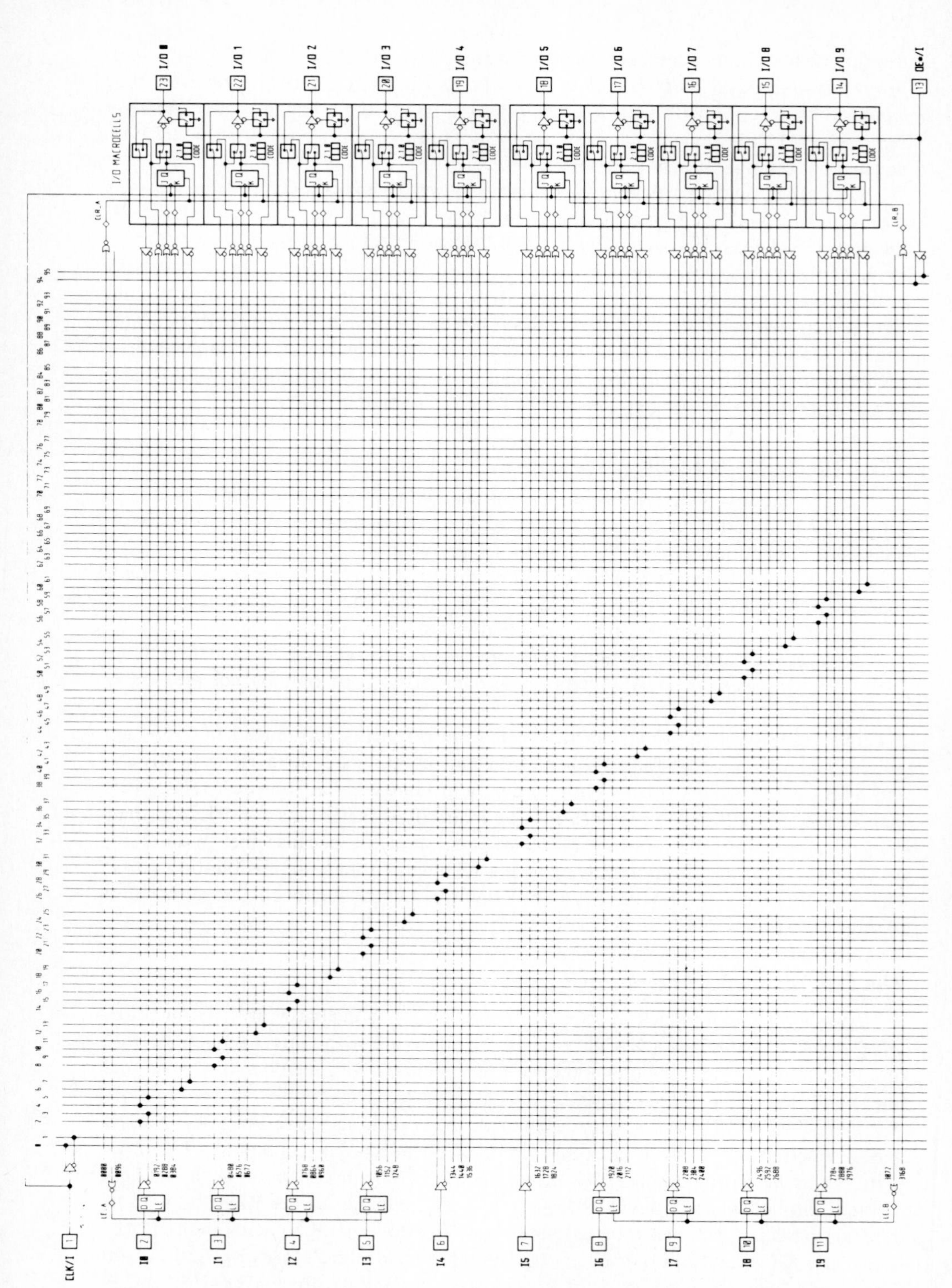
I/O MACROCELLS
23 I/O 0
22 I/O 1
21 I/O 2
20 I/O 3
19 I/O 4
18 I/O 5
17 I/O 6
16 I/O 7
15 I/O 8
14 I/O 9
13 OE•/I
CLK/I 1
I0 2
I1 3
I2 4
I3 5
I4 6
I5 7
I6 8
I7 9
I8 10
I9 11

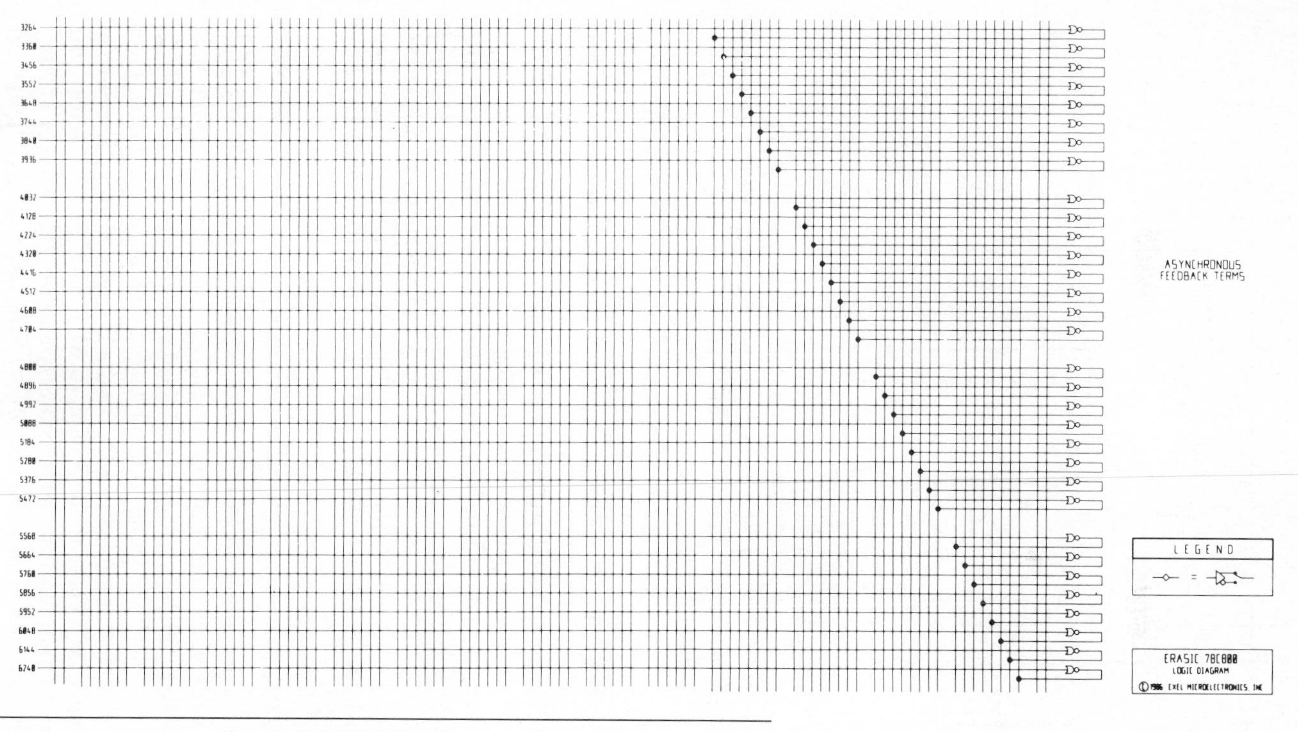

Figure 10.58 The logic diagram of the ERASIC XL78C800 with its folded-NOR array. (Courtesy EXEL Microelectronics, San Jose, California)

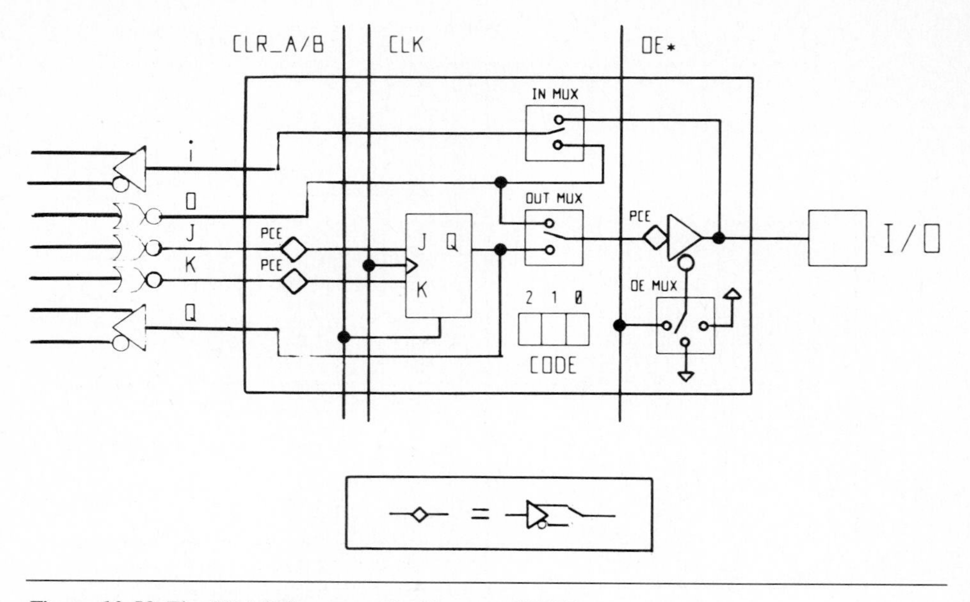

Figure 10.59 The ERASIC macrocell. (Courtesy EXEL Microelectronics, San Jose, California)

specified as using a maximum active power supply current of 35 mA at a 10 MHz clock rate.

The first product of the Signetics PML family of PLDs, introduced in late 1987, is the bipolar PLHS501, with the equivalent of 1300 gates in a 52-pin J-leaded chip-carrier package. Its logic diagram is shown in Figure 10.63. Larger than the ERASIC, the 32 × 72 × 24 folded-NAND array provides 72 levels of logic and has 16 buried registers with 24 dedicated inputs and 24 outputs, 8 of which function as I/O buffers. The CMOS version of the architecture has been in development and will extend densities up to at least 10,000 gates. A registered version, the PLHS502 with 16 buried registers, is also available.

The Logic-Cell Array

The current pace-setter in high-density PLDs is Xilinx, Inc. In 1986, the company introduced a family of the so-called *logic-cell array* (*LCA*) devices that use an uncommitted architecture with no prefixed logic structures such as macrocells or registers and are capable of densities of up to 9000 gates using a 1.2-μm double-layer metal CMOS process. Unlike traditional PLDs, they are based on an architecture in which the internal logic, as illustrated in Figure 10.64 for the first LCA part, the XC2064, is essentially an array of uncommitted *configurable logic blocks* (*CLBs*) that are surrounded by uncommitted programmable *input/output blocks* (*IOBs*), all of which are linked by programmable interconnect lines controlled by a 12,038-bit configuration program stored in an internally distributed array of static memory cells.

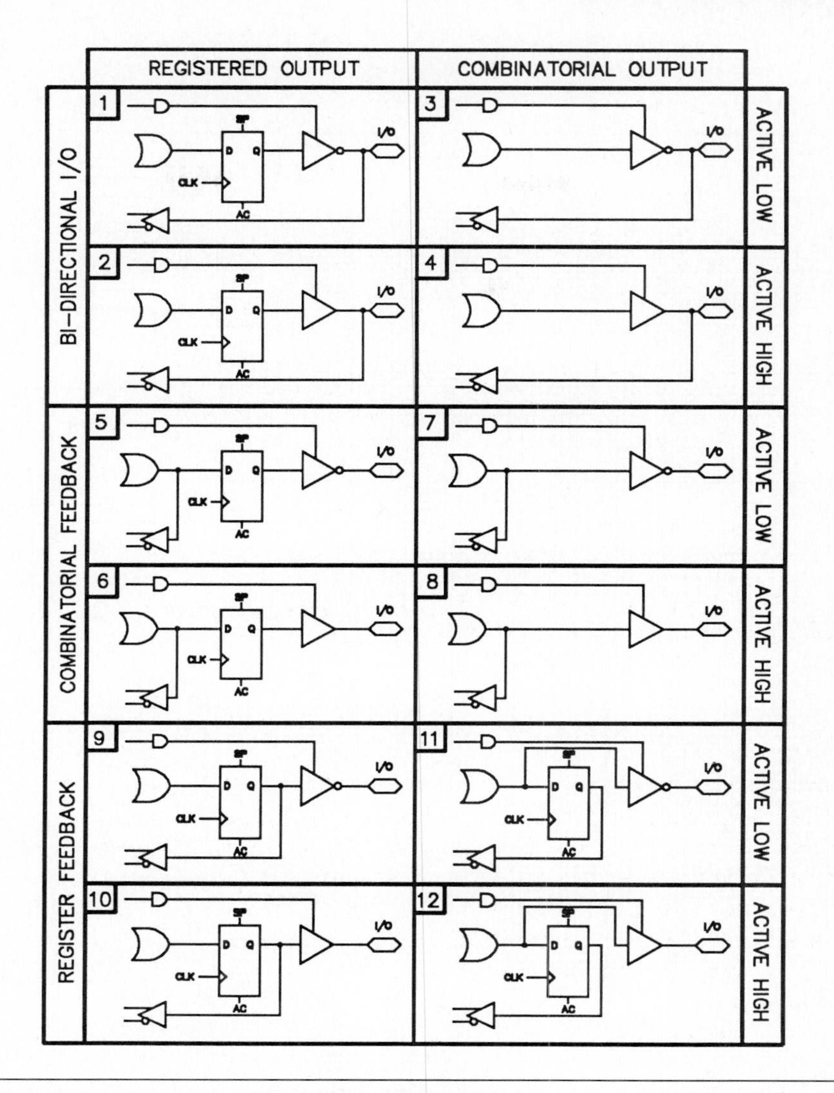

Figure 10.60 Twelve configurations of the PEEL I/O macrocell. (Reprinted with permission from International CMOS Technology, Inc.)

The 58 IOBs provide an interface between the array and the package pins by being individually configured as inputs, outputs, or bidirectional I/Os. Each also includes a latch for holding input values. The 64 CLBs perform user-specified logic functions, and the interconnects are programmed to form networks that carry signals among blocks, analogous to the printed circuit board traces that connect ICs. The logic diagrams of the IOB and the CLB are shown in Figure 10.65. The XC2064

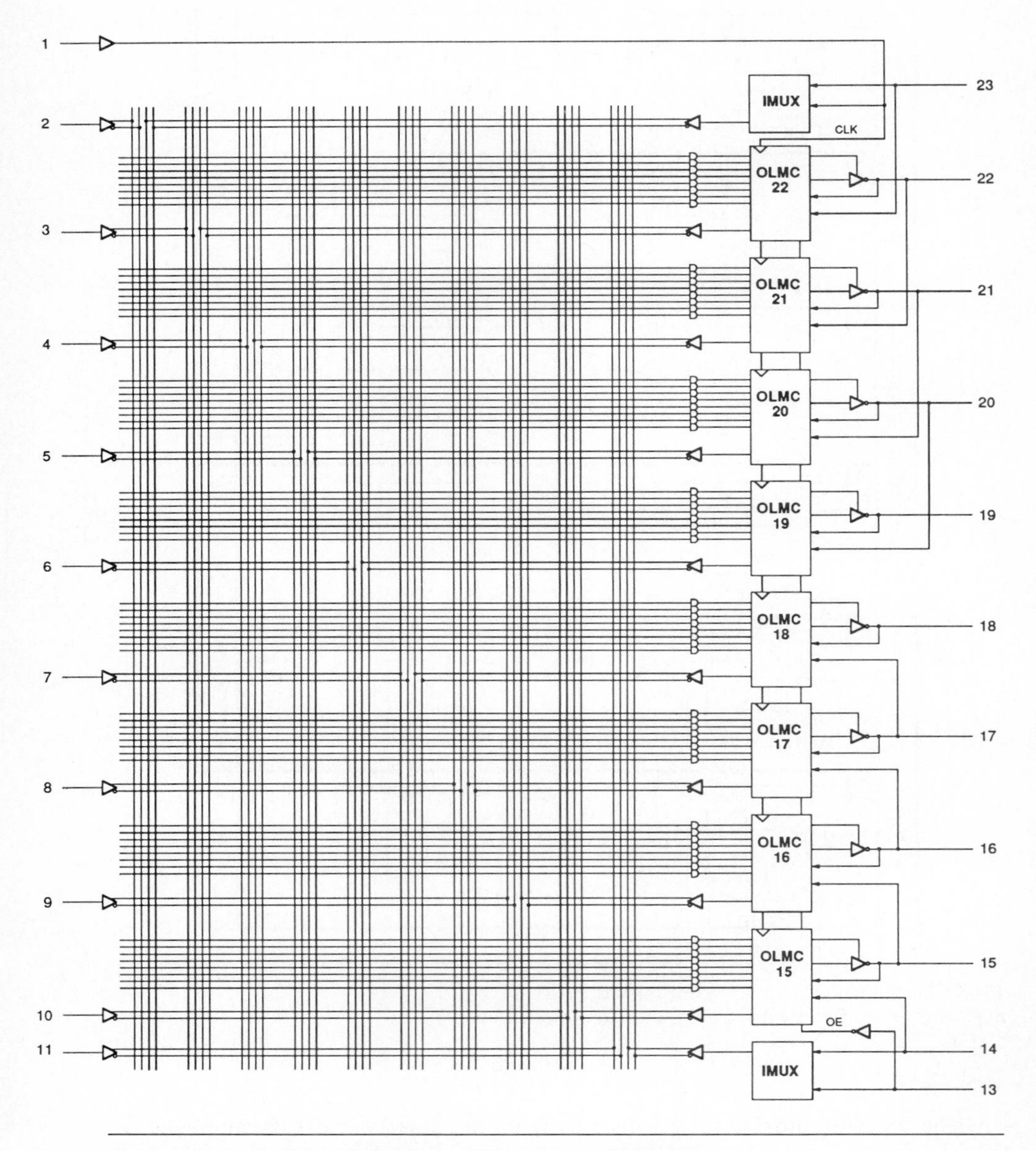

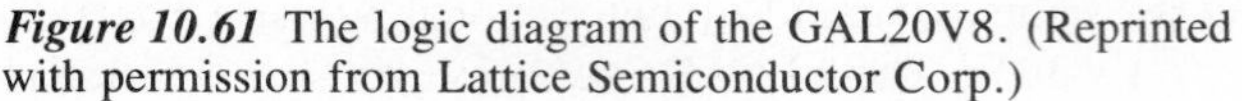
Figure 10.61 The logic diagram of the GAL20V8. (Reprinted with permission from Lattice Semiconductor Corp.)

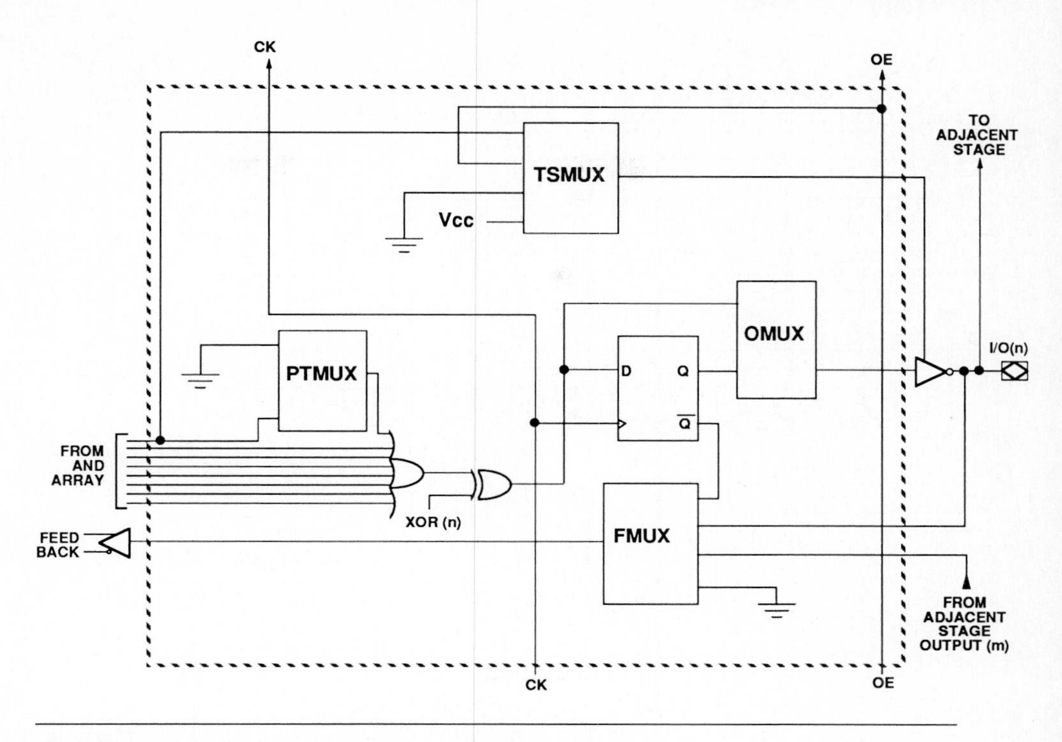

Figure 10.62 The GAL macrocell. (Reprinted with permission from Lattice Semiconductor Corp.)

offers the logical equivalent of 1200 usable gates, with the typical clock rate specified at 30 MHz. Since it is SRAM-based, it must be dynamically configured on power-up from a nonvolatile source.

The company is trying to get users to stop associating LCAs with conventional PLDs because, generally, the latter are not thought of as alternatives to gate arrays. Hence, Xilinx is turning from the proprietary name LCA to the more descriptive *programmable gate array*. With its 9000 equivalent gates, the XCA3090, introduced in late 1987, raises PLDs to a functional level that lets them go head-to-head with large gate arrays while still offering reprogrammability and *off-the-shelf* supply.

Summary

Semiconductor memories are arranged in arrays of cells, and each cell stores one bit of information. The time required to retrieve data in random access memories is independent of its location. RAM can be classified as either read/write or read-only. A characteristic of R/W memories is volatility; that is, removal of power results in the loss of data. CMOS R/W memories are divided into two groups: dynamic and static. Dynamic memory cells consist of a storage capacitor and one transistor. Therefore,

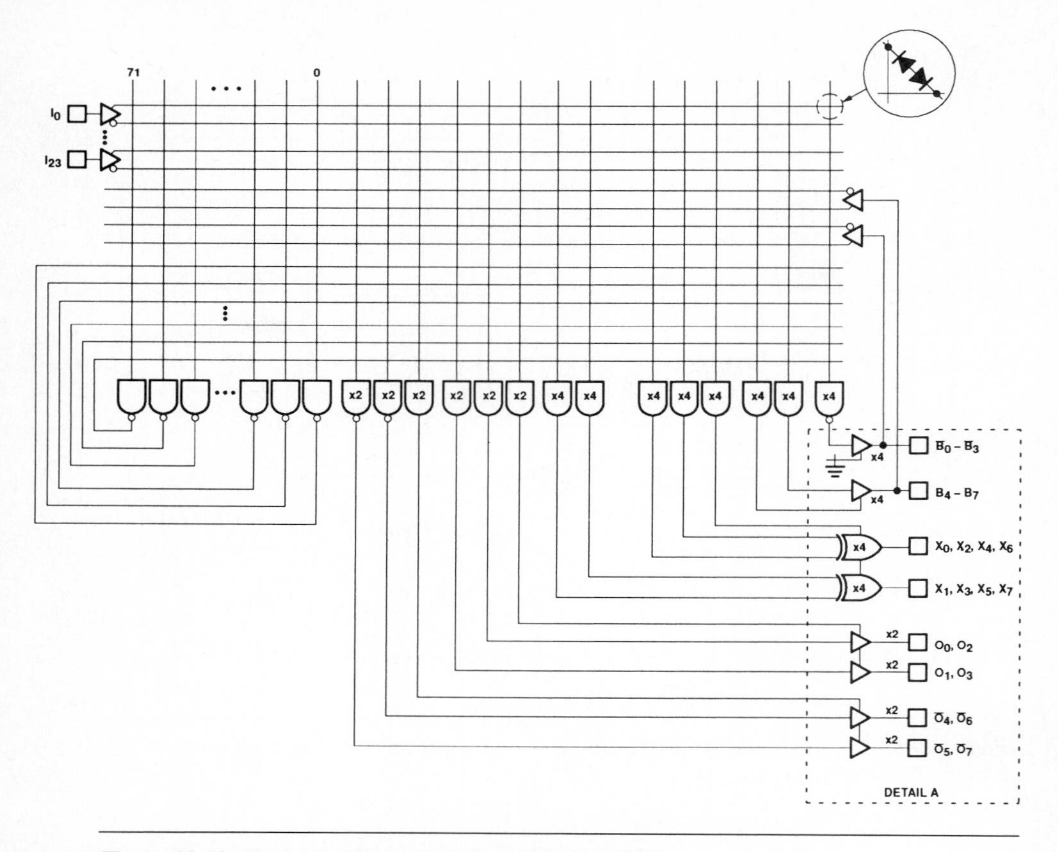

Figure 10.63 The logic diagram of the PLHS501 with its folded-NAND array. (Courtesy of the copyright owner, North American Philips Corp.)

dynamic memories lend themselves to very high circuit density. The disadvantage of dynamic memories is that the data stored as a voltage on a very small capacitor needs to be refreshed frequently because the charge leaks off with time. Static memory cells are flip-flops that require at least four transistors per cell. They also require more power and cost more per bit than dynamic memories, but they are very fast and do not require any extra circuitry for refresh.

Read-only memories are nonvolatile memories. A ROM is said to be mask-programmed when it is programmed by the manufacturer during fabrication. A more flexible approach provides the user with the means to program the ROM. This is achieved by selectively blowing fusible links in order to store data permanently; thus, the procedure is irreversible. EPROMs, on the other hand, can be reprogrammed many times; any stored data can be erased by exciting the trapped electrons on a

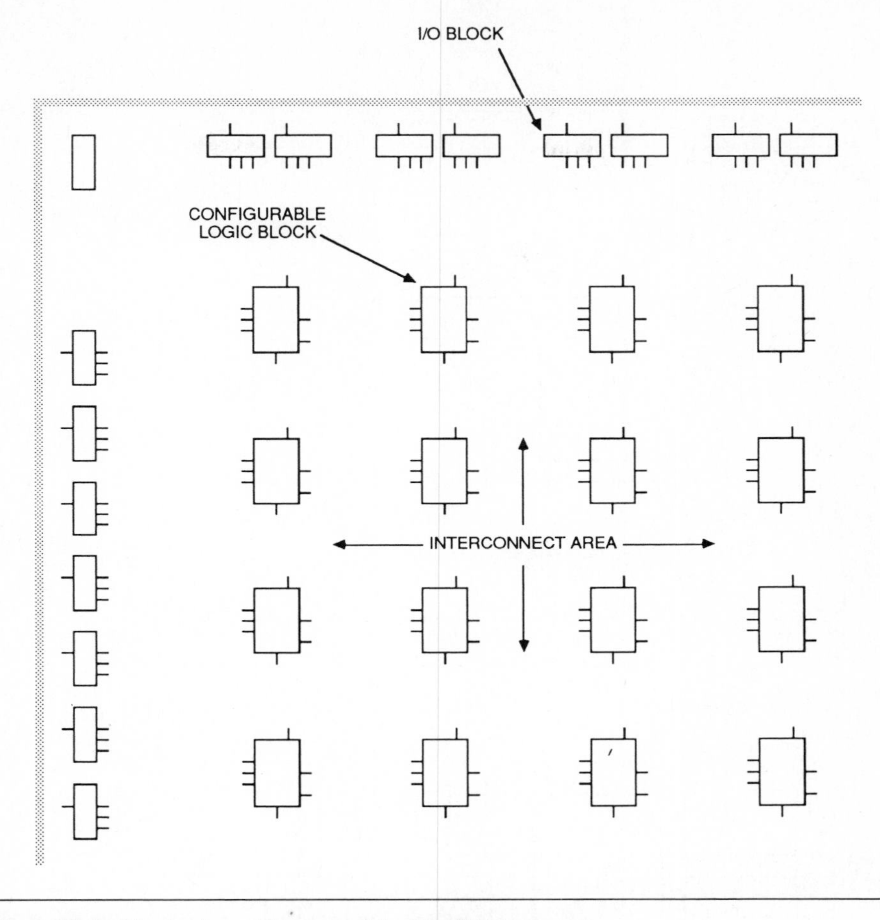

Figure 10.64 The basic architecture of the XC2064 logic-cell array. (Courtesy Xilinx, Inc.)

floating gate to an energy level at which they can escape from the gate. This is accomplished by exposing the chip to intense ultraviolet radiation. The most versatile ROM is the E^2PROM in which the erasure and reprogramming are done in the circuit by applying electrical pulses.

PLAs were developed as an extension of bipolar PROMs. In the latter, only the OR array is programmable, with the input address decoder acting as a fixed AND plane. In the PLA, on the other hand, the address decoder is replaced by an AND array, and both AND and OR matrices are programmable. The PAL has a programmable AND matrix and a fixed OR array. In recent years, various features have been added, especially to high-density CMOS PLDs, to make their highly structured architecture more flexible.

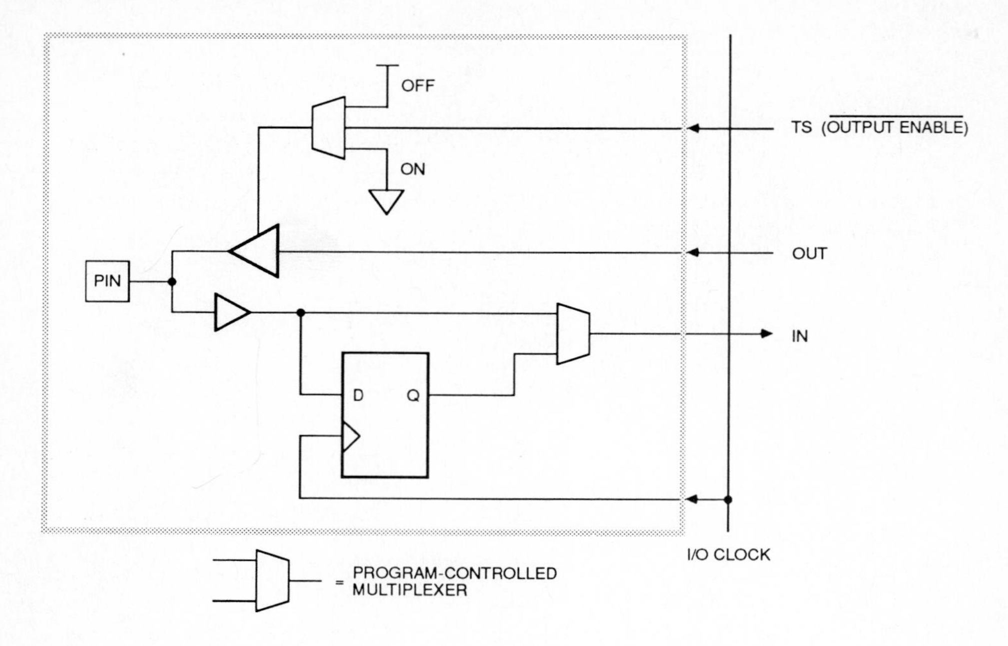

(*a*)

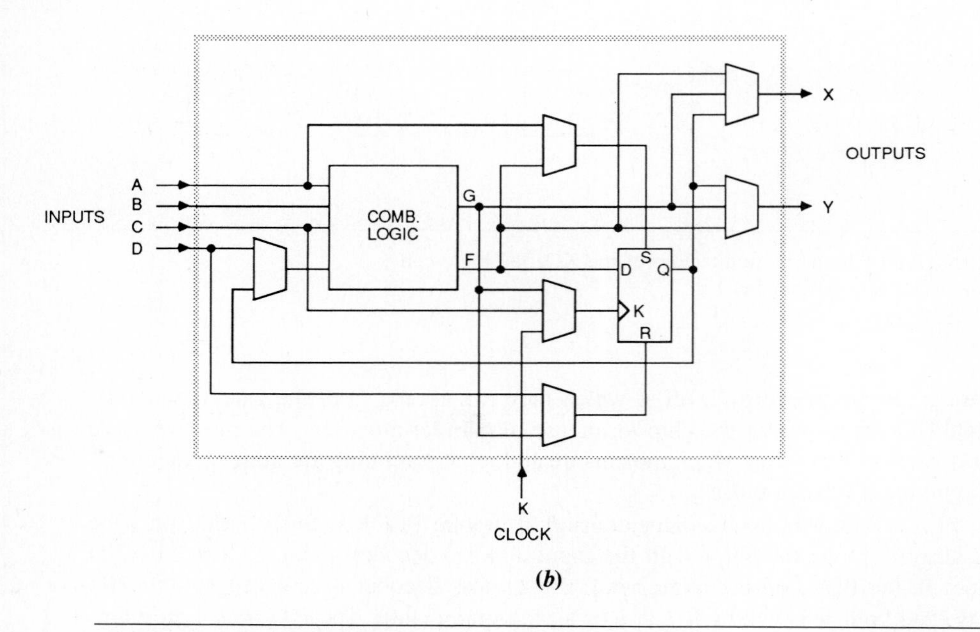

(*b*)

Figure 10.65 Logic diagrams of the (a) I/O block, and (b) configurable logic block for the XC2064. (Courtesy Xilinx, Inc.)

References

1. Editorial Staff. "Thick Oxide Beats Thin Film in Building Big E²PROMs." *Electronics*, 12 May 1986.
2. H. A. R. Wegener. "Endurance of Xicor E²PROMs and NOVRAMs." *Xicor Databook.* Xicor, Milpitas, CA: 1985.
3. R. Palm. "Determining System Reliability From E²PROM Endurance Data." *Xicor Databook.* Xicor, Milpitas, CA: 1985.
4. J. M. Caywood and B. L. Prickett. "Radiation-Induced Soft Errors and Floating Gate Memories." *Xicor Databook.* Xicor, Milpitas, CA: 1985.
5. H. A. R. Wegener. "Endurance Model for Textured-Poly Floating Gate Memories." *Xicor Databook.* Xicor, Milpitas, CA: 1985.
6. A. S. Sedra and K. C. Smith. *Microelectronic Circuits.* Holt, Rinehart & Winston, New York: 1982.
7. R. Proebsting. "Dynamic MOS RAMs, Technology." *Memory Databook and Designers Guide.* MOSTEK, Carrollton, TX: 1979.
8. D. Coker. "An In-Depth Look at MOSTEK's High Performance MK4027." *Memory Databook and Designers Guide.* MOSTEK, Carrollton, TX: 1979.
9. D. Coker. "16K—The New Generation Dynamic RAM." *Memory Databook and Designers Guide.* MOSTEK, Carrollton, TX: 1979.
10. P. K. Chatterjee et al. "A Survey of High-Density Dynamic RAM Cell Concepts." *IEEE Trans. Electron Devices*, vol. ED-26 (June 1979): 827–39.
11. R. R. Troutman. "VLSI Limitations from Drain-Induced Barrier Lowering." *IEEE Trans. Electron Devices*, vol. ED-26 (April 1979): 461–69.
12. T. C. May and M. H. Woods. "Alpha-Particle-Induced Soft Errors in Dynamic Memories." *IEEE Trans. Electron Devices*, vol. ED-26 (January 1979): 2–9.
13. *IEEE Journal of Solid-State Circuits.* The October issue of each year since 1970 has been devoted to semiconductor memory and logic circuits.

PROBLEMS

For Problems 1–5, assume 1-T cell DRAMS with a cell-storage capacitance of 50 fF, $V_{CC} = 5$ V and $V_T = 1$ V unless otherwise stated.

10.1 Suppose that the word line is raised from 0 to 6 V to maximize the logical **1** level on the storage capacitor.

a. What value of gain constant k_n is required for $C = 50$ fF to be charged to 4 V in 10 ns?

b. Verify your answer in (*a*) with SPICE.

10.2 In a certain memory with a 64 × 64 array, the bit-line capacitance is 40 fF, and the input capacitance of the sense amplifier is .5 pF. If the minimum allowable signal is 6 V on the cell capacitance, what will be the signal available at the sense-amplifier input after the access transistor is turned on?

10.3 For a certain DRAM, the sense amplifier allows the stored charge to decay to $1/e$ of its original value before refresh is required. The maximum permitted refresh interval is 2 ms. Determine the value of the smallest equivalent resistor that can shunt the storage capacitor. What is the worst-case leakage current that the node can tolerate?

10.4 A DRAM operates with a minimum refresh time of 5 ms. The storage capacitor in each cell is 5 μm^2 and is fully charged at V_{CC}.

a. Calculate the number of electrons stored in each cell.
b. Estimate the worst-case leakage current that can be tolerated.

10.5 What are the stored charge and number of electrons on a fully-charged storage capacitor with an area of 10 μm^2, and a dielectric of 250Å-thick SiO_2?

10.6 Raw endurance data in thousands of write cycles for a certain 20-piece lot of E^2PROM devices is given as follows:

450, 315, 940, 365, 680, 572, 620, 297, 992, 800, 400, 1400, 820, 355, 952, 410, 650, 620, 450, 510

a. Use the data to generate the corresponding extreme value distribution on log paper.
b. What is the mode of the distribution, that is, the predicted most-probable endurance of devices in this lot?
c. Extrapolate the graph to determine the percentage of all devices that will fail before 200,000 cycles.

10.7 Consider the following requirements for a system utilizing the E^2PROMs of Problem 10.6:

1. The life expectancy of the system is five years.
2. Over the lifetime of the system, each device will perform nonvolatile writes roughly uniformly and will experience 250,000 write cycles.

Determine the reliability of the system in terms of the percentage of failures per 1000 hours.

11. NOISE IN DIGITAL CIRCUITS

As the overall system size decreases, signal lines are brought in closer proximity to one another as well as to power supply and ground lines, hindering the ability of the system to provide good low-impedance grounds. An advance in switching speeds has been coupled with new packaging restrictions. Since the cross-coupling between adjacent signal, power supply, and ground lines is proportional to the speed of the transitions on these lines, and since an order of magnitude of speed improvement frequently accompanies the denser packaging schemes, it is not unusual that *crosstalk* experienced in digital systems is becoming increasingly greater. Furthermore, in some circumstances, digital circuits are continually being used in close proximity to non-digital equipment, being exposed to *radio frequency (RF) interference* in the surrounding atmosphere as well as on interconnecting ground and power supply lines. In view of these considerations, it is appropriate to discuss some of the factors contributing to noise generation in digital circuits.

There are always three elements involved in a noise problem: a *noise source*, such as line transients, relays, and magnetic fields; a *coupling medium*, for example, capacitance or mutual inductance; and a *receiver*, that is, a circuit susceptible to noise. Different noise problems require different solutions. The types of noise encountered in digital systems can be classified as follows:

1. *Transmission line* reflections from unterminated transmission lines that cause ringing and overshoot
2. *Crosstalk*, that is, the coupling of signals from one line in the system to another through mutual inductance and capacitance between them
3. Supply-current spikes caused by switching several loads simultaneously
4. *Electromagnetic interference* (*EMI*), namely electrical noise from external circuits as *radiated* into the system or coupled in from other electrical apparatus in the system or its vicinity, such as switching power supplies, radio transmitters, motor brushes, circuit breakers, arcing relay contacts, fluorescent lights, switch contacts, and electrostatic discharge
5. Power-line noise that is coupled through ac or dc power distribution system

Intrinsic noise, such as thermally-generated noise and shot noise, originating within the devices that constitute a circuit, will not be treated in this chapter primarily because it does not cause a problem for digital circuits due to their built-in noise margins.

11.1 High-Speed Logic System Interconnections

Interconnections from one board to another are important to preserve the best possible system performance. When designing system interconnections with high-speed logic families having edge speeds around 1 ns, the following parameters must be taken into account:

1. Attenuation of the line
2. Propagation delay per unit length of line
3. Reflections due to mismatched impedance between the line and line termination
4. Crosstalk between lines

The attenuation is a characteristic of the line that increases with frequency due to higher impedance in the line. It first appears as a degradation in edge rate, then as a loss of signal amplitude on long lines at high frequencies. Fortunately, for the great majority of interconnections in digital systems, the impedance of conductors is much less than the input and output impedance of the circuits. Similarly, the insulating materials have very good dielectric properties. These circumstances permit such factors as attenuation, phase distortion, and bandwidth limitations to be ignored. With these simplifications, interconnections can be dealt with in terms of the characteristic impedance and propagation delay only.

The propagation delay of the line is critical because unequal delays in parallel lines give rise to timing errors. Furthermore, the total delay time on long lines will have a forbidding effect on the system speed.

Reflections due to mismatched lines also result in loss of noise immunity. Successful termination of a line depends on the uniformity of the line impedance. A *coaxial cable*, for example, is easier to terminate than an open wire because of its constant impedance.

Crosstalk is the coupling of a signal from one signal path to a nearby path. A coupled pulse in the direction of undershoot leads to a degeneration of noise immunity. This type of noise will be treated in detail in Section 11.4.

Two general types of lines are mainly used between logic cards, card panels, or for other system interconnections. They are *single-ended lines* and *differential twisted-pair lines*.

Single-Ended Lines

Single-ended lines are interconnections such as a coaxial cable or other single-path transmission lines as opposed to a twisted pair of lines over which a differential signal is sent. The well-defined and uniform characteristic impedance of the coaxial cable allows easy matching of the transmission line, and the internal ground shield reduces the crosstalk. Moreover, low attenuation at high frequencies lets the cable transmit the edge rates associated with high-speed logic families.

Multiple-conductor *parallel-wire cables* are not used in high-speed logic families unless individual shields on each wire are employed because of the crosstalk that is due to the capacitive and inductive coupling of signals between parallel lines. Such a cable is also susceptible to external signals coupling to the entire cable.

Systems requiring a large number of board-to-board interconnections may utilize a multiconductor *ribbon cable*, whose cross-section is shown in Figure 11.1. The side-by-side arrangement of signal lines establishes a defined characteristic impedance because of the presence of alternate ground wires.

Differential Twisted-Pair Lines

Twisted-pair lines that are differentially driven into a line receiver furnish the maximum noise immunity. Any noise coupled into a twisted-pair appears equally on both wires. Since the receiver senses only the differential voltage between the lines, the crosstalk noise has no adverse effect on the signal up to the common-mode rejection limit of the receiver.

As a specific example, consider the arrangement in Figure 11.2 in which the difference between the OR and NOR outputs of the basic ECL gate is transmitted over a twisted pair of wires that is matched at its receiving end to a difference amplifier generally referred to as a receiver. The difference voltage is twice as large in magnitude as the signal available from either output. The twisting of the transmission wires keeps them together and regularly reverses their relative positions so that any signal path that might have induced a signal from one of the wires in the pair may be expected to have an equal and opposite signal induced by the other wire. As a result, crosstalk from the twisted pair to other signal paths is reduced. Similarly, crosstalk of other signals to the twisted pair will be introduced into the difference amplifier as a common-mode signal and hence will be restricted from appearing at the single-ended output of the receiver.

Even though we will not be discussing reflections due to mismatched lines, the reader should know that they come about not only from mismatched load and source impedances, as will be seen in the next section, but also from changes in the line impedance that can be caused by an unshielded twisted pair in contact with metal, mismatch between *printed circuit board* (*PCB*) traces and backplane wiring, or bends in the coaxial cable.

Figure 11.1 Cross section of a multiconductor ribbon cable.

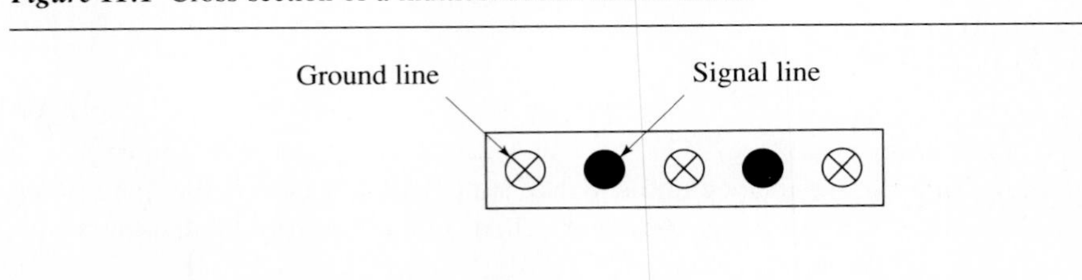

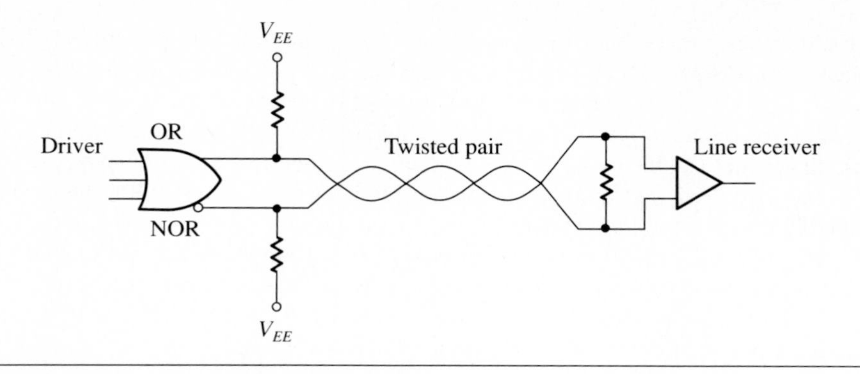

Figure 11.2 Twisted-pair transmission in ECL.

11.2 Transmission-Line Effects in Digital Circuits

In this section, a brief review of the basic concepts of transmission-line characteristics is presented, and simplified methods of analysis as well as SPICE simulations are employed to examine situations commonly encountered in high-speed digital systems. Since the principles and methods apply to any type of logic circuit, normalized pulse amplitudes are used in most of the calculations.

When information on a signal line changes, a finite amount of time is necessary for it to travel from the transmitting end to the receiving end. As the circuit speed becomes faster and clock rates increase, the dynamic behavior of the interconnection line becomes increasingly important. The edge speeds, that is, rise and fall times of the logic elements, fan-out, delay times of signal paths, and some other transient characteristics all influence the reliable operation of the system.

A transmission line, as used in high-speed logic families, is a signal path such as a *microstrip* or a *stripline* on a PCB that exhibits a *characteristic impedance* Z_o similar to coaxial cables and twisted-pair lines. The transmission-line nature of an interconnection manifests itself when the signal edge speeds are comparable to the propagation delay along the line. When the transition times are long in comparison to the time of propagation, the line can be approximated by lumped-circuit elements.

Signal-Line Considerations

To determine when a pair of connecting wires may be viewed as a transmission line, consider the source V of Figure 11.3, which is connected through a switch S and a pair of uniform wires of length ℓ to a load Z. When the switch is closed at time $t = 0$, its effect is not felt everywhere immediately but propagates along the line with a finite velocity u given by

$$u \equiv \frac{1}{\sqrt{LC}} \tag{11.1}$$

where L and C are the inductance and capacitance per unit length of the line. Even though both L and C depend on the geometry, their product is constant and the prop-

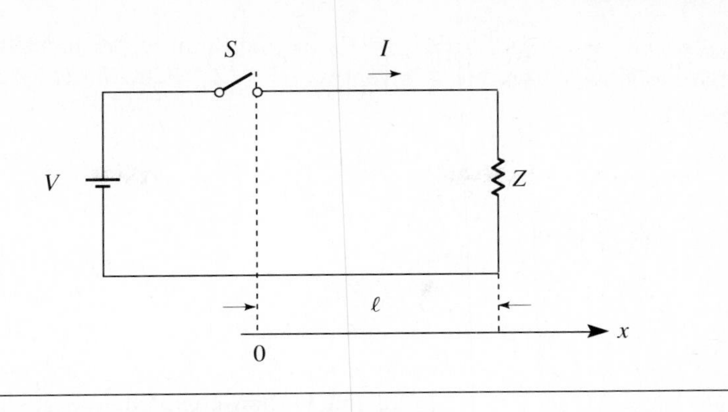

Figure 11.3 A source connected to a load Z through a transmission line of length ℓ.

agation velocity has the value equal to the velocity of light (i.e., 3×10^8 m/s), when the surrounding medium is air. This fact shows that the power is actually not carried by the wires but is instead transferred through the free space surrounding the wires, and the latter serve only as guides to direct the power flow.

As the voltage front moves a distance dx down the line, the additional capacitance that is charged to voltage V is Cdx, and the charge required is $dQ = V(Cdx)$. Therefore, we find an expression for the magnitude of the accompanying current necessary for this charging as

$$I = \frac{dQ}{dt} = VC\frac{dx}{dt} = VCu = \frac{V}{Z_o} \tag{11.2}$$

where

$$Z_o \equiv \sqrt{\frac{L}{C}} \tag{11.3}$$

is called the characteristic impedance of the line that is seen by the source looking into the line. Note that to the right of the voltage front, the current is still zero.

If $Z = Z_o$ at the line's termination, then this termination appears as an infinite extension of the line, so that there will be no *reflection* when the front reaches the termination. If, on the other hand, $Z \neq Z_o$, then a reflection develops when the front arrives at the end of the line due to the potential inconsistency at the termination. This reflected front starts moving toward the left. Its amplitude and polarity are such that the total voltage V_ℓ, that is, the sum of the incident and reflected voltages at $x = \ell$, and the total current I_ℓ are related by

$$\frac{V_\ell}{I_\ell} = Z \tag{11.4}$$

This implies that if the incident voltage front has an amplitude V, then the reflected voltage will have an amplitude ΓV where the parameter Γ, which lies in the range

± 1, is called the *reflection coefficient*. Furthermore, since the incident current is V/Z_o while the reflected current is given by $-\Gamma V/Z_o$, Equation (11.4) can be expressed as

$$\frac{V + \Gamma V}{V/Z_o - \Gamma V/Z_o} = Z$$

so that solving for Γ yields

$$\Gamma = \frac{Z - Z_o}{Z + Z_o} \tag{11.5}$$

When $Z = Z_o$, the reflection coefficient is zero. It is equal to 1 when the end of the line is open-circuited, and -1 when the line is short-circuited.

Example 11.1

For a given .8-V logic swing on a 50-Ω line and using a 51-Ω carbon resistor load whose impedance is given as $(51.4 + j5.6)$ Ω at a certain frequency, the reflected voltage is calculated to be

$$V_r = \left|\frac{51.4 + j5.6 - 50}{51.4 + j5.6 + 50}\right| \times .8 = .055 \times .8 = 44 \text{ mV}$$

Consider the connection of Figure 11.4a, in which a driver with a low output impedance R_s of around 10 Ω drives a load whose input impedance R is on the order of Kohms. Suppose that the characteristic impedance R_o of the interconnecting wire is in the range of tens of ohms, so that $R_s << R_o << R$.

If the input V_i makes a sudden transition between voltage levels such that the rise time $t_r >> t_d$, the one-way delay of the wire, then the output V_o will have the damped oscillatory waveform, as depicted in Figure 11.4b. Therefore, the single transition at the driver output between two logic states can be interpreted by the load as several transitions. If the input waveform makes a transition slower in comparison with t_d, then the output will be able to follow the input more closely, as shown in Figure 11.4c.

In general, then, the transmission-line character of an interconnection is evident when edge speeds are very fast. On the other hand, when signals change over long intervals when compared to the line delay, the lines may be substituted by lumped circuit equivalents. In ECL, advanced Schottky TTL subfamilies, and more recently in ACL CMOS, the rise times are in the order of few nanoseconds, so that even connecting wires of a few centimeters in length become transmission lines.

The aforementioned oscillations can be suppressed by terminating the line at its receiving end in its characteristic impedance. If the line is not terminated, the oscillation becomes more pronounced as the lines get longer for an input logic swing of a fixed rise time. The allowable unterminated line length depends on the edge rate, the fan-out, Z_o, and the propagation delay per unit length of line.

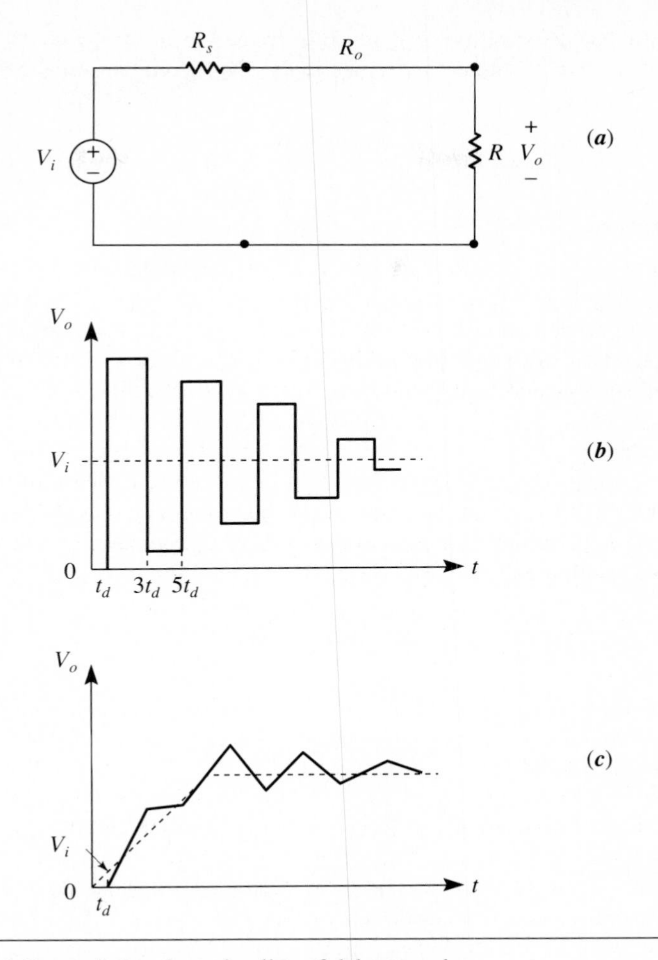

Figure 11.4 (a) Transmission through a line of delay t_d and characteristic impedance R_o with $R_s \ll R_o \ll R$; (b) response at the output for a step input; and (c) response for a slowly rising ramp input.

The propagation delay per unit length for an unloaded transmission line is given by

$$\delta \equiv \frac{t_d}{\ell} = \sqrt{LC} \tag{11.6}$$

For a homogeneous medium the propagation delay is also equal to

$$\delta = \sqrt{\mu\epsilon} \tag{11.7}$$

where $\mu \equiv \mu_r\mu_o$ is the permeability and $\epsilon \equiv \epsilon_r\epsilon_o$ is the permittivity of the medium. In transmission lines used in digital circuits, the relative permeability μ_r is unity, and the permeability and permittivity of the free space are $\mu_o = 4\pi \times 10^{-7}$ Henry/m and $\epsilon_o = 8.85 \times 10^{-12}$ Farad/m, respectively. Therefore, in terms of relative permittivity,

$$\delta = 33.35\sqrt{\epsilon_r} \text{ ps/cm} \tag{11.8}$$

Lattice Diagram

Multiple reflections develop on a transmission line when neither the signal source impedance nor the load impedance matches the line impedance. When the source and load reflection coefficients, denoted by Γ_s and Γ_ℓ, respectively, are of opposite polarity, the reflections alternate in polarity, giving rise to the oscillation of the signal about its final steady state value, known as *ringing*. When the signal rise time is long compared to the line delay, the signal shape is distorted because the individual reflections overlap in time.

In the presence of multiple reflections, keeping track of incremental waves on the line and the net voltage at the ends becomes cumbersome. A convenient method of indicating the conditions, which combines magnitude, polarity, and time, employs a graphic construction called a *lattice diagram*.

Example 11.2

Consider the circuit of Figure 11.5a with a line mismatched on both ends. The source is a step function of 1 V in amplitude occurring at $t = t_o$. The initial value of V_i' starting down the line is found to be

$$V_i' = \left(\frac{Z_o}{Z_o + R_s}\right)V_i = \frac{93}{124} \times 1 = .75 \text{ V}$$

Since neither end of the line is terminated in its characteristic impedance, multiple reflections develop with reflection coefficients

$$\Gamma_\ell = \frac{\infty - 93}{\infty + 93} = +1$$

and

$$\Gamma_s = \frac{31 - 93}{31 + 93} = -.5$$

A lattice diagram for this circuit is shown in Figure 11.5b. The vertical lines symbolize the discontinuity points at the generator and the load, and a time scale is marked off on each line in increments of $2t_d$, starting at t_o for V_i' and t_d for V_o. The diagonal lines show the incremental voltages traveling down the plot from the top and between the ends of the line, with solid lines used for positive voltages and dashed lines for negative.

At $t = t_o$, an incident voltage, $V_i' = .75$ V, travels down the line from t_o to t_d. Next, a line is drawn from $t = t_d$ to $t = 2t_d$, with the reflected voltage value $\Gamma_\ell V_i' = .75$ V indicated. To find the net voltage at either end at $t = nt_d$ for any n, all the incremental voltages arriving at and leaving from this point up until that time are summed. Therefore, for $t = t_d$, $V_o = (1 + \Gamma_\ell)V_i' = 1.5$ V. The reflected voltage on the line between $t = 2t_d$

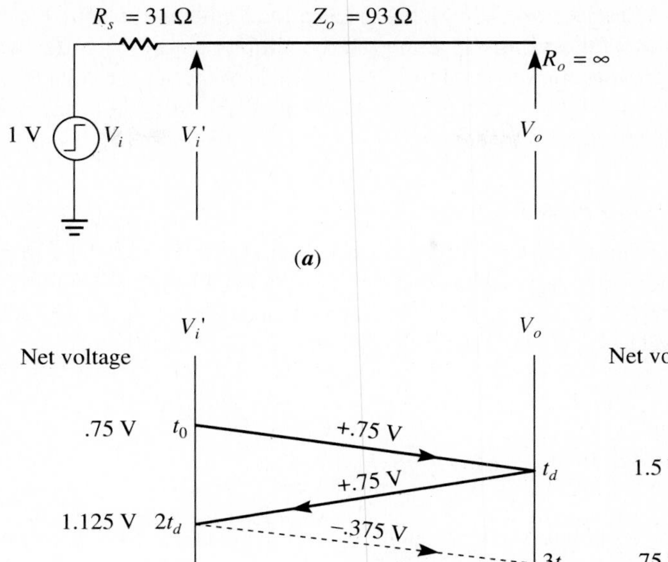

(*a*)

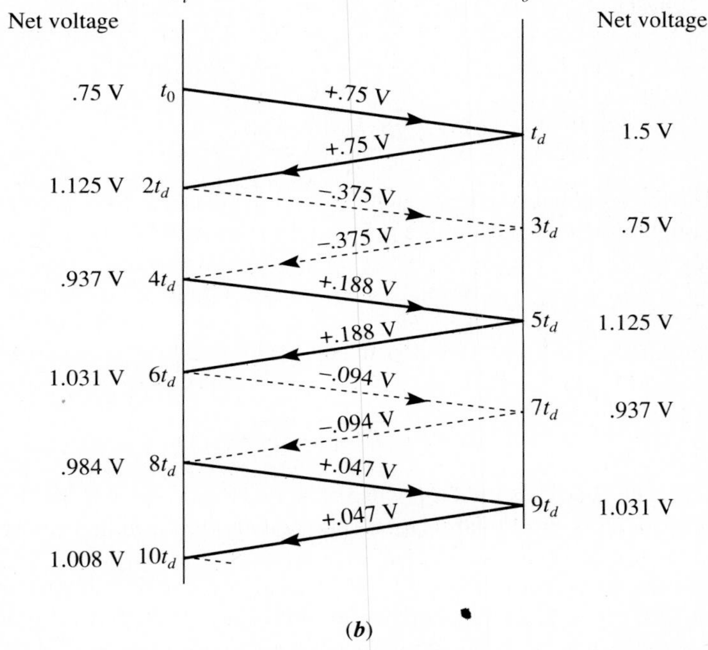

(*b*)

Figure 11.5 (a) Multiple reflections due to mismatch at the load and source for an open-ended line driven by a 1-V source, and (b) the lattice diagram.

and $3t_d$ is then found to be $\Gamma_\ell \Gamma_s V_i' = -.375$ V. Thus, for $t = 2t_d$, $V_i' = [1 + \Gamma_\ell(1 + \Gamma_s)](.75) = 1.125$ V at the source end of the line. Similarly, the output voltage at $t = 3t_d$ is found as $V_o = (1 + \Gamma_\ell)(1 + \Gamma_\ell\Gamma_s)(.75) = .75$ V. Note that as the time progresses, successively higher order reflection coefficient terms come into play. Subsequent terms may be positive or negative depending on the resulting sign, so that damped ringing can take place. The process continues until the voltage at the receiving end asymptotically approaches the steady-state condition, that is, 1 V in this example.

The open-ended line of Example 11.2 has a load reflection coefficient of +1, so that the successive reflections cause the line conditions to approach the steady-state conditions of zero line current and a line voltage equal to the source voltage. A shorted line will have a reflection coefficient of −1, eventually leading to a zero voltage and a line current determined by the source voltage and resistance.

Signal Traces on Printed Circuit Boards

Signal interconnections on a two-sided or multilayer PCB can be grouped into two general categories: *microstrip lines* and *strip lines*. The microstrip line of Figure 11.6 consists of a signal conductor separated from a ground plane by a dielectric insulating material. Its characteristic line impedance is given by

$$Z_o = \frac{87 \ln\left(\dfrac{5.98h}{.8w + t}\right)}{\sqrt{\epsilon_r + 1.41}} \tag{11.9}$$

so that if the trace thickness t, trace width w, and the distance from the ground plane h are controlled, the line will exhibit a predictable characteristic impedance depending on the relative dielectric constant ϵ_r of the board material. Equation (11.9) is accurate for ratios of width to height between .1 and 3, and for dielectric constants between 1 and 15.

The line propagation delay is dependent only on the dielectric constant and is given by

$$\delta = 33.35 \sqrt{.475\epsilon_r + .67} \ \text{ps/cm} \tag{11.10}$$

For G-10 fiberglass epoxy boards for which $\epsilon_r \approx 5$, δ is calculated to be 58.28 ps/cm. The microstrip offers easier fabrication and higher propagation velocity than the strip line.

A strip line, as shown in Figure 11.7, consists of a copper ribbon centered in a dielectric insulating medium between two ground planes. This type of line is used in multilayer boards when operating at top speed in high-density packaging applications. Its characteristic impedance depends on the thickness t and width w of the line as

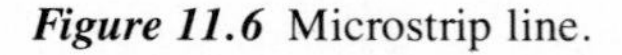
Figure 11.6 Microstrip line.

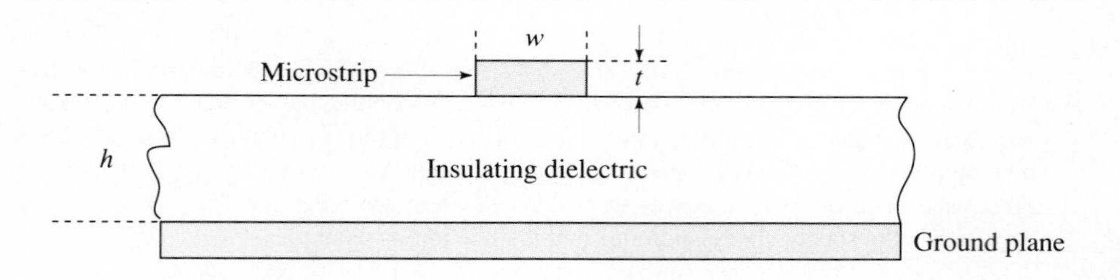

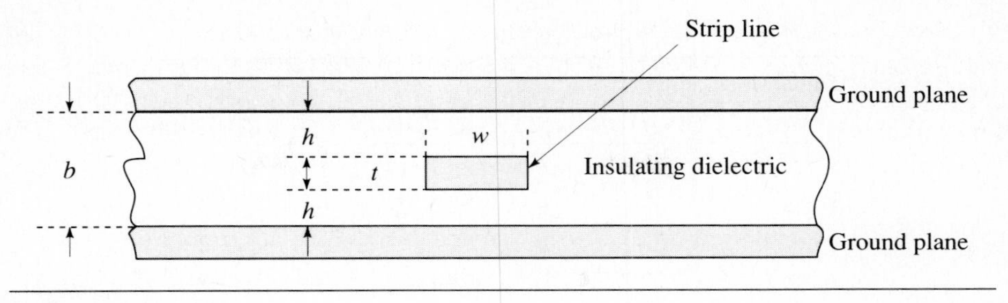

Figure 11.7 Strip line.

well as on the dielectric constant of the medium ϵ_r and the distance b between the ground planes, and is expressed as

$$Z_o = \frac{60}{\sqrt{\epsilon_r}} \ln \left[\frac{4b}{.67\pi(.8w + t)} \right] \tag{11.11}$$

which proves to be accurate for $w/(b - t) < .35$ and $t/b < .25$.

The propagation delay per unit length of the line depends only on the dielectric constant, as in the case of the microstrip, and is given by Equation (11.8). For G-10 fiberglass epoxy boards, it is found to be 74.68 ps/cm.

The important feature of both PCB lines is that their impedances are highly predictable and can be closely controlled, as explained above. Typical impedances of these conductors with respect to their physical size and relative spacings are shown in Tables 11.1 and 11.2.

Both microstrip and strip lines may be treated as operating in the *transverse elec-*

Table 11.1 Typical Impedance of Microstrip Lines, $\epsilon_r \approx 5$

Dimensions, mms			
h	w	Z_o, Ω	C, pF/cm
.15	.5	35	1.3
.15	.375	40	1.15
.375	.5	56	1
.375	.375	66	.85
.75	.5	80	.66
.75	.375	89	.59
1.5	.5	105	.52
1.5	.375	114	.46
2.5	.5	124	.43
2.5	.375	132	.39

Table 11.2 Typical Impedance of Strip Lines, $\epsilon_r \approx 5$

Dimensions, mms			
h	w	Z_o, Ω	C, pF/cm
.15	.5	27	2.63
.15	.375	32	2.3
.25	.5	34	2.2
.25	.375	40	1.84
.3	.5	37	1.87
.3	.375	43	1.58
.5	.5	44	1.58
.5	.375	51	1.38
.75	.5	55	1.28
.75	.375	61	1.15

tromagnetic mode (*TEM*) for all practical purposes, even though the microstrip propagation is not purely TEM because of nonuniform dielectrics.

Transmission-Line Termination Techniques for ECL Gates

We already know that in order to prevent reflections and hence ringing, transmission lines should be properly terminated. The two most frequently utilized fan-out arrangements for ECL gates are called *parallel termination* and *series termination*.

Two ways of placing gates on a parallel-terminated transmission line are *lumped* and *distributed* loadings. Figure 11.8 depicts a parallel-terminated interconnection between ECL 10K series gates with lumped fan-out at the end of the line. A transmission line will have a reflection coefficient Γ_ℓ of zero when driving a load impedance equal to its characteristic impedance. Thus, the ECL 10K series can source current for driving a 50-Ω line with the line terminated by 50 Ω to an auxiliary supply of −2 volts. Note that in Figure 11.8, one end of the termination is connected to the emitter of the output emitter follower of the driver. The other end cannot return to $V_{EE} = -5.2$ V because such a connection, with the terminating resistor in parallel with the emitter resistor of the driver, would use excessive current. The disadvantage of the connection in Figure 11.8 is, of course, the requirement for an additional supply.

A full logic swing that is available all along a parallel-terminated transmission line allows distributed loading to be placed anywhere along the line. Figure 11.9 illustrates the parallel termination with distributed loads, in which case gate inputs appear as high-impedance stubs to the transmission line and hence should be as short as possible. No matter where the loads appear along the line, the terminating resistor should always be at the end of the line. In this configuration, as the fan-out increases,

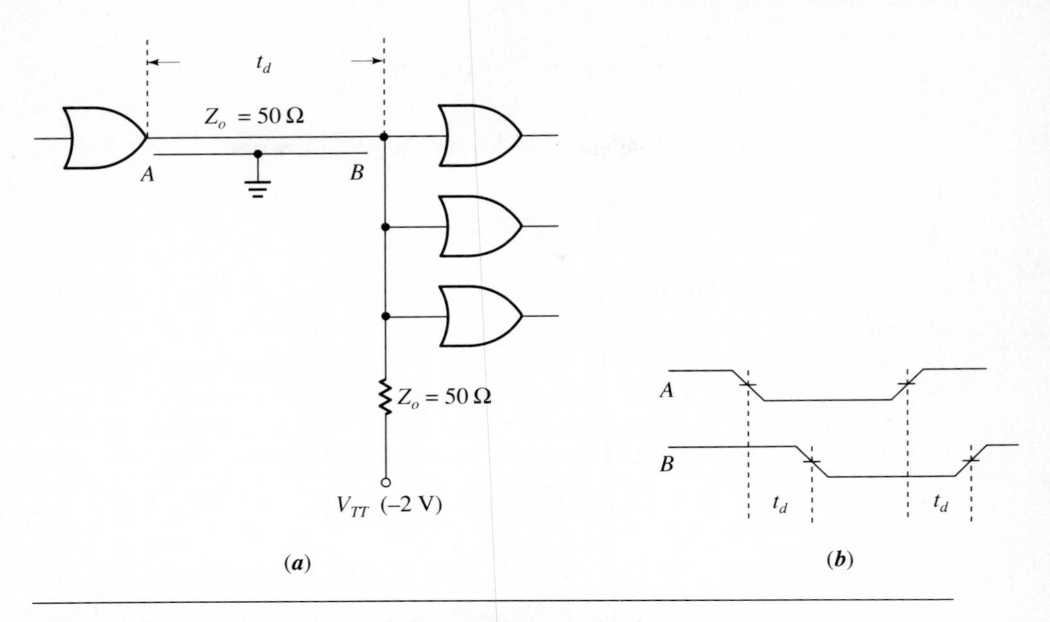

Figure 11.8 (a) Parallel-terminated line for ECL gates with the loads lumped at the end of a matched line, and (b) the waveforms.

the edge of the waveform slows down due to an increase in the capacitive load, although the waveform itself is undistorted along the full length of the line.

The input impedance of logic gates can usually be assumed to be purely capacitive as far as reflections are concerned. If a distributed capacitive load C_d is placed along a parallel-terminated line, then the propagation delay will be modified as

Figure 11.9 Parallel-terminated line with distributed fan-out.

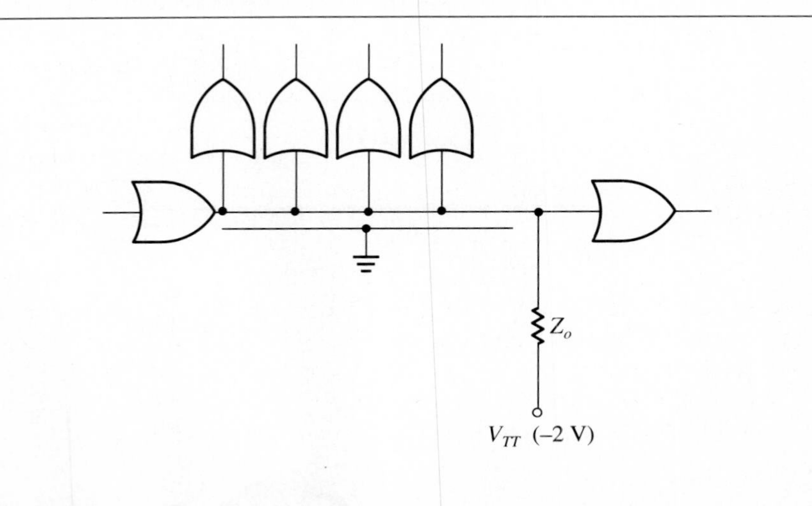

$$\delta' = \sqrt{L(C + C_d)} = \sqrt{1 + \frac{C_d}{C}}\,\delta \qquad (11.12)$$

Distributed capacitive loads also change the characteristic impedance of the line as

$$Z_o' = \sqrt{\frac{L}{C + C_d}} = \left(\frac{1}{\sqrt{1 + C_d/C}}\right)Z_o \qquad (11.13)$$

Example 11.3

Four identical ECL loads with a total capacitance of 13 pF and spaced equally at 5-cm intervals along a 68-Ω, 20-cm microstrip line on a glass epoxy board with $\epsilon_r \approx 5$ are to be driven. To determine a value for a parallel terminating impedance that will eliminate reflections at the end of the line, we first find the line capacitance as

$$C = \frac{\delta}{Z_o} = \frac{58.28}{68} = .86 \text{ pF/cm}$$

Since $C_d = 13/20 = .65$ pF/cm, (11.13) yields the value of the termination impedance

$$Z_o' = 51.3\ \Omega$$

The resulting circuit is shown in Figure 11.10.

Another approach to utilizing two power supplies is to use two resistors, as depicted in Figure 11.11. The Thévenin equivalent of the resistive network is a resistor equal to the characteristic impedance of the line, returned to −2 V. The values of R_1 and R_2 may be calculated as

$$R_2 = 2.6Z_o$$

and

Figure 11.10 Distributed loading with four identical ECL loads.

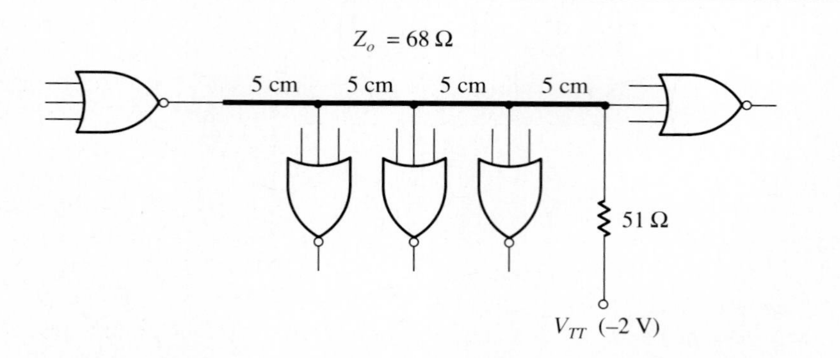

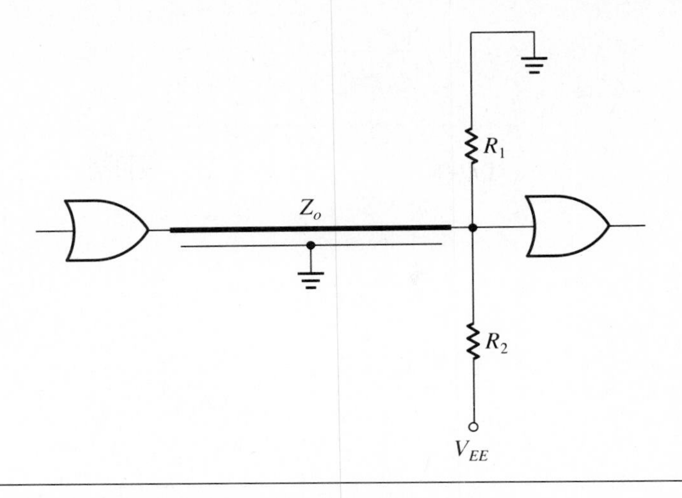

Figure 11.11 Parallel termination using Thévenin equivalent resistor network.

$$R_1 = \frac{R_2}{1.6}$$

The advantage of this approach is that only one power supply is required even though at the expense of more overall power.

A series-terminated line eliminates reflections at the transmitting end of the line. This type of termination is attained by inserting a resistor in series with the output of the driver, as illustrated in Figure 11.12a. Series termination is especially useful when driving an open-ended line. Thus, the input impedance of the load must be much greater than the characteristic impedance of the transmission line, a condition easily satisfied by the high-impedance inputs of ECL circuits.

The series resistor value plus the circuit output impedance is matched to the transmission-line impedance to control overshoot and ringing on long lines. The advantages of using series-terminated lines include the employment of a single power supply, as in the Thévenin equivalent parallel-termination technique, hence low power consumption, and low crosstalk between adjacent lines.

A disadvantage of this technique, on the other hand, is that distributed loading cannot be used because of the ½ logic swing propagating down the line. Lumped loads may be placed at the end of the terminated line as far as reflection at the receiving end is concerned because a full logic swing is observed at this point, and all subsequent reflections will be absorbed at the source since $\Gamma_s = 0$. The waveform for a series-terminated line is shown in Figure 11.12b.

The aforementioned drawback and slower propagation delay of a series-terminated line can be eliminated at the expense of more lines. This is diagrammed in Figure 11.13, where n transmission lines are connected in parallel.

A low-impedance parallel-terminated line has a shorter propagation delay (as explained below) than a series-terminated line with equivalent fan-out. However,

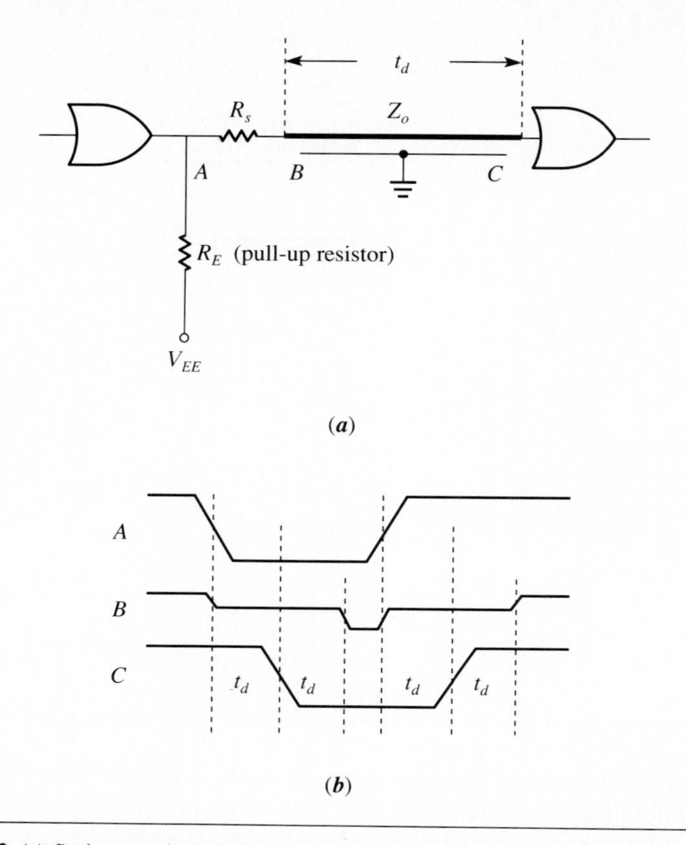

Figure 11.12 (a) Series-terminated line for ECL gates, and (b) the waveforms.

multiple series-terminated lines driven from a single gate output will lower fan-out per line and thus will show shorter delay times than a single parallel-terminated line with an equivalent total fan-out.

The effect of the capacitive load at the end of the line on the output waveform is shown in Figure 11.14. With no load, the delay between the 50% points of the input and output is only the line delay t_d. A load capacitance causes an extra delay Δt_d due to the increase in the rise time of the output signal. This extra delay can be calculated by using a ramp approximation for the incident voltage and characterizing the circuit as a fixed impedance in series with the load capacitance C_L, as depicted in Figure 11.15 for the series- and parallel-terminated cases, respectively. Also shown are the corresponding Thévenin equivalent circuits, with Z_{Th} signifying the Thévenin equivalent impedance and $\tau \equiv C_L Z_{Th}$ the time constant. The characteristic impedance of the series-terminated line is twice as large as that for the parallel-terminated line. Note that by definition the input rise time $t_{ri} \equiv .8\alpha$ where α is the ramp time. In the following discussion, we assume that the line is *long* (that is, $t_d >> t_{ri}/2$).

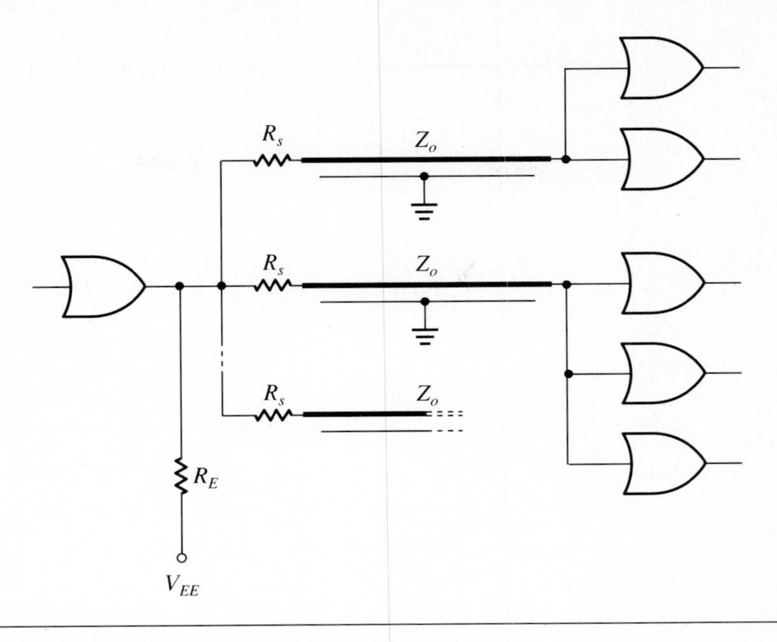

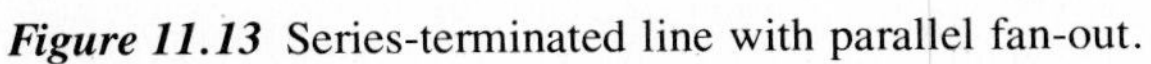

Figure 11.13 Series-terminated line with parallel fan-out.

The input voltage can then be expressed as

$$v_{in}(t) = \left(\frac{V}{\alpha}\right)\left[tu(t) - (t - \alpha)u(t - \alpha)\right] \tag{11.14}$$

where

$$u(t - \alpha) \equiv \begin{cases} 0 \text{ for } t < \alpha \\ 1 \text{ for } t > \alpha \end{cases}$$

is the unit step function applied at $t = \alpha$. Taking the Laplace transform of both sides in Equation (11.14), we get

Figure 11.14 Effect of capacitive loading on the output rise time.

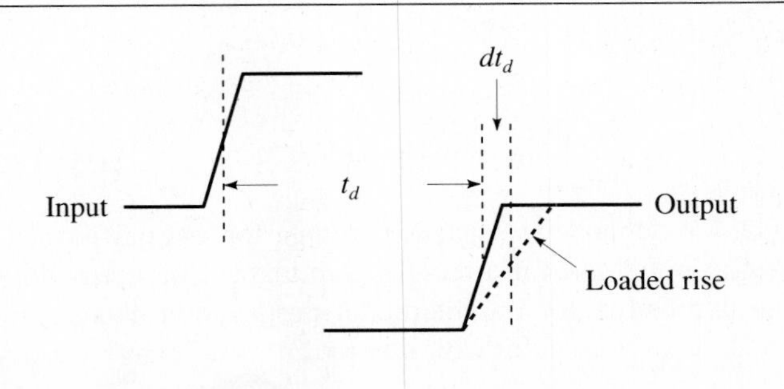

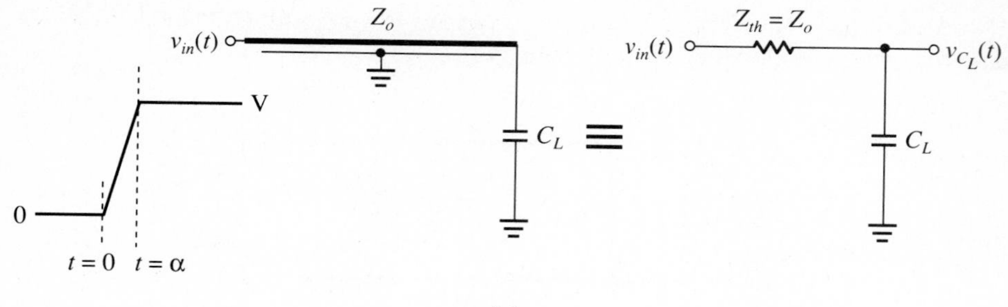

(*a*)

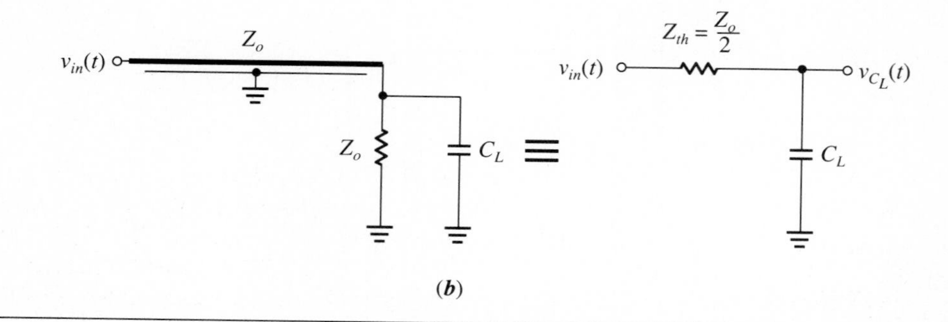

(*b*)

Figure 11.15 (a) Series termination with a capacitive load and its Thévenin equivalent, and (b) parallel termination with a capacitive load and its Thévenin equivalent.

$$V_{in}(s) = \left(\frac{V}{\alpha s^2}\right)(1 - e^{-\alpha s}) \tag{11.15}$$

The Laplace transform of the voltage drop across the capacitive load yields

$$V_{C_L}(s) = \left(\frac{V}{\alpha \tau}\right)\frac{1}{s^2(s + 1/\tau)}(1 - e^{-\alpha s}) \tag{11.16}$$

with the Laplacian impedance of the equivalent circuit being $Z_{Th} + sC_L$. Taking the inverse Laplace transformation of Equation (11.16) produces an expression for the waveform at the end of a line with termination capacitance as

$$v_{C_L}(t) = \left(\frac{V}{\alpha}\right)[t - \tau(1 - e^{-t/\tau})]u(t)$$
$$- \left(\frac{V}{\alpha}\right)[(t - \alpha) - \tau(1 - e^{-(t-\alpha)/\tau})]u(t - \alpha) \tag{11.17}$$

Equation (11.17) defines the capacitor voltage for a series-terminated transmission line when $Z_{Th} = Z_o$, and for a parallel-terminated line when $Z_{Th} = Z_o/2$. The derivation of an expression for the additional propagation delay due to capacitive loading at the end of the line is difficult due to the complexity of Equation (11.17).

However, it becomes simpler if the input is assumed to be a step function, in which case the capacitive voltage is found to be

$$v_{C_L}(t) = V(1 - e^{-t/\tau}) \tag{11.18a}$$

for a series-terminated line and

$$v_{C_L}(t) = V(1 - e^{-2t/\tau}) \tag{11.18b}$$

for a parallel-terminated line.

Now, to find the additional delay time due to gate loading, let $v_{C_L}(t) = V/2$ and solve (11.18) for Δt to obtain

$$\Delta t_d = \tau \ln 2 \tag{11.19a}$$

for series termination and

$$\Delta t_d = \frac{1}{2} \tau \ln 2 \tag{11.19b}$$

for parallel termination. These quantities should be added to the intrinsic delay, as expressed in Equation (11.6), to obtain the value of the total system line delay, which will be very close to the value that can be calculated using Equation (11.12) in the case of parallel termination. Note from Equation Set (11.19) that the propagation-delay increase is twice as much for a series-terminated line as for a parallel-terminated line.

Next, to find an expression for the output rise time t_{ro}, we use the second term on the right-hand side of Equation (11.17) which reduces to

$$v_{C_L}(t) = \left(\frac{V_T}{a}\right)(1 - e^{-(t-\alpha)/\tau}) + V \tag{11.20}$$

for $t > \alpha$. Recall that the rise time is defined as the time it takes for the voltage to go from 10% to 90% of its final value. Also note that $\alpha = 1.25 t_{ri}$. Therefore, using Equation (11.20), rearranging, and taking the natural log of both sides, the rise time of the voltage waveform at the end of the transmission line is obtained as

$$t_{ro} = \left(\frac{\tau}{m}\right) \ln \left[\frac{\tau}{.18 m t_{ri}} (e^{1.25 m t_{ri}/\tau} - 1)\right] \tag{11.21}$$

where $m = 1$ for a series termination and $m = 2$ for a parallel termination.

11.3 SPICE Transmission-Line Model

SPICE considers the transmission line an ideal *lossless* delay line. This is understandable for a circuit analysis program with IC emphasis. Transmission lines within an integrated circuit are very short, electrically as well as physically, so that the loss is negligible for most purposes. In SPICE, loss can be added only by using lumped resistive elements. A long and lossy line can be simulated by breaking it up into shorter sections and incorporating a small series resistance in each section to represent the distributed conductor loss. However, in transient analysis, SPICE will make the computing interval less than or equal to one-half the minimum transmission delay of the shortest transmission line. Thus, if short transmission lines are employed, a transient analysis may take a very long time. Moreover, lumped-resistance simulation of the line loss is valid only at a certain frequency because a frequency-dependent resistor cannot be specified. It is worth noting at this point that the transmission-line model is the only SPICE model that allows only one connection to a node. Thus, open-circuited lines may be simulated.

The length of the line may be expressed in either of the following two forms:

T ⟨*name*⟩ ⟨*+A port node*⟩ ⟨*−A port node*⟩ ⟨*+B port node*⟩ ⟨*−B port node*⟩

\+ **ZO** = ⟨*value*⟩ **TD** = ⟨*value*⟩

or

T ⟨*name*⟩ ⟨*+A port node*⟩ ⟨*−A port node*⟩ ⟨*+B port node*⟩ ⟨*−B port node*⟩

\+ **ZO** = ⟨*value*⟩ **F** = ⟨*value*⟩ [**NL** = ⟨*value*⟩]

The line has two ports, *A* and *B* with (+) and (−) nodes defining the polarity of a positive voltage at a port. The first form above specifies the transmission delay **TD** in seconds. The delay can be found from

$$\mathbf{TD} = \text{physical length}/(\text{velocity factor}) \cdot c$$

where c is the free-space velocity of 3×10^8 m/s, and (velocity factor) ≤ 1.

The second form specifies the frequency **F** and the dimensionless normalized electric length **NL** of the line with respect to the wavelength in the line at that frequency. The latter is determined by

$$\mathbf{NL} = \mathbf{TD} \cdot \mathbf{F}$$

and defaults to 1/4. The frequency is then the quarter-wave frequency. Notice that only one propagating mode is modeled by this element. If all four nodes are actually distinct, then two modes may be excited in simulation by employing two transmission-line elements.

Example 11.4

Figure 11.16a illustrates a pulse voltage source with a 50-Ω output impedance that is connected to a 50-Ω transmission line terminated in a matched load. The line has a delay of 2 ns. The following SPICE input file produces a plot of voltages at the input and output of the transmission line:

```
A SHORT TRANSMISSION LINE
VIN 1 0 PULSE 0 5 0 0 5N
RIN 1 2 50
T   2 0 3 0 ZO 50 TD 2N
RL  3 0 50
.TRAN .2N 10N
.PROBE
.END
```

The resulting waveforms are shown in Figure 11.16b.

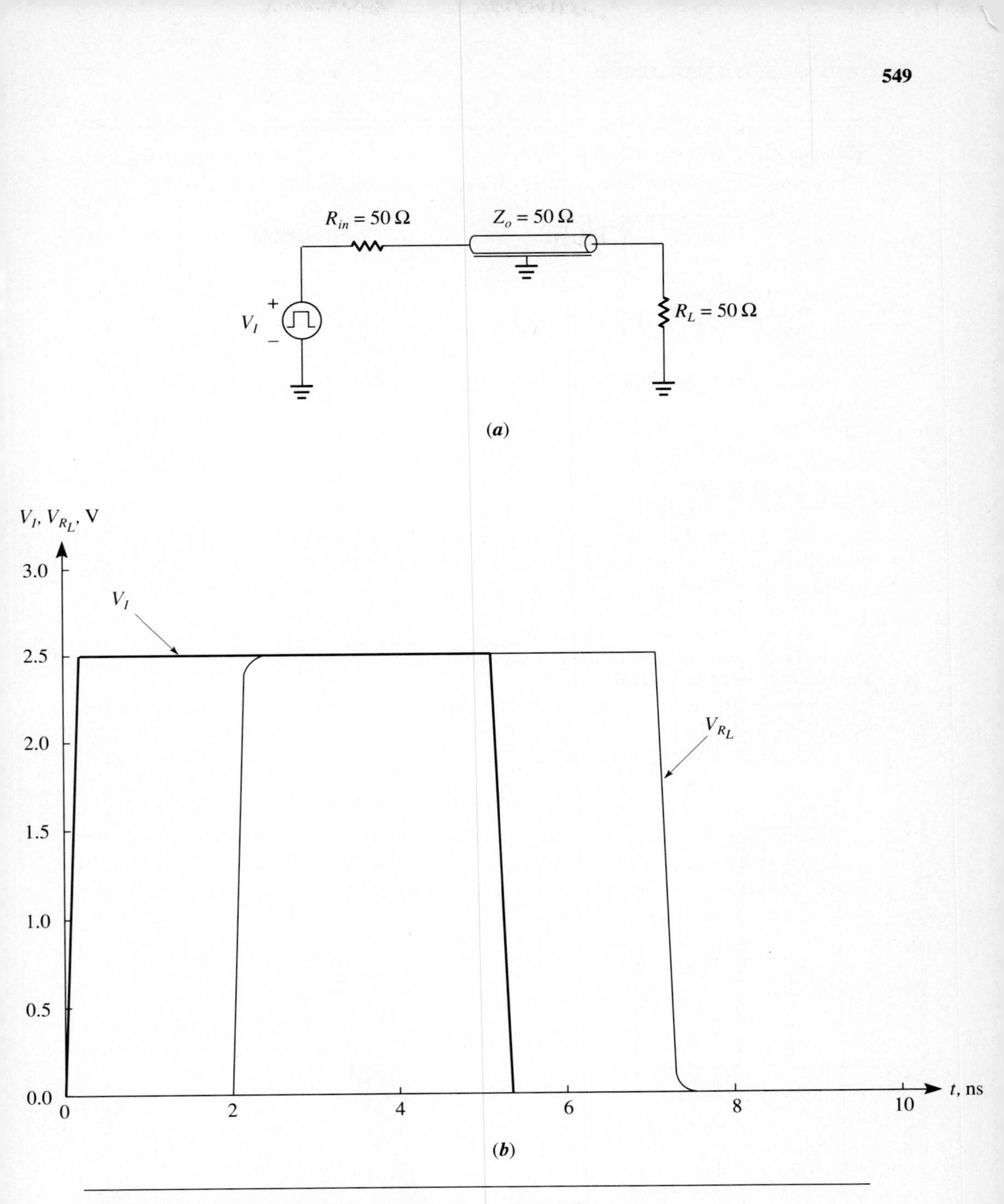

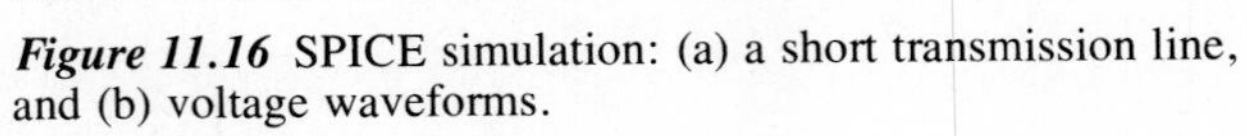

Figure 11.16 SPICE simulation: (a) a short transmission line, and (b) voltage waveforms.

Example 11.5

The following SPICE file simulates the circuit of Example 11.2:

```
A MISMATCHED TRANSMISSION LINE
VIN 1 0 PULSE 0 1 0 0 0 30N
RIN 1 2 31
T    2 0 3 0 ZO 93 TD 2N
RL   3 0 25MEG
.TRAN  .2N 30N
.PROBE
.END
```

The resulting waveforms at the input and output are shown in Figure 11.17.

11.4 Crosstalk

Crosstalk is defined as the coupling of signals from one line in the system to another through capacitance and mutual inductance between them. When two conductors such as wires, ground shields, or printed wiring are adjacent, they geometrically form a

Figure 11.17 Input and output voltage waveforms for the mismatched circuit of Example 11.2.

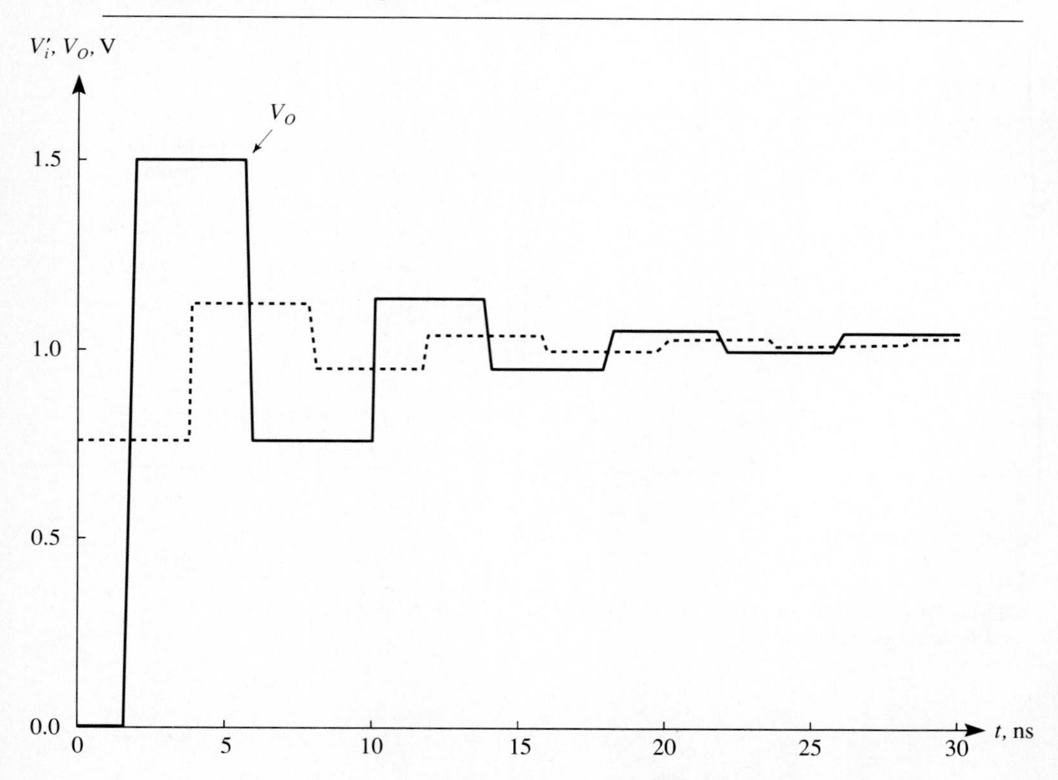

capacitance: the conductors form the plates, and the insulation between them becomes the dielectric. Consequently, between any two conductors there exists capacitance distributed along their parallel lengths that can transmit a signal from one line to another. A source on one line, distributed capacitance, and a load resistance on the other line form a differentiating circuit. Since a step input to a differentiating circuit yields a spike across the load, the distributed capacitance causes crosstalk.

A coupled noise spike is limited in width by the time constant of the differentiating circuit. A low-impedance circuit connected to the line will reduce this time constant and hence narrow the spike.

The distributed capacitance between a line and ground decreases the bandwidth of the line and increases the rise time of an imposed signal or noise spike. If the spike is narrow enough, the increase in rise time will in turn reduce the amplitude. This means that the spike will be partially filtered out by the distributed shunt capacitance, as depicted in Figure 11.18. Sufficient reduction in amplitude will confine the spike within the noise margin. Even if it surpasses the noise margin, it has to remain a sufficient time to produce an undesirable result. While a high-speed circuit responds to very short pulses, a slower circuit may not have a chance to respond to a short spike.

Similarly, if two conductors with distributed inductance are in close proximity, the magnetic flux around each conductor will link the other to form a transformer with very light mutual coupling. Thus, a change in current through one conductor will induce a small undesirable voltage pulse along the other.

While designing circuits that are immune to noise spikes below a certain amplitude is a way to reduce the effect of noise, another way is to minimize the noise itself by either employing low-impedance circuitry or avoiding high-speed circuitry, when and if possible.

Naturally, the most rational way of interconnecting digital circuits is to use lines that are as short as possible to minimize the interconnection delay. An added advantage of doing this is that if the interconnection delay is much less than the rise and fall times of the signals on the lines, then the lines need not be terminated with their characteristic impedance; thus, the crosstalk can be estimated using a lumped-parameter approach. In this case, the maximum unmatched line length is given as

$$\ell_{max} = t_r u / k \tag{11.22}$$

where t_r is the signal rise time, u is the velocity of propagation, and k is an empirical constant that depends on the particular circuit design and the wiring scheme. For

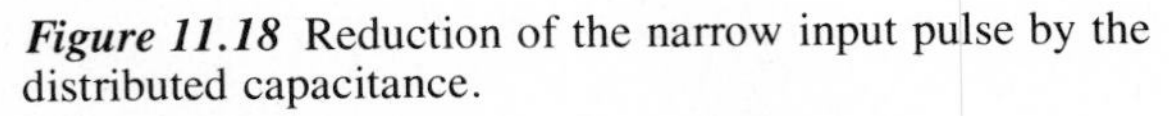

Figure 11.18 Reduction of the narrow input pulse by the distributed capacitance.

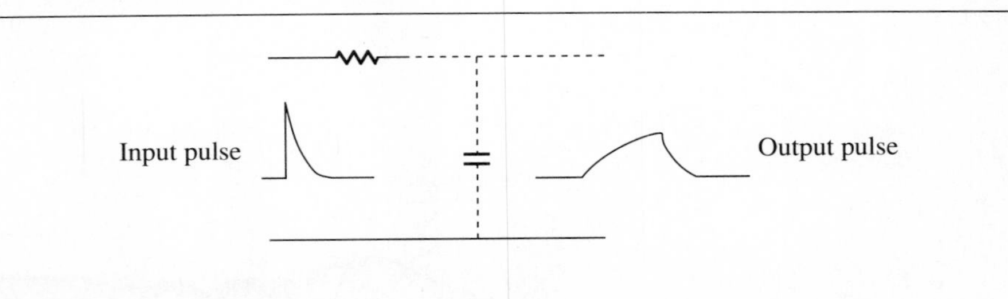

typical values of $k = 4$ and $u = 2 \times 10^8$ m/s, and with $t_r = 1$ ns, ℓ_{max} is found to be 5 cm. The restriction on line length imposed by Equation (11.22) is an indication as to when consideration of a distributed analysis is warranted.

Lumped Parameter Analysis

Assuming $\ell < \ell_{max}$, consider the circuit of Figure 11.19 using the lumped-parameter analysis to determine the effect of coupling between signal lines. A driver with a pulse input is tied to a load by a short line. A second pair of gates is connected by an identical line that runs parallel to the first line. Assume that the second line is in a quiescent state. Nevertheless, there will be a time-varying signal on this line due to capacitive and inductive coupling between the two lines. The coupling mechanism is represented by the per-length mutual capacitance C_m and the per-length mutual inductance L_m. The equivalent circuit of the quiescent line is shown in Figure 11.20 where C_M and L_M signify the total mutual capacitance and inductance, defined as ℓC_m and ℓL_m, respectively. Note that the driver and load for this line are replaced by their equivalent output and input resistances, omitting the reactive portion of their impedances. Neglecting line impedances and denoting the current and voltage on the active line as $i_a(t)$ and $v_a(t)$, we find the voltages resulting from crosstalk as

$$v_{no}(t) = \frac{R_o}{R_i + R_o}\left[R_i C_M \frac{dv_a(t)}{dt} + L_M \frac{di_a(t)}{dt}\right] \tag{11.23}$$

and

$$v_{ni}(t) = \frac{R_i}{R_i + R_o}\left[R_o C_M \frac{dv_a(t)}{dt} - L_M \frac{di_a(t)}{dt}\right] \tag{11.24}$$

Note from Figure 11.19 that

Figure 11.19 Active and quiescent lines with $\ell_{max} > \ell$.

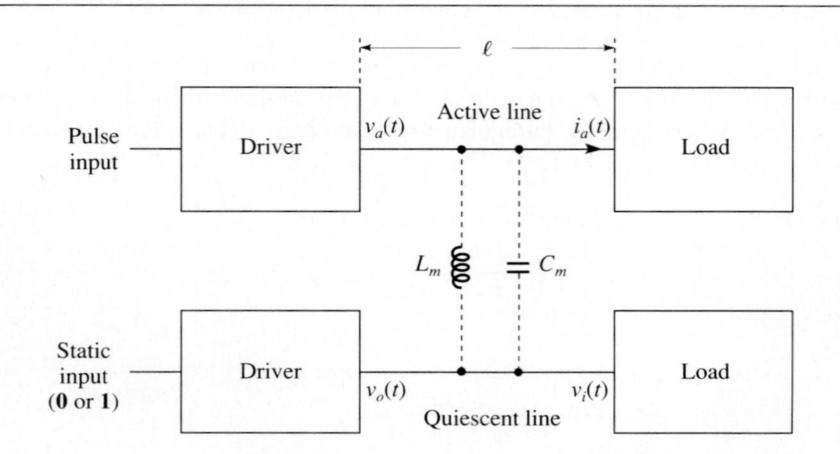

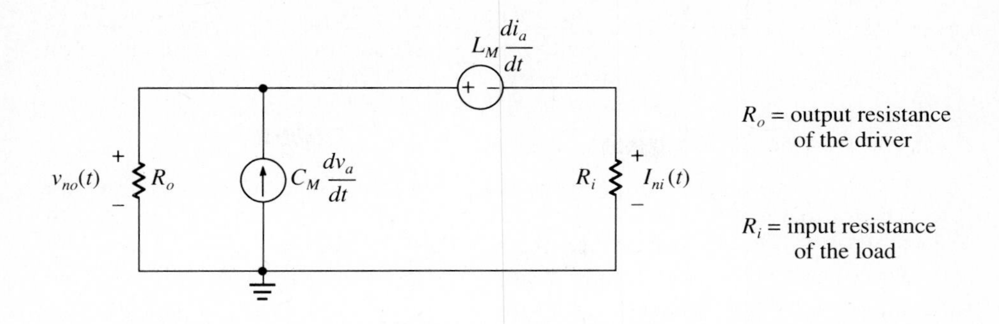

Figure 11.20 Equivalent circuit of the quiescent line for the lumped parameter analysis.

$$v_o(t) = V + v_{no}(t)$$

$$v_i(t) = V + v_{ni}(t)$$

where V is dc output voltage of the quiescent line driver, and is either V_{OL} or V_{OH}. From Equations (11.23) and (11.24) we see that the noise voltages on the quiescent line are proportional to the time rate of change of the voltage and current on the active line.

Distributed Analysis

A distributed analysis is employed to find the variation of current and voltage when the edge rates of the pulse on the active line are not substantially greater than the propagation delay down the line. Consider the two lossless and identical lines of Figure 11.21 that are coupled together by a distributed mutual capacitance C_m and mutual inductance L_m per unit length. The lines have the same values of capacitance C and inductance L per unit length. The active line is driven by a voltage source $v(t)$ and is terminated by its characteristic impedance Z_o.

In our analysis, we will assume TEM waves, which are the most likely to be encountered in a digital environment. In actual practice, however, the system may not be able to support a pure TEM mode because of the inhomogeneous dielectric, such as the air above the lines and glass epoxy for the printed circuit board. It can be shown that the voltage on the active line is expressed by

$$v_a(x, t) = v(t - x/u) \tag{11.25}$$

where $v(t) = v_a(0, t)$ is the source voltage applied at $t = 0$ and u is the velocity of propagation, as defined in Equation (11.1). Therefore, voltage on the active line at any point x and any time t is equal to the source voltage at the time $t - x/u$, and it propagates unchanged along the active line without any reflection at the end.

The quiescent line voltage v_Q, then, is expressed by a second-order partial differential equation as

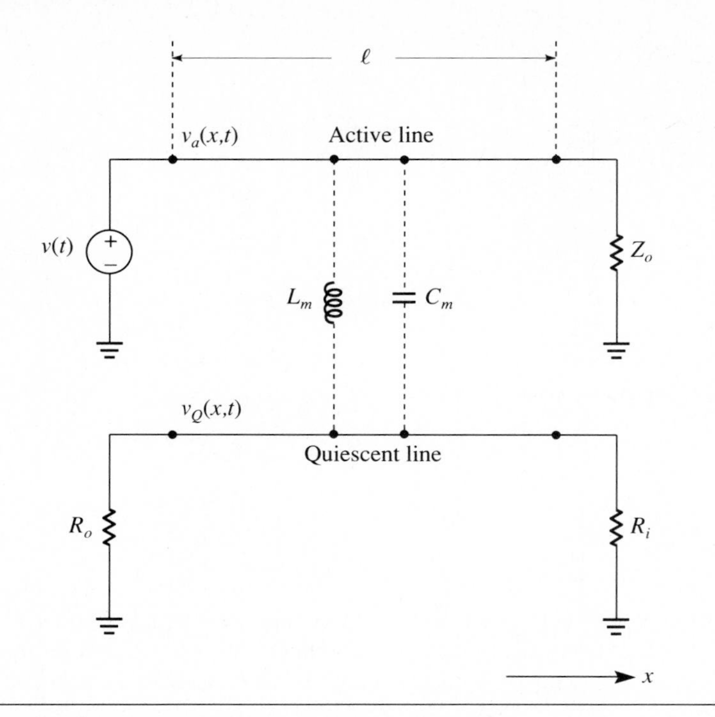

Figure 11.21 Active and quiescent lines for the distributed analysis.

$$u^2 \frac{\partial^2 v_Q}{\partial x^2} = \frac{\partial^2 v_Q}{\partial t^2} + \left(\frac{L_m}{L} - \frac{C_m}{C} \right) \frac{\partial^2 v_a}{\partial t^2} \tag{11.26}$$

Using Laplace transform techniques with zero initial conditions, we obtain the solution to Equation (11.26) as

$$V_Q = Ae^{-sx/u} + Be^{sx/u} - \frac{1}{2}\phi(\kappa - 1)(x/u)e^{-sx/u}(sV) \tag{11.27}$$

where V is the Laplace transform of the source voltage $v(t)$, $\phi \equiv C_m/C$, $\kappa \equiv L_mC/LC_m$, and A and B are constants to be determined from the boundary conditions on the quiescent line. Next, we are going to examine six common line terminations that are found in digital circuits. In all cases, we assume that the active line is properly terminated at both ends and the signal on this line is a step voltage. Figure 11.22 illustrates the waveforms for the quiescent line voltage for all cases.

Case 1. Solving Equation (11.27) when the quiescent line is terminated at both ends by Z_o yields the quiescent line voltage V_Q as

$$V_Q = \frac{1}{4}\phi(\kappa + 1)\{e^{-sx/u} - e^{-s(2\ell - x)/u}\}V - \frac{1}{2}\phi(\kappa - 1)(x/u)e^{-sx/u}(sV) \tag{11.28}$$

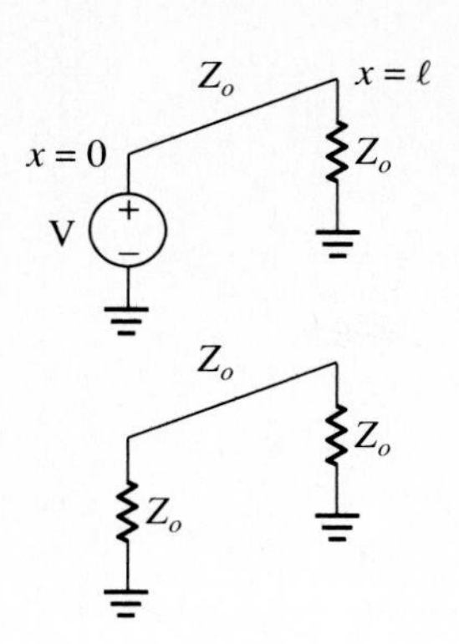
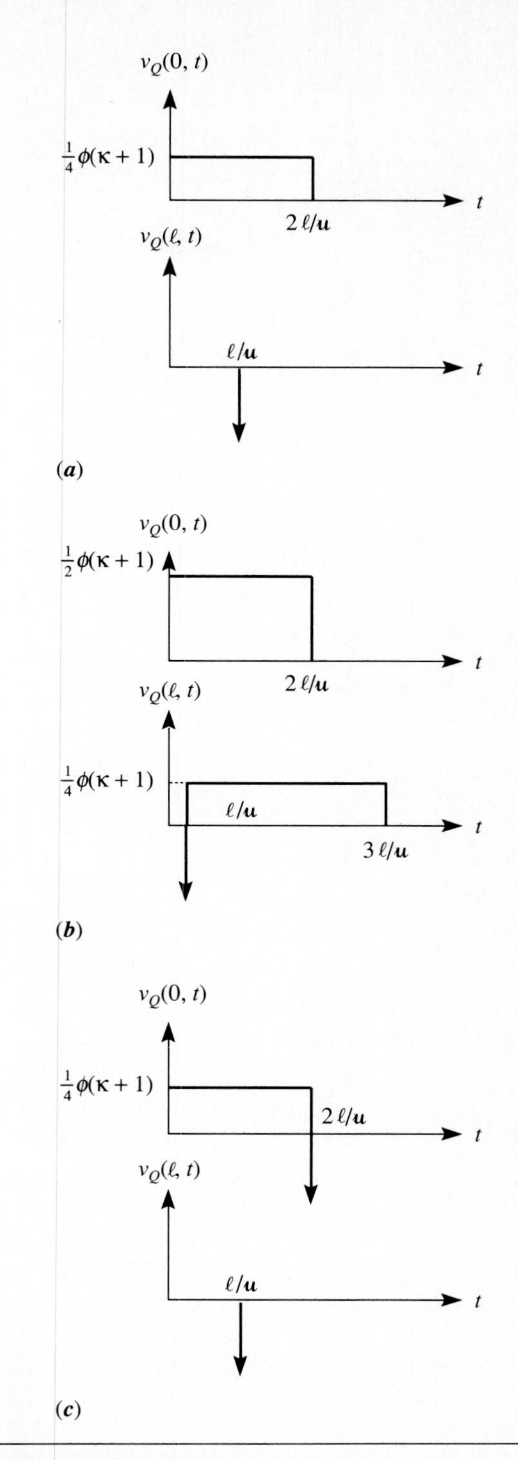

Figure 11.22 Waveforms for the quiescent line voltage for the six common cases with a unit step voltage on the active line: quiescent line is (a) terminated at both ends, (b) open at $x = 0$ and terminated at $x = \ell$, (c) open at $x = \ell$ and terminated at $x = 0$.

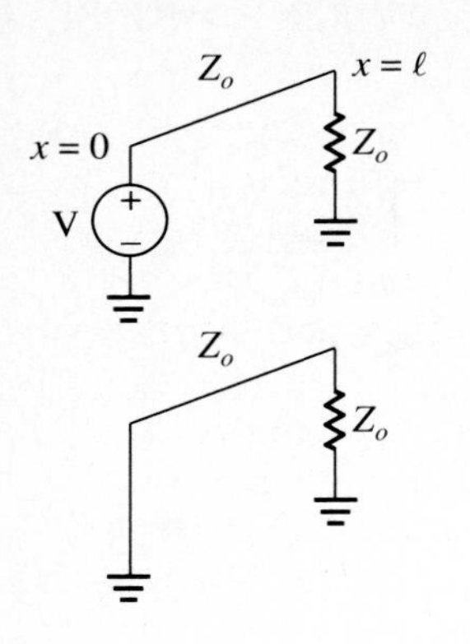

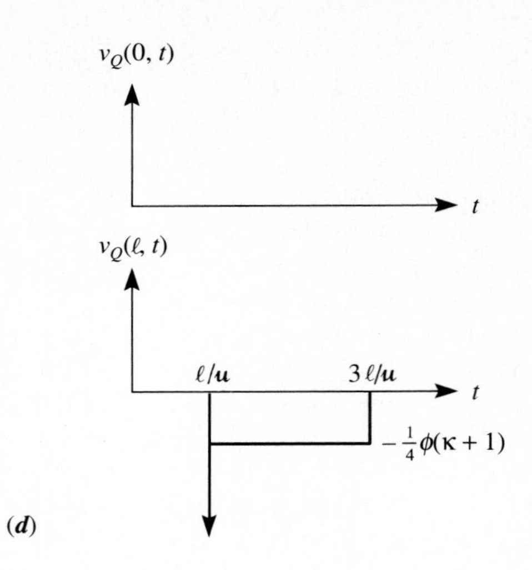

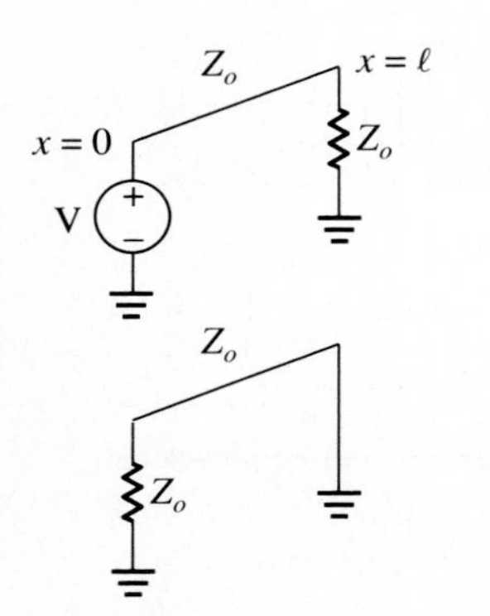

(*d*)

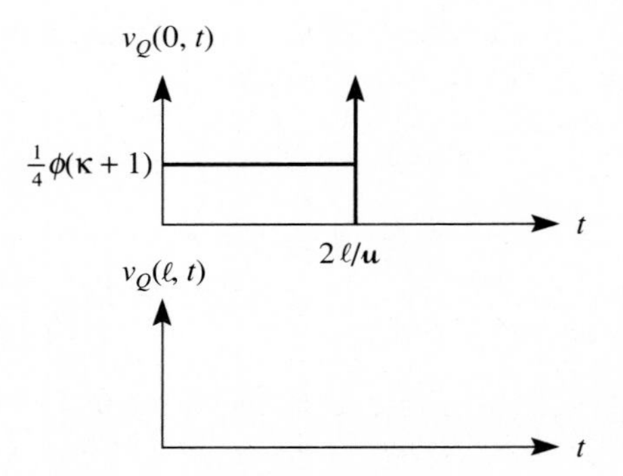

(*e*)

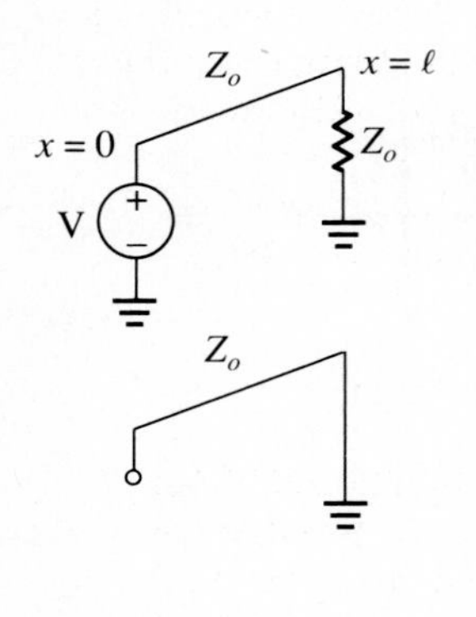

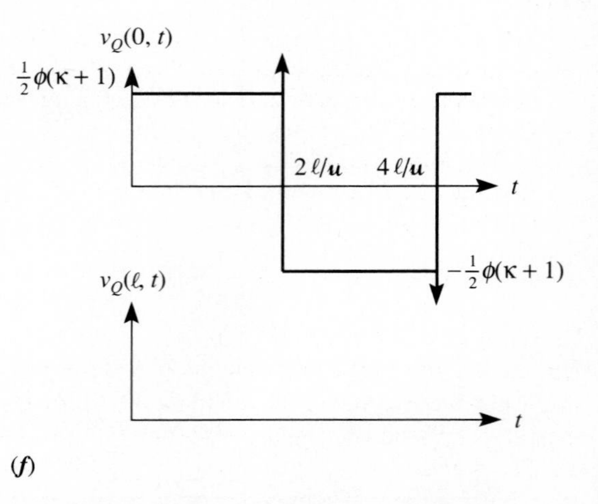

(*f*)

Figure 11.22 (*continued*) (d) shorted at $x = 0$ and terminated at $x = \ell$, (e) terminated at $x = 0$ and shorted at $x = \ell$, and (f) open at $x = 0$ and shorted at $x = \ell$.

Note from Equation (11.28) that the quiescent voltage is made up of three components. The first term on the right-hand side corresponds in the time domain to a wave having the same shape as the source voltage but reduced in amplitude by a factor of $\phi(\kappa + 1)/4$ and traveling in the positive x direction with a velocity u. The second term corresponds to a wave moving in the direction of decreasing x such that at $x = \ell$, the sum of the first two terms is zero.

Consider for the time being only the first two components on the right side of Equation (11.28) and suppose that $v(t)$ is a unit step function. As this voltage travels along the active line, according to the first term, the quiescent line is charged up to a voltage of $\phi(\kappa + 1)/4$ volts. When the step reaches the end of the active line, a negative step is produced in the quiescent line, represented by the second term. From this time on, the voltage on the active line does not change since it is correctly terminated and decoupled from the quiescent line. Meanwhile, though, the negative step on the quiescent line travels backward with a velocity u, resulting in the cancellation of the voltage on the active line. The net result at $x = 0$, then, is a pulse of duration $2\ell/u$. As we move farther down the line, the pulse decreases in width and eventually disappears at $x = \ell$. Consequently, the effect of these terms can be thought of as a backward wave; the induced wave travels in the opposite direction to the inducing wave.

The last term in Equation (11.28) represents a forward wave whose shape is determined by the time rate of change of the source voltage. Also note that its amplitude is proportional to x; this component can thus be crucial in determining the effects of crosstalk for large-amplitude pulses with fast edge rates that are coupled for an appreciable distance. On the other hand, however, this term is eliminated when all the electric and magnetic fields produced by the lines are confined to a homogeneous medium, such as the air above a ground plane because, in that case, $\kappa = 1$.

Example 11.6

Suppose that the propagation delay down a 3-meter active line that is terminated in its characteristic impedance is longer than the 1-ns rise time of the 5-V voltage step from a generator, and an otherwise identical 2.1-m quiescent line runs parallel to the former. Given $u = 17.68$ cm/ns, $C = .48$ pF/cm, $L = 6.91$ nH/cm, $C_m = .12$ pF/cm, and $L_m = 1.9$ nH/cm. With $\kappa = 1.12$ and $\phi = .25$, taking the inverse Laplace transform of Equation (11.28), we find the pulse width and amplitude of the quiescent voltage at $x = 0$ to be 24 ns and .66 V.

Case 2. Since logic gates display extremes in input and output impedance depending on the quiescent state of the gate, we cannot expect in practice that the quiescent line is properly terminated at both ends. Consider that the quiescent line is open at $x = 0$ and is correctly terminated at the other end of the line. Then, Equation (11.27) yields

$$V_Q\big|_{x=0} = \frac{1}{2}\phi(\kappa + 1)\{1 - e^{-2s\ell/u}\}V \tag{11.29a}$$

$$V_Q\Big|_{x=\ell} = \frac{1}{4}\phi(\kappa+1)\{e^{-s\ell/u} - e^{-3s\ell/u}\}V - \frac{1}{2}\phi(\kappa-1)(\ell/u)e^{-s\ell/u}(sV) \tag{11.29b}$$

Therefore, the pulse at $x = 0$ is twice the amplitude of that found for the previous case. At $x = \ell$, Equation (11.29b) predicts that the waveform consists of a pulse and a negative spike occurring at the leading end of the pulse with the actual shape depending on the fall time of the impulse instead of its rise time.

Case 3. Next, consider the case where the quiescent line is open at $x = \ell$ and correctly terminated at $x = 0$. Then, the quiescent voltage is expressed as

$$V_Q\Big|_{x=0} = \frac{1}{4}\phi(\kappa+1)\{1 - e^{-2s\ell/u}\}V - \frac{1}{2}\phi(\kappa-1)(\ell/u)e^{-2s\ell/u}(sV) \tag{11.30a}$$

$$V_Q\Big|_{x=\ell} = -\phi(\kappa-1)(\ell/u)e^{-s\ell/u}(sV) \tag{11.30b}$$

Equation (11.30a) reveals that the pulse and spike have the same magnitudes as in the previous case, but the spike now occurs at the trailing edge of the pulse. Equation (11.30b), on the other hand, suggests that only the spike is present at $x = \ell$, and its amplitude is twice that of the previous cases.

Case 4. Now, assume that, at $x = 0$, the quiescent line is connected to the saturated output stage of a driver, and the line is properly terminated at the other end. Approximating the resistance of the saturated transistor as zero, the voltages at either end are found to be

$$V_Q\Big|_{x=0} = 0 \tag{11.31a}$$

$$V_Q\Big|_{x=\ell} = \frac{1}{4}\phi(\kappa+1)\{e^{-3s\ell/u} - e^{-s\ell/u}\}V - \frac{1}{2}\phi(\kappa-1)(\ell/u)e^{-s\ell/u}(sV) \tag{11.31b}$$

Case 5. When the line is terminated in its characteristic impedance at $x = 0$, and is connected to a saturated load, we obtain the following quiescent voltage expressions:

$$V_Q\Big|_{x=0} = \frac{1}{4}\phi(\kappa+1)\{1 - e^{-2s\ell/u}\}V + \frac{1}{2}\phi(\kappa-1)(\ell/u)e^{-2s\ell/u}(sV) \tag{11.32a}$$

$$V_Q\Big|_{x=\ell} = 0 \tag{11.32b}$$

Case 6. Finally, if, instead of being correctly terminated at least at one end of the line, as in all previous cases, the quiescent line is shorted at $x = \ell$ and open at $x = 0$, then the voltage at $x = 0$ becomes

$$V_Q\Big|_{x=0} = \frac{1}{2}\phi(\kappa+1)\left(1 - 2\sum_{n=1}^{\infty}(-1)^n e^{-2ns\ell/u}\right)V + \phi(\kappa-1)\left(\frac{\ell}{u}\right)\left(\sum_{n=1}^{\infty}(-1)^{n+1}e^{-2ns\ell/u}\right)(sV) \tag{11.33a}$$

while at $x = \ell$, it is obviously

$$V_Q\big|_{x=\ell} = 0 \tag{11.33b}$$

Therefore, the waveform at $x = 0$ is periodic. With a unit step voltage on the active line at $t = 0$, there is a corresponding voltage step on the quiescent line with a magnitude of $\phi(\kappa + 1)/2$. Then, at $t = 2\ell/u$, a negative step of twice the magnitude occurs as well as a positive spike from the second term in Equation (11.33a). At $t = 4\ell/u$, a positive step and a negative impulse take place to nullify the preceding negative step and positive spike. Even though the pattern repeats itself theoretically, the waveform in actual practice eventually dies out due to the loss associated with the PCB traces.

Coaxial cable effectively eliminates crosstalk but is necessary only in the noisiest environments. Differential twisted-pair line connections are adequate for most applications and are typically less expensive and easier to use.

Twisted-pair cables offer several advantages. They avoid crosstalk problems by virtue of the common-mode rejection of line receivers. Twisting of the wires insures a homogeneous distribution of capacitances. Both capacitances to ground and to extraneous sources are balanced. This is effective in reducing capacitive coupling while maintaining high common-mode rejection.

11.5 Transient Switching Currents and Decoupling

Another type of noise that is encountered in digital circuits arises during transitions between two logical states. As the devices change state, current levels change because of the different device current requirements in each state, the transients caused by charging and discharging external capacitive loads, and the conduction overlap in the output stage.

The majority of current flow in a CMOS system, for example, is transient by nature, occurring on the waveform edges or transitions where instantaneous demand for current occurs. These current transients emanate from charging and discharging of the capacitive loads and can cause noise on power supply lines.

In addition, there is a brief period, on the order of a nanosecond, during which both output transistors are on simultaneously, so that the device draws a substantial supply current, giving rise to a current spike on the V_{CC} and ground leads to the gate. This spike, having di/dt as high as 5000 A/s, will react with the distributed inductance of the supply wiring as well as with the inductance in the ground and power supply leads of the package to produce significant voltage transients on V_{CC} and ground unless adequate supply *decoupling* is provided.

Similarly, a characteristic common to all TTL totem-pole output stages gives rise to a current transient when the output changes from a logical **0** to logical **1**. This spike is caused by the overlap in conduction of the output transistors. Due to the active pull-down circuitry employed in advanced Schottky subfamilies, the low-level driver will be only slightly in the linear region during conduction overlap, and the spike will be less compared to those seen in other TTL subfamilies.

When the transition is from low to high at the device output, the current needed to charge the load capacitance C_L is supplied by the supply voltage. When the output goes from high to low, however, C_L is shorted to ground by the low-level driver and has no effect on the supply current I_{CC}.

The total supply current switching transient is then a combination of three major effects: the difference in high-level and low-level supply currents, the charging of the capacitive load, and the conduction overlap. The charging of the load capacitance overshadows the other two effects as far as the noise produced on V_{CC} by switching current transients is concerned. Therefore, one precaution that can be taken by the system designer is to avoid the unnecessary stray capacitance in circuit wiring.

The magnitude of current transients is not the only factor in determining the size of supply voltage variations. If the time interval over which this current is switched is relatively long, the transient effect on the power supply voltage is minimal; switching the same amount of current more quickly, however, will have a greater impact on V_{CC} due to the parasitic inductance that results from system interconnections, decoupling capacitor leads, and device package contributions.

Thus, the actual transition time determines the amplitude and frequency spectrum of the generated signal at the higher harmonics. Application of the Fourier integral to waveforms in high-speed logic circuits reveals frequency components of appreciable amplitude that exceed 100 MHz. Because of the frequency spectrum generated when high-speed devices switch, a system using them must consider problems caused by RF even if the repetition rates are only a few MHz.

Noise at high frequencies of 100 MHz and above comes mainly from two sources. First, it results from fast switching times, as explained above. The greater part of the switching current from a low-to-high transition shows up in supply current surges; the bulk of the switching current from a high-to-low transition gives rise to surges in ground current. High-frequency noise is also transmitted through the changing magnetic fields that result from the changing electric fields in a switching line and are picked up on adjacent signal paths. Low-frequency noise, on the other hand, is caused by changes in the supply current demand as devices switch between states. Therefore, it is not only the switching frequency that generates the noise but also the frequency of the signal's slew rate. Typical noise characteristics of various logic subfamilies are given in Table 11.3.

For an output driving a capacitive load, the transient current is given by

$$i_t(t) = C_L \frac{dv_o(t)}{dt} \tag{11.34}$$

where $v_o(t)$ is the output voltage across the load, while the induced transient voltage is related to the transient current flow by

$$v_t(t) = -L \frac{di_t(t)}{dt} \tag{11.35}$$

where L is the total inductance, which is the sum of the line inductance and parasitic inductance due to each package lead and its bonding wire. These voltage spikes can appear at the output as false signals, thereby limiting the usefulness of the faster devices. The effect is additive as more outputs of an IC switch simultaneously. Even

Table 11.3 Noise Characteristics of Logic Subfamilies*

Subfamily	dV/dt (V/ns)	V_{CC}/Gnd Spike (pC)	ℓ_{max} (cm)
74	2.00	26	12
74H	2.75	12	9
74L	.47	98	52
74S	1.88	9.2	12
74LS	.67	31	37
74AS	1.24	5.5	16
74F	2.20	9.2	11
74ALS	1.18	12	21
74C	.02	440	1667
74HC	.63	40	60
74HCT	.87	15	42
74AC	2.29	5	16
74ACT	5.38	6.5	6
10K	.4	.8	15
100K	1.00	.25	5

*$V_{EE} = -5.2$ V for the 10K and 100K series, and $V_{CC} = 5$ V for all others

quiescent outputs may have unacceptable spikes due to crosstalk. To reduce the amplitude of the voltage spike it is necessary to consider two components, namely, the current slew rate and L.

Large values of current slew rate naturally lead to larger voltage spikes. Figure 11.23 shows the current through a transistor as it turns on and the resulting di/dt value over the same period. Note that the slew rate falls back to zero as the transistor current reaches a steady-state value. To reduce the di/dt peak, the output transistor can be made smaller so that the steady-state value of the current is lowered, which in turn slows the rate at which it reaches this new value. However, compromising the current-handling capability of the device unfavorably affects fan-out specifications, propagation delay, and the ability to drive transmission lines at high frequencies.

Consider as an example the CMOS output structure for a high-to-low transition that can be modeled as an RC network as shown in Figure 11.24, where r_{ds} is the finite on-resistance of the conducting channel. The transient current to charge or discharge the load is then given by

$$i_t(t) = \left(\frac{1}{r_{ds}}\right) e^{-t/r_{ds}C_L} V_i \qquad (11.36)$$

where the initial voltage across the capacitor V_i is either V_{OH} or V_{OL}. Since the transient currents flow through parasitic inductances between the supply and ground nodes,

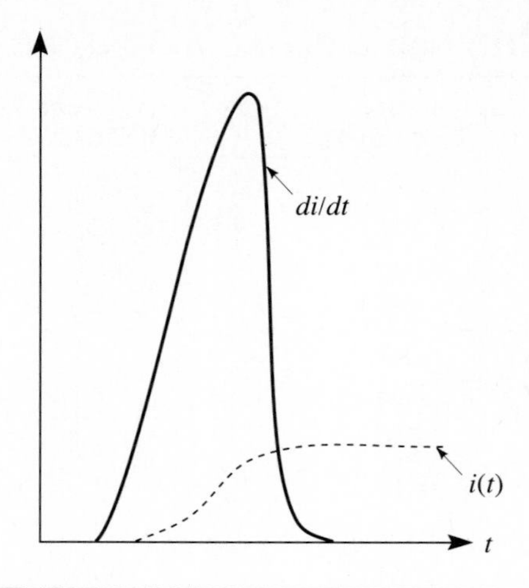

Figure 11.23 Current through a switching transistor and the corresponding slew rate.

the induced voltage spikes can be expressed, using Equations (11.35) and (11.36), as

$$v_t(t) = \left(\frac{L}{r_{ds}^2 C_L}\right) e^{-t/r_{ds}C_L} V_i \tag{11.37}$$

Decoupling

The simplest and most common method of tackling this type of noise problem is to decouple the supply line by using capacitors that filter out unwanted frequency components in the supply current. The main concern with power supply distribution lines is to reduce the areas of *current loops*. The significance of these lines is that they have access to every PCB in the system, so that the power supply current loop is a very large one susceptible to a lot of noise pickup.

Figure 11.25a shows a load circuit that draws current spikes from a supply volt-

Figure 11.24 Equivalent circuit of a CMOS output during a high-to-low transition.

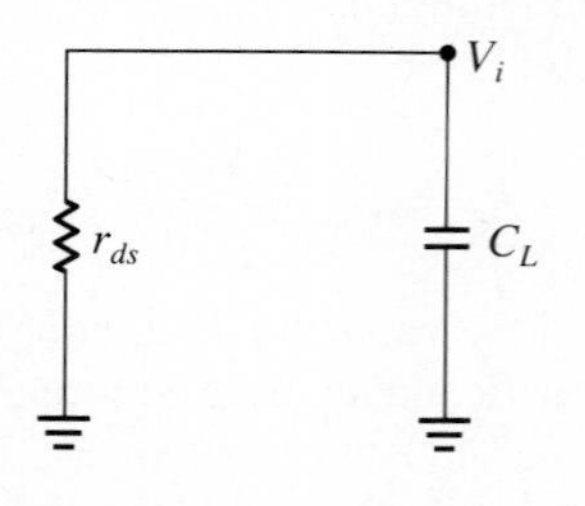

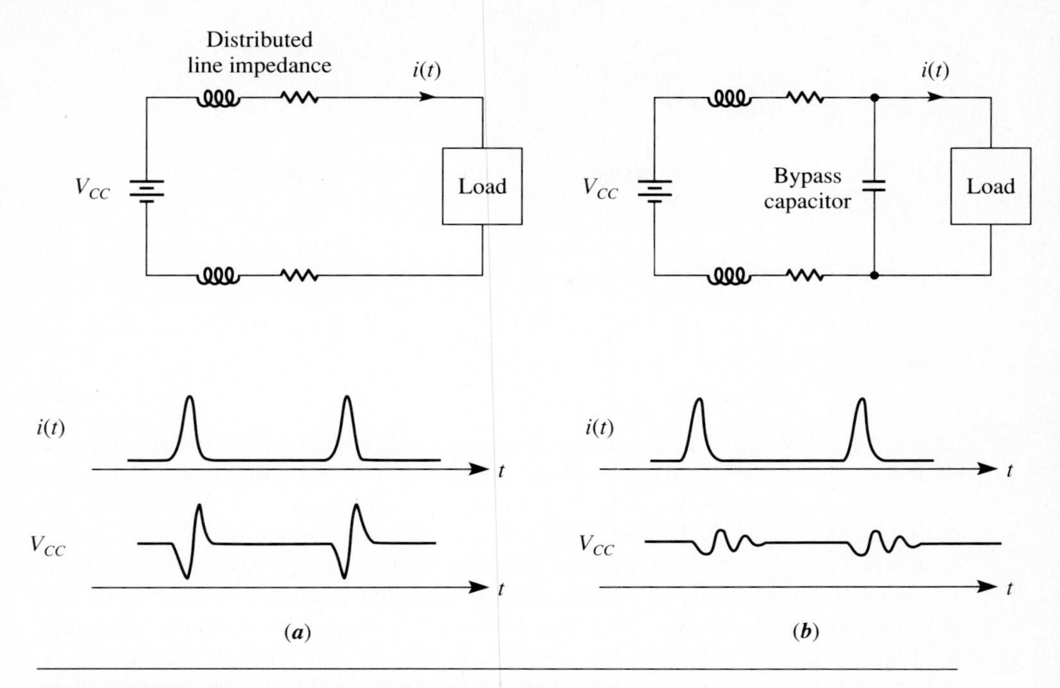

Figure 11.25 Load draws current spikes (a) from the supply, and (b) from the bypass capacitor.

age through the line impedance. The effect of the inductive coupling associated with a large loop area on the supply waveform is also shown. Connection of a *bypass capacitor*, as depicted in Figure 11.25b, acts as a nearby source of charge to supply current spikes through a smaller line impedance. It also defines a much smaller loop area for the higher frequency components of the noise. Between current spikes it recovers via the line impedance. For the bypass capacitor to be able to supply the current spikes required by the load, the inductance of this current loop must be kept small.

For effective filtering and decoupling, the capacitors must be able to supply the change in current for a period of time greater than the pulse width of this current. Since the problem is essentially one of dc level changes due to transitions between logic states coupled with high-frequency transients associated with the changes, two different values of time constant must be taken into account. Capacitors having both high capacitance for long periods of time and low series reactance for fast transients are prohibitive in cost and size. A good compromise is the arrangement shown in Figure 11.26, in which a low-frequency decoupling capacitor and an RF bypass capacitor are utilized in parallel. A typical value for the high-frequency bypass capacitor C_1 may be found from

$$C_1 = \frac{I_{CC}}{\Delta V_{CC}/\Delta T}$$

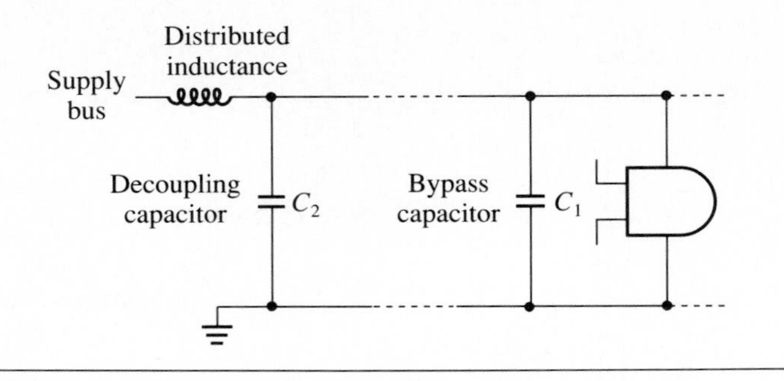

Figure 11.26 Decoupling of the power distribution.

where ΔT is the worst-case transient time. For example, with a 3-ns rise time, a 75-mA supply current, and a maximum supply droop of .1 V, we get $C_1 = .03\ \mu\text{F}$.

The same method may be used to find the value of the low-frequency board decoupling capacitor C_2. However, the factor ΔT becomes somewhat ambiguous, and an analysis of the current cycling on a statistical basis is the best method to determine its value. It will normally be a 10 to 100 μF electrolytic capacitor placed where the power supply enters the PCB. Its purpose is to accommodate the continually changing I_{CC} requirements of the supply bus line and thereby provide enough energy storage to prevent supply droop while refreshing the charge on the bypass capacitors that are placed near the ICs. The latter are what actually provide the current spikes to the chips. A bypass capacitor will normally be a .01 to 1 μF capacitor connected to the IC by traces that minimize the area of the loop formed by the capacitor and the IC. The decoupling path is the trace distance from a power pin through a bypass capacitor and to package ground. The impedance of this path is determined by the line inductance and the series impedance of the bypass capacitor. Since the current transients usually have significant harmonic content over 100 MHz, the line inductance is one of the most critical factors. It can be minimized by providing a *power plane*.

Proper and improper placements of bypass capacitors are shown in Figure 11.27 while Figure 11.28a illustrates a circuit for testing the effectiveness of decoupling. In this circuit, the supply and ground connections consist of two parallel copper strips on an epoxy-glass circuit board. Figure 11.28b depicts the increase in V_{CC} transients as a .01-μF bypass capacitor between the supply and ground is physically moved away from the load. The results picture the correct procedure to obtain adequate coupling. As evident from Figure 11.28b, bypass capacitors should be located as close as possible to the IC in order to maximize noise margins.

Most capacitors, because of their leads and the nature of their dielectrics, tend to become inductive or lossy at higher frequencies. This is especially true of electrolytic capacitors; mica, glass, ceramic, and polystyrene dielectrics work well to several hundred MHz. Bypass capacitors for high-speed logic circuits should have low *equivalent series impedance*, which is primarily made up of series inductance and resistance internal to the capacitor.

It is now clear that the capacitor with its lead inductance forms a series *LC* circuit. Below the *resonant frequency*, the net impedance of the combination is capacitive.

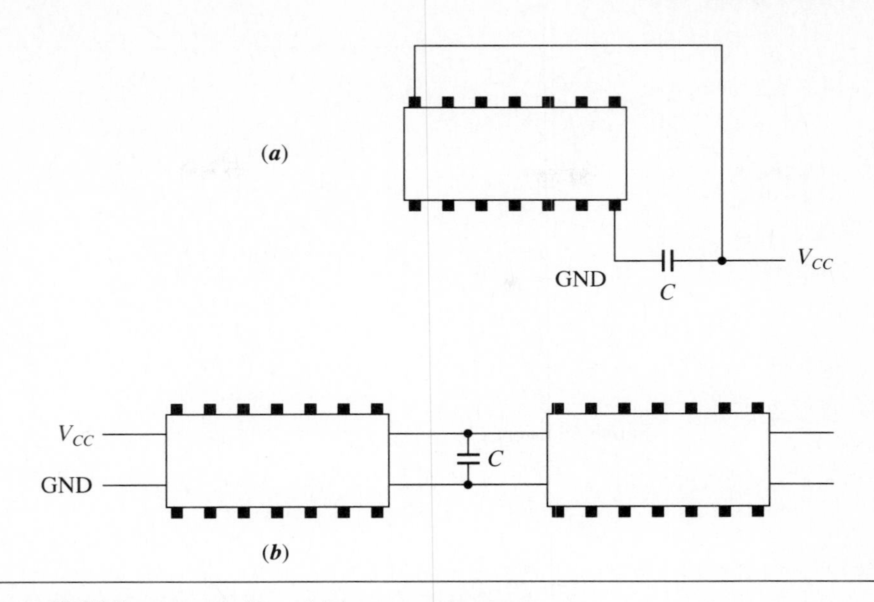

Figure 11.27 Bypass capacitors: (*a*) improper placement; (*b*) proper placement.

Above that frequency, it is inductive. Thus, a bypass capacitor is capacitive only below the resonant frequency as given by

$$f_o \equiv \frac{1}{2\pi \sqrt{LC}} \tag{11.38}$$

where L is the lead inductance between the bypass capacitor C and the chip. On a PCB, this inductance is determined by the layout and is the same whether the value of the capacitor is .01 μF or 1 μF. Thus, increasing the capacitance lowers the series resonant frequency. Since a length of wire has no inductance at all, figures quoted by manufacturers on the inductance of a length of wire are usually based on a presumably very large loop area such that the magnetic field produced by the return current has no cancellation effect on the field produced by the current in the given length of wire. Such a loop geometry is of course not the case with the decoupling loop.

Supply-line glitches are not always picked up in the distribution networks but can come from the power supply circuit itself. In this case, a well-designed distribution network faithfully delivers the glitch throughout the system. In case the usual board decoupling techniques are not effective in removing it, employment of an on-board voltage regulator can be used. An additional advantage of this approach is the alleviation of the requirements on the heat sinking at the supply circuit. Its disadvantage is the possibility that different boards would be operating at slightly different supply levels due to the tolerance in the regulator ICs, which in turn would lead to different logic levels from board to board, giving rise to implications that may vary from nothing to latchups, depending on the types of logic families being used.

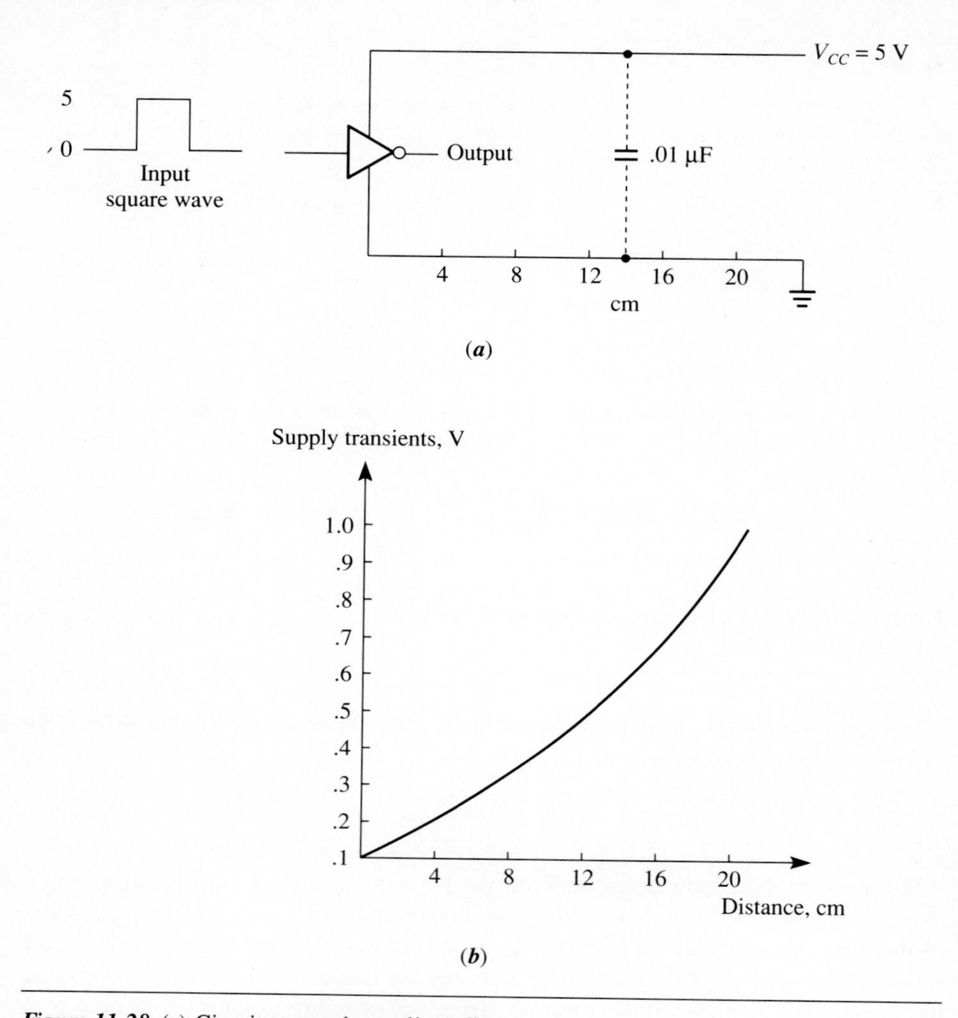

Figure 11.28 (a) Circuit to test decoupling effects, and (b) supply transients as a function of bypass capacitor distance from the IC.

Unused inputs on TTL devices float at the transition region, anywhere from 1.1 V to 1.5 V, depending upon the IC and the logic subfamily. While this usually produces a logical **1**, many application problems can be traced to floating inputs because they are susceptible to induced noise transmitted from other lines and can easily switch the state of the device. Therefore, a good design rule is to tie unused inputs to a logic level. They are tied to V_{CC} through a resistor-to-supply I_{IH} current which is several orders of magnitude smaller than I_{IL}. The resistor is used to protect the input against possible supply voltage surges.

Effect of Output Impedance on Noise Margins

The ability of a logic element to operate in a noisy environment involves more than the noise margins discussed in previous chapters. The amount of noise needed to develop a given voltage is a function of the circuit impedance. Even though the low output impedance of TTL circuits improves their noise immunity, fast-operating systems, for example those using advanced Schottky TTL subfamilies, are still susceptible to noise.

To show how and under what circumstances the low impedance of an active TTL output rejects noise spikes, consider the circuit of Figure 11.29a where the outputs of *G1* and *G2* are coupled simply by the stray capacitance. In practice, the coupling is usually more complex but resolvable into *RLC* series coupling elements. Moreover, the driving source impedance is ignored because the source effect is in a direction to improve rather than degrade the noise rejection, so that our analysis will result in a worst-case type of response reaction.

Figure 11.29b illustrates the simple *RC* equivalent circuit, where $v_i(t)$ is the voltage waveform at the *G1* output that can be considered as a stimulated noise pulse from *G2* and *G3*'s points of view, $v_o(t)$ is the coupled voltage at the *G2* output, *C* is the coupling stray capacitance, and *R* represents the nominal output impedance of *G2*. Assuming that a fast-rising waveform at the *G1* output is essentially a ramp, as shown in Figure 11.29c, we have

$$v_i(t) = \frac{V_{max} t}{t_r}$$

where V_{max} is the maximum value of the *G1* output voltage, and t_r is its rise time. It is easy to verify that the output pulse will be given by

$$v_o(t) = \left(\frac{RCV_{max}}{t_r}\right)\left[1 - \exp\frac{-t/t_r}{RC/t_r}\right] \qquad (11.39)$$

Now, define the normalized values of the time constant, output pulse time, and output voltage as

$$\tau \equiv \frac{RC}{t_r}$$

$$\iota \equiv \frac{t}{t_r}$$

$$\nu(\iota) \equiv \frac{v_o(\iota)}{V_{max}}$$

respectively. Then Equation (11.39) can be rewritten in terms of these normalized values as

$$\nu(\iota) = \tau(1 - e^{-\iota/\tau}) \qquad (11.40)$$

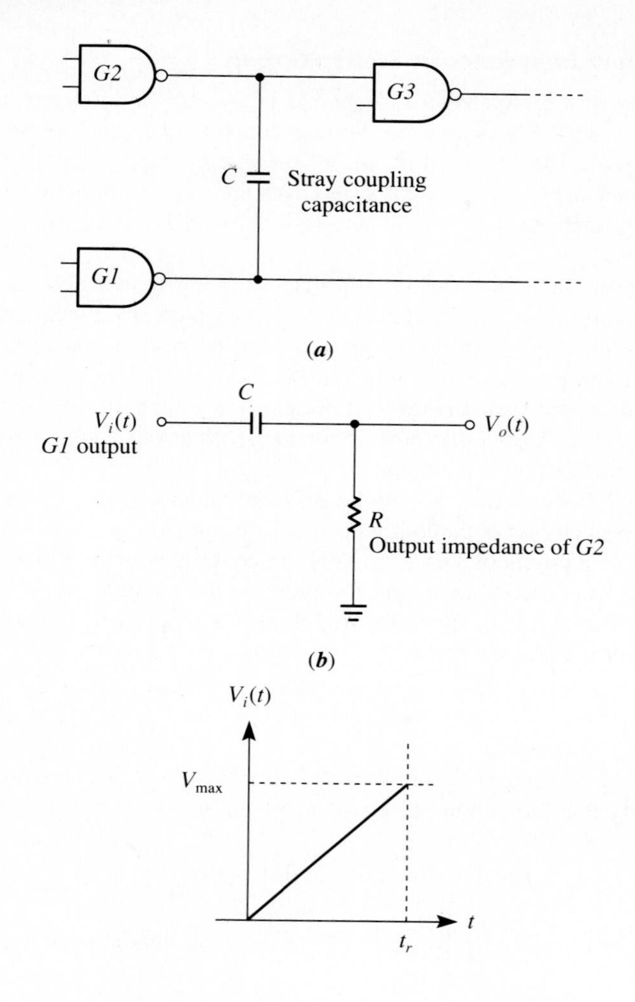

Figure 11.29 (a) Capacitance coupling between two TTL outputs, (b) equivalent circuit, and (c) voltage waveform at *G1* output.

which holds for unit time. Thus, the voltage decays exponentially with a time constant τ. This is plotted, for various values of τ, in Figure 11.30, from which the pulse width and amplitude of the coupled noise can be estimated.

Example 11.7

Suppose that we utilize ALS gates and the *G1* output changes state from 0 to 3 V, rising at 1 V/ns, while *G2* is at a logical **1**. Assuming a coupling capacitance of 10 pF and an

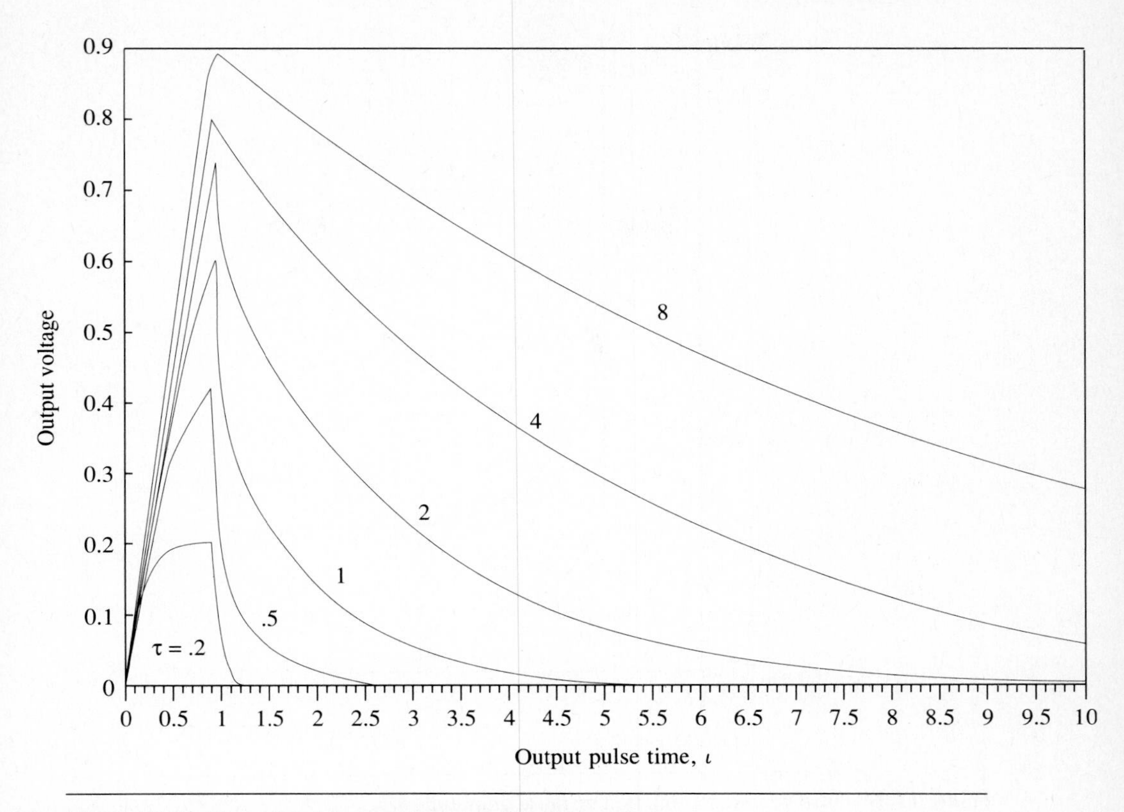

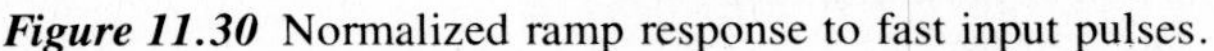

Figure 11.30 Normalized ramp response to fast input pulses.

output impedance of 58 Ω for *G2*, we find the rise time to be

$$t_r = \frac{3 \text{ V}}{1 \text{ V/ns}} = 3 \text{ ns}$$

so that

$$\tau = \frac{RC}{t_r} = 58 \times 10 \times \frac{10^{-12}}{3} = 193 \text{ ps}$$

Using the $\tau = .2$ ns curve in Figure 11.30 yields a peak voltage v_o of .57 V and a pulse width of 3 ns at the 50% points. Now, in order to determine if this pulse causes an interference at the *G2* output and is propagated by *G3*, we consider the graph of Figure 11.31, which shows the input noise immunity for AS and ALS gates in terms of the pulse amplitude and width. It is clear that for (.57 V, 3 ns), the pulse will be rejected, and the gates should not be affected.

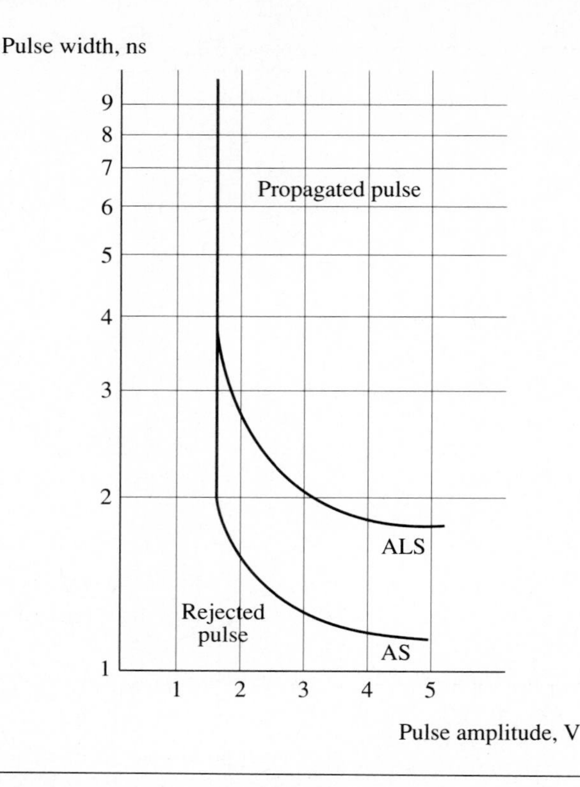

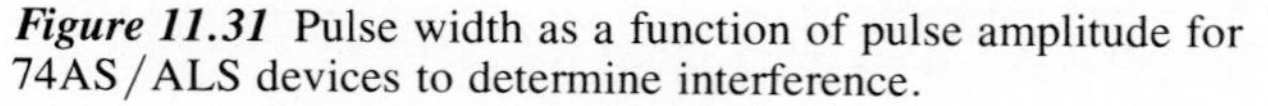

Figure 11.31 Pulse width as a function of pulse amplitude for 74AS/ALS devices to determine interference.

11.6 Electromagnetic Interference

In the case of noise entering from external sources, the interconnections between circuits act as antennas. Motors, power switches, fluorescent lights, and electrostatic discharge (ESD) are all sources of EMI. This type of noise is coupled capacitively or inductively. It can also be directly *conducted* into digital systems when power supply lines must be shared with other circuits. Thus, decoupling and filtering of these lines should be a standard design procedure, as already mentioned in the previous section. In order to prevent *radiated* EMI from entering the equipment, *shielding* is utilized.

The analogous relationship between circuits coupled capacitively and inductively is summarized in Table 11.4, and depicted in Figure 11.32. When the noise is coupled inductively, voltage noise $v_n(t)$ appears in series with the receiver circuit; in the case of capacitively coupled noise, the voltage noise across the receiver is caused by the noise current $i_n(t)$ flowing through the receiver. Also note that reducing the receiver impedance Z will reduce the capacitively coupled noise. This may not be the case in the inductively coupled circuits.

Table 11.4 Characteristics of Capacitive and Inductive Coupling

	Capacitive coupling	Inductive coupling
Noise source	Voltage slew rate	Current slew rate
Coupling medium	Mutual capacitance	Mutual inductance
Coupled noise	Current	Voltage

Shielding against Electric Fields

Since capacitive coupling is caused by electric fields, shielding against it corresponds to shielding against the latter. The way to prevent capacitive coupling is to enclose the circuit or conductor in a grounded metal known as a *Faraday shield* to shunt the interference current to ground. This is illustrated in Figure 11.33. A shield works because a charge cannot exist on the interior of a closed conducting surface.

Shielding against Magnetic Fields

Strong magnetic fields are found where cables carry current or where ac power is distributed, and near machinery, power transformers, fans, and so forth. The physical mechanism involved in inductive coupling is the linkage of a magnetic flux Φ from an external source with a current loop in the circuit and the resultant voltage in the loop in accordance with *Faraday's Law*, which states that the induced electromotive force is directly proportional to the time rate of change of magnetic flux through the circuit or

$$emf = -\frac{d\Phi}{dt} \tag{11.41}$$

which can be expressed in terms of the area A bounded by the current loop as

$$emf = -A\left(\frac{dB}{dt}\right)\cos\theta \tag{11.42}$$

Figure 11.32 Comparison of capacitive and inductive noise coupling.

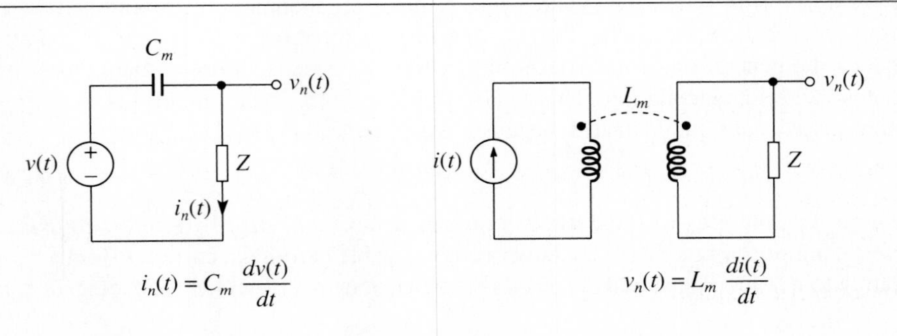

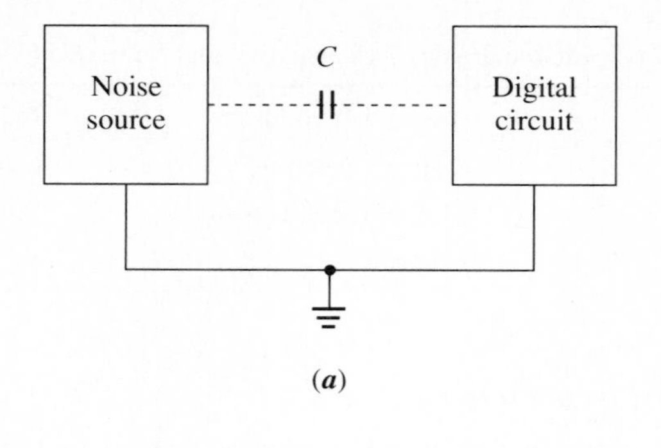

(*a*)

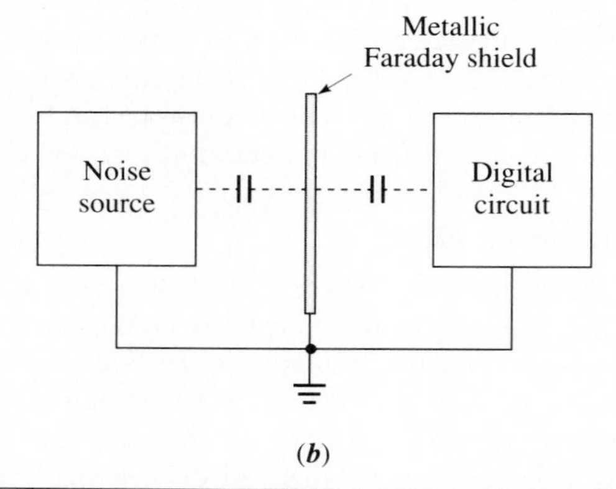

(*b*)

Figure 11.33 Employment of a Faraday shield to prevent capacitive coupling.

where B is the magnetic flux density and θ is the angle B makes with the normal to the loop.

Any flow of current generates a magnetic field whose intensity varies inversely with the distance from the wire that carries the current. Consider, for example, two parallel wires carrying currents $\pm I$, as in signal and return lines. These lines will generate a nonzero magnetic field near the wires where the distance from a given point to one wire is noticeably different from the distance to the other wire. However, farther away, where distances from a given point to either wire are the same, the fields from both wires will cancel out. Therefore, maintaining proximity between signal feed and return paths, that is, minimizing the current loop area, is a way to reduce the generation of and susceptibility to EMI because holding them closer promotes the field cancellation. This is also obvious from Equation (11.42). It also implies a small loop inductance L because, from Equation (11.41),

$$\Phi \equiv LI \tag{11.43}$$

Minimizing the area of the current loop can be accomplished by employing coaxial cable shielding. Figure 11.34a shows a coaxial cable carrying a current I from a signal source to a load. The shield carries the same current as the center conductor. Outside

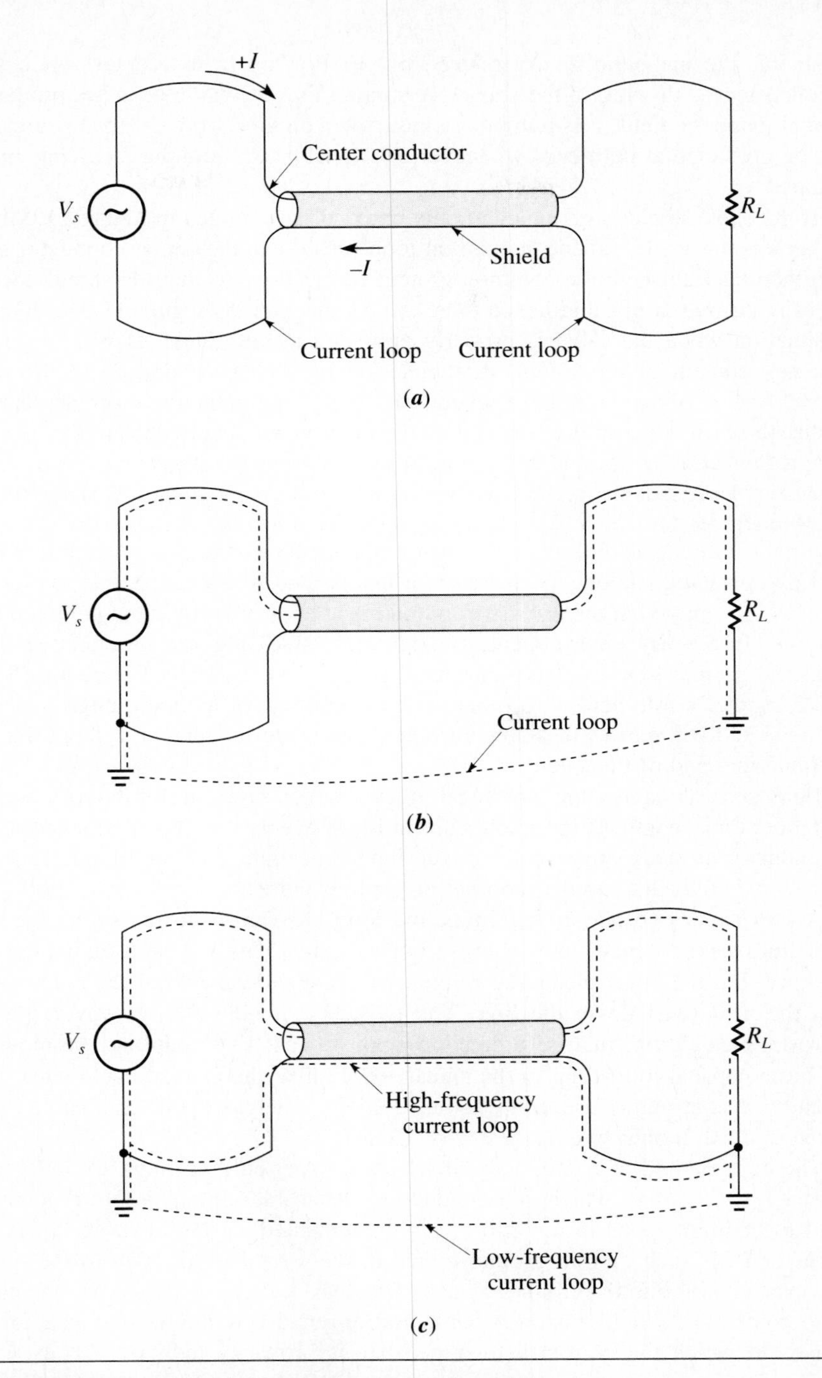

Figure 11.34 (a) A coaxial cable carrying current from the source to the load; (b) grounded at only one end, the shield has no effect; and (c) cable grounded at both ends.

the shield, the magnetic field produced by $+I$ flowing in the center conductor is cancelled by $-I$ flowing in the shield. Assuming that the cable does not produce an external magnetic field, it is immune to inductive coupling from external sources and must be grounded at both ends whenever the signal source and the receiving end are grounded.

If the cable shield is grounded at only one end, as depicted in Figure 11.34b, the loop area is not well-defined; the current loop runs down the center conductor of the cable then back through the common ground connection, so that the shield does not carry any current at all, and hence there is no cancellation. Figure 11.34c diagrams the situation when the cable is properly grounded at both ends. Depending on the frequency content of the signal, the shield carries at least a portion of the return current. At low frequencies up to several kHz where the inductive reactance is insignificant, the current will follow the path of least resistance, that is, the ground, whereas above a few kHz, as the inductive reactance predominates, the current will follow the path of least inductance, that is, the path of minimum loop area. Therefore, at high frequencies the shield carries in the opposite direction virtually the current of the same magnitude as the center conductor and is effective against reception of EMI.

Any unwanted and unexpected current in a ground line is referred to as a *ground loop* problem, in which the true earth-ground is not really at the same potential in all locations. In a noisy environment, coaxial cable shielding can be effective if one breaks the ground loop to insert an optical coupler, as shown in Figure 11.35. The optical coupler's role here is to redefine the signal source as being ungrounded, so that the shield still carries the same current as the center conductor without the need to ground that end of the cable.

Real coaxial cables are not ideal: they do not have uniform cross sections throughout their length. If the shield current is not evenly distributed around the center conductor at every cross section, then field cancellation is not complete, so that the effective area added to the loop by the cable is not zero.

A less expensive way to minimize the loop area is to run the signal feed and return lines next to each other using a twisted pair. This not only maintains their proximity, but the noise picked up in one twist tends to cancel out the noise picked up in the next twist down the line. The twisted pair does not, however, provide electrostatic shielding, that is, protection against capacitive coupling. Moreover, it adds more capacitive loading to the signal source than the coaxial cable does. Consequently, it is normally useful up to only 1 MHz, whereas the coaxial cable can be utilized at much higher frequencies up to 1 GHz.

The best method to reduce loop areas when many current loops are involved is to use a *ground plane*, which is a conducting surface serving as a return conductor for all the current loops in a circuit. It is implemented as one or more layers of a multilayer PCB such that all ground points in the circuit go directly to the ground plane instead of going to a grounded trace. This leaves each current loop in the circuit free to complete itself in whatever configuration yields minimum loop area for frequencies in which the ground path impedance is primarily inductive. Thus, if the signal path for a given signal meanders across the PCB, the return path for this particular signal is free to roam right along beneath it on the ground plane in such a way as to minimize the energy stored in the magnetic field produced by this current loop,

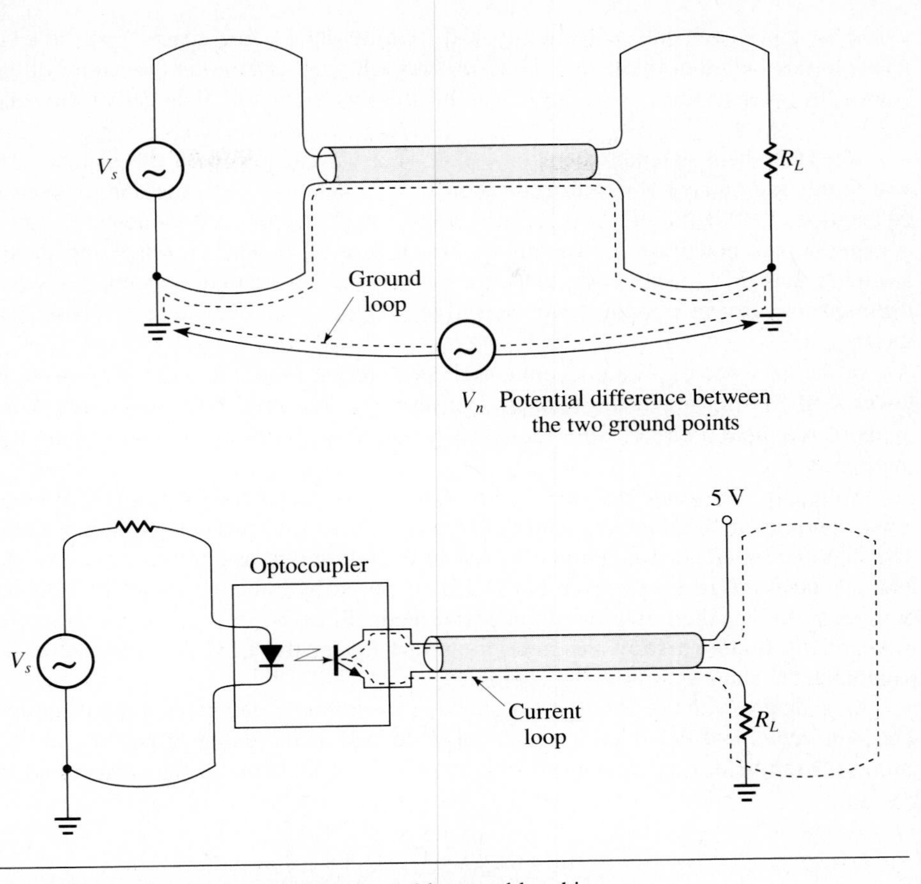

Figure 11.35 The illustration of the ground loop and breaking it using an optocoupler.

resulting in minimal susceptibility to inductive coupling. Note that the important thing here is to let the ground currents distribute themselves around the entire area of the board as freely as possible, so the current loops can reduce their own magnetic fields.

Shielding against Time-Varying Electromagnetic Fields

A time-varying electric field generates a time-varying magnetic field, and vice versa. Far from the source of a time-varying *electromagnetic field* (*EM*), the *wave impedance*, that is, the ratio of the amplitudes of the electric and magnetic fields is given as

$$\frac{E}{H} \equiv \sqrt{\frac{\mu_o}{\epsilon_o}} = 377\ \Omega \tag{11.44}$$

where μ_o and ϵ_o are the permeability and permittivity of free space, respectively. Close to the source of the fields, however, this ratio can depend on the nature of the source. Anywhere the ratio is significantly different from 377 Ω is called the *near field*.

The near field extends about one-sixth of a wavelength from the source, and within this region an RF interference problem can be almost entirely due to electric or magnetic field coupling. This fact can affect the choice of an RF shield.

In the presence of a whip antenna such as a wire-wrap post, the wave impedance is higher than 377 Ω, implying that it is mainly an electric field generator, so that techniques to protect a circuit from capacitive coupling would be effective against RF interference.

In the presence of a loop antenna such as a current loop, the wave impedance is lower than 377 Ω, suggesting that it is primarily a magnetic field generator; thus, methods to shield a circuit from inductive coupling would be successful against RF interference.

A metallic RF shield may be required in certain cases. Time-varying EM fields induce currents in the shielding material, which in turn dissipates energy in two ways. The first one is I^2R, that is, ohmic, losses and is referred to as an *absorption loss A*. The second is called a *reflection loss B* and is caused by radiation losses as induced currents re-radiate their own electromagnetic fields. Since the energy for both of these mechanisms is drawn from the interfering EM fields, the EMI is weakened as it penetrates the shield.

As a magnetic field penetrates a shield, its amplitude decreases exponentially. The *skin depth* δ of the shielding material is defined as the depth of penetration required for the field to be attenuated to $e^{-1} = 37\%$ of its value in free space, and is given by

$$\delta = \sqrt{\frac{2}{\omega\mu\sigma}}$$

Table 11.5 lists typical values of δ for several materials at various frequencies. Note that steel yields at least an order of magnitude more effective shielding at any frequency than copper or aluminum.

Absorption loss is the primary shielding mechanism for magnetic fields, and reflection loss is the main protective mechanism for electric fields. Both loss mechanisms are dependent on the frequency $\omega = 2\pi f$ of the impinging EMI field as well as on the permeability μ and conductivity σ of the shielding material. Furthermore, absorption loss is a surface phenomenon and also dependent on the thickness t of the shielding material. These losses can be expressed as

$$A = 8.69\left(\frac{t}{\delta}\right)\text{ dB} \tag{11.45}$$

$$B = 168 - 10\log\left(\frac{\mu f}{\sigma}\right)\text{ dB} \tag{11.46}$$

Equation (11.45) indicates that the magnetic field shielding is more effective at high frequencies and with material that has both high conductivity and permeability. Fig-

Table 11.5 The Skin Depth, δ, versus Frequency

	Skin depth, mm		
Frequency, KHz	Copper	Aluminum	Steel
.06	8.51	10.9	.86
.1	6.6	8.46	.66
1	2.08	2.67	.2
10	.66	.846	.066
100	.208	.267	.02
1000	.066	.0846	.0066

ure 11.36 shows the absorption loss as a function of frequency for steel and copper with different values of thickness.

From Equation (11.46), we see that the electric field shielding is more effective if the material is highly conductive, and that low-frequency fields are easier to block than high-frequency fields. Figure 11.37 illustrates the dependence of the reflection loss on the frequency for various metals.

In practice, the choice of a shielding material is less important than the seams, joints, and holes present in the physical structure of the enclosure, which may cause RF leakage. The current must be allowed to flow freely without having to swerve around them in order for the shield not to lose much of its effectiveness.

11.7 Voltage Surges and Overcurrent

Most line-powered electronic products and systems are protected by service power line circuit breakers and fuses. However, complete circuit protection demands protection against *voltage surges* and *overcurrent*, and is being encouraged by the higher

Figure 11.36 Absorption loss as a function of frequency.

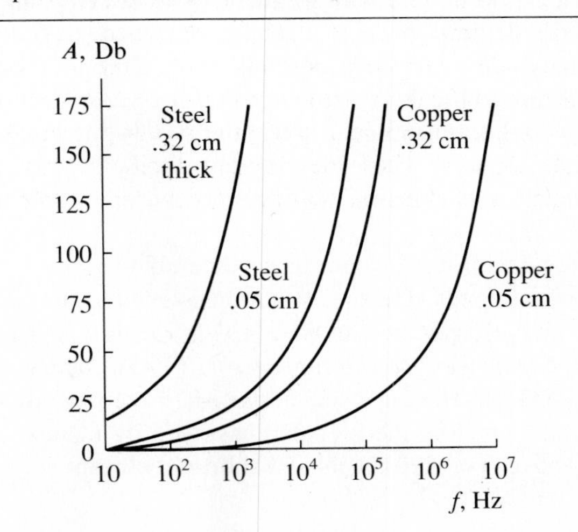

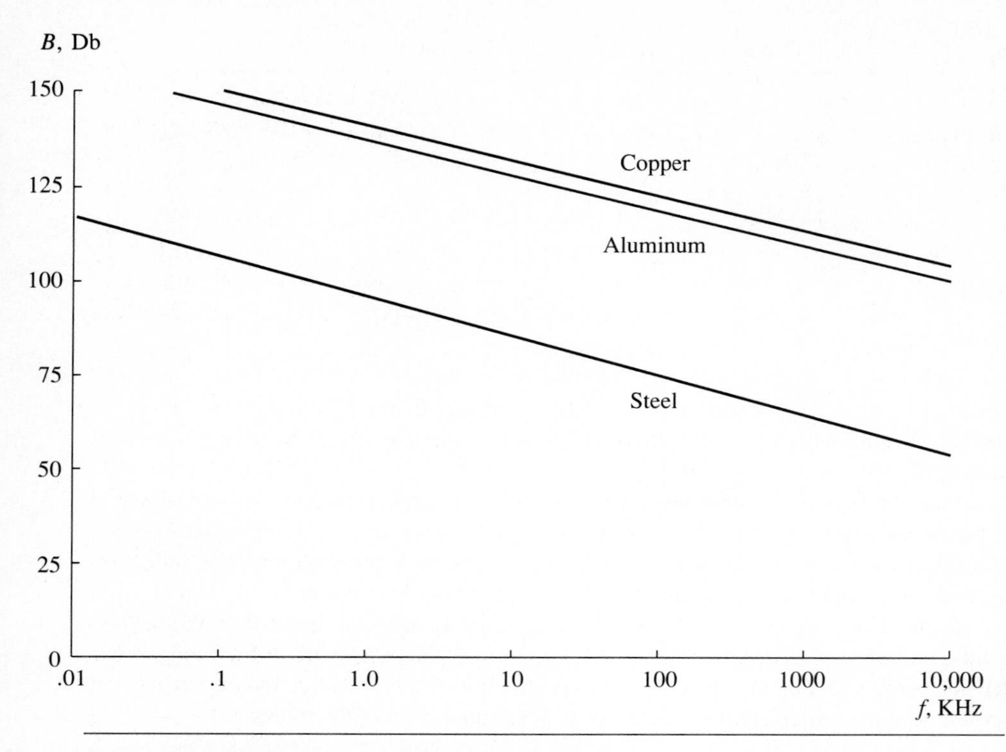

Figure 11.37 Reflection loss as a function of frequency.

performance of the products being protected and the growing vulnerability of VLSI circuits to voltage transients, current surges, ESD, and EMI.

One source of power line noise is due to high-voltage transients in inductive circuits such as relays, solenoids, and motors when they are turned on or off. When devices having high self-inductance are turned off, the collapsing fields can generate transients in the order of kilovolts, with frequencies from .1 Hz up to a few MHz.

Circuit-protection devices prevent damage or destruction to electronic circuits that can be caused by excessive currents and voltages. *Fuses* and miniature magnetic or thermal circuit breakers, called *circuit protectors*, are good at handling current surges that would be insufficient to trip the power line breaker but that could destroy more delicate circuits. These devices are similar to those used to protect household, commercial, and industrial wiring, motors, and appliances from excess currents.

Sensitive semiconductor devices are also vulnerable to fast-acting voltage transients occurring in picoseconds. Devices that are designed to protect semiconductors against high-speed overvoltages include *metal-oxide variable resistors* (*MOVs*), silicon Zener diode type *transient voltage suppressors* (*TVSs*), and gas discharge *surge voltage protectors* (*SVPs*). Having been designed to recover after the passage of transient voltages and to restore themselves unassisted by manual action, these devices clip voltages above specified levels and short out current surges.

Protection against Overcurrent

Fuses are overcurrent protection devices placed in series with the electronic circuits they are protecting. When the current exceeds the fuse's rated value, the conductive element in the fuse melts or blows, thereby opening the circuit and isolating the component's downstream. Their slow response time is in the order of milliseconds, and they offer no protection against high-speed voltage transients.

There are no clear distinctions between the kinds of fuses used in electronics equipment and those used in electrical equipment. They are expendable, one-time devices that must be replaced after they are blown. By contrast, circuit protectors must be manually restored or reset after contact tripping.

Building-service main-power circuit-breaker design has recently been adapted to electronic components for direct mounting on electronic equipment. These miniature circuit breakers have been termed *circuit protectors* in order to distinguish them from the larger heavy-duty products. As in the case of fuses, they guard against overcurrent that may result from catastrophic short circuits or failures in critical components, which can build to destructive levels of thousands of amperes.

Circuit protectors are small, light-duty circuit breakers that are designed with contacts that open up during overcurrent surges. They are manually resettable, and remain reset only if the electrical conditions causing the overcurrent have been cleared. They are intended to replace fuses and employ either a magnetic protection mechanism that trips in the presence of the field produced by the overcurrent, or a thermal protection based on a bimetal element that moves to open the circuit when exposed to heating caused by the overcurrent.

The *magnetic* circuit protector is favored in applications in which it is important to differentiate between nondestructive *nuisance* spikes due to the presence of inductive elements in the circuit and the catastrophic overcurrent that is caused by short circuits. The hydraulic *dashpot* in the magnetic protector senses time relationships in overcurrents, allowing the device to recognize truly destructive overcurrents. A typical magnetic circuit protector is programmed in such a way that the time delay dashpot will respond to an overcurrent 25% over its rating but will not trip, permitting it to distinguish between normal high inrush and overload; thus, the unit may ignore momentary spikes larger than the rated current to avoid false tripping yet react quickly to an actual overload six to eight times the rated current.

This mechanism utilizes a solenoid with a clapper-type armature linked by collapsible couplings to electrical contacts. When a current surge is in excess of the protector's rated value, the solenoid is actuated to create a magnetic field that attracts a hinged armature similar to that of an electromagnetic relay. However, in contrast to the relay design, the protector armature closure does not directly open the electrical contacts but collapses a train of linkage within the protector to permit fast-acting springs to snap open the contacts without arcing or sticking. When the conditions that cause the overcurrent are cleared, the linkage can be reset manually with a toggle or lever. Thus, protectors can be dual-purpose components, serving both as switches and protectors.

The time delay dashpot, included in the protector to minimize nuisance tripping by permitting the contacts to remain closed during routine nondestructive overcurrents, is a ferrous metal core. It slides freely within a silicon-oil-filled tube and is mounted within a solenoid coil. With an overcurrent, the solenoid's magnetic field will pull up the core against an internal spring. If the overcurrent is sustained, the core will move to the end of the armature's dashpot, tightly compressing the spring. The core reinforces the magnetic field until it is great enough to pull in the armature, collapsing the linkage and actuating the contact-opening springs.

The time delay can be changed by the user to meet specific circuit-protection requirements. It may slightly be affected by the ambient temperature due to changes in the viscosity of the silicon oil. However, the magnetic protector is essentially unaffected by the thermal effects of the overcurrent.

Thermal protectors can also be manually reset, but there may be a slight delay until the element cools enough to permit it to reopen the contacts. Unlike the magnetic protector, the thermal unit has a slower response time and no provision for adjusting delays related to time and current surge. Its external contacts are tripped when the heating effects of the overcurrent are sufficient to cause its bimetal element to bend.

Protection against Voltage Surges

MOVs have been available now for over a decade. A MOV is a zinc-oxide nonlinear resistor whose resistance changes as a function of applied voltage. It provides very reliable protection against repeated high-voltage transients such as those produced by lightning, switching surges, and noise spikes. Since it furnishes circuit-clamping protection in either direction, it is mostly employed in ac circuits. One disadvantage of a MOV is that it can wear out with long-term use; its ability to clamp voltages degrades with each pulse, even when the pulsed ratings are not exceeded.

The smaller MOV units are used to protect semiconductor devices, including thyristors, against locally produced relay surges; the large-bulk units can suppress voltage transients in industrial motor-control and electrical power distribution applications. With the increasing use of electronics in automobiles, a new market has opened recently for these devices to protect sensors, transistors, and diodes in ignition and fuel-injection systems.

Being nonlinear resistors with resistance changes that are a function of the applied voltage, MOV devices have bilateral and symmetrical characteristic curves, as a result of which the circuit protection is in either direction. With a voltage above the device's rating, its resistance drops sharply, and the MOV effectively becomes a short circuit to bypass the voltage transient while absorbing the flowing current.

These protective devices are made up of a material formed from powdered metal oxides, particularly zinc-oxide, mixed with binders and pressed into disks, blocks or cylinders. When fired at temperatures above 1000°C, the pressed slugs become a matrix of conductive metallic oxide grains separated by highly resistive boundaries. The amount of energy absorbed is limited by slug volume, but they have a high short-term absorption capability.

Silicon Zener diodes, when used as protective devices, are called transient voltage suppressors, or TVSs. They have large junction areas that can withstand high pulse power. Among their advantages are their precise voltage clamping and their lack of *wear-out* mechanisms.

A silicon Zener diode has inherent protective qualities because of its reverse-bias voltage-clamping characteristic. The diode will break down and become a short circuit when the applied voltage exceeds its rated avalanche level. When the applied reverse-bias voltage is reduced below the breakdown level, the current is restored to its normal saturation level.

TVSs are Zener diodes optimized for this protective function. They have better surge-handling characteristics than conventional Zener diodes, with response times measurable in picoseconds, and are mainly intended for use in direct current circuits. Compared to MOVs, they have sharper current-voltage characteristic curves, resulting in more precise breakdown values with limits that can be specified with greater accuracy.

The SVP, or spark-gap discharge tube, is employed to protect circuits against high-speed, high-voltage transients by providing low-resistance paths for excessive voltage transients when their voltage ratings are exceeded.

An SVP is a gas-filled metal or ceramic cell with metal end caps designed to switch current at a preset breakdown voltage. Its electrical characteristics are determined by the geometry of the gas-filled chamber, the electrode spacing, and the selection of gas and pressure. The excess voltage ionizes the internal gas, causing the SVP to switch from a nonconducting to a conducting state in which an arc is formed to short out the SVP. After the voltage transient has passed, the gas deionizes extremely fast, and the device is reset. The SVP has a slower response time, however, of a few microseconds as compared to either a TVS or a MOV.

Summary

In the past, slew rates were low, so that crosstalk did not pose a problem, and rise times were so slow that ringing could settle down before a logic device could switch states. Thus, the assumptions of lumped-element circuit theory worked out quite well. Today's high-speed logic circuitry, however, with slew rates on the order of 1 to 2 V/ns and subnanosecond rise times, demand that the PCB layout conform with the results of distributed-element theory to avoid ringing, crosstalk, and transmission-line effects.

High-speed logic systems require the employment of ground and power planes for reliable board performance, proper termination of lines to minimize reflections, control of conductor spacings among all high-speed parallel signal lines to eliminate crosstalk, and extensive use of decoupling capacitors to reduce noise on the power plane.

References

1. A. Rich. "Shielding and Guarding." *Analog Dialogue* (Analog Devices technical magazine), vol. 17-1. Norwood, MA: 1983.
2. A. Rich. "Understanding Interference-Type Noise." *Analog Dialogue* (Analog Devices technical magazine), vol. 16-3. Norwood, MA: 1982.

3. Advanced Bipolar Division. *F100K ECL User's Handbook.* Fairchild Camera and Instrument Corp., Mountain View, CA: 1982.
4. C. S. Meyer et al., eds. *Analysis and Design of Integrated Circuits.* McGraw-Hill, New York: 1968.
5. T. Williamson. ''Designing Microcontroller Systems for Electrically Noisy Environments.'' *Application Note AP-125.* Intel Corp., Santa Clara, CA: November 1986.
6. Technical Engineering Staff. ''Advanced Schottky Family (ALS/AS) Application.'' Texas Instruments, Inc., Dallas, TX: 1986.

PROBLEMS

11.1 a. For the following circuit diagram, construct the lattice diagram and sketch the waveforms at points A and B.
b. Verify (*a*) with SPICE.

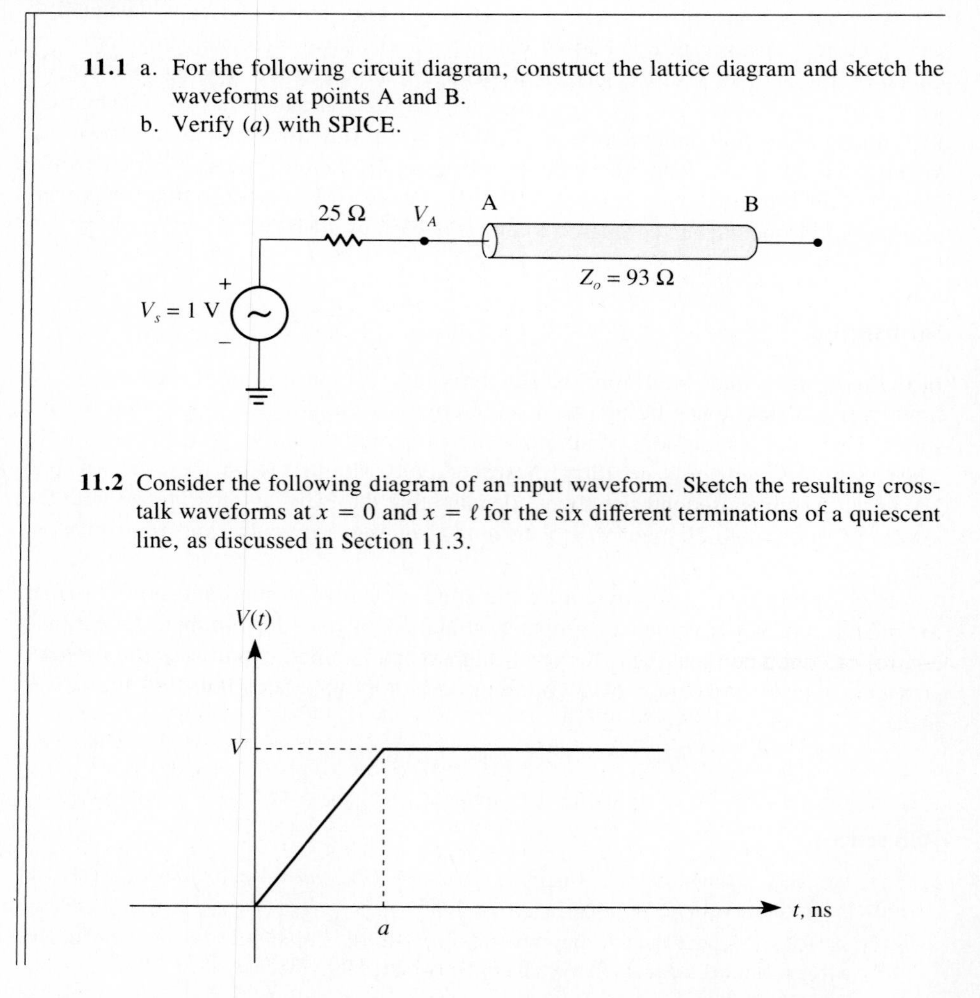

11.2 Consider the following diagram of an input waveform. Sketch the resulting crosstalk waveforms at $x = 0$ and $x = \ell$ for the six different terminations of a quiescent line, as discussed in Section 11.3.

11.3 Calculate the near field for EMI sources with the following frequencies:
a. 1 MHz
b. 10 MHz
c. 100 MHz

11.4 Consider the diagram of two 30-cm conductors, separated by 2.5 cm in a 10-Gauss, 60-Hz magnetic field as found in power wiring or transformers. Assuming that the field is parallel to the area A, find the voltage V_n induced in the closed loop. What happens to V_n if the conductors are placed 2.5 mm apart?

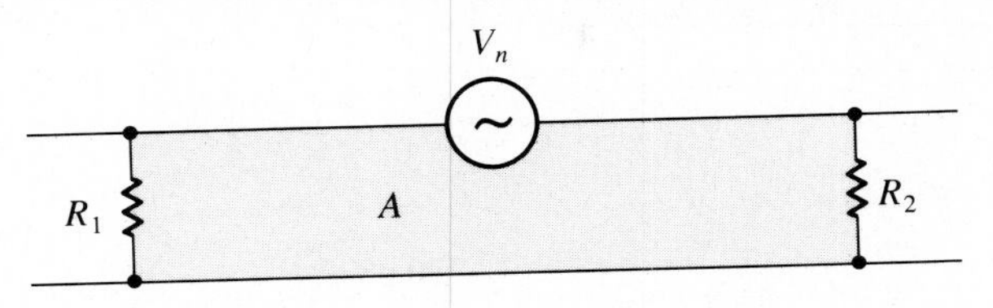

11.5 The rms-induced voltage V_n in a conductor in parallel with a second conductor that is carrying a current I at an angular frequency $\omega = 2\pi f$ is expressed as

$$V_n = \omega MI$$

where M is the mutual inductance between the conductors. Consider a 30-m shielded cable grounded at both the source and destination and used to carry a 10-V low-impedance signal to a 12-bit data converter (see diagram). The shield has a series resistance of .033 Ω/m and $M = 2\ \mu$H/m. Between the two ground points, a potential of 1 V at 60 Hz exists.
a. Determine the induced voltage in the conductor.
b. How is the resolution of the data converter affected?

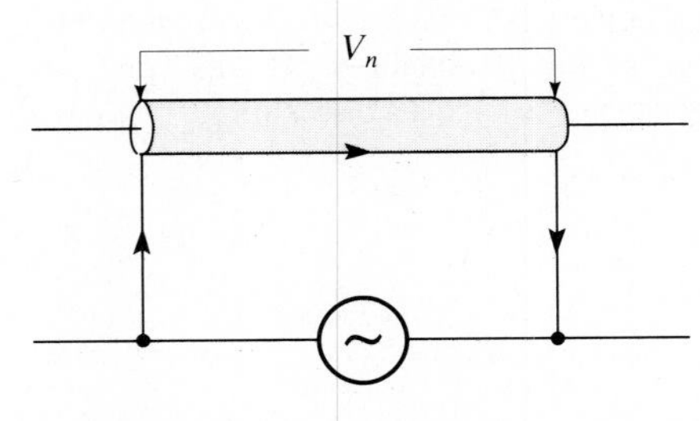

11.6 Consider the improperly configured shield system shown in the diagram, in which a precision voltage source V_s and a digital logic gate share a common shield connection. A 5-V step from the logic gate output couples capacitively to its shield, creating a current in the shield return. This, in turn, develops a shield voltage $v_n(t)$ common to both analog and digital shields. The equivalent circuit is also illustrated, where R_o represents the output impedance of the gate, C is the capacitor from the shield to the center conductor of the shielded cable, and R_n and L_n are the .1-Ω resistance and 1-μH inductance of the 60-cm wire connecting the shield to the system ground.
a. Use SPICE simulation analysis to compute the transient response $v_n(t)$ that is coupled to the analog system input.
b. What are the resonant frequency, and damping time constant?

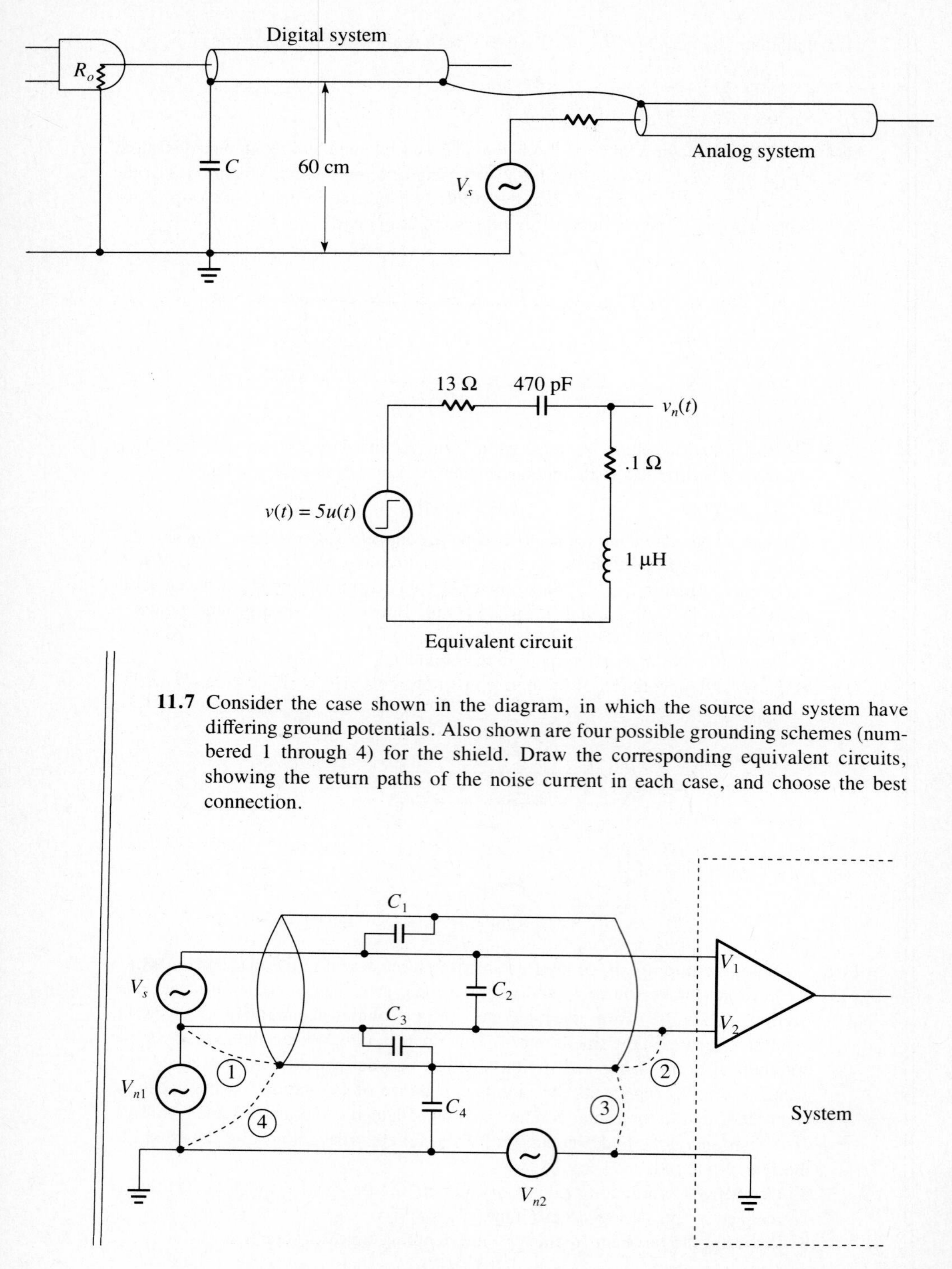

11.7 Consider the case shown in the diagram, in which the source and system have differing ground potentials. Also shown are four possible grounding schemes (numbered 1 through 4) for the shield. Draw the corresponding equivalent circuits, showing the return paths of the noise current in each case, and choose the best connection.

11.8 A transmission line with $Z_o = 50\ \Omega$ and a velocity factor of 2/3 is 10 m long.
a. Determine **TD**.
b. What is the normalized electric length at 50 MHz?
c. Write the element line for SPICE input in two different ways.

11.9 The series stray capacitance between two TTL gates is given as 20 pF. The logic swing is 3 V and the gate output resistance is 150 Ω.
a. Determine the noise transient for a step input.
b. If the gates were of standard type, would the transient interfere with their operation?
c. Repeat (*b*) for Schottky gates.

11.10 Repeat Problem 11.9 for an input signal source with a nonzero rise time as shown in the diagram.

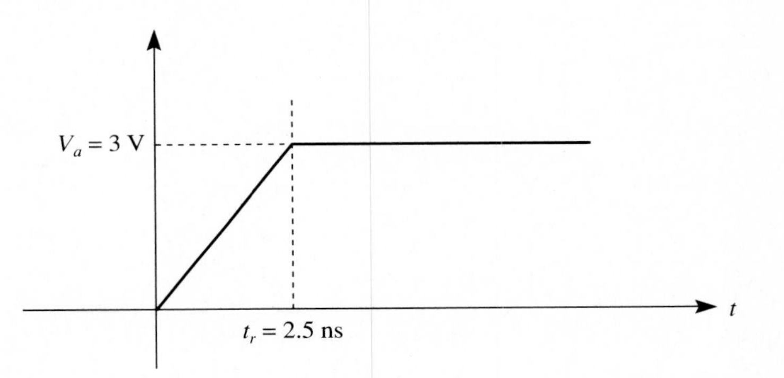

11.11 One way to eliminate the effect of oscillations on improperly terminated data inputs of synchronous digital circuits is to delay the clock input to the device. A stable input is read only when the clock is not pulsed until the oscillations on the data input have dampened. The relationship

$$t_c > k\tau$$

gives the minimum clock period t_c that satisfies this condition in terms of the signal one-way propagation time τ. Here, the odd integer k signifies the number of round trips required for the originally transmitted signal to reach steady state at the data input.

For $k = 3$, a signal propagation time of .17 ns/m, and a data-line length of 50 cm, what clock frequency allows a stable data input?

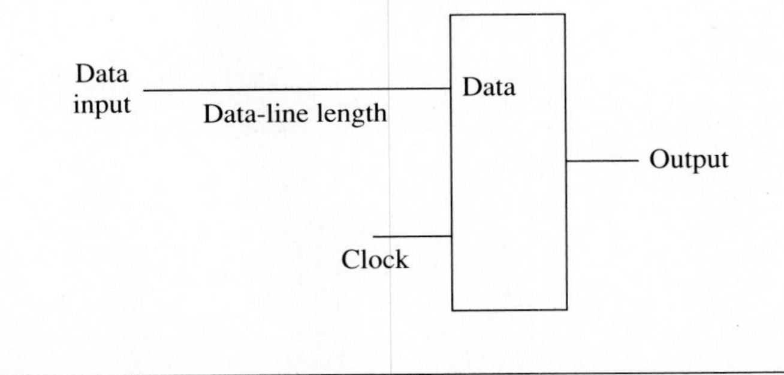

APPENDIX A: MANUFACTURERS' DATA SHEETS

The following is a condensed explanation of the information available in typical TTL and CMOS logic data sheets of commercial and military ICs as provided by the manufacturers.

The six major sections contained in a data sheet are as follows:

1. General description
2. Absolute maximum ratings
3. Recommended operating conditions
4. Electrical characteristics
5. Timing requirements
6. Switching characteristics

General Description

The first section of a data sheet, as shown in Figure A.1 for the 54/74ALS00 and 54/74AS00 quadruple two-input positive NAND gates, embodies all of the general information for that particular device. This includes the following:

1. Device number and title
2. Main features
3. Package options and pinouts
4. Functional description
5. Truth or function table
6. Logic symbol and sometimes the logic diagram

Absolute Maximum Ratings

The absolute maximum ratings table of 54/74xx is shown in Figure A.2. It specifies the stress levels that, when exceeded, may lead to permanent damage to the device. In addition, exposure to these conditions for an extended period of time may adversely affect device reliability. Therefore, manufacturers do not guarantee reliable operation of the device beyond these values.

SN54ALS00A, SN54AS00, SN74ALS00A, SN74AS00
QUADRUPLE 2-INPUT POSITIVE-NAND GATES

D2661, APRIL 1982—REVISED SEPTEMBER 1987

- **Package Options Include Plastic "Small Outline" Packages, Ceramic Chip Carriers, and Standard Plastic and Ceramic 300-mil DIPs**
- **Dependable Texas Instruments Quality and Reliability**

SN54ALS00A, SN54AS00 . . . J PACKAGE
SN74ALS00A, SN74AS00 . . . D OR N PACKAGE
(TOP VIEW)

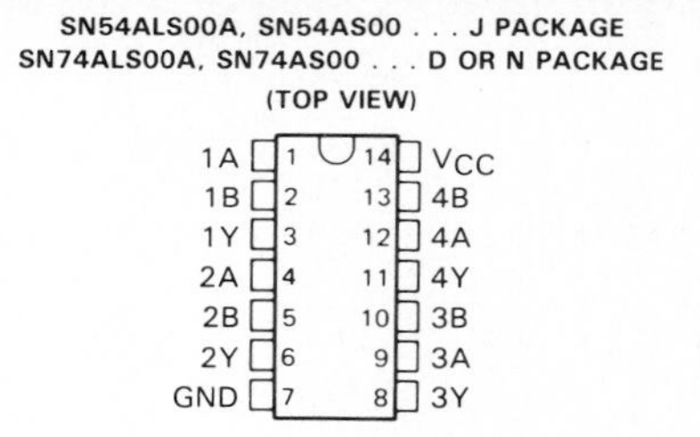

description

These devices contain four independent 2-input NAND gates. They perform the Boolean functions $Y = \overline{A \cdot B}$ or $Y = \overline{A} + \overline{B}$ in positive logic.

The SN54ALS00A and SN54AS00 are characterized for operation over the full military temperature range of $-55\,°C$ to $125\,°C$. The SN74ALS00A and SN74AS00 are characterized for operation from $0\,°C$ to $70\,°C$.

SN54ALS00A, SN54AS00 . . . FK PACKAGE
(TOP VIEW)

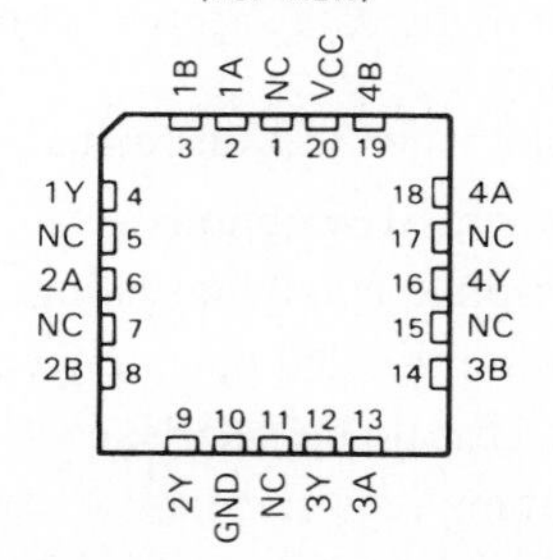

NC—No internal connection

FUNCTION TABLE (each gate)

INPUTS		OUTPUT
A	B	Y
H	H	L
L	X	H
X	L	H

logic symbol†

1A (1), 1B (2) → & → (3) 1Y
2A (4), 2B (5) → (6) 2Y
3A (9), 3B (10) → (8) 3Y
4A (12), 4B (13) → (11) 4Y

†This symbol is in accordance with ANSI/IEEE Std 91-1984 and IEC Publication 617-12.
Pin numbers shown are for D, J, and N packages.

logic diagram (positive logic)

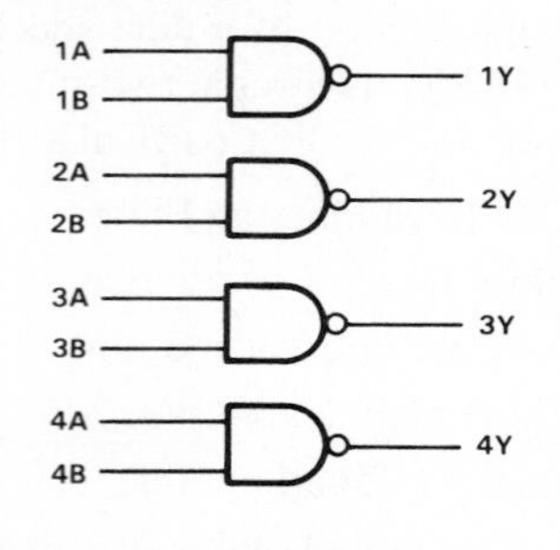

Figure A.1 The general description section of the data sheet for the 54/74ALS00 and 54/74AS00. (Reproduced by permission of Texas Instruments. Copyright © Texas Instruments.)

SN5400, SN54LS00, SN54S00, SN7400, SN74LS00, SN74S00 QUADRUPLE 2-INPUT POSITIVE-NAND GATES

absolute maximum ratings over operating free-air temperature range (unless otherwise noted)

Supply voltage, V_{CC} (see Note 1) . . . . 7 V
Input voltage: '00, 'S00 . . . . 5.5 V
'LS00 . . . . 7 V
Operating free-air temperature range: SN54' . . . . −55°C to 125°C
SN74' . . . . 0°C to 70°C
Storage temperature range . . . . −65°C to 150°C

NOTE 1: Voltage values are with respect to network ground terminal.

SN5400, SN7400 QUADRUPLE 2-INPUT POSITIVE-NAND GATES

recommended operating conditions

		SN5400			SN7400			UNIT
		MIN	NOM	MAX	MIN	NOM	MAX	
V_{CC}	Supply voltage	4.5	5	5.5	4.75	5	5.25	V
V_{IH}	High-level input voltage	2			2			V
V_{IL}	Low-level input voltage			0.8			0.8	V
I_{OH}	High-level output current			−0.4			−0.4	mA
I_{OL}	Low-level output current			16			16	mA
T_A	Operating free-air temperature	−55		125	0		70	°C

electrical characteristics over recommended operating free-air temperature range (unless otherwise noted)

PARAMETER	TEST CONDITIONS †	SN5400			SN7400			UNIT
		MIN	TYP‡	MAX	MIN	TYP‡	MAX	
V_{IK}	V_{CC} = MIN, I_I = −12 mA			−1.5			−1.5	V
V_{OH}	V_{CC} = MIN, V_{IL} = 0.8 V, I_{OH} = −0.4 mA	2.4	3.4		2.4	3.4		V
V_{OL}	V_{CC} = MIN, V_{IH} = 2 V, I_{OL} = 16 mA		0.2	0.4		0.2	0.4	V
I_I	V_{CC} = MAX, V_I = 5.5 V			1			1	mA
I_{IH}	V_{CC} = MAX, V_I = 2.4 V			40			40	μA
I_{IL}	V_{CC} = MAX, V_I = 0.4 V			−1.6			−1.6	mA
I_{OS}§	V_{CC} = MAX	−20		−55	−18		−55	mA
I_{CCH}	V_{CC} = MAX, V_I = 0 V		4	8		4	8	mA
I_{CCL}	V_{CC} = MAX, V_I = 4.5 V		12	22		12	22	mA

† For conditions shown as MIN or MAX, use the appropriate value specified under recommended operating conditions.
‡ All typical values are at V_{CC} = 5 V, T_A = 25°C.
§ Not more than one output should be shorted at a time.

switching characteristics, V_{CC} = 5 V, T_A = 25°C (see note 2)

PARAMETER	FROM (INPUT)	TO (OUTPUT)	TEST CONDITIONS	MIN	TYP	MAX	UNIT
t_{PLH}	A or B	Y	R_L = 400 Ω, C_L = 15 pF		11	22	ns
t_{PHL}					7	15	ns

NOTE 2: Load circuits and voltage waveforms are shown in Section 1.

Figure A.2 Data sheets for the two-input NAND gate IC for various TTL and CMOS subfamilies. (Reproduced by permission of Texas Instruments. Copyright © Texas Instruments.)

TYPES SN54H00, SN74H00
QUADRUPLE 2-INPUT POSITIVE-NAND GATES

recommended operating conditions

		SN54H00			SN74H00			UNIT
		MIN	NOM	MAX	MIN	NOM	MAX	
V_{CC}	Supply voltage	4.5	5	5.5	4.75	5	5.25	V
V_{IH}	High-level input voltage	2			2			V
V_{IL}	Low-level input voltage			0.8			0.8	V
I_{OH}	High-level output current			– 0.5			– 0.5	mA
I_{OL}	Low-level output current			20			20	mA
T_A	Operating free-air temperature	– 55		125	0		70	°C

electrical characteristics over recommended operating free-air temperature range (unless otherwise noted)

PARAMETER	TEST CONDITIONS †			MIN	TYP ‡	MAX	UNIT
V_{IK}	V_{CC} = MIN,	$I_I = -8$ mA				– 1.5	V
V_{OH}	V_{CC} = MIN,	V_{IL} = 0.8 V,	$I_{OH} = -0.5$ mA	2.4	3.5		V
V_{OL}	V_{CC} = MIN,	V_{IH} = 2 V,	I_{OL} = 20 mA		0.2	0.4	V
I_I	V_{CC} = MAX,	V_I = 5.5 V				1	mA
I_{IH}	V_{CC} = MAX,	V_I = 2.4 V				50	μA
I_{IL}	V_{CC} = MAX,	V_I = 0.4 V				– 2	mA
I_{OS} §	V_{CC} = MAX			– 40		– 100	mA
I_{CCH}	V_{CC} = MAX,	V_I = 0 V			10	16.8	mA
I_{CCL}	V_{CC} = MAX,	V_I = 4.5 V			26	40	mA

† For conditions shown as MIN or MAX, use the appropriate value specified under recommended operating conditions.
‡ All typical values are at V_{CC} = 5 V, T_A = 25°C.
§ Not more than one output should be shorted at a time, and the duration of the short-circuit should not exceed one second.

switching characteristics, V_{CC} = 5 V, T_A = 25°C (see note 2)

PARAMETER	FROM (INPUT)	TO (OUTPUT)	TEST CONDITIONS		MIN	TYP	MAX	UNIT
t_{PLH}	A or B	Y	R_L = 280 Ω,	C_L = 25 pF		5.9	10	ns
t_{PHL}						6.2	10	ns

NOTE 2: See General Information Section for load circuits and voltage waveforms.

Figure A.2 (*Continued*) (Reproduced by permission of Texas Instruments. Copyright © Texas Instruments.)

TYPE SN54L00
QUADRUPLE 2-INPUT POSITIVE-NAND GATES

recommended operating conditions

		SN54L00			UNIT
		MIN	NOM	MAX	
V_{CC}	Supply voltage	4.5	5	5.5	V
V_{IH}	High-level input voltage	2			V
V_{IL}	Low-level input voltage			0.7	V
I_{OH}	High-level output current			−0.1	mA
I_{OL}	Low-level output current			2	mA
T_A	Operating free-air temperature	−55		125	°C

electrical characteristics over recommended operating free-air temperature range (unless otherwise noted)

PARAMETER	TEST CONDITIONS †	SN54L00			UNIT
		MIN	TYP ‡	MAX	
V_{OH}	V_{CC} = MIN, V_{IL} = 0.7 V, I_{OH} = −0.1 mA	2.4	3.3		V
V_{OL}	V_{CC} = MIN, V_{IH} = 2 V, I_{OL} = 2 mA		0.15	0.3	V
I_I	V_{CC} = MAX, V_I = 5.5 V			0.1	mA
I_{IH}	V_{CC} = MAX, V_I = 2.4 V			10	μA
I_{IL}	V_{CC} = MAX, V_I = 0.3 V			−0.18	mA
I_{OS}§	V_{CC} = MAX	−3		−15	mA
I_{CCH}	V_{CC} = MAX, V_I = 0 V		0.44	0.8	mA
I_{CCL}	V_{CC} = MAX, V_I = 4.5 V		1.16	2.04	mA

† For conditions shown as MIN or MAX, use the appropriate value specified under recommended operating conditions.
‡ All typical values are at V_{CC} = 5 V, T_A = 25°C.
§ Not more than one output should be shorted at a time.

switching characteristics, V_{CC} = 5 V, T_A = 25°C (see note 2)

PARAMETER	FROM (INPUT)	TO (OUTPUT)	TEST CONDITIONS	MIN	TYP	MAX	UNIT
t_{PLH}	A or B	Y	R_L = 4 kΩ, C_L = 50 pF		35	60	ns
t_{PHL}					31	60	ns

Figure A.2 (*Continued*) (Reproduced by permission of Texas Instruments. Copyright © Texas Instruments.)

SN54S00, SN74S00
QUADRUPLE 2-INPUT POSITIVE-NAND GATES

recommended operating conditions

		SN54S00			SN74S00			UNIT
		MIN	NOM	MAX	MIN	NOM	MAX	
V_{CC}	Supply voltage	4.5	5	5.5	4.75	5	5.25	V
V_{IH}	High-level input voltage	2			2			V
V_{IL}	Low-level input voltage			0.8			0.8	V
I_{OH}	High-level output current			– 1			– 1	mA
I_{OL}	Low-level output current			20			20	mA
T_A	Operating free-air temperature	– 55		125	0		70	°C

electrical characteristics over recommended operating free-air temperature range (unless otherwise noted)

PARAMETER	TEST CONDITIONS †	SN54S00			SN74S00			UNIT
		MIN	TYP‡	MAX	MIN	TYP‡	MAX	
V_{IK}	V_{CC} = MIN, I_I = –18 mA			–1.2			–1.2	V
V_{OH}	V_{CC} = MIN, V_{IL} = 0.8 V, I_{OH} = – 1 mA	2.5	3.4		2.7	3.4		V
V_{OL}	V_{CC} = MIN, V_{IH} = 2 V, I_{OL} = 20 mA			0.5			0.5	V
I_I	V_{CC} = MAX, V_I = 5.5 V			1			1	mA
I_{IH}	V_{CC} = MAX, V_I = 2.7 V			50			50	μA
I_{IL}	V_{CC} = MAX, V_I = 0.5 V			–2			–2	mA
I_{OS}§	V_{CC} = MAX	–40		–100	–40		–100	mA
I_{CCH}	V_{CC} = MAX, V_I = 0 V		10	16		10	16	mA
I_{CCL}	V_{CC} = MAX, V_I = 4.5 V		20	36		20	36	mA

† For conditions shown as MIN or MAX, use the appropriate value specified under recommended operating conditions.
‡ All typical values are at V_{CC} = 5 V, T_A = 25°C.
§ Not more than one output should be shorted at a time, and the duration of the short-circuit should not exceed one second.

switching characteristics, V_{CC} = 5 V, T_A = 25°C (see note 2)

PARAMETER	FROM (INPUT)	TO (OUTPUT)	TEST CONDITIONS	MIN	TYP	MAX	UNIT
t_{PLH}	A or B	Y	R_L = 280 Ω, C_L = 15 pF		3	4.5	ns
t_{PHL}					3	5	ns
t_{PLH}			R_L = 280 Ω, C_L = 50 pF		4.5		ns
t_{PHL}					5		ns

Figure A.2 (*Continued*) (Reproduced by permission of Texas Instruments. Copyright © Texas Instruments.)

SN54LS00, SN74LS00
QUADRUPLE 2-INPUT POSITIVE-NAND GATES

recommended operating conditions

		SN54LS00			SN74LS00			UNIT
		MIN	NOM	MAX	MIN	NOM	MAX	
V_{CC}	Supply voltage	4.5	5	5.5	4.75	5	5.25	V
V_{IH}	High-level input voltage	2			2			V
V_{IL}	Low-level input voltage			0.7			0.8	V
I_{OH}	High-level output current			− 0.4			− 0.4	mA
I_{OL}	Low-level output current			4			8	mA
T_A	Operating free-air temperature	− 55		125	0		70	°C

electrical characteristics over recommended operating free-air temperature range (unless otherwise noted)

PARAMETER	TEST CONDITIONS †			SN54LS00			SN74LS00			UNIT
				MIN	TYP‡	MAX	MIN	TYP‡	MAX	
V_{IK}	V_{CC} = MIN,	I_I = − 18 mA				− 1.5			− 1.5	V
V_{OH}	V_{CC} = MIN,	V_{IL} = MAX,	I_{OH} = − 0.4 mA	2.5	3.4		2.7	3.4		V
V_{OL}	V_{CC} = MIN,	V_{IH} = 2 V,	I_{OL} = 4 mA		0.25	0.4		0.25	0.4	V
	V_{CC} = MIN,	V_{IH} = 2 V,	I_{OL} = 8 mA					0.35	0.5	
I_I	V_{CC} = MAX,	V_I = 7 V				0.1			0.1	mA
I_{IH}	V_{CC} = MAX,	V_I = 2.7 V				20			20	μA
I_{IL}	V_{CC} = MAX,	V_I = 0.4 V				− 0.4			− 0.4	mA
I_{OS}§	V_{CC} = MAX			− 20		− 100	− 20		− 100	mA
I_{CCH}	V_{CC} = MAX,	V_I = 0 V			0.8	1.6		0.8	1.6	mA
I_{CCL}	V_{CC} = MAX,	V_I = 4.5 V			2.4	4.4		2.4	4.4	mA

† For conditions shown as MIN or MAX, use the appropriate value specified under recommended operating conditions.
‡ All typical values are at V_{CC} = 5 V, T_A = 25°C
§ Not more than one output should be shorted at a time, and the duration of the short-circuit should not exceed one second.

switching characteristics, V_{CC} = 5 V, T_A = 25°C (see note 2)

PARAMETER	FROM (INPUT)	TO (OUTPUT)	TEST CONDITIONS		MIN	TYP	MAX	UNIT
t_{PLH}	A or B	Y	R_L = 2 kΩ,	C_L = 15 pF		9	15	ns
t_{PHL}						10	15	ns

Figure A.2 (*Continued*) (Reproduced by permission of Texas Instruments. Copyright © Texas Instruments.)

SN54AS00, SN74AS00
QUADRUPLE 2-INPUT POSITIVE-NAND GATES

absolute maximum ratings over operating free-air temperature range (unless otherwise noted)

Supply voltage, V_{CC} . . . . 7 V
Input voltage . . . . 7 V
Operating free-air temperature range: SN54AS00 . . . . −55 °C to 125 °C
SN74AS00 . . . . 0 °C to 70 °C
Storage temperature range . . . . −65 °C to 150 °C

recommended operating conditions

		SN54AS00			SN74AS00			UNIT
		MIN	NOM	MAX	MIN	NOM	MAX	
V_{CC}	Supply voltage	4.5	5	5.5	4.5	5	5.5	V
V_{IH}	High-level input voltage	2			2			V
V_{IL}	Low-level input voltage			0.8			0.8	V
I_{OH}	High-level output current			−2			−2	mA
I_{OL}	Low-level output current			20			20	mA
T_A	Operating free-air temperature	−55		125	0		70	°C

electrical characteristics over recommended operating free-air temperature range (unless otherwise noted)

PARAMETER	TEST CONDITIONS		SN54AS00			SN74AS00			UNIT
			MIN	TYP†	MAX	MIN	TYP†	MAX	
V_{IK}	V_{CC} = 4.5 V,	I_I = −18 mA			−1.2			−1.2	V
V_{OH}	V_{CC} = 4.5 V to 5.5 V,	I_{OH} = −2 mA	$V_{CC}-2$			$V_{CC}-2$			V
V_{OL}	V_{CC} = 4.5 V,	I_{OL} = 20 mA		0.35	0.5		0.35	0.5	V
I_I	V_{CC} = 5.5 V,	V_I = 7 V			0.1			0.1	mA
I_{IH}	V_{CC} = 5.5 V,	V_I = 2.7 V			20			20	μA
I_{IL}	V_{CC} = 5.5 V,	V_I = 0.4 V			−0.5			−0.5	mA
I_O‡	V_{CC} = 5.5 V,	V_O = 2.25 V	−30		−112	−30		−112	mA
I_{CCH}	V_{CC} = 5.5 V,	V_I = 0 V		2	3.2		2	3.2	mA
I_{CCL}	V_{CC} = 5.5 V,	V_I = 4.5 V		10.8	17.4		10.8	17.4	mA

†All typical values are at V_{CC} = 5 V, T_A = 25 °C.
‡The output conditions have been chosen to produce a current that closely approximates one half of the true short-circuit output current, I_{OS}.

switching characteristics (see Note 1)

PARAMETER	FROM (INPUT)	TO (OUTPUT)	V_{CC} = 4.5 V to 5.5 V, C_L = 50 pF, R_L = 50 Ω, T_A = MIN to MAX				UNIT
			SN54AS00		SN74AS00		
			MIN	MAX	MIN	MAX	
t_{PLH}	A or B	Y	1	5	1	4.5	ns
t_{PHL}	A or B	Y	1	5	1	4	

Figure A.2 (*Continued*) (Reproduced by permission of Texas Instruments. Copyright © Texas Instruments.)

SN54F00, SN74F00
QUADRUPLE 2-INPUT POSITIVE-NAND GATES

absolute maximum ratings over operating free-air temperature range (unless otherwise noted)

Supply voltage, V_{CC} . . . . −0.5 V to 7 V
Input voltage† . . . . −1.2 V to 7 V
Input current . . . . −30 mA to 5 mA
Voltage applied to any output in the high state . . . . −0.5 V to V_{CC}
Current into any output in the low state . . . . 40 mA
Operating free-air temperature range: SN54F00 . . . . −55 °C to 125 °C
SN74F00 . . . . 0 °C to 70 °C
Storage temperature range . . . . −65 °C to 150 °C

†The input voltage ratings may be exceeded provided the input current ratings are observed.

recommended operating conditions

		SN54F00			SN74F00			UNIT
		MIN	NOM	MAX	MIN	NOM	MAX	
V_{CC}	Supply voltage	4.5	5	5.5	4.5	5	5.5	V
V_{IH}	High-level input voltage	2			2			V
V_{IL}	Low-level input voltage			0.8			0.8	V
I_{IK}	Input clamp current			−18			−18	mA
I_{OH}	High-level output current			−1			−1	mA
I_{OL}	Low-level output current			20			20	mA
T_A	Operating free-air temperature	−55		125	0		70	°C

electrical characteristics over recommended operating free-air temperature range (unless otherwise noted)

PARAMETER	TEST CONDITIONS		SN54F00			SN74F00			UNIT
			MIN	TYP‡	MAX	MIN	TYP‡	MAX	
V_{IK}	V_{CC} = 4.5 V,	I_I = −18 mA			−1.2			−1.2	V
V_{OH}	V_{CC} = 4.5 V,	I_{OH} = −1 mA	2.5	3.4		2.5	3.4		V
	V_{CC} = 4.75 V,	I_{OH} = −1 mA				2.7			
V_{OL}	V_{CC} = 4.5 V,	I_{OL} = 20 mA		0.30	0.5		0.30	0.5	V
I_I	V_{CC} = 5.5 V,	V_I = 7 V			0.1			0.1	mA
I_{IH}	V_{CC} = 5.5 V,	V_I = 2.7 V			20			20	μA
I_{IL}	V_{CC} = 5.5 V,	V_I = 0.5 V			−0.6			−0.6	mA
I_{OS}§	V_{CC} = 5.5 V,	V_O = 0	−60		−150	−60		−150	mA
I_{CCH}	V_{CC} = 5.5 V,	V_I = 0		1.9	2.8		1.9	2.8	mA
I_{CCL}	V_{CC} = 5.5 V,	V_I = 4.5 V		6.8	10.2		6.8	10.2	mA

switching characteristics (see Note 1)

PARAMETER	FROM (INPUT)	TO (OUTPUT)	V_{CC} = 5 V, C_L = 50 pF, R_L = 500 Ω, T_A = 25°C			V_{CC} = 4.5 V to 5.5 V, C_L = 50 pF, R_L = 500 Ω, T_A = MIN to MAX¶				UNIT
			'F00			SN54F00		SN74F00		
			MIN	TYP	MAX	MIN	MAX	MIN	MAX	
t_{PLH}	A or B	Y	1.6	3.3	5	1.2	7	1.6	6	ns
t_{PHL}	A or B	Y	1	2.8	4.3	1	6.5	1	5.3	ns

‡All typical values are at V_{CC} = 5 V, T_A = 25 °C.
§Not more than one output should be shorted at a time and the duration of the short circuit should not exceed one second.
¶For conditions shown as MIN or MAX, use the appropriate value specified under Recommended Operating Conditions.

Figure A.2 (*Continued*) (Reproduced by permission of Texas Instruments.

SN54ALS00A, SN74ALS00A
QUADRUPLE 2-INPUT POSITIVE-NAND GATES

absolute maximum ratings over operating free-air temperature range (unless otherwise noted)

Supply voltage, V_{CC} .... 7 V
Input voltage .... 7 V
Operating free-air temperature range: SN54ALS00A .... −55 °C to 125 °C
SN74ALS00A .... 0 °C to 70 °C
Storage temperature range .... −65 °C to 150 °C

recommended operating conditions

		SN54ALS00A			SN74ALS00A			UNIT
		MIN	NOM	MAX	MIN	NOM	MAX	
V_{CC}	Supply voltage	4.5	5	5.5	4.5	5	5.5	V
V_{IH}	High-level input voltage	2			2			V
V_{IL}	Low-level input voltage						0.8	V
				0.8†				
				0.7‡				
I_{OH}	High-level output current			−0.4			−0.4	mA
I_{OL}	Low--level output current			4			8	mA
T_A	Operating free-air temperature	−55		125	0		70	°C

†Tested at −55 °C to 70 °C.
‡Tested at 70 °C to 125 °C, per MIL-STD-833, method 5005, sub-group 1, 2, and 3. Static test is performed at 25 °C, 125 °C, and −55 °C.

electrical characteristics over recommended operating free-air temperature range (unless otherwise noted)

PARAMETER	TEST CONDITIONS		SN54ALS00A			SN74ALS00A			UNIT
			MIN	TYP§	MAX	MIN	TYP§	MAX	
V_{IK}	V_{CC} = 4.5 V,	I_I = −18 mA			−1.5			−1.5	V
V_{OH}	V_{CC} = 4.5 V to 5.5 V,	I_{OH} = −0.4 mA	V_{CC}−2						V
V_{OL}	V_{CC} = 4.5 V,	I_{OL} = 4 mA		0.25	0.4		0.25	0.4	V
	V_{CC} = 4.5 V,	I_{OL} = 8 mA					0.35	0.5	
I_I	V_{CC} = 5.5 V,	V_I = 7 V			0.1			0.1	mA
I_{IH}	V_{CC} = 5.5 V,	V_I = 2.7 V			20			20	μA
I_{IL}	V_{CC} = 5.5 V,	V_I = 0.4 V			−0.1			−0.1	mA
I_O¶	V_{CC} = 5.5 V,	V_O = 2.25 V	−30		−112	−30		−112	mA
I_{CCH}	V_{CC} = 5.5 V,	V_I = 0 V		0.5	0.85		0.5	0.85	mA
I_{CCL}	V_{CC} = 5.5 V,	V_I = 4.5 V		1.5	3		1.5	3	mA

§All typical values are at V_{CC} = 5 V, T_A = 25 °C.
¶The output conditions have been chosen to produce a current that closely approximates one half of the true short-circuit output current, I_{OS}.

switching characteristics (see Note 1)

PARAMETER	FROM (INPUT)	TO (OUTPUT)	V_{CC} = 5 V, C_L = 50 pF, R_L = 500 Ω, T_A = 25 °C	V_{CC} = 4.5 V to 5.5 V, C_L = 50 pF, R_L = 500 Ω, T_A = MIN to MAX				UNIT
			'ALS00A	SN54ALS00A		SN74ALS00A		
			TYP	MIN	MAX	MIN	MAX	
t_{PLH}	A or B	Y	7	3	15	3	11	ns
t_{PHL}	A or B	Y	5	2	9	2	8	ns

Figure A.2 (*Continued*) (Reproduced by permission of Texas Instruments. Copyright © Texas Instruments.)

switching characteristics over recommended operating free-air temperature range (unless otherwise noted), C_L = 50 pF (see Note 1)

PARAMETER	FROM (INPUT)	TO (OUTPUT)	V_{CC}	T_A = 25°C MIN	TYP	MAX	SN54HC00 MIN	MAX	SN74HC00 MIN	MAX	UNIT
t_{pd}	A or B	Y	2 V		45	90		135		115	ns
			4.5 V		9	18		27		23	
			6 V		8	15		23		20	
t_t		Y	2 V		38	75		110		95	ns
			4.5 V		8	15		22		19	
			6 V		6	13		19		16	

C_{pd}	Power dissipation capacitance per gate	No load, T_A = 25°C	20 pF typ

absolute maximum ratings over operating free-air temperature range†

Supply voltage, V_{CC} . . . . −0.5 V to 7 V
Input clamp current, I_{IK} ($V_I < 0$ or $V_I > V_{CC}$) . . . . ±20 mA
Output clamp current, I_{OK} ($V_O < 0$ or $V_O > V_{CC}$ . . . . ±20 mA
Continuous output current, I_O (V_O = 0 to V_{CC}) . . . . ±25 mA
Continuous current through V_{CC} or GND pins . . . . ±50 mA
Lead temperature 1,6 mm (1/16 in) from case for 60 s: FK or J package . . . . 300°C
Lead temperature 1,6 mm (1/16 in) from case for 10 s: D or N package . . . . 260°C
Storage temperature range . . . . −65°C to 150°C

† Stresses beyond those listed under "absolute maximum ratings" may cause permanent damage to the device. These are stress ratings only, and functional operation of the device at these or any other conditions beyond those indicated under "recommended operating conditions" is not implied. Exposure to absolute-maximum-rated conditions for extended periods may affect device reliability.

recommended operating conditions

		SN54HC00 MIN	NOM	MAX	SN74HC00 MIN	NOM	MAX	UNIT
V_{CC} Supply voltage		2	5	6	2	5	6	V
V_{IH} High-level input voltage	V_{CC} = 2 V	1.5			1.5			V
	V_{CC} = 4.5 V	3.15			3.15			
	V_{CC} = 6 V	4.2			4.2			
V_{IL} Low-level input voltage	V_{CC} = 2 V	0		0.3	0		0.3	V
	V_{CC} = 4.5 V	0		0.9	0		0.9	
	V_{CC} = 6 V	0		1.2	0		1.2	
V_I Input voltage		0		V_{CC}	0		V_{CC}	V
V_O Output voltage		0		V_{CC}	0		V_{CC}	V
t_t Input transition (rise and fall) times	V_{CC} = 2 V	0		1000	0		1000	ns
	V_{CC} = 4.5 V	0		500	0		500	
	V_{CC} = 6 V	0		400	0		400	
T_A Operating free-air temperature		−55		125	−40		85	°C

electrical characteristics over recommended operating free-air temperature range (unless otherwise noted)

PARAMETER	TEST CONDITIONS	V_{CC}	T_A = 25°C MIN	TYP	MAX	SN54HC00 MIN	MAX	SN74HC00 MIN	MAX	UNIT
V_{OH}	$V_I = V_{IH}$ or V_{IL}, I_{OH} = −20 µA	2 V	1.9	1.998		1.9		1.9		V
		4.5 V	4.4	4.499		4.4		4.4		
		6 V	5.9	5.999		5.9		5.9		
	$V_I = V_{IH}$ or V_{IL}, I_{OH} = −4 mA	4.5 V	3.98	4.30		3.7		3.84		
	$V_I = V_{IH}$ or V_{IL}, I_{OH} = −5.2 mA	6 V	5.48	5.80		5.2		5.34		
V_{OL}	$V_I = V_{IH}$ or V_{IL}, I_{OL} = 20 µA	2 V		0.002	0.1		0.1		0.1	V
		4.5 V		0.001	0.1		0.1		0.1	
		6 V		0.001	0.1		0.1		0.1	
	$V_I = V_{IH}$ or V_{IL}, I_{OL} = 4 mA	4.5 V		0.17	0.26		0.4		0.33	
	$V_I = V_{IH}$ or V_{IL}, I_{OL} = 5.2 mA	6 V		0.15	0.26		0.4		0.33	
I_I	$V_I = V_{CC}$ or 0	6 V		±0.1	±100		±1000		±1000	nA
I_{CC}	$V_I = V_{CC}$ or 0, I_O = 0	6 V			2		40		20	µA
C_i		2 to 6 V		3	10		10		10	pF

Figure A.2 (*Continued*) (Reproduced by permission of Texas Instruments. Copyright © Texas Instruments.)

54AC11000, 74AC11000
QUADRUPLE 2-INPUT POSITIVE-NAND GATES

absolute maximum ratings over operating free-air temperature range (unless otherwise noted)†

Supply voltage range, V_{CC} . . . . -0.5 V to 6 V
Input voltage range, V_I (see Note 1) . . . . -0.5 V to $V_{CC} + 0.5$ V
Output voltage range, V_O (see Note 1) . . . . -0.5 V to $V_{CC} + 0.5$ V
Input clamp current, I_{IK} ($V_I < 0$ or $V_I > V_{CC}$) . . . . ± 20 mA
Output clamp current, I_{OK} ($V_O < 0$ or $V_O > V_{CC}$) . . . . ± 50 mA
Continuous output current, I_O ($V_O = 0$ to V_{CC}) . . . . ± 50 mA
Continuous current through V_{CC} or GND pins . . . . ± 100 mA
Storage temperature range . . . . -65 °C to 150 °C

†Stresses beyond those listed under "absolute maximum ratings" may cause permanent damage to the device. These are stress ratings only and functional operation of the device at these or any other conditions beyond those indicated under "recommended operating conditions" is not implied. Exposure to absolute-maximum-rated conditions for extended periods may affect device reliability.
NOTE 1: The input and output voltage ratings may be exceeded if the input and output current ratings are observed.

recommended operating conditions

			54AC11000			74AC11000			UNIT
			MIN	NOM	MAX	MIN	NOM	MAX	
V_{CC}	Supply voltage		3	5	5.5	3	5	5.5	V
V_{IH}	High-level input voltage	V_{CC} = 3 V	2.1			2.1			V
		V_{CC} = 4.5 V	3.15			3.15			
		V_{CC} = 5.5 V	3.85			3.85			
V_{IL}	Low-level input voltage	V_{CC} = 3 V			0.9			0.9	V
		V_{CC} = 4.5 V			1.35			1.35	
		V_{CC} = 5.5 V			1.65			1.65	
I_{OH}	High-level output current	V_{CC} = 3 V			−4			−4	mA
		V_{CC} = 4.5 V			−24			−24	
		V_{CC} = 5.5 V			−24			−24	
I_{OL}	Low-level output current	V_{CC} = 3 V			12			12	mA
		V_{CC} = 4.5 V			24			24	
		V_{CC} = 5.5 V			24			24	
V_I	Input voltage		0		V_{CC}	0		V_{CC}	V
V_O	Output voltage		0		V_{CC}	0		V_{CC}	V
dt/dv	Input transition rise or fall rate		0		10	0		10	ns/V
T_A	Operating free-air temperature		−55		125	−40		85	°C

Figure A.2 (*Continued*) (Reproduced by permission of Texas Instruments. Copyright © Texas Instruments.)

electrical characteristics over recommended operating free-air temperature range (unless otherwise noted)

PARAMETER	TEST CONDITIONS	V_{CC}	$T_A = 25°C$ MIN	$T_A = 25°C$ TYP	$T_A = 25°C$ MAX	54AC11000 MIN	54AC11000 MAX	74AC11000 MIN	74AC11000 MAX	UNIT
V_{OH}	$I_{OH} = -50\ \mu A$	3 V	2.9			2.9		2.9		V
		4.5 V	4.4			4.4		4.4		
		5.5 V	5.4			5.4		5.4		
	$I_{OH} = -4$ mA	3 V	2.58			2.4		2.48		
	$I_{OH} = -24$ mA	4.5 V	3.94			3.7		3.8		
		5.5 V	4.94			4.7		4.8		
	$I_{OH} = -50$ mA†	5.5 V				3.85				
	$I_{OH} = -75$ mA†	5.5 V						3.85		
V_{OL}	$I_{OL} = 50\ \mu A$	3 V			0.1		0.1		0.1	V
		4.5 V			0.1		0.1		0.1	
		5.5 V			0.1		0.1		0.1	
	$I_{OL} = 12$ mA	3 V			0.36		0.5		0.44	
	$I_{OL} = 24$ mA	4.5 V			0.36		0.5		0.44	
		5.5 V			0.36		0.5		0.44	
	$I_{OL} = 50$ mA†	5.5 V					1.65			
	$I_{OL} = 75$ mA†	5.5 V							1.65	
I_I	$V_I = V_{CC}$ or GND	5.5 V			±0.1		±1		±1	μA
I_{CC}	$V_I = V_{CC}$ or GND, $I_O = 0$	5.5 V			4		80		40	μA
C_I	$V_I = V_{CC}$ or GND	5 V		3.5						pF

†Not more than one output should be tested at a time, and the duration of the test should not exceed 10 ms.

switching characteristics over recommended operating free-air temperature range, $V_{CC} = 3.3\ V \pm 0.3\ V$ (unless otherwise noted) (see Figure 1)

PARAMETER	FROM (INPUT)	TO (OUTPUT)	$T_A = 25°C$ MIN	$T_A = 25°C$ TYP	$T_A = 25°C$ MAX	54AC11000 MIN	54AC11000 MAX	74AC11000 MIN	74AC11000 MAX	UNIT
t_{PLH}	A or B	Y	1.5	7.2	9.8	1.5	11.9	1.5	11.1	ns
t_{PHL}			1.5	5.8	8.6	1.5	10.2	1.5	9.6	

switching characteristics over recommended operating free-air temperature range, $V_{CC} = 5\ V \pm 0.5\ V$ (unless otherwise noted) (see Figure 1)

PARAMETER	FROM (INPUT)	TO (OUTPUT)	$T_A = 25°C$ MIN	$T_A = 25°C$ TYP	$T_A = 25°C$ MAX	54AC11000 MIN	54AC11000 MAX	74AC11000 MIN	74AC11000 MAX	UNIT
t_{PLH}	A or B	Y	1.5	5	6.5	1.5	8.1	1.5	7.4	ns
t_{PHL}			1.5	4.4	6.1	1.5	7.3	1.5	6.8	

operating characteristics, $V_{CC} = 5$ V, $T_A = 25°C$

PARAMETER		TEST CONDTIONS	TYP	UNIT
C_{pd}	Power dissipation capacitance per gate	$C_L = 50$ pF, f = 1 MHz	33	pF

Figure A.2 (*Continued*) (Reproduced by permission of Texas Instruments. Copyright © Texas Instruments.)

absolute maximum ratings over operating free-air temperature range (unless otherwise noted)†

Supply voltage range, V_{CC} . . . −0.5 V to 6 V
Input voltage range, V_I (see Note 1) . . . −0.5 V to V_{CC} +0.5 V
Output voltage range, V_O (see Note 1) . . . −0.5 V to V_{CC} +0.5 V
Input clamp current, I_{IK} ($V_I < 0$ or $V_I > V_{CC}$) . . . ±20 mA
Output clamp current, I_{OK} ($V_O < 0$ or $V_O > V_{CC}$) . . . ±50 mA
Continuous output current, I_O ($V_O = 0$ to V_{CC}) . . . ±50 mA
Continuous current through V_{CC} or GND pins . . . ±100 mA
Storage temperature range . . . −65°C to 150°C

†Stresses beyond those listed under "absolute maximum ratings" may cause permanent damage to the device. These are stress ratings only and functional operation of the device at these or any other conditions beyond those indicated under "recommended operating conditions" is not implied. Exposure to absolute-maximum-rated conditions for extended periods may affect device reliability.
NOTE 1: The input and output voltage ratings may be exceeded if the input and output current ratings are observed.

recommended operating conditions

		54ACT11000		74ACT11000		UNIT
		MIN	MAX	MIN	MAX	
V_{CC}	Supply voltage	4.5	5.5	4.5	5.5	V
V_{IH}	High-level input voltage	2		2		V
V_{IL}	Low-level input voltage		0.8		0.8	V
V_I	Input voltage	0	V_{CC}	0	V_{CC}	V
V_O	Output voltage	0	V_{CC}	0	V_{CC}	V
I_{OH}	High-level output current		−24		−24	mA
I_{OL}	Low-level output current		24		24	mA
$\Delta t/\Delta v$	Input transition rise or fall rate	0	10	0	10	ns/V
T_A	Operating free-air temperature	−55	125	−40	85	°C

electrical characteristics over recommended operating free-air temperature range (unless otherwise noted)

PARAMETER	TEST CONDITIONS	V_{CC}	T_A = 25°C			54ACT11000		74ACT11000		UNIT
			MIN	TYP	MAX	MIN	MAX	MIN	MAX	
V_{OH}	I_{OH} = −50 μA	4.5 V	4.4			4.4		4.4		V
		5.5 V	5.4			5.4		5.4		
	I_{OH} = −24 mA	4.5 V	3.94			3.7		3.8		
		5.5 V	4.94			4.7		4.8		
	I_{OH} = −50 mA†	5.5 V				3.85				
	I_{OH} = −75 mA†	5.5 V						3.85		
V_{OL}	I_{OL} = 50 μA	4.5 V			0.1		0.1		0.1	V
		5.5 V			0.1		0.1		0.1	
	I_{OL} = 24 mA	4.5 V			0.36		0.5		0.44	
		5.5 V			0.36		0.5		0.44	
	I_{OL} = 50 mA†	5.5 V					1.65			
	I_{OL} = 75 mA†	5.5 V							1.65	
I_I	$V_I = V_{CC}$ or GND	5.5 V			±0.1		±1		±1	μA
I_{CC}	$V_I = V_{CC}$ or GND, $I_O = 0$	5.5 V			4		80		40	μA
ΔI_{CC}‡	One input at 3.4 V, Other inputs at GND or V_{CC}	5.5 V			0.9		1		1	mA
C_i	$V_I = V_{CC}$ or GND	5 V		3.5						pF

†Not more than one output should be tested at a time, and the duration of the test should not exceed 10 ms.
‡This is the increase in supply current for each input that is at one of the specified TTL voltage levels rather than 0 V or V_{CC}.

switching characteristics, V_{CC} = 5 V ±0.5 V (see Figure 1)

PARAMETER	FROM (INPUT)	TO (OUTPUT)	T_A = 25°C			54ACT11000		74ACT11000		UNIT
			MIN	TYP	MAX	MIN	MAX	MIN	MAX	
t_{PLH}	A or B	Y	1.5	7.2	10.9	1.5	13.3	1.5	12.3	ns
t_{PHL}			1.5	5.8	8	1.5	9.5	1.5	8.8	

operating characteristics, V_{CC} = 5 V, T_A = 25°C

PARAMETER		TEST CONDITIONS	TYP	UNIT
C_{pd}	Power dissipation capacitance per gate	C_L = 50 pF, f = 1 MHz	23	pF

Figure A.2 (*Continued*) (Reproduced by permission of Texas Instruments. Copyright © Texas Instruments.)

In data sheets, any current that flows out of a terminal of the IC is considered to be a negative quantity. In the convention used to specify limits, *maximum* refers to the greater magnitude limit of a range of values having the same sign. If the range spans both positive and negative values, then both limit values are maximums. The following is the definition of each of the parameters in this table.

The supply voltage V_{CC} defines the maximum voltage that can safely be applied to this terminal with respect to the ground of the device.

The input voltage V_I indicates the maximum voltage that can safely be applied to an input terminal with respect to the ground of the device. Actually, this specification may be exceeded as long as the input clamp current rating, as described below, is observed.

The output voltage V_O shows the maximum voltage that can safely be applied to an output terminal with respect to the ground of the device. This specification may also be exceeded as long as the output clamp current rating, as described below, is observed.

The input clamp current I_{IK} is the maximum current that can safely flow into or out of an input terminal of the device at voltages below or above the normal operating range.

The output clamp current I_{OK} is the maximum current that can safely flow into or out of an output terminal of the device at voltages below or above the normal operating range.

The continuous output current I_O is the maximum output sink or source current that can safely flow into or out of an output terminal of the device at voltages within the normal operating range.

The maximum current that can safely flow into or out of the V_{CC} and ground (GND) terminals of the IC is defined as the continuous current through those terminals.

Finally, the range of temperatures over which the device can be stored without causing excessive degradation of its performance characteristics is specified.

Recommended Operating Conditions

The conditions over which the manufacturer guarantees device operation are set in this section. The limits for the parameters that appear in this section are employed as test conditions for the limits that appear in the subsequent sections.

The first parameter defines the range of supply voltages for which the operation of the logic device within specification limits is guaranteed.

The high-level input voltage V_{IH} defines the more positive of the two ranges of values used to represent the logic states. For positive logic, a voltage within this range corresponds to logical **1**. The specified value is the least positive value of the input voltage for which the device is guaranteed to recognize this signal as a **1**.

The low-level input voltage V_{IL} defines the less positive of the two ranges of values used to represent the logic states. For positive logic, a voltage within this range corresponds to logical **0**. The specified value is the most positive value of the input voltage for which the device is guaranteed to recognize that signal as a **0**.

The high-level output current I_{OH} is the maximum source current out of the output terminal with input conditions that will establish a **1** at the output.

The low-level output current I_{OL} is the maximum sink current into the output terminal with input conditions that will establish a **0** at the output.

Defined for CMOS devices, the input voltage V_I is the range of input voltage levels over which the logic element is designed to operate.

Also defined for CMOS devices, the output voltage V_O is the range of output voltage levels over which the logic element is designed to operate.

Defined for HCMOS devices, the input transition time t_t is the time interval between two reference points, usually 10% and 90%, on the input waveform that is changing from **0** to **1** or vice versa. Therefore, it refers to rise and fall times for three specific supply-voltage values.

Defined for ACL devices, the input transition rise and fall rate dt/dv points to the input voltage edge rates.

The operating free-air temperature T_A gives the range of temperatures over which the device is designed to operate.

Electrical Characteristics

This section of the data sheet provides the electrical characteristic limits of the logic device as guaranteed by the manufacturer and when tested under the conditions furnished in the previous section. For any device specification in this section, a test is defined by a pair of test conditions. The *forcing functions* appear under the column labeled *Test Conditions* and define the external operating constraints placed upon the device tested. The *test limit*, on the other hand, defines how well the device responds to these constraints.

The input clamp voltage V_{IK} specification for bipolar logic families tests the quality of the input clamp diode used for limiting the voltage swing by damping out the ringing during transients. It is measured with all but one input tied high and the input clamp current I_{IK} forced on the remaining inputs while V_{CC} is set to the minimum.

The high-level output voltage V_{OH} is the minimum voltage at an output terminal with input conditions that will establish a **1** at the output. The output is forced to source the required current, as defined in the previous section, and the voltage is measured to check if it is greater than V_{OH}.

The low-level output voltage V_{OL} is the maximum voltage at an output terminal with input conditions that will establish a **0** at the output. The output is forced to sink the required current, as defined in the previous section, and the voltage is measured to check if it is less than V_{OL}.

The off-state output current I_{OZ} is defined for advanced CMOS devices as the current flowing into an output having the three-state capability with input conditions that will establish the high-impedance state at the output. This characteristic is verified employing two tests, I_{OZH} and I_{OZL}, which guarantee that the device will not excessively load a bus line when its output is put into the high-impedance mode.

I_{OZH} is the sink current flowing into a three-state output with a high-level voltage applied at the output and with conditions to the output-control input establishing a high-impedance state at the output while the data inputs are such that they would cause the output to be at **0** if it were enabled. Similarly, I_{OZL} is the source current flowing out of a three-state output with a low-level voltage applied at the output and with conditions to the output-control input establishing a high-impedance state at the output while the data inputs are such that they would cause the output to be at **1** if it were enabled.

The input current I_I is the maximum current flowing into an input. This test is employed to guarantee the minimum reverse breakdown voltage of the input structure. For TTL, it is tested when that input has the maximum input voltage specified for the subfamily. For CMOS devices, it is tested with a V_{CC} voltage level and then a GND level applied to the input, so that the specification guarantees the maximum input current for any input voltage within the normal range of operation.

Defined for TTL subfamilies, the high-level input current I_{IH} is the maximum current flowing into an input when a high-level voltage is applied to that input. The test set-up consists of all inputs tied to maximum V_{CC} with the exception of the one under test, upon which a high-level voltage value is forced. The resultant current is measured to test the input leakage by checking for the emitter-to-collector action for multi-emitter inputs, and reverse-bias action for diode and *pnp* inputs. For multi-emitter inputs an additional set-up checks for emitter-to-emitter transistor action by tying the other inputs to ground.

Defined for TTL subfamilies, the low-level input current, I_{IL}, is the maximum current flowing out of an input when a low-level voltage is applied to that input. The associated test is intended to measure the value of the base pull-up resistor on a multi-emitter or a diode input, and to guarantee the maximum input load an IC presents, that is, the specified fan-in of the subfamily. One input at a time is tested by connecting all the other inputs to a logical **1** value, setting V_{CC} to its maximum, forcing a low-level voltage upon that input, and measuring the corresponding current.

The output short-circuit I_{OS} is also defined for TTL only. It is the current into an output when that output is short-circuited to ground potential with input conditions that would otherwise establish a **1** at the output.

The supply current I_{CC} is the current into the supply terminal under static no-load conditions. When all of the outputs are at the high level, the corresponding supply current is denoted by I_{CCH}, while I_{CCL} signifies the supply current when all outputs are at the low level.

The supply current change ΔI_{CC} is specified for ACT devices only and indicates the increase in supply current for each input that is at one of the specified TTL voltage levels other than 0 V or V_{CC}. If n inputs are at voltages other than these two values, then the increase in supply current will be $n\Delta I_{CC}$. This parameter is tested by applying the specified V_{CC} level, setting one input at 3.4 V and the other inputs at 0 V or V_{CC}, and then measuring the current into the device under no-load conditions.

For CMOS devices, the typical values of the input capacitance, C_i—in the case of AC also the output capacitance C_o—are given. These parameters are not tested values but are defined by the design and process of the device.

For advanced CMOS devices, there is only one parameter specified under *operating characteristics* in addition to those already mentioned. That is the power-dissipation capacitance C_{PD} which is the equivalent capacitance used in calculating the dynamic power dissipation, as explained in Chapter 4. The values given are typical values and not tested.

Timing Requirements

Timing requirements are specified in a different section of the data sheet for the advanced CMOS devices. For the other subfamilies, they are included in the *Recommended Operating Conditions* section. These input parameters apply only to sequential devices such as latches, flip-flops, and registers.

The clock frequency f_{clock} defines the range of clock frequencies over which a bistable element can be operated while maintaining stable transitions between the logic states at the outputs.

The pulse duration t_w signifies the shortest time interval between specified reference points on the rising and falling edges of the input pulse waveform for which the correct operation of the logic element is guaranteed.

The setup time t_{su} defines the shortest time interval between the application of a stable signal that is maintained at a specified input data terminal and a subsequent active transition at the clock input. It may have a negative value, in which case it defines the longest interval between the aforementioned transitions for which the correct operation of the logic element is guaranteed.

The hold time t_h denotes the shortest time interval during which a signal is maintained at a specified input data terminal after the active transition of the clock signal.

Switching Characteristics

This is usually the last section of the data sheet. It consists of those parameters that specify the speed of the outputs to signal changes at the inputs under specified conditions of the supply voltage and temperature.

The maximum clock frequency f_{max} is the upper limit of the f_{clock} specification defined above. This parameter applies only to sequential parts.

The propagation delay time t_{PHL} is the time between the 50% points on the input and output voltage waveforms with the output changing from the defined high level to the defined low level.

The propagation delay time t_{PLH} is the time between the 50% points on the input and output voltage waveforms with the output changing from the defined low level to the defined high level.

Data sheets for the two-input NAND gate in all subfamilies of the TTL and CMOS families are shown in Figure A.2.

APPENDIX B: CMOS AND BIPOLAR FABRICATION TECHNIQUES

This appendix briefly discusses the techniques of fabricating CMOS and bipolar ICs.

Basic Concepts in CMOS Fabrication

Being the starting material for ICs, silicon is the most important semiconductor in both digital and analog electronics. Currently, around 95% of all semiconductor devices sold worldwide are based on silicon. It occurs naturally in the form of silica and silicates. The material is grown as a single crystal. A solid, steel-gray cylinder of up to 20 cm in diameter and 1 m in length, the crystal is sawed to produce 500-μm thick wafers. This thickness is needed to provide mechanical strength; only 10 μm or less is actually necessary to meet the electronic requirements. The surface of each wafer is then polished to a mirror finish. For typical CMOS processes, the uniform wafer doping is either *n*- or *p*-type with $N_a = 3 \times 10^{15}$ cm^{-3} or $N_d = 10^{15}$ cm^{-3}.

The IC fabrication starts with the *layout* process. From the circuit schematic of the IC to be fabricated, geometrical shapes necessary to produce all the devices in the circuit are defined. Then they are *laid out* on paper or screen in greatly magnified form, treating each physical *layer* such as metal, polysilicon, and so forth separately, and strictly following a set of geometrical constraints called *design rules*. The rules specify such parameters as the minimum permissible dimensions of geometrical patterns in each physical layer, minimum distance between such patterns, minimum overlap between patterns of different layers, and so on. Each layout layer corresponds to one *mask*, which in turn is used for one fabrication step. A mask is a transparent glass plate on one surface of which is deposited an opaque film patterned in geometrical shapes. It is used to transfer the geometric patterns onto the surface of a silicon wafer through a series of fabrication processing steps some of which are repeated several times. The fabrication steps can be classified as *deposition*, *lithography*, *etching*, *ion implantation*, *diffusion*, and *epitaxy*.

At several stages during fabrication, the entire wafer is covered by films of materials, which are then selectively removed except in certain areas where they are used to form the various layers on a chip. Deposited films provide conducting regions within the device, electrical insulation between metals, and protection from the environment. The most widely used materials for film deposition are polycrystalline silicon (*polysilicon* or *poly*), silicon dioxide (SiO_2), silicon nitride (Si_3N_4), and metal.

The thin *gate oxide* between the poly gate and silicon channel and the thick *field oxide* to isolate the MOS transistors are usually *grown* by *thermal oxidation*. In this process, the wafer is heated to temperatures up to 1300°C in the presence of oxygen before an SiO_2 layer is formed.

The polysilicon is used as the gate electrode material in MOS devices, as a conducting material for multilevel metallization, and as a contact material for devices with shallow junctions. It is usually deposited without impurities. The latter are added later by diffusion or ion implantation to reduce the resistivity of the poly. A metal or metal silicide such as tungsten or tantalum silicide may be deposited over the poly gate to increase the electrical conductivity.

The resistance of poly is often reduced by doping with boron, phosphorus, or arsenic. Since the gate electrode is over a thin oxide of 150 to 500 Å, it is important that the impurity atoms in the poly film not diffuse through the gate oxide to cause degradation of the gate oxide. To minimize this problem, the film is deposited at a low temperature without doping elements. After the gate region is defined, the film is doped by diffusion from a doped-oxide or chemical source or by ion implantation.

Poly film is also used as the diffusion source for emitters in high frequency bipolar ICs. In this process, shallow emitters with junction depths less than .5 μm are formed by introducing emitter dopants into the poly film.

Dielectric materials, in general, are used for insulation between conducting layers, for diffusion and ion implantation masks, for diffusion from doped oxides, and for passivation to protect devices from impurities, moisture, and scratches.

SiO_2 is grown as an insulating layer on top of the wafer surface or other films already on that surface. It is used to separate transistor gates from the channel, to selectively protect silicon surface areas against ion implantation, and to isolate connecting layers from one another. Undoped SiO_2 is used as an insulating layer between multilevel metallizations, ion implantations or diffusion masks. Phosphorus-doped SiO_2 is used as an insulator between metal layers, and as a final passivation over devices.

Si_3N_4 is employed to protect selected areas against oxidation. It is an extremely good barrier to sodium and water diffusion. These impurities cause device metallization to corrode or devices to become unstable.

Metal, usually aluminum, is used to form interconnections. Initially, it was also employed to form the gates of MOS transistors. The most common technique for depositing metal onto the surface of a substrate is called *sputtering*, in which argon ions are accelerated before they bombard the *target*, consisting of the material to be deposited. Atoms near the target surface are released and deposited on the substrate.

Geometrical patterns are transferred from a mask onto the wafer surface or onto the surface of a film covering the wafer by means of a *lithographical* process. First, the surface is covered by a layer of polymer called the *resist*, and the patterns are transferred onto it from a mask. Since the currently dominant method for this transfer makes use of the photographic technique, the process is called *photolithography*, while the polymer layer is known as the *photoresist*.

The photoresist in liquid form is applied uniformly on the surface and is hardened. The patterns are then transferred onto it by using UV light that casts a precise shadow of the mask. The exposure to UV light causes the molecular structure of the photoresist to be changed selectively below the transparent areas of the mask. Finally, the photoresist is developed by using an organic solvent. If the photoresist is positive, which is the case in the VLSI process with its better resolution, the development process leaves the unexposed areas intact and removes the rest. Otherwise (i.e., if it is negative), the opposite happens.

Following the lithography step, selected areas of the deposited film are removed from the wafer by the *etching* process. An etching ambient etches away the film areas not protected by the resist. Later, the photoresist is also removed.

Ion implantation is the principal process for changing the doping concentration. It is primarily used for source and drain formations, and for V_T adjustment. It consists of shooting uniformly over the entire target surface precise amounts of ionized atoms of the desired impurity. The *ion dose*, the number of implanted ions per unit area, can be accurately controlled.

Initially, dopant was diffused into the wafer from a surface source such as a doped glass. Over the past two decades, ion implantation has become the method of choice for IC fabrication because the dopant introduction is not only controllable and reproducible but free from undesirable side effects as well.

During ion implantation, impurity atoms, directed at the substrate and accelerated to gain sufficient energy, enter the crystal lattice where not prevented by the resist, SiO_2, Si_3N_4, polysilicon or a combination of them, collide with silicon atoms, and gradually lose energy, eventually coming to rest at some depth within the lattice. Implantation energies range from 1 keV to 1 MeV, resulting in ion distributions with average depths ranging from 100 Å to 10 μm. Doses range from 10^{12} ions/cm^2 for threshold adjustment to 10^{18} ions/cm^2 for buried isolators.

Ion implantation damages the silicon lattice due to the numerous collisions of ions with the lattice atoms, which occur until the ions stop. The damage *annealing*, that is, repair, and impurity atom *activation*, that is, placement of the impurity atoms on lattice sites where they can be ionized, are achieved by exposing the wafer to an elevated temperature.

Dopant atoms deposited near the surface of Si by ion implantation are driven deeper into the Si by *diffusion*. With the exception of deep *well* formation in CMOS, the diffusion of dopants must be minimized. Therefore, diffusion is not an independent step but occurs naturally during thermal oxidation and annealing because all these processes need similar temperatures and times.

Epitaxy is used only once at the beginning of the fabrication sequence. It consists of growing a film of Si on top of a silicon wafer that can be doped at levels and/or type different from those of the underlying substrate.

Silicon epitaxy was first developed to enhance the performance of discrete BJTs. At that time, resistivity of the bulk wafer was used to determine the collector voltage breakdown. High breakdown voltage requires high-resistivity material. However, this fact and the thickness of the wafer result in excessive collector resistance, which in turn hinders high-frequency response and increases power consumption. The solution was the epitaxial growth of a high-resistivity layer on a low-resistivity substrate.

One fundamental advantage of epitaxial wafers over bulk wafers is that the layer on a substrate offers the device designer a means of controlling the doping profile and hence the electrical parameters in a device structure beyond that available with ion implantation.

Fabrication Steps for a CMOS Inverter

The fabrication steps for a typical double-metal *p*-well CMOS inverter are shown in Figure B.1 and enumerated in Table B.1. The starting material is a silicon wafer on which an *n*-type *epitaxial layer* (*n*-epi) and an SiO_2 layer are grown. The silicon substrate is shown only in Figure B.1a because the transistors are formed in the epitaxial layer.

First, a hole is opened in the SiO_2 to form a *p*-well in the *n*-epi where the NMOS transistor will be fabricated. Upon formation of the *p*-well, the SiO_2 is removed, and a layer of Si_3N_4 is deposited. Subsequent etching of this layer acts as the mask to form the field-oxide regions.

Then the nitride mask is removed by etching, and a second thermal oxidation leads to the creation of a thin gate oxide that covers those regions where transistors will be fabricated. Next, polysilicon is deposited across the entire wafer, and the gate mask and lithography are employed to pattern the polysilicon. Both the polysilicon and gate oxide are etched to create the gate structure of both the NMOS and PMOS.

The entire wafer is then coated with resist to initialize the formation of the n^+ source and drain regions of the *n*-channel device by utilizing the n^+ mask and lithography. A phosphorus ion implantation with the gate poly acting as the mask creates the source and drain regions in the *p*-well. Similarly, the p^+ mask and lithography and subsequent implantation of boron ions form the source and drain regions of the *p*-channel device in the *n*-epi.

Next, SiO_2 is deposited over the entire wafer to act as an insulator. Contact holes are opened using lithography and etching. The first-level metal is sputtered over the entire wafer and then patterned by lithography and etching to form the necessary interconnections. Following the deposition of a second SiO_2 layer, *via* holes are opened using lithography and etching. The second-level metal is sputtered. The metal 2 mask patterns the metal by lithography and etching. Finally, phosphorus-doped SiO_2, deposited across the entire wafer, acts as the passivation layer to seal the top.

Fabrication of a Bipolar Junction Transistor

In order to avoid current conduction between them, devices on the same substrate in a bipolar IC are isolated from each other. Two ways to provide an isolation are the employment of (1) reverse-biased *pn* junctions, and (2) dielectric materials.

1. In the traditional bipolar technology, *junction isolation* uses reverse-biased *pn* junctions by employing impurity diffusion to produce *n*-type *islands* surrounded by *p*-type materials. The standard fabrication procedure begins with a *p*-type Si substrate of approximately 200-μm in thickness upon which an *n*-type epitaxial layer of around 10 μm is grown. A series of oxidation and

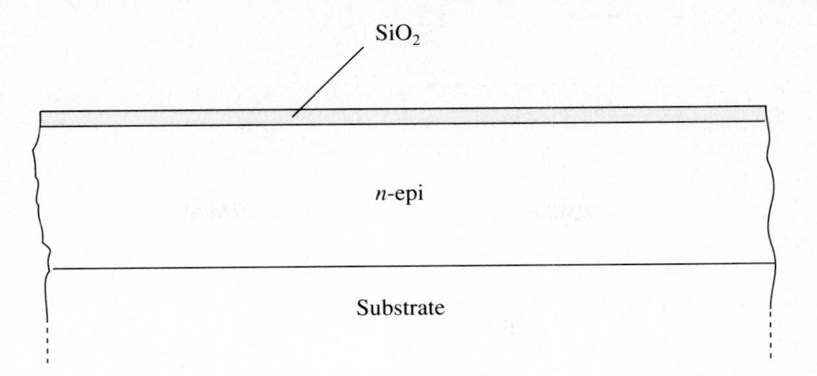

(*a*) Starting material

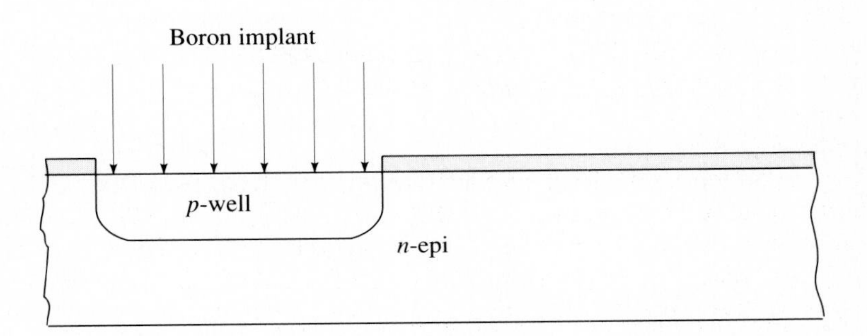

(*b*) #1: *p*-well mask

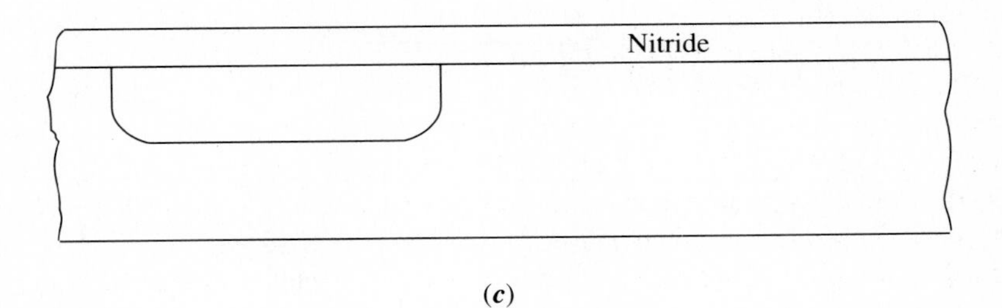

(*c*)

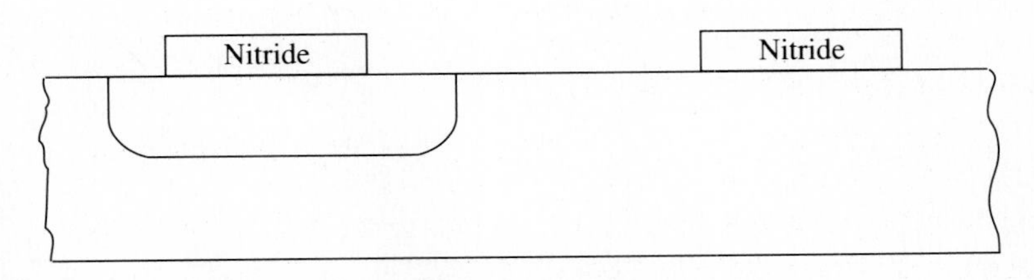

(*d*) #2: Field-oxide mask

Figure B.1 Fabrication steps for a double-metal *p*-well CMOS inverter.

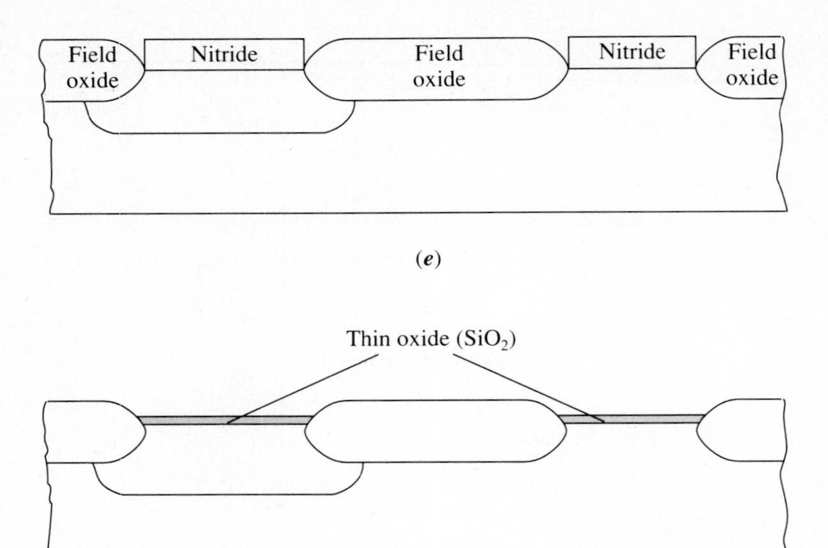

(*e*)

(*f*)

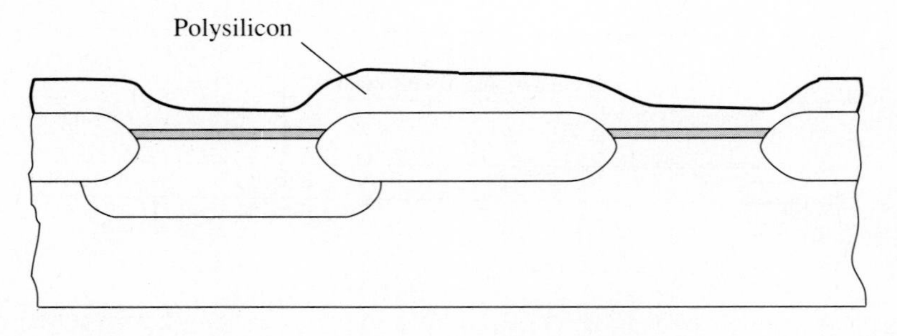

(*g*)

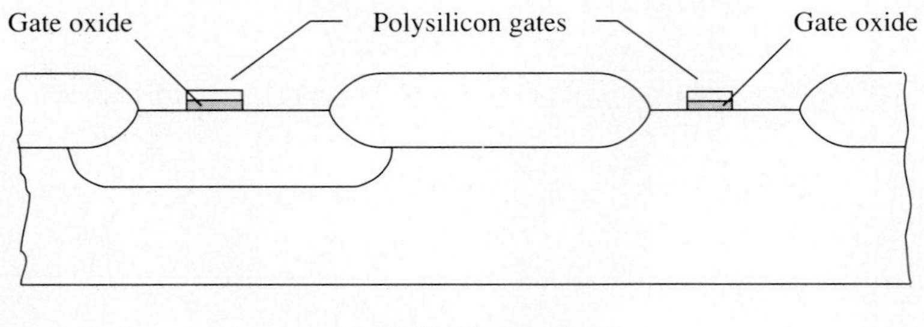

(*h*) #3: Gate mask

Figure B.1 (*Continued*)

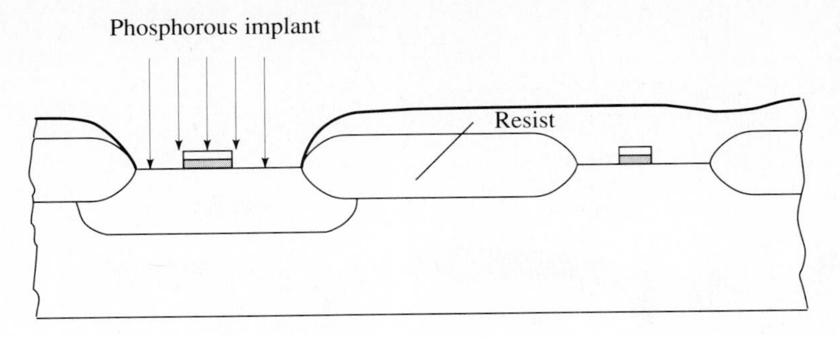

(i) #4: n^+ mask

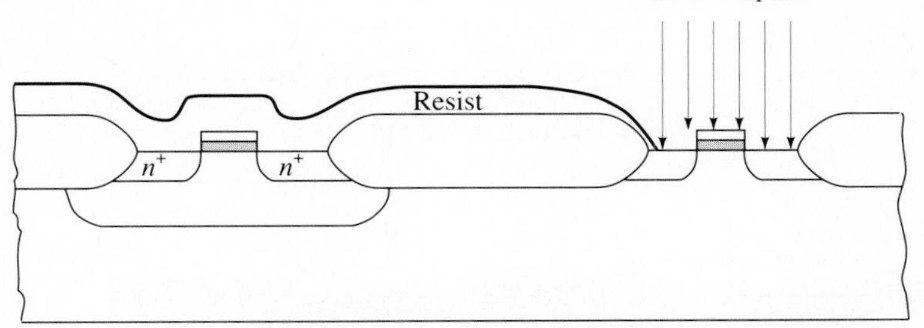

(j) #5: p^+ mask

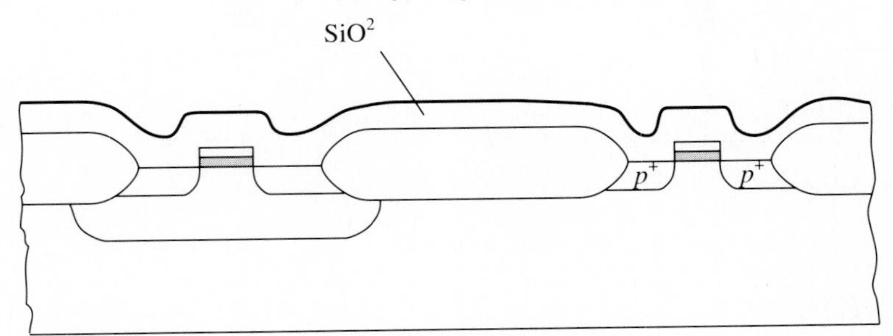

(k)

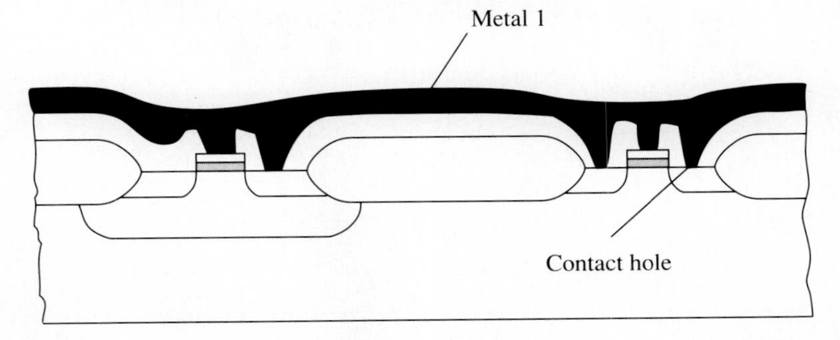

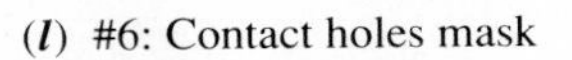
(l) #6: Contact holes mask

Figure B.1 (*Continued*)

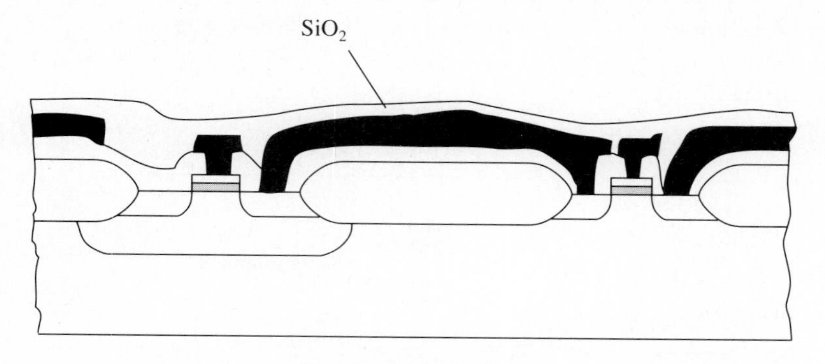

(*m*) #7: Metal 1 mask

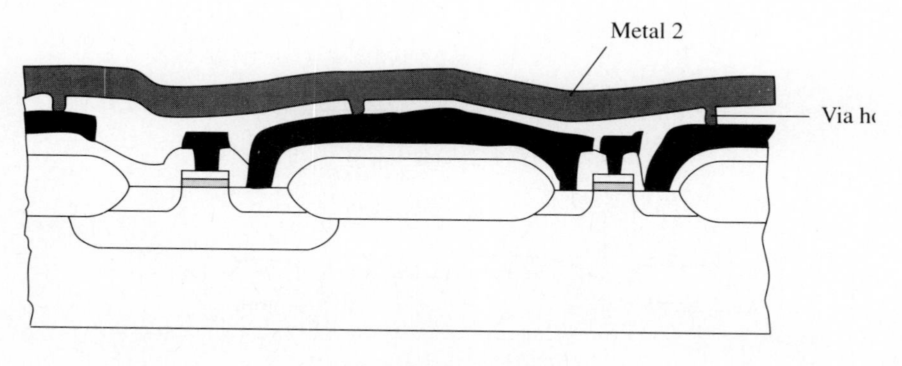

(*n*) #8: Via hole mask

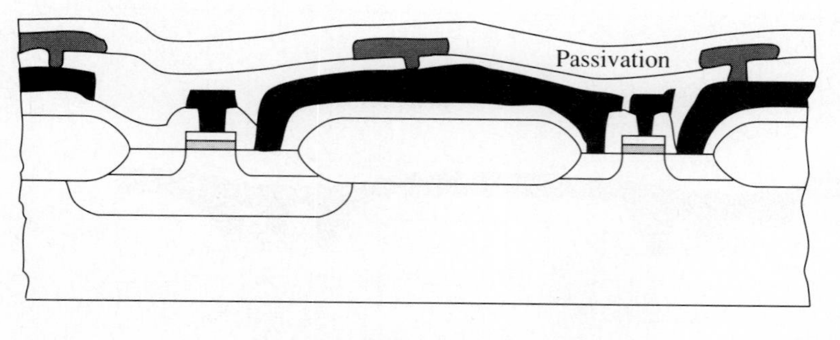

(*o*) #9: Metal 2 mask

Figure B.1 (*Continued*)

Table B.1 The CMOS Inverter Fabrication Steps

Mask	Steps
1. *p*-well	Lithography
	Etch SiO_2
	p-well ion implant
	Annealing
	Remove SiO_2
	Deposit Si_3N_4
2. Field-oxide	Lithography
	Etch Si_3N_4
	Thermal oxidation
	Remove Si_3N_4
	Thermal oxidation
	Deposit polysilicon
3. Gate	Lithography
	Etch polysilicon
	Etch SiO_2
4. n^+	Lithography
	n^+ ion implant
	Annealing
5. p^+	Lithography
	p^+ ion implant
	Annealing
	SiO_2 deposition
6. Contact hole	Lithography
	Etching
	Metal 1 sputtering
7. Metal 1	Lithography
	Etching
	SiO_2 deposition
8. Via hole	Lithography
	Etching
	Metal 2 sputtering
9. Metal 2	Lithography
	Etching
	Passivation deposition

photoresist steps define the n-type islands. Acceptor impurities are diffused into regions between the islands until the entire epitaxial layer has been penetrated so that the islands are bounded on all sides by p-type silicon and hence are electrically isolated. In these newly diffused isolation regions, the doping density is higher than the donor concentration in the epitaxial layer.

The island itself is used as the collector region. Then, an n^+ diffusion follows a p diffusion to form the emitter and base regions, respectively. During the emitter diffusion, additional donor impurities are diffused into the n-type island to provide ohmic contacts. Note that the final structure of Figure B.2a has a high collector series resistance due to the horizontal current through the lightly doped collector. To overcome this, an n^+ diffusion, called a *buried layer*, is made before growing the epitaxial layer to provide a low-resistance connection from the collector contacts on the top surface to the active portion of the transistor under the center of the completed device. This is shown in Figure B.2b.

2. The active area of the junction-isolated *npn* transistor is the area beneath the emitter mask, which is about 5% of the total area. The remaining area is

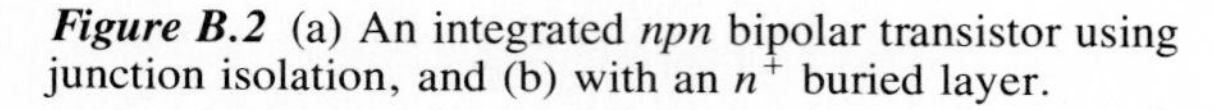

Figure B.2 (a) An integrated *npn* bipolar transistor using junction isolation, and (b) with an n^+ buried layer.

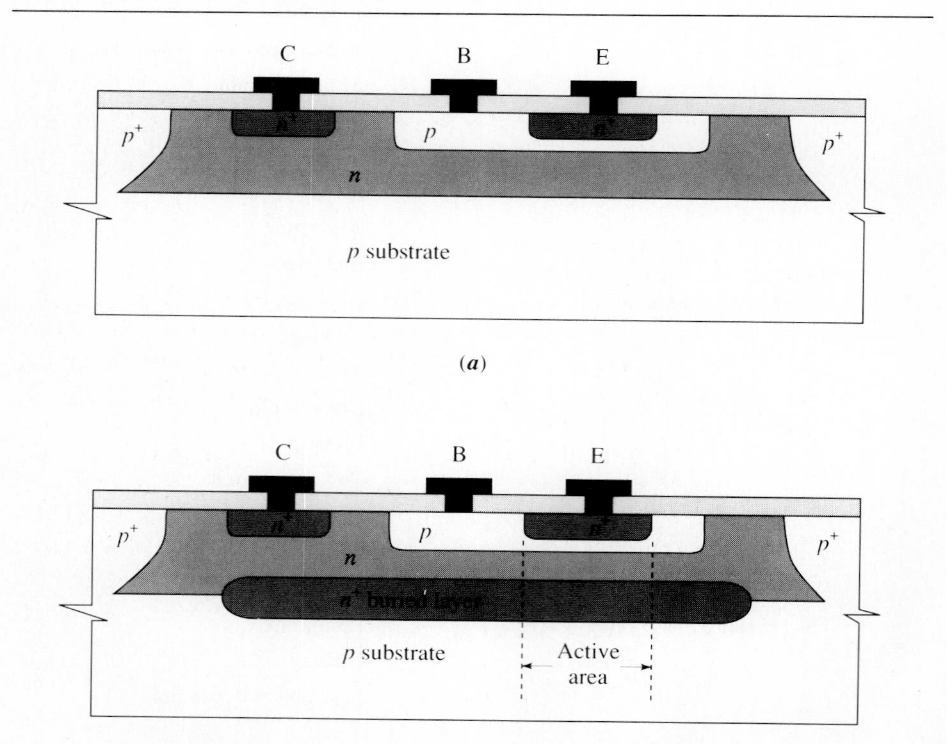

required for contacts and overlaps. In addition, the electronically inactive isolation regions take up a large area and hence introduce parasitic capacitances, as a result of which the chip density and circuit speed suffer. These disadvantages can be significantly reduced when the devices are isolated by the so-called *field oxide*.

Oxide isolation allows for a decrease in the device size and leads to an increase in both the density and speed of bipolar ICs. In this process, the transistor is completely surrounded by the oxidized silicon; thus, the p-type base region no longer has to be enclosed by the n-type collector region and hence can abut the isolation region, allowing for a large increase in the scale of integration. Moreover, a significant decrease in collector parasitic capacitance results in an increase in the switching speed.

In the oxide-isolated process, a thin oxide of around 50 nm and then a patterned film of about 75-nm silicon nitride produced by photolithography are deposited on a wafer with an n^+ buried layer, which is used, as explained before, to minimize resistance between the collector contact and the BC junction. The thin oxide is needed to release the stress at the Si_3N_4-Si interface, thereby preventing defect formation during the following high-temperature cycling. Next, an isolation mask is used to remove the nitride from the regions where the field-oxide insulation will be created. Then an etching step using the nitride as the mask is employed and is continued until the etched grooves are half the depth of the epitaxial layer.

It turns out that a net positive charge exists within the thermally grown oxide that induces negative charge in the silicon, as a result of which electrical isolation may be destroyed by the connection of the n^+ buried layers through the undesirable formation of an n layer just beneath the surface in the silicon. Thus, a boron implant is performed to increase the concentration of p-type impurities, thereby producing p^+ *channel stoppers* in those regions that will eventually be underneath the field oxide. In this way, creation of a conducting channel is prevented.

The wafer is then placed in a high-temperature furnace for thermal oxidation. A thick oxide layer grows downward in the grooves in the silicon and expands upward to fill up the etched space. Consequently, n-type islands, which are isolated from one another by SiO_2 walls and from the substrate by the pn^+ junction, are created.

The nitride is removed, and boron ions are again implanted, this time to form the p-type base. A contact mask is used to remove SiO_2 from the regions in which contacts will be created. Finally, a mask, a layer of photoresist, and a shallow n^+ phosphorous implant form the emitter region and collector contact before the metallization. The final structure is depicted in Figure B.3.

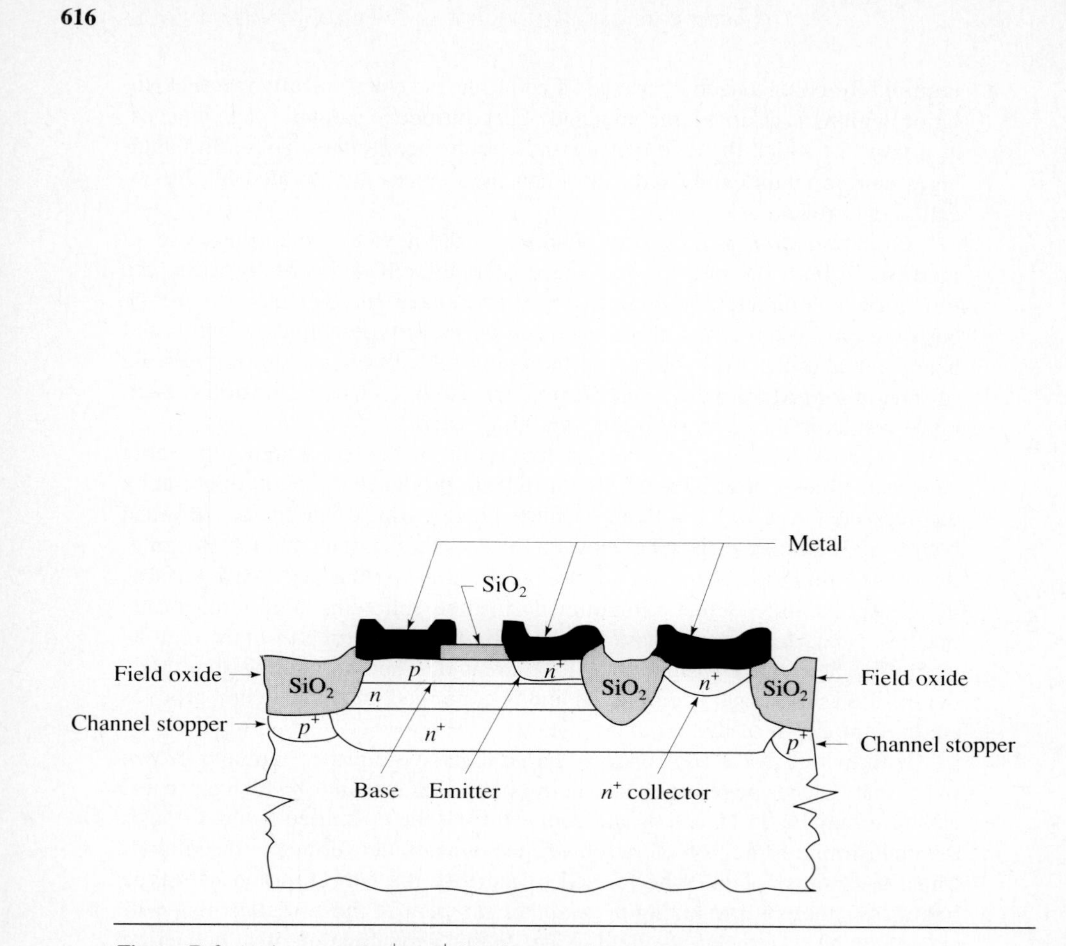

Figure B.3 An integrated n^+pnn^+ bipolar transistor using oxide isolation.

APPENDIX C

Some Important Physical Constants

Quantity	Symbol	Magnitude
Boltzmann's constant	k	1.38066×10^{-23} J/K
Electronic charge	q	1.60218×10^{-19} C
Electron volt	eV	1.60218×10^{-19} J
Electron rest mass	m_e	$.91093897 \times 10^{-30}$ kg
Permittivity of free space	ϵ_o	8.85418×10^{-12} F/m
Thermal voltage at 300 K	ϕ_T	.025860 V

Properties of Silicon at 300 K

Dielectric constant	11.8
Diffusion coefficient, cm^2/s	
electrons	37.5
holes	13
Effective density of energy states, cm^{-3}	
conduction band	2.8×10^{19}
valence band	1.04×10^{19}
Electron affinity, V	4.05
Energy gap, eV	1.12
Intrinsic carrier concentration, cm^{-3}	1.5×10^{10}
Mobility, $cm^2/V \cdot s$	
electron	1200–1350
holes	400–500

INDEX